INSTRUCTOR'S SOLUTIONS MANUAL

to accompany

Physics Sixth Edition

John D. Cutnell
Kenneth W. Johnson

Southern Illinois University at Carbondale

Volume 1

Chapters 1 – 17

JOHN WILEY & SONS, INC.

To order books or for customer service call 1-800-CALL-WILEY (225-5945).

ISBN 0-471-23126-6

Printed in the United States of America

10 9 8 7 6 5 4 3 2 1

Printed and bound by Victor Graphics, Inc.

PREFACE for VOLUME 1

This volume contains the complete solutions to the problems and conceptual questions for chapters 1 through 17. These chapters include mechanics (including fluids), thermal physics, and wave motion. The solutions for chapters 18 through 32 (electricity and magnetism, light and optics, and modern physics) are contained in Volume 2.

The manual is organized so that the solutions to the problems are in the front part of the book (part A), while those for the conceptual questions are in the back (part B).

An electronic version of this manual is also available on the Instructor's Resource CD. The files are available in two formats: Microsoft Word, and PDF files of each individual solution.

There are two types of icons at the beginning of some problems. The [SSM] and [WWW] icons indicate, respectively, that the solution is also available in the Student Solutions Manual and on the World Wide Web at www.wiley.com/college/cutnell.

Note to adopters regarding The Instructor's Solutions Manual to accompany *Physics,* 6e by Cutnell & Johnson

CONTENTS for VOLUME 1

PART A: PROBLEMS

PART B: CONCEPTUAL QUESTIONS

CHAPTER 1 | INTRODUCTION AND MATHEMATICAL CONCEPTS

PROBLEMS

---

1. SSM ***REASONING*** When converting between units, we write down the units explicitly in the calculations and treat them like any algebraic quantity. We construct the appropriate conversion factor (equal to unity) so that the final result has the desired units.

SOLUTION
a. Since 1.0×10^3 grams = 1.0 kilogram, it follows that the appropriate conversion factor is $(1.0 \times 10^3 \text{ g})/(1.0 \text{ kg}) = 1$. Therefore,

$$\left(5\times10^{-6}\ \text{kg}\right)\left(\frac{1.0\times10^{3}\ \text{g}}{1.0\ \text{kg}}\right)=\boxed{5\times10^{-3}\ \text{g}}$$

b. Since 1.0×10^3 milligrams = 1.0 gram,

$$\left(5\times10^{-3}\ \text{g}\right)\left(\frac{1.0\times10^{3}\ \text{mg}}{1.0\ \text{g}}\right)=\boxed{5\ \text{mg}}$$

c. Since 1.0×10^6 micrograms = 1.0 gram,

$$\left(5\times10^{-3}\ \text{g}\right)\left(\frac{1.0\times10^{6}\ \mu\text{g}}{1.0\ \text{g}}\right)=\boxed{5\times10^{3}\ \mu\text{g}}$$

---

2. ***REASONING*** We use the facts that 1 mi = 5280 ft, 1 m = 3.281 ft, and 1 yd = 3 ft. With these facts we construct three conversion factors: (5280 ft)/(1 mi) = 1, (1 m)/(3.281 ft) = 1, and (3 ft)/(1 yd) = 1.

SOLUTION By multiplying by the given distance d of the fall by the appropriate conversion factors we find that

$$d=\left(6\ \text{mi}\right)\left(\frac{5280\text{ft}}{1\ \text{mi}}\right)\left(\frac{1\ \text{m}}{3.281\ \text{ft}}\right)+\left(551\ \text{yd}\right)\left(\frac{3\ \text{ft}}{1\ \text{yd}}\right)\left(\frac{1\ \text{m}}{3.281\ \text{ft}}\right)=\boxed{10\ 159\ \text{m}}$$

---

3. ***REASONING AND SOLUTION***

a. 1 minute = 60 seconds, 1 hour = 3600 seconds.
(35 minutes) [60 seconds/(1 minute)] = 2100 seconds.
Hence, 1 hour 35 minutes = 3600 seconds + 2100 seconds = $\boxed{5700 \text{ s}}$

b. 1 day = 24 hours, 1 hour = 3600 s
1 day = (24 hours) [3600 seconds/(1 hour)] = $\boxed{86\,400 \text{ s}}$

---

4. ***REASONING***

a. To convert the speed from miles per hour (mi/h) to kilometers per hour (km/h), we need to convert miles to kilometers. This conversion is achieved by using the relation 1.609 km = 1 mi (see the page facing the inside of the front cover of the text).

b. To convert the speed from miles per hour (mi/h) to meters per second (m/s), we must convert miles to meters and hours to seconds. This is accomplished by using the conversions 1 mi = 1609 m and 1 h = 3600.

SOLUTION a. Multiplying the speed of 34.0 mi/h by a factor of unity, (1.609 km)/(1 mi) = 1, we find the speed of the bicyclists is

$$\text{Speed} = \left(34.0\frac{\text{mi}}{\text{h}}\right)(1) = \left(34.0\frac{\cancel{\text{mi}}}{\text{h}}\right)\left(\frac{1.609\,\text{km}}{1\,\cancel{\text{mi}}}\right) = \boxed{54.7\frac{\text{km}}{\text{h}}}$$

b. Multiplying the speed of 34.0 mi/h by two factors of unity, (1609 m)/(1 mi) = 1 and (1 h)/(3600 s) = 1, the speed of the bicyclists is

$$\text{Speed} = \left(34.0\frac{\text{mi}}{\text{h}}\right)(1)(1) = \left(34.0\frac{\cancel{\text{mi}}}{\cancel{\text{h}}}\right)\left(\frac{1609\,\text{m}}{1\,\cancel{\text{mi}}}\right)\left(\frac{1\,\cancel{\text{h}}}{3600\,\text{s}}\right) = \boxed{15.2\frac{\text{m}}{\text{s}}}$$

---

5. SSM ***REASONING AND SOLUTION*** We use the fact that 0.200 g = 1 carat and that, under the conditions stated, 1000 g has a weight of 2.205 lb to construct the two conversion factors: (0.200 g)/(1 carat) = 1 and (2.205 lb)/(1000 g) = 1. Then,

$$(3106 \text{ carats})\left(\frac{0.200\text{ g}}{1\text{ carat}}\right)\left(\frac{2.205\text{ lb}}{1000\text{ g}}\right) = \boxed{1.37\text{ lb}}$$

---

6. ***REASONING*** This problem involves using unit conversions to determine the number of magnums in one jeroboam. The necessary relationships are

$$1.0 \text{ magnum} = 1.5 \text{ liters}$$
$$1.0 \text{ jeroboam} = 0.792 \text{ U. S. gallons}$$
$$1.00 \text{ U. S. gallon} = 3.785 \times 10^{-3} \text{ m}^3 = 3.785 \text{ liters}$$

These relationships may be used to construct the appropriate conversion factors.

SOLUTION By multiplying one jeroboam by the appropriate conversion factors we can determine the number of magnums in a jeroboam as shown below:

$$(1.0 \text{ jeroboam})\left(\frac{0.792 \text{ gallons}}{1.0 \text{ jeroboam}}\right)\left(\frac{3.785 \text{ liters}}{1.0 \text{ gallon}}\right)\left(\frac{1.0 \text{ magnum}}{1.5 \text{ liters}}\right) = \boxed{2.0 \text{ magnums}}$$

---

7. ***REASONING AND SOLUTION***

a. $F = [\text{M}][\text{L}]/[\text{T}]^2$; $ma = [\text{M}][\text{L}]/[\text{T}]^2 = [\text{M}][\text{L}]/[\text{T}]^2$ so $F = ma$ $\boxed{\text{is dimensionally correct}}$.

b. $x = [\text{L}]$; $at^3 = ([\text{L}]/[\text{T}]^2)[\text{T}]^3 = [\text{L}][\text{T}]$ so $x = (1/2)at^3$ $\boxed{\text{is } \textit{not} \text{ dimensionally correct}}$.

c. $E = [\text{M}][\text{L}]^2/[\text{T}]^2$; $mv = [\text{M}][\text{L}]/[\text{T}]$ so $E = (1/2)mv$ $\boxed{\text{is } \textit{not} \text{ dimensionally correct}}$.

d. $E = [\text{M}][\text{L}]^2/[\text{T}]^2$; $max = [\text{M}]([\text{L}]/[\text{T}]^2)[\text{L}] = [\text{M}][\text{L}]^2/[\text{T}]^2$ so $E = max$ $\boxed{\text{is dimensionally correct}}$.

e. $v = [\text{L}]/[\text{T}]$; $(Fx/m)^{1/2} = \{([\text{M}][\text{L}]/[\text{T}]^2)([\text{L}]/[\text{M}])\}^{1/2} = \{[\text{L}]^2/[\text{T}]^2\}^{1/2} = [\text{L}]/[\text{T}]$ so $v = (Fx/m)^{1/2}$ $\boxed{\text{is dimensionally correct}}$.

---

8. ***REASONING AND SOLUTION*** x has the dimensions of [L], v has the dimensions of [L]/[T], and a has the dimensions of $[\text{L}]/[\text{T}]^2$. The equation under consideration is $v^n = 2ax$.

The dimensions of the right hand side are $\frac{[\text{L}]}{[\text{T}]^2}[\text{L}] = \frac{[\text{L}]^2}{[\text{T}]^2}$, while the dimensions of the left hand side are $\left(\frac{[\text{L}]}{[\text{T}]}\right)^n = \frac{[\text{L}]^n}{[\text{T}]^n}$. The right side will equal the left side only when $\boxed{n = 2}$.

---

9. SSM ***REASONING*** The volume of water at a depth d beneath the rectangle is equal to the area of the rectangle multiplied by d. The area of the rectangle = (1.20 nautical miles) × (2.60 nautical miles) = 3.12 (nautical miles)^2. Since 6076 ft = 1 nautical mile and 0.3048 m = 1 ft, the conversion factor between nautical miles and meters is

$$\left(\frac{6076\text{ ft}}{1\text{ nautical mile}}\right)\left(\frac{0.3048\text{ m}}{1\text{ ft}}\right)=\frac{1.852\times10^3\text{ m}}{1\text{ nautical mile}}$$

SOLUTION The area of the rectangle of water in m^2 is, therefore,

$$\left[3.12\text{ (nautical miles)}^2\right]\left(\frac{1.852\times10^3\text{ m}}{1\text{ nautical mile}}\right)^2=1.07\times10^7\text{ m}^2$$

Since 1 fathom = 6 ft, and 1 ft = 0.3048 m, the depth d in meters is

$$(16.0\text{ fathoms})\left(\frac{6\text{ ft}}{1\text{ fathom}}\right)\left(\frac{0.3048\text{ m}}{1\text{ ft}}\right)=2.93\times10^1\text{ m}$$

The volume of water beneath the rectangle is

$$(1.07\times10^7\text{ m}^2)(2.93\times10^1\text{ m})=\boxed{3.13\times10^8\text{ m}^3}$$

---

10. ***REASONING*** The dimension of the spring constant k can be determined by first solving the equation $T=2\pi\sqrt{m/k}$ for k in terms of the time T and the mass m. Then, the dimensions of T and m can be substituted into this expression to yield the dimension of k.

SOLUTION Algebraically solving the expression above for k gives $k=4\pi^2m/T^2$. The term $4\pi^2$ is a numerical factor that does not have a dimension, so it can be ignored in this analysis. Since the dimension for mass is [M] and that for time is [T], the dimension of k is

$$\text{Dimension of } k=\boxed{\frac{[\text{M}]}{[\text{T}]^2}}$$

---

11. ***REASONING AND SOLUTION*** The following figure (*not* drawn to scale) shows the geometry of the situation, when the observer is a distance r from the base of the arch.

The angle θ is related to r and h by $\tan\theta = h/r$. Solving for r, we find

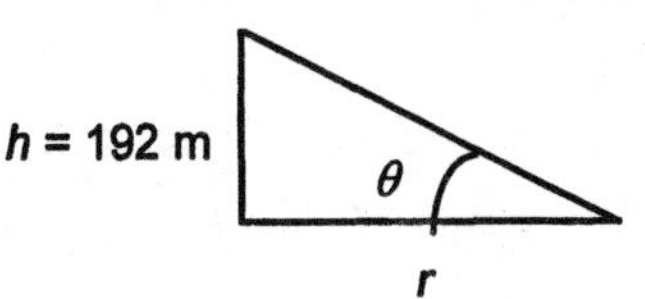

$$r = \frac{h}{\tan\theta} = \frac{192 \text{ m}}{\tan 2.0°} = 5.5 \times 10^3 \text{ m} = \boxed{5.5 \text{ km}}$$

---

12. ***REASONING AND SOLUTION*** The observer's eye, the treetop and a point on the tree trunk 1.83 m above the ground are the vertices of a right triangle. The height of the tree, H is

$$H = 1.83 \text{ m} + (32.0 \text{ m}) \tan 20.0° = \boxed{13.5 \text{ m}}$$

---

13 SSM WWW ***REASONING*** The shortest distance between the two towns is along the line that joins them. This distance, h, is the hypotenuse of a right triangle whose other sides are $h_o = 35.0$ km and $h_a = 72.0$ km, as shown in the figure below.

SOLUTION The angle θ is given by $\tan\theta = h_o/h_a$ so that

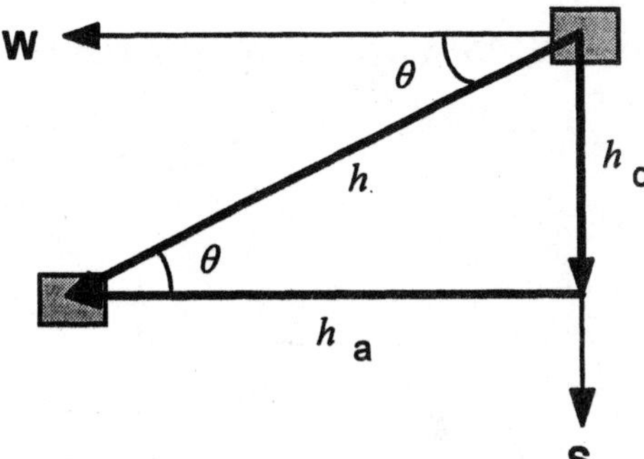

$$\theta = \tan^{-1}\left(\frac{35.0 \text{ km}}{72.0 \text{ km}}\right) = \boxed{25.9° \text{ S of W}}$$

We can then use the Pythagorean theorem to find h.

$$h = \sqrt{h_o^2 + h_a^2} = \sqrt{(35.0 \text{ km})^2 + (72.0 \text{ km})^2} = \boxed{80.1 \text{ km}}$$

---

14. ***REASONING*** The drawing shows the heights of the two balloonists and the horizontal distance x between them. Also shown in dashed lines is a right triangle, one angle of which is 13.3°. Note that the side adjacent to the 13.3° angle is the horizontal distance x, while the side opposite the angle is the distance between the two heights, 61.0 m − 48.2 m. Since we know the angle and the length of one side of the right triangle, we can use trigonometry to find the length of the other side.

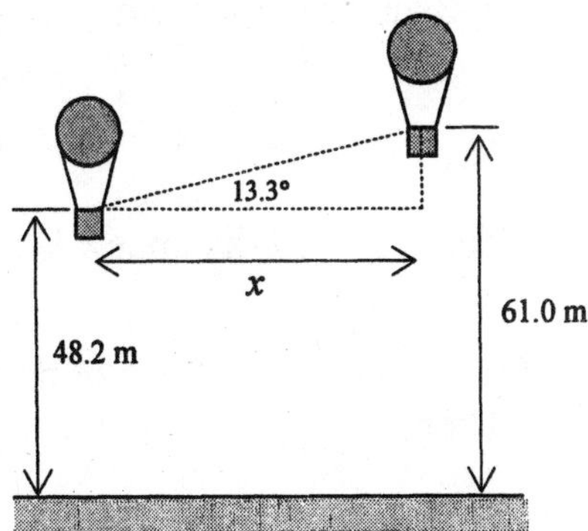

SOLUTION The definition of the tangent function, Equation 1.3, can be used to find the horizontal distance x, since the angle and the length of the opposite side are known:

$$\tan 13.3° = \frac{\text{length of opposite side}}{\text{length of adjacent side } (= x)}$$

Solving for x gives

$$x = \frac{\text{length of opposite side}}{\tan 13.3°} = \frac{61.0\ \text{m} - 48.2\ \text{m}}{\tan 13.3°} = \boxed{54.1\ \text{m}}$$

---

15. ***REASONING AND SOLUTION*** One half of the tree forms a right triangle as shown.

$$H = \frac{1.00\ \text{m}}{\tan 15.0°} = \boxed{3.73\ \text{m}}$$

15.0°
H
1.00 m

---

16. ***REASONING AND SOLUTION*** Consider the following views of the cube.

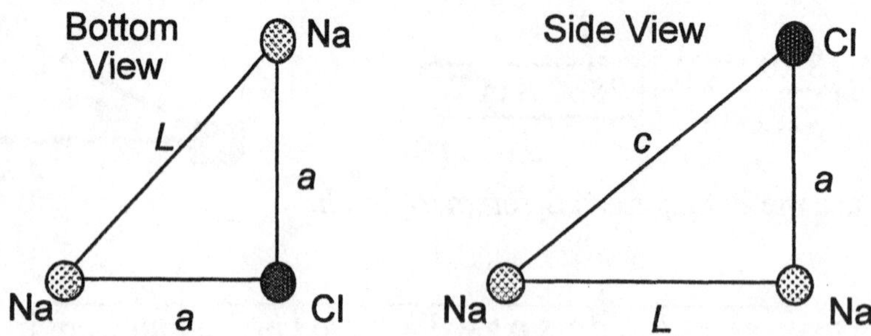

The length, L, of the diagonal of the bottom face of the cube can be found using the Pythagorean theorem to be

$$L^2 = a^2 + a^2 = 2(0.281\ \text{nm})^2 = 0.158\ \text{nm}^2 \quad \text{or} \quad L = 0.397\ \text{nm}$$

The required distance c is also found using the Pythagorean theorem.

$$c^2 = L^2 + a^2 = (0.397\ \text{nm})^2 + (0.281\ \text{nm})^2 = 0.237\ \text{nm}^2$$

Then,

$$c = \boxed{0.487\ \text{nm}}$$

---

17. ***REASONING*** Note from the drawing that the shaded right triangle contains the angle θ, the side opposite the angle (length = 0.281 nm), and the side adjacent to the angle (length = L). If the length L can be determined, we can use trigonometry to find θ. The bottom face of the cube is a square whose diagonal has a length L. This length can be found from the Pythagorean theorem, since the lengths of the two sides of the square are known.

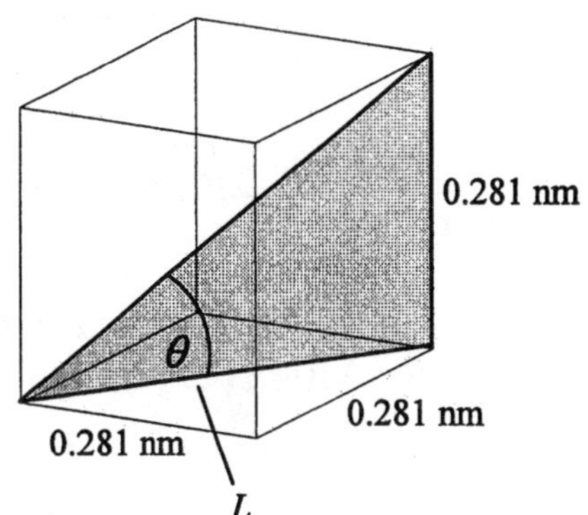

SOLUTION The angle can be obtained from the inverse tangent function, Equation 1.6, as $\theta = \tan^{-1}\left[(0.281\text{ nm})/L\right]$. Since L is the length of the hypotenuse of a right triangle whose sides have lengths of 0.281 nm, its value can be determined from the Pythagorean theorem:

$$L = \sqrt{(0.281\text{ nm})^2 + (0.281\text{ nm})^2} = 0.397\text{ nm}$$

Thus, the angle is

$$\theta = \tan^{-1}\left(\frac{0.281\text{ nm}}{L}\right) = \tan^{-1}\left(\frac{0.281\text{ nm}}{0.397\text{ nm}}\right) = \boxed{35.3^\circ}$$

---

18. ***REASONING***

a. The drawing shows the person standing on the earth and looking at the horizon. Notice the right triangle, the sides of which are R, the radius of the earth, and d, the distance from the person's eyes to the horizon. The length of the hypotenuse is $R + h$, where h is the height of the person's eyes above the water. Since we know the lengths of two sides of the triangle, the Pythagorean theorem can be employed to find the length of the third side.

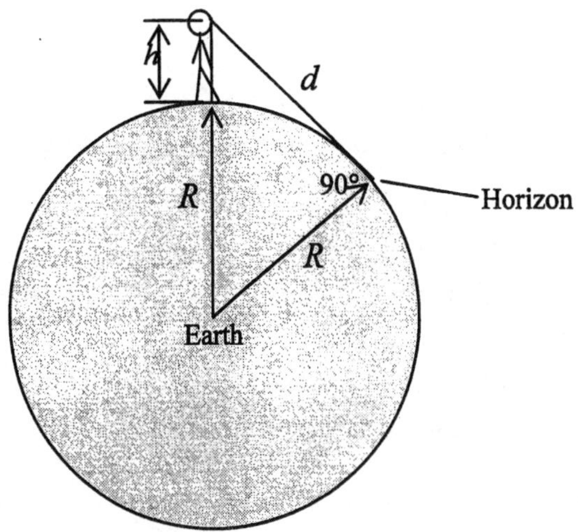

b. To convert the distance from meters to miles, we use the relation 1609 m = 1 mi (see the page facing the inside of the front cover of the text).

SOLUTION

a. The Pythagorean theorem (Equation 1.7) states that the square of the hypotenuse is equal to the sum of the squares of the sides, or $(R+h)^2 = d^2 + R^2$. Solving this equation for d yields

$$d=\sqrt{(R+h)^2-R^2}=\sqrt{R^2+2Rh+h^2-R^2}$$

$$=\sqrt{2Rh+h^2}=\sqrt{2\left(6.38\times10^6\text{ m}\right)(1.6\text{ m})+(1.6\text{ m})^2}=\boxed{4500\text{ m}}$$

b. Multiplying the distance of 4500 m by a factor of unity, (1 mi)/(1609 m) = 1, the distance (in miles) from the person's eyes to the horizon is

$$d=(4500\text{ m})(1)=\left(4500\not{\text{m}}\right)\left(\frac{1\text{ mi}}{1609\not{\text{m}}}\right)=\boxed{2.8\text{ mi}}$$

---

19. **SSM** ***REASONING*** The law of cosines is

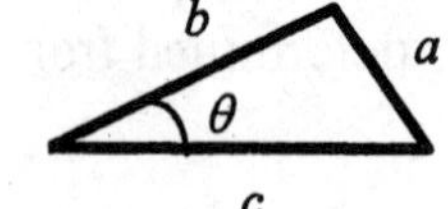

$$a^2=b^2+c^2-2bc\cos\theta$$

where a is the side opposite angle θ, and b and c are the other two sides. Solving for θ, we have

$$\theta=\cos^{-1}\left[(b^2+c^2-a^2)/(2bc)\right]$$

SOLUTION For a = 190 cm, b = 95 cm, c = 150 cm,

$$\theta=\cos^{-1}\left[\frac{(95\text{ cm})^2+(150\text{ cm})^2-(190\text{ cm})^2}{2(95\text{ cm})(150\text{ cm})}\right]=99°$$

Thus, the angle opposite the side of length 190 cm is $\boxed{99\text{ degrees}}$.

Similarly, when a = 95 cm, b =190 cm, c = 150 cm, we find that the angle opposite the side of length 95 cm is $\boxed{\theta=3.0\times10^1\text{ degrees}}$.

Finally, when a = 150 cm, b = 190 cm, c = 95 cm, we find that the angle opposite the side of length 150 cm is $\boxed{\theta=51\text{ degrees}}$.

---

20. ***REASONING AND SOLUTION*** Consider the bottom face of a tetrahedron which is an equilateral triangle.

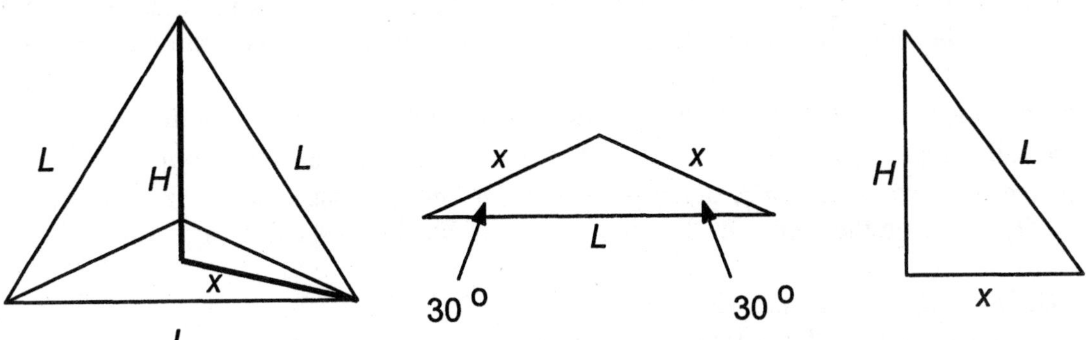

The distance x from a vertex to the intersection of the perpendicular bisectors is

$$x = \frac{\frac{1}{2}L}{\cos 30°} = \frac{L}{\sqrt{3}}$$

Now consider any right triangle in the tetrahedron whose sides are H, L, and x. The Pythagorean theorem gives

$$L^2 = H^2 + x^2 = H^2 + \tfrac{1}{3}L^2$$

Then

$$H^2 = L^2 - \tfrac{1}{3}L^2 = \tfrac{2}{3}L^2 \qquad \text{so that} \qquad H/L = \boxed{\sqrt{2/3}}$$

---

21. ***REASONING AND SOLUTION*** Since the initial force and the resultant force point along the east/west line, the second force must also point along the east/west line. The direction of the second force is not specified; it could point either due east or due west.

If the second force points $\boxed{\text{due east}}$, both forces point in the same direction and the magnitude of the resultant force is the sum of the two magnitudes: $F_1 + F_2 = F_R$. Therefore,

$$F_2 = F_R - F_1 = 400\text{ N} - 200\text{ N} = \boxed{200\text{ N}}$$

If the second force points $\boxed{\text{due west}}$, the two forces point in opposite directions, and the magnitude of the resultant force is the difference of the two magnitudes: $F_2 - F_1 = F_R$. Therefore,

$$F_2 = F_R + F_1 = 400\text{ N} + 200\text{ N} = \boxed{600\text{ N}}$$

---

22. ***REASONING*** a. Since the two force vectors **A** and **B** have directions due west and due north, they are perpendicular. Therefore, the resultant vector **F** = **A** + **B** has a magnitude given by the Pythagorean theorem: $F^2 = A^2 + B^2$. Knowing the magnitudes of **A** and **B**, we can calculate the magnitude of **F**. The direction of the resultant can be obtained using trigonometry.
b. For the vector **F′** = **A** – **B** we note that the subtraction can be regarded as an addition in the following sense: **F′** = **A** + (–**B**). The vector –**B** points due south, opposite the vector **B**, so the two vectors are once again perpendicular and the magnitude of **F′** again is given by the Pythagorean theorem. The direction again can be obtained using trigonometry.

SOLUTION a. The drawing shows the two vectors and the resultant vector. According to the Pythagorean theorem, we have

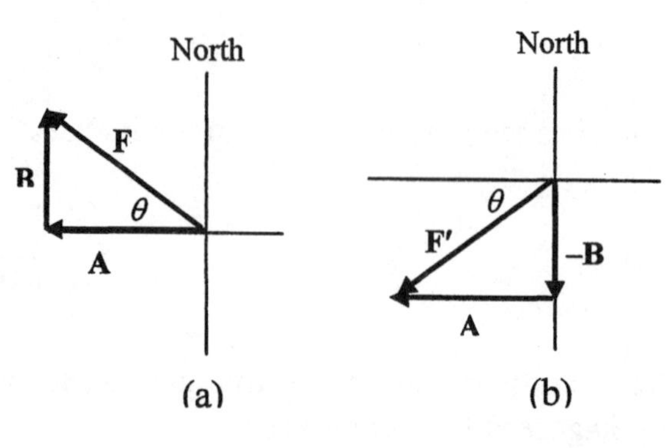

$$F^2 = A^2 + B^2$$

$$F = \sqrt{A^2 + B^2}$$

$$F = \sqrt{(445\text{ N})^2 + (325\text{ N})^2}$$

$$= \boxed{551\text{ N}}$$

Using trigonometry, we can see that the direction of the resultant is

$$\tan\theta = \frac{B}{A} \quad \text{or} \quad \theta = \tan^{-1}\left(\frac{325\text{ N}}{445\text{ N}}\right) = \boxed{36.1°\text{ north of west}}$$

b. Referring to the drawing and following the same procedure as in part a, we find

$$F'^2 = A^2 + (-B)^2 \quad \text{or} \quad F' = \sqrt{A^2 + (-B)^2} = \sqrt{(445\text{ N})^2 + (-325\text{ N})^2} = \boxed{551\text{ N}}$$

$$\tan\theta = \frac{B}{A} \quad \text{or} \quad \theta = \tan^{-1}\left(\frac{325\text{ N}}{445\text{ N}}\right) = \boxed{36.1°\text{ south of west}}$$

---

23. SSM ***REASONING*** Using the component method, we find the components of the resultant **R** that are due east and due north. The magnitude and direction of the resultant **R** can be determined from its components, the Pythagorean theorem, and the tangent function.

SOLUTION The first four rows of the table below give the components of the vectors **A**, **B**, **C**, and **D**. Note that east and north have been taken as the positive directions; hence vectors pointing due west and due south will appear with a negative sign.

Vector	*East/West Component*	*North/South Component*
A	+ 2.00 km	0
B	0	+ 3.75 km
C	– 2.50 km	0
D	0	–3.00 km
R = **A** + **B** + **C** + **D**	– 0.50 km	+ 0.75 km

The fifth row in the table gives the components of **R**. The magnitude of **R** is given by the Pythagorean theorem as

$$R = \sqrt{(-0.50\text{ km})^2 + (+0.75\text{ km})^2} = \boxed{0.90\text{ km}}$$

The angle θ that **R** makes with the direction due west is

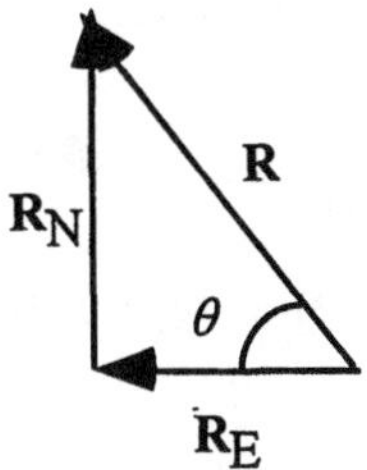

$$\theta = \tan^{-1}\left(\frac{0.75\text{ km}}{0.50\text{ km}}\right) = \boxed{56°\text{ north of west}}$$

---

24. ***REASONING*** The triple jump consists of a double jump (assumed to end on a square that we label C) followed by a single jump. The single jump is perpendicular to the double jump, so that the length ℓ_{AC} of the double jump, the length ℓ_{CB} of the single jump, and the magnitude ℓ_{AB} of the total displacement form a right triangle. Thus, we have

$$\ell_{AB} = \sqrt{\ell_{AC}^2 + \ell_{CB}^2}$$

SOLUTION The diagonal of one square on the checkerboard has a length d of

$$d = \sqrt{(4.0\text{ cm})^2 + (4.0\text{ cm})^2} = 5.66\text{ cm}$$

Since $\ell_{AC} = 4\,d$ and $\ell_{CB} = 2\,d$, it follows that

$$\ell_{AB} = \sqrt{(4d)^2 + (2d)^2} = d\sqrt{20} = (5.66\text{ cm})\sqrt{20} = \boxed{25\text{ cm}}$$

---

25. SSM WWW ***REASONING*** When a vector is multiplied by – 1, the magnitude of the vector remains the same, but the direction is reversed. Vector subtraction is carried out in the same manner as vector addition except that one of the vectors has been multiplied by –1.

SOLUTION Since both vectors point north, they are colinear. Therefore their magnitudes may be added by the rules of ordinary algebra.

a. Taking north as the positive direction, we have

$$\mathbf{A}-\mathbf{B}=\mathbf{A}+(-\mathbf{B})=+2.43 \text{ km} + (-7.74 \text{ km}) = -5.31 \text{ km}$$

The minus sign in the answer for **A** – **B** indicates that the direction is south so that

$$\boxed{\mathbf{A}-\mathbf{B}=5.31 \text{ km, south}}$$

b. Similarly,

$$\mathbf{B}-\mathbf{A}=\mathbf{B}+(-\mathbf{A})=+7.74 \text{ km} + (-2.43 \text{ km}) = +5.31 \text{ km}$$

The plus sign in the answer for **B** – **A** indicates that the direction is north, so that

$$\boxed{\mathbf{B}-\mathbf{A}=5.31 \text{ km, north}}$$

---

26. ***REASONING*** At the turning point, the distance to the campground is labeled d in the drawing. Note that d is the length of the hypotenuse of a right triangle. Since we know the lengths of the other two sides of the triangle, the Pythagorean theorem can be used to find d. The direction that cyclist #2 must head during the last part of the trip is given by the angle θ. It can be determined by using the inverse tangent function.

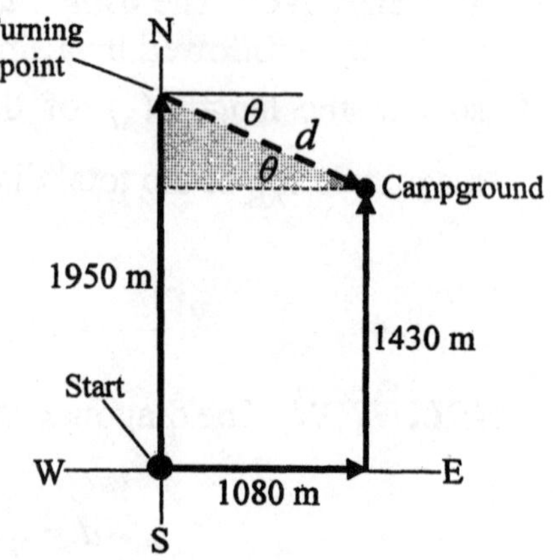

SOLUTION
a. The two sides of the triangle have lengths of 1080 m and 520 m (1950 m – 1430 m = 520 m). The length d of the hypotenuse can be determined from the Pythagorean theorem, Equation (1.7), as

$$d=\sqrt{(1080 \text{ m})^2+(520 \text{ m})^2}=\boxed{1200 \text{ m}}$$

b. Since the lengths of the sides opposite and adjacent to the angle θ are known, the inverse tangent function (Equation 1.6) can be used to find θ.

$$\theta = \tan^{-1}\left(\frac{520 \text{ m}}{1080 \text{ m}}\right) = \boxed{26° \text{ south of east}}$$

---

27. **SSM** **WWW** ***REASONING AND SOLUTION*** The single force needed to produce the same effect is equal to the resultant of the forces provided by the two ropes. The figure below shows the force vectors drawn to scale and arranged tail to head. The magnitude and direction of the resultant can be found by direct measurement using the scale factor shown in the figure.

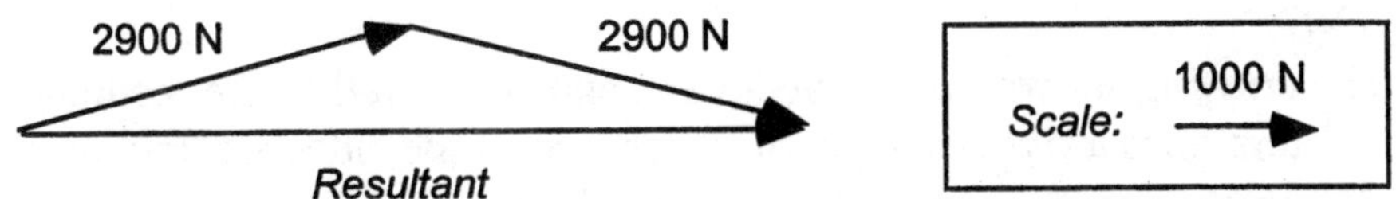

a. From the figure, the magnitude of the resultant is $\boxed{5600 \text{ N}}$.

b. The single rope should be directed $\boxed{\text{along the dashed line}}$ in the text drawing.

---

28. ***REASONING*** The displacement **Q** needed for the bear to return to its starting point is the vector that is equal in magnitude and opposite in direction to the actual displacement **R** of the bear.

SOLUTION The first two rows of the following table gives components of the individual displacements, **A** and **B**, of the bear. The third row gives the components of $\mathbf{R} = \mathbf{A} + \mathbf{B}$. The directions due east and due north have been chosen to be positive.

Displacement	*East/West Component*	*North/South Component*
A	–1563 m	0
B	(– 3348 m) cos 32°= –2839 m	(– 3348 m) sin 32°= 1774 m
$\mathbf{R} = \mathbf{A} + \mathbf{B}$	–4402 m	1774 m

Thus, the components of the displacement **Q** needed for the bear to *return to its starting point* are an eastward component of +4402 m and a southward component of –1774 m.

a. From the Pythagorean theorem, we have

$$Q = \sqrt{(4402\text{ m})^2 + (-1774\text{ m})^2} = \boxed{4746\text{ m}}$$

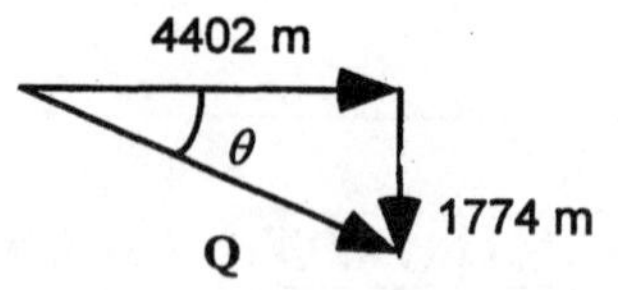

b. The angle θ is given by

$$\theta = \tan^{-1}\left(\frac{1774\text{ m}}{4402\text{ m}}\right) = \boxed{21.9°,\text{ south of east}}$$

---

29. ***REASONING***

a. and b. The drawing shows the two vectors **A** and **B**, as well as the resultant vector **A** + **B**. The three vectors form a right triangle, of which two of the sides are known. We can employ the Pythagorean theorem, Equation 1.7, to find the length of the third side. The angle θ in the drawing can be determined by using the inverse cosine function, Equation 1.5, since the side adjacent to θ and the length of the hypotenuse are known.

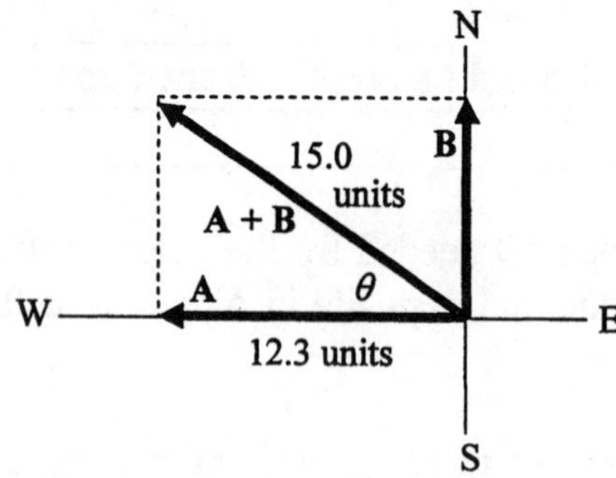

c. and d. The drawing illustrates the two vectors **A** and **–B**, as well as the resultant vector **A** – **B.** The three vectors form a right triangle, which is identical to the one above, except for the orientation. Therefore, the lengths of the hypotenuses and the angles are equal.

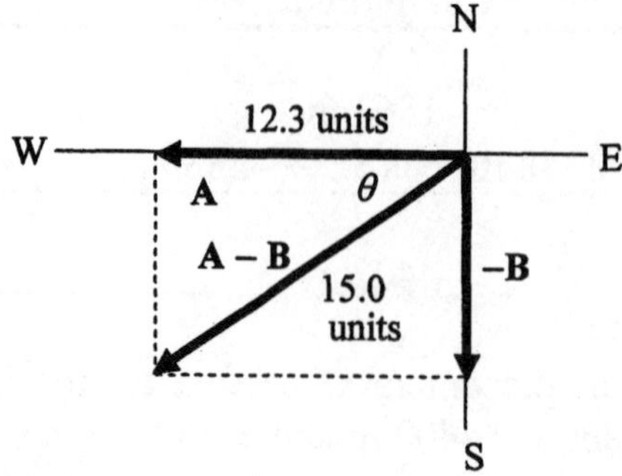

SOLUTION

a. Let **R** = **A** + **B**. The Pythagorean theorem (Equation 1.7) states that the square of the hypotenuse is equal to the sum of the squares of the sides, so that $R^2 = A^2 + B^2$. Solving for B yields

$$B = \sqrt{R^2 - A^2} = \sqrt{(15.0 \text{ units})^2 - (12.3 \text{ units})^2} = \boxed{8.6 \text{ units}}$$

b. The angle θ can be found from the inverse cosine function, Equation 1.5:

$$\theta = \cos^{-1}\left(\frac{12.3 \text{ units}}{15.0 \text{ units}}\right) = \boxed{34.9° \text{ north of west}}$$

c. Except for orientation, the triangles in the two drawings are the same. Thus, the value for B is the same as that determined in part (a) above: $B = \boxed{8.6 \text{ units}}$

d. The angle θ is the same as that found in part (a), except the resultant vector points south of west, rather than north of west: $\theta = \boxed{34.9° \text{ south of west}}$

---

30. ***REASONING AND SOLUTION*** The following figure is a scale diagram of the forces drawn tail-to-head. The scale factor is shown in the figure. The head of $\mathbf{F}_3$ touches the tail of $\mathbf{F}_1$, because the resultant of the three forces is zero.

a. From the figure, $\mathbf{F}_3$ must have a magnitude of $\boxed{78 \text{ N}}$ if the resultant force acting on the ball is zero.

b. Measurement with a protractor indicates that the angle $\boxed{\theta = 34°}$.

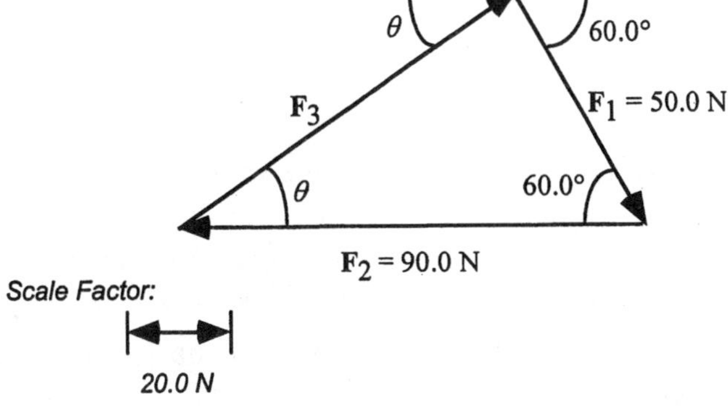

---

31. SSM ***REASONING*** The ostrich's velocity vector **v** and the desired components are shown in the figure at the right. The components of the velocity in the directions due west and due north are $\mathbf{v}_W$ and $\mathbf{v}_N$, respectively. The sine and cosine functions can be used to find the components.

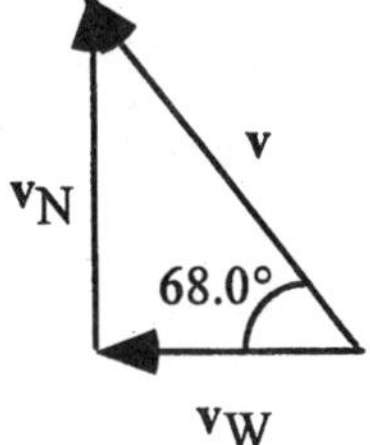

SOLUTION

a. According to the definition of the sine function, we have for the vectors in the figure

$$\sin\,\theta = \frac{v_N}{v} \quad \text{or} \quad v_N = v \sin\,\theta = (17.0 \text{ m/s}) \sin 68° = \boxed{15.8 \text{ m/s}}$$

b. Similarly,

$$\cos\theta = \frac{v_W}{v} \quad \text{or} \quad v_W = v\cos\theta = (17.0\text{ m/s})\cos 68.0° = \boxed{6.37\text{ m/s}}$$

---

32. ***REASONING AND SOLUTION***

a. From the Pythagorean theorem, we have

$$F = \sqrt{(150\text{ N})^2 + (130\text{ N})^2} = \boxed{2.0 \times 10^2\text{ N}}$$

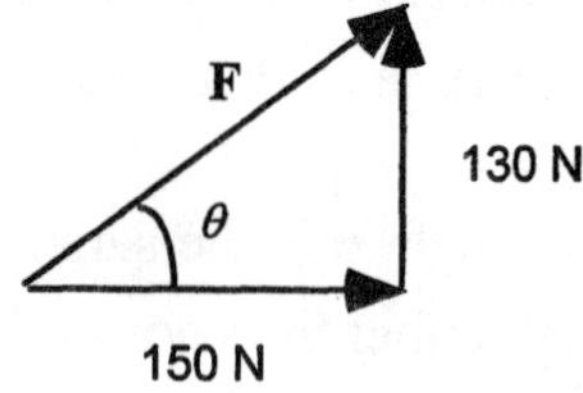

b. The angle θ is given by

$$\theta = \tan^{-1}\left(\frac{130\text{ N}}{150\text{ N}}\right) = \boxed{41°}$$

---

33. ***REASONING AND SOLUTION*** The east and north components are, respectively

a. $$A_e = A\cos\theta = (155\text{ km})\cos 18.0° = \boxed{147\text{ km}}$$

b. $$A_n = A\sin\theta = (155\text{ km})\sin 18.0° = \boxed{47.9\text{ km}}$$

---

34. ***REASONING*** The triangle in the drawing is a right triangle. We know one of its angles is 30.0°, and the length of the hypotenuse is 8.6 m. Therefore, the sine and cosine functions can be used to find the magnitudes of $\mathbf{A_x}$ and $\mathbf{A_y}$. The directions of these vectors can be found by examining the diagram.

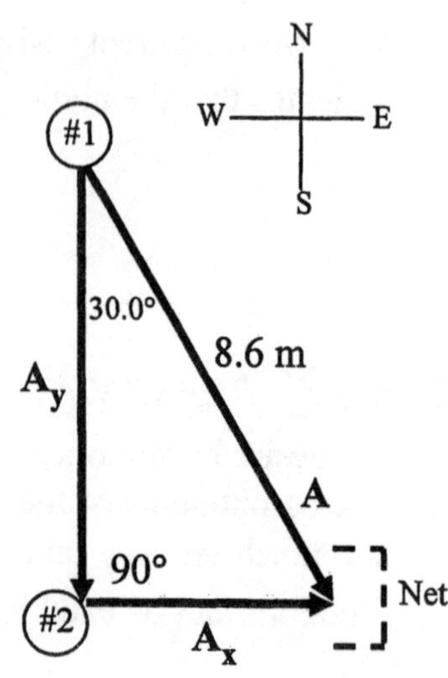

SOLUTION

a. The magnitude A_x of the displacement vector $\mathbf{A_x}$ is related to the length of the hypotenuse and the 30.0° angle by the sine function (Equation 1.1). The drawing shows that the direction of $\mathbf{A_x}$ is due east.

$$A_x = A\sin 30.0° = (8.6\text{ m})\sin 30.0° = \boxed{4.3\text{ m, due east}}$$

b. In a similar manner, the magnitude A_y of $\mathbf{A_y}$ can be found by using the cosine function (Equation 1.2). Its direction is due south.

$$A_y = A\cos 30.0° = (8.6\text{ m})\cos 30.0° = \boxed{7.4\text{ m, due south}}$$

---

35. SSM ***REASONING AND SOLUTION*** A single rope must supply the resultant of the two forces. Since the forces are perpendicular, the magnitude of the resultant can be found from the Pythagorean theorem.

a. Applying the Pythagorean theorem,

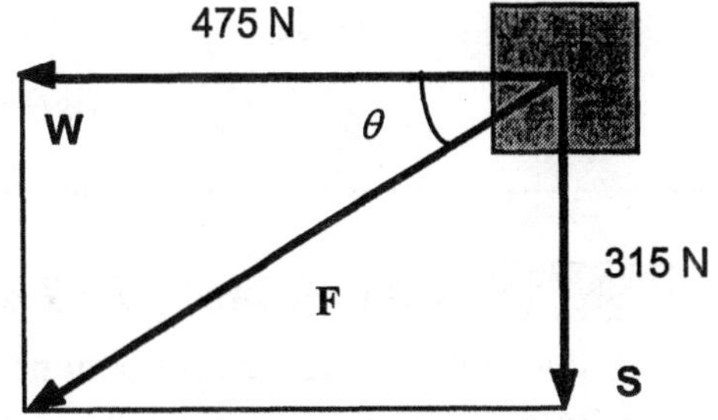

$$F = \sqrt{(475\text{ N})^2 + (315\text{ N})^2} = \boxed{5.70 \times 10^2\text{ N}}$$

b. The angle θ that the resultant makes with the westward direction is

$$\theta = \tan^{-1}\left(\frac{315\text{ N}}{475\text{ N}}\right) = 33.6°$$

Thus, the rope must make an angle of $\boxed{33.6° \text{ south of west}}$.

---

36. ***REASONING AND SOLUTION*** The horizontal component of the plane's velocity is

$$v_x = v\cos\theta = (180\text{ m/s})\cos 34° = \boxed{150\text{ m/s}}$$

---

37. ***REASONING*** The force **F** and its two components form a right triangle. The hypotenuse is 82.3 newtons, and the side parallel to the +x axis is F_x = 74.6 newtons. Therefore, we can use the trigonometric cosine and sine functions to determine the angle of **F** relative to the +x axis and the component F_y of **F** along the +y axis.

SOLUTION

a. The direction of **F** relative to the +x axis is specified by the angle θ as

$$\theta = \cos^{-1}\left(\frac{74.6\text{ newtons}}{82.3\text{ newtons}}\right) = \boxed{25.0°} \qquad (1.5)$$

b. The component of **F** along the +y axis is

$$F_y = F\sin 25.0° = (82.3\text{ newtons})\sin 25.0° = \boxed{34.8\text{ newtons}} \qquad (1.4)$$

---

38. ***REASONING*** The force vector **F** points at an angle of θ above the +x axis. Therefore, its x and y components are given by $F_x = F\cos\theta$ and $F_y = F\sin\theta$.

SOLUTION a. The magnitude of the vector can be obtained from the *y* component as follows:

$$F_y = F \sin\theta \quad \text{or} \quad F = \frac{F_y}{\sin\theta} = \frac{290\text{ N}}{\sin 52°} = \boxed{370\text{ N}}$$

b. Now that the magnitude of the vector is known, the *x* component of the vector can be calculated as follows:

$$F_x = F\cos\theta = (370\text{ N})\cos 52° = \boxed{+230\text{ N}}$$

---

39. SSM WWW ***REASONING AND SOLUTION*** The force **F** can be first resolved into two components; the *z* component F_z and the projection onto the *x-y* plane, F_p as shown in the figure below on the left. According to that figure,

$$F_p = F \sin 54.0° = (475\text{ N}) \sin 54.0° = 384\text{ N}$$

The projection onto the *x-y* plane, F_p, can then be resolved into *x* and *y* components.

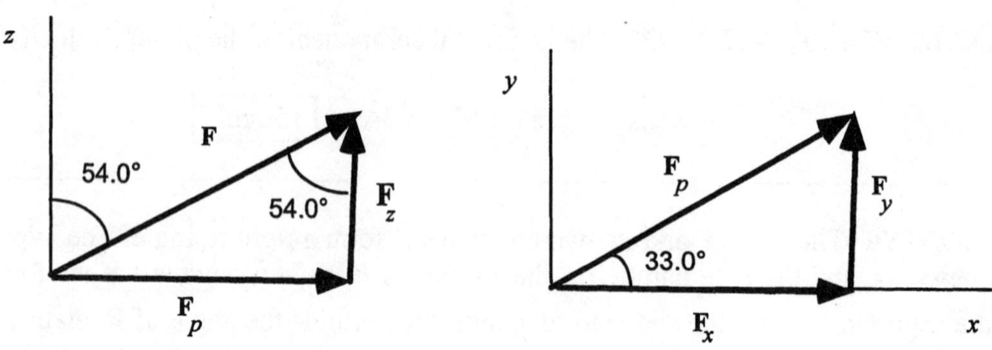

a. From the figure on the right,

$$F_x = F_p \cos 33.0° = (384\text{ N}) \cos 33.0° = \boxed{322\text{ N}}$$

b. Also from the figure on the right,

$$F_y = F_p \sin 33.0° = (384\text{ N}) \sin 33.0° = \boxed{209\text{ N}}$$

c. From the figure on the left,

$$F_z = F \cos 54.0° = (475\text{ N}) \cos 54.0° = \boxed{279\text{ N}}$$

---

40. ***REASONING*** Using the component method for vector addition, we will find the x component of the resultant force vector by adding the x components of the individual vectors. Then we will find the y component of the resultant vector by adding the y components of the individual vectors. Once the x and y components of the resultant are known, we will use the Pythagorean theorem to find the magnitude of the resultant and trigonometry to find its direction. We will take east as the $+x$ direction and north as the $+y$ direction.

SOLUTION The x component of the resultant force **F** is

$$F_x = \underbrace{(2240\text{ N})\cos 34.0°}_{F_{Ax}} + \underbrace{(3160\text{ N})\cos 90.0°}_{F_{Bx}} = (2240\text{ N})\cos 34.0°$$

The y component of the resultant force **F** is

$$F_y = \underbrace{-(2240\text{ N})\sin 34.0°}_{F_{Ay}} + \underbrace{(-3160\text{ N})}_{F_{By}}$$

Using the Pythagorean theorem, we find that the magnitude of the resultant force is

$$F = \sqrt{F_x^2 + F_y^2} = \sqrt{[(2240\text{ N})\cos 34.0°]^2 + [-(2240\text{ N})\sin 34.0° - 3160\text{ N}]^2} = \boxed{4790\text{ N}}$$

Using trigonometry, we find that the direction of the resultant force is

$$\theta = \tan^{-1}\left[\frac{(2240\text{ N})\sin 34.0° + 3160\text{ N}}{(2240\text{ N})\cos 34.0°}\right] = \boxed{67.2°\text{ south of east}}$$

---

41. SSM ***REASONING*** The individual displacements of the golf ball, **A**, **B**, and **C** are shown in the figure. Their resultant, **R**, is the displacement that would have been needed to "hole the ball" on the very first putt. We will use the component method to find **R**.

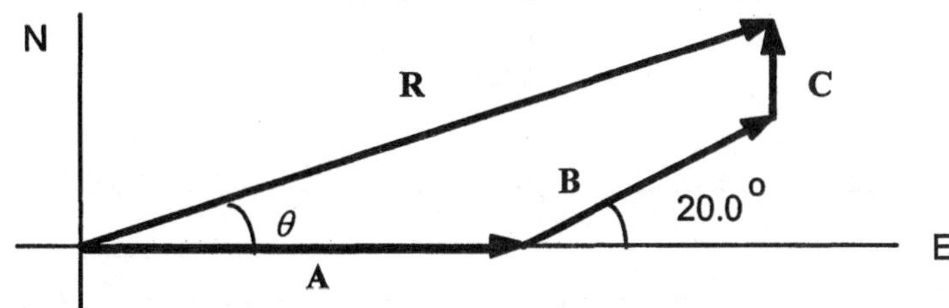

SOLUTION The components of each displacement vector are given in the table below.

Vector	***x* Components**	***y* Components**
A	(5.0 m) cos 0° = 5.0 m	(5.0 m) sin 0° = 0
B	(2.1 m) cos 20.0° = 2.0 m	(2.1 m) sin 20.0° = 0.72 m
C	(0.50 m) cos 90.0° = 0	(0.50 m) sin 90.0° = 0.50 m
R = **A** + **B** + **C**	7.0 m	1.22 m

The resultant vector **R** has magnitude

$$R = \sqrt{(7.0\text{ m})^2 + (1.22\text{ m})^2} = \boxed{7.1\text{ m}}$$

and the angle θ is

$$\theta = \tan^{-1}\left(\frac{1.22\text{ m}}{7.0\text{ m}}\right) = 9.9°$$

Thus, the required direction is $\boxed{9.9° \text{ north of east}}$.

---

42. ***REASONING AND SOLUTION*** The first three rows of the following table give the components of each of the three individual displacements. The fourth row gives the components of the resultant displacement. The directions due east and due north have been taken as the positive directions.

Displacement	*East/West Component*	*North/South Component*
A	–52 paces	0
B	–(42 paces) cos 30.0° = –36 paces	(42 paces) sin 30.0° = 21 paces
C	0	25 paces
R = **A** + **B** + **C**	–88 paces	46 paces

a. Therefore, the magnitude of the displacement in the direction due north is $\boxed{46 \text{ paces}}$.

b. Similarly, the magnitude of the displacement in the direction due west is $\boxed{88 \text{ paces}}$.

---

43. ***REASONING AND SOLUTION*** The x-components of the vectors are

$$+A_x = A\ \cos(180° - 20.0°) = (5.00\text{ m})\cos 160° = -4.70\text{ m}$$

$$+B_x = B\ \cos(60.0°) = (5.00\text{ m})\cos 60.0° = 2.50\text{ m}$$
$$+C_x = C\ \cos(270°) = 0.00\text{ m}$$

The x-component of the resultant vector is then

$$+R_x = -4.70\text{ m} + 2.50\text{ m} + 0.00\text{ m} = -2.20\text{ m}$$

Similarly, the y-components of the vectors are

$$+A_y = A\ \sin 160° = 1.71\text{ m}$$
$$+B_y = B\ \sin 60.0° = 4.33\text{ m}$$
$$+C_y = C\ \sin 270° = -4.00\text{ m}$$

The y-component of the resultant vector is then

$$R_y = 1.71\text{ m} + 4.33\text{ m} - 4.00\text{ m} = 2.04\text{ m}$$

Therefore,

$$R = \sqrt{R_x^2 + R_y^2} = \sqrt{(-2.20\text{ m})^2 + (2.04\text{ m})^2} = \boxed{3.00\text{ m}}$$

$$\theta = \tan^{-1}(R_y/R_x) = \boxed{42.8°\text{ above the negative } x \text{ axis}}$$

---

44. ***REASONING*** The magnitude of the resultant force in part (b) of the drawing is given by

$$F_A + F_B\cos 20.0° + F_C\cos 20.0° = F_A + 2F\cos 20.0°$$

where, since $F_B \sin 20.0° = -F_C \sin 20.0°$, the y components add to zero. As stated in the problem, the magnitude of the resultant force acting on the elephant in part (b) of the drawing is twice that in part (a); therefore,

$$F_A + 2F\cos 20.0° = 2F_A \qquad \text{or} \qquad 2F\cos 20.0° = F_A$$

SOLUTION Solving for the ratio F/F_A, we have

$$\frac{F}{F_A} = \frac{1}{2(\cos 20.0°)} = \boxed{0.532}$$

---

45. ***REASONING*** If we let the directions due east and due north be the positive directions, then the desired displacement **A** has components

$$A_E = (4.8 \text{ km}) \cos 42° = 3.57 \text{ km}$$

$$A_N = (4.8 \text{ km}) \sin 42° = 3.21 \text{ km}$$

while the actual displacement **B** has components

$$B_E = (2.4 \text{ km}) \cos 22° = 2.23 \text{ km}$$

$$B_N = (2.4 \text{ km}) \sin 22° = 0.90 \text{ km}$$

Therefore, to reach the research station, the research team must go

$$3.57 \text{ km} - 2.23 \text{ km} = 1.34 \text{ km, eastward}$$

and

$$3.21 \text{ km} - 0.90 \text{ km} = 2.31 \text{ km, northward}$$

SOLUTION
a. From the Pythagorean theorem, we find that the magnitude of the displacement vector required to bring the team to the research station is

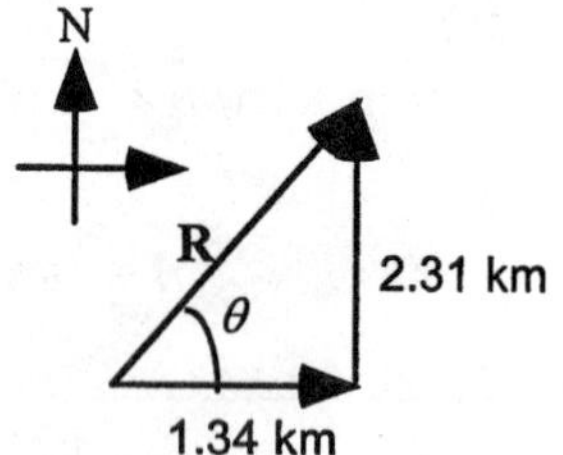

$$R = \sqrt{(1.34 \text{ km})^2 + (2.31 \text{ km})^2} = \boxed{2.7 \text{ km}}$$

b. The angle θ is given by

$$\theta = \tan^{-1}\left(\frac{2.31 \text{ km}}{1.34 \text{ km}}\right) = \boxed{6.0 \times 10^1 \text{ degrees, north of east}}$$

---

46. ***REASONING*** We are given that the vector sum of the three forces is zero, so $\mathbf{F_1} + \mathbf{F_2} + \mathbf{F_3} = 0$ N. Since $\mathbf{F_1}$ and $\mathbf{F_2}$ are known, $\mathbf{F_3}$ can be found from the relation $\mathbf{F_3} = -(\mathbf{F_1} + \mathbf{F_2})$. We will use the x- and y-components of this equation to find the magnitude and direction of $\mathbf{F_3}$.

SOLUTION The x- and y-components of the equation $\mathbf{F_3} = -(\mathbf{F_1} + \mathbf{F_2})$ are:

x-component $$F_{3x} = -(F_{1x} + F_{2x}) \qquad (1)$$

y-component $$F_{3y} = -(F_{1y} + F_{2y}) \qquad (2)$$

The table below gives the x- and y-components of $\mathbf{F_1}$ and $\mathbf{F_2}$:

Vector	*x* component	*y* component
$\mathbf{F}_1$	$F_{1x} = -(21.0\text{ N})\sin 30.0° = -10.5\text{ N}$	$F_{1y} = -(21.0\text{ N})\cos 30.0° = +18.2\text{ N}$
$\mathbf{F}_2$	$F_{2x} = +15.0\text{ N}$	$F_{2y} = 0\text{ N}$

Substituting the values for F_{1x} and F_{2x} into Equation (1) gives

$$F_{3x} = -\left(F_{1x} + F_{2x}\right) = -\left(-10.5\text{ N} + 15.2\text{ N}\right) = -4.5\text{ N}$$

Substituting F_{1y} and F_{2y} into Equation (2) gives

$$F_{3y} = -\left(F_{1y} + F_{2y}\right) = -\left(+18.2\text{ N} + 0\text{ N}\right) = -18.2\text{ N}$$

The magnitude of $\mathbf{F}_3$ can now be obtained by employing the Pythagorean theorem:

$$F_3 = \sqrt{F_{3x}^2 + F_{3y}^2} = \sqrt{(-4.5\text{ N})^2 + (-18.2\text{ N})^2} = \boxed{18.7\text{ N}}$$

The angle θ that $\mathbf{F}_3$ makes with respect to the $-x$ axis can be determined from the inverse tangent function (Equation 1.6),

$$\theta = \tan^{-1}\left(\frac{F_{3y}}{F_{3x}}\right) = \tan^{-1}\left(\frac{-18.2\text{ N}}{-4.5\text{ N}}\right) = \boxed{76°}$$

---

47. SSM ***REASONING AND SOLUTION*** We take due north to be the direction of the $+y$ axis. Vectors **A** and **B** are the components of the resultant, **C**. The angle that **C** makes with the x axis is then $\theta = \tan^{-1}(B/A)$. The symbol u denotes the units of the vectors.

a. Solving for B gives

$$B = A\tan\theta = (6.00\text{ u})\tan 60.0° = \boxed{10.4\text{ u}}$$

b. The magnitude of **C** is

$$C = \sqrt{A^2 + B^2} = \sqrt{(6.00\text{ u})^2 + (10.4\text{ u})^2} = \boxed{12.0\text{ u}}$$

---

48. ***REASONING*** We know that the three displacement vectors have a resultant of zero, so that $\mathbf{A} + \mathbf{B} + \mathbf{C} = \mathbf{0}$. This means that the sum of the x components of the vectors and the sum of the y components of the vectors are separately equal to zero. From these two equations we

will be able to determine the magnitudes of vectors **B** and **C**. The directions east and north are, respectively, the $+x$ and $+y$ directions.

SOLUTION Setting the sum of the x components of the vectors and the sum of the y components of the vectors separately equal to zero, we have

$$\underbrace{(1550\text{ m})\cos 25.0°}_{A_x} + \underbrace{B\sin 41.0°}_{B_x} + \underbrace{(-C\cos 35.0°)}_{C_x} = 0$$

$$\underbrace{(1550\text{ m})\sin 25.0°}_{A_y} + \underbrace{(-B\cos 41.0°)}_{B_y} + \underbrace{C\sin 35.0°}_{C_y} = 0$$

These two equations contain two unknown variables, B and C. They can be solved simultaneously to show that

a. $B = \boxed{5550\text{ m}}$ and b. $C = \boxed{6160\text{ m}}$

---

49. ***REASONING*** The drawing shows the vectors **A**, **B**, and **C**. Since these vectors add to give a resultant that is zero, we can write that **A** + **B** + **C** = 0. This addition will be carried out by the component method. This means that the x-component of this equation must be zero ($A_x + B_x + C_x = 0$) and the y-component must be zero ($A_y + B_y + C_y = 0$). These two equations will allow us to find the magnitudes of **B** and **C**.

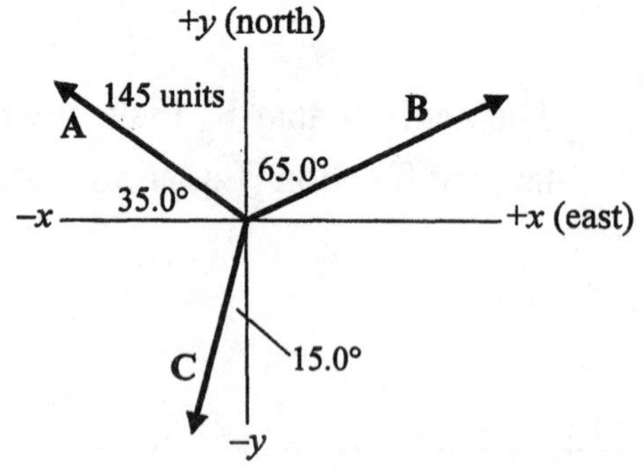

SOLUTION The x- and y-components of **A**, **B**, and **C** are given in the table below. The plus and minus signs indicate whether the components point along the positive or negative axes.

Vector	*x Component*	*y Component*
A	−(145 units) cos 35.0° = −119 units	+(145 units) sin 35.0° = +83.2 units
B	$+B$ sin 65.0° = $+B$ (0.906)	$+B$ cos 65.0° = $+B$ (0.423)
C	$-C$ sin 15.0° = $-C$ (0.259)	$-C$ cos 15.0° = $-C$ (0.966)
A + B + C	−119 units + B (0.906) − C (0.259)	+83.2 units + B (0.423) − C (0.966)

Setting the separate x- and y- components of **A + B + C** equal to zero gives

x-component $(-119\text{ units}) + B\,(0.906) - C\,(0.259) = 0$

y-component $(+83.2\text{ units}) + B\,(0.423) - C\,(0.966) = 0$

Solving these two equations simultaneously, we find that

a. $B =$ $\boxed{178 \text{ units}}$ b. $C =$ $\boxed{164 \text{ units}}$

---

50. ***REASONING*** The following table shows the components of the individual displacements and the components of the resultant. The directions due east and due north are taken as the positive directions.

Displacement	*East/West Component*	*North/South Component*
(1)	–27.0 cm	0
(2)	$-(23.0 \text{ cm}) \cos 35.0° = -18.84 \text{ cm}$	$-(23.0 \text{ cm}) \sin 35.0° = -13.19 \text{ cm}$
(3)	$(28.0 \text{ cm}) \cos 55.0° = 16.06 \text{ cm}$	$-(28.0 \text{ cm}) \sin 55.0° = -22.94 \text{ cm}$
(4)	$(35.0 \text{ cm}) \cos 63.0° = 15.89 \text{ cm}$	$(35.0 \text{ cm}) \sin 63.0° = 31.19 \text{ cm}$
Resultant	–13.89 cm	–4.94 cm

SOLUTION

a. From the Pythagorean theorem, we find that the magnitude of the resultant displacement vector is

$$R = \sqrt{(13.89 \text{ cm})^2 + (4.94 \text{ cm})^2} = \boxed{14.7 \text{ cm}}$$

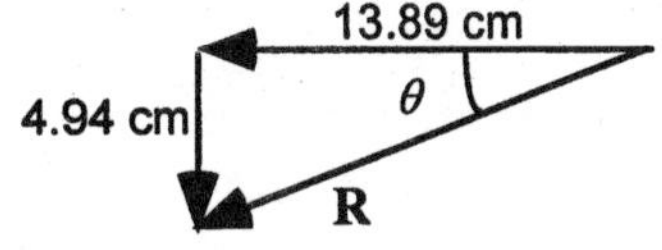

b. The angle θ is given by

$$\theta = \tan^{-1}\left(\frac{4.94 \text{ cm}}{13.89 \text{ cm}}\right) = \boxed{19.6°, \text{ south of west}}$$

---

51. ***REASONING*** The shortest distance between the tree and the termite mound is equal to the magnitude of the chimpanzee's displacement **r**.

SOLUTION

a. From the Pythagorean theorem, we have

$$r = \sqrt{(51 \text{ m})^2 + (39 \text{ m})^2} = \boxed{64 \text{ m}}$$

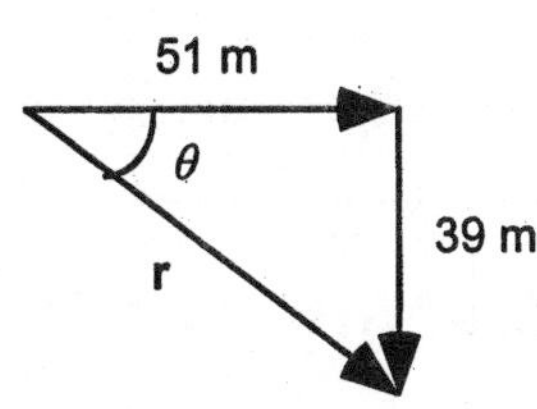

b. The angle θ is given by

$$\theta = \tan^{-1}\left(\frac{39 \text{ m}}{51 \text{ m}}\right) = \boxed{37° \text{ south of east}}$$

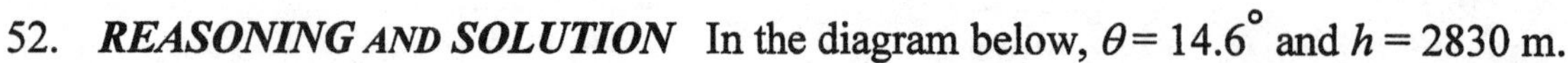

52. ***REASONING AND SOLUTION*** In the diagram below, $\theta = 14.6^\circ$ and $h = 2830$ m.

We know that $\sin\theta = H/h$ and, therefore,

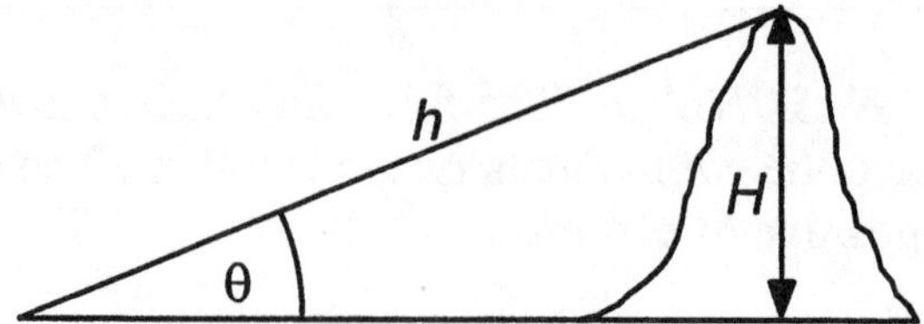

$$H = h\sin\theta$$

$$= (2830\text{ m})\sin 14.6^\circ = \boxed{713\text{ m}}$$

---

53. **SSM** ***REASONING AND SOLUTION*** In order to determine which vector has the largest *x* and *y* components, we calculate the magnitude of the *x* and *y* components explicitly and compare them. In the calculations, the symbol u denotes the units of the vectors.

$A_x = (100.0\text{ u})\cos 90.0^\circ = 0.00\text{ u}$ $\quad$ $A_y = (100.0\text{ u})\sin 90.0^\circ = 1.00\times 10^2\text{ u}$

$B_x = (200.0\text{ u})\cos 60.0^\circ = 1.00\times 10^2\text{ u}$ $\quad$ $B_y = (200.0\text{ u})\sin 60.0^\circ = 173\text{ u}$

$C_x = (150.0\text{ u})\cos 0.00^\circ = 150.0\text{ u}$ $\quad$ $C_y = (150.0\text{ u})\sin 0.00^\circ = 0.00\text{ u}$

a. **C** has the largest *x* component.

b. **B** has the largest *y* component.

---

54. ***REASONING AND SOLUTION*** Since $v = \frac{1}{3}zxt^2$, it follows that $z = \dfrac{3v}{xt^2}$. Since $x = [\text{L}]$, $v = [\text{L}]/[\text{T}]$, and $t = [\text{T}]$, we have

$$\frac{v}{xt^2} = \frac{[\text{L}]/[\text{T}]}{[\text{L}][\text{T}]^2} = \boxed{\frac{1}{[\text{T}]^3}}$$

Note that the numerical factor "3" does not contribute to the dimensions.

---

55. ***REASONING AND SOLUTION*** The horizontal component of the resultant vector is

$$R_\text{h} = A_\text{h} + B_\text{h} + C_\text{h} = (0.00\text{ m}) + (15.0\text{ m}) + (18.0\text{ m})\cos 35.0^\circ = 29.7\text{ m}$$

Similarly, the vertical component is

$$R_V = A_V + B_V + C_V = (5.00 \text{ m}) + (0.00 \text{ m}) + (-18.0 \text{ m}) \sin 35.0° = -5.32 \text{ m}$$

The magnitude of the resultant vector is

$$R = \sqrt{R_h^2 + R_v^2} = \sqrt{(29.7 \text{ m})^2 + (-5.32 \text{ m})^2} = \boxed{30.2 \text{ m}}$$

The angle θ is obtained from

$$\theta = \tan^{-1}[(5.32 \text{ m})/(29.7 \text{ m})] = \boxed{10.2°}$$

---

56. ***REASONING*** The performer walks out on the wire a distance *d*, and the vertical distance to the net is *h*. Since these two distances are perpendicular, the magnitude of the displacement is given by the Pythagorean theorem as $s = \sqrt{d^2 + h^2}$. Values for *s* and *h* are given, so we can solve this expression for the distance *d*. The angle that the performer's displacement makes below the horizontal can be found using trigonometry.

SOLUTION

a. Using the Pythagorean theorem, we find that

$$s = \sqrt{d^2 + h^2} \quad \text{or} \quad d = \sqrt{s^2 - h^2} = \sqrt{(26.7 \text{ ft})^2 - (25.0 \text{ ft})^2} = \boxed{9.4 \text{ ft}}$$

b. The angle θ that the performer's displacement makes below the horizontal is given by

$$\tan\theta = \frac{h}{d} \quad \text{or} \quad \theta = \tan^{-1}\left(\frac{h}{d}\right) = \tan^{-1}\left(\frac{25.0 \text{ ft}}{9.4 \text{ ft}}\right) = \boxed{69°}$$

---

57. SSM ***REASONING*** The *x* and *y* components of **r** are mutually perpendicular; therefore, the magnitude of **r** can be found using the Pythagorean theorem. The direction of **r** can be found using the definition of the tangent function.

SOLUTION According to the Pythagorean theorem, we have

$$r = \sqrt{x^2 + y^2} = \sqrt{(-125 \text{ m})^2 + (-184 \text{ m})^2} = \boxed{222 \text{ m}}$$

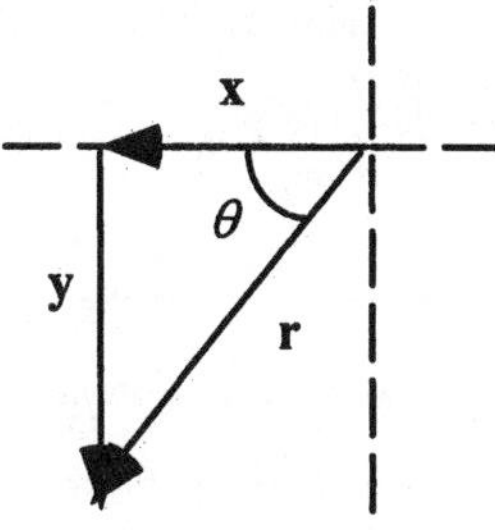

The angle θ is

$$\theta = \tan^{-1}\left(\frac{184 \text{ m}}{125 \text{ m}}\right) = \boxed{55.8°}$$

---

58. ***REASONING AND SOLUTION***

a. $A_x = A \cos 30.0° = (750 \text{ units})(0.866) = \boxed{650 \text{ units}}$

$A_y = A \sin 30.0° = (750 \text{ units})(0.500) = \boxed{380 \text{ units}}$

b. $A_{x'} = A \cos 40.0° = (750 \text{ units})(0.643) = \boxed{570 \text{ units}}$

$A_{y'} = -A \sin 40.0° = -(750 \text{ units})(0.766) = \boxed{-480 \text{ units}}$

---

59. SSM ***REASONING*** Since the finish line is coincident with the starting line, the net displacement of the sailboat is zero. Hence the sum of the components of the displacement vectors of the individual legs must be zero. In the drawing in the text, the directions to the right and upward are taken as positive.

SOLUTION In the horizontal direction $R_h = A_h + B_h + C_h + D_h = 0$

$$R_h = (3.20 \text{ km}) \cos 40.0° - (5.10 \text{ km}) \cos 35.0° - (4.80 \text{ km}) \cos 23.0° + D \cos \theta = 0$$

$$D \cos \theta = 6.14 \text{ km}. \qquad (1)$$

In the vertical direction $R_v = A_v + B_v + C_v + D_v = 0$.

$$R_v = (3.20 \text{ km}) \sin 40.0° + (5.10 \text{ km}) \sin 35.0° - (4.80 \text{ km}) \sin 23.0° - D \sin \theta = 0.$$

$$D \sin \theta = 3.11 \text{ km} \qquad (2)$$

Dividing (2) by (1) gives

$$\tan \theta = (3.11 \text{ km})/(6.14 \text{ km}) \quad \text{or} \quad \theta = \boxed{26.9°}$$

Solving (1) gives

$$D = (6.14 \text{ km})/\cos 26.9° = \boxed{6.88 \text{ km}}$$

---

60. ***REASONING*** a. Since the two displacement vectors **A** and **B** have directions due south and due east, they are perpendicular. Therefore, the resultant vector **R** = **A** + **B** has a magnitude given by the Pythagorean theorem: $R^2 = A^2 + B^2$. Knowing the magnitudes of **R** and **A**, we can calculate the magnitude of **B**. The direction of the resultant can be obtained using trigonometry.

b. For the vector **R′** = **A** – **B** we note that the subtraction can be regarded as an addition in the following sense: **R′** = **A** + (–**B**). The vector –**B** points due west, opposite the vector **B**, so the two vectors are once again perpendicular and the magnitude of **R′** again is given by the Pythagorean theorem. The direction again can be obtained using trigonometry.

SOLUTION a. The drawing shows the two vectors and the resultant vector. According to the Pythagorean theorem, we have

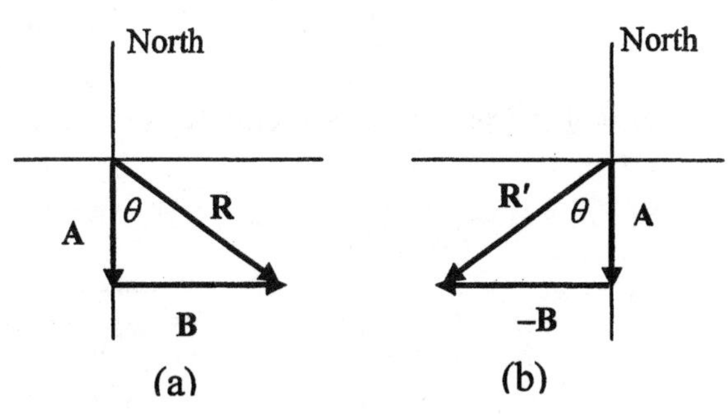

$$R^2 = A^2 + B^2$$

$$B = \sqrt{R^2 - A^2}$$

$$B = \sqrt{(3.75 \text{ km})^2 - (2.50 \text{ km})^2}$$

$$= \boxed{2.8 \text{ km}}$$

Using trigonometry, we can see that the direction of the resultant is

$$\tan\theta = \frac{B}{A} \quad \text{or} \quad \theta = \tan^{-1}\left(\frac{2.8 \text{ km}}{2.50 \text{ km}}\right) = \boxed{48° \text{ east of south}}$$

b. Referring to the drawing and following the same procedure as in part a, we find

$$R'^2 = A^2 + (-B)^2 \quad \text{or} \quad B = \sqrt{R'^2 - A^2} = \sqrt{(3.75 \text{ km})^2 - (2.50 \text{ km})^2} = \boxed{2.8 \text{ km}}$$

$$\tan\theta = \frac{B}{A} \quad \text{or} \quad \theta = \tan^{-1}\left(\frac{2.8 \text{ km}}{2.50\text{km}}\right) = \boxed{48° \text{ west of south}}$$

---

61. ***REASONING*** The drawing at the right shows the location of each deer A, B, and C. From the problem statement it follows that

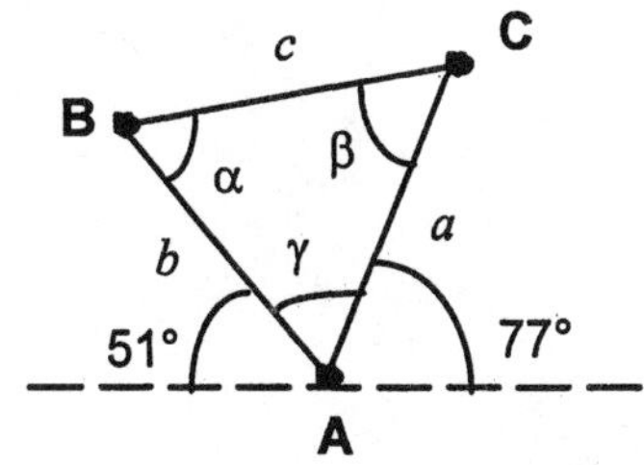

$$b = 62 \text{ m}$$

$$c = 95 \text{ m}$$

$$\gamma = 180° - 51° - 77° = 52°$$

Applying the law of cosines (given in Appendix E) to the geometry in the figure, we have

$$a^2 - 2ab\cos\gamma + (b^2 - c^2) = 0$$

which is an expression that is quadratic in *a*. It can be simplified to $Aa^2 + Ba + C = 0$, with

$$A = 1$$

$$B = -2b \cos \gamma = -2(62 \text{ m}) \cos 52° = -76 \text{ m}$$

$$C = (b^2 - c^2) = (62 \text{ m})^2 - (95 \text{ m})^2 = -5181 \text{ m}^2$$

This quadratic equation can be solved for the desired quantity *a*.

SOLUTION Suppressing units, we obtain from the quadratic formula

$$a = \frac{-(-76) \pm \sqrt{(-76)^2 - 4(1)(-5181)}}{2(1)} = 1.2 \times 10^2 \text{ m and } -43 \text{ m}$$

Discarding the negative root, which has no physical significance, we conclude that the distance between deer A and C is $\boxed{1.2 \times 10^2 \text{ m}}$.

---

62. ***REASONING AND SOLUTION*** The figures below are scale diagrams of the forces drawn tail-to-head. The scale factor is shown in the figure.

a. From the figure on the left, we see that $\boxed{F_A - F_B = 142 \text{ N}, \ \theta = 67° \text{ south of east}}$.

b. Similarly, from the figure on the right, $\boxed{F_B - F_A = 142 \text{ N}, \ \theta = 67° \text{ north of west}}$.

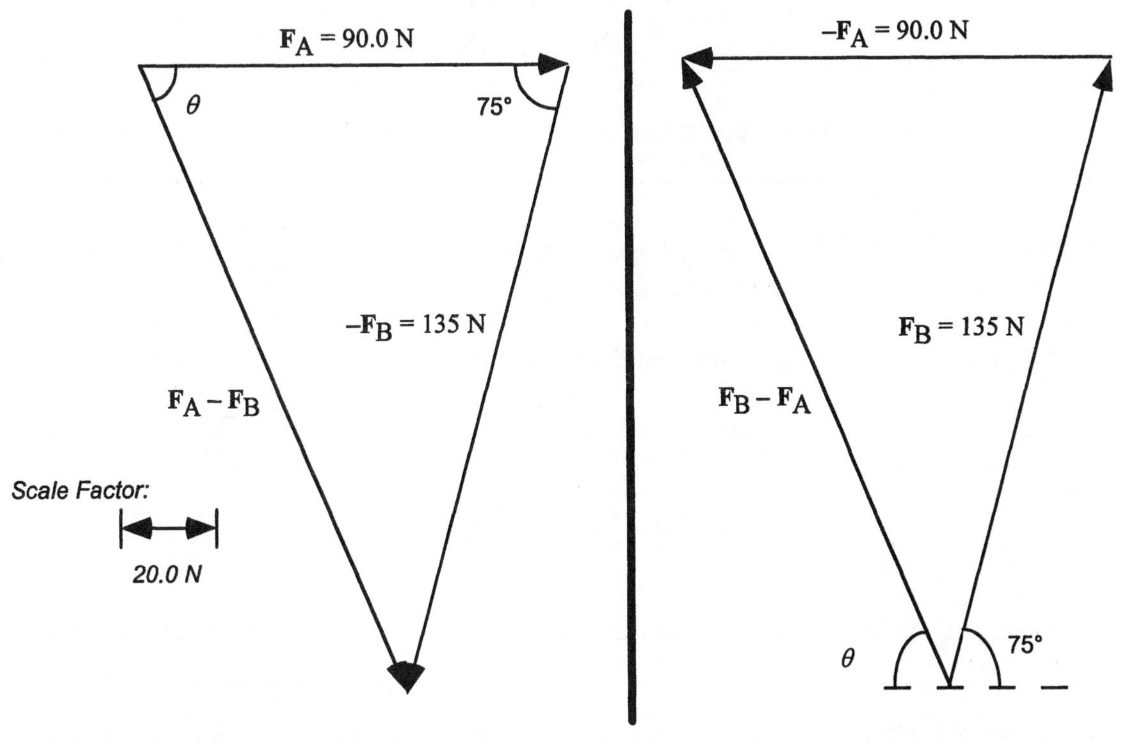

63. ***REASONING AND SOLUTION*** If **D** is the unknown vector, then **A** + **B** + **C** + **D** = 0 requires that $D_E = -(A_E + B_E + C_E)$ or

$$D_E = (113\text{ u})\cos 60.0° - (222\text{ u})\cos 35.0° - (177\text{ u})\cos 23.0° = \boxed{-288\text{ units}}$$

The minus sign indicates that D_E has a direction of due west.

Also, $D_N = -(A_N + B_N + C_N)$ or

$$D_N = (113\text{ u})\sin 60.0° + (222\text{ u})\sin 35.0° - (177\text{ u})\sin 23.0° = \boxed{156\text{ units}}$$

---

64. ***CONCEPT QUESTIONS*** a. Since there are 3.28 ft in 1 m, the larger unit for measuring area is 1 m^2.

b. The apartment has less than 1330 m^2 of area, because the same physical area measured in larger units will contain a smaller number of those units.

SOLUTION We use the fact that 1 m = 3.28 ft to form the following conversion factor: (1 m)/(3.28 ft) =1. To convert ft^2 into m^2, we apply this conversion factor twice:

$$\text{area} = (1330 \text{ ft}^2)\left(\frac{1 \text{ m}}{3.28 \text{ ft}}\right)\left(\frac{1 \text{ m}}{3.28 \text{ ft}}\right) = \boxed{124 \text{ m}^2}$$

As expected, the area of the apartment is less than 1330 m^2.

---

65. ***CONCEPT QUESTION*** The Pythagorean states that $D = \sqrt{L^2 + L^2} = \sqrt{2}L$. Therefore, D is larger than L, or L is smaller than D.

SOLUTION Using the Pythagorean theorem, we find

$$D = \sqrt{L^2 + L^2} = \sqrt{2}L$$

$$L = \frac{D}{\sqrt{2}} = \frac{0.35 \text{ m}}{\sqrt{2}} = \boxed{0.25 \text{ m}}$$

As expected, the length of the side of the square is less than the diameter of the circle.

---

66. ***CONCEPT QUESTION*** The magnitude of either component is given by the magnitude of the vector times the sine function or the cosine function. Since the sine and cosine functions are both less than or equal to one, the magnitudes of the components can never be greater than the magnitude of the vector itself.

SOLUTION The x and y scalar components are

a. $F_x = (575 \text{ newtons}) \cos 36.0° = \boxed{465 \text{ newtons}}$

b. $F_y = -(575 \text{ newtons}) \sin 36.0° = \boxed{-338 \text{ newtons}}$

As expected, the magnitudes of the components are less than the magnitude of the vector itself.

---

67. ***CONCEPT QUESTIONS*** a. Increasing the component A_x, while holding the component A_y constant, causes the vector **A** to rotate toward the $+x$ axis, thus decreasing the angle θ.

b. Increasing the component A_y, while holding the component A_x constant, causes the vector **A** to rotate toward the $+y$ axis, thus increasing the angle θ.

SOLUTION Using trigonometry, we can determine the angle θ from the relation $\tan \theta = A_y/A_x$:

a. $$\theta = \tan^{-1}\left(\frac{A_y}{A_x}\right) = \tan^{-1}\left(\frac{12\text{ m}}{12\text{ m}}\right) = \boxed{45°}$$

b. $$\theta = \tan^{-1}\left(\frac{A_y}{A_x}\right) = \tan^{-1}\left(\frac{12\text{ m}}{17\text{ m}}\right) = \boxed{35°}$$

c. $$\theta = \tan^{-1}\left(\frac{A_y}{A_x}\right) = \tan^{-1}\left(\frac{17\text{ m}}{12\text{ m}}\right) = \boxed{55°}$$

As expected, θ decreases with increasing A_x and increases with increasing A_y.

---

68. ***CONCEPT QUESTIONS*** a. Arranging the vectors in tail-to-head fashion, we can see that the vector **A** gives the resultant a westerly direction and vector **B** gives the resultant a southerly direction. Therefore, the resultant **A** + **B** points south of west.

b. Arranging the vectors in tail-to-head fashion, we can see that the vector **A** gives the resultant a westerly direction and vector **–B** gives the resultant a northerly direction. Therefore, the resultant **A** + (**–B)** points north of west.

SOLUTION Using the Pythagorean theorem to find the magnitude of the resultant and trigonometry to find the direction, we obtain the following results:

a. magnitude of $\mathbf{A}+\mathbf{B} = \sqrt{(63\text{ units})^2 + (63\text{ units})^2} = \boxed{89\text{ units}}$

$$\theta = \tan^{-1}\left(\frac{63\text{ units}}{63\text{ units}}\right) = \boxed{45°\text{ south of west}}$$

b. magnitude of $\mathbf{A}-\mathbf{B} = \sqrt{(63\text{ units})^2 + (63\text{ units})^2} = \boxed{89\text{ units}}$

$$\theta = \tan^{-1}\left(\frac{63\text{ units}}{63\text{ units}}\right) = \boxed{45°\text{ north of west}}$$

As expected, the direction of A + B south of west, while the direction of A – B is north of west.

---

69. ***CONCEPT QUESTIONS***

a. The magnitude of **R** can be seen from the drawing to be smaller than the magnitude of **A**, that is, smaller than 244 km.

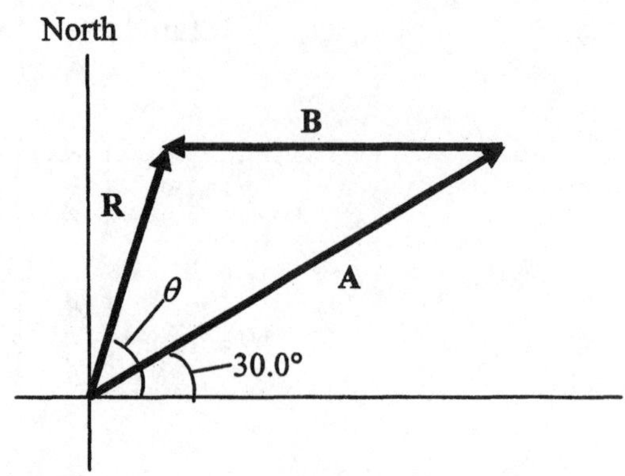

b. The magnitude of **R** can be seen from the drawing to be smaller than the magnitude of **B**, that is, smaller than 175 km.

c. The angle θ can be seen to be greater

SOLUTION According to the component method for vector addition, the x component of the resultant is the sum of the x component of **A** and the x component of **B**. Similarly, the y component of the resultant is the sum of the y component of **A** and the y component of **B**. The magnitude of the resultant can be obtained from the x and y components of the resultant by using the Pythagorean theorem. The directional angle can be obtained using trigonometry. We find the following results:

$$R_x = \underbrace{(244\text{ km})\cos 30.0°}_{A_x} + \underbrace{(-175\text{ km})}_{B_x} = 36\text{ km}$$

$$R_y = \underbrace{(244\text{ km})\sin 30.0°}_{A_y} + \underbrace{(0\text{ km})}_{B_y} = 122\text{ km}$$

$$R = \sqrt{R_x^2 + R_y^2} = \sqrt{(36\text{ km})^2 + (122\text{ km})^2} = \boxed{127\text{ km}}$$

$$\theta = \tan^{-1}\left(\frac{R_y}{R_x}\right) = \tan^{-1}\left(\frac{122\text{ km}}{36\text{ km}}\right) = \boxed{74°}$$

As expected, the magnitude of the resultant is less than 175 km, and the directional angle is greater than 30.0°.

CHAPTER 2 | *KINEMATICS IN ONE DIMENSION*

PROBLEMS

---

1. SSM ***REASONING AND SOLUTION*** The moving plane can be seen through the window of the stationary plane for the time that it takes the moving plane to travel its own length. To find the time, we write Equation 2.1 as

$$\text{Elapsed time} = \frac{\text{Distance}}{\text{Average speed}} = \frac{36 \text{ m}}{45 \text{ m/s}} = \boxed{0.80 \text{ s}}$$

---

2. ***REASONING AND SOLUTION***

a. The distance traveled is equal to three-fourths of the circumference of the circular lake.

$$\frac{3}{4} \times \text{circumference of lake} = \frac{3}{4}(2\pi r) = \frac{3}{2}\pi(1.50 \text{ km}) = \boxed{7.07 \text{ km}}$$

b. The displacement R of the couple is shown in the figure at the right. The displacement forms the hypotenuse of a right triangle with sides equal to the radius of the circle. From the Pythagorean theorem,

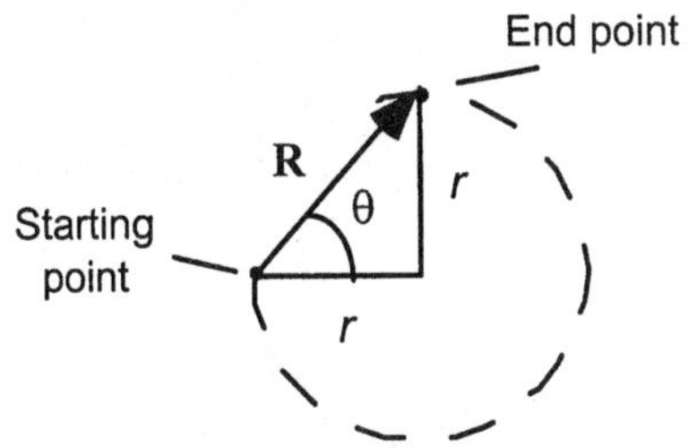

$$R = \sqrt{r^2 + r^2} = \sqrt{2r^2} = \sqrt{2(1.50 \text{ km})^2} = \boxed{2.12 \text{ km}}$$

The angle θ that R makes with due east is

$$\theta = \tan^{-1}\left(\frac{r}{r}\right) = \tan^{-1}(1) = \boxed{45.0° \text{ north of east}}$$

---

3. SSM ***REASONING AND SOLUTION***

a. The total distance traveled is found by adding the distances traveled during each segment of the trip.

$$6.9 \text{ km} + 1.8 \text{ km} + 3.7 \text{ km} = \boxed{12.4 \text{ km}}$$

b. All three segments of the trip lie along the east-west line. Taking east as the positive direction, the individual displacements can then be added to yield the resultant displacement.

$$6.9 \text{ km} + (-1.8 \text{ km}) + 3.7 \text{ km} = +8.8 \text{ km}$$

The displacement is positive, indicating that it points due east. Therefore,

$$\text{Displacement of the whale} = \boxed{8.8 \text{ km, due east}}$$

---

4. ***REASONING*** The distance traveled by the Space Shuttle is equal to its speed multiplied by the time. The number of football fields is equal to this distance divided by the length L of one football field.

SOLUTION The number of football fields is

$$\text{Number} = \frac{x}{L} = \frac{vt}{L} = \frac{\left(7.6\times10^{3}\text{ m/s}\right)\left(110\times10^{-3}\text{ s}\right)}{91.4\text{ m}} = \boxed{9.1}$$

---

5. ***REASONING*** Equation 2.1 gives the average speed as the distance traveled divided by the elapsed time. The distance is the circumference of the circular path, $2\pi r$, where r is the radius of the earth

SOLUTION
a. The average speed (in m/s) of the person standing on the equator is

$$\text{Average speed} = \frac{\text{Distance}}{\text{Elapsed time}} = \frac{2\pi r}{t} = \frac{2\pi\left(6.38\times10^{6}\text{ m}\right)}{\left(24\text{ h}\right)\left(\dfrac{3600\text{ s}}{1\text{ h}}\right)} = \boxed{464\text{ m/s}}$$

b. Since 1 mi/h = 0.447 m/s, the average speed (in mi/h) is

$$\text{Average speed} = \left(464\text{ m/s}\right)\left(\frac{1\text{ mi/h}}{0.447\text{ m/s}}\right) = \boxed{1040\text{ mi/h}}$$

---

6. ***REASONING*** In a race against el-Guerrouj, Bannister would run a distance given by his average speed times the time duration of the race (see Equation 2.1). The time duration of the race would be el-Guerrouj's winning time of 3:43.13 (223.13 s). The difference between Bannister's distance and the length of the race is el-Guerrouj's winning margin.

SOLUTION From the table of conversion factors on the page facing the front cover, we find that one mile corresponds to 1609 m. According to Equation 2.1, Bannister's average speed is

$$\text{Average speed} = \frac{\text{Distance}}{\text{Elapsed time}} = \frac{1609\text{ m}}{239.4\text{ s}}$$

Had he run against el-Guerrouj at this average speed for the 223.13-s duration of the race, he would have traveled a distance of

$$\text{Distance} = \text{Average speed} \times \text{time} = \left(\frac{1609\ \text{m}}{239.4\ \text{s}}\right)(223.13\ \text{s})$$

while el-Guerrouj traveled 1609 m. Thus, el-Guerrouj would have won by a distance of

$$1609\ \text{m} - \left(\frac{1609\ \text{m}}{239.4\ \text{s}}\right)(223.13\ \text{s}) = \boxed{109\ \text{m}}$$

---

7. ***REASONING*** In order for the bear to catch the tourist over the distance d, the bear must reach the car at the same time as the tourist. During the time t that it takes for the tourist to reach the car, the bear must travel a total distance of d + 26 m. From Equation 2.1,

$$v_{\text{tourist}} = \frac{d}{t} \quad (1) \qquad \text{and} \qquad v_{\text{bear}} = \frac{d + 26\ \text{m}}{t} \quad (2)$$

Equations (1) and (2) can be solved simultaneously to find d.

SOLUTION Solving Equation (1) for t and substituting into Equation (2), we find

$$v_{\text{bear}} = \frac{d + 26\ \text{m}}{d / v_{\text{tourist}}} = \frac{(d + 26\ \text{m})v_{\text{tourist}}}{d}$$

$$v_{\text{bear}} = \left(1 + \frac{26\ \text{m}}{d}\right)v_{\text{tourist}}$$

Solving for d yields:

$$d = \frac{26\ \text{m}}{\dfrac{v_{\text{bear}}}{v_{\text{tourist}}} - 1} = \frac{26\ \text{m}}{\dfrac{6.0\ \text{m/s}}{4.0\ \text{m/s}} - 1} = \boxed{52\ \text{m}}$$

---

8. ***REASONING AND SOLUTION*** Let west be the positive direction. The average velocity of the backpacker is

$$v = \frac{x_{\text{w}} + x_{\text{e}}}{t_{\text{w}} + t_{\text{e}}} \quad \text{where} \quad t_{\text{w}} = \frac{x_{\text{w}}}{v_{\text{w}}} \quad \text{and} \quad t_{\text{e}} = \frac{x_{\text{e}}}{v_{\text{e}}}$$

Combining these equations and solving for x_{e} (suppressing the units) gives

$$x_e = \frac{-(1 - v/v_w)x_w}{(1 - v/v_e)} = \frac{-[1 - (1.34 \text{ m/s})/(2.68 \text{ m/s})](6.44 \text{ km})}{1 - (1.34 \text{ m/s})/(0.447 \text{ m/s})} = -0.81 \text{ km}$$

The distance traveled is the magnitude of x_e, or $\boxed{0.81 \text{ km}}$.

---

9. SSM WWW ***REASONING*** Since the woman runs for a known distance at a known constant speed, we can find the time it takes for her to reach the water from Equation 2.1. We can then use Equation 2.1 to determine the total distance traveled by the dog in this time.

SOLUTION The time required for the woman to reach the water is

$$\text{Elapsed time} = \frac{d_{\text{woman}}}{v_{\text{woman}}} = \left(\frac{4.0 \text{ km}}{2.5 \text{ m/s}}\right)\left(\frac{1000 \text{ m}}{1.0 \text{ km}}\right) = 1600 \text{ s}$$

In 1600 s, the dog travels a total distance of

$$d_{\text{dog}} = v_{\text{dog}}t = (4.5 \text{ m/s})(1600 \text{ s}) = \boxed{7.2 \times 10^3 \text{ m}}$$

---

10. ***REASONING*** The definition of average velocity is given by Equation 2.2 as Average velocity = Displacement/(Elapsed time). The displacement in this expression is the total displacement, which is the sum of the displacements for each part of the trip. Displacement is a vector quantity, and we must be careful to account for the fact that the displacement in the first part of the trip is north, while the displacement in the second part is south.

SOLUTION According to Equation 2.2, the displacement for each part of the trip is the average velocity for that part times the corresponding elapsed time. Designating north as the positive direction, we find for the total displacement that

$$\text{Displacement} = \underbrace{(27 \text{ m/s})t_{\text{North}}}_{\text{Northward}} + \underbrace{(-17 \text{ m/s})t_{\text{South}}}_{\text{Southward}}$$

where t_{North} and t_{South} denote, respectively, the times for each part of the trip. Note that the minus sign indicates a direction due south. Noting that the total elapsed time is $t_{\text{North}} + t_{\text{South}}$, we can use Equation 2.2 to find the average velocity for the entire trip as follows:

$$\text{Average velocity} = \frac{\text{Displacement}}{\text{Elapsed time}} = \frac{(27\text{ m/s})t_{\text{North}} + (-17\text{ m/s})t_{\text{South}}}{t_{\text{North}} + t_{\text{South}}}$$

$$= (27\text{ m/s})\left(\frac{t_{\text{North}}}{t_{\text{North}} + t_{\text{South}}}\right) + (-17\text{ m/s})\left(\frac{t_{\text{South}}}{t_{\text{North}} + t_{\text{South}}}\right)$$

But $\left(\frac{t_{\text{North}}}{t_{\text{North}} + t_{\text{South}}}\right) = \frac{3}{4}$ and $\left(\frac{t_{\text{South}}}{t_{\text{North}} + t_{\text{South}}}\right) = \frac{1}{4}$. Therefore, we have that

$$\text{Average velocity} = (27\text{ m/s})\left(\frac{3}{4}\right) + (-17\text{ m/s})\left(\frac{1}{4}\right) = \boxed{+16\text{ m/s}}$$

The plus sign indicates that the average velocity for the entire trip points north.

---

11. ***REASONING AND SOLUTION*** The upper edge of the wall will disappear after the train has traveled the distance d in the figure below.

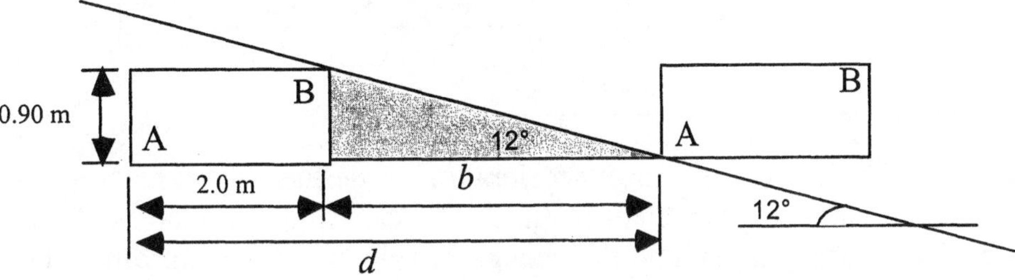

The distance d is equal to the length of the window plus the base of the 12° right triangle of height 0.90 m.

The base of the triangle is given by

$$b = \frac{0.90\text{ m}}{\tan 12°} = 4.2\text{ m}$$

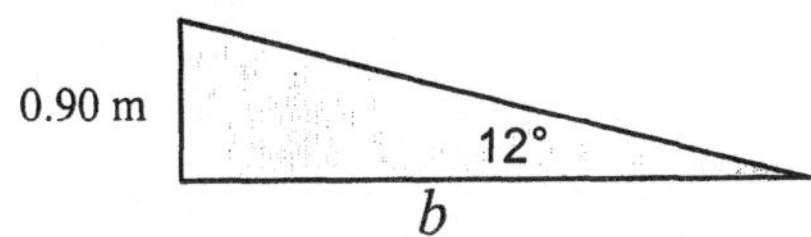

Thus, $d = 2.0\text{ m} + 4.2\text{ m} = 6.2\text{ m}$.

The time required for the train to travel 6.2 m is, from the definition of average speed,

$$t = \frac{x}{v} = \frac{6.2\text{ m}}{3.0\text{ m/s}} = \boxed{2.1\text{ s}}$$

---

12. ***REASONING AND SOLUTION*** The magnitude of the car's acceleration can be found from Equation 2.4 ($v = v_0 + at$) as

$$a = \frac{v - v_0}{t} = \frac{26.8\text{ m/s} - 0\text{ m/s}}{3.275\text{ s}} = \boxed{8.18\text{ m/s}^2}$$

---

13. SSM ***REASONING*** Since the velocity and acceleration of the motorcycle point in the same direction, their numerical values will have the same algebraic sign. For convenience, we will choose them to be positive. The velocity, acceleration, and the time are related by Equation 2.4: $v = v_0 + at$.

SOLUTION

a. Solving Equation 2.4 for t we have

$$t = \frac{v - v_0}{a} = \frac{(+31\text{ m/s}) - (+21\text{ m/s})}{+2.5\text{ m/s}^2} = \boxed{4.0\text{ s}}$$

b. Similarly,

$$t = \frac{v - v_0}{a} = \frac{(+61\text{ m/s}) - (+51\text{ m/s})}{+2.5\text{ m/s}^2} = \boxed{4.0\text{ s}}$$

---

14. ***REASONING*** The average acceleration is defined by Equation 2.4 as the change in velocity divided by the elapsed time. We can find the elapsed time from this relation because the acceleration and the change in velocity are given. Since the acceleration of the spacecraft is constant, it is equal to the average acceleration.

SOLUTION

a. The time Δt that it takes for the spacecraft to change its velocity by an amount $\Delta v = +2700$ m/s is

$$\Delta t = \frac{\Delta v}{a} = \frac{+2700\text{ m/s}}{+9.0\,\dfrac{\text{m/s}}{\text{day}}} = \boxed{3.0\times10^2\text{ days}}$$

b. Since 24 hr = 1 day and 3600 s = 1 hr, the acceleration of the spacecraft (in m/s^2) is

$$a = \frac{\Delta v}{t} = \frac{+9.0\text{ m/s}}{(1\text{ day})\left(\dfrac{24\text{ hr}}{1\text{ day}}\right)\left(\dfrac{3600\text{ s}}{1\text{ hr}}\right)} = \boxed{+1.04\times10^{-4}\text{ m/s}^2}$$

---

15. SSM ***REASONING AND SOLUTION*** The initial velocity of the runner can be found by solving Equation 2.4 ($v = v_0 + at$) for v_0. Taking west as the positive direction, we have

$$v_0 = v - at = (+5.36 \text{ m/s}) - (+0.640 \text{ m/s}^2)(3.00 \text{ s}) = +3.44 \text{ m/s}$$

Therefore, the initial velocity of the runner is $\boxed{3.44 \text{ m/s, due west}}$.

---

16. ***REASONING*** The fact that the emu is slowing down tells us that the acceleration and the velocity have opposite directions. Furthermore, since the acceleration remains the same in both parts of the motion, we can determine its value from the first part of the motion and then use it in the second part to determine the bird's final velocity at the end of the total 7.0-s time interval.

SOLUTION
a. The initial velocity of the emu is directed due south. Since the bird is slowing down, its acceleration must point in the opposite direction, or $\boxed{\text{due north}}$.

b. We assume that due south is the positive direction. With the data given for the first part of the motion, Equation 2.4 shows that the average acceleration is

$$\overline{\mathbf{a}} = \frac{\mathbf{v} - \mathbf{v}_0}{t - t_0} = \frac{(11.0 \text{ m/s}) - (13.9 \text{ m/s})}{3.0 \text{ s} - 0 \text{ s}} = -0.97 \text{ m/s}^2$$

Solving Equation 2.4 for the final velocity gives

$$\mathbf{v} = \mathbf{v}_0 + \overline{\mathbf{a}}\left(t - t_0\right) = +11.0 \text{ m/s} + \left(-0.97 \text{ m/s}^2\right)\left(4.0 \text{ s} - 0 \text{ s}\right) = +7.1 \text{ m/s}$$

Thus, the bird's velocity after an additional 4.0 s is $\boxed{7.1 \text{ m/s, due south}}$.

---

17. ***REASONING*** According to Equation 2.4, the average acceleration of the car for the first twelve seconds after the engine cuts out is

$$\overline{a}_1 = \frac{v_{1\text{f}} - v_{10}}{\Delta t_1} \tag{1}$$

and the average acceleration of the car during the next six seconds is

$$\overline{a}_2 = \frac{v_{2\text{f}} - v_{20}}{\Delta t_2} = \frac{v_{2\text{f}} - v_{1\text{f}}}{\Delta t_2} \tag{2}$$

The velocity v_{1f} of the car at the end of the initial twelve-second interval can be found by solving Equations (1) and (2) simultaneously.

SOLUTION Dividing Equation (1) by Equation (2), we have

$$\frac{\bar{a}_1}{\bar{a}_2} = \frac{(v_{1f} - v_{10})/\Delta t_1}{(v_{2f} - v_{1f})/\Delta t_2} = \frac{(v_{1f} - v_{10})\Delta t_2}{(v_{2f} - v_{1f})\Delta t_1}$$

Solving for v_{1f}, we obtain

$$v_{1f} = \frac{\bar{a}_1 \Delta t_1 v_{2f} + \bar{a}_2 \Delta t_2 v_{10}}{\bar{a}_1 \Delta t_1 + \bar{a}_2 \Delta t_2} = \frac{(\bar{a}_1/\bar{a}_2)\Delta t_1 v_{2f} + \Delta t_2 v_{10}}{(\bar{a}_1/\bar{a}_2)\Delta t_1 + \Delta t_2}$$

$$v_{1f} = \frac{1.50(12.0\text{ s})(+28.0\text{ m/s}) + (6.0\text{ s})(+36.0\text{ m/s})}{1.50(12.0\text{ s}) + 6.0\text{ s}} = \boxed{+30.0\text{ m/s}}$$

---

18. ***REASONING AND SOLUTION*** Both motorcycles have the same velocity v at the end of the four second interval. Now

$$v = v_{0A} + a_A t$$

for motorcycle A and

$$v = v_{0B} + a_B t$$

for motorcycle B. Subtraction of these equations and rearrangement gives

$$v_{0A} - v_{0B} = (4.0\text{ m/s}^2 - 2.0\text{ m/s}^2)(4\text{ s}) = \boxed{+8.0\text{ m/s}}$$

The positive result indicates that motorcycle A was initially traveling faster.

---

19. ***REASONING AND SOLUTION*** The average acceleration of the basketball player is $\bar{a} = v/t$, so

$$x = \frac{1}{2}\bar{a}t^2 = \frac{1}{2}\left(\frac{6.0\text{ m/s}}{1.5\text{ s}}\right)(1.5\text{ s})^2 = \boxed{4.5\text{ m}}$$

---

20. ***REASONING AND SOLUTION*** Since $v = v_0 + at$, the acceleration is given by $a = (v - v_0)/t$. Since the direction of travel is in the negative direction throughout the problem, all velocities will be negative.

a. $$a = \frac{(-29.0\text{ m/s}) - (-27.0\text{ m/s})}{5.0\text{ s}} = \boxed{-0.40\text{ m/s}^2}$$

Since the acceleration is negative, it is in the same direction as the velocity and the car is speeding up.

b.
$$a = \frac{(-23.0\text{ m/s}) - (-27.0\text{ m/s})}{5.0\text{ s}} = \boxed{+0.80\text{ m/s}^2}$$

Since the acceleration is positive, it is in the opposite direction to the velocity and the car is slowing down or decelerating.

---

21. SSM ***REASONING*** The average acceleration is defined by Equation 2.4 as the change in velocity divided by the elapsed time. We can find the elapsed time from this relation because the acceleration and the change in velocity are given.

SOLUTION

a. The time Δt that it takes for the VW Beetle to change its velocity by an amount $\Delta v = v - v_0$ is (and noting that 0.4470 m/s = 1 mi/h)

$$\Delta t = \frac{v - v_0}{a} = \frac{(60.0\text{ mi/h})\left(\dfrac{0.4470\text{ m/s}}{1\text{ mi/h}}\right) - 0\text{ m/s}}{2.35\text{ m/s}^2} = \boxed{11.4\text{ s}}$$

b. From Equation 2.4, the acceleration (in m/s^2) of the dragster is

$$a = \frac{v - v_0}{t - t_0} = \frac{(60.0\text{ mi/h})\left(\dfrac{0.4470\text{ m/s}}{1\text{ mi/h}}\right) - 0\text{ m/s}}{0.600\text{ s} - 0\text{ s}} = \boxed{44.7\text{ m/s}^2}$$

---

22. ***REASONING AND SOLUTION***

a. From Equation 2.4, the definition of average acceleration, the magnitude of the average acceleration of the skier is

$$\bar{a} = \frac{v - v_0}{t - t_0} = \frac{8.0\text{ m/s} - 0\text{ m/s}}{5.0\text{ s}} = \boxed{1.6\text{ m/s}^2}$$

b. With x representing the displacement traveled along the slope, Equation 2.7 gives:

$$x = \tfrac{1}{2}(v_0 + v)t = \tfrac{1}{2}(8.0\text{ m/s} + 0\text{ m/s})(5.0\text{ s}) = \boxed{2.0 \times 10^1\text{ m}}$$

---

23. ***REASONING*** We know the initial and final velocities of the blood, as well as its displacement. Therefore, Equation 2.9 $\left(v^2 = v_0^2 + 2ax\right)$ can be used to find the acceleration of the blood. The time it takes for the blood to reach it final velocity can be found by using Equation 2.7 $\left[t = \dfrac{x}{\frac{1}{2}(v_0 + v)}\right]$.

SOLUTION
a. The acceleration of the blood is

$$a = \frac{v^2 - v_0^2}{2x} = \frac{(26\text{ cm/s})^2 - (0\text{ cm/s})^2}{2(2.0\text{ cm})} = \boxed{1.7\times10^2\text{ cm/s}^2}$$

b. The time it takes for the blood, starting from 0 cm/s, to reach a final velocity of +26 cm/s is

$$t = \frac{x}{\frac{1}{2}(v_0 + v)} = \frac{2.0\text{ cm}}{\frac{1}{2}(0\text{ cm/s} + 26\text{ cm/s})} = \boxed{0.15\text{ s}}$$

---

24. ***REASONING*** The cheetah and its prey run the same distance. The prey runs at a constant velocity, so that its distance is the magnitude of its displacement, which is given by Equation 2.2 as the product of velocity and time. The distance for the cheetah can be expressed using Equation 2.8, since the cheetah's initial velocity (zero, since it starts from rest) and the time are given, and we wish to determine the acceleration. The two expressions for the distance can be equated and solved for the acceleration.

SOLUTION We begin by using Equation 2.2 and assuming that the initial position of the prey is $x_0 = 0$ m. The distance run by the prey is

$$\Delta x = x - x_0 = x = v_{\text{Prey}}t$$

The distance run by the cheetah is given by Equation 2.8 as

$$x = v_{0,\text{ Cheetah}}t + \tfrac{1}{2}a_{\text{Cheetah}}t^2$$

Equating the two expressions for x and using the fact that $v_{0,\text{ Cheetah}} = 0$ m/s, we find that

$$v_{\text{Prey}}t = \tfrac{1}{2}a_{\text{Cheetah}}t^2$$

Solving for the acceleration gives

$$a_{\text{Cheetah}} = \frac{2v_{\text{Prey}}}{t} = \frac{2(+9.0\text{ m/s})}{3.0\text{ s}} = \boxed{+6.0\text{ m/s}^2}$$

25. SSM ***REASONING AND SOLUTION*** The average acceleration of the plane can be found by solving Equation 2.9 $\left(v^2 = v_0^2 + 2ax\right)$ for a. Taking the direction of motion as positive, we have

$$a = \frac{v^2 - v_0^2}{2x} = \frac{(+6.1\ \text{m/s})^2 - (+69\ \text{m/s})^2}{2(+750\ \text{m})} = \boxed{-3.1\ \text{m/s}^2}$$

The minus sign indicates that the direction of the acceleration is opposite to the direction of motion, and the plane is slowing down.

26. ***REASONING*** At a constant velocity the time required for Secretariat to run the final mile is given by Equation 2.2 as the displacement (+1609 m) divided by the velocity. The actual time required for Secretariat to run the final mile can be determined from Equation 2.8, since the initial velocity, the acceleration, and the displacement are given. It is the difference between these two results for the time that we seek.

SOLUTION According to Equation 2.2, with the assumption that the initial time is $t_0 = 0$ s, the run time at a constant velocity is

$$\Delta t = t - t_0 = t = \frac{\Delta x}{v} = \frac{+1609\ \text{m}}{+16.58\ \text{m/s}} = 97.04\ \text{s}$$

Solving Equation 2.8 $\left(x = v_0 t + \frac{1}{2}at^2\right)$ for the time shows that

$$t = \frac{-v_0 \pm \sqrt{v_0^2 - 4\left(\frac{1}{2}a\right)(-x)}}{2\left(\frac{1}{2}a\right)}$$

$$= \frac{-16.58\ \text{m/s} \pm \sqrt{(+16.58\ \text{m/s})^2 - 4\left(\frac{1}{2}\right)\left(+0.0105\ \text{m/s}^2\right)(-1609\ \text{m})}}{2\left(\frac{1}{2}\right)\left(+0.0105\ \text{m/s}^2\right)} = 94.2\ \text{s}$$

We have ignored the negative root as being unphysical. The acceleration allowed Secretariat to run the last mile in a time that was faster by

$$97.04\ \text{s} - 94.2\ \text{s} = \boxed{2.8\ \text{s}}$$

27. SSM WWW ***REASONING*** Since the belt is moving with constant velocity, the displacement ($x_0 = 0$ m) covered by the belt in a time t_{belt} is giving by Equation 2.2 (with x_0 assumed to be zero) as

$$x = v_{belt} t_{belt} \qquad (1)$$

Since Clifford moves with constant acceleration, the displacement covered by Clifford in a time t_{Cliff} is, from Equation 2.8,

$$x = v_0 t_{Cliff} + \tfrac{1}{2} a t_{Cliff}^2 = \tfrac{1}{2} a t_{Cliff}^2 \qquad (2)$$

The speed v_{belt} with which the belt of the ramp is moving can be found by eliminating x between Equations (1) and (2).

SOLUTION Equating the right hand sides of Equations (1) and (2), and noting that $t_{Cliff} = \tfrac{1}{4} t_{belt}$, we have

$$v_{belt} t_{belt} = \tfrac{1}{2} a (\tfrac{1}{4} t_{belt})^2$$

$$v_{belt} = \tfrac{1}{32} a\, t_{belt} = \tfrac{1}{32}(0.37 \text{ m/s}^2)(64 \text{ s}) = \boxed{0.74 \text{ m/s}}$$

---

28. ***REASONING AND SOLUTION*** The speed of the car at the end of the first (402 m) phase can be obtained as follows:

$$v_1^2 = v_0^2 + 2a_1 x_1$$

$$v_1 = \sqrt{2(17.0 \text{ m/s}^2)(402 \text{ m})}$$

The speed after the second phase (3.50×10^2 m) can be obtained in a similar fashion.

$$v_2^2 = v_{02}^2 + 2a_2 x_2$$

$$v_2 = \sqrt{v_1^2 + 2(-6.10 \text{ m/s}^2)(3.50 \times 10^2 \text{ m})}$$

$$v_2 = \boxed{96.9 \text{ m/s}}$$

---

29. ***REASONING*** The stopping distance is the sum of two parts. First, there is the distance the car travels at 20.0 m/s before the brakes are applied. According to Equation 2.2, this distance is the magnitude of the displacement and is the magnitude of the velocity times the time. Second, there is the distance the car travels while it decelerates as the brakes are applied. This distance is given by Equation 2.9, since the initial velocity, the acceleration, and the final velocity (0 m/s when the car comes to a stop) are given.

SOLUTION With the assumption that the initial position of the car is $x_0 = 0$ m, Equation 2.2 gives the first contribution to the stopping distance as

$$\Delta x_1 = x_1 = vt_1 = (20.0\text{ m/s})(0.530\text{ s})$$

Solving Equation 2.9 $(v^2 = v_0^2 + 2ax)$ for x shows that the second part of the stopping distance is

$$x_2 = \frac{v^2 - v_0^2}{2a} = \frac{(0\text{ m/s})^2 - (20.0\text{ m/s})^2}{2(-7.00\text{ m/s}^2)}$$

Here, the acceleration is assigned a negative value, because we have assumed that the car is traveling in the positive direction, and it is decelerating. Since it is decelerating, its acceleration points opposite to its velocity. The stopping distance, then, is

$$x_{\text{Stopping}} = x_1 + x_2 = (20.0\text{ m/s})(0.530\text{ s}) + \frac{(0\text{ m/s})^2 - (20.0\text{ m/s})^2}{2(-7.00\text{ m/s}^2)} = \boxed{39.2\text{ m}}$$

---

30. ***REASONING AND SOLUTION***

a. The velocity at the end of the first (7.00 s) period is

$$v_1 = v_0 + a_1 = (2.01\text{ m/s}^2)(7.00\text{ s})$$

At the end of the second period the velocity is

$$v_2 = v_1 + a_2t_2 = v_1 + (0.518\text{ m/s}^2)(6.00\text{ s})$$

And the velocity at the end of the third (8.00 s) period is

$$v_3 = v_2 + a_3t_3 = v_2 + (-1.49\text{ m/s}^2)(8.00\text{ s}) = \boxed{5.26\text{ m/s}}$$

b. The displacement for the first time period is found from

$$x_1 = v_0t_1 + 1/2\ a_1t_1^{\ 2}$$
$$x_1 = (0\text{ m/s})(7.00\text{ s}) + (1/2)(2.01\text{ m/s}^2)(7.00\text{ s})^2 = 49.2\text{ m}$$

Similarly, $x_2 = 93.7$ m and $x_3 = 89.7$ m, so the total displacement of the boat is

$$x = x_1 + x_2 + x_3 = \boxed{233\text{ m}}$$

---

31. ***REASONING*** At a constant velocity the time required for the first car to travel to the next exit is given by Equation 2.2 as the magnitude of the displacement (2.5×10^3 m) divided by the magnitude of the velocity. This is also the travel time for the second car to reach the next exit. The acceleration for the second car can be determined from Equation 2.8, since the initial velocity, the displacement, and the time are known. This equation applies, because the acceleration is constant.

SOLUTION According to Equation 2.2, with the assumption that the initial time is $t_0 = 0$ s, the time for the first car to reach the next exit at a constant velocity is

$$\Delta t = t - t_0 = t = \frac{\Delta x}{v} = \frac{2.5 \times 10^3 \text{ m}}{33 \text{ m/s}} = 76 \text{ s}$$

Remembering that the initial velocity v_0 of the second car is zero, we can solve Equation 2.8 $\left(x = v_0 t + \frac{1}{2}at^2 = \frac{1}{2}at^2\right)$ for the acceleration to show that

$$a = \frac{2x}{t^2} = \frac{2(2.5 \times 10^3 \text{ m})}{(76 \text{ s})^2} = \boxed{0.87 \text{ m/s}^2}$$

Since the second car's speed is increasing, this acceleration must be $\boxed{\text{in the same direction as the velocity}}$.

---

32. ***REASONING AND SOLUTION*** The distance covered by the cab driver during the two phases of the trip must satisfy the relation

$$x_1 + x_2 = 2.00 \text{ km} \qquad (1)$$

where x_1 and x_2 are the displacements of the acceleration and deceleration phases of the trip, respectively. The quantities x_1 and x_2 can be determined from Equation 2.9 $\left(v^2 = v_0^2 + 2ax\right)$:

$$x_1 = \frac{v_1^2 - (0 \text{ m/s})^2}{2a_1} = \frac{v_1^2}{2a_1} \quad \text{and} \quad x_2 = \frac{(0 \text{ m/s})^2 - v_{02}^2}{2a_2} = -\frac{v_{02}^2}{2a_2}$$

with $v_{02} = v_1$ and $a_2 = -3a_1$. Thus,

$$\frac{x_1}{x_2} = \frac{v_1^2/(2a_1)}{-v_1^2/(-6a_1)} = 3$$

so that

$$x_1 = 3x_2 \qquad (2)$$

Combining (1) and (2), we have,

$$3x_2 + x_2 = 2.00 \text{ km}$$

Therefore, $x_2 = 0.50$ km, and from Equation (1), $x_1 = 1.50$ km. Thus, the length of the acceleration phase of the trip is $x_1 = \boxed{1.50 \text{ km}}$, while the length of the deceleration phase is $x_2 = \boxed{0.50 \text{ km}}$.

---

33. ***REASONING*** Let the total distance between the first and third sign be equal to $2d$. Then, the time t_A is given by

$$t_A = \frac{d}{v_{55}} + \frac{d}{v_{35}} = \frac{d(v_{35} + v_{55})}{v_{55}v_{35}} \qquad (1)$$

Equation 2.7 $\left[x = \frac{1}{2}(v_0 + v)t\right]$ can be written as $t = 2x/(v_0 + v)$, so that

$$t_B = \frac{2d}{v_{55} + v_{35}} + \frac{2d}{v_{35} + v_{25}} = \frac{2d\left[(v_{35} + v_{25}) + (v_{55} + v_{35})\right]}{(v_{55} + v_{35})(v_{35} + v_{25})} \qquad (2)$$

SOLUTION Dividing Equation (2) by Equation (1) and suppressing units, we obtain

$$\frac{t_B}{t_A} = \frac{2v_{55}v_{35}\left[(v_{35} + v_{25}) + (v_{55} + v_{35})\right]}{(v_{55} + v_{35})^2 (v_{35} + v_{25})} = \frac{2(55)(35)[(35+25)+(55+35)]}{(55+35)^2\,(35+25)} = \boxed{1.2}$$

---

34. ***REASONING AND SOLUTION*** As the plane decelerates through the intersection, it covers a total distance equal to the length of the plane plus the width of the intersection, so

$$x = 59.7 \text{ m} + 25.0 \text{ m} = 84.7 \text{ m}$$

The speed of the plane as it enters the intersection can be found from Equation 2.9. Solving Equation 2.9 for v_0 gives

$$v_0 = \sqrt{v^2 - 2ax} = \sqrt{(45.0 \text{ m})^2 - 2(-5.70 \text{ m/s}^2)(84.7 \text{ m})} = 54.7 \text{ m/s}$$

The time required to traverse the intersection can then be found from Equation 2.4. Solving Equation 2.4 for t gives

$$t = \frac{v - v_0}{a} = \frac{45.0 \text{ m/s} - 54.7 \text{ m/s}}{-5.70 \text{ m/s}^2} = \boxed{1.7 \text{ s}}$$

---

35. SSM ***REASONING*** Since the car is moving with a constant velocity, the displacement of the car in a time t can be found from Equation 2.8 with $a = 0$ m/s^2 and v_0 equal to the velocity of the car: $x_{car} = v_{car}t$. Since the train starts from rest with a constant acceleration, the displacement of the train in a time t is given by Equation 2.8 with $v_0 = 0$ m/s:

$$x_{train} = \frac{1}{2}a_{train}t^2$$

At a time t_1, when the car just reaches the front of the train, $x_{car} = L_{train} + x_{train}$, where L_{train} is the length of the train. Thus, at time t_1,

$$v_{car}t_1 = L_{train} + \frac{1}{2}a_{train}t_1^2 \qquad (1)$$

At a time t_2, when the car is again at the rear of the train, $x_{car} = x_{train}$. Thus, at time t_2

$$v_{car}t_2 = \frac{1}{2}a_{train}t_2^2 \qquad (2)$$

Equations (1) and (2) can be solved simultaneously for the speed of the car v_{car} and the acceleration of the train a_{train}.

SOLUTION
a. Solving Equation (2) for a_{train} we have

$$a_{train} = \frac{2v_{car}}{t_2} \qquad (3)$$

Substituting this expression for a_{train} into Equation (1) and solving for v_{car}, we have

$$v_{car} = \frac{L_{train}}{t_1\left(1 - \dfrac{t_1}{t_2}\right)} = \frac{92\text{ m}}{(14\text{ s})\left(1 - \dfrac{14\text{ s}}{28\text{ s}}\right)} = \boxed{13\text{ m/s}}$$

b. Direct substitution into Equation (3) gives the acceleration of the train:

$$a_{train} = \frac{2v_{car}}{t_2} = \frac{2\,(13\text{ m/s})}{28\text{ s}} = \boxed{0.93\text{ m/s}^2}$$

---

36. ***REASONING AND SOLUTION*** During the acceleration phase of the motion, the sprinter runs a distance

$$x_1 = (1/2)\,at_1^2 \qquad (1)$$

acquiring a final speed

$$v = at_1 \qquad (2)$$

During the rest of the race he runs a distance

$$x_2 = vt_2 \qquad (3)$$

Adding Equations (1) and (3) gives

$$x = (1/2)\,at_1^2 + vt_2$$

where $x = x_1 + x_2$.

Substituting from Equation (2) and noting that the total time is $t = t_1 + t_2$, we have

$$(1/2)\,at_1^2 - att_1 + x = 0$$

or (suppressing the units)

$$1.34t_1^2 - 32.2t_1 + 100.0 = 0$$

Solving for t_1 gives 20 s and 3.7 s. The first is obviously not a physically realistic solution, since it is larger than the total time for the race so, t_1 = 3.7 s. Using this value in Equation (1) gives

$$x_1 = (1/2)(2.68\text{ m/s}^2)(3.7\text{ s})^2 = \boxed{18\text{ m}}$$

---

37. SSM ***REASONING AND SOLUTION*** The speed of the penny as it hits the ground can be determined from Equation 2.9: $v^2 = v_0^2 + 2ay$. Since the penny is dropped from rest, $v_0 = 0$ m/s. Solving for v, with downward taken as the positive direction, we have

$$v = \sqrt{2(9.80\text{ m/s}^2)(427\text{ m})} = \boxed{91.5\text{ m/s}}$$

---

38. ***REASONING AND SOLUTION***

a. Once the pebble has left the slingshot, it is subject only to the acceleration due to gravity. Since the downward direction is negative, the acceleration of the pebble is $\boxed{-9.80\text{ m/s}^2}$. The pebble is not decelerating. Since its velocity and acceleration both point downward, the magnitude of the pebble's velocity is increasing, not decreasing.

b. The displacement y traveled by the pebble as a function of the time t can be found from Equation 2.8. Using Equation 2.8, we have

$$y = v_0 t + \tfrac{1}{2} a_y t^2 = (-9.0\text{ m/s})(0.50\text{ s}) + \tfrac{1}{2}\left[(-9.80\text{ m/s}^2)(0.50\text{ s})^2\right] = -5.7\text{ m}$$

Thus, after 0.50 s, the pebble is $\boxed{5.7\text{ m}}$ beneath the cliff-top.

---

39. ***REASONING*** The initial velocity and the elapsed time are given in the problem. Since the rock returns to the same place from which it was thrown, its displacement is zero ($y = 0$ m). Using this information, we can employ Equation 2.8 $\left(y = v_0 t + \tfrac{1}{2} a t^2\right)$ to determine the acceleration a due to gravity.

SOLUTION Solving Equation 2.8 for the acceleration yields

$$a = \frac{2(y - v_0 t)}{t^2} = \frac{2\left[0\text{ m} - (+15\text{ m/s})(20.0\text{ s})\right]}{(20.0\text{ s})^2} = \boxed{-1.5\text{ m/s}^2}$$

---

40. ***REASONING AND SOLUTION*** In a time t the card will undergo a vertical displacement y given by

$$y = \tfrac{1}{2}\ at^2$$

where $a = -9.80\text{ m/s}^2$. When $t = 60.0\text{ ms} = 6.0 \times 10^{-2}$ s, the displacement of the card is 0.018 m, and the distance is the magnitude of this value or $\boxed{d_1 = 0.018\text{ m}}$.

Similarly, when t = 120 ms, $\boxed{d_2 = 0.071\text{ m}}$, and when t = 180 ms, $\boxed{d_3 = 0.16\text{ m}}$.

---

41. SSM ***REASONING AND SOLUTION*** Since the balloon is released from rest, its initial velocity is zero. The time required to fall through a vertical displacement y can be found from Equation 2.8 $\left(y = v_0 t + \frac{1}{2} at^2\right)$ with $v_0 = 0$ m/s. Assuming upward to be the positive direction, we find

$$t = \sqrt{\frac{2y}{a}} = \sqrt{\frac{2(-6.0\text{ m})}{-9.80\text{ m/s}^2}} = \boxed{1.1\text{ s}}$$

---

42. ***REASONING*** The minimum time that a player must wait before touching the basketball is the time required for the ball to reach its maximum height. The initial and final velocities are known, as well as the acceleration due to gravity, so Equation 2.4 ($v = v_0 + at$) can be used to find the time.

SOLUTION Solving Equation 2.4 for the time yields

$$t = \frac{v - v_0}{a} = \frac{0 \text{ m/s} - 4.6 \text{ m/s}}{-9.8 \text{ m/s}^2} = \boxed{0.47 \text{ s}}$$

---

43. ***REASONING AND SOLUTION*** The figure at the right shows the paths taken by the pellets fired from gun **A** and gun **B**. The two paths differ by the extra distance covered by the pellet from gun **A** as it rises and falls back to the edge of the cliff. When it falls back to the edge of the cliff, the pellet from gun A will have the same speed as the pellet fired from gun B, as Conceptual Example 15 discusses. Therefore, the flight time of pellet **A** will be greater than that of **B** by the amount of time that it takes for pellet **A** to cover the extra distance.

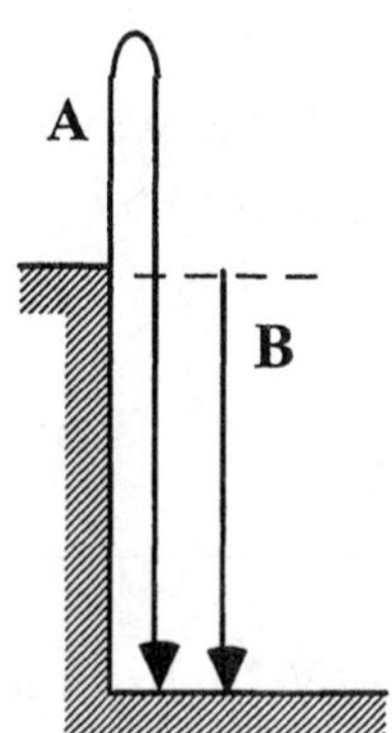

The time required for pellet **A** to return to the cliff edge after being fired can be found from Equation 2.4: $v = v_0 + at$.

If "up" is taken as the positive direction then $v_0 = +30.0$ m/s and $v = -30.0$ m/s. Solving Equation 2.4 for t gives

$$t = \frac{v - v_0}{a} = \frac{(-30.0 \text{ m/s}) - (+30.0 \text{ m/s})}{-9.80 \text{ m/s}^2} = \boxed{6.12 \text{ s}}$$

Notice that this result is *independent* of the height of the cliff.

---

44. ***REASONING AND SOLUTION***

a. $$v^2 = v_0^2 + 2ay$$

$$v = \pm\sqrt{(1.8 \text{ m/s})^2 + 2\left(-9.80 \text{ m/s}^2\right)\left(-3.0 \text{ m}\right)} = \pm 7.9 \text{ m/s}$$

The minus is chosen, since the diver is now moving down. Hence, $\boxed{v = -7.9 \text{ m/s}}$.

b. The diver's velocity is zero at his highest point. The position of the diver relative to the board is

$$y = -\frac{v_0^2}{2a} = -\frac{(1.8 \text{ m/s})^2}{2\left(-9.80 \text{ m/s}^2\right)} = 0.17 \text{ m}$$

The position above the water is 3.0 m + 0.17 m = $\boxed{3.2 \text{ m}}$.

---

45. SSM ***REASONING AND SOLUTION*** Equation 2.8 can be used to determine the displacement that the ball covers as it falls halfway to the ground. Since the ball falls from rest, its initial velocity is zero. Taking down to be the negative direction, we have

$$y = v_0 t + \frac{1}{2}at^2 = \frac{1}{2}at^2 = \frac{1}{2}(-9.80 \text{ m/s}^2)(1.2 \text{ s})^2 = -7.1 \text{ m}$$

In falling all the way to the ground, the ball has a displacement of $y = -14.2$ m. Solving Equation 2.8 with this displacement then yields the time

$$t = \sqrt{\frac{2y}{a}} = \sqrt{\frac{2(-14.2 \text{ m})}{-9.80 \text{ m/s}^2}} = \boxed{1.7 \text{ s}}$$

---

46. ***REASONING*** Equation 2.9 $\left(v^2 = v_0^2 + 2ay\right)$ can be used to find out how far above the cliff's edge the pellet would have gone if the gun had been fired straight upward, provided that we can determine the initial speed imparted to the pellet by the gun. This initial speed can be found by applying Equation 2.9 to the downward motion of the pellet described in the problem statement.

SOLUTION If we assume that upward is the positive direction, the initial speed of the pellet is, from Equation 2.9,

$$v_0 = \sqrt{v^2 - 2ay} = \sqrt{(-27 \text{ m/s})^2 - 2(-9.80 \text{ m/s}^2)(-15 \text{ m})} = 20.9 \text{ m/s}$$

Equation 2.9 can again be used to find the maximum height of the pellet if it were fired straight up. At its maximum height, $v = 0$ m/s, and Equation 2.9 gives

$$y = \frac{-v_0^2}{2a} = \frac{-(20.9 \text{ m/s})^2}{2(-9.80 \text{ m/s}^2)} = \boxed{22 \text{ m}}$$

---

47. ***REASONING*** The initial speed of the ball can be determined from Equation 2.9 $\left(v^2 = v_0^2 + 2ay\right)$. Once the initial speed of the ball is known, Equation 2.9 can be used a second time to determine the height above the launch point when the speed of the ball has decreased to one half of its initial value.

SOLUTION When the ball has reached its maximum height, its velocity is zero. If we take upward as the positive direction, we have from Equation 2.9

$$v_0 = \sqrt{v^2 - 2ay} = \sqrt{(0 \text{ m/s})^2 - 2(-9.80 \text{ m/s}^2)(12.0 \text{ m})} = +15.3 \text{ m/s}$$

When the speed of the ball has decreased to one half of its initial value, $v = \frac{1}{2}v_0$, and Equation 2.9 gives

$$y = \frac{v^2 - v_0^2}{2a} = \frac{(\frac{1}{2}v_0)^2 - v_0^2}{2a} = \frac{v_0^2}{2a}\left(\frac{1}{4} - 1\right) = \frac{(+15.3 \text{ m/s})^2}{2(-9.80 \text{ m/s}^2)}\left(\frac{1}{4} - 1\right) = \boxed{+8.96 \text{ m}}$$

---

48. ***REASONING AND SOLUTION*** The time required for the first arrow to reach its maximum height can be determined from Equation 2.4 ($v = v_0 + at$). Taking upward as the positive direction, we have

$$t = \frac{v - v_0}{a} = \frac{0 \text{ m/s} - 25.0 \text{ m/s}}{-9.80 \text{ m/s}^2} = 2.55 \text{ s}$$

Since both arrows reach their maximum height at the same time, the second arrow reaches its maximum height

$$2.55 \text{ s} - 1.20 \text{ s} = 1.35 \text{ s}$$

after being fired. The initial speed of the second arrow can then be found from Equation 2.4:

$$v_0 = v - at = 0 \text{ m/s} - (-9.80 \text{ m/s}^2)(1.35 \text{ s}) = \boxed{13.2 \text{ m/s}}$$

---

49. ***REASONING*** To calculate the speed of the raft, it is necessary to determine the distance it travels and the time interval over which the motion occurs. The speed is the distance divided by the time, according to Equation 2.1. The distance is 7.00 m – 4.00 m = 3.00 m. The time is the time it takes for the stone to fall, which can be obtained from Equation 2.8 $\left(y = v_0 t + \frac{1}{2}at^2\right)$, since the displacement y, the initial velocity v_0, and the acceleration a are known.

SOLUTION During the time t that it takes the stone to fall, the raft travels a distance of 7.00 m – 4.00 m = 3.00 m, and according to Equation 2.1, its speed is

$$\text{speed} = \frac{3.00 \text{ m}}{t}$$

The stone falls downward for a distance of 75.0 m, so its displacement is $y = -75.0$ m, where the downward direction is taken to be the negative direction. Equation 2.8 can be used to find the time of fall. Setting $v_0 = 0$ m/s, and solving Equation 2.8 for the time t, we have

$$t = \sqrt{\frac{2y}{a}} = \sqrt{\frac{2(-75.0 \text{ m})}{-9.80 \text{ m/s}^2}} = 3.91 \text{ s}$$

Therefore, the speed of the raft is

$$\text{speed} = \frac{3.00 \text{ m}}{3.91 \text{ s}} = \boxed{0.767 \text{ m/s}}$$

---

50. ***REASONING AND SOLUTION*** The initial speed of either student can be found from Equation (2.9): $v^2 = v_0^2 + 2ay$. Since the speed of either student is zero when the student is at her highest point, $v = 0$ m/s and

$$v_0 = \sqrt{-2ay}$$

Since Anne bounces twice as high as Joan, $y_A = 2y_J$, and $v_{0A} = \sqrt{2}\, v_{0J}$.

The time it takes for either student to reach the highest point in her trajectory can be found from Equation 2.4: $v = at + v_0$. Solving for t with $v = 0$ m/s gives

$$t = \frac{-v_o}{a}$$

The total time in the air is twice this value. Therefore,

$$\frac{t_A}{t_J} = \frac{-2v_{0A}/a}{-2v_{0J}/a} = \frac{v_{0A}}{v_{0J}} = \frac{\sqrt{2}v_{0J}}{v_{0J}} = \boxed{\sqrt{2}}$$

---

51. SSM ***REASONING AND SOLUTION*** The stone will reach the water (and hence the log) after falling for a time *t*, where *t* can be determined from Equation 2.8: $y = v_0 t + \frac{1}{2}at^2$. Since the stone is dropped from rest, $v_0 = 0$ m/s. Assuming that downward is positive and solving for *t*, we have

$$t = \sqrt{\frac{2y}{a}} = \sqrt{\frac{2(75 \text{ m})}{9.80 \text{ m/s}^2}} = 3.9 \text{ s}$$

During that time, the displacement of the log can be found from Equation 2.8. Since the log moves with constant velocity, $a = 0 \text{ m/s}^2$, and v_0 is equal to the velocity of the log.

$$x = v_0 t = (5.0 \text{ m/s})(3.9 \text{ s}) = 2.0 \times 10^1 \text{ m}$$

Therefore, the horizontal distance between the log and the bridge when the stone is released is $\boxed{2.0 \times 10^1 \text{ m}}$.

---

52. ***REASONING AND SOLUTION***

a. We can use Equation 2.9 to obtain the speed acquired as she falls through the distance *H*. Taking downward as the positive direction, we find

$$v^2 = v_0^2 + 2ay = (0\text{ m/s})^2 + 2aH \qquad \text{or} \qquad v = \sqrt{2aH}$$

To acquire a speed of twice this value or $2\sqrt{2aH}$, she must fall an additional distance H'. According to Equation 2.9 $\left(v^2 = v_0^2 + 2ay\right)$, we have

$$\left(2\sqrt{2aH}\right)^2 = \left(\sqrt{2aH}\right)^2 + 2aH' \qquad \text{or} \qquad 4(2aH) = 2aH + 2aH'$$

The acceleration due to gravity *a* can be eliminated algebraically from this result, giving

$$4H = H + H' \qquad \text{or} \qquad \boxed{H' = 3H}$$

b. In the previous calculation the acceleration due to gravity was eliminated algebraically. Thus, a value other than 9.80 m/s^2 would [not have affected the answer to part (a)] .

---

53. [SSM] [WWW] ***REASONING AND SOLUTION*** The stone requires a time, t_1, to reach the bottom of the hole, a distance *y* below the ground. Assuming downward to be the positive direction, the variables are related by Equation 2.8 with $v_0 = 0$ m/s:

$$y = \tfrac{1}{2}at_1^2 \tag{1}$$

The sound travels the distance *y* from the bottom to the top of the hole in a time t_2. Since the sound does not experience any acceleration, the variables *y* and t_2 are related by Equation 2.8 with $a = 0\ \text{m/s}^2$ and v_{sound} denoting the speed of sound:

$$y = v_{\text{sound}}\, t_2 \tag{2}$$

Equating the right hand sides of Equations (1) and (2) and using the fact that the total elapsed time is $t = t_1 + t_2$, we have

$$\tfrac{1}{2}at_1^2 = v_{\text{sound}}t_2 \qquad \text{or} \qquad \tfrac{1}{2}at_1^2 = v_{\text{sound}}(t - t_1)$$

Rearranging gives

$$\tfrac{1}{2}at_1^2 + v_{\text{sound}}t_1 - v_{\text{sound}}t = 0$$

Substituting values and suppressing units for brevity, we obtain the following quadratic equation for t_1:

$$4.90t_1^2 + 343t_1 - 514 = 0$$

From the quadratic formula, we obtain

$$t_1 = \frac{-343 \pm \sqrt{(343)^2 - 4(4.90)(-514)}}{2(4.90)} = 1.47 \text{ s} \quad \text{or} \quad -71.5 \text{ s}$$

The negative time corresponds to a nonphysical result and is rejected. The depth of the hole is then found using Equation 2.8 with the value of t_1 obtained above:

$$y = v_0 t_1 + \tfrac{1}{2} a t_1^2 = (0 \text{ m/s})(1.47 \text{ s}) + \tfrac{1}{2}(9.80 \text{ m/s}^2)(1.47 \text{ s})^2 = \boxed{10.6 \text{ m}}$$

---

54. ***REASONING AND SOLUTION*** The position of the ball relative to the top of the building at any time t is

$$y = v_0 t + \tfrac{1}{2} a t^2 = v_0 t + \tfrac{1}{2}(-9.80 \text{ m/s}^2) t^2$$

For the particular time that the ball arrives at the bottom of the building, $y = -25.0$ m. The equation becomes

$$4.90t^2 - 12.0t - 25.0 = 0$$

The quadratic equation yields solutions $t = 3.79$ s and -1.34 s. The negative solution is rejected as being non-physical. During time t the person has run a distance $x = vt$ so

$$v = \frac{x}{t} = \frac{31.0 \text{ m}}{3.79 \text{ s}} = \boxed{8.18 \text{ m/s}}$$

---

55. ***REASONING AND SOLUTION*** The balls pass at a time t when both are at a position y above the ground. Applying Equation 2.8 to the ball that is dropped from rest, we have

$$y = 24 \text{ m} + v_{01} t + \tfrac{1}{2} a t^2 = 24 \text{ m} + (0 \text{ m/s}) t + \tfrac{1}{2} a t^2 \qquad (1)$$

Note that we have taken into account the fact that $y = 24$ m when $t = 0$ s in Equation (1). For the second ball that is thrown straight upward,

$$y = v_{02} t + \frac{1}{2} a t^2 \qquad (2)$$

Equating Equations (1) and (2) for y yields

$$24 \text{ m} + \frac{1}{2} a t^2 = v_{02} t + \frac{1}{2} a t^2 \qquad \text{or} \qquad 24 \text{ m} = v_{02} t$$

Thus, the two balls pass at a time t, where

$$t = \frac{24\text{ m}}{v_{02}}$$

The initial speed v_{02} of the second ball is exactly the same as that with which the first ball hits the ground. To find the speed with which the first ball hits the ground, we take upward as the positive direction and use Equation 2.9 $\left(v^2 = v_0^2 + 2ay\right)$. Since the first ball is dropped from rest, we find that

$$v_{02} = v = \sqrt{2ay} = \sqrt{2(-9.80\text{ m/s}^2)(-24\text{ m})} = 21.7\text{ m/s}$$

Thus, the balls pass after a time

$$t = \frac{24\text{ m}}{21.7\text{ m/s}} = 1.11\text{ s}$$

At a time $t = 1.11$ s, the position of the first ball according to Equation (1) is

$$y = 24\text{ m} + \tfrac{1}{2}(-9.80\text{ m/s}^2)(1.11\text{ s})^2 = 24\text{ m} - 6.0\text{ m}$$

which is $\boxed{\text{6.0 m below the top of the cliff}}$.

---

56. ***REASONING AND SOLUTION*** We measure the positions of the balloon and the pellet relative to the ground and assume up to be positive. The balloon has no acceleration, since it travels at a constant velocity v_B, so its displacement in time t is $v_B t$. Its position above the ground, therefore, is

$$y_B = H_0 + v_B t$$

where H_0 = 12 m. The pellet moves under the influence of gravity (a = –9.80 m/s^2), so its position above the ground is given by Equation 2.8 as

$$y_P = v_0 t + \frac{1}{2}at^2$$

But $y_P = y_B$ at time t, so that

$$v_0 t + \frac{1}{2}at^2 = H_0 + v_B t$$

Rearranging this result and suppressing the units gives

$$\tfrac{1}{2}at^2+(v_0-v_B)t-H_0=\tfrac{1}{2}(-9.80)t^2+(30.0-7.0)t-12.0=0$$

$$4.90t^2-23.0t+12.0=0$$

$$t=\frac{23.0\pm\sqrt{23.0^2-4(4.90)(12.0)}}{2(4.90)}=4.09\text{ s}\quad\text{or}\quad 0.602\text{ s}$$

Substituting each of these values in the expression for y_B gives

$$y_B=12.0\text{ m}+(7.0\text{ m/s})(4.09\text{ s})=\boxed{41\text{ m}}$$

$$y_B=12.0\text{ m}+(7.0\text{ m/s})(0.602\text{ s})=\boxed{16\text{ m}}$$

---

57. SSM ***REASONING*** In order to construct the graph, the time for each segment of the trip must be determined.

SOLUTION From the definition of average velocity

$$\Delta t=\frac{\Delta x}{v}$$

Therefore,

$$\Delta t_1=\left(\frac{10.0\text{ km}}{15.0\text{ km/h}}\right)\left(\frac{60\text{ min}}{1.0\text{ h}}\right)=40\text{ min}$$

$$\Delta t_2=\left(\frac{15.0\text{ km}}{10.0\text{ km/h}}\right)\left(\frac{60\text{ min}}{1.0\text{ h}}\right)=90\text{ min}$$

Thus, the second time interval ends 40 min + 90 min = 130 min after the trip begins.

$$\Delta t_3=\left(\frac{15.0\text{ km}}{5.0\text{ km/h}}\right)\left(\frac{60\text{ min}}{1.0\text{ h}}\right)=180\text{ min}$$

and the third time interval ends 130 min + 180 min = 310 min after the trip begins.

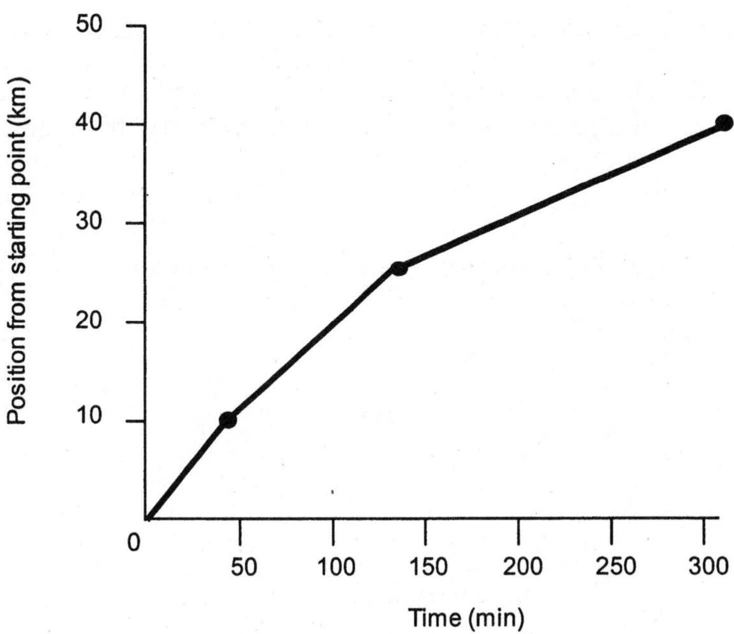

Note that the slope of each segment of the graph gives the average velocity during that interval.

---

58. ***REASONING*** The average velocity for each segment is the slope of the line for that segment.

SOLUTION Taking the direction of motion as positive, we have from the graph for segments *A*, *B*, and *C*,

$$v_A = \frac{10.0 \text{ km} - 40.0 \text{ km}}{1.5 \text{ h} - 0.0 \text{ h}} = \boxed{-2.0 \times 10^1 \text{ km/h}}$$

$$v_B = \frac{20.0 \text{ km} - 10.0 \text{ km}}{2.5 \text{ h} - 1.5 \text{ h}} = \boxed{1.0 \times 10^1 \text{ km/h}}$$

$$v_C = \frac{40.0 \text{ km} - 20.0 \text{ km}}{3.0 \text{ h} - 2.5 \text{ h}} = \boxed{40 \text{ km/h}}$$

---

59. ***REASONING AND SOLUTION*** The average acceleration for each segment is the slope of that segment.

$$a_A = \frac{40 \text{ m/s} - 0 \text{ m/s}}{21 \text{ s} - 0 \text{ s}} = \boxed{1.9 \text{ m/s}^2}$$

$$a_B = \frac{40 \text{ m/s} - 40 \text{ m/s}}{48 \text{ s} - 21 \text{ s}} = \boxed{0 \text{ m/s}^2}$$

$$a_C = \frac{80 \text{ m/s} - 40 \text{ m/s}}{60 \text{ s} - 48 \text{ s}} = \boxed{3.3 \text{ m/s}^2}$$

---

60. ***REASONING*** The slope of a straight-line segment in a position-versus-time graph is the average velocity. The algebraic sign of the average velocity, therefore, corresponds to the sign of the slope.

SOLUTION
a. The slope, and hence the average velocity, is *positive* for segments A and C, *negative* for segment B, and zero for segment D.

b.

$$v_A = \frac{1.25\text{ km} - 0\text{ km}}{0.20\text{ h} - 0\text{ h}} = \boxed{+6.3\text{ km/h}}$$

$$v_B = \frac{0.50\text{ km} - 1.25\text{ km}}{0.40\text{ h} - 0.20\text{ h}} = \boxed{-3.8\text{ km/h}}$$

$$v_C = \frac{0.75\text{ km} - 0.50\text{ km}}{0.80\text{ h} - 0.40\text{ h}} = \boxed{+0.63\text{ km/h}}$$

$$v_D = \frac{0.75\text{ km} - 0.75\text{ km}}{1.00\text{ h} - 0.80\text{ h}} = \boxed{0\text{ km/h}}$$

---

61. SSM ***REASONING*** The average acceleration is given by Equation 2.4: $\bar{a} = (v_C - v_A)/\Delta t$. The velocities v_A and v_C can be found from the slopes of the position-time graph for segments A and C.

SOLUTION The average velocities in the segments A and C are

$$v_A = \frac{24\text{ km} - 0\text{ km}}{1.0\text{ h} - 0\text{ h}} = 24\text{ km/h}$$

$$v_C = \frac{27\text{ km} - 33\text{ km}}{3.5\text{ h} - 2.2\text{ h}} = -5\text{ km/h}$$

From the definition of average acceleration,

$$\bar{a} = \frac{\Delta v}{\Delta t} = \frac{v_C - v_A}{\Delta t} = \frac{(-5\text{ km/h}) - (24\text{ km/h})}{3.5\text{ h} - 0\text{ h}} = \boxed{-8.3\text{ km/h}^2}$$

---

62. ***REASONING AND SOLUTION*** The runner is at the position $x = 0$ m when time $t = 0$ s; the finish line is 100 m away. During each ten-second segment, the runner has a constant velocity and runs half the remaining distance to the finish line. Therefore, from $t = 0$ s to

$t = 10.0$ s, the position of the runner changes from $x = 0$ m to $x = 50.0$ m. From $t = 10.0$ s to $t = 20.0$ s, the position of the runner changes from $x = 50.0$ m to $x = 50.0\text{ m} + 25.0\text{ m} = 75.0$ m. From $t = 20.0$ s to $t = 30.0$ s, the position of the runner changes from $x = 75.0$ m to $x = 75.0\text{ m} + 12.5\text{ m} = 87.5$ m. Finally, from $t = 30.0$ s to $t = 40.0$ s, the position of the runner changes from $x = 87.5$ m to $x = 87.5\text{ m} + 6.25\text{ m} = 93.8$ m. This data can be used to construct the position-time graph. Since the runner has a constant velocity during each ten-second segment, we can find the velocity during each segment from the slope of the position-time graph for that segment.

a. The following figure shows the position-time graph for the first forty seconds.

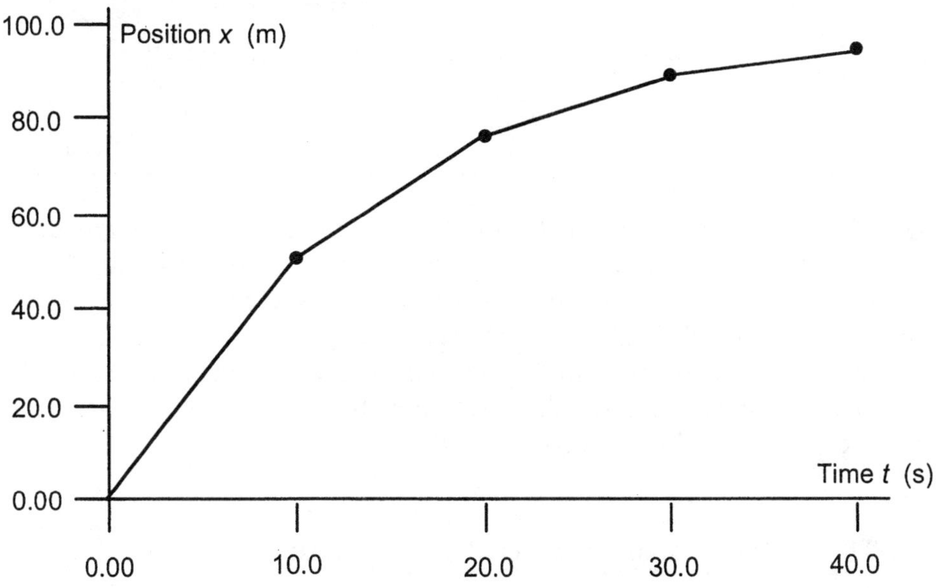

b. The slope of each segment of the position-time graph is calculated as follows:

$$[0.00\text{ s to }10.0\text{ s}] \qquad v = \frac{\Delta x}{\Delta t} = \frac{50.0\text{ m} - 0.00\text{ m}}{10.0\text{ s} - 0\text{ s}} = 5.00\text{ m/s}$$

$$[10.0\text{ s to }20.0\text{ s}] \qquad v = \frac{\Delta x}{\Delta t} = \frac{75.0\text{ m} - 50.0\text{ m}}{20.0\text{ s} - 10.0\text{ s}} = 2.50\text{ m/s}$$

$$[20.0\text{ s to }30.0\text{ s}] \qquad v = \frac{\Delta x}{\Delta t} = \frac{87.5\text{ m} - 75.0\text{ m}}{30.0\text{ s} - 20.0\text{ s}} = 1.25\text{ m/s}$$

$$[30.0\text{ s to }40.0\text{ s}] \qquad v = \frac{\Delta x}{\Delta t} = \frac{93.8\text{ m} - 87.5\text{ m}}{40.0\text{ s} - 30.0\text{ s}} = 0.625\text{ m/s}$$

Therefore, the velocity-time graph is:

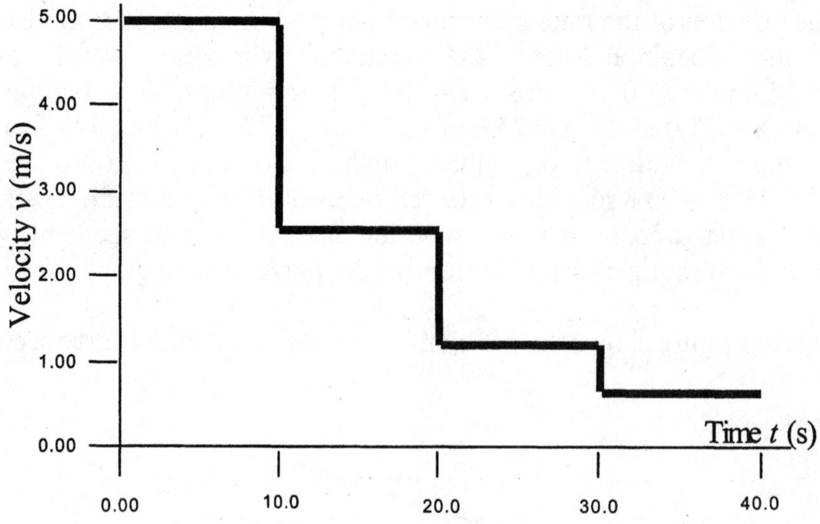

---

63. ***REASONING*** The two runners start one hundred meters apart and run toward each other. As stated in the problem, each runs ten meters during the first second and, during each second thereafter, each runner runs ninety percent of the distance he ran in the previous second. While the velocity of each runner changes from second to second, it remains constant during any one second. The following table shows the distance covered during each second for one of the runners, and the position at the end of each second (assuming that he begins at the origin) for the first eight seconds.

Time t (s)	Distance covered (m)	Position x (m)
0.00		0.00
1.00	10.00	10.00
2.00	9.00	19.00
3.00	8.10	27.10
4.00	7.29	34.39
5.00	6.56	40.95
6.00	5.90	46.85
7.00	5.31	52.16
8.00	4.78	56.94

The following graph is the position-time graph constructed from the data in the table above.

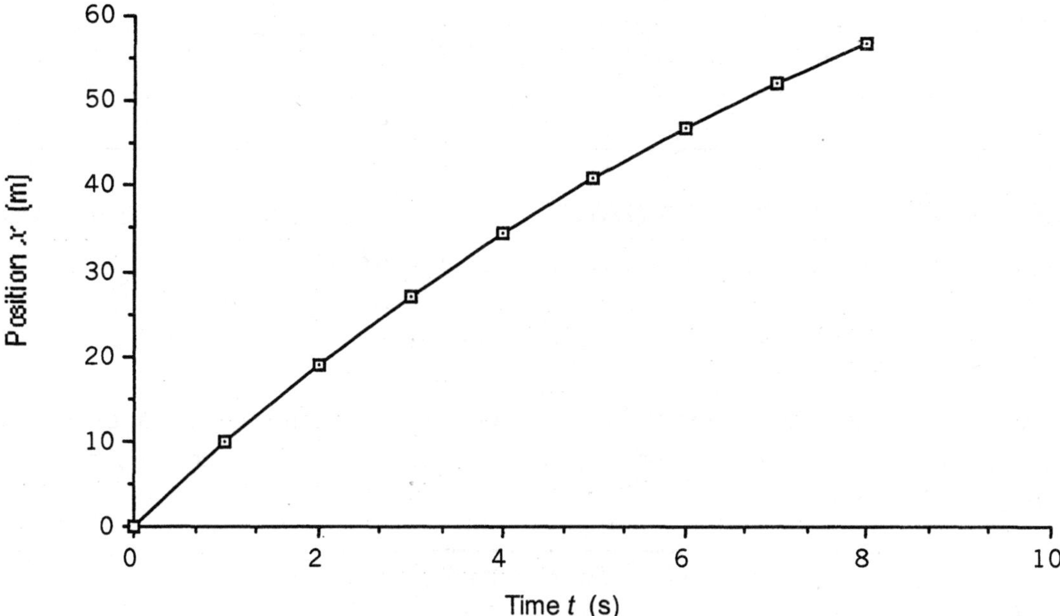

a. Since the two runners are running toward each other in exactly the same way, they will meet halfway between their respective starting points. That is, they will meet at $x = 50.0$ m. According to the graph, therefore, this position corresponds to a time of $\boxed{6.6 \text{ s}}$.

b. Since the runners collide during the seventh second, the speed at the instant of collision can be found by taking the slope of the position-time graph for the seventh second. The speed of either runner in the interval from $t = 6.00$ s to $t = 7.00$ s is

$$v = \frac{\Delta x}{\Delta t} = \frac{52.16 \text{ m} - 46.85 \text{ m}}{7.00 \text{ s} - 6.00 \text{ s}} = 5.3 \text{ m/s}$$

Therefore, at the moment of collision, the speed of either runner is $\boxed{5.3 \text{ m/s}}$.

---

64. ***REASONING*** The cart has an initial velocity of $v_0 = +5.0$ m/s, so initially it is moving to the right, which is the positive direction. It eventually reaches a point where the displacement is $x = +12.5$ m, and it begins to move to the left. This must mean that the cart comes to a momentary halt at this point (final velocity is $v = 0$ m/s), before beginning to move to the left. In other words, the cart is decelerating, and its acceleration must point opposite to the velocity, or to the left. Thus, the acceleration is negative. Since the initial velocity, the final velocity, and the displacement are known, Equation 2.9 $\left(v^2 = v_0^2 + 2ax\right)$ can be used to determine the acceleration.

SOLUTION Solving Equation 2.9 for the acceleration a shows that

$$a=\frac{v^2-v_0^2}{2x}=\frac{(0\text{ m/s})^2-(+5.0\text{ m/s})^2}{2(+12.5\text{ m})}=\boxed{-1.0\text{ m/s}^2}$$

---

65. SSM ***REASONING AND SOLUTION*** When air resistance is neglected, free fall conditions are applicable. The final speed can be found from Equation 2.9;

$$v^2=v_0^2+2ay$$

where v_0 is zero since the stunt man falls from rest. If the origin is chosen at the top of the hotel and the upward direction is positive, then the displacement is y = – 99.4 m. Solving for v, we have

$$v=-\sqrt{2ay}=-\sqrt{2(-9.80\text{ m/s}^2)(-99.4\text{ m})}=-44.1\text{ m/s}$$

The speed at impact is the magnitude of this result or $\boxed{44.1\text{ m/s}}$.

---

66. ***REASONING AND SOLUTION*** In 12 minutes the sloth travels a distance of

$$x_s=v_st=(0.037\text{ m/s})(12\text{ min})\left(\frac{60\text{ s}}{1\text{ min}}\right)=27\text{ m}$$

while the tortoise travels a distance of

$$x_t=v_t\text{t}=(0.076\text{ m/s})(12\text{ min})\left(\frac{60\text{ s}}{1\text{ min}}\right)=55\text{ m}$$

The tortoise goes farther than the sloth by an amount that equals 55 m – 27 m = $\boxed{28\text{ m}}$

---

67. SSM WWW ***REASONING AND SOLUTION***

a. The magnitude of the acceleration can be found from Equation 2.4 ($v=v_0+at$) as

$$a=\frac{v-v_0}{t}=\frac{3.0\text{ m/s}-0\text{ m/s}}{2.0\text{ s}}=\boxed{1.5\text{ m/s}^2}$$

b. Similarly the magnitude of the acceleration of the car is

$$a = \frac{v - v_0}{t} = \frac{41.0 \text{ m/s} - 38.0 \text{ m/s}}{2.0 \text{ s}} = \boxed{1.5 \text{ m/s}^2}$$

c. Assuming that the acceleration is constant, the displacement covered by the car can be found from Equation 2.9 ($v^2 = v_0^2 + 2ax$):

$$x = \frac{v^2 - v_0^2}{2a} = \frac{(41.0 \text{ m/s})^2 - (38.0 \text{ m/s})^2}{2(1.5 \text{ m/s}^2)} = 79 \text{ m}$$

Similarly, the displacement traveled by the jogger is

$$x = \frac{v^2 - v_0^2}{2a} = \frac{(3.0 \text{ m/s})^2 - (0 \text{ m/s})^2}{2(1.5 \text{ m/s}^2)} = 3.0 \text{ m}$$

Therefore, the car travels 79 m – 3.0 m = $\boxed{76 \text{ m}}$ further than the jogger.

---

68. ***REASONING*** The initial velocity of the compass is +2.50 m/s. The initial position of the compass is 3.00 m and its final position is 0 m when it strikes the ground. The displacement of the compass is the final position minus the initial position, or $y = -3.00$ m. As the compass falls to the ground, its acceleration is the acceleration due to gravity, $a = -9.80 \text{ m/s}^2$. Equation 2.8 $\left(y = v_0 t + \frac{1}{2} a t^2\right)$ can be used to find how much time elapses before the compass hits the ground.

SOLUTION Starting with Equation 2.8, we use the quadratic equation to find the elapsed time.

$$t = \frac{-v_0 \pm \sqrt{v_0^2 - 4\left(\frac{1}{2}a\right)(-y)}}{2\left(\frac{1}{2}a\right)} = \frac{-(2.50 \text{ m/s}) \pm \sqrt{(2.50 \text{ m/s})^2 - 4\left(-4.90 \text{ m/s}^2\right)\left[-(-3.00 \text{ m})\right]}}{2\left(-4.90 \text{ m/s}^2\right)}$$

There are two solutions to this quadratic equation, $t_1 = \boxed{1.08 \text{ s}}$ and $t_2 = -0.568$ s. The second solution, being a negative time, is discarded.

---

69. ***REASONING*** The average speed is the distance traveled divided by the elapsed time (Equation 2.1). Since the average speed and distance are known, we can use this relation to find the time.

SOLUTION The time it takes for the continents to drift apart by 1500 m is

$$\text{Elapsed time} = \frac{\text{Distance}}{\text{Average speed}} = \frac{1500\,\text{m}}{\left(3\dfrac{\text{cm}}{\text{y}}\right)\left(\dfrac{1\text{ m}}{100\text{ cm}}\right)} = \boxed{5\times10^4\text{ y}}$$

---

70. ***REASONING AND SOLUTION*** The velocity of the automobile for each stage is given by Equation 2.4: $v = v_0 + at$. Therefore,

$$v_1 = v_0 + a_1 t = 0\text{ m/s} + a_1 t \quad \text{and} \quad v_2 = v_1 + a_2 t$$

Since the magnitude of the car's velocity at the end of stage 2 is 2.5 times greater than it is at the end of stage 1, $v_2 = 2.5v_1$. Thus, rearranging the result for v_2, we find.

$$a_2 = \frac{v_2 - v_1}{t} = \frac{2.5v_1 - v_1}{t} = \frac{1.5v_1}{t} = \frac{1.5(a_1 t)}{t} = 1.5a_1 = 1.5(3.0\text{ m/s}^2) = \boxed{4.5\text{ m/s}^2}$$

---

71. ***REASONING*** At time t both rockets return to their starting points and have a displacement of zero. This occurs, because each rocket is decelerating during the first half of its journey. However, rocket A has a smaller initial velocity than rocket B. Therefore, in order for rocket B to decelerate and return to its point of origin in the same time as rocket A, rocket B must have a deceleration with a greater magnitude than that for rocket A. Since we know that the displacement of each rocket is zero at time t, since both initial velocities are given, and since we seek information about the acceleration, we begin our solution with Equation 2.8, for it contains just these variables.

SOLUTION Applying Equation 2.8 to each rocket gives

$$x_A = v_{0A}t + \tfrac{1}{2}a_A t^2 \qquad x_B = v_{0B}t + \tfrac{1}{2}a_B t^2$$

$$0 = v_{0A}t + \tfrac{1}{2}a_A t^2 \qquad 0 = v_{0B}t + \tfrac{1}{2}a_B t^2$$

$$0 = v_{0A} + \tfrac{1}{2}a_A t \qquad 0 = v_{0B} + \tfrac{1}{2}a_B t$$

$$t = \frac{-2v_{0A}}{a_A} \qquad t = \frac{-2v_{0B}}{a_B}$$

The time for each rocket is the same, so that we can equate the two expressions for t, with the result that

$$\frac{-2v_{0A}}{a_A} = \frac{-2v_{0B}}{a_B} \quad \text{or} \quad \frac{v_{0A}}{a_A} = \frac{v_{0B}}{a_B}$$

Solving for a_B gives

$$a_B = \frac{a_A}{v_{0A}} v_{0B} = \frac{-15\text{ m/s}^2}{5800\text{ m/s}}(8600\text{ m/s}) = \boxed{-22\text{ m/s}^2}$$

As expected, the magnitude of the acceleration for rocket B is greater than that for rocket A.

---

72. ***REASONING AND SOLUTION*** The car enters the speedway with a speed of

$$v_{01} = a_1 t_1 = (6.0\text{ m/s}^2)(4.0\text{ s}) = 24\text{ m/s}$$

After an additional time, t, it will have traveled a distance of

$$x = v_{01}t + a_1 t^2/2$$

to overtake the other car. This second car travels the same distance $x = v_{02}t$. Equating and solving for t yields

$$t = \frac{2(v_{02} - v_{01})}{a_1} = \frac{2(70.0\text{ m/s} - 24\text{ m/s})}{6.0\text{ m/s}^2} = \boxed{15\text{ s}}$$

---

73. SSM ***REASONING AND SOLUTION***

a. The total displacement traveled by the bicyclist for the entire trip is equal to the sum of the displacements traveled during each part of the trip. The displacement traveled during each part of the trip is given by Equation 2.2: $\Delta x = \bar{v}\Delta t$. Therefore,

$$\Delta x_1 = (7.2\text{ m/s})(22\text{ min})\left(\frac{60\text{ s}}{1\text{ min}}\right) = 9500\text{ m}$$

$$\Delta x_2 = (5.1\text{ m/s})(36\text{ min})\left(\frac{60\text{ s}}{1\text{ min}}\right) = 11\ 000\text{ m}$$

$$\Delta x_3 = (13\text{ m/s})(8.0\text{ min})\left(\frac{60\text{ s}}{1\text{ min}}\right) = 6200\text{ m}$$

The total displacement traveled by the bicyclist during the entire trip is then

$$\Delta x = 9500\text{ m} + 11\ 000\text{ m} + 6200\text{ m} = \boxed{2.67\times 10^4\text{ m}}$$

b. The average velocity can be found from Equation 2.2.

$$\bar{v} = \frac{\Delta x}{\Delta t} = \frac{2.67 \times 10^4 \text{ m}}{(22 \text{ min} + 36 \text{ min} + 8.0 \text{ min})}\left(\frac{1 \text{min}}{60 \text{ s}}\right) = \boxed{6.74 \text{ m/s, due north}}$$

---

74. ***REASONING AND SOLUTION*** Her average speed is the total distance she falls divided by the total time of the fall:

$$v = \frac{x}{t} = \frac{625 \text{ m} + 356 \text{ m}}{15.0 \text{ s} + 142 \text{ s}} = \boxed{6.25 \text{ m/s}}$$

The direction of the velocity is $\boxed{\text{downward}}$.

---

75. SSM ***REASONING*** Once the man sees the block, the man must get out of the way in the time it takes for the block to fall through an additional 12.0 m. The velocity of the block at the instant that the man looks up can be determined from Equation 2.9. Once the velocity is known at that instant, Equation 2.8 can be used to find the time required for the block to fall through the additional distance.

SOLUTION When the man first notices the block, it is 14.0 m above the ground and its displacement from the starting point is $y = 14.0 \text{ m} - 53.0 \text{ m}$. Its velocity is given by Equation 2.9 $\left(v^2 = v_0^2 + 2ay\right)$. Since the block is moving down, its velocity has a negative value,

$$v = -\sqrt{v_0 + 2ay} = -\sqrt{(0 \text{ m/s})^2 + 2(-9.80 \text{ m/s}^2)(14.0 \text{ m} - 53.0 \text{ m})} = -27.7 \text{ m/s}$$

The block then falls the additional 12.0 m to the level of the man's head in a time t which satisfies Equation 2.8:

$$y = v_0 t + \frac{1}{2} a t^2$$

where $y = -12.0$ m and $v_0 = -27.7$ m/s. Thus, t is the solution to the quadratic equation

$$4.90 t^2 + 27.7 t - 12.0 = 0$$

where the units have been suppressed for brevity. From the quadratic formula, we obtain

$$t = \frac{-27.7 \pm \sqrt{(27.7)^2 - 4(4.90)(-12.0)}}{2(4.90)} = 0.40 \text{ s} \quad \text{or} \quad -6.1 \text{ s}$$

The negative solution can be rejected as nonphysical, and the time it takes for the block to reach the level of the man is $\boxed{0.40 \text{ s}}$.

---

76. ***REASONING*** We choose due north as the positive direction. Our solution is based on the fact that when the police car catches up, both cars will have the same displacement, relative to the point where the speeder passed the police car. The displacement of the speeder can be obtained from the definition of average velocity given in Equation 2.2, since the speeder is moving at a constant velocity. During the 0.800-s reaction time of the policeman, the police car is also moving at a constant velocity. Once the police car begins to accelerate, its displacement can be expressed as in Equation 2.8 $\left(x = v_0 t + \frac{1}{2}at^2\right)$, because the initial velocity v_0 and the acceleration a are known and it is the time t that we seek. We will set the displacements of the speeder and the police car equal and solve the resulting equation for the time t.

SOLUTION Let t equal the time during the accelerated motion of the police car. Relative to the point where he passed the police car, the speeder then travels a time of $t + 0.800$ s before the police car catches up. During this time, according to the definition of average velocity given in Equation 2.2, his displacement is

$$x_{\text{Speeder}} = v_{\text{Speeder}}\left(t + 0.800 \text{ s}\right) = \left(42.0 \text{ m/s}\right)\left(t + 0.800 \text{ s}\right)$$

The displacement of the police car consists of two contributions, the part due to the constant-velocity motion during the reaction time and the part due to the accelerated motion. Using Equation 2.2 for the contribution from the constant-velocity motion and Equation 2.9 for the contribution from the accelerated motion, we obtain

$$x_{\text{Police car}} = \underbrace{v_{\text{0, Police car}}\left(0.800 \text{ s}\right)}_{\substack{\text{Constant velocity motion,} \\ \text{Equation 2.2}}} + \underbrace{v_{\text{0, Police car}}t + \tfrac{1}{2}at^2}_{\substack{\text{Accelerated motion,} \\ \text{Equation 2.8}}}$$

$$= \left(18.0 \text{ m/s}\right)\left(0.800 \text{ s}\right) + \left(18.0 \text{ m/s}\right)t + \tfrac{1}{2}\left(5.00 \text{ m/s}^2\right)t^2$$

Setting the two displacements equal we obtain

$$\underbrace{\left(42.0 \text{ m/s}\right)\left(t + 0.800 \text{ s}\right)}_{\text{Displacement of speeder}} = \underbrace{\left(18.0 \text{ m/s}\right)\left(0.800 \text{ s}\right) + \left(18.0 \text{ m/s}\right)t + \tfrac{1}{2}\left(5.00 \text{ m/s}^2\right)t^2}_{\text{Displacement of police car}}$$

Rearranging and combining terms gives this result in the standard form of a quadratic equation:

$$\left(2.50 \text{ m/s}^2\right)t^2 - \left(24.0 \text{ m/s}\right)t - 19.2 \text{ m} = 0$$

Solving for t shows that

$$t = \frac{-(-24.0\text{ m/s}) \pm \sqrt{(-24.0\text{ m/s})^2 - 4(2.50\text{ m/s}^2)(-19.2\text{ m})}}{2(2.50\text{ m/s}^2)} = 10.3\text{ s}$$

We have ignored the negative root, because it leads to a negative value for the time, which is unphysical. The total time for the police car to catch up, including the reaction time, is

$$0.800\text{ s} + 10.3\text{ s} = \boxed{11.1\text{ s}}$$

---

77. ***REASONING*** The players collide when they have the same x coordinate relative to a common origin. For convenience, we will place the origin at the starting point of the first player. From Equation 2.8, the x coordinate of each player is given by

$$x_1 = v_{01}t_1 + \tfrac{1}{2}a_1t_1^2 = \tfrac{1}{2}a_1t_1^2 \qquad (1)$$

and

$$x_2 = d + v_{02}t_2 + \tfrac{1}{2}a_2t_2^2 = d + \tfrac{1}{2}a_2t_2^2 \qquad (2)$$

where d = +48 m is the initial position of the second player. When $x_1 = x_2$, the players collide at time $t = t_1 = t_2$.

SOLUTION
a. Equating Equations (1) and (2) when $t_1 = t_2 = t$, we have

$$\tfrac{1}{2}a_1t^2 = d + \tfrac{1}{2}a_2t^2$$

We note that a_1 = +0.50 m/s^2, while a_2 = –0.30 m/s^2, since the first player accelerates in the +x direction and the second player in the –x direction. Solving for t, we have

$$t = \sqrt{\frac{2d}{(a_1 - a_2)}} = \sqrt{\frac{2(48\text{ m})}{(0.50\text{ m/s}^2) - (-0.30\text{ m/s}^2)}} = \boxed{11\text{ s}}$$

b. From Equation (1),

$$x_1 = \tfrac{1}{2}a_1t_1^2 = \tfrac{1}{2}(0.50\text{ m/s}^2)(11\text{ s})^2 = \boxed{3.0 \times 10^1\text{ m}}$$

---

78. ***REASONING AND SOLUTION*** Assuming that down is positive, the distance that the tile falls in going from the roof top to the top of the window is found to be

$$y = \frac{v^2}{2a}$$

where v is the velocity of the tile at the top of the window. The tile travels an additional distance h in traversing the window in time t.

$$h = v_0 t + (1/2)\, at^2$$

Solving for v_0 yields,

$$v_0 = \frac{h}{t} - (1/2)at = \frac{1.6\text{ m}}{0.20\text{ s}} - (1/2)(9.80\text{ m/s}^2)(0.20\text{ s}) = 7.0\text{ m/s}.$$

Then,

$$y = \frac{(7.0\text{ m/s})^2}{2\left(9.80\text{ m/s}^2\right)} = \boxed{2.5\text{ m}}$$

---

79. SSM ***REASONING*** As the train passes through the crossing, its motion is described by Equations 2.4 ($v = v_0 + at$) and 2.7 $\left[x = \frac{1}{2}(v + v_0)t\right]$, which can be rearranged to give

$$v - v_0 = at \qquad \text{and} \qquad v + v_0 = \frac{2x}{t}$$

These can be solved simultaneously to obtain the speed v when the train reaches the end of the crossing. Once v is known, Equation 2.4 can be used to find the time required for the train to reach a speed of 32 m/s.

SOLUTION Adding the above equations and solving for v, we obtain

$$v = \frac{1}{2}\left(at + \frac{2x}{t}\right) = \frac{1}{2}\left[(1.6\text{ m/s}^2)(2.4\text{ s}) + \frac{2(20.0\text{ m})}{2.4\text{ s}}\right] = 1.0 \times 10^1\text{ m/s}$$

The motion from the end of the crossing until the locomotive reaches a speed of 32 m/s requires a time

$$t = \frac{v - v_0}{a} = \frac{32\text{ m/s} - 1.0 \times 10^1\text{ m/s}}{1.6\text{ m/s}^2} = \boxed{14\text{ s}}$$

---

80. ***REASONING AND SOLUTION*** During the first phase of the acceleration,

$$a_1 = \frac{v}{t_1}$$

During the second phase of the acceleration,

$$v = (6.4 \text{ m/s}) - (1.1 \text{ m/s}^2)(2.0 \text{ s}) = 4.2 \text{ m/s}$$

Then

$$a_1 = \frac{4.2 \text{ m/s}}{3.0 \text{ s}} = \boxed{1.4 \text{ m/s}^2}$$

---

81. ***CONCEPT QUESTION*** The displacement is a vector that points from an object's initial position to its final position. If the final position is greater than the initial position, the displacement is positive. On the other hand, if the final position is less than the initial position, the displacement is negative

a. The final position is greater than the initial position, so the displacement is $\boxed{\text{positive}}$.

b. The final position is less than the initial position, so the displacement is $\boxed{\text{negative}}$.

c. The final position is greater than the initial position, so the displacement is $\boxed{\text{positive}}$.

SOLUTION The displacement is defined as Displacement = $x - x_0$, where x is the final position and x_0 is the initial position. The displacements for the three cases are:

a. Displacement = 6 m – 2 m = $\boxed{+4 \text{ m}}$

b. Displacement = 2 m – 6 m = $\boxed{-4 \text{ m}}$

c. Displacement = 7 m – (–3 m) = $\boxed{+10 \text{ m}}$

---

82. ***CONCEPT QUESTION*** According to Equation 2.2, the direction of the car's average velocity is the same as its displacement. Therefore, the answers to this Concept Question are the same as those to the Concept Question in problem 81. The average velocities are: (a) $\boxed{\text{positive}}$, (b) $\boxed{\text{negative}}$, (c) $\boxed{\text{positive}}$.

SOLUTION The average velocity is equal to the displacement divided by the elapsed time (Equation 2.2), where the displacement is equal to the final position minus the initial position;

$$\bar{v} = \frac{x - x_0}{t - t_0}$$

The average velocities for the three cases are:

a. Average velocity = (6 m – 2 m)/(0.5 s) = $\boxed{+8 \text{ m/s}}$

b. Average velocity = (2 m – 6 m)/(0.5 s) = $\boxed{-8 \text{ m/s}}$

c. Average velocity = [7 m – (–3 m)]/(0.5 s) = $\boxed{+20 \text{ m/s}}$

---

83. ***CONCEPT QUESTIONS*** The average acceleration is defined by Equation 2.4 as the change in velocity divided by the elapsed time. The change in velocity is equal to the final velocity minus the initial velocity. Therefore, the change in velocity, and hence the acceleration, is positive if the final velocity is greater than the initial velocity. The acceleration is negative if the final velocity is less than the initial velocity.

a. The final velocity is greater than the initial velocity, so the acceleration is $\boxed{\text{positive}}$.

b. The final velocity is less than the initial velocity, so the acceleration is $\boxed{\text{negative}}$.

c. The final velocity is greater than the initial velocity (–3.0 m/s is greater than –6.0 m/s), so the acceleration is $\boxed{\text{positive}}$.

d. The final velocity is less than the initial velocity, so the acceleration is $\boxed{\text{negative}}$.

SOLUTION The average acceleration is equal to the change in velocity divided by the elapsed time (Equation 2.4), where the change in velocity is equal to the final velocity minus the initial velocity;

$$\bar{a} = \frac{v - v_0}{t - t_0}$$

The average accelerations are:

a. $\bar{a} = (5.0\text{ m/s} - 2.0\text{ m/s})/(2.0\text{ s}) = \boxed{+1.5\text{ m/s}^2}$

b. $\bar{a} = (2.0\text{ m/s} - 5.0\text{ m/s})/(2.0\text{ s}) = \boxed{-1.5\text{ m/s}^2}$

c. $\bar{a} = [-3.0\text{ m/s} - (-6.0\text{ m/s})]/(2.0\text{ s}) = \boxed{+1.5\text{ m/s}^2}$

d. $\bar{a} = (-4.0\text{ m/s} - 4.0\text{ m/s})/(2.0\text{ s}) = \boxed{-4.0\text{ m/s}^2}$

---

84. ***CONCEPT QUESTIONS*** When the velocity and acceleration vectors are in the same direction, the speed of the object increases in time. When the velocity and acceleration vectors are in opposite directions, the speed of the object decreases in time.

a. The initial velocity and acceleration are in the same direction, so the speed is $\boxed{\text{increasing}}$.

b. The initial velocity and acceleration are in opposite directions, so the speed is $\boxed{\text{decreasing}}$.

c. The initial velocity and acceleration are in opposite directions, so the speed is $\boxed{\text{decreasing}}$.

d. The initial velocity and acceleration are in the same direction, so the speed is $\boxed{\text{increasing}}$.

SOLUTION The final velocity v is related to the initial velocity v_0, the acceleration a, and the elapsed time t through Equation 2.4, $v = v_0 + a\,t$. The final velocities and speeds for the four moving objects are:

a. $v = 12\text{ m/s} + (3.0\text{ m/s}^2)(2.0\text{ s}) = 18\text{ m/s}$. The final speed is $\boxed{18\text{ m/s}}$.

b. $v = 12\text{ m/s} + (-3.0\text{ m/s}^2)(2.0\text{ s}) = 6.0\text{ m/s}$. The final speed is $\boxed{6.0\text{ m/s}}$.

c. $v = -12\text{ m/s} + (3.0\text{ m/s}^2)(2.0\text{ s}) = -6.0\text{ m/s}$. The final speed is $\boxed{6.0\text{ m/s}}$.

d. $v = -12\text{ m/s} + (-3.0\text{ m/s}^2)(2.0\text{ s}) = -18\text{ m/s}$. The final speed is $\boxed{18\text{ m/s}}$.

---

85. ***CONCEPT QUESTIONS***

a. No. A zero acceleration means that the velocity is constant, but not necessarily zero.

b. No, because the direction of the car changes and so the final velocity is not equal to the initial velocity. The change in the velocity is not zero, so the average acceleration is not zero.

SOLUTION

The average acceleration is equal to the change in velocity divided by the elapsed time,

$$\bar{a} = \frac{v - v_0}{t - t_0} \qquad (2.4)$$

a. The initial and final velocities are +82 m/s and +82 m/s. The average acceleration is

$$\bar{a} = (82\text{ m/s} - 82\text{ m/s})/(t - t_0) = \boxed{0\text{ m/s}^2}$$

b. The initial velocity is + 82 m/s and the final velocity is –82 m/s. The average acceleration is

$$\bar{a} = (-82\text{ m/s} - 82\text{ m/s})/(12\text{ s}) = \boxed{-14\text{ m/s}^2}$$

---

86. ***CONCEPT QUESTIONS***

a. The acceleration of the ball does not reverse direction on the downward part of the trip. The acceleration is the same for both the upward and downward parts, namely -9.80 m/s^2.

b. The displacement is $y = 0$ m, since the final and initial positions of the ball are the same.

SOLUTION The displacement of the ball, the acceleration due to gravity, and the elapsed time are known. We may use Equation 2.8, $y = v_0\,t + \tfrac{1}{2}at^2$, to find the initial velocity of the ball. Solving this equation for the initial velocity gives

$$v_0 = \frac{y - \frac{1}{2}at^2}{t} = \frac{0\text{ m} - \frac{1}{2}\left(-9.80\text{ m/s}^2\right)\left(8.0\text{ s}\right)^2}{8.0\text{ s}} = \boxed{+39\text{ m/s}}$$

---

87. ***CONCEPT QUESTIONS***

a. Runner B's final velocity is greater than runner A's constant velocity. This must be so, because B starts from rest and runs more slowly than A at the beginning. The only way in which B can cover the same distance in the same time as A and run more slowly at the beginning is to run more quickly at the end.

b. Runner B's average velocity is the same as runner A's constant velocity. This follows from the definition of average velocity given in Equation 2.2 as the displacement divided by the elapsed time. When the velocity is constant, as it is for runner A, the average velocity is the same as the constant velocity. Since both displacement and time are the same for each runner, this equation gives the same value for runner B's average velocity and runner A's constant velocity.

c. Runner B's constant acceleration can be calculated from Equation 2.4 ($v_B = v_{B0} + a_B t$), which is one of the equations of kinematics and gives the acceleration as [$a_B = (v_B - v_{B0})/t$]. Since runner B starts from rest, we know that $v_{B0} = 0$ m/s. Furthermore, t is given. Therefore, calculation of the acceleration a_B requires that we first determine the final velocity v_B.

SOLUTION

a. According to Equation 2.2, the velocity of runner A is the displacement L divided by the time t. Thus, we obtain

$$v_A = \frac{L}{t} = \frac{460\text{ m}}{210\text{ s}} = \boxed{2.2\text{ m/s}}$$

b. Since the acceleration of runner B is constant, we know that his average velocity is given by Equation 2.6 as $\bar{v}_B = \frac{1}{2}\left(v_B + v_{B0}\right)$, where v_B is the final velocity and v_{B0} is the initial velocity. Solving for the final velocity and using the fact that runner B starts from rest ($v_{B0} = 0$ m/s) gives

$$v_B = 2\bar{v}_B - v_{B0} = 2\bar{v}_B \qquad (1)$$

As discussed in the answer to Concept Question (b), the average velocity of runner B is equal to the constant velocity of runner A. Substituting this result into Equation (1), we find that

$$v_B = 2\bar{v}_B = 2v_A = 2\left(2.2\text{ m/s}\right) = \boxed{4.4\text{ m/s}}$$

As expected, runner B's final velocity is greater than runner A's constant velocity.

c. Solving Equation 2.4 ($v_B = v_{B0} + a_B t$) for the acceleration shows that

$$a_B = \frac{v_B - v_{B0}}{t} = \frac{4.4\text{ m/s} - 0\text{ m/s}}{210\text{ s}} = \boxed{0.021\text{ m/s}^2}$$

---

88. ***CONCEPT QUESTIONS***

a. The stone that is thrown upward loses speed on the way up. The initial velocity points upward, while the acceleration due to gravity points downward. Under these circumstances, the stone decelerates on the way up. In other words, it loses speed.

b. The stone that is thrown downward gains speed on the way down. The initial velocity points downward, in the same direction as the acceleration due to gravity. Under these circumstances, the stone accelerates on the way down and gains speed.

c. The stones cross paths below the point that corresponds to half the height of the cliff. To see why, consider where they would cross paths if they each maintained their initial speed as the moved. Then, they would cross paths exactly at the halfway point. However, the stone traveling upward begins immediately to lose speed, while the stone traveling downward immediately gains speed. Thus, the upward moving stone travels more slowly than the downward moving stone. Consequently, the stone thrown downward has traveled farther when it reaches the crossing point than the stone thrown upward. The crossing point, then, is below the halfway point.

SOLUTION The initial velocity v_0 is known for both stones, as is the acceleration a due to gravity. In addition, we know that at the crossing point the stones are at the same place at the same time t. Furthermore, the position of each stone is specified by its displacement y from its starting point. The equation of kinematics that relates the variables v_0, a, t and y is Equation 2.8 $\left(y = v_0 t + \frac{1}{2}at^2\right)$, and we will use it in our solution. In using this equation, we will assume upward to be the positive direction. Applying Equation 2.8 to each stone, we have

$$\underbrace{y_{\text{up}} = v_0^{\text{up}} t + \tfrac{1}{2}at^2}_{\text{Upward moving stone}} \quad \text{and} \quad \underbrace{y_{\text{down}} = v_0^{\text{down}} t + \tfrac{1}{2}at^2}_{\text{Downward moving stone}}$$

In these expressions t is the time it takes for either stone to reach the crossing point, and a is the acceleration due to gravity. Note that y_{up} is the displacement of the upward moving stone above the base of the cliff, y_{down} is the displacement of the downward moving stone below the top of the cliff, and H is the displacement of the cliff-top above the base of the cliff, as the drawing shows. The distances above and below the crossing point must add to equal the height of the cliff, so we have

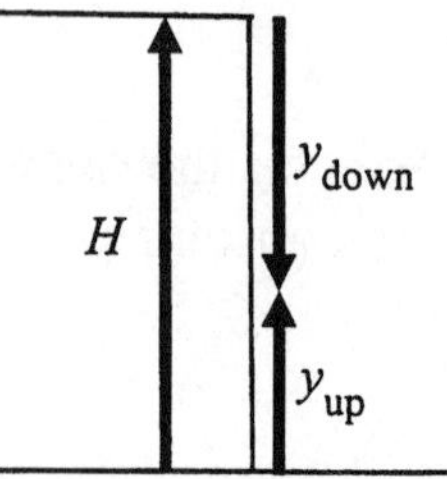

$$y_{\text{up}} - y_{\text{down}} = H$$

where the minus sign appears because the displacement y_{down} points in the negative direction. Substituting the two expressions for y_{up} and y_{down} into this equation gives

$$v_0^{\text{up}} t + \tfrac{1}{2}at^2 - \left(v_0^{\text{down}} t + \tfrac{1}{2}at^2\right) = H$$

This equation can be solved for t to show that the travel time to the crossing point is

$$t = \frac{H}{v_0^{\text{up}} - v_0^{\text{down}}}$$

Substituting this result into the expression from Equation 2.8 for y_{up} gives

$$y_{\text{up}} = v_0^{\text{up}} t + \tfrac{1}{2}at^2 = v_0^{\text{up}}\left(\frac{H}{v_0^{\text{up}} - v_0^{\text{down}}}\right) + \tfrac{1}{2}a\left(\frac{H}{v_0^{\text{up}} - v_0^{\text{down}}}\right)^2$$

$$= (9.00\text{ m/s})\left[\frac{6.00\text{ m}}{9.00\text{ m/s} - (-9.00\text{ m/s})}\right] + \tfrac{1}{2}\left(-9.80\text{ m/s}^2\right)\left[\frac{6.00\text{ m}}{9.00\text{ m/s} - (-9.00\text{ m/s})}\right]^2$$

$$= 2.46\text{ m}$$

Thus, the crossing is located a distance of $\boxed{2.46\text{ m}}$ above the base of the cliff, which is below the halfway point of 3.00 m, as expected.

CHAPTER 3 | *KINEMATICS IN TWO DIMENSIONS*

PROBLEMS

---

1. SSM ***REASONING*** The displacement of the elephant seal has two components; 460 m due east and 750 m downward. These components are mutually perpendicular; hence, the Pythagorean theorem can be used to determine their resultant.

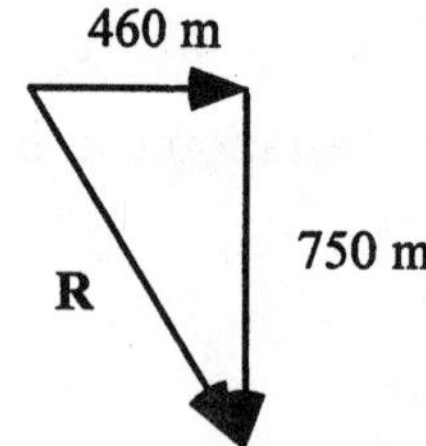

SOLUTION From the Pythagorean theorem,

$$R^2 = (460\text{ m})^2 + (750\text{ m})^2$$

Therefore,

$$R = \sqrt{(460\text{ m})^2 + (750\text{ m})^2} = \boxed{8.8 \times 10^2\text{ m}}$$

---

2. ***REASONING AND SOLUTION*** The batter ends up at third base, 27.4 m from home plate, where he started. Therefore, the magnitude of his displacement is simply $\boxed{27.4\text{ m}}$.

---

3. ***REASONING AND SOLUTION*** The horizontal displacement is

$$x = 19\,600\text{ m} - 11\,200\text{ m} = 8400\text{ m}$$

The vertical displacement is

$$y = 4900\text{ m} - 3200\text{ m} = 1700\text{ m}$$

The magnitude of the displacement is therefore,

$$\Delta r = \sqrt{x^2 + y^2} = \sqrt{(8400\text{ m})^2 + (1700\text{ m})^2} = \boxed{8600\text{ m}}$$

---

4. ***REASONING AND SOLUTION*** The increase in altitude represents $v_y = 6.80$ m/s. The movement of the shadow represents $v_x = 15.5$ m/s. The magnitude of the glider's velocity is therefore

$$v = \sqrt{v_x^2 + v_y^2} = \sqrt{(15.5\text{ m/s})^2 + (6.80\text{ m/s})^2} = \boxed{16.9\text{ m/s}}$$

---

5. SSM ***REASONING AND SOLUTION*** The horizontal and vertical components of the plane's velocity are related to the speed of the plane by the Pythagorean theorem: $v^2 = v_\text{h}^2 + v_\text{v}^2$. Solving for v_h we have

$$v_\text{h} = \sqrt{v^2 - v_\text{v}^2} = \sqrt{(245 \text{ m/s})^2 - (40.6 \text{ m/s})^2} = \boxed{242 \text{ m/s}}$$

---

6. ***REASONING*** To determine the horizontal and vertical components of the launch velocity, we will use trigonometry. To do so, however, we need to know both the launch angle and the magnitude of the launch velocity. The launch angle is given. The magnitude of the launch velocity can be determined from the given acceleration and the definition of acceleration given in Equation 3.2.

SOLUTION According to Equation 3.2, we have

$$a = \frac{v - v_0}{t - t_0} \quad \text{or} \quad 340 \text{ m/s}^2 = \frac{v - 0 \text{ m/s}}{0.050 \text{ s}} \quad \text{or} \quad v = \left(340 \text{ m/s}^2\right)\left(0.050 \text{ s}\right)$$

Using trigonometry, we find the components to be

$$v_x = v \cos 51° = \left(340 \text{ m/s}^2\right)\left(0.050 \text{ s}\right)\cos 51° = \boxed{11 \text{ m/s}}$$

$$v_y = v \sin 51° = \left(340 \text{ m/s}^2\right)\left(0.050 \text{ s}\right)\sin 51° = \boxed{13 \text{ m/s}}$$

---

7. ***REASONING*** Trigonometry indicates that the x and y components of the dolphin's velocity are related to the launch angle θ according to $\tan \theta = v_y/v_x$.

SOLUTION Using trigonometry, we find that the y component of the dolphin's velocity is

$$v_y = v_x \tan \theta = v_x \tan 35° = (7.7 \text{ m/s})\tan 35° = \boxed{5.4 \text{ m/s}}$$

---

8. ***REASONING***
a. We designate the direction down and parallel to the ramp as the $+x$ direction, and the table shows the variables that are known. Since three of the five kinematic variables have values, one of the equations of kinematics can be employed to find the acceleration a_x.

x-Direction Data

x	a_x	v_x	v_{0x}	t
+12.0 m	?	+7.70 m/s	0 m/s	

b. The acceleration vector points down and parallel to the ramp, and the angle of the ramp is 25.0° relative to the ground (see the drawing). Therefore, trigonometry can be used to determine the component a_{parallel} of the acceleration that is parallel to the ground.

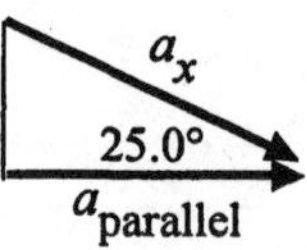

SOLUTION

a. Equation 3.6a $\left(v_x^2 = v_{0x}^2 + 2a_x x\right)$ can be used to find the acceleration in terms of the three known variables. Solving this equation for a_x gives

$$a_x = \frac{v_x^2 - v_{0x}^2}{2x} = \frac{(+7.70 \text{ m/s})^2 - (0 \text{ m/s})^2}{2(+12.0 \text{ m})} = \boxed{2.47 \text{ m/s}^2}$$

b. The drawing shows that the acceleration vector is oriented 25.0° relative to the ground. The component a_{parallel} of the acceleration that is parallel to the ground is

$$a_{\text{parallel}} = a_x \cos 25.0° = \left(2.47 \text{ m/s}^2\right)\cos 25.0° = \boxed{2.24 \text{ m/s}^2}$$

9\. SSM WWW ***REASONING AND SOLUTION*** The escalator is the hypotenuse of a right triangle formed by the lower floor and the vertical distance between floors, as shown in the figure. The angle θ at which the escalator is inclined above the horizontal is related to the length L of the escalator and the vertical distance between the floors by the sine function:

$$\sin\theta = \frac{6.00 \text{ m}}{L} \qquad (1)$$

The length L of the escalator can be found from the right triangle formed by the components of the shopper's displacement up the escalator and to the right of the escalator.

From the Pythagorean theorem, we have

$$L^2 + (9.00 \text{ m})^2 = (16.0 \text{ m})^2$$

so that

$$L = \sqrt{(16.0 \text{ m})^2 - (9.00 \text{ m})^2} = 13.2 \text{ m}$$

From Equation (1) we have

$$\theta = \sin^{-1}\left(\frac{6.00 \text{ m}}{13.2 \text{ m}}\right) = \boxed{27.0°}$$

---

10. ***REASONING*** The component method can be used to determine the magnitude and direction of the bird watcher's displacement. Once the displacement is known, Equation 3.1 can be used to find the average velocity.

SOLUTION The following table gives the components of the individual displacements of the bird watcher. The last entry gives the components of the bird watcher's resultant displacement. Due east and due north have been chosen as the positive directions.

Displacement	*East/West Component*	*North/South Component*
A	0.50 km	0
B	0	–0.75 km
C	–(2.15 km) cos 35.0° = –1.76 km	(2.15 km) sin 35.0° = 1.23 km
$\mathbf{r} = \mathbf{A} + \mathbf{B} + \mathbf{C}$	–1.26 km	0.48 km

a. From the Pythagorean theorem, we have

$$r = \sqrt{(-1.26 \text{ km})^2 + (0.48 \text{ km})^2} = \boxed{1.35 \text{ km}}$$

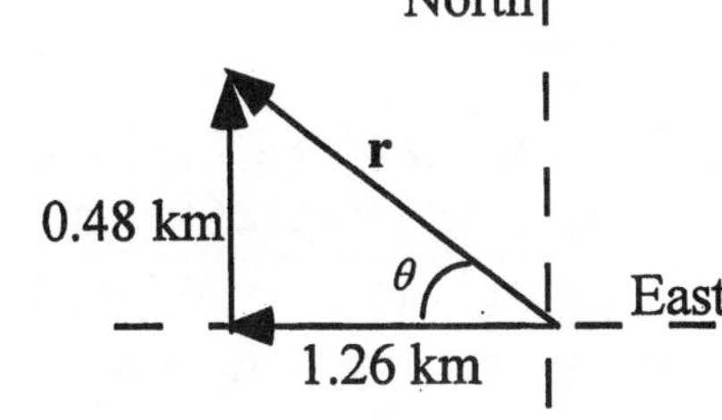

The angle θ is given by

$$\theta = \tan^{-1}\left(\frac{0.48 \text{ km}}{1.26 \text{ km}}\right) = \boxed{21°\text{, north of west}}$$

b. From Equation 3.1, the average velocity is

$$\bar{\mathbf{v}} = \frac{\Delta \mathbf{r}}{\Delta t} = \frac{1.35 \text{ km}}{2.50 \text{ h}} = \boxed{0.540 \text{ km/h, } 21° \text{ north of west}}$$

Note that the direction of the average velocity is, by definition, the same as the direction of the displacement.

---

11. ***REASONING AND SOLUTION***

a. Average speed is defined as the total distance d covered divided by the time Δt required to cover the distance. The total distance covered by the earth is one-fourth the circumference of its circular orbit around the sun:

$$d = \frac{1}{4} \times 2\pi\,(1.50 \times 10^{11}\text{ m}) = 2.36 \times 10^{11}\text{ m}$$

$$\bar{v} = \frac{d}{\Delta t} = \frac{2.36 \times 10^{11}\text{ m}}{7.89 \times 10^{6}\text{ s}} = \boxed{2.99 \times 10^{4}\text{ m/s}}$$

b. The average velocity is defined as the displacement divided by the elapsed time.

In moving one-fourth of the distance around the sun, the earth completes the displacement shown in the figure at the right. From the Pythagorean theorem, the magnitude of this displacement is

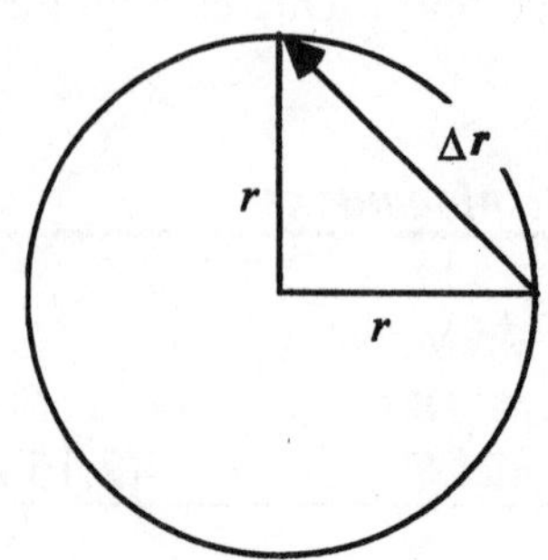

$$\Delta r = \sqrt{r^2 + r^2} = \sqrt{2}\,r$$

Thus, the magnitude of the average velocity is

$$\bar{v} = \frac{\Delta r}{\Delta t} = \frac{\sqrt{2} \times 1.50 \times 10^{11}\text{ m}}{7.89 \times 10^{6}\text{ s}} = \boxed{2.69 \times 10^{4}\text{ m/s}}$$

---

12. ***REASONING AND SOLUTION***

a. The x component of velocity is

$$v_x = v_{0x} + a_x t = 5480\text{ m/s} + \left(1.20\text{ m/s}^2\right)(842\text{ s}) = \boxed{6490\text{ m/s}}$$

b. For the y component

$$v_y = v_{0y} + a_y t = 0\text{ m/s} + \left(8.40\text{ m/s}^2\right)(842\text{ s}) = \boxed{7070\text{ m/s}}$$

---

13. ***REASONING AND SOLUTION*** Use the information concerning the x motion to find the time of flight of the ball

$$x = v_{ox}t \quad \text{or} \quad t = x/v_{ox} = (19.6\text{ m})/(28.0\text{ m/s}) = 0.700\text{ s}$$

The motion in the y direction is therefore subject to

$$y = v_{oy}t - (1/2)gt^2 = 0 - (1/2)(9.80\text{ m/s}^2)(0.700\text{ s})^2 = -2.40\text{ m}$$

The height of the tennis ball is $\boxed{2.40 \text{ m}}$.

---

14. ***REASONING*** The vertical component of the ball's velocity $\mathbf{v}_0$ changes as the ball approaches the opposing player. It changes due to the acceleration of gravity. However, the horizontal component does not change, assuming that air resistance can be neglected. Hence, the horizontal component of the ball's velocity when the opposing player fields the ball is the same as it was initially.

SOLUTION Using trigonometry, we find that the horizontal component is

$$v_x = v_0 \cos\theta = (15 \text{ m/s})\cos 55° = \boxed{8.6 \text{ m/s}}$$

---

15. SSM ***REASONING*** The time that the ball spends in the air is determined by its vertical motion. The time required for the ball to reach the lake can be found by solving Equation 3.5b for *t*. The motion of the golf ball is characterized by constant velocity in the *x* direction and accelerated motion (due to gravity) in the *y* direction. Thus, the *x* component of the velocity of the golf ball is constant, and the *y* component of the velocity at any time *t* can be found from Equation 3.3b. Once the *x* and *y* components of the velocity are known for a particular time *t*, the speed can be obtained from $v = \sqrt{v_x^2 + v_y^2}$.

SOLUTION
a. Since the ball rolls off the cliff horizontally, $v_{0y} = 0$. If the origin is chosen at top of the cliff and upward is assumed to be the positive direction, then the vertical component of the ball's displacement is $y = -15.5$ m. Thus, Equation 3.5b gives

$$t = \sqrt{\frac{2y}{a_y}} = \sqrt{\frac{2(-15.5 \text{ m})}{(-9.80 \text{ m/s}^2)}} = \boxed{1.78 \text{ s}}$$

b. Since there is no acceleration in the *x* direction, $v_x = v_{0x} = 11.4$ m/s. The *y* component of the velocity of the ball just before it strikes the water is, according to Equation 3.3b,

$$v_y = v_{0y} + a_y t = \left[0 + (-9.80 \text{ m/s}^2)(1.78 \text{ s})\right] = -17.4 \text{ m/s}$$

The speed of the ball just before it strikes the water is, therefore,

$$v = \sqrt{v_x^2 + v_y^2} = \sqrt{(11.4 \text{ m/s})^2 + (-17.4 \text{ m/s})^2} = \boxed{20.8 \text{ m/s}}$$

---

16. ***REASONING AND SOLUTION*** Using $v_y = 0$ and

$$v_{0y} = v_0 \sin\theta = (11 \text{ m/s}) \sin 65^\circ = 1.0 \times 10^1 \text{ m/s}$$

and $v_y{}^2 = v_{0y}{}^2 + 2a_y y$, we have

$$y = \frac{-v_{0y}^2}{2a_y} = \frac{-(1.0\times10^1 \text{ m/s})^2}{2(-9.80 \text{ m/s}^2)} = \boxed{5.1 \text{ m}}$$

---

17. SSM WWW ***REASONING*** Once the diver is airborne, he moves in the x direction with constant velocity while his motion in the y direction is accelerated (at the acceleration due to gravity). Therefore, the magnitude of the x component of his velocity remains constant at 1.20 m/s for all times t. The magnitude of the y component of the diver's velocity after he has fallen through a vertical displacement y can be determined from Equation 3.6b: $v_y^2 = v_{0y}^2 + 2a_y y$. Since the diver runs off the platform horizontally, $v_{0y} = 0$. Once the x and y components of the velocity are known for a particular vertical displacement y, the speed of the diver can be obtained from $v = \sqrt{v_x^2 + v_y^2}$.

SOLUTION For convenience, we will take downward as the positive y direction. After the diver has fallen 10.0 m, the y component of his velocity is, from Equation 3.6b,

$$v_y = \sqrt{v_{0y}^2 + 2a_y y} = \sqrt{0^2 + 2(9.80 \text{ m/s}^2)(10.0 \text{ m})} = 14.0 \text{ m/s}$$

Therefore,

$$v = \sqrt{v_x^2 + v_y^2} = \sqrt{(1.20 \text{ m/s})^2 + (14.0 \text{ m/s})^2} = \boxed{14.1 \text{ m/s}}$$

---

18. ***REASONING***

a. The maximum possible distance that the ball can travel occurs when it is launched at an angle of 45.0°. When the ball lands on the green, it is at the same elevation as the tee, so the vertical component (or y component) of the ball's displacement is zero. The time of flight is given by the y variables, which are listed in the table below. We designate "up" as the $+y$ direction.

y-Direction Data

y	a_y	v_y	v_{0y}	t
0 m	-9.80 m/s^2		$+(30.3 \text{ m/s}) \sin 45.0° = +21.4 \text{ m/s}$	?

Since three of the five kinematic variables are known, we can employ one of the equations of kinematics to find the time t that the ball is in the air.

b. The longest hole in one that the golfer can make is equal to the range R of the ball. This distance is given by the x variables and the time of flight, as determined in part (a). Once again, three variables are known, so an equation of kinematics can be used to find the range of the ball. The $+x$ direction is taken to be from the tee to the green.

x-Direction Data

x	a_x	v_x	v_{0x}	t
$R = ?$	$0\ \text{m/s}^2$		$+(30.3\ \text{m/s})\cos 45.0° = +21.4\ \text{m/s}$	from part a

SOLUTION

a. We will use Equation 3.5b to find the time, since this equation involves the three known variables in the y direction:

$$y = v_{0y}t + \tfrac{1}{2}a_y t^2 = \left(v_{0y} + \tfrac{1}{2}a_y t\right)t$$

$$0\ \text{m} = \left[+21.4\ \text{m/s} + \tfrac{1}{2}\left(-9.80\ \text{m/s}^2\right)t\right]t$$

Solving this quadratic equation yields two solutions, $t = 0$ s and $t = 4.37$ s. The first solution represents the situation when the golf ball just begins its flight, so we discard this one. Therefore, $t = \boxed{4.37\ \text{s}}$.

b. With the knowledge that $t = 4.37$ s and the values for a_x and v_{0x} (see the x-direction data table above), we can use Equation 3.5a to obtain the range R of the golf ball.

$$\underbrace{x}_{=R} = v_{0x}t + \tfrac{1}{2}a_x t^2 = (+21.4\ \text{m/s})(4.37\ \text{s}) + \tfrac{1}{2}\left(0\ \text{m/s}^2\right)(4.37\ \text{s})^2 = \boxed{93.5\ \text{m}}$$

---

19. ***REASONING AND SOLUTION*** The components of the initial velocity are

$$v_{0x} = v_0 \cos\theta = (22\ \text{m/s})\cos 40.0° = 17\ \text{m/s}$$
$$v_{0y} = v_0 \sin\theta = (22\ \text{m/s})\sin 40.0° = 14\ \text{m/s}$$

a. Solving Equation 3.6b for y gives

$$y = \frac{v_y^2 - v_{0y}^2}{2a_y}$$

When the football is at the maximum height $y = H$, and the football is momentarily at rest, so $v_y = 0$. Thus,

$$H = \frac{0 - v_{0y}^2}{2a_y} = \frac{0-(14\text{ m/s})^2}{2(-1.62\text{ m/s}^2)} = \boxed{6.0 \times 10^1\text{ m}}$$

b. When the ball strikes the ground, $y = 0$; therefore, the time of flight can be determined from Equation 3.5b with $y = 0$.

$$y = v_{0y}t + \tfrac{1}{2}a_y t^2$$

or

$$0 = [(14\text{ m/s}) + \tfrac{1}{2}(-1.62\text{ m/s}^2)t]\,t$$

$$t = 17\text{ s}$$

The range is

$$x = R = v_{0x}t = (17\text{ m/s})(17\text{ s}) = \boxed{290\text{ m}}$$

---

20. ***REASONING AND SOLUTION*** The maximum vertical displacement y attained by a projectile is given by Equation 3.6b ($v_y^2 = v_{0y}^2 + 2a_y y$) with $v_y = 0$:

$$y = -\frac{v_{0y}^2}{2a_y}$$

In order to use Equation 3.6b, we must first estimate his initial speed v_{0y}. When Jordan has reached his maximum vertical displacement, $v_y = 0$, and $t = 1.00$ s. Therefore, according to Equation 3.3b ($v_y = v_{0y} + a_y t$), with upward taken as positive, we find that

$$v_{0y} = -a_y t = -\,(-9.80\text{ m/s}^2)\,(1.00\text{ s}) = 9.80\text{ m/s}$$

Therefore, Jordan's maximum jump height is

$$y = -\frac{(9.80\text{ m/s})^2}{2(-9.80\text{ m/s}^2)} = \boxed{4.90\text{ m}}$$

This result far exceeds Jordan's maximum jump height, so the claim that he can remain in the air for two full seconds is false.

---

21. SSM ***REASONING AND SOLUTION*** The water exhibits projectile motion. The x component of the motion has zero acceleration while the y component is subject to the acceleration due to gravity. In order to reach the highest possible fire, the displacement of the hose from the building is x, where, according to Equation 3.5a (with $a_x = 0$),

$$x = v_{0x}t = (v_0 \cos\theta)t$$

with t equal to the time required for the water the reach its maximum vertical displacement. The time t can be found by considering the vertical motion. From Equation 3.3b,

$$v_y = v_{0y} + a_y t$$

When the water has reached its maximum vertical displacement, $v_y = 0$. Taking up and to the right as the positive directions, we find that

$$t = \frac{-v_{0y}}{a_y} = \frac{-v_0 \sin\theta}{a_y}$$

and

$$x = (v_0 \cos\theta)\left(\frac{-v_0 \sin\theta}{a_y}\right)$$

Therefore, we have

$$x = -\frac{v_0^2 \cos\theta \ \sin\theta}{a_y} = -\frac{(25.0 \text{ m/s})^2 \cos 35.0° \ \sin 35.0°}{-9.80 \text{ m/s}^2} = \boxed{30.0 \text{ m}}$$

---

22. ***REASONING AND SOLUTION*** The time required for the car to fall to the ground is given by

$$t = \sqrt{\frac{-2y}{g}} = \sqrt{\frac{-2(-54 \text{ m})}{9.80 \text{ m/s}^2}} = 3.3 \text{ s}$$

During this time, the car traveled a horizontal distance of 130 m. Using $a_x = 0 \text{ m/s}^2$ gives

$$v_{0x} = x/t = (130 \text{ m})/(3.3 \text{ s}) = \boxed{39 \text{ m/s}}$$

---

23. ***REASONING*** We begin by considering the flight time of the ball on the distant planet. Once the flight time is known, we can determine the maximum height and the range of the ball.

The range of a projectile is proportional to the time that the projectile is in the air. Therefore, the flight time on the distant plant 3.5 times larger than on earth. The flight time can be found from Equation 3.3b ($v_y = v_{0y} + a_y t$). When the ball lands, it is at the same

level as the tee; therefore, from the symmetry of the motion $v_y = -v_{0y}$. Taking upward and to the right as the positive directions, we find that the flight time on earth would be

$$t = \frac{v_y - v_{0y}}{a_y} = \frac{-2v_{0y}}{a_y} = \frac{-2v_0 \sin\theta}{a_y} = \frac{-2(45 \text{ m/s}) \sin 29°}{-9.80 \text{ m/s}^2} = 4.45 \text{ s}$$

Therefore, the flight time on the distant planet is 3.5 × (4.45 s) = 15.6 s. From the symmetry of the problem, we know that this is twice the amount of time required for the ball to reach its maximum height, which, consequently, is 7.80 s.

SOLUTION

a. The height y of the ball at any instant is given by Equation 3.4b as the product of the average velocity component in the y direction $\frac{1}{2}(v_{0y} + v_y)$ and the time t: $y = \frac{1}{2}(v_{0y} + v_y)t$. Since the maximum height H is reached when the final velocity component in the y direction is zero ($v_y = 0$), we find that

$$H = \tfrac{1}{2}v_{0y}t = \tfrac{1}{2}v_0 (\sin 29°)t = \tfrac{1}{2}(45 \text{ m/s}) (\sin 29°) (7.80 \text{ s}) = \boxed{85 \text{ m}}$$

b. The range of the ball on the distant planet is

$$x = v_{0x}t = v_0 (\cos 29°)t = (45 \text{ m/s}) (\cos 29°) (15.6 \text{ s}) = \boxed{610 \text{ m}}$$

---

24. ***REASONING*** When the skier leaves the ramp, she exhibits projectile motion. Since we know the maximum height attained by the skier, we can find her launch speed v_0 using Equation 3.6b, $v_y^2 = v_{0y}^2 + 2a_y y$, where $v_{0y} = v_0 \sin 63°$.

SOLUTION At the highest point in her trajectory, $v_y = 0$. Solving Equation 3.6b for v_{0y} we obtain, taking upward as the positive direction,

$$v_{0y} = v_0 \sin 63° = \sqrt{-2a_y y} \quad \text{or} \quad v_0 = \frac{\sqrt{-2a_y y}}{\sin 63°} = \frac{\sqrt{-2(-9.80 \text{ m/s}^2)(13 \text{ m})}}{\sin 63°} = \boxed{18 \text{ m/s}}$$

---

25. **SSM** ***REASONING*** The speed of the fish at any time t is given by $v = \sqrt{v_x^2 + v_y^2}$, where v_x and v_y are the x and y components of the velocity at that instant. Since the horizontal motion of the fish has zero acceleration, $v_x = v_{0x}$ for all times t. Since the fish is dropped by the eagle, v_{0x} is equal to the horizontal speed of the eagle and $v_{0y} = 0$. The y component of

the velocity of the fish for any time t is given by Equation 3.3b with $v_{0y} = 0$. Thus, the speed at any time t is given by $v = \sqrt{v_{0x}^2 + (a_y t)^2}$.

SOLUTION

a. The initial speed of the fish is $v_0 = \sqrt{v_{0x}^2 + v_{0y}^2} = \sqrt{v_{0x}^2 + 0^2} = v_{0x}$. When the fish's speed doubles, $v = 2v_{0x}$. Therefore,

$$2v_{0x} = \sqrt{v_{0x}^2 + (a_y t)^2} \quad \text{or} \quad 4v_{0x}^2 = v_{0x}^2 + (a_y t)^2$$

Assuming that downward is positive and solving for t, we have

$$t = \sqrt{3}\frac{v_{0x}}{a_y} = \sqrt{3}\left(\frac{6.0 \text{ m/s}}{9.80 \text{ m/s}^2}\right) = \boxed{1.1 \text{ s}}$$

b. When the fish's speed doubles again, $v = 4v_{0x}$. Therefore,

$$4v_{0x} = \sqrt{v_{0x}^2 + (a_y t)^2} \quad \text{or} \quad 16v_{0x}^2 = v_{0x}^2 + (a_y t)^2$$

Solving for t, we have

$$t = \sqrt{15}\frac{v_{0x}}{a_y} = \sqrt{15}\left(\frac{6.0 \text{ m/s}}{9.80 \text{ m/s}^2}\right) = 2.37 \text{ s}$$

Therefore, the additional time for the speed to double again is $(2.4 \text{ s}) - (1.1 \text{ s}) = \boxed{1.3 \text{ s}}$.

---

26. ***REASONING*** The vertical displacement y of the ball depends on the time that it is in the air before being caught. These variables depend on the y-direction data, as indicated in the table, where the $+y$ direction is "up."

y-Direction Data

y	a_y	v_y	v_{0y}	t
?	-9.80 m/s^2		0 m/s	?

Since only two variables in the y direction are known, we cannot determine y at this point. Therefore, we examine the data in the x direction, where $+x$ is taken to be the direction from the pitcher to the catcher.

x-Direction Data

x	a_x	v_x	v_{0x}	t
+17.0 m	0 m/s^2		+41.0 m/s	?

Since this table contains three known variables, the time t can be evaluated by using an equation of kinematics. Once the time is known, it can then be used with the y-direction data, along with the appropriate equation of kinematics, to find the vertical displacement y.

SOLUTION Using the x-direction data, Equation 3.5a can be employed to find the time t that the baseball is in the air:

$$x = v_{0x}t + \tfrac{1}{2}a_x t^2 = v_{0x}t \quad \left(\text{since } a_x = 0 \text{ m/s}^2\right)$$

Solving for t gives

$$t = \frac{x}{v_{0x}} = \frac{+17.0 \text{ m}}{+41.0 \text{ m/s}} = 0.415 \text{ s}$$

The displacement in the y direction can now be evaluated by using the y-direction data table and the value of $t = 0.415$ s. Using Equation 3.5b, we have

$$y = v_{0y}t + \tfrac{1}{2}a_y t^2 = (0 \text{ m/s})(0.415 \text{ s}) + \tfrac{1}{2}\left(-9.80 \text{ m/s}^2\right)(0.415 \text{ s})^2 = -0.844 \text{ m}$$

The distance that the ball drops is given by the magnitude of this result, so Distance = $\boxed{0.844 \text{ m}}$.

---

27. ***REASONING AND SOLUTION*** The time of flight of the motorcycle is given by

$$t = \frac{2v_0 \sin\theta_0}{g} = \frac{2(33.5 \text{ m/s})\sin 18.0°}{9.80 \text{ m/s}^2} = 2.11 \text{ s}$$

The horizontal distance traveled by the motorcycle is then

$$x = v_0 \cos\theta_0 \, t = (33.5 \text{ m/s})(\cos 18.0°)(2.11 \text{ s}) = 67.2 \text{ m}$$

The daredevil can jump over (67.2 m)/(2.74 m/bus) = 24.5 buses. In even numbers, this means $\boxed{24 \text{ buses}}$.

---

28. ***REASONING*** The drawing at the right shows the velocity vector **v** of the water at a point below the top of the falls. The components of the velocity are also shown. The angle θ is given by $\tan\theta = v_y / v_x$, so that

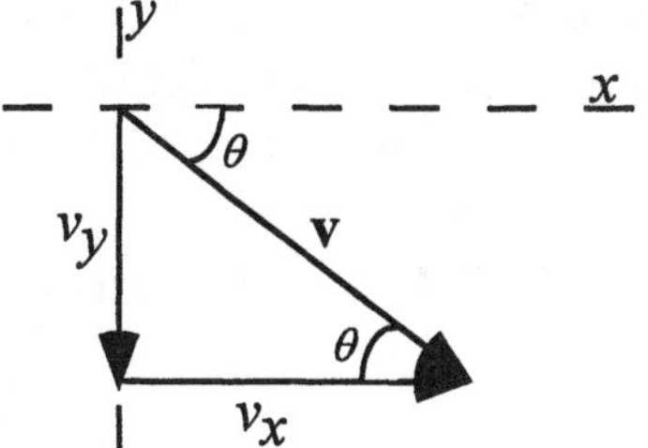

$$v_y = -v_x \tan\theta = -(2.7 \text{ m/s}) \tan 75°$$

Here we have used the fact that the horizontal velocity component v_x remains unchanged at its initial value of 2.7 m/s as the water falls. Knowing the y component of the velocity, we can use Equation 3.6b, ($v_y^2 = v_{0y}^2 + 2a_y y$) to find the vertical distance y.

SOLUTION Taking $v_{0y} = 0$, and taking upward as the positive direction, we have from Equation 3.5b that

$$y = \frac{v_y^2}{2a_y} = \frac{[-(2.7 \text{ m/s}) \tan 75°]^2}{2(-9.80 \text{ m/s}^2)} = -5.2 \text{ m}$$

Therefore, the velocity vector of the water points downward at a 75° angle below the horizontal at a vertical distance of $\boxed{5.2 \text{ m}}$ below the edge.

---

29. $\boxed{\text{SSM}}$ ***REASONING*** The upward direction is chosen as positive. Since the ballast bag is released from rest relative to the balloon, its initial velocity relative to the ground is equal to the velocity of the balloon relative to the ground, so that $v_{0y} = 3.0 \text{ m/s}$. Time required for the ballast to reach the ground can be found by solving Equation 3.5b for t.

SOLUTION Using Equation 3.5b, we have

$$\tfrac{1}{2} a_y t^2 + v_{0y} t - y = 0 \quad \text{or} \quad \tfrac{1}{2}(-9.80 \text{ m/s}^2) t^2 + (3.0 \text{ m/s}) t - (-9.5 \text{ m}) = 0$$

This equation is quadratic in t, and t may be found from the quadratic formula. Using the quadratic formula, suppressing the units, and discarding the negative root, we find

$$t = \frac{-3.0 \pm \sqrt{(3.0)^2 - 4(-4.90)(9.5)}}{2(-4.90)} = \boxed{1.7 \text{ s}}$$

---

30. ***REASONING*** The rocket will clear the top of the wall by an amount that is the height of the rocket as it passes over the wall minus the height of the wall. To find the height of the rocket as it passes over the wall, we separate the rocket's projectile motion into its horizontal and vertical parts and treat each one separately. From the horizontal part we will obtain the

time of flight until the rocket reaches the location of the wall. Then, we will use this time along with the acceleration due to gravity in the equations of kinematics to determine the height of the rocket as it passes over the wall.

SOLUTION We begin by finding the horizontal and vertical components of the launch velocity

$$v_{0x} = v_0 \cos 60.0° = (75.0 \text{ m/s})\cos 60.0°$$

$$v_{0y} = v_0 \sin 60.0° = (75.0 \text{ m/s})\sin 60.0°$$

Using v_{0x}, we can obtain the time of flight, since the distance to the wall is known to be 27.0 m:

$$t = \frac{27.0 \text{ m}}{v_{0x}} = \frac{27.0 \text{ m}}{(75.0 \text{ m/s})\cos 60.0°} = 0.720 \text{ s}$$

The height of the rocket as it clears the wall can be obtained from Equation 3.5b, in which we take upward to be the positive direction. The amount by which the rocket clears the wall can then be obtained:

$$y = v_{0y}t + \tfrac{1}{2}a_y t^2$$

$$y = (75.0 \text{ m/s})(\sin 60.0°)(0.720 \text{ s}) + \tfrac{1}{2}(-9.80 \text{ m/s}^2)(0.720 \text{ s})^2 = 44.2 \text{ m}$$

$$\text{clearance} = 44.2 \text{ m} - 11.0 \text{ m} = \boxed{33.2 \text{ m}}$$

---

31. ***REASONING AND SOLUTION*** The coordinates of the bullet when it hits the target are $y = -(1/2)gt^2$ and $x = v_0 t$. The first of these yields the time of flight

$$t = \sqrt{\frac{-2y}{g}} = \sqrt{\frac{-2(-0.025 \text{ m})}{9.80 \text{ m/s}^2}} = 0.071 \text{ s}$$

The horizontal distance traveled is then

$$x = (670 \text{ m/s})(0.071 \text{ s}) = \boxed{48 \text{ m}}$$

---

32. ***REASONING***

a. The drawing shows the initial velocity $\mathbf{v}_0$ of the package when it is released. The initial speed of the package is 97.5 m/s. The component of its displacement along the ground is labeled as x. The data for the x direction are indicated in the data table below.

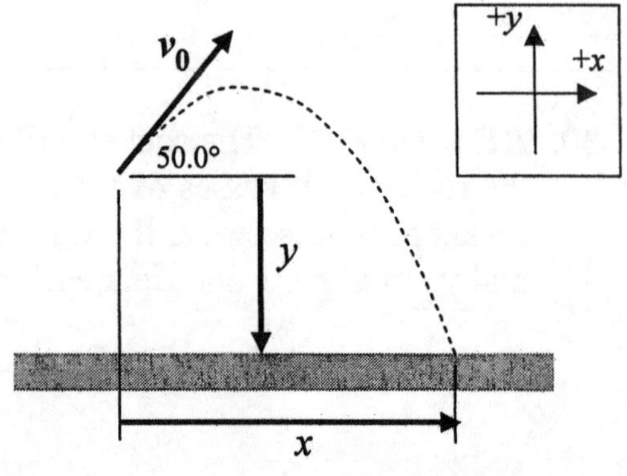

x-Direction Data

x	a_x	v_x	v_{0x}	t
?	$0\ m/s^2$		$+(97.5\ m/s)\cos 50.0° = +62.7\ m/s$	

Since only two variables are known, it is not possible to determine x from the data in this table. A value for a third variable is needed. We know that the time of flight t is the same for both the x and y motions, so let's now look at the data in the y direction.

y-Direction Data

y	a_y	v_y	v_{0y}	t
$-732\ m$	$-9.80\ m/s^2$		$+(97.5\ m/s)\sin 50.0° = +74.7\ m/s$	?

Note that the displacement y of the package points from its initial position toward the ground, so its value is negative, i.e., $y = -732$ m. The data in this table, along with the appropriate equation of kinematics, can be used to find the time of flight t. This value for t can, in turn, be used in conjunction with the x-direction data to determine x.

b. The drawing at the right shows the velocity of the package just before impact. The angle that the velocity makes with respect to the ground can be found from the inverse tangent function as $\theta = \tan^{-1}\left(v_y / v_x\right)$. Once the time has been found in part (a), the values of v_y and v_x can be determined from the data in the tables and the appropriate equations of kinematics.

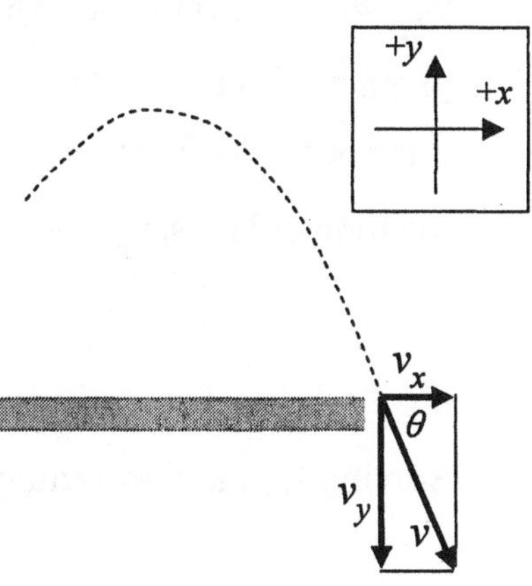

SOLUTION

a. To determine the time that the package is in the air, we will use Equation 3.5b $\left(y = v_{0y}t + \frac{1}{2}a_y t^2\right)$ and the data in the y-direction data table. Solving this quadratic equation for the time yields

$$t = \frac{-v_{0y} \pm \sqrt{v_{0y}^2 - 4\left(\frac{1}{2}a_y\right)(-y)}}{2\left(\frac{1}{2}a_y\right)}$$

$$t = \frac{-(74.7\ \text{m/s}) \pm \sqrt{(74.7\ \text{m/s})^2 - 4\left(\frac{1}{2}\right)\left(-9.80\ \text{m/s}^2\right)(732\ \text{m})}}{2\left(\frac{1}{2}\right)\left(-9.80\ \text{m/s}^2\right)} = -6.78\ \text{s} \text{ and } 22.0\ \text{s}$$

We discard the first solution, since it is a negative value and, hence, unrealistic. The displacement x can be found using t = 22.0 s, the data in the x-direction data table, and Equation 3.5a:

$$x = v_{0x}t + \tfrac{1}{2}a_x t^2 = (+62.7\text{ m/s})(22.0\text{ s}) + \underbrace{\tfrac{1}{2}(0\text{ m/s}^2)(22.0\text{ s})^2}_{=0} = \boxed{+1380\text{ m}}$$

b. The angle θ that the velocity of the package makes with respect to the ground is given by $\theta = \tan^{-1}(v_y / v_x)$. Since there is no acceleration in the x direction (a_x = 0 m/s^2), v_x is the same as v_{0x}, so that v_x = v_{0x} = +62.7 m/s. Equation 3.3b can be employed with the y-direction data to find v_y:

$$v_y = v_{0y} + a_y t = +74.7\text{ m/s} + (-9.80\text{ m/s}^2)(22.0\text{ s}) = -141\text{ m/s}$$

Therefore,

$$\theta = \tan^{-1}\left(\frac{v_y}{v_x}\right) = \tan^{-1}\left(\frac{-141\text{ m/s}}{+62.7\text{ m/s}}\right) = -66.0°$$

where the minus sign indicates that the angle is $\boxed{66.0° \text{ below the horizontal}}$.

---

33. SSM ***REASONING AND SOLUTION*** The horizontal displacement of the jumper is given by Equation 3.5a with $a_x = 0$: $x = v_{0x}t = (v_0 \cos\theta)t$. The vertical component of the jumper's velocity at any time t is given by Equation 3.3b: $v_y = v_{0y} + a_y t$. At the instant that the jumper lands, $v_y = -v_{0y}$. Therefore, the vertical component of the jumper's velocity is

$$-v_{0y} = v_{0y} + a_y t$$

Solving for t and assuming that upward is the positive direction gives

$$t = \frac{-2v_{0y}}{a_y} = \frac{-2v_0 \sin\theta}{a_y}$$

Substituting this expression for t into the expression for x gives

$$x = (v_0 \cos\theta)t = v_0 \cos\theta \left(\frac{-2v_0 \sin\theta}{a_y}\right)$$

or

$$x = \frac{-2v_0^2 \sin\theta \cos\theta}{a_y}$$

Solving for v_0 gives

$$v_0 = \sqrt{\frac{-x\,a_y}{2\cos\theta \sin\theta}} = \sqrt{\frac{-(8.7\text{ m})(-9.80\text{ m/s}^2)}{2\cos 23° \sin 23°}} = \boxed{11\text{ m/s}}$$

---

34. ***REASONING*** As discussed in Conceptual Example 4, the horizontal velocity component of the bullet does not change from its initial value and is equal to the horizontal velocity of the car. The same thing is true here for the tomato. In other words, regardless of its vertical position relative to the ground, the tomato always remains above you as you travel in the convertible. From the symmetry of free fall motion, we know that when you catch the tomato, its velocity will be 11 m/s straight downward. The time t required to catch the tomato can be found by solving Equation 3.3b $(v_y = v_{0y} + a_y t)$ with $v_y = -v_{0y}$. Once t is known, the distance that the car moved can be found from $x = v_x t$.

SOLUTION Taking upward as the positive direction, we find the flight time of the tomato to be

$$t = \frac{v_y - v_{0y}}{a_y} = \frac{-2v_{0y}}{a_y} = \frac{-2(11\text{ m/s})}{-9.80\text{ m/s}^2} = 2.24\text{ s}$$

Thus, the car moves through a distance of

$$x = v_x t = (25\text{ m/s})(2.24\text{ s}) = \boxed{56\text{ m}}$$

---

35. ***REASONING*** The speed v of the soccer ball just before the goalie catches it is given by $v = \sqrt{v_x^2 + v_y^2}$, where v_x and v_x are the x and y components of the final velocity of the ball. The data for this problem are (the $+x$ direction is from the kicker to the goalie, and the $+y$ direction is the "up" direction):

x-Direction Data

x	a_x	v_x	v_{0x}	t
+16.8 m	0 m/s^2	?	+(16.0 m/s) cos 28.0° = +14.1 m/s	

y-Direction Data

y	a_y	v_y	v_{0y}	t
	−9.80 m/s^2	?	+(16.0 m/s) sin 28.0° = +7.51 m/s	

Since there is no acceleration in the *x* direction ($a_x = 0$ m/s^2), v_x remains the same as v_{0x}, so $v_x = v_{0x} = +14.1$ m/s. The time *t* that the soccer ball is in the air can be found from the *x*-direction data, since three of the variables are known. With this value for the time and the *y*-direction data, the *y* component of the final velocity can be determined.

SOLUTION Since $a_x = 0$ m/s^2, the time can be calculated from Equation 3.5a as $t = \dfrac{x}{v_{0x}} = \dfrac{+16.8\text{ m}}{+14.1\text{ m/s}} = 1.19\text{ s}$. The value for v_y can now be found by using Equation 3.3b with this value of the time and the *y*-direction data:

$$v_y = v_{0y} + a_y t = +7.51\text{ m/s} + \left(-9.80\text{ m/s}^2\right)(1.19\text{ s}) = -4.15\text{ m/s}$$

The speed of the ball just as it reaches the goalie is

$$v = \sqrt{v_x^2 + v_y^2} = \sqrt{(+14.1\text{ m/s})^2 + (-4.15\text{ m/s})^2} = \boxed{14.7\text{ m/s}}$$

---

36. ***REASONING*** As shown in the drawing, the angle that the velocity vector makes with the horizontal is given by

$$\tan\theta = \frac{v_y}{v_x}$$

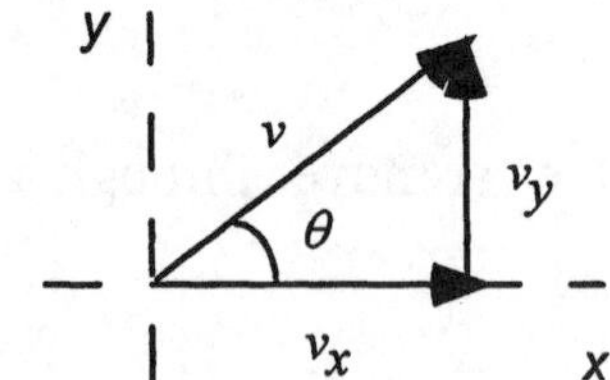

where, from Equation 3.3b,

$$v_y = v_{0y} + a_y t = v_0 \sin\theta_0 + a_y t$$

and, from Equation 3.3a (since $a_x = 0$),

$$v_x = v_{0x} = v_0 \cos\theta_0$$

Therefore,

$$\tan\theta = \frac{v_y}{v_x} = \frac{v_0 \sin\theta_0 + a_y t}{v_0 \cos\theta_0}$$

SOLUTION Solving for *t*, we find

$$t = \frac{v_0\left(\cos\theta_0\ \tan\theta - \sin\theta_0\right)}{a_y} = \frac{(29\text{ m/s})(\cos 36°\ \tan 18° - \sin 36°)}{-9.80\text{ m/s}^2} = \boxed{0.96\text{ s}}$$

---

37. SSM WWW ***REASONING*** The horizontal distance covered by stone **1** is equal to the distance covered by stone **2** after it passes point **P** in the following diagram. Thus, the distance Δx between the points where the stones strike the ground is equal to x_2, the horizontal distance covered by stone **2** when it reaches **P**. In the diagram, we assume up and to the right are positive.

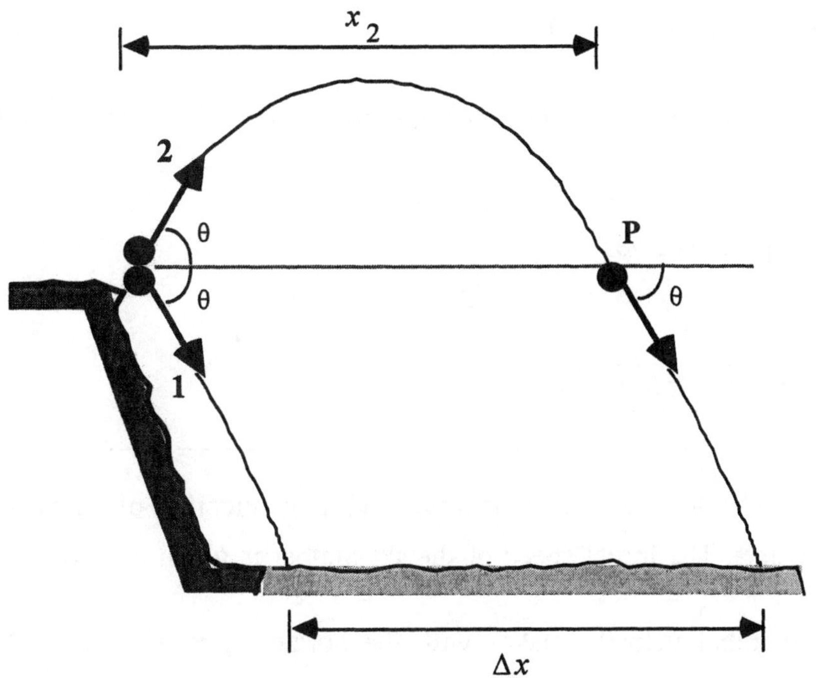

SOLUTION If t_P is the time required for stone **2** to reach **P**, then

$$x_2 = v_{0x}t_P = (v_0 \cos\theta)t_P$$

For the vertical motion of stone **2**, $v_y = v_0 \sin\theta + a_y t$. Solving for t gives

$$t = \frac{v_y - v_0 \sin\theta}{a_y}$$

When stone **2** reaches **P**, $v_y = -v_0 \sin\theta$, so the time required to reach **P** is

$$t_P = \frac{-2v_0 \sin\theta}{a_y}$$

Then,

$$x_2 = v_{0x}t_P = (v_0 \cos\theta)\left(\frac{-2v_0 \sin\theta}{a_y}\right)$$

$$x_2 = \frac{-2v_0^2 \sin\theta \cos\theta}{a_y} = \frac{-2(13.0 \text{ m/s})^2 \sin 30.0° \cos 30.0°}{-9.80 \text{ m/s}^2} = \boxed{14.9 \text{ m}}$$

---

38. ***REASONING AND SOLUTION*** The horizontal position of the impact point is $x_1 = v_{0x}t_1$ for one stone and $x_2 = v_{0x}t_2$ for the other stone.

The vertical positions are $y_1 = -(1/2)gt_1^2$ and $y_2 = -(1/2)gt_2^2$. Dividing the x equations gives

$$x_2/x_1 = t_2/t_1 = 2$$

Dividing the y equations gives

$$y_2/y_1 = t_2^2/t_1^2 = 4$$

So one building is $\boxed{\text{four times}}$ as tall as the other.

---

39. ***REASONING*** The drawings show the initial and final velocities of the ski jumper and their scalar components. The initial speed of the ski jumper is given by $v_0 = \sqrt{v_{0x}^2 + v_{0y}^2}$, and the angle that the initial velocity makes with the horizontal is $\theta = \tan^{-1}\left(\frac{v_{0y}}{v_{0x}}\right)$. The scalar components v_{0x} and v_{0y} can be determined by using the equations of kinematics and the data in the following tables. (The $+x$ direction is in the direction of the horizontal displacement of the skier, and the $+y$ direction is "up.")

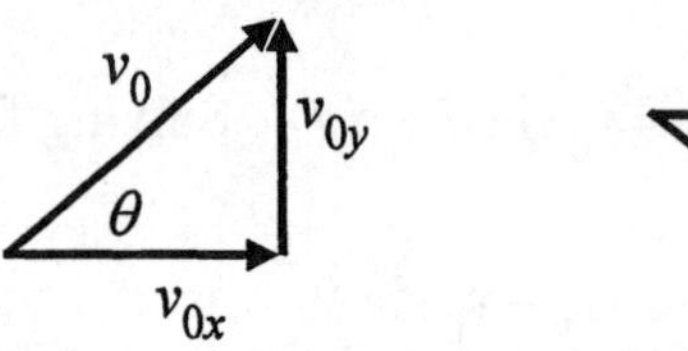

Initial velocity — Final velocity

***x*-Direction Data**

x	a_x	v_x	v_{0x}	t
+51.0 m	0 m/s^2	+(23.0 m/s) cos 43.0° = +16.8 m/s	?	

***y*-Direction Data**

y	a_y	v_y	v_{0y}	t
	−9.80 m/s^2	−(23.0 m/s) sin 43.0° = −15.7 m/s	?	

Since there is no acceleration in the x direction ($a_x = 0$ m/s^2), v_{0x} is the same as v_x, so $v_{0x} = v_x = +16.8$ m/s. The time that the skier is in the air can be found from the x-direction data, since three of the variables are known. With the value for the time and the y-direction data, the y component of the initial velocity can be determined.

SOLUTION Since $a_x = 0$ m/s^2, the time can be determined from Equation 3.5a as $t = \dfrac{x}{v_{0x}} = \dfrac{+51.0 \text{ m}}{+16.8 \text{ m/s}} = 3.04 \text{ s}$. The value for v_{0y} can now be found by using Equation 3.3b with this value of the time and the y-direction data:

$$v_{0y} = v_y - a_y t = -15.7 \text{ m/s} - \left(-9.80 \text{ m/s}^2\right)\left(3.04 \text{ s}\right) = +14.1 \text{ m/s}$$

The speed of the skier when he leaves the end of the ramp is

$$v_0 = \sqrt{v_{0x}^2 + v_{0y}^2} = \sqrt{(+16.8 \text{ m/s})^2 + (+14.1 \text{ m/s})^2} = \boxed{21.9 \text{ m/s}}$$

The angle that the initial velocity makes with respect to the horizontal is

$$\theta = \tan^{-1}\left(\frac{v_{0y}}{v_{0x}}\right) = \tan^{-1}\left(\frac{+14.1 \text{ m/s}}{+16.8 \text{ m/s}}\right) = \boxed{40.0^\circ}$$

---

40. ***REASONING*** Using the data given in the problem, we can find the maximum flight time t of the ball using Equation 3.5b ($y = v_{0y}t + \frac{1}{2}a_y t^2$). Once the flight time is known, we can use the definition of average velocity to find the minimum speed required to cover the distance x in that time.

SOLUTION Equation 3.5b is quadratic in t and can be solved for t using the quadratic formula. According to Equation 3.5b, the maximum flight time is (with upward taken as the positive direction)

$$t = \frac{-v_{0y} \pm \sqrt{v_{0y}^2 - 4\left(\frac{1}{2}\right)a_y(-y)}}{2\left(\frac{1}{2}\right)a_y} = \frac{-v_{0y} \pm \sqrt{v_{0y}^2 + 2a_y y}}{a_y}$$

$$= \frac{-(15.0 \text{ m/s}) \sin 50.0^\circ \pm \sqrt{\left[(15.0 \text{ m/s}) \sin 50.0^\circ\right]^2 + 2(-9.80 \text{ m/s}^2)(2.10 \text{ m})}}{-9.80 \text{ m/s}^2}$$

$$= 0.200 \text{ s} \quad \text{and} \quad 2.145 \text{ s}$$

where the first root corresponds to the time required for the ball to reach a vertical displacement of $y = +2.10$ m as it travels upward, and the second root corresponds to the time required for the ball to have a vertical displacement of $y = +2.10$ m as the ball travels upward and then downward. The desired flight time *t* is 2.145 s.

During the 2.145 s, the horizontal distance traveled by the ball is

$$x = v_x t = (v_0 \cos\theta)t = \left[(15.0 \text{ m/s}) \cos 50.0°\right] (2.145 \text{ s}) = 20.68 \text{ m}$$

Thus, the opponent must move $20.68 \text{ m} - 10.0 \text{ m} = 10.68 \text{ m}$ in $2.145 \text{ s} - 0.30 \text{ s} = 1.845 \text{ s}$. The opponent must, therefore, move with a minimum average speed of

$$\bar{v}_{\text{min}} = \frac{10.68 \text{ m}}{1.845 \text{ s}} = \boxed{5.79 \text{ m/s}}$$

---

41. SSM ***REASONING AND SOLUTION*** In the absence of air resistance, the bullet exhibits projectile motion. The *x* component of the motion has zero acceleration while the *y* component of the motion is subject to the acceleration due to gravity. The horizontal distance traveled by the bullet is given by Equation 3.5a (with $a_x = 0$):

$$x = v_{0x}t = (v_0 \cos\theta)t$$

with *t* equal to the time required for the bullet to reach the target. The time *t* can be found by considering the vertical motion. From Equation 3.3b,

$$v_y = v_{0y} + a_y t$$

When the bullet reaches the target, $v_y = -v_{0y}$. Assuming that up and to the right are the positive directions, we have

$$t = \frac{-2v_{0y}}{a_y} = \frac{-2v_0 \sin\theta}{a_y} \qquad \text{and} \qquad x = (v_0 \cos\theta)\left(\frac{-2v_0 \sin\theta}{a_y}\right)$$

Using the fact that $2\sin\theta \cos\theta = \sin 2\theta$, we have

$$x = -\frac{2v_0^2 \cos\theta \sin\theta}{a_y} = -\frac{v_0^2 \sin 2\theta}{a_y}$$

Thus, we find that

$$\sin 2\theta = -\frac{x\,a_y}{v_0^2} = -\frac{(91.4\text{ m})\,(-9.80\text{ m/s}^2)}{(427\text{ m/s})^2} = 4.91\times10^{-3}$$

and

$$2\theta = 0.281° \quad \text{or} \quad 2\theta = 180.000° - 0.281° = 179.719°$$

Therefore,

$$\theta = \boxed{0.141° \text{ and } 89.860°}$$

---

42. ***REASONING*** The drawing shows the trajectory of the ball, along with its initial speed v_0, horizontal displacement x, and vertical displacement y. The angle that the initial velocity of the ball makes with the horizontal is θ. The known data are shown in the tables below:

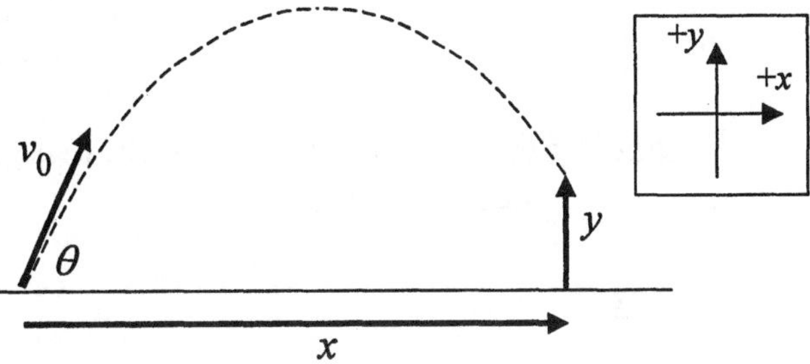

x-Direction Data

x	a_x	v_x	v_{0x}	t
+26.9 m	0 m/s^2		+(19.8 m/s) cos θ	

y-Direction Data

y	a_y	v_y	v_{0y}	t
+2.74 m	−9.80 m/s^2		+(19.8 m/s) sin θ	

There are only two known variables in each table, so we cannot directly use the equations of kinematics to find the angle θ. Our approach will be to first use the x direction data and obtain an expression for the time of flight t in terms of x and v_{0x}. We will then enter this expression for t into the y-direction data table. The four variables in this table, y, a_y, v_{0y}, and t, can be related by using the appropriate equation of kinematics. This equation can then be solved for the angle θ.

SOLUTION Using the x-direction data, Equation 3.5a can be employed to find the time t that the ball is in the air:

$$x = v_{0x}t + \tfrac{1}{2}a_x t^2 = v_{0x}t \qquad \left(\text{since } a_x = 0\text{ m/s}^2\right)$$

Solving for t gives

$$t = \frac{x}{v_{0x}} = \frac{+26.9\text{ m}}{+(19.8\text{ m/s})\cos\theta}$$

Using the expression above for the time t and the data in the y-direction data table, the displacement in the y direction can be written with the aid of Equation 3.5b:

$$y = v_{0y}t + \tfrac{1}{2}a_y t^2$$

$$+2.74\text{ m} = \left[\left(\cancel{19.8\text{ m/s}}\right)\sin\theta\right]\underbrace{\left(\frac{+26.9\text{ m}}{\left(\cancel{19.8\text{ m/s}}\right)\cos\theta}\right)}_{=t} + \tfrac{1}{2}\left(-9.80\text{ m/s}^2\right)\underbrace{\left(\frac{+26.9\text{ m}}{(19.8\text{ m/s})\cos\theta}\right)^2}_{=t^2}$$

Evaluating the numerical factors and using the fact that sin θ /cos θ= tan θ, the equation above becomes

$$+2.74\text{ m} = (+26.9\text{ m})\tan\theta + \frac{(-9.04\text{ m})}{\cos^2\theta}$$

Using $\frac{1}{\cos^2\theta} = 1 + \tan^2\theta$, this equation can be rearranged and placed into a quadratic form:

$$(-9.04\text{ m})\tan^2\theta + (26.9\text{ m})\tan\theta - 11.8\text{ m} = 0$$

The solutions to this quadratic equation are

$$\tan\theta = \frac{-26.9\text{ m} \pm \sqrt{(26.9\text{ m})^2 - 4(-9.04\text{ m})(-11.8\text{ m})}}{2(-9.04\text{ m})} = 0.535 \quad\text{and}\quad 2.44$$

The two angles are $\theta_1 = \tan^{-1} 0.535 = \boxed{28.1^\circ}$ and $\theta_2 = \tan^{-1} 2.44 = \boxed{67.7^\circ}$

---

43. SSM ***REASONING*** Since the horizontal motion is not accelerated, we know that the x component of the velocity remains constant at 340 m/s. Thus, we can use Equation 3.5a (with $a_x = 0$) to determine the time that the bullet spends in the building before it is embedded in the wall. Since we know the vertical displacement of the bullet after it enters the building, we can use the flight time in the building and Equation 3.5b to find the y component of the velocity of the bullet as it enters the window. Then, Equation 3.6b can be used (with $v_{0y} = 0$) to determine the vertical displacement y of the bullet as it passes between the buildings. We can determine the distance H by adding the magnitude of y to the vertical distance of 0.50 m within the building.

Once we know the vertical displacement of the bullet as it passes between the buildings, we can determine the time t_1 required for the bullet to reach the window using Equation 3.4b. Since the motion in the x direction is not accelerated, the distance D can then be found from $D = v_{0x} t_1$.

SOLUTION Assuming that the direction to the right is positive, we find that the time that the bullet spends in the building is (according to Equation 3.5a)

$$t = \frac{x}{v_{0x}} = \frac{6.9 \text{ m}}{340 \text{ m/s}} = 0.0203 \text{ s}$$

The vertical displacement of the bullet after it enters the building is, taking down as the negative direction, equal to –0.50 m. Therefore, the vertical component of the velocity of the bullet as it passes through the window is, from Equation 3.5b,

$$v_{0y(\text{window})} = \frac{y - \frac{1}{2} a_y t^2}{t} = \frac{y}{t} - \frac{1}{2} a_y t = \frac{-0.50 \text{ m}}{0.0203 \text{ s}} - \frac{1}{2}(-9.80 \text{ m/s}^2)(0.0203 \text{ s}) = -24.5 \text{ m/s}$$

The vertical displacement of the bullet as it travels between the buildings is (according to Equation 3.6b with $v_{0y} = 0$)

$$y = \frac{v_y^2}{2a_y} = \frac{(-24.5 \text{ m/s})^2}{2(-9.80 \text{ m/s}^2)} = -30.6 \text{ m}$$

Therefore, the distance H is

$$H = 30.6 \text{ m} + 0.50 \text{ m} = \boxed{31 \text{ m}}$$

The time for the bullet to reach the window, according to Equation 3.4b, is

$$t_1 = \frac{2y}{v_{0y} + v_y} = \frac{2y}{v_y} = \frac{2(-30.6 \text{ m})}{(-24.5 \text{ m/s})} = 2.50 \text{ s}$$

Hence, the distance D is given by

$$D = v_{0x} t_1 = (340 \text{ m/s})(2.50 \text{ s}) = \boxed{850 \text{ m}}$$

---

44. ***REASONING AND SOLUTION*** The ball is caught at a position of $y = +\,0.914$ m and $x = v_F t + 1.10 \times 10^2$ m where t is the time of flight and v_F is the speed of the center fielder.

$$v_F = \frac{x - 1.10\times10^2 \text{ m}}{t} = v_{0x} - \frac{1.10\times10^2 \text{ m}}{t}$$

The time t can be found from

$$y = v_{0y}t - (1/2)\,gt^2 = (v_0 \sin\theta_0)t - (1/2)\,gt^2$$

or $$0.914 \text{ m} = (36.6 \text{ m/s})(\sin 50.0°)t - (4.9 \text{ m/s}^2)t^2$$

Rearrangement gives

$$4.90\,t^2 - 28.0t + 0.914 = 0,$$ where the units have been suppressed.

The quadratic formula gives $t = 5.68$ s. Hence

$$v_F = (36.6 \text{ m/s})\cos 50.0^\circ - (1.10\times10^2 \text{ m})/(5.68 \text{ s}) = \boxed{4.2 \text{ m/s}}$$

---

45. ***REASONING*** We can use information about the motion of clown A and the collision to determine the initial velocity components for clown B. Once the initial velocity components are known, the launch speed v_{0B} and the launch angle θ_B for clown B can be determined.

SOLUTION From Equation 3.5b $\left(y = v_{0y}t + \tfrac{1}{2}a_y t^2\right)$ we can find the time of flight until the collision. Taking upward as positive and noting for clown A that $v_{0y} = (9.00 \text{ m/s})\sin 75.0° = 8.693$ m/s, we have

$$1.00 \text{ m} = (8.693 \text{ m/s})t + \tfrac{1}{2}(-9.80 \text{ m/s}^2)t^2$$

Rearranging this result and suppressing the units gives

$$4.90t^2 - 8.693t + 1.00 = 0$$

The quadratic equation reveals that

$$t = \frac{8.693 \pm \sqrt{(-8.693)^2 - 4(4.90)(1.00)}}{2(4.90)} = 1.650 \text{ s} \quad \text{or} \quad 0.1237 \text{ s}$$

Using these values for t with the magnitudes v_{0xA} and v_{0xB} of the horizontal velocity components for each clown, we note that the horizontal distances traveled by each clown must add up to 6.00 m. Thus,

$$v_{0xA}t + v_{0xB}t = 6.00 \text{ m} \quad \text{or} \quad v_{0xB} = \frac{6.00 \text{ m}}{t} - v_{0xA}$$

Using $v_{0xA} = (9.00 \text{ m/s}) \cos 75.0° = 2.329 \text{ m/s}$, we find

$$v_{0xB} = \frac{6.00 \text{ m}}{1.650 \text{ s}} - 2.329 \text{ m/s} = 1.307 \text{ m/s} \quad \text{or} \quad v_{0xB} = \frac{6.00 \text{ m}}{0.1237 \text{ s}} - 2.329 \text{ m/s} = 46.175 \text{ m/s}$$

The vertical component of clown B's velocity is $v_{0y\text{B}}$ and must be the same as that for clown A, since each clown has the same vertical displacement of 1.00 m at the same time. Hence, $v_{0y\text{B}} = 8.693$ m/s (see above). The launch speed of clown B, finally, is $v_{0B} = \sqrt{v_{0xB}^2 + v_{0yB}^2}$. Thus, we find

$$v_{0B} = \sqrt{(1.307 \text{ m/s})^2 + (8.693 \text{ m/s})^2} = 8.79 \text{ m/s}$$

or

$$v_{0B} = \sqrt{(46.175 \text{ m/s})^2 + (8.693 \text{ m/s})^2} = 47.0 \text{ m/s}$$

For these two possible launch speeds, we find the corresponding launch angles using the following drawings, neither of which is to scale:

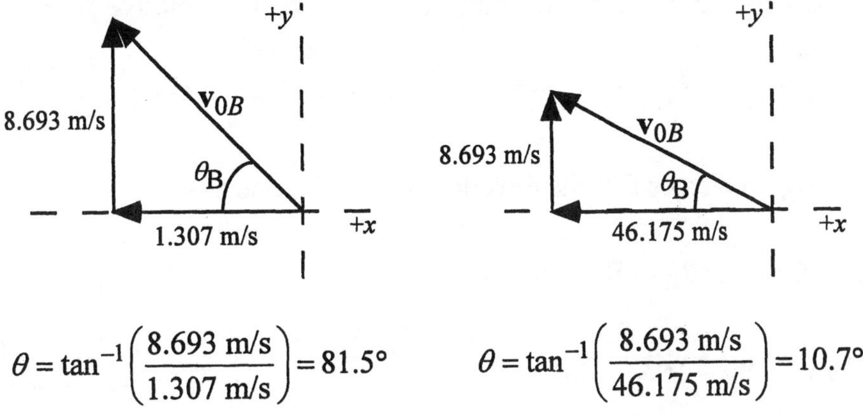

$$\theta = \tan^{-1}\left(\frac{8.693 \text{ m/s}}{1.307 \text{ m/s}}\right) = 81.5° \qquad \theta = \tan^{-1}\left(\frac{8.693 \text{ m/s}}{46.175 \text{ m/s}}\right) = 10.7°$$

Since the problem states that $\theta_\text{B} > 45°$, the solution is $\boxed{v_{0B} = 8.79 \text{ m/s} \text{ and } \theta_\text{B} = 81.5°}$.

---

46. ***REASONING AND SOLUTION*** Let H be the initial height of the can above the muzzle of the rifle. Relative to its initial position the vertical coordinate of the bullet at time, t, is

$$y = v_{0y}t - \tfrac{1}{2}gt^2$$

Relative to the can's initial position the vertical coordinate of the can at the *same* time is

$y' = (1/2)gt^2$.

NOTE: $y = H - y'$ if the bullet hits the can. Then $y = v_{oy}t - y'$. The horizontal distance traveled by the bullet in time t is $x = v_{ox}t$.

Solving for t and substituting gives

$$y = (v_{oy}/v_{ox})x - y'$$

Now $v_{oy}/v_{ox} = \tan\theta$ and it is seen from the figure that $H = x\tan\theta$ if the bullet is to hit the can so $y = H - y'$. Hence, both objects are at the same place at the same time, and the bullet will always strike the can.

---

47. SSM ***REASONING*** The velocity $\mathbf{v}_{AB}$ of train **A** relative to train **B** is the vector sum of the velocity $\mathbf{v}_{AG}$ of train **A** relative to the ground and the velocity $\mathbf{v}_{GB}$ of the ground relative to train **B**, as indicated by Equation 3.7: $\mathbf{v}_{AB} = \mathbf{v}_{AG} + \mathbf{v}_{GB}$. The values of $\mathbf{v}_{AG}$ and $\mathbf{v}_{BG}$ are given in the statement of the problem. We must also make use of the fact that $\mathbf{v}_{GB} = -\mathbf{v}_{BG}$.

SOLUTION

a. Taking east as the positive direction, the velocity of **A** relative to **B** is, according to Equation 3.7,

$$\mathbf{v}_{AB} = \mathbf{v}_{AG} + \mathbf{v}_{GB} = \mathbf{v}_{AG} - \mathbf{v}_{BG} = (+13\text{ m/s}) - (-28\text{ m/s}) = \boxed{+41\text{ m/s}}$$

The positive sign indicates that the direction of $\mathbf{v}_{AB}$ is $\boxed{\text{due east}}$.

b. Similarly, the velocity of **B** relative to **A** is

$$\mathbf{v}_{BA} = \mathbf{v}_{BG} + \mathbf{v}_{GA} = \mathbf{v}_{BG} - \mathbf{v}_{AG} = (-28\text{ m/s}) - (+13\text{ m/s}) = \boxed{-41\text{ m/s}}$$

The negative sign indicates that the direction of $\mathbf{v}_{BA}$ is $\boxed{\text{due west}}$.

---

48. ***REASONING*** Since car A is moving faster, it will eventually catch up with car B. Each car is traveling at a constant velocity, so the time t it takes for A to catch up with B is equal to the displacement between the two cars ($\mathbf{x}$ = +186 m) divided by the velocity $\mathbf{v}_{AB}$ of A relative to B. (If the relative velocity were zero, A would never catch up with B). We can find the velocity of A relative to B by using the subscripting technique developed in Section 3.4 of the text.

$\mathbf{v_{AB}}$ = velocity of car **A** relative to car **B**
$\mathbf{v_{AG}}$ = velocity of car **A** relative to the **G**round = +24.4 m/s
$\mathbf{v_{BG}}$ = velocity of car **B** relative to the **G**round = +18.6 m/s

We have chosen the positive direction for the displacement and velocities to be the direction in which the cars are moving. The velocities are related by

$$\mathbf{v_{AB}} = \mathbf{v_{AG}} + \mathbf{v_{GB}}$$

SOLUTION The velocity of car A relative to car B is

$$\mathbf{v_{AB}} = \mathbf{v_{AG}} + \mathbf{v_{GB}} = +24.4 \text{ m/s} + (-18.6 \text{ m/s}) = +5.8 \text{ m/s},$$

where we have used the fact that $\mathbf{v_{GB}} = -\,\mathbf{v_{BG}} = -18.6$ m/s. The time it takes for car A to catch car B is

$$t = \frac{\mathbf{x}}{\mathbf{v_{AB}}} = \frac{+186 \text{ m}}{+5.8 \text{ m/s}} = \boxed{32.1 \text{ s}}$$

---

49. ***REASONING AND SOLUTION*** The speed of a person relative to the ground is

$$v = x/t = (105 \text{ m})/(75 \text{ s}) = 1.4 \text{ m/s}$$

If the person walks at this rate relative to the conveyor belt, his speed relative to the ground is

$$v' = 1.4 \text{ m/s} + 2.0 \text{ m/s} = 3.4 \text{ m/s}$$

Thus,

$$t' = x/v' = (105 \text{ m})/(3.4 \text{ m/s}) = \boxed{31 \text{ s}}$$

---

50. ***REASONING*** As indicated by Equation 3.7, the velocity of the student relative to the ground is

$$\mathbf{v_{SG}} = \mathbf{v_{SE}} + \mathbf{v_{EG}}$$

where $\mathbf{v_{SE}}$ is the velocity of the student relative to the escalator and $\mathbf{v_{EG}}$ is the velocity of the escalator relative to the ground. Thus, we have

$$\mathbf{v_{SE}} = \mathbf{v_{SG}} - \mathbf{v_{EG}}$$

SOLUTION Taking the direction in which the student runs as the positive direction, we find that

$$v_{SE} = v_{SG} - v_{EG} = \frac{30.0 \text{ m}}{11 \text{ s}} - (-1.8 \text{ m/s}) = \boxed{4.5 \text{ m/s}}$$

The student must exceed this speed of 4.5 m/s to beat the record.

---

51. SSM ***REASONING*** The velocity $\mathbf{v}_{SG}$ of the swimmer relative to the ground is the vector sum of the velocity $\mathbf{v}_{SW}$ of the swimmer relative to the water and the velocity $\mathbf{v}_{WG}$ of the water relative to the ground as shown at the right: $\mathbf{v}_{SG} = \mathbf{v}_{SW} + \mathbf{v}_{WG}$.

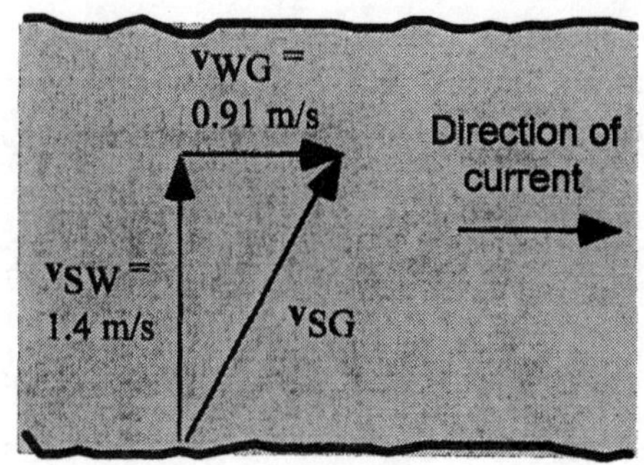

The component of $\mathbf{v}_{SG}$ that is parallel to the width of the river determines how fast the swimmer is moving across the river; this parallel component is v_{SW}. The time for the swimmer to cross the river is equal to the width of the river divided by the magnitude of this velocity component.

The component of $\mathbf{v}_{SG}$ that is parallel to the direction of the current determines how far the swimmer is carried down stream; this component is v_{WG}. Since the motion occurs with constant velocity, the distance that the swimmer is carried downstream while crossing the river is equal to the magnitude of v_{WG} multiplied by the time it takes for the swimmer to cross the river.

SOLUTION

a. The time t for the swimmer to cross the river is

$$t = \frac{\text{width}}{v_{SW}} = \frac{2.8 \times 10^3 \text{ m}}{1.4 \text{ m/s}} = \boxed{2.0 \times 10^3 \text{ s}}$$

b. The distance x that the swimmer is carried downstream while crossing the river is

$$x = v_{WG} t = (0.91 \text{ m/s})(2.0 \times 10^3 \text{ s}) = \boxed{1.8 \times 10^3 \text{ m}}$$

---

52. ***REASONING*** The time it takes for the passenger to walk the distance on the boat is the distance divided by the passenger's speed v_{PB} relative to the boat. The time it takes for the passenger to cover the distance on the water is the distance divided by the passenger's speed

v_{PW} relative to the water. The passenger's velocity relative to the boat is given. However, we need to determine the passenger's velocity relative to the water.

SOLUTION a. In determining the velocity of the passenger relative to the water, we define the following symbols:

$\mathbf{v_{PW}}$ = Passenger's velocity relative to the water
$\mathbf{v_{PB}}$ = Passenger's velocity relative to the boat
$\mathbf{v_{BW}}$ = Boat's velocity relative to the water

The passenger's velocity relative to the water is

$$\mathbf{v}_{\mathbf{PW}} = \mathbf{v}_{\mathbf{PB}} + \mathbf{v}_{\mathbf{BW}} = (1.5 \text{ m/s, north}) + (5.0 \text{ m/s, south}) = \boxed{3.5 \text{ m/s, south}}$$

b. The time it takes for the passenger to walk a distance of 27 m on the boat is

$$t = \frac{27 \text{ m}}{v_{\text{PB}}} = \frac{27 \text{ m}}{1.5 \text{ m/s}} = \boxed{18 \text{ s}}$$

c. The time it takes for the passenger to cover a distance of 27 m on the water is

$$t = \frac{27 \text{ m}}{v_{\text{PW}}} = \frac{27 \text{ m}}{3.5 \text{ m/s}} = \boxed{7.7 \text{ s}}$$

---

53. ***REASONING*** Let $\mathbf{v}_{\text{HB}}$ represent the velocity of the hawk relative to the balloon and $\mathbf{v}_{\text{BG}}$ represent the velocity of the balloon relative to the ground. Then, as indicated by Equation 3.7, the velocity of the hawk relative to the ground is $\mathbf{v}_{\text{HG}} = \mathbf{v}_{\text{HB}} + \mathbf{v}_{\text{BG}}$. Since the vectors $\mathbf{v}_{\text{HB}}$ and $\mathbf{v}_{\text{BG}}$ are at right angles to each other, the vector addition can be carried out using the Pythagorean theorem.

SOLUTION Using the drawing at the right, we have from the Pythagorean theorem,

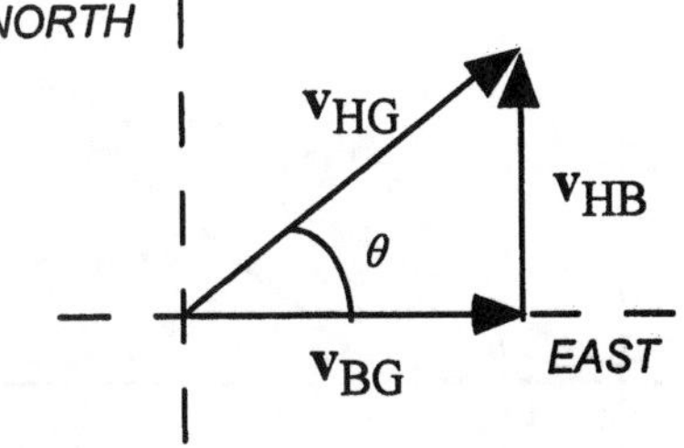

$$v_{\text{HG}} = \sqrt{v_{\text{HB}}^2 + v_{\text{BG}}^2}$$

$$= \sqrt{(2.0 \text{ m/s})^2 + (6.0 \text{ m/s})^2} = \boxed{6.3 \text{ m/s}}$$

The angle θ is

$$\theta = \tan^{-1}\left(\frac{v_{\text{HB}}}{v_{\text{BG}}}\right) = \tan^{-1}\left(\frac{2.0 \text{ m/s}}{6.0 \text{ m/s}}\right) = \boxed{18°\text{, north of east}}$$

---

54. ***REASONING*** The relative velocities in this problem are:

$\mathbf{v_{PG}}$ = velocity of the **P**lane relative to the **G**round (direction is due west)
$\mathbf{v_{PA}}$ = velocity of the **P**lane relative to the **A**ir (magnitude = 245 m/s)
$\mathbf{v_{AG}}$ = velocity of the **A**ir relative to the **G**round (38.0 m/s, due north)

These velocities are related by the subscripting method discussed in Section 3.4: $\mathbf{v_{PG}} = \mathbf{v_{PA}} + \mathbf{v_{AG}}$. This vector sum is shown in the drawing in the drawing on the left.

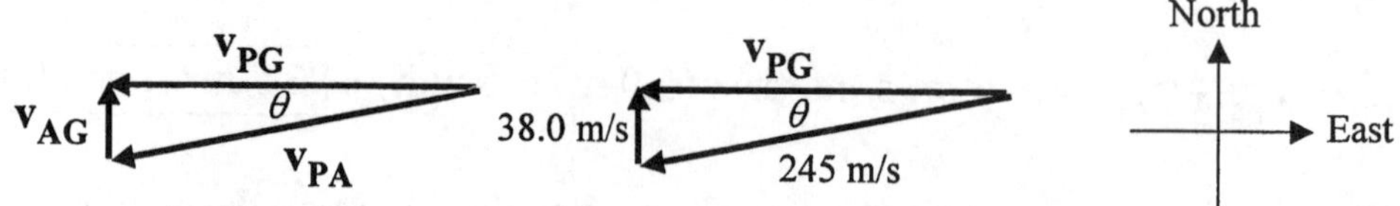

SOLUTION Since the triangles in the drawings are right triangles, the angle θ can be found by using the inverse sine function:

$$\theta = \sin^{-1}\left(\frac{v_{AG}}{v_{PA}}\right) = \sin^{-1}\left(\frac{38.0 \text{ m/s}}{245 \text{ m/s}}\right) = \boxed{8.92°\text{, south of west}}$$

---

55. ***REASONING*** The relative velocities in this problem are:

$\mathbf{v_{PW}}$ = velocity of the **P**assenger relative to the **W**ater
$\mathbf{v_{PB}}$ = velocity of the **P**assenger relative to the **B**oat (2.50 m/s, due east)
$\mathbf{v_{BW}}$ = velocity of the **B**oat relative to the **W**ater (5.50 m/s, at 38.0° north of east)

The velocities are shown in the drawing are related by the subscripting method discussed in Section 3.4:

$$\mathbf{v_{PW}} = \mathbf{v_{PB}} + \mathbf{v_{BW}}$$

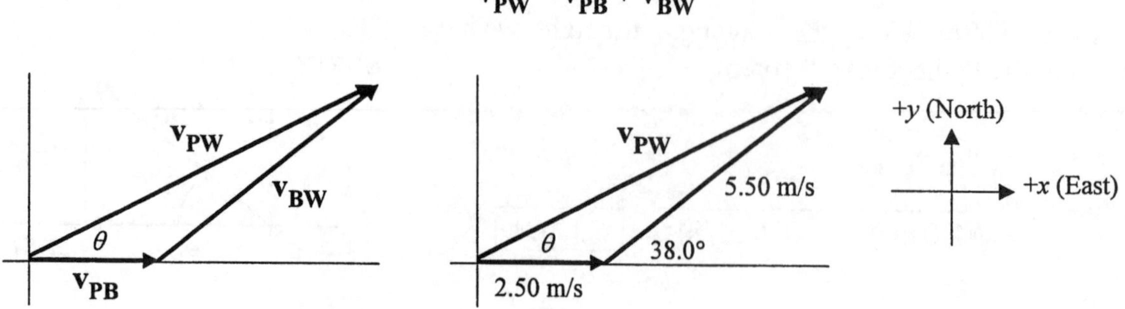

We will determine the magnitude and direction of $\mathbf{v_{PW}}$ from the equation above by using the method of scalar components.

SOLUTION The table below lists the scalar components of the three vectors.

Vector	*x Component*	*y Component*
$\mathbf{v}_{\mathbf{PB}}$	+2.50 m/s	0 m/s
$\mathbf{v}_{\mathbf{BW}}$	+(5.50 m/s) cos 38.0° = +4.33 m/s	+(5.50 m/s) sin 38.0° = +3.39 m/s
$\mathbf{v}_{\mathbf{PW}} = \mathbf{v}_{\mathbf{PB}} + \mathbf{v}_{\mathbf{BW}}$	+2.50 m/s + 4.33 m/s = +6.83 m/s	0 m/s + 3.39 m/s = +3.39 m/s

The magnitude of $\mathbf{v}_{\mathbf{PW}}$ can be found by applying the Pythagorean theorem to the x and y components:

$$v_{\mathrm{PW}} = \sqrt{(6.83\text{ m/s})^2 + (3.39\text{ m/s})^2} = \boxed{7.63\text{ m/s}}$$

The angle θ (see the drawings) that $\mathbf{v}_{\mathbf{PW}}$ makes with due east is

$$\theta = \tan^{-1}\left(\frac{+3.39\text{ m/s}}{+6.83\text{ m/s}}\right) = \boxed{26.4°\text{ north of east}}$$

---

56. ***REASONING AND SOLUTION*** The velocity of the raindrops relative to the train is given by

$$v_{\mathrm{RT}} = v_{\mathrm{RG}} + v_{\mathrm{GT}}$$

where v_{RG} is the velocity of the raindrops relative to the ground and v_{GT} is the velocity of the ground relative to the train.

Since the train moves horizontally, and the rain falls vertically, the velocity vectors are related as shown in the figure at the right. Then

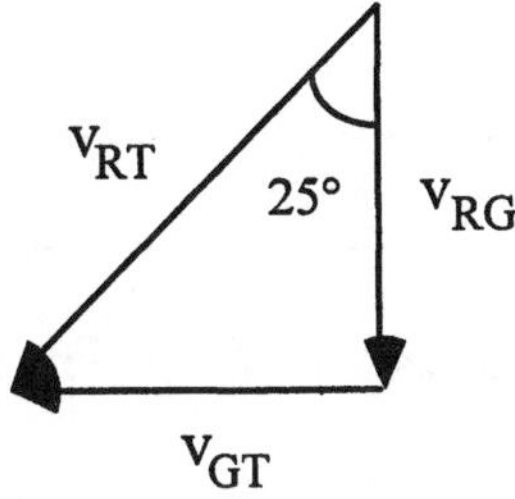

$$v_{\mathrm{GT}} = v_{\mathrm{RG}} \tan\theta = (5.0\text{ m/s})(\tan 25°) = 2.3\text{ m/s}$$

The train is moving at a speed of $\boxed{2.3\text{ m/s}}$

---

57. **SSM** ***REASONING*** The velocity $\mathbf{v}_{\mathbf{PG}}$ of the plane relative to the ground is the vector sum of the velocity $\mathbf{v}_{\mathbf{PA}}$ of the plane relative to the air and the velocity $\mathbf{v}_{\mathbf{AG}}$ of the air relative to the ground, as indicated by Equation 3.7: $\mathbf{v}_{\mathbf{PG}} = \mathbf{v}_{\mathbf{PA}} + \mathbf{v}_{\mathbf{AG}}$, or $\mathbf{v}_{\mathbf{AG}} = \mathbf{v}_{\mathbf{PG}} - \mathbf{v}_{\mathbf{PA}}$. We are given $\mathbf{v}_{\mathbf{PA}}$ and can find $\mathbf{v}_{\mathbf{PG}}$ from the given displacement and travel time. Thus,

$$\mathbf{v}_{PG} = \frac{81.0\times10^{3}\ \text{m}}{9.00\times10^{2}\ \text{s}} = 90.0\ \text{m/s},\ 45°\ \text{west of south}$$

SOLUTION The first two rows of the following table give the east/west and north/south components of the vectors $\mathbf{v}_{PG}$ and $-\mathbf{v}_{PA}$. The third row gives the components of $\mathbf{v}_{AG} = \mathbf{v}_{PG} - \mathbf{v}_{PA}$. Due east and due north have been chosen as the positive directions.

Vector	*East/West Component*	*North/South Component*
$\mathbf{v}_{PG}$	$-(90.0\ \text{m/s})\sin 45° = -63.6\ \text{m/s}$	$-(90.0\ \text{m/s})\cos 45° = -63.6\ \text{m/s}$
$-\mathbf{v}_{PA}$	0 m/s	+57.8 m/s
$\mathbf{v}_{AG} = \mathbf{v}_{PG} - \mathbf{v}_{PA}$	−63.6 m/s	−5.8 m/s

The magnitude of the velocity of the wind with respect to the ground can be obtained using the Pythagorean theorem:

$$v_{AG} = \sqrt{(-63.6\ \text{m/s})^2 + (-5.8\ \text{m/s})^2} = \boxed{63.9\ \text{m/s}}$$

The direction of $\mathbf{v}_{AG}$ is found from

$$\phi = \tan^{-1}\left(\frac{63.6\ \text{m/s}}{5.8\ \text{m/s}}\right) = \boxed{85°\ \text{west of south}}$$

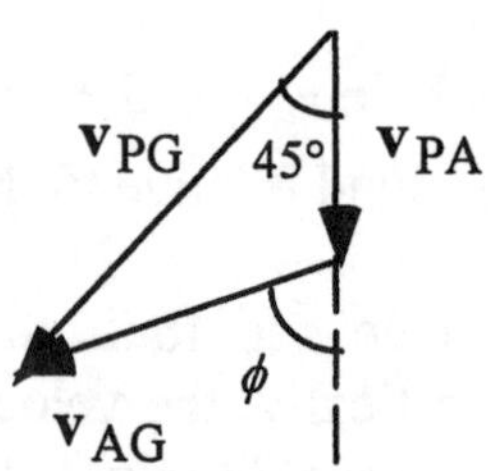

58. ***REASONING*** The relative velocities in this problem are:

$\mathbf{v}_{PS}$ = velocity of the **P**assenger relative to the **S**hore
$\mathbf{v}_{P2}$ = velocity of the **P**assenger relative to Boat **2** (1.20 m/s, due east)
$\mathbf{v}_{2S}$ = velocity of Boat **2** relative to the **S**hore
$\mathbf{v}_{21}$ = velocity of Boat **2** relative to Boat **1** (1.60 m/s, at 30.0° north of east)
$\mathbf{v}_{1S}$ = velocity of Boat **1** relative to the **S**hore (3.00 m/s, due north)

The velocity $\mathbf{v}_{PS}$ of the passenger relative to the shore is related to $\mathbf{v}_{P2}$ and $\mathbf{v}_{2S}$ by (see the method of subscripting discussed in Section 3.4):

$$\mathbf{v}_{PS} = \mathbf{v}_{P2} + \mathbf{v}_{2S}$$

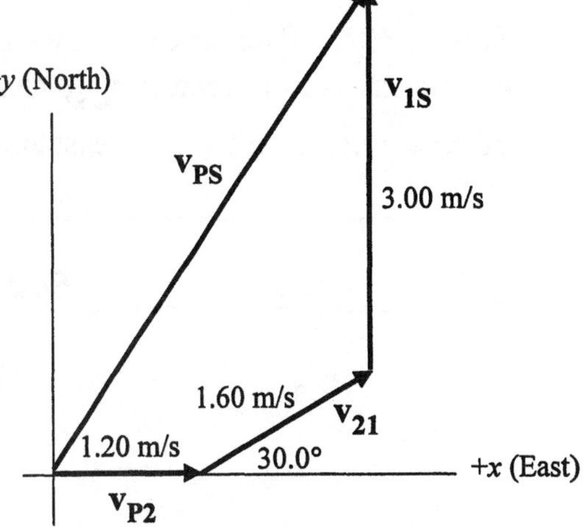

But $\mathbf{v}_{2S}$, the velocity of Boat 2 relative to the shore, is related to $\mathbf{v}_{21}$ and $\mathbf{v}_{1S}$ by

$$\mathbf{v}_{2S} = \mathbf{v}_{21} + \mathbf{v}_{1S}$$

Substituting this expression for $\mathbf{v}_{2S}$ into the first equation yields

$$\mathbf{v}_{PS} = \mathbf{v}_{P2} + \mathbf{v}_{21} + \mathbf{v}_{1S}$$

This vector sum is shown in the diagram. We will determine the magnitude of $\mathbf{v}_{PS}$ from the equation above by using the method of scalar components.

SOLUTION The table below lists the scalar components of the four vectors in the drawing.

Vector	*x Component*	*y Component*
$\mathbf{v}_{P2}$	+1.20 m/s	0 m/s
$\mathbf{v}_{21}$	+(1.60 m/s) cos 30.0° = +1.39 m/s	+(1.60 m/s) sin 30.0° = +0.80 m/s
$\mathbf{v}_{1S}$	0 m/s	+3.00 m/s
$\mathbf{v}_{PS} = \mathbf{v}_{P2} + \mathbf{v}_{21} + \mathbf{v}_{1S}$	+1.20 m/s + 1.39 m/s = +2.59 m/s	+0.80 m/s +3.00 m/s = +3.80 m/s

The magnitude of $\mathbf{v}_{PS}$ can be found by applying the Pythagorean theorem to its x and y components:

$$v_{PS} = \sqrt{(2.59 \text{ m/s})^2 + (3.80 \text{ m/s})^2} = \boxed{4.60 \text{ m/s}}$$

---

59. SSM WWW ***REASONING*** The velocity $\mathbf{v}_{OW}$ of the object relative to the water is the vector sum of the velocity $\mathbf{v}_{OS}$ of the object relative to the ship and the velocity $\mathbf{v}_{SW}$ of the ship relative to the water, as indicated by Equation 3.7: $\mathbf{v}_{OW} = \mathbf{v}_{OS} + \mathbf{v}_{SW}$. The value of $\mathbf{v}_{SW}$ is given in the statement of the problem. We can find the value of $\mathbf{v}_{OS}$ from the fact that we know the position of the object relative to the ship at two different times. The initial position is $\mathbf{r}_{OS1}$, and the final position is $\mathbf{r}_{OS2}$. Since the object moves with constant velocity,

$$\mathbf{v}_{OS} = \frac{\Delta \mathbf{r}_{OS}}{\Delta t} = \frac{\mathbf{r}_{OS2} - \mathbf{r}_{OS1}}{\Delta t} \quad (1)$$

SOLUTION The first two rows of the following table give the east/west and north/south components of the vectors $\mathbf{r}_{OS2}$ and $-\mathbf{r}_{OS1}$. The third row of the table gives the components of $\Delta\mathbf{r}_{OS} = \mathbf{r}_{OS2} - \mathbf{r}_{OS1}$. Due east and due north have been taken as positive.

Vector	*East/West Component*	*North/South Component*
$\mathbf{r}_{OS2}$	$-(1120\text{ m})\cos 57.0° = -6.10\times10^2\text{ m}$	$-(1120\text{ m})\sin 57.0° = -9.39\times10^2\text{ m}$
$-\mathbf{r}_{OS1}$	$-(2310\text{ m})\cos 32.0° = -1.96\times10^3\text{ m}$	$+(2310\text{ m})\sin 32.0° = 1.22\times10^3\text{ m}$
$\Delta\mathbf{r}_{OS} = \mathbf{r}_{OS2} - \mathbf{r}_{OS1}$	$-2.57\times10^3\text{ m}$	$2.81\times10^2\text{ m}$

Now that the components of $\Delta\mathbf{r}_{OS}$ are known, the Pythagorean theorem can be used to find the magnitude.

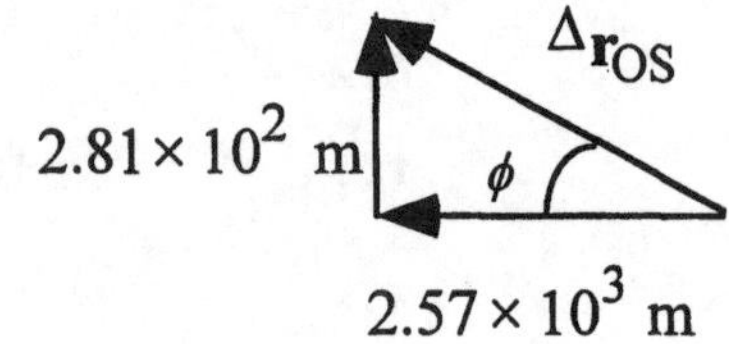

$$\Delta r_{OS} = \sqrt{(-2.57\times10^3\text{ m})^2 + (2.81\times10^2\text{ m})^2} = 2.59\times10^3\text{ m}$$

The direction of $\Delta\mathbf{r}_{OS}$ is found from

$$\phi = \tan^{-1}\left(\frac{2.81\times10^2\text{ m}}{2.57\times10^3\text{ m}}\right) = 6.24°$$

Therefore, from Equation (1),

$$\mathbf{v}_{OS} = \frac{\Delta\mathbf{r}_{OS}}{\Delta t} = \frac{\mathbf{r}_{OS2} - \mathbf{r}_{OS1}}{\Delta t} = \frac{2.59\times10^3\text{ m}}{360\text{ s}} = 7.19\text{ m/s, } 6.24°\text{ north of west}$$

Now that $\mathbf{v}_{OS}$ is known, we can find $\mathbf{v}_{OW}$, as indicated by Equation 3.7: $\mathbf{v}_{OW} = \mathbf{v}_{OS} + \mathbf{v}_{SW}$. The following table summarizes the vector addition:

Vector	*East/West Component*	*North/South Component*
$\mathbf{v}_{OS}$	$-(7.19\text{ m/s})\cos 6.24° = -7.15\text{ m/s}$	$(7.19\text{ m/s})\sin 6.24° = 0.782\text{ m/s}$

$\mathbf{v}_{SW}$	+4.20 m/s	0 m/s
$\mathbf{v}_{OW} = \mathbf{v}_{OS} + \mathbf{v}_{SW}$	–2.95 m/s	0.782 m/s

Now that the components of $\mathbf{v}_{OW}$ are known, the Pythagorean theorem can be used to find the magnitude.

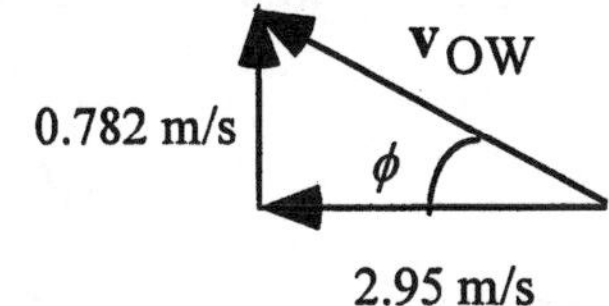

$$v_{OW} = \sqrt{(-2.95\ \text{m/s})^2 + (0.782\ \text{m/s})^2} = \boxed{3.05\ \text{m/s}}$$

The direction of $\mathbf{v}_{OW}$ is found from

$$\phi = \tan^{-1}\left(\frac{0.782\ \text{m/s}}{2.95\ \text{m/s}}\right) = \boxed{14.8°\ \text{north of west}}$$

---

60. ***REASONING AND SOLUTION***

$$x = r\cos\theta = (162\ \text{km})\cos 62.3° = \boxed{75.3\ \text{km}}$$

$$y = r\sin\theta = (162\ \text{km})\sin 62.3° = \boxed{143\ \text{km}}$$

---

61. ***REASONING AND SOLUTION*** The vertical motion consists of the ball rising for a time t stopping and returning to the ground in another time t. For the upward portion

$$t = \frac{v_{0y} - v_y}{g} = \frac{v_{0y}}{g}$$

Note: $v_y = 0$ m/s since the ball stops at the top. Now

$$v_{0y} = v_0 \sin\theta_0 = (25.0\ \text{m/s})\sin 60.0° = 21.7\ \text{m/s}$$

$$t = (21.7\ \text{m/s})/(9.80\ \text{m/s}^2) = 2.21\ \text{s}$$

The required "hang time" is $2t = \boxed{4.42\ \text{s}}$.

---

62. ***REASONING AND SOLUTION***
a. The bullet strikes the ground when its y coordinate is – 1.6 m. We know

$$v_{oy} = v_o \sin 0° = 0$$

$$t = \sqrt{\frac{-2y}{g}} = \sqrt{\frac{-2(-1.6 \text{ m})}{9.80 \text{ m/s}^2}} = \boxed{0.57 \text{ s}}$$

b. During this time the bullet travels a horizontal distance x with $a_x = 0$.

$$x = v_{ox}t = (v_o \cos \theta_o)t = (1100 \text{ m/s})(\cos 0°)(0.57 \text{ s}) = \boxed{630 \text{ m}}$$

---

63. SSM ***REASONING AND SOLUTION*** As shown in Example 2, the time required for the package to hit the ground is given by $t = \sqrt{2y/a_y}$ and is independent of the plane's horizontal velocity. Thus, the time needed for the package to hit the ground is still $\boxed{14.6 \text{ s}}$.

---

64. ***REASONING*** The magnitude and direction of the initial velocity $\mathbf{v_0}$ can be obtained using the Pythagorean theorem and trigonometry, once the x and y components of the initial velocity v_{0x} and v_{0y} are known. These components can be calculated using Equations 3.3a and 3.3b.

SOLUTION Using Equations 3.3a and 3.3b, we obtain the following results for the velocity components:

$$v_{0x} = v_x - a_x t = 3775 \text{ m/s} - (5.10 \text{ m/s}^2)(565 \text{ s}) = 893.5 \text{ m/s}$$

$$v_{0y} = v_y - a_y t = 4816 \text{ m/s} - (7.30 \text{ m/s}^2)(565 \text{ s}) = 691.5 \text{ m/s}$$

Using the Pythagorean theorem and trigonometry, we find

$$v_0 = \sqrt{v_{0x}^2 + v_{0y}^2} = \sqrt{(893.5 \text{ m/s})^2 + (691.5 \text{ m/s})^2} = \boxed{1130 \text{ m/s}}$$

$$\theta = \tan^{-1}\left(\frac{v_{0y}}{v_{0x}}\right) = \tan^{-1}\left(\frac{691.5 \text{ m/s}}{893.5 \text{ m/s}}\right) = \boxed{37.7°}$$

---

65. SSM ***REASONING*** Since the magnitude of the velocity of the fuel tank is given by $v = \sqrt{v_x^2 + v_y^2}$, it is necessary to know the velocity components v_x and v_y just before impact. At the instant of release, the empty fuel tank has the same velocity as that of the plane. Therefore, the magnitudes of the initial velocity components of the fuel tank are

given by $v_{0x} = v_0 \cos\theta$ and $v_{0y} = v_0 \sin\theta$, where v_0 is the speed of the plane at the instant of release. Since the x motion has zero acceleration, the x component of the velocity of the plane remains equal to v_{0x} for all later times while the tank is airborne. The y component of the velocity of the tank after it has undergone a vertical displacement y is given by Equation 3.6b.

SOLUTION

a. Taking up as the positive direction, the velocity components of the fuel tank just before it hits the ground are

$$v_x = v_{0x} = v\cos\theta = (135 \text{ m/s}) \cos 15° = 1.30\times10^2 \text{ m/s}$$

From Equation 3.6b, we have

$$v_y = -\sqrt{v_{0y}^2 + 2a_y y} = -\sqrt{(v_0 \sin\theta)^2 + 2a_y y}$$

$$= -\sqrt{[(135 \text{ m/s}) \sin 15.0°]^2 + [2(-9.80 \text{ m/s}^2)(-2.00\times10^3 \text{ m})]} = -201 \text{ m/s}$$

Therefore, the magnitude of the velocity of the fuel tank just before impact is

$$v = \sqrt{v_x^2 + v_y^2} = \sqrt{(1.30\times10^2 \text{ m/s})^2 + (201 \text{ m/s})^2} = \boxed{239 \text{ m/s}}$$

The velocity vector just before impact is inclined at an angle ϕ above the horizontal. This angle is

$$\phi = \tan^{-1}\left(\frac{201 \text{ m/s}}{1.30\times10^2 \text{ m/s}}\right) = \boxed{57.1°}$$

b. As shown in Conceptual Example 9, once the fuel tank in part a rises and falls to the same altitude at which it was released, its motion is identical to the fuel tank in part b. Therefore, the velocity of the fuel tank in part b just before impact is $\boxed{\text{239 m/s at an angle of 57.1° above the horizontal}}$.

---

66. ***REASONING*** Since we know the launch angle $\theta = 15.0°$, the launch speed v_0 can be obtained using trigonometry, which gives the y component of the launch velocity as $v_{0y} = v_0 \sin\theta$. Solving this equation for v_0 requires a value for v_{0y}, which we can obtain from the vertical height of $y = 13.5$ m by using Equation 3.6b from the equations of kinematics.

SOLUTION From Equation 3.6b we have

$$v_y^2 = v_{0y}^2 + 2a_y y$$

$$(0\text{ m/s})^2 = v_{0y}^2 + 2(-9.80\text{ m/s}^2)(13.5\text{ m}) \quad \text{or} \quad v_{0y} = \sqrt{2(9.80\text{ m/s}^2)(13.5\text{ m})}$$

Using trigonometry, we find

$$v_0 = \frac{v_{0y}}{\sin 15.0°} = \frac{\sqrt{2(9.80\text{ m/s}^2)(13.5\text{ m})}}{\sin 15.0°} = \boxed{62.8\text{ m/s}}$$

---

67. ***REASONING*** The drawing shows the trajectory of the ball, along with its initial speed v_0 and vertical displacement y. The angle that the initial velocity of the ball makes with the ground is 35.0°. The known data are shown in the table below:

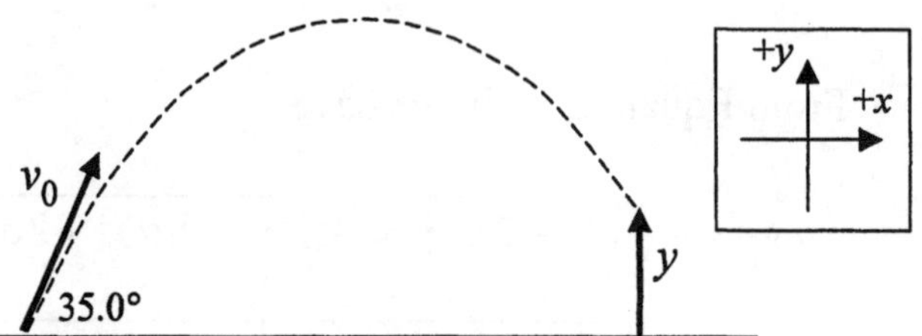

***y*-Direction Data**

y	a_y	v_y	v_{0y}	t
+5.50 m	−9.80 m/s^2		+(46.0 m/s) sin 35.0° = +26.4 m/s	?

Since three of the kinematic variables are known, we will employ the appropriate equation of kinematics to determine the time of flight.

SOLUTION Equation 3.5b $\left(y = v_{0y}t + \frac{1}{2}a_y t^2\right)$ relates the time t to the three known variables. The terms in this equation can be rearranged to as to place it in a standard quadratic form: $\frac{1}{2}a_y t^2 + v_{0y}t - y = 0$. The solution of this quadratic equation is

$$t = \frac{-v_{0y} \pm \sqrt{v_{0y}^2 - 4\left(\frac{1}{2}a_y\right)(-y)}}{2\left(\frac{1}{2}a_y\right)}$$

$$t = \frac{-(+26.4\text{ m/s}) \pm \sqrt{(+26.4\text{ m/s})^2 - 4\left(\frac{1}{2}\right)(-9.80\text{ m/s}^2)(-5.50\text{ m})}}{-9.80\text{ m/s}^2} = 0.217\text{ s or } 5.17\text{ s}$$

The first solution (t = 0.217 s) corresponds to the situation where the ball is moving upward and has a displacement of y = +5.50 m. The second solution represents the later time when the ball is moving downward and its displacement is also y = +5.50 m (see the drawing). This is the solution we seek, so $t = \boxed{5.17\text{ s}}$.

---

68. ***REASONING AND SOLUTION*** On impact

$$v_x = v\cos 75.0° = (8.90\text{ m/s})\cos 75.0° = 2.30\text{ m/s}$$

and

$$v_{0y}^2 = v_y^2 + 2gy = (8.90\text{ m/s})^2 \sin^2 75.0° + 2(9.80\text{ m/s}^2)(-3.00\text{ m})$$

so

$$v_{0y} = 3.89\text{ m/s}$$

The magnitude of the diver's initial velocity is

$$v_0 = \sqrt{(2.30\text{ m/s})^2 + (3.89\text{ m/s})^2} = \boxed{4.52\text{ m/s}}$$

The angle the initial velocity vector makes with the horizontal is

$$\theta_0 = \tan^{-1}(v_{oy}/v_{ox}) = \boxed{59.4°}$$

---

69. SSM ***REASONING*** The velocity $\mathbf{v}_{PM}$ of the puck relative to Mario is the vector sum of the velocity $\mathbf{v}_{PI}$ of the puck relative to the ice and the velocity $\mathbf{v}_{IM}$ of the ice relative to Mario as indicated by Equation 3.7: $\mathbf{v}_{PM} = \mathbf{v}_{PI} + \mathbf{v}_{IM}$. The values of $\mathbf{v}_{MI}$ and $\mathbf{v}_{PI}$ are given in the statement of the problem. In order to use the data, we must make use of the fact that $\mathbf{v}_{IM} = -\mathbf{v}_{MI}$, with the result that $\mathbf{v}_{PM} = \mathbf{v}_{PI} - \mathbf{v}_{MI}$.

SOLUTION The first two rows of the following table give the east/west and north/south components of the vectors $\mathbf{v}_{PI}$ and $-\mathbf{v}_{MI}$. The third row gives the components of their resultant $\mathbf{v}_{PM} = \mathbf{v}_{PI} - \mathbf{v}_{MI}$. Due east and due north have been taken as positive.

Vector	*East/West Component*	*North/South Component*
$\mathbf{v}_{PI}$	$-(11.0\text{ m/s})\sin 22° = -4.1\text{ m/s}$	$-(11.0\text{ m/s})\cos 22° = -10.2\text{ m/s}$
$-\mathbf{v}_{MI}$	0	+7.0 m/s
$\mathbf{v}_{PM} = \mathbf{v}_{PI} - \mathbf{v}_{MI}$	−4.1 m/s	−3.2 m/s

Now that the components of $\mathbf{v}_{PM}$ are known, the Pythagorean theorem can be used to find the magnitude.

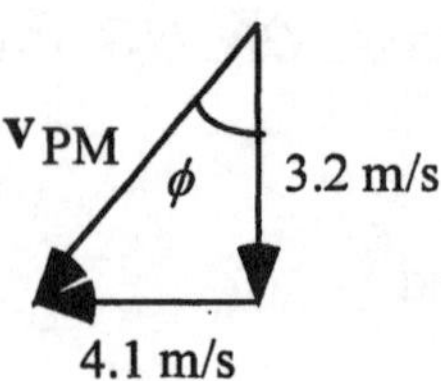

$$v_{PM} = \sqrt{(-4.1 \text{ m/s})^2 + (-3.2 \text{ m/s})^2} = \boxed{5.2 \text{ m/s}}$$

The direction of $\mathbf{v}_{PM}$ is found from

$$\phi = \tan^{-1}\left(\frac{4.1 \text{ m/s}}{3.2 \text{ m/s}}\right) = \boxed{52° \text{ west of south}}$$

---

70. ***REASONING AND SOLUTION*** In the book's notation $\mathbf{v}_{SW} = \mathbf{v}_{SP} + \mathbf{v}_{PW}$. The vectors form the sides of a triangle for which the law of cosines gives

$$v_{SP}{}^2 = v_{SW}{}^2 + v_{PW}{}^2 - 2v_{SW}v_{PW} \cos 45°$$

$$= (1.0 \text{ m/s})^2 + (8.5 \text{ m/s})^2 - 2(1.0 \text{ m/s})(8.5 \text{ m/s}) \cos 45°, \quad \text{so} \quad v_{SP} = \boxed{7.8 \text{ m/s}}$$

The law of sines gives the directional angle to be

$$\sin\theta = (v_{SW}/v_{SP}) \sin 45° \quad \text{or} \quad \theta = 5.2° \text{ east of south}$$

This corresponds to $180.0° - 5.2° = \boxed{174.8° \text{ east of north}}$.

---

71. **SSM** ***REASONING*** The angle θ can be found from

$$\theta = \tan^{-1}\left(\frac{2400 \text{ m}}{x}\right) \tag{1}$$

where x is the horizontal displacement of the flare. Since $a_x = 0$, it follows that $x = (v_0 \cos 30.0°)t$. The flight time t is determined by the vertical motion. In particular, the time t can be found from Equation 3.5b. Once the time is known, x can be calculated.

SOLUTION From Equation 3.5b, assuming upward is the positive direction, we have

$$y = -(v_0 \sin 30.0°)t + \tfrac{1}{2}a_y t^2$$

which can be rearranged to give the following equation that is quadratic in t:

$$\tfrac{1}{2}a_y t^2 - (v_0 \sin 30.0°)t - y = 0$$

Using $y = -2400$ m and $a_y = -9.80$ m/s^2 and suppressing the units, we obtain the quadratic equation

$$4.9t^2 + 120t - 2400 = 0$$

Using the quadratic formula, we obtain $t = 13$ s. Therefore, we find that

$$x = (v_0 \cos 30.0°)t = (240 \text{ m/s})(\cos 30.0°)(13 \text{ s}) = 2700 \text{ m}$$

Equation (1) then gives

$$\theta = \tan^{-1}\left(\frac{2400 \text{ m}}{2700 \text{ m}}\right) = \boxed{42°}$$

---

72. ***REASONING AND SOLUTION*** While flying west, the airplane has a ground speed of

$$v_{PG} = 2.40 \times 10^2 \text{ m/s} - 57.8 \text{ m/s} = 182 \text{ m/s}$$

and requires time $t_W = x/(182 \text{ m/s})$ to reach the turn-around point. While flying east the airplane has a ground speed of

$$v_{PG} = 2.40 \times 10^2 \text{ m/s} + 57.8 \text{ m/s} = 298 \text{ m/s}$$

and requires $t_E = x/(298 \text{ m/s})$ to return home. Now the total time for the trip is $t = t_W + t_E = 6.00 \text{ h} = 2.16 \times 10^4$ s, so

$$x/182 + x/298 = 2.16 \times 10^4 \quad \text{or} \quad x = \boxed{2.44 \times 10^6 \text{ m} = 2440 \text{ km}}$$

---

73. ***REASONING*** We will treat the horizontal and vertical parts of the motion separately. The range R is the product of the horizontal component of the initial velocity v_{0x} and the time of flight t. The time of flight can be obtained from the vertical part of the motion by using Equation 3.5b $\left(y = v_{0y}t + \tfrac{1}{2}a_y t^2\right)$ and the fact that the displacement y in the vertical direction is zero, since the projectile is launched from and returns to ground level. The expression for the range obtained in this way can then be applied to obtain the desired launch angle for doubling the range.

SOLUTION The range of the projectile is

$$R = v_{0x}t = (v_0 \cos\theta)t$$

Using Equation 3.5b, we obtain the time of flight as

$$y = 0 = v_{0y}t + \tfrac{1}{2}a_y t^2 \quad \text{or} \quad t = -\frac{2v_{0y}}{a_y} = -\frac{2v_0 \sin\theta}{a_y}$$

Substituting this expression for *t* into the range expression gives

$$R = (v_0 \cos\theta)\left(\frac{-2v_0 \sin\theta}{a_y}\right) = -\frac{2v_0^2 \cos\theta \sin\theta}{a_y} = -\frac{v_0^2 \sin 2\theta}{a_y}$$

where we have used the fact that 2 cos θ sin θ = sin 2θ. We can now apply this expression for the range to the initial range R_1 for θ_1 and the range R_2 for θ_2:

$$R_1 = -\frac{v_0^2 \sin 2\theta_1}{a_y} \quad \text{or} \quad R_2 = -\frac{v_0^2 \sin 2\theta_2}{a_y}$$

Dividing these two expressions gives

$$\frac{R_2}{R_1} = \frac{-(v_0^2 \sin 2\theta_2)/a_y}{-(v_0^2 \sin 2\theta_1)/a_y} = \frac{\sin 2\theta_2}{\sin 2\theta_1} = 2$$

where we have used the fact that $R_2/R_1 = 2$. Since $\theta_1 = 12.0°$, we find that

$$\sin 2\theta_2 = 2.00 \sin 2\theta_1 = 2.00 \sin 2(12.0°) = 0.813$$

$$\theta_2 = \frac{\sin^{-1} 0.813}{2.00} = \boxed{27.2°}$$

---

74. ***CONCEPT QUESTIONS*** a. Since both v_{0x} and a_x are positive, the *x* component of the initial velocity and the *x* component of the acceleration have the same direction. In such a case, the magnitude of the velocity component increases in time.

b. Since v_{0y} is positive and a_y is negative, the *y* component of the initial velocity and the *y* component of the acceleration have opposite directions. In such a case, the magnitude of the velocity component decreases in time.

SOLUTION Using Equations 3.3a and 3.3b, we find

$$v_x = v_{0x} + a_x t = +1.0 \text{ m/s} + \left(2.0 \text{ m/s}^2\right)\left(0.50 \text{ s}\right) = +2.0 \text{ m/s}$$

$$v_y = v_{0y} + a_y t = +2.0 \text{ m/s} + \left(-2.0 \text{ m/s}^2\right)\left(0.50 \text{ s}\right) = +1.0 \text{ m/s}$$

$$v = \sqrt{\left(2.0 \text{ m/s}\right)^2 + \left(1.0 \text{ m/s}\right)^2} = \boxed{2.2 \text{ m/s}}$$

$$\theta = \tan^{-1}\left(\frac{1.0 \text{ m/s}}{2.0 \text{ m/s}}\right) = \boxed{27° \text{ above the } +x \text{ axis}}$$

---

75. ***CONCEPT QUESTIONS*** a. The speed at the peak of the trajectory is v_{0x}, because at this point the vertical component of the velocity has decreased to zero and in the absence of air resistance the horizontal component remains constant at its launch value.

b. When the projectile lands at the same vertical level from which it was launched, it has the same speed v_0 as it did initially, provided there is no air resistance. The initial horizontal component of the velocity never changes during the flight, while the vertical component decreases to zero as the projectile rises to the peak of the trajectory and then increases on the way back down. At the point of landing, the vertical component has regained exactly its initial magnitude, but now points downward, instead of upward. Since the magnitudes of both the horizontal and vertical velocity components at the end are the same as they were initially, the magnitude of the final velocity or the final speed is equal to the initial speed.

c. At this intermediate point the projectile has a speed between v_{0x} and v_0. This is because the speed changes gradually from point to point along the trajectory, and the speed at the top is v_{0x}, while the speed at the end is v_0.

SOLUTION Using trigonometry to find the x component v_{0x} of the initial velocity, using Equation 3.6b for the y component v_y, and using the Pythagorean theorem, we find:

$$v_x = v_{0x} = v_0 \cos 40.0° = (14.0 \text{ m/s})\cos 40.0° = 10.7 \text{ m/s}$$

$$v_y^2 = v_{0y}^2 + 2a_y y = \left(v_0 \sin 40.0°\right)^2 + 2a_y y$$

$$v = \sqrt{v_x^2 + v_y^2} = \sqrt{\left(v_0 \cos 40.0°\right)^2 + \left(v_0 \sin 40.0°\right)^2 + 2a_y y} = \sqrt{v_0^2 + 2a_y y}$$

$$v = \sqrt{(14.0 \text{ m/s})^2 + 2\left(-9.80 \text{ m/s}^2\right)(3.0 \text{ m})} = \boxed{11.7 \text{ m/s}}$$

---

76. ***CONCEPT QUESTIONS*** a. In the absence of air resistance, the horizontal velocity component never changes from its initial value of v_{0x}.

b. Yes. Since the horizontal velocity component never changes from its initial value of v_{0x}, the horizontal distance traveled after launch can be calculated simply as velocity times the travel time, or $D = v_{0x}t$.

c. Yes. In calculating the fall-time, the vertical part of the motion is exactly like that of a ball dropped from rest. This is because a horizontally-launched projectile has an initial velocity component of $v_{0y} = 0$ m/s in the vertical direction.

SOLUTION The horizontal distance traveled after launch is

$$D = v_{0x}t = (5.3 \text{ m/s})t$$

Using Equation 3.5b to find the fall-time, we find

$$y = v_{0y}t + \tfrac{1}{2}a_y t^2 \quad \text{or} \quad -2.0 \text{ m} = (0 \text{ m/s})t + \tfrac{1}{2}(-9.80 \text{ m/s}^2)t^2$$

$$t = \sqrt{\frac{2(-2.0 \text{ m})}{-9.80 \text{ m/s}^2}}$$

$$D = (5.3 \text{ m/s})\sqrt{\frac{2(-2.0 \text{ m})}{-9.80 \text{ m/s}^2}} = \boxed{3.4 \text{ m}}$$

---

77. ***CONCEPT QUESTIONS*** a. The vertical component of the launch velocity determines the maximum height attained by the projectile. The horizontal component determines the range of the projectile, not the height.

b. The magnitudes of the components are smaller than v_0. The reason is that these components are given by v_0 times the trigonometric sine or cosine function, and these functions are always less than one.

c. When thrown straight upward at speed v_0, the projectile attains a greater height than when launched at an angle. The reason is that when the projectile is thrown straight upward, the entire velocity is directed upward, rather than just component of it. With more the velocity directed upward, the projectile naturally goes higher.

SOLUTION Using Equation 3.6b, we can find the y component of the initial velocity:

$$v_y^2 = v_{0y}^2 + 2a_y y \quad \text{or} \quad (0 \text{ m/s})^2 = v_{0y}^2 + 2(-9.80 \text{ m/s}^2)(7.5 \text{ m})$$

$$v_{0y} = \sqrt{2(9.80 \text{ m/s}^2)(7.5 \text{ m})}$$

Using trigonometry, we can use this value for v_{0y} to determine the initial velocity v_0:

$$v_0 = \frac{v_{0y}}{\sin\theta} = \frac{\sqrt{2(9.80 \text{ m/s}^2)(7.5 \text{ m})}}{\sin 52°}$$

With this value for v_0 and the fact that v_{0y} = 0 m/s at the top of the trajectory, we can use Equation 3.6b to determine the desired height:

$$v_y^2 = v_{0y}^2 + 2a_y y$$

$$(0 \text{ m/s})^2 = \left[\frac{\sqrt{2(9.80 \text{ m/s}^2)(7.5 \text{ m})}}{\sin 52°}\right]^2 + 2(-9.80 \text{ m/s}^2)y$$

$$y = \frac{\left[\dfrac{\sqrt{2(9.80 \text{ m/s}^2)(7.5 \text{ m})}}{\sin 52°}\right]^2}{2(9.80 \text{ m/s}^2)} = \boxed{12 \text{ m}}$$

---

78. ***CONCEPT QUESTIONS*** a. The horizontal component v_{0x} of the launch velocity is proportional to the launch speed v_0, because $v_{0x} = v_0 \cos\theta$, where θ is the launch angle.

b. The time of flight t is also proportional to the launch speed v_0. The greater the launch speed, the longer the projectile is in flight. More exactly, we know that, for a projectile that is launched from and returns to ground level, the vertical displacement is $y = 0$ m. Using Equation 3.5b, we have

$$y = v_{0y}t + \tfrac{1}{2}a_y t^2$$

$$0 \text{ m} = v_{0y}t + \tfrac{1}{2}a_y t^2 \quad \text{or} \quad 0 \text{ m} = v_{0y} + \tfrac{1}{2}a_y t \quad \text{or} \quad t = \frac{-2v_{0y}}{a_y}$$

The flight time is proportional to the vertical component of the launch velocity v_{0y}, which, in turn, is proportional to the launch speed v_0.

c. Since the range is given by $R = v_{0x}t$ and since both v_{0x} and t are proportional to v_0, the range R is proportional to v_0^2.

SOLUTION The given range is 23 m. When the launch speed doubles, the range increases by a factor of $2^2 = 4$, since the range is proportional to the square of the speed. Thus, the new range is

$$R = 4(23 \text{ m}) = \boxed{92 \text{ m}}$$

---

79. ***CONCEPT QUESTIONS*** a. Barbara's motion and Neil's motion are perpendicular. Therefore, relative to herself, she sees him moving west.

b. Relative to the ground, Neil is not moving in the north/south direction. However, because she is moving due south relative to the ground, she sees Neil coming closer to her in the north/south direction. Therefore, relative to herself, she sees Neil moving north.

c. Relative to herself, Barbara sees Neil as moving toward the west and north.

SOLUTION Using B, G, and N to denote Barbara, the ground, and Neil, we have the following velocities:

$$\mathbf{v}_{\mathbf{BG}} = 4.0 \text{ m/s, due south}$$

$$\mathbf{v}_{\mathbf{NG}} = 3.2 \text{ m/s, due west}$$

$$\mathbf{v}_{\mathbf{NB}} = \text{Neil's velocity as seen by Barbara}$$

Following the procedure outlined in Section 3.4 and remembering that $\mathbf{v}_{\mathbf{GB}} = -\mathbf{v}_{\mathbf{BG}}$, we find

$$\mathbf{v}_{\mathbf{NB}} = \mathbf{v}_{\mathbf{NG}} + \mathbf{v}_{\mathbf{GB}} = \mathbf{v}_{\mathbf{NG}} + (-\mathbf{v}_{\mathbf{BG}})$$

Using the Pythagorean theorem and trigonometry, we find

$$v_{NB} = \sqrt{(3.2 \text{ m/s})^2 + (4.0 \text{ m/s})^2} = \boxed{5.1 \text{ m/s}}$$

$$\theta = \tan^{-1}\left(\frac{4.0 \text{ m/s}}{3.2 \text{ m/s}}\right) = \boxed{51° \text{ north of west}}$$

CHAPTER 4 | *FORCES AND NEWTON'S LAWS OF MOTION*

PROBLEMS

---

1. ***REASONING AND SOLUTION*** According to Newton's second law, the acceleration is $a = \Sigma F/m$. Since the pilot and the plane have the same acceleration, we can write

$$\left(\frac{\Sigma F}{m}\right)_{\text{PILOT}} = \left(\frac{\Sigma F}{m}\right)_{\text{PLANE}} \quad \text{or} \quad (\Sigma F)_{\text{PILOT}} = m_{\text{PILOT}}\left(\frac{\Sigma F}{m}\right)_{\text{PLANE}}$$

Therefore, we find

$$(\Sigma F)_{\text{PILOT}} = (78\text{ kg})\left(\frac{3.7\times10^4\text{ N}}{3.1\times10^4\text{ kg}}\right) = \boxed{93\text{ N}}$$

---

2. ***REASONING*** Newton's second law of motion gives the relationship between the net force ΣF and the acceleration a that it causes for an object of mass m. The net force is the vector sum of all the external forces that act on the object. Here the external forces are the drive force, the force due to the wind, and the resistive force of the water.

SOLUTION We choose the direction of the drive force (due west) as the positive direction. Solving Newton's second law $(\Sigma F = ma)$ for the acceleration gives

$$a = \frac{\Sigma F}{m} = \frac{+4100\text{ N} - 800\text{ N} - 1200\text{ N}}{6800\text{ kg}} = \boxed{+0.31\text{ m/s}^2}$$

The positive sign for the acceleration indicates that its direction is $\boxed{\text{due west}}$.

---

3. ***REASONING*** According to Newton's second law, Equation 4.1, the average net force $\Sigma\bar{F}$ is equal to the product of the object's mass m and the average acceleration $\bar{a}$. The average acceleration is equal to the change in velocity divided by the elapsed time (Equation 2.4), where the change in velocity is the final velocity v minus the initial velocity v_0.

SOLUTION The average net force exerted on the car and riders is

$$\Sigma\bar{F} = m\bar{a} = m\frac{v - v_0}{t - t_0} = \left(5.5\times10^3\text{ kg}\right)\frac{45\text{ m/s} - 0\text{ m/s}}{7.0\text{ s}} = \boxed{3.5\times10^4\text{ N}}$$

---

4. ***REASONING AND SOLUTION*** Using Equation 2.4 and assuming that $t_0 = 0$ s, we have for the required time that

$$t = \frac{v - v_0}{a}$$

Since $\Sigma F = ma$, it follows that

$$t = \frac{v - v_0}{\Sigma F / m} = \frac{m(v - v_0)}{\Sigma F} = \frac{(5.0\text{ kg})\left[(4.0 \times 10^3\text{ m/s}) - (0\text{ m/s})\right]}{4.9 \times 10^5\text{ N}} = \boxed{4.1 \times 10^{-2}\text{ s}}$$

---

5. SSM ***REASONING*** The net force acting on the ball can be calculated using Newton's second law. Before we can use Newton's second law, however, we must use Equation 2.9 from the equations of kinematics to determine the acceleration of the ball.

SOLUTION According to Equation 2.9, the acceleration of the ball is given by

$$a = \frac{v^2 - v_0^2}{2x}$$

Thus, the magnitude of the net force on the ball is given by

$$\Sigma F = ma = m\left(\frac{v^2 - v_0^2}{2x}\right) = (0.058\text{ kg})\left[\frac{(45\text{ m/s})^2 - (0\text{ m/s})^2}{2(0.44\text{ m})}\right] = \boxed{130\text{ N}}$$

---

6. ***REASONING AND SOLUTION*** The acceleration required is

$$\text{a} = \frac{v^2 - v_0^2}{2\text{x}} = \frac{-(15.0\text{ m/s})^2}{2(50.0\text{ m})} = -2.25\text{ m/s}^2$$

Newton's second law then gives the magnitude of the net force as

$$\text{F} = \text{ma} = (1580\text{ kg})(2.25\text{ m/s}^2) = \boxed{3560\text{ N}}$$

---

7. SSM ***REASONING*** According to Newton's second law of motion, the net force applied to the fist is equal to the mass of the fist multiplied by its acceleration. The data in the problem gives the final velocity of the fist and the time it takes to acquire that velocity. The average acceleration can be obtained directly from these data using the definition of average acceleration given in Equation 2.4.

SOLUTION The magnitude of the average net force applied to the fist is, therefore,

$$\Sigma \overline{F} = ma = m\left(\frac{\Delta v}{\Delta t}\right) = (0.70 \text{ kg})\left(\frac{8.0 \text{ m/s} - 0 \text{ m/s}}{0.15 \text{ s}}\right) = \boxed{37 \text{ N}}$$

---

8. ***REASONING AND SOLUTION*** From Equation 2.9,

$$v^2 = v_0^2 + 2ax$$

Since the arrow starts from rest, $v_0 = 0$ m/s. In both cases, x is the same so

$$\frac{v_1^2}{v_2^2} = \frac{2a_1x}{2a_2x} = \frac{a_1}{a_2} \quad \text{or} \quad \frac{v_1}{v_2} = \sqrt{\frac{a_1}{a_2}}$$

Since F = ma, it follows that a = F/m. The mass of the arrow is unchanged, and

$$\frac{v_1}{v_2} = \sqrt{\frac{F_1}{F_2}}$$

or

$$v_2 = v_1\sqrt{\frac{F_2}{F_1}} = v_1\sqrt{\frac{2F_1}{F_1}} = (25.0 \text{ m/s})\sqrt{2} = \boxed{35.4 \text{ m/s}}$$

---

9. SSM WWW ***REASONING*** Let due east be chosen as the positive direction. Then, when both forces point due east, Newton's second law gives

$$\underbrace{F_A + F_B}_{\Sigma F} = ma_1 \qquad (1)$$

where $a_1 = 0.50 \text{ m/s}^2$. When F_A points due east and F_B points due west, Newton's second law gives

$$\underbrace{F_A - F_B}_{\Sigma F} = ma_2 \qquad (2)$$

where $a_2 = 0.40 \text{ m/s}^2$. These two equations can be used to find the magnitude of each force.

SOLUTION

a. Adding Equations 1 and 2 gives

$$F_A = \frac{m(a_1 + a_2)}{2} = \frac{(8.0\text{ kg})(0.50\text{ m/s}^2 + 0.40\text{ m/s}^2)}{2} = \boxed{3.6\text{ N}}$$

b. Subtracting Equation 2 from Equation 1 gives

$$F_B = \frac{m(a_1 - a_2)}{2} = \frac{(8.0\text{ kg})(0.50\text{ m/s}^2 - 0.40\text{ m/s}^2)}{2} = \boxed{0.40\text{ N}}$$

---

10. ***REASONING AND SOLUTION***

$$F_E = F\cos\theta = (720\text{ N})\cos 38° = \boxed{570\text{ N}}$$

$$F_N = F\sin\theta = (720\text{ N})\sin 38° = \boxed{440\text{ N}}$$

---

11. ***REASONING*** Newton's second law gives the acceleration as $\mathbf{a} = (\Sigma\mathbf{F})/m$. Since we seek only the horizontal acceleration, it is the x component of this equation that we will use; $a_x = (\Sigma F_x)/m$. For completeness, however, the free-body diagram will include the vertical forces also (the normal force F_N and the weight W).

SOLUTION The free body diagram is shown at the right, where

$$F_1 = 59.0\text{ N}$$
$$F_2 = 33.0\text{ N}$$
$$\theta = 70.0°$$

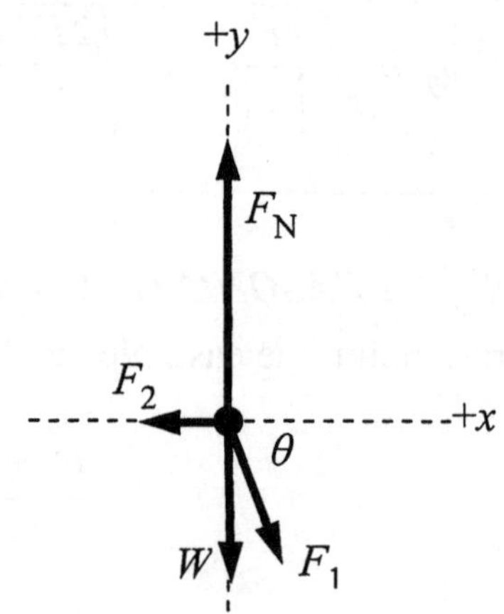

When F_1 is replaced by its x and y components, we obtain the free body diagram in the second drawing.

Choosing right to be the positive direction, we have

$$a_x = \frac{\Sigma F_x}{m} = \frac{F_1\cos\theta - F_2}{m}$$

$$a_x = \frac{(59.0\text{ N})\cos 70.0° - (33.0\text{ N})}{7.00\text{ kg}} = \boxed{-1.83\text{ m/s}^2}$$

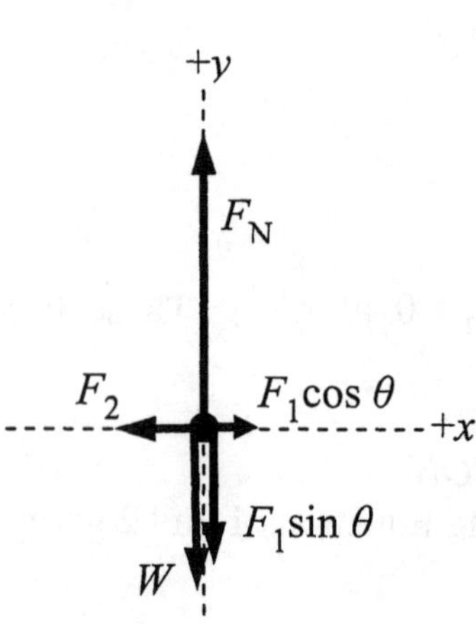

The minus sign indicates that the horizontal acceleration points to the $\boxed{\text{left}}$.

---

12. ***REASONING AND SOLUTION*** Newton's second law, $\Sigma\mathbf{F} = m\mathbf{a}$, implies that the acceleration **a** and the net force are in the same direction. This is $\boxed{64° \text{ N of E}}$. The magnitude of the net force is

$$F = ma = (350 \text{ kg})(0.62 \text{ m/s}^2) = \boxed{220 \text{ N}}$$

---

13. $\boxed{\text{SSM}}$ ***REASONING*** According to Newton's second law ($\Sigma\mathbf{F} = m\mathbf{a}$), the acceleration of the object is given by $\mathbf{a} = \Sigma\mathbf{F}/m$, where $\Sigma\mathbf{F}$ is the net force that acts on the object. We must first find the net force that acts on the object, and then determine the acceleration using Newton's second law.

SOLUTION The following table gives the *x* and *y* components of the two forces that act on the object. The third row of that table gives the components of the net force.

Force	*x-Component*	*y-Component*
$\mathbf{F}_1$	40.0 N	0 N
$\mathbf{F}_2$	(60.0 N) cos 45.0° = 42.4 N	(60.0 N) sin 45.0° = 42.4 N
$\Sigma\mathbf{F} = \mathbf{F}_1 + \mathbf{F}_2$	82.4 N	42.4 N

The magnitude of $\Sigma\mathbf{F}$ is given by the Pythagorean theorem as

$$\Sigma F = \sqrt{(82.4 \text{ N})^2 + (42.4)^2} = 92.7 \text{ N}$$

The angle θ that $\Sigma\mathbf{F}$ makes with the $+x$ axis is

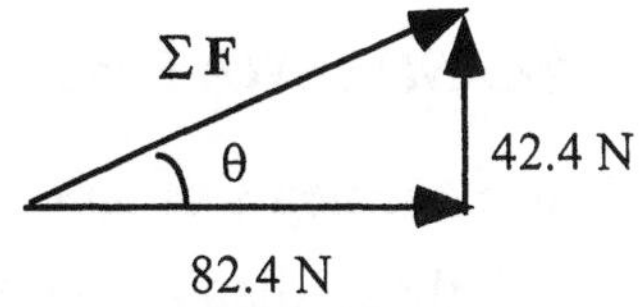

$$\theta = \tan^{-1}\left(\frac{42.4 \text{ N}}{82.4 \text{ N}}\right) = 27.2°$$

According to Newton's second law, the magnitude of the acceleration of the object is

$$a = \frac{\Sigma F}{m} = \frac{92.7 \text{ N}}{3.00 \text{ kg}} = \boxed{30.9 \text{ m/s}^2}$$

Since Newton's second law is a vector equation, we know that the direction of the right hand side must be equal to the direction of the left hand side. In other words, the direction of the acceleration a is the same as the direction of the net force $\Sigma\mathbf{F}$. Therefore, the direction of the acceleration of the object is $\boxed{27.2° \text{ above the } +x \text{ axis}}$.

---

14. ***REASONING*** Newton's second law, $\Sigma\mathbf{F} = m\mathbf{a}$, states that a net force of $\Sigma\mathbf{F}$ must act on an object of mass m in order to impart an acceleration $\mathbf{a}$ to the object. In the impact shock test the box is subjected to a large deceleration and, hence, a correspondingly large net force. To determine the net force we will determine the deceleration in a kinematics calculation and use it in Newton's second law.

SOLUTION According to Newton's second law, the net force is $\Sigma\mathbf{F} = m\mathbf{a}$, where the acceleration can be determined with the aid of Equation 2.4 ($v = v_0 + at$). According to this equation

$$a = \frac{v - v_0}{t}$$

Substituting this result for the acceleration into the second law gives

$$\Sigma F = ma = m\left(\frac{v - v_0}{t}\right)$$

Since the initial velocity ($v_0 = +220$ m/s), final velocity ($v = 0$ m/s), and the duration of the collision ($t = 6.5 \times 10^{-3}$ s) are known, we find

$$\Sigma F = m\left(\frac{v - v_0}{t}\right) = (41\text{ kg})\left(\frac{0\text{ m/s} - 220\text{ m/s}}{6.5 \times 10^{-3}\text{ s}}\right) = -1.39 \times 10^6\text{ N}$$

The minus sign indicates that the net force points opposite to the direction in which the box is thrown, which has been assumed to be the positive direction. The magnitude of the net force is $\boxed{1.39 \times 10^6\text{ N}}$, which is over three hundred thousand pounds.

---

15. ***REASONING*** Equations 3.5a $\left(x = v_{0x}t + \frac{1}{2}a_x t^2\right)$ and 3.5b $\left(y = v_{0y}t + \frac{1}{2}a_y t^2\right)$ give the displacements of an object under the influence of constant accelerations a_x and a_y. We can add these displacements as vectors to find the magnitude and direction of the resultant displacement. To use Equations 3.5a and 3.5b, however, we must have values for a_x and a_y. We can obtain these values from Newton's second law, provided that we combine the given forces to calculate the x and y components of the net force acting on the duck, and it is here that our solution begins.

SOLUTION Let the directions due east and due north, respectively, be the $+x$ and $+y$ directions. Then, the components of the net force are

$$\Sigma F_x = 0.10 \text{ N} + (0.20 \text{ N})\cos 52° = 0.2231 \text{ N}$$

$$\Sigma F_y = -(0.20 \text{ N})\sin 52° = -0.1576 \text{ N}$$

According to Newton's second law, the components of the acceleration are

$$a_x = \frac{\Sigma F_x}{m} = \frac{0.2231 \text{ N}}{2.5 \text{ kg}} = 0.08924 \text{ m/s}^2$$

$$a_y = \frac{\Sigma F_y}{m} = \frac{-0.1576 \text{ N}}{2.5 \text{ kg}} = -0.06304 \text{ m/s}^2$$

From Equations 3.5a and 3.5b, we now obtain the displacements in the x and y directions:

$$x = v_{0x}t + \frac{1}{2}a_x t^2 = (0.11 \text{ m/s})(3.0 \text{ s}) + \frac{1}{2}\left(0.08924 \text{ m/s}^2\right)(3.0 \text{ s})^2 = 0.7316 \text{ m}$$

$$y = v_{0y}t + \frac{1}{2}a_y t^2 = (0 \text{ m/s})(3.0 \text{ s}) + \frac{1}{2}\left(-0.06304 \text{ m/s}^2\right)(3.0 \text{ s})^2 = -0.2837 \text{ m}$$

The magnitude of the resultant displacement is

$$r = \sqrt{x^2 + y^2} = \sqrt{(0.7316 \text{ m})^2 + (-0.2837 \text{ m})^2} = \boxed{0.78 \text{ m}}$$

The direction of the resultant displacement is

$$\theta = \tan^{-1}\left(\frac{0.2837 \text{ m}}{0.7316 \text{ m}}\right) = \boxed{21° \text{ south of east}}$$

---

16. ***REASONING*** For both the tug and the asteroid, Equation 2.8 $\left(x = v_0 t + \frac{1}{2}at^2\right)$ applies with $v_0 = 0$ m/s, since both are initially at rest. In applying this equation, we must be careful and use the proper acceleration for each object. Newton's second law indicates that the acceleration is given by $a = \Sigma F/m$. In this expression, we note that the magnitudes of the net forces acting on the tug and the asteroid are the same, according to Newton's action-reaction law. The masses of the tug and the asteroid are different, however. Thus, the distance traveled for either object is given by, where we use for ΣF only the magnitude of the pulling force

$$x = v_0 t + \frac{1}{2}at^2 = \frac{1}{2}\left(\frac{\Sigma F}{m}\right)t^2$$

SOLUTION Let L be the initial distance between the tug and the asteroid. When the two objects meet, the distances that each has traveled must add up to equal L. Therefore,

$$L = x_T + x_A = \frac{1}{2}a_T t^2 + \frac{1}{2}a_A t^2$$

$$L = \frac{1}{2}\left(\frac{\Sigma F}{m_T}\right)t^2 + \frac{1}{2}\left(\frac{\Sigma F}{m_A}\right)t^2 = \frac{1}{2}\Sigma F\left(\frac{1}{m_T} + \frac{1}{m_A}\right)t^2$$

Solving for the time t gives

$$t = \sqrt{\frac{2L}{\Sigma F\left(\frac{1}{m_T} + \frac{1}{m_A}\right)}} = \sqrt{\frac{2(450 \text{ m})}{(490 \text{ N})\left(\frac{1}{3500 \text{ kg}} + \frac{1}{6200 \text{ kg}}\right)}} = \boxed{64 \text{ s}}$$

---

17. SSM WWW ***REASONING*** We first determine the acceleration of the boat. Then, using Newton's second law, we can find the net force $\Sigma\mathbf{F}$ that acts on the boat. Since two of the three forces are known, we can solve for the unknown force $\mathbf{F}_W$ once the net force $\Sigma\mathbf{F}$ is known.

SOLUTION Let the direction due east be the positive x direction and the direction due north be the positive y direction. The x and y components of the initial velocity of the boat are then

$$v_{0x} = (2.00 \text{ m/s}) \cos 15.0° = 1.93 \text{ m/s}$$

$$v_{0y} = (2.00 \text{ m/s}) \sin 15.0° = 0.518 \text{ m/s}$$

Thirty seconds later, the x and y velocity components of the boat are

$$v_x = (4.00 \text{ m/s}) \cos 35.0° = 3.28 \text{ m/s}$$

$$v_y = (4.00 \text{ m/s}) \sin 35.0° = 2.29 \text{ m/s}$$

Therefore, according to Equations 3.3a and 3.3b, the x and y components of the acceleration of the boat are

$$a_x = \frac{v_x - v_{0x}}{t} = \frac{3.28 \text{ m/s} - 1.93 \text{ m/s}}{30.0 \text{ s}} = 4.50 \times 10^{-2} \text{ m/s}^2$$

$$a_y = \frac{v_y - v_{0y}}{t} = \frac{2.29 \text{ m/s} - 0.518 \text{ m/s}}{30.0 \text{ s}} = 5.91 \times 10^{-2} \text{ m/s}^2$$

Thus, the x and y components of the net force that act on the boat are

$$\Sigma F_x = ma_x = (325 \text{ kg})(4.50 \times 10^{-2} \text{ m/s}^2) = 14.6 \text{ N}$$

$$\sum F_y = ma_y = (325\text{ kg})\,(5.91\times10^{-2}\text{ m/s}^2) = 19.2\text{ N}$$

The following table gives the x and y components of the net force $\sum \mathbf{F}$ and the two known forces that act on the boat. The fourth row of that table gives the components of the unknown force $\mathbf{F}_\text{W}$.

Force	*x-Component*	*y-Component*
$\sum \mathbf{F}$	14.6 N	19.2 N
$\mathbf{F}_1$	(31.0 N) cos 15.0° = 29.9 N	(31.0 N) sin 15.0° = 8.02 N
$\mathbf{F}_2$	–(23.0 N) cos 15.0° = –22.2 N	–(23.0 N) sin 15.0° = –5.95 N
$\mathbf{F}_\text{W} = \sum \mathbf{F} - \mathbf{F}_1 - \mathbf{F}_2$	14.6 N – 29.9 N + 22.2 N = 6.9 N	19.2 N – 8.02 N + 5.95 N = 17.1 N

The magnitude of $\mathbf{F}_\text{W}$ is given by the Pythagorean theorem as

$$F_\text{W} = \sqrt{(6.9\text{ N})^2 + (17.1\text{N})^2} = \boxed{18.4\text{ N}}$$

The angle θ that $\mathbf{F}_\text{W}$ makes with the x axis is

$$\theta = \tan^{-1}\left(\frac{17.1\text{ N}}{6.9\text{ N}}\right) = 68°$$

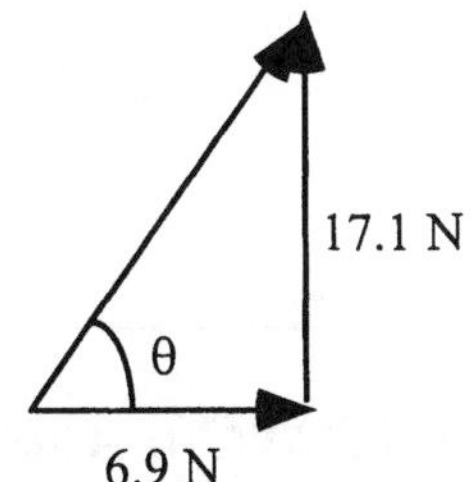

Therefore, the direction of $\mathbf{F}_\text{W}$ is $\boxed{68°\text{, north of east}}$.

---

18. ***REASONING*** Newton's law of universal gravitation indicates that the gravitational force that each uniform sphere exerts on the other has a magnitude that is inversely proportional to the square of the distance between the centers of the spheres. Therefore, the maximum gravitational force between two uniform spheres occurs when the centers of the spheres are as close together as possible, and this occurs when the surfaces of the spheres are touching. Then, the distance between the centers of the spheres is the sum of the two radii.

    SOLUTION When the bowling ball and the billiard ball are touching, the distance between their centers is $r = r_\text{Bowling} + r_\text{Billiard}$. Using this expression in Newton's law of universal gravitation gives

$$F = \frac{Gm_{\text{Bowling}}m_{\text{Billiard}}}{r^2} = \frac{Gm_{\text{Bowling}}m_{\text{Billiard}}}{\left(r_{\text{Bowling}} + r_{\text{Billiard}}\right)^2}$$

$$= \frac{\left(6.67\times10^{-11}\ \text{N}\cdot\text{m}^2/\text{kg}^2\right)\left(7.2\ \text{kg}\right)\left(0.38\ \text{kg}\right)}{\left(0.11\ \text{m} + 0.028\ \text{m}\right)^2} = \boxed{9.6\times10^{-9}\ \text{N}}$$

---

19. ***REASONING*** The magnitude of the gravitational force that each part exerts on the other is given by Newton's law of gravitation as $F = Gm_1m_2/r^2$. To use this expression, we need the masses m_1 and m_2 of the parts, whereas the problem statement gives the weights W_1 and W_2. However, the weight is related to the mass by $W = mg$, so that for each part we know that $m = W/g$.

SOLUTION The gravitational force that each part exerts on the other is

$$F = \frac{Gm_1m_2}{r^2} = \frac{G\left(W_1/g\right)\left(W_2/g\right)}{r^2}$$

$$= \frac{\left(6.67\times10^{-11}\ \text{N}\cdot\text{m}^2/\text{kg}^2\right)\left(11\,000\ \text{N}\right)\left(3400\ \text{N}\right)}{\left(9.80\ \text{m/s}^2\right)^2\left(12\ \text{m}\right)^2} = \boxed{1.8\times10^{-7}\ \text{N}}$$

---

20. ***REASONING AND SOLUTION*** The forces that act on the rock are shown at the right. Newton's second law (with the direction of motion as positive) is

$$\Sigma F = mg - R = ma$$

Solving for the acceleration a gives

$$a = \frac{mg - R}{m} = \frac{\left(45\ \text{kg}\right)\left(9.80\ \text{m/s}^2\right) - \left(250\ \text{N}\right)}{45\ \text{kg}} = \boxed{4.2\ \text{m/s}^2}$$

---

21. SSM ***REASONING AND SOLUTION***

a. According to Equation 4.5, the weight of the space traveler of mass $\boxed{m = 115\ \text{kg}}$ on earth is

$$W = mg = (115\ \text{kg})\ (9.80\ \text{m/s}^2) = \boxed{1.13\times10^3\ \text{N}}$$

b. In interplanetary space where there are no nearby planetary objects, the gravitational force exerted on the space traveler is zero and $g = 0\text{ m/s}^2$. Therefore, the weight is $\boxed{W = 0\text{ N}}$. Since the mass of an object is an intrinsic property of the object and is independent of its location in the universe, the mass of the space traveler is still $\boxed{m = 115\text{ kg}}$.

---

22. ***REASONING*** As discussed in Conceptual Example 7, the same net force is required on the moon as on the earth. This net force is given by Newton's second law as $\Sigma F = ma$, where the mass m is the same in both places. Thus, from the given mass and acceleration, we can calculate the net force. On the moon, the net force comes about due to the drive force and the opposing frictional force. Since the drive force is given, we can find the frictional force.

SOLUTION Newton's second law, with the direction of motion taken as positive, gives

$$\Sigma F = ma \qquad \text{or} \qquad (1430\text{ N}) - f = \left(5.90 \times 10^3\text{ kg}\right)\left(0.220\text{ m/s}^2\right)$$

Solving for the frictional force f, we find

$$f = (1430\text{ N}) - \left(5.90 \times 10^3\text{ kg}\right)\left(0.220\text{ m/s}^2\right) = \boxed{130\text{N}}$$

---

23. SSM ***REASONING AND SOLUTION*** According to Equations 4.4 and 4.5, the weight of an object of mass m at a distance r from the *center* of the earth is

$$mg = \frac{GM_E m}{r^2}$$

In a circular orbit that is $3.59 \times 10^7\text{ m}$ above the surface of the earth (radius $= 6.38 \times 10^6\text{ m}$, mass $= 5.98 \times 10^{24}\text{ kg}$), the total distance from the center of the earth is $r = 3.59 \times 10^7\text{ m} + 6.38 \times 10^6\text{ m}$. Thus the acceleration g due to gravity is

$$g = \frac{GM_E}{r^2} = \frac{(6.67 \times 10^{-11}\text{ N}\cdot\text{m}^2/\text{kg}^2)(5.98 \times 10^{24}\text{ kg})}{(3.59 \times 10^7\text{ m} + 6.38 \times 10^6\text{ m})^2} = \boxed{0.223\text{ m/s}^2}$$

---

24. ***REASONING AND SOLUTION*** The magnitude of the net force acting on the moon is found by the Pythagorean theorem to be

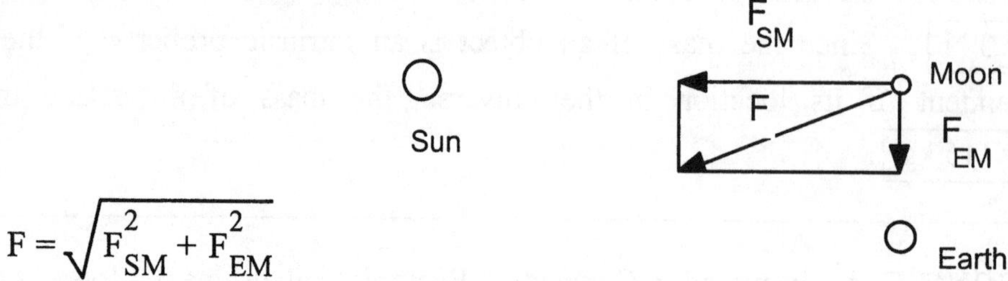

$$F = \sqrt{F_{SM}^2 + F_{EM}^2}$$

Newton's law of gravitation applied to the sun-moon (the units have been suppressed)

$$F_{SM} = G\frac{m_S m_M}{r_{SM}^2} = (6.67 \times 10^{-11})\frac{(1.99 \times 10^{30})(7.35 \times 10^{22})}{(1.50 \times 10^{11}\text{ m})^2} = 4.34 \times 10^{20}\text{ N.}$$

A similar application to the earth-moon gives

$$F_{EM} = G\frac{m_E m_M}{r_{EM}^2} = (6.67 \times 10^{-11})\frac{(5.98 \times 10^{24})(7.35 \times 10^{22})}{(3.85 \times 10^{8}\text{ m})^2} = 1.98 \times 10^{20}\text{ N.}$$

The net force on the moon is then

$$F = \sqrt{(4.34 \times 10^{20}\text{ N})^2 + (1.98 \times 10^{20}\text{ N})^2} = \boxed{4.77 \times 10^{20}\text{ N}}$$

---

25. ***REASONING*** According to Equation 4.4, the weights of an object of mass m on the surfaces of planet A (mass = M_A, radius = R) and planet B (mass = M_B , radius = R) are

$$W_A = \frac{GM_A m}{R^2} \quad \text{and} \quad W_B = \frac{GM_B m}{R^2}$$

The difference between these weights is given in the problem.

SOLUTION The difference in weights is

$$W_A - W_B = \frac{GM_A m}{R^2} - \frac{GM_B m}{R^2} = \frac{Gm}{R^2}(M_A - M_B)$$

Rearranging this result, we find

$$M_A - M_B = \frac{(W_A - W_B)R^2}{Gm} = \frac{(3620\text{ N})(1.33 \times 10^7\text{ m})^2}{(6.67 \times 10^{-11}\text{ N} \cdot \text{m}^2/\text{kg}^2)(5450\text{ kg})} = \boxed{1.76 \times 10^{24}\text{ kg}}$$

---

26. ***REASONING*** Newton's law of gravitation shows how the weight W of an object of mass m is related to the mass M and radius r of the planet on which the object is located: $W = GMm/r^2$. In this expression G is the universal gravitational constant. Using the law of gravitation, we can express the weight of the object on each planet, set the two weights equal, and obtain the desired ratio.

SOLUTION According to Newton's law of gravitation, we have

$$\underbrace{\frac{GM_{\text{A}}m}{r_{\text{A}}^2}}_{\text{Weight on planet A}} = \underbrace{\frac{GM_{\text{B}}m}{r_{\text{B}}^2}}_{\text{Weight on planet B}}$$

The mass m of the object, being an intrinsic property, is the same on both planets and can be eliminated algebraically from this equation. The universal gravitational constant can likewise be eliminated algebraically. As a result, we find that

$$\frac{M_{\text{A}}}{r_{\text{A}}^2} = \frac{M_{\text{B}}}{r_{\text{B}}^2} \quad \text{or} \quad \frac{M_{\text{A}}}{M_{\text{B}}} = \frac{r_{\text{A}}^2}{r_{\text{B}}^2}$$

$$\frac{r_{\text{A}}}{r_{\text{B}}} = \sqrt{\frac{M_{\text{A}}}{M_{\text{B}}}} = \sqrt{0.60} = \boxed{0.77}$$

---

27. SSM ***REASONING AND SOLUTION***

a. According to Equation 4.4, the weight of an object of mass m on the surface of Mars would be given by

$$W = \frac{GM_{\text{M}}m}{R_{\text{M}}^2}$$

where M_{M} is the mass of Mars and R_{M} is the radius of Mars. On the surface of Mars, the weight of the object can be given as $W = mg$ (see Equation 4.5), so

$$mg = \frac{GM_{\text{M}}m}{R_{\text{M}}^2} \quad \text{or} \quad g = \frac{GM_{\text{M}}}{R_{\text{M}}^2}$$

Substituting values, we have

$$g = \frac{(6.67\times10^{-11}\text{N}\cdot\text{m}^2/\text{kg}^2)(6.46\times10^{23}\text{ kg})}{(3.39\times10^6\text{ m})^2} = \boxed{3.75\text{ m/s}^2}$$

b. According to Equation 4.5,

$$W = mg = (65\text{ kg})(3.75\text{ m/s}^2) = \boxed{2.4\times10^2\text{ N}}$$

---

28. ***REASONING AND SOLUTION*** The figure at the right shows the three spheres with sphere 3 being the sphere of unknown mass. Sphere 3 feels a force $\mathbf{F}_{31}$ due to the presence of sphere 1, and a force $\mathbf{F}_{32}$ due to the presence of sphere 2. The net force on sphere 3 is the resultant of $\mathbf{F}_{31}$ and $\mathbf{F}_{32}$.

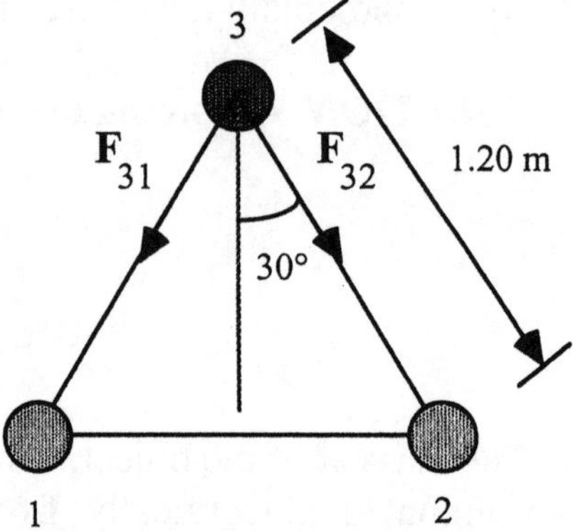

Note that since the spheres form an equilateral triangle, each interior angle is 60°. Therefore, both $\mathbf{F}_{31}$ and $\mathbf{F}_{32}$ make a 30° angle with the vertical line as shown.

Furthermore, $\mathbf{F}_{31}$ and $\mathbf{F}_{32}$ have the same magnitude given by

$$F = \frac{GMm_3}{r^2}$$

where M is the mass of either sphere 1 or 2 and m_3 is the mass of sphere 3. The components of the two forces are shown in the following drawings:

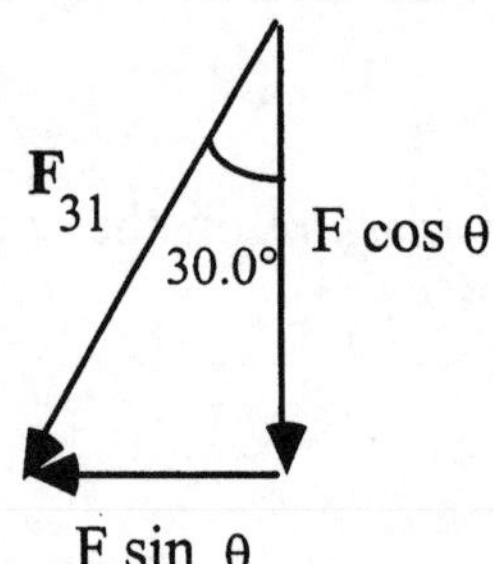

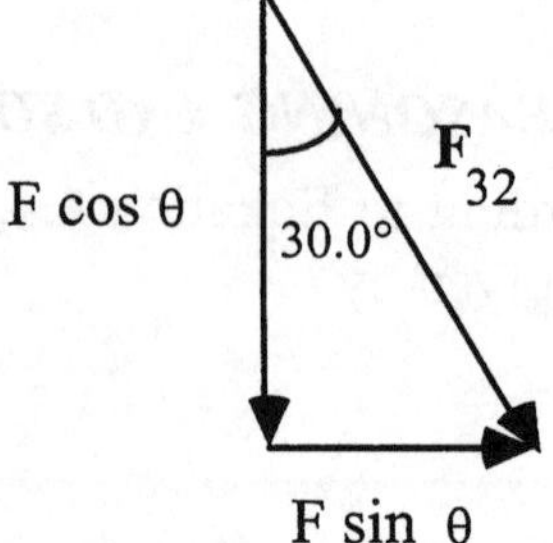

Clearly, the horizontal components of the two forces add to zero. Thus, the net force on sphere 3 is the resultant of the vertical components of $\mathbf{F}_{31}$ and $\mathbf{F}_{32}$:

$$F_3 = 2F\cos\theta = 2\frac{GMm_3}{r^2}\cos\theta.$$

The acceleration of sphere 3 is given by Newton's second law:

$$a_3 = \frac{F_3}{m_3} = 2\frac{GM}{r^2}\cos\theta = 2\frac{(6.67\times10^{-11}\ \text{N}\cdot\text{m}^2/\text{kg}^2)(2.80\ \text{kg})}{(1.20\ \text{m})^2}\cos 30.0°$$

$$= \boxed{2.25\times10^{-10}\ \text{m/s}^2}$$

---

29. SSM WWW ***REASONING AND SOLUTION*** There are two forces that act on the balloon; they are, the combined weight of the balloon and its load, Mg, and the upward buoyant force F_B. If we take upward as the positive direction, then, initially when the balloon is motionless, Newton's second law gives $F_B - Mg = 0$. If an amount of mass m is dropped overboard so that the balloon has an upward acceleration, Newton's second law for this situation is

$$F_B - (M-m)g = (M-m)a$$

But $F_B = mg$, so that

$$Mg - (M-m)g = mg = (M-m)a$$

Solving for the mass m that should be dropped overboard, we obtain

$$m = \frac{Ma}{g+a} = \frac{(310\ \text{kg})(0.15\ \text{m/s}^2)}{9.80\ \text{m/s}^2 + 0.15\ \text{m/s}^2} = \boxed{4.7\ \text{kg}}$$

---

30. ***REASONING*** The gravitational force that the sun exerts on a person standing on the earth is given by Equation 4.3 as $F_{\text{sun}} = GM_{\text{sun}}m/r^2_{\text{sun-earth}}$, where M_{sun} is the mass of the sun, m is the mass of the person, and $r_{\text{sun-earth}}$ is the distance from the sun to the earth. Likewise, the gravitational force that the moon exerts on a person standing on the earth is given by $F_{\text{moon}} = GM_{\text{moon}}m/r^2_{\text{moon-earth}}$, where M_{moon} is the mass of the moon and $r_{\text{moon-earth}}$ is the distance from the moon to the earth. These relations will allow us to determine whether the sun or the moon exerts the greater gravitational force on the person.

SOLUTION Taking the ratio of F_{sun} to F_{moon}, and using the mass and distance data from the inside of the text's front cover, we find

$$\frac{F_{\text{sun}}}{F_{\text{moon}}} = \frac{\dfrac{GM_{\text{sun}}m}{r^2_{\text{sun-earth}}}}{\dfrac{GM_{\text{moon}}m}{r^2_{\text{moon-earth}}}} = \left(\frac{M_{\text{sun}}}{M_{\text{moon}}}\right)\left(\frac{r_{\text{moon-earth}}}{r_{\text{sun-earth}}}\right)^2$$

$$= \left(\frac{1.99\times10^{30}\ \text{kg}}{7.35\times10^{22}\ \text{kg}}\right)\left(\frac{3.85\times10^{8}\ \text{m}}{1.50\times10^{11}\ \text{m}}\right)^2 = \boxed{178}$$

Therefore, the sun exerts the greater gravitational force.

---

31. ***REASONING*** According to Equation 4.4, the weights of an object of mass m on the surface of a planet (mass = M, radius = R) and at a height H above the surface are

$$\underbrace{W=\frac{GMm}{R^2}}_{\text{On surface}} \quad \text{and} \quad \underbrace{W_H=\frac{GMm}{(R+H)^2}}_{\text{At height } H \text{ above surface}}$$

The fact that W is one percent less than W_H tells us that the $W_H/W = 0.9900$, which is the starting point for our solution.

SOLUTION The ratio W_H/W is

$$\frac{W_H}{W}=\frac{\dfrac{GMm}{(R+H)^2}}{\dfrac{GMm}{R^2}}=\frac{R^2}{(R+H)^2}=\frac{1}{(1+H/R)^2}=0.9900$$

Solving for H/R gives

$$1+\frac{H}{R}=\sqrt{\frac{1}{0.9900}} \quad \text{or} \quad \frac{H}{R}=\boxed{0.0050}$$

---

32. ***REASONING AND SOLUTION*** Since both motions are characterized by constant acceleration, it follows that

$$\frac{y_J}{y_E}=\frac{\frac{1}{2}a_J t_J^2}{\frac{1}{2}a_E t_E^2}$$

where the subscripts designate those quantities that pertain to Jupiter and Earth. Since both objects fall the same distance, the above ratio is equal to unity. Solving for the ratio of the times yields

$$\frac{t_J}{t_E}=\sqrt{\frac{a_E}{a_J}}=\sqrt{\frac{GM_E/R_E^2}{GM_J/R_J^2}}=\frac{R_J}{R_E}\sqrt{\frac{M_E}{M_J}}=(11.2)\sqrt{\frac{1}{318}}=\boxed{0.628}$$

---

33. ***REASONING*** We place the third particle (mass = m_3) as shown in the following drawing:

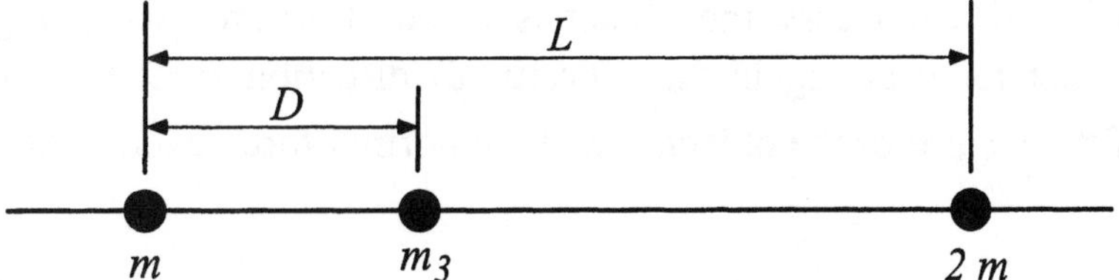

The magnitude of the gravitational force that one particle exerts on another is given by Newton's law of gravitation as $F = Gm_1m_2/r^2$. Before the third particle is in place, this law indicates that the force on each particle has a magnitude $F_{\text{before}} = Gm2m/L^2$. After the third particle is in place, each of the first two particles experiences a greater net force, because the third particle also exerts a gravitational force on them.

SOLUTION For the particle of mass m, we have

$$\frac{F_{\text{after}}}{F_{\text{before}}} = \frac{\dfrac{Gmm_3}{D^2} + \dfrac{Gm2m}{L^2}}{\dfrac{Gm2m}{L^2}} = \frac{L^2 m_3}{2mD^2} + 1$$

For the particle of mass $2m$, we have

$$\frac{F_{\text{after}}}{F_{\text{before}}} = \frac{\dfrac{G2mm_3}{(L-D)^2} + \dfrac{Gm2m}{L^2}}{\dfrac{Gm2m}{L^2}} = \frac{L^2 m_3}{m(L-D)^2} + 1$$

Since $F_{\text{after}}/F_{\text{before}} = 2$ for both particles, we have

$$\frac{L^2 m_3}{2mD^2} + 1 = \frac{L^2 m_3}{m(L-D)^2} + 1 \quad \text{or} \quad 2D^2 = (L-D)^2$$

Expanding and rearranging this result gives $D^2 + 2LD - L^2 = 0$, which can be solved for D using the quadratic formula:

$$D = \frac{-2L \pm \sqrt{(2L)^2 - 4(1)(-L^2)}}{2(1)} = 0.414L \quad \text{or} \quad -2.414L$$

The negative solution is discarded because the third particle lies on the $+x$ axis between m and $2m$. Thus, $D = \boxed{0.414L}$.

34. ***REASONING*** In each case the object is in equilibrium. According to Equation 4.9b, $\Sigma F_y = 0$, the net force acting in the y (vertical) direction must be zero. The net force is composed of the weight of the object(s) and the normal force exerted on them.

SOLUTION
a. There are three vertical forces acting on the crate: an upward normal force $+F_N$ that the floor exerts, the weight $-m_1g$ of the crate, and the weight $-m_2g$ of the person standing on the crate. Since the weights act downward, they are assigned negative numbers. Setting the sum of these forces equal to zero gives

$$\underbrace{F_N + (-m_1g) + (-m_2g)}_{\Sigma F_y} = 0$$

The magnitude of the normal force is

$$F_N = m_1g + m_2g = (35\text{ kg} + 65\text{ kg})(9.80\text{ m/s}^2) = \boxed{980\text{ N}}$$

b. There are only two vertical forces acting on the person: an upward normal force $+F_N$ that the crate exerts and the weight $-m_2g$ of the person. Setting the sum of these forces equal to zero gives

$$\underbrace{F_N + (-m_2g)}_{\Sigma F_y} = 0$$

The magnitude of the normal force is

$$F_N = m_2g = (65\text{ kg})(9.80\text{ m/s}^2) = \boxed{640\text{ N}}$$

---

35. SSM ***REASONING AND SOLUTION*** According to Equation 3.3b, the acceleration of the astronaut is $a_y = (v_y - v_{0y})/t = v_y/t$. The apparent weight and the true weight of the astronaut are related according to Equation 4.6. Direct substitution gives

$$\underbrace{F_N}_{\text{Apparent weight}} = \underbrace{mg}_{\text{True weight}} + ma_y = m(g + a_y) = m\left(g + \frac{v_y}{t}\right)$$

$$= (57\text{ kg})\left(9.80\text{ m/s}^2 + \frac{45\text{ m/s}}{15\text{ s}}\right) = \boxed{7.3\times10^2\text{ N}}$$

---

36. ***REASONING AND SOLUTION*** The apparent weight is

$$F_N = m_w(g + a)$$

We need to find the acceleration, a. Let T represent the force applied by the hoisting cable. Newton's second law applied to the elevator gives

$$T - (m_w + m_e)g = (m_w + m_e)a$$

Solving for a gives

$$a = \frac{T}{m_w + m_e} - g = \frac{9410 \text{ N}}{60.0 \text{ kg} + 815 \text{ kg}} - 9.80 \text{ m/s}^2 = 0.954 \text{ m/s}^2.$$

Now the apparent weight is

$$F_N = 60.0 \text{ kg}(9.80 \text{ m/s}^2 + 0.954 \text{ m/s}^2) = \boxed{645 \text{ N}}$$

---

37. ***REASONING AND SOLUTION*** The block will move only if the applied force is greater than the maximum static frictional force acting on the block. That is, if

$$F > \mu_s F_N = \mu_s mg = (0.650)(45.0 \text{ N}) = 29.2 \text{ N}$$

The applied force is given to be F = 36.0 N which is greater than the maximum static frictional force, so $\boxed{\text{the block will move}}$.

The block's acceleration is found from Newton's second law.

$$a = \frac{\Sigma F}{m} = \frac{F - f_k}{m} = \frac{F - \mu_k mg}{m} = \boxed{3.72 \text{ m/s}^2}$$

---

38. ***REASONING*** It is the static friction force that accelerates the cup when the plane accelerates. The maximum magnitude that this force can have will determine the maximum acceleration, according to Newton's second law.

SOLUTION According to Newton's second law, we have

$$\Sigma F = f_s^{MAX} = \mu_s F_N = \mu_s mg = ma$$

In this result, we have used the fact that the magnitude of the normal force is $F_N = mg$, since the plane is flying horizontally and the normal force acting on the cup balances the cup's weight. Solving for the acceleration gives

$$a = \mu_s g = (0.30)\left(9.80 \text{ m/s}^2\right) = \boxed{2.9 \text{ m/s}^2}$$

---

39. SSM ***REASONING AND SOLUTION*** Four forces act on the sled. They are the pulling force P, the force of kinetic friction $\mathbf{f}_k$, the weight mg of the sled, and the normal force $\mathbf{F}_N$ exerted on the sled by the surface on which it slides. The following figures show free-body diagrams for the sled. In the diagram on the right, the forces have been resolved into their x and y components.

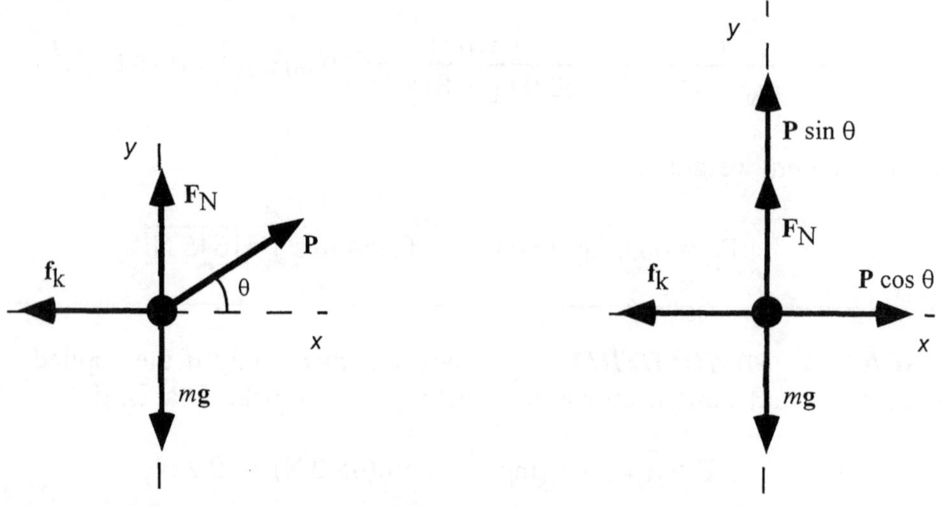

Since the sled is pulled at constant velocity, its acceleration is zero, and Newton's second law in the direction of motion is (with right chosen as the positive direction)

$$\Sigma F_x = P\cos\theta - f_k = ma_x = 0$$

From Equation 4.8, we know that $f_k = \mu_k F_N$, so that the above expression becomes

$$P\cos\theta - \mu_k F_N = 0 \qquad (1)$$

In the vertical direction,

$$\Sigma F_y = P\sin\theta + F_N - mg = ma_y = 0 \qquad (2)$$

Solving Equation (2) for the normal force, and substituting into Equation (1), we obtain

$$P\cos\theta - \mu_k\left(mg - P\sin\theta\right) = 0$$

Solving for μ_k, the coefficient of kinetic friction, we find

$$\mu_k = \frac{P\cos\theta}{mg - P\sin\theta} = \frac{(80.0\text{ N})\cos 30.0°}{(20.0\text{ kg})(9.80\text{ m/s}^2) - (80.0\text{N})\sin 30.0°} = \boxed{0.444}$$

40. ***REASONING*** In each of the three cases under consideration the kinetic frictional force is given by $f_k = \mu_k F_N$. However, the normal force F_N varies from case to case. To determine the normal force, we use Equation 4.6 ($F_N = mg + ma$) and thereby take into account the acceleration of the elevator. The normal force is greatest when the elevator accelerates upward (a positive) and smallest when the elevator accelerates downward (a negative).

SOLUTION

a. When the elevator is stationary, its acceleration is $a = 0$ m/s^2. Using Equation 4.6, we can express the kinetic frictional force as

$$f_k = \mu_k F_N = \mu_k (mg + ma) = \mu_k m(g + a)$$

$$= (0.360)(6.00 \text{ kg})\left[\left(9.80 \text{ m/s}^2\right) + \left(0 \text{ m/s}^2\right)\right] = \boxed{21.2 \text{ N}}$$

b. When the elevator accelerates upward, $a = +1.20$ m/s^2. Then,

$$f_k = \mu_k F_N = \mu_k (mg + ma) = \mu_k m(g + a)$$

$$= (0.360)(6.00 \text{ kg})\left[\left(9.80 \text{ m/s}^2\right) + \left(1.20 \text{ m/s}^2\right)\right] = \boxed{23.8 \text{ N}}$$

c. When the elevator accelerates downward, $a = -1.20$ m/s^2. Then,

$$f_k = \mu_k F_N = \mu_k (mg + ma) = \mu_k m(g + a)$$

$$= (0.360)(6.00 \text{ kg})\left[\left(9.80 \text{ m/s}^2\right) + \left(-1.20 \text{ m/s}^2\right)\right] = \boxed{18.6 \text{ N}}$$

---

41. ***REASONING*** The magnitude of the kinetic frictional force is given by Equation 4.8 as the coefficient of kinetic friction times the magnitude of the normal force. Since the slide into second base is horizontal, the normal force is vertical. It can be evaluated by noting that there is no acceleration in the vertical direction and, therefore, the normal force must balance the weight.

To find the player's initial velocity v_0, we will use kinematics. The time interval for the slide into second base is given as $t = 1.6$ s. Since the player comes to rest at the end of the slide, his final velocity is $v = 0$ m/s. The player's acceleration a can be obtained from Newton's second law, since the net force is the kinetic frictional force, which is known from part (a), and the mass is given. Since t, v, and a are known and we seek v_0, the appropriate kinematics equation is Equation 2.4 ($v = v_0 + at$).

SOLUTION

a. Since the normal force F_N balances the weight mg, we know that $F_N = mg$. Using this fact and Equation 4.8, we find that the magnitude of the kinetic frictional force is

$$f_k = \mu_k F_N = \mu_k mg = (0.49)(81 \text{ kg})(9.8 \text{ m/s}^2) = \boxed{390 \text{ N}}$$

b. Solving Equation 2.4 ($v = v_0 + at$) for v_0 gives $v_0 = v - at$. Taking the direction of the player's slide to be the positive direction, we use Newton's second law and Equation 4.8 for the kinetic frictional force to write the acceleration a as follows:

$$a = \frac{\Sigma F}{m} = \frac{-\mu_k mg}{m} = -\mu_k g$$

The acceleration is negative, because it points opposite to the player's velocity, since the player slows down during the slide. Thus, we find for the initial velocity that

$$v_0 = v - (-\mu_k g)t = 0 \text{ m/s} - \left[-(0.49)(9.8 \text{ m/s}^2)\right](1.6 \text{ s}) = \boxed{+7.7 \text{ m/s}}$$

---

42. ***REASONING*** Static friction determines the magnitude of the applied force at which either the upper or lower block begins to slide. For the upper block the static frictional force is applied only by the lower block. For the lower block, however, separate static frictional forces are applied by the upper block and by the horizontal surface. The maximum magnitude of any of the individual frictional forces is given by Equation 4.7 as the coefficient of static friction times the magnitude of the normal force.

SOLUTION We begin by drawing the free-body diagram for the lower block.

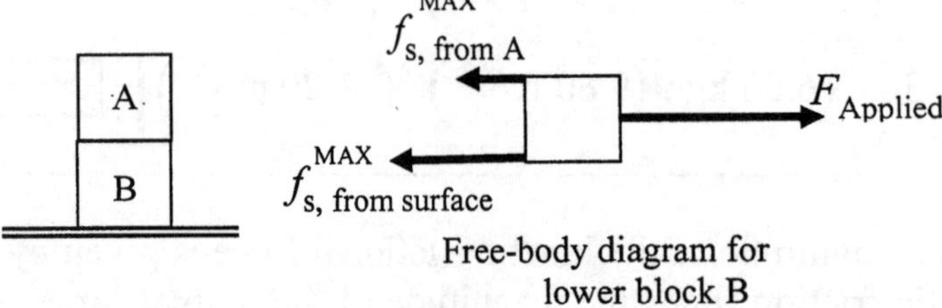

Free-body diagram for lower block B

This diagram shows that three horizontal forces act on the lower block, the applied force, and the two maximum static frictional forces, one from the upper block and one from the horizontal surface. At the instant that the lower block just begins to slide, the blocks are in equilibrium and the applied force is balanced by the two frictional forces, with the result that

$$F_{\text{Applied}} = f_{\text{s, from A}}^{\text{MAX}} + f_{\text{s, from surface}}^{\text{MAX}} \quad (1)$$

According to Equation 4.7, the magnitude of the maximum frictional force from the surface is

$$f_{\text{s, from surface}}^{\text{MAX}} = \mu_s F_N = \mu_s 2mg \quad (2)$$

Here, we have recognized that the normal force F_N from the horizontal surface must balance the weight $2mg$ of both blocks.

It remains now to determine the magnitude of the maximum frictional force $f_{\text{s, from A}}^{\text{MAX}}$ from the upper block. To this end, we draw the free-body diagram for the upper block at the instant that it just begins to slip due to the 47.0-N applied force. At this instant the block is in equilibrium, so that the frictional force from the lower block B balances the 47.0-N force. Thus, $f_{\text{s, from B}}^{\text{MAX}} = 47.0\text{ N}$, and according to Equation 4.7, we have

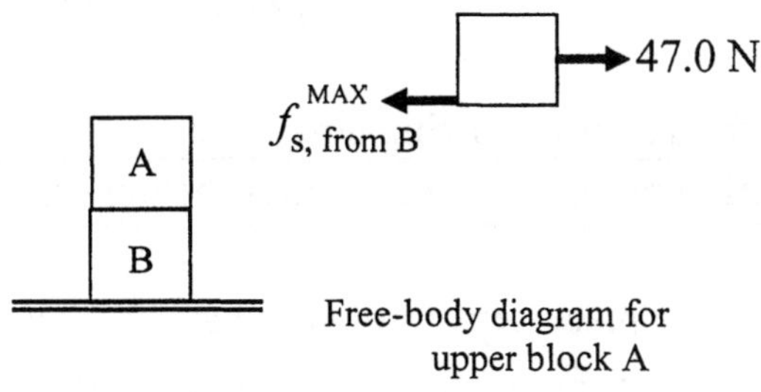

Free-body diagram for upper block A

$$f_{\text{s, from B}}^{\text{MAX}} = \mu_s F_N = \mu_s mg = 47.0\text{ N}$$

Here, we have recognized that the normal force F_N from the lower block must balance the weight mg of only the upper block. This result tells us that $\mu_s mg = 47.0$ N. To determine $f_{\text{s, from A}}^{\text{MAX}}$ we invoke Newton's third law to conclude that the magnitudes of the frictional forces at the A-B interface are equal, since they are action-reaction forces. Thus, $f_{\text{s, from A}}^{\text{MAX}} = \mu_s mg$. Substituting this result and Equation (2) into Equation (1) gives

$$F_{\text{Applied}} = f_{\text{s, from A}}^{\text{MAX}} + f_{\text{s, from surface}}^{\text{MAX}} = \mu_s mg + \mu_s 2mg = 3(47.0\text{ N}) = \boxed{141\text{ N}}$$

---

43. SSM ***REASONING*** If we assume that kinetic friction is the only horizontal force that acts on the skater, then, since kinetic friction is a resistive force, it acts opposite to the direction of motion and the skater slows down. According to Newton's second law ($\Sigma\mathbf{F} = m\mathbf{a}$), the magnitude of the deceleration is $a = f_k / m$.

The magnitude of the frictional force that acts on the skater is, according to Equation 4.8, $f_k = \mu_k F_N$ where μ_k is the coefficient of kinetic friction between the ice and the skate blades. There are only two vertical forces that act on the skater; they are the upward normal force $\mathbf{F}_N$ and the downward pull of gravity (the weight) mg. Since the skater has no vertical acceleration, Newton's second law in the vertical direction gives (if we take upward as the positive direction) $F_N - mg = 0$. Therefore, the magnitude of the normal force is $F_N = mg$ and the magnitude of the deceleration is given by

$$a = \frac{f_k}{m} = \frac{\mu_k F_N}{m} = \frac{\mu_k mg}{m} = \mu_k g$$

SOLUTION

a. Direct substitution into the previous expression gives

$$a = \mu_k g = (0.100)(9.80\ \text{m/s}^2) = \boxed{0.980\ \text{m/s}^2}$$

Since the skater is slowing down, the $\boxed{\text{direction of the acceleration must be opposite to the direction of motion}}$.

b. The displacement through which the skater will slide before he comes to rest can be obtained from Equation 2.9 ($v^2 - v_0^2 = 2ax$). Since the skater comes to rest, $v = 0$ m/s. If we take the direction of motion of the skater as the positive direction, then, solving for x, we obtain

$$x = \frac{-v_0^2}{2a} = \frac{-(7.60\ \text{m/s})^2}{2(-0.980\ \text{m/s}^2)} = \boxed{29.5\ \text{m}}$$

---

44. ***REASONING*** The free-body diagrams for the large cube (mass = M) and the small cube (mass = m) are shown in the following drawings. In the case of the large cube, we have omitted the weight and the normal force from the surface, since the play no role in the solution (although they do balance).

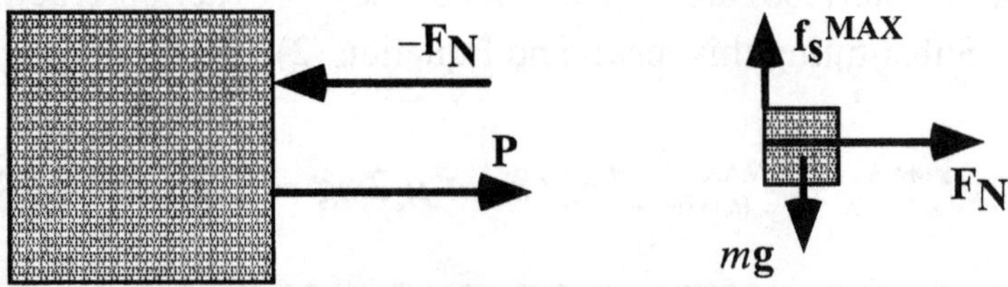

In these diagrams, note that the two blocks exert a normal force on each other; the large block exerts the force F_N on the smaller block, while the smaller block exerts the force $-F_N$ on the larger block. In accord with Newton's third law these forces have opposite directions and equal magnitudes F_N. Under the influence of the forces shown, the two blocks have the same acceleration a. We begin our solution by applying Newton's second law to each one.

SOLUTION According to Newton's second law, we have

$$\underbrace{\Sigma F = P - F_N = Ma}_{\text{Large block}} \qquad \underbrace{F_N = ma}_{\text{Small block}}$$

Substituting $F_N = ma$ into the large-block expression and solving for P gives

$$P = (M + m)\,a$$

For the smaller block to remain in place against the larger block, the static frictional force must balance the weight of the smaller block, so that $f_s^{MAX} = mg$. But f_s^{MAX} is given by

$f_s^{MAX} = \mu_s F_N$, where, from the Newton's second law, we know that $F_N = ma$. Thus, we have $\mu_s ma = mg$ or $a = g/\mu_s$. Using this result in the expression for P gives

$$P = (M+m)a = \frac{(M+m)g}{\mu_s} = \frac{(25\text{ kg} + 4.0\text{ kg})(9.80\text{ m/s}^2)}{0.71} = \boxed{4.0 \times 10^2\text{ N}}$$

---

45. ***REASONING*** The free-body diagram for the box is shown in the following drawing on the left. On the right the same drawing is repeated, except that the pushing force P is resolved into its horizontal and vertical components.

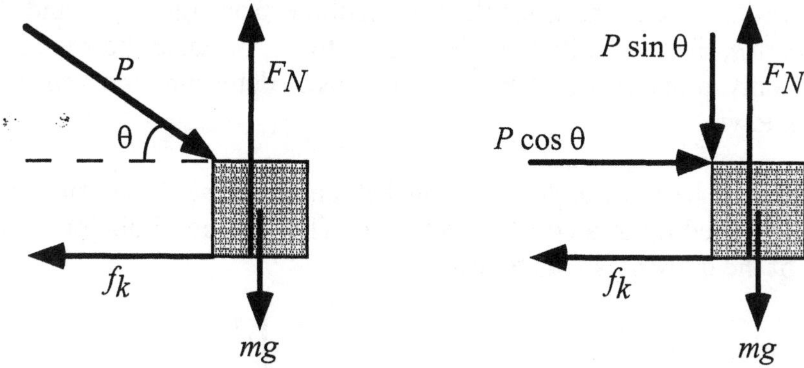

Since the block is moving at a constant velocity, it has no acceleration, and Newton's second law indicates that the net vertical and net horizontal forces must separately be zero.

SOLUTION Taking upward and to the right as the positive directions, we write the zero net vertical and horizontal forces as follows:

$$\underbrace{F_N - mg - P\sin\theta = 0}_{\text{Vertical}} \qquad \underbrace{P\cos\theta - f_k = 0}_{\text{Horizontal}}$$

From the equation for the horizontal forces, we have $P\cos\theta = f_k$. But the kinetic frictional force is $f_k = \mu_k F_N$. Furthermore, from the equation for the vertical forces, we have $F_N = mg + P\sin\theta$. With these substitutions, we obtain

$$P\cos\theta = f_k = \mu_k F_N = \mu_k(mg + P\sin\theta)$$

Solving for P gives

$$P = \frac{\mu_k mg}{\cos\theta - \mu_k \sin\theta}$$

The necessary pushing force becomes infinitely large when the denominator in this expression is zero. Hence, we find that $\cos\theta - \mu_k \sin\theta = 0$, which can be rearranged to show that

$$\frac{\sin\theta}{\cos\theta} = \tan\theta = \frac{1}{\mu_k} \quad \text{or} \quad \theta = \tan^{-1}\left(\frac{1}{0.41}\right) = \boxed{68°}$$

---

46. ***REASONING*** At first glance there seems to be very little information given. However, it is enough. In part *a* of the drawing the bucket is hanging stationary and, therefore, is in equilibrium. The forces acting on it are its weight and the two tension forces from the rope. There are two tension forces from the rope, because the rope is attached to the bucket handle at two places. These three forces must balance, which will allow us to determine the weight of the bucket. In part *b* of the drawing, the bucket is again in equilibrium, since it is traveling at a constant velocity and, therefore, has no acceleration. The forces acting on the bucket now are its weight and a single tension force from the rope, and they again must balance. In part *b*, there is only a single tension force, because the rope is attached to the bucket handle only at one place. This will allow us to determine the tension in part *b*, since the weight is known.

SOLUTION Let *W* be the weight of the bucket, and let *T* be the tension in the rope as the bucket is being pulled up at a constant velocity. The free-body diagrams for the bucket in parts *a* and *b* of the drawing are as follows:

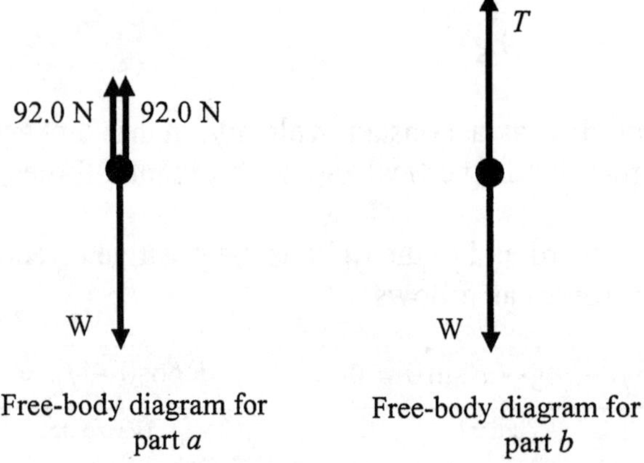

Free-body diagram for part *a* Free-body diagram for part *b*

Since the bucket in part *a* is in equilibrium, the net force acting on it is zero. Taking upward to be the positive direction, we have

$$\Sigma F = 92.0\text{ N} + 92.0\text{ N} - W = 0 \quad \text{or} \quad W = 184\text{ N}$$

Similarly, in part *b* we have

$$\Sigma F = T - W = 0 \quad \text{or} \quad T = W = \boxed{184\text{ N}}$$

---

47. ***REASONING AND SOLUTION***

a. In the horizontal direction the thrust, F, is balanced by the resistive force, f_r, of the water. That is,

$$\Sigma F_x = 0$$

or

$$f_r = F = \boxed{7.40 \times 10^5 \text{ N}}$$

b. In the vertical direction, the weight, mg, is balanced by the buoyant force, F_b. So

$$\Sigma F_y = 0$$

gives

$$F_b = mg = (1.70 \times 10^8 \text{ kg})(9.80 \text{ m/s}^2) = \boxed{1.67 \times 10^9 \text{ N}}$$

---

48. ***REASONING AND SOLUTION*** Newton's second law applied in the vertical and horizontal directions gives

$$L \cos 21.0° - W = 0 \qquad (1)$$
$$L \sin 21.0° - R = 0 \qquad (2)$$

a. Equation (1) gives

$$L = \frac{W}{\cos 21.0°} = \frac{53\ 800 \text{ N}}{\cos 21.0°} = \boxed{57\ 600 \text{ N}}$$

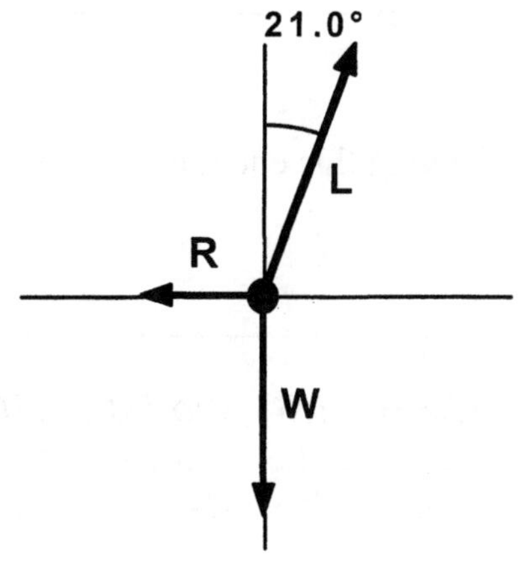

b. Equation (2) gives

$$R = L \sin 21.0° = (57\ 600 \text{ N}) \sin 21.0° = \boxed{20\ 600 \text{ N}}$$

---

49. **SSM** ***REASONING AND SOLUTION*** The figure at the right shows the forces that act on the wine bottle. Newton's second law applied in the horizontal and vertical directions gives

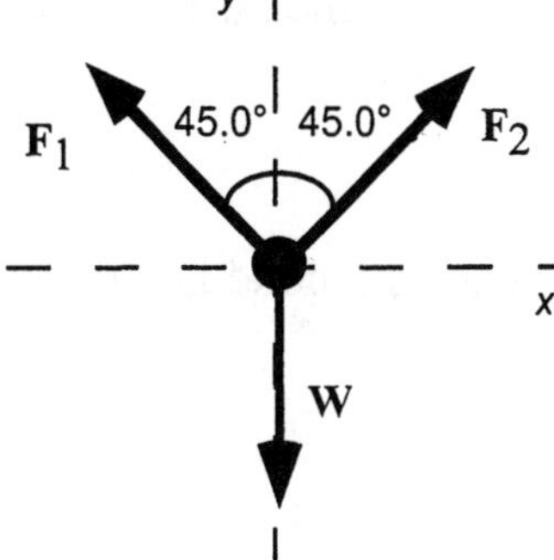

$$\Sigma F_y = F_1 \cos 45.0° + F_2 \cos 45.0° - W = 0 \qquad (1)$$

$$\Sigma F_x = F_2 \sin 45.0° - F_1 \sin 45.0° = 0 \qquad (2)$$

From Equation (2), we see that $F_1 = F_2$. According to Equation (1), we have

$$F_1 = \frac{W}{2 \cos 45.0°} = \frac{mg}{2 \cos 45.0°}$$

Therefore,

$$F_1 = F_2 = \frac{(1.40\text{ kg})\,(9.80\text{ m/s}^2)}{2\cos 45.0°} = \boxed{9.70\text{ N}}$$

---

50. ***REASONING*** The drawing shows the I-beam and the three forces that act on it, its weight **W** and the tension **T** in each of the cables. Since the I-beam is moving upward at a constant velocity, its acceleration is zero and it is in vertical equilibrium. According to Equation 4.9b, $\sum F_y = 0$, the net force in the vertical (or *y*) direction must be zero. This relation will allow us to find the magnitude of the tension.

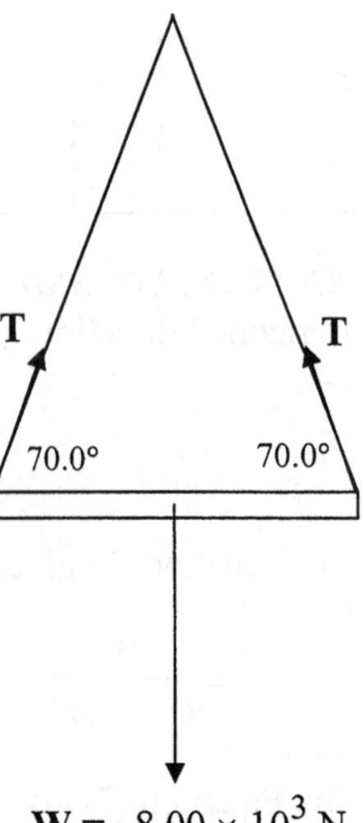

SOLUTION Taking up as the +*y* direction, Equation 4.9b becomes

$$\underbrace{+T\sin 70.0° + T\sin 70.0° - 8.00\times 10^3\text{ N}}_{\Sigma F_y} = 0$$

Solving this equation for the tension gives $T = \boxed{4260\text{ N}}$.

---

51. ***REASONING AND SOLUTION*** Consider a free body diagram for the stuntman with the x-axis parallel to the ground and the +y-axis vertically upward. The motion is along the +x-axis. Newton's second law written for no motion along the y-axis is $\Sigma F_y = 0$ or

$$F_N - mg = 0$$

This gives the normal force to be

$$F_N = mg = (109\text{ kg})(9.80\text{ m/s}^2)$$

Newton's second law for uniform motion in the x direction is $\Sigma F_x = 0$ or

$$T - f_k = 0$$

Then

$$T = f_k = \mu_k F_N = (0.870)(109\text{ kg})(9.80\text{ m/s}^2) = \boxed{929\text{ N}}$$

---

52. ***REASONING AND SOLUTION*** The free body diagram for the plane is shown below to the left. The figure at the right shows the forces resolved into components parallel to and perpendicular to the line of motion of the plane.

If the plane is to continue at constant velocity, the resultant force must still be zero after the fuel is jettisoned. Therefore (using the directions of T and L to define the positive directions),

$$T - R - W(\sin\theta) = 0 \qquad (1)$$
$$L - W(\cos\theta) = 0 \qquad (2)$$

From Example 13, before the fuel is jettisoned, the weight of the plane is 86 500 N, the thrust is 103 000 N, and the lift is 74 900 N. The force of air resistance is the same before and after the fuel is jettisoned and is given in Example 13 as R = 59 800 N.

After the fuel is jettisoned, $W = 86\ 500\ \text{N} - 2800\ \text{N} = 83\ 700\ \text{N}$

From Equation (1) above, the thrust after the fuel is jettisoned is

$$T = R + W(\sin\theta) = [(59\ 800\ \text{N}) + (83\ 700\ \text{N})(\sin 30.0°)] = 101\ 600\ \text{N}$$

From Equation (2), the lift after the fuel is jettisoned is

$$L = W(\cos\theta) = (83\ 700\ \text{N})(\cos 30.0°) = 72\ 500\ \text{N}$$

a. The pilot must, therefore, reduce the thrust by

$$103\ 000\ \text{N} - 101\ 600\ \text{N} = \boxed{1400\ \text{N}}$$

b. The pilot must reduce the lift by

$$74\ 900\ \text{N} - 72\ 500\ \text{N} = \boxed{2400\ \text{N}}$$

---

53. SSM ***REASONING*** In order for the object to move with constant velocity, the net force on the object must be zero. Therefore, the north/south component of the third force must be equal in magnitude and opposite in direction to the 80.0 N force, while the east/west component of the third force must be equal in magnitude and opposite in direction to the

60.0 N force. Therefore, the third force has components: 80.0 N due south and 60.0 N due east. We can use the Pythagorean theorem and trigonometry to find the magnitude and direction of this third force.

SOLUTION The magnitude of the third force is

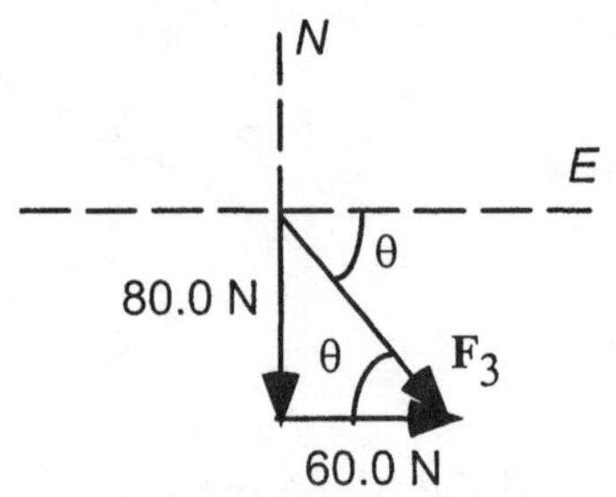

$$F_3 = \sqrt{(80.0\text{ N})^2 + (60.0\text{ N})^2} = \boxed{1.00 \times 10^2\text{ N}}$$

The direction of F_3 is specified by the angle θ where

$$\theta = \tan^{-1}\left(\frac{80.0\text{ N}}{60.0\text{ N}}\right) = \boxed{53.1°,\text{ south of east}}$$

---

54. ***REASONING*** The free-body diagram in the drawing at the right shows the forces that act on the clown (weight = W). In this drawing, note that P denotes the pulling force. Since the rope passes around three pulleys, forces of magnitude P are applied both to the clown's hands and his feet. The normal force due to the floor is F_N, and the maximum static frictional force is f_s^{MAX}. At the instant just before the clown's feet move, the net vertical and net horizontal forces are zero, according to Newton's second law, since there is no acceleration at this instant.

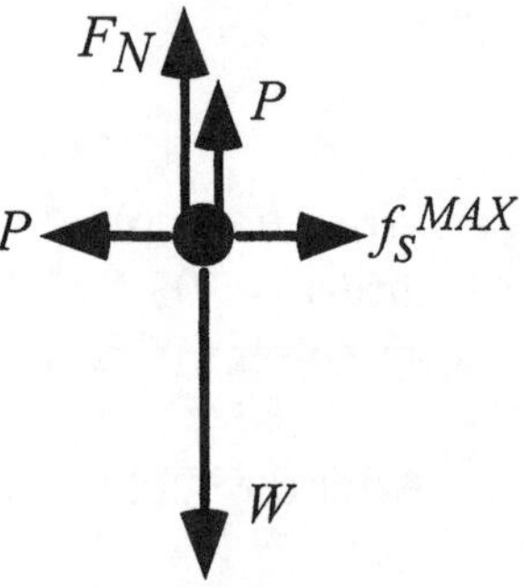

SOLUTION According to Newton's second law, with upward and to the right chosen as the positive directions, we have

$$\underbrace{F_N + P - W = 0}_{\text{Vertical forces}} \quad \text{and} \quad \underbrace{f_s^{MAX} - P = 0}_{\text{Horizontal forces}}$$

From the horizontal-force equation we find $P = f_s^{MAX}$. But $f_s^{MAX} = \mu_s F_N$. From the vertical-force equation, the normal force is $F_N = W - P$. With these substitutions, it follows that

$$P = f_s^{MAX} = \mu_s F_N = \mu_s (W - P)$$

Solving for P gives

$$P = \frac{\mu_s W}{1 + \mu_s} = \frac{(0.53)(890\text{ N})}{1 + 0.53} = \boxed{310\text{ N}}$$

---

55. ***REASONING*** Since the boxes are at rest, they are in equilibrium. According to Equation 4.9b, the net force in the vertical, or y, direction is zero, $\Sigma F_y = 0$. There are two unknowns in this problem, the normal force that the table exerts on box 1 and the tension in the rope that connects boxes 2 and 3. To determine these unknowns we will apply the relation $\Sigma F_y = 0$ twice, once to the boxes on the left of the pulley and once to the box on the right.

SOLUTION There are four forces acting on the two boxes on the left. The boxes are in equilibrium, so that the net force must be zero. Choosing the $+y$ direction as being the upward direction, we have that

$$\underbrace{-W_1 - W_2 + F_N + T}_{\Sigma F_y} = 0 \qquad (1)$$

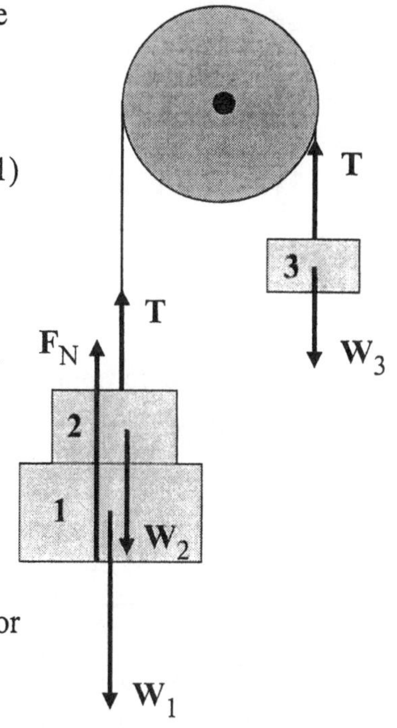

where W_1 and W_2 are the magnitudes of the weights of the boxes, F_N is the magnitude of the normal force that the table exerts on box 1, and T is the magnitude of the tension in the rope. We know the weights. To find the unknown tension, note that the box 3 is also in equilibrium, so that the net force acting on it must be zero.

$$\underbrace{-W_3 + T}_{\Sigma F_y} = 0 \qquad \text{so that} \qquad T = W_3$$

Substituting this expression for T into Equation (1) and solving for the normal force gives

$$F_N = W_1 + W_2 - W_3 = 55 \text{ N} + 35 \text{ N} - 28 \text{ N} = \boxed{62 \text{ N}}$$

---

56. ***REASONING*** Since the wire beneath the limb is at rest, it is in equilibrium and the net force acting on it must be zero. Three forces comprise the net force, the 151-N force from the limb, the 447-N tension force from the left section of the wire, and the tension force **T** from the right section of the wire. We will resolve the forces into components and set the sum of the x components and the sum of the y components separately equal to zero. In so doing we will obtain two equations containing the unknown quantities, which are the horizontal and vertical components of the tension force **T**. These two equations will be solved simultaneously to give values for the two unknowns. Knowing the components of the tension force, we can determine its magnitude and direction.

SOLUTION Let T_x and T_y be the horizontal and vertical components of the tension force. The free-body diagram for the wire beneath the limb is as follows:

Taking upward and to the right as the positive directions, we find for the x components of the forces that

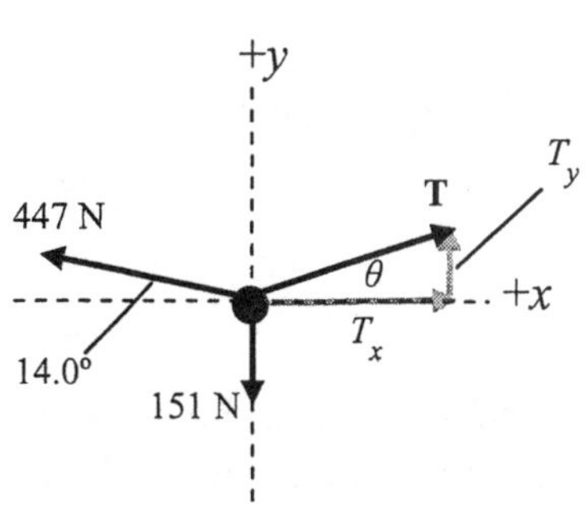

$$\Sigma F_x = T_x - (447 \text{ N})\cos 14.0° = 0$$

$$T_x = (447 \text{ N})\cos 14.0° = 434 \text{ N}$$

For the y components of the forces we have

$$\Sigma F_y = T_y + (447\text{ N})\sin 14.0° - 151\text{ N} = 0$$

$$T_y = -(447\text{ N})\sin 14.0° + 151\text{ N} = 43\text{ N}$$

The magnitude of the tension force is

$$T = \sqrt{T_x^2 + T_y^2} = \sqrt{(434\text{ N})^2 + (43\text{ N})^2} = \boxed{436\text{ N}}$$

Since the components of the tension force and the angle θ are related by $\tan\theta = T_y / T_x$, we find that

$$\theta = \tan^{-1}\left(\frac{T_y}{T_x}\right) = \tan^{-1}\left(\frac{43\text{ N}}{434\text{ N}}\right) = \boxed{5.7°}$$

---

57. SSM ***REASONING AND SOLUTION*** The free-body diagram is shown at the right. The forces that act on the picture are the pressing force P, the normal force $\mathbf{F_N}$ exerted on the picture by the wall, the weight mg of the picture, and the force of static friction $\mathbf{f_s^{MAX}}$. The maximum magnitude for the frictional force is given by Equation 4.7: $f_s^{MAX} = \mu_s F_N$. The picture is in equilibrium, and, if we take the directions to the right and up as positive, we have in the x direction

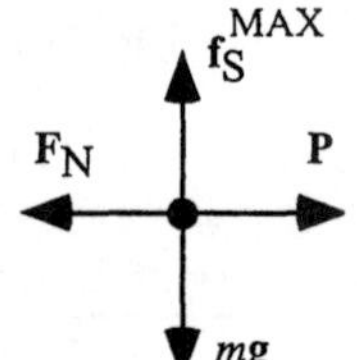

$$\Sigma F_x = P - F_N = 0 \qquad \text{or} \qquad P = F_N$$

and in the y direction

$$\Sigma F_y = f_s^{MAX} - mg = 0 \qquad \text{or} \qquad f_s^{MAX} = mg$$

Therefore,

$$f_s^{MAX} = \mu_s F_N = mg$$

But since $F_N = P$, we have

$$\mu_s P = mg$$

Solving for P, we have

$$P = \frac{mg}{\mu_s} = \frac{(1.10\text{ kg})(9.80\text{ m/s}^2)}{0.660} = \boxed{16.3\text{ N}}$$

---

58. ***REASONING*** Since the mountain climber is at rest, she is in equilibrium and the net force acting on her must be zero. Three forces comprise the net force, her weight, and the tension forces from the left and right sides of the rope. We will resolve the forces into components and set the sum of the x components and the sum of the y components separately equal to

zero. In so doing we will obtain two equations containing the unknown quantities, the tension T_L in the left side of the rope and the tension T_R in the right side. These two equations will be solved simultaneously to give values for the two unknowns.

SOLUTION Using W to denote the weight of the mountain climber and choosing right and upward to be the positive directions, we have the following free-body diagram for the climber:

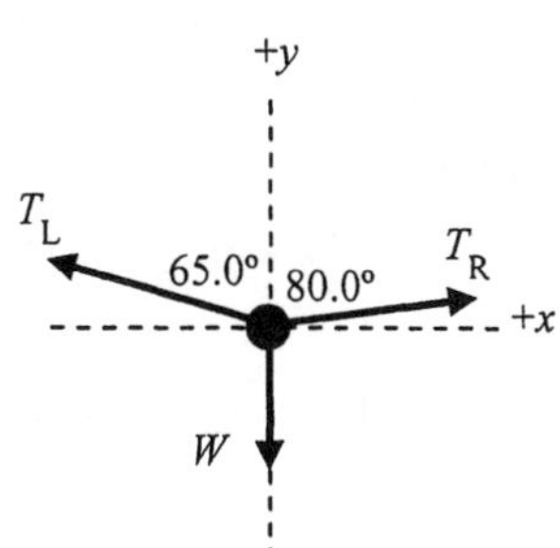

For the x components of the forces we have

$$\Sigma F_x = T_R \sin 80.0° - T_L \sin 65.0° = 0$$

For the y components of the forces we have

$$\Sigma F_y = T_R \cos 80.0° + T_L \cos 65.0° - W = 0$$

Solving the first of these equations for T_R, we find that

$$T_R = T_L \frac{\sin 65.0°}{\sin 80.0°}$$

Substituting this result into the second equation gives

$$T_L \frac{\sin 65.0°}{\sin 80.0°} \cos 80.0° + T_L \cos 65.0° - W = 0 \quad \text{or} \quad T_L = 1.717\, W$$

Using this result in the expression for T_R reveals that

$$T_R = T_L \frac{\sin 65.0°}{\sin 80.0°} = (1.717W) \frac{\sin 65.0°}{\sin 80.0°} = 1.580\, W$$

Since the weight of the climber is $W = 535$ N, we find that

$$T_L = 1.717\, W = 1.717(535 \text{ N}) = \boxed{919 \text{ N}}$$

$$T_R = 1.580\, W = 1.580(535 \text{ N}) = \boxed{845 \text{ N}}$$

59. ***REASONING AND SOLUTION*** If the +x axis is taken in the direction of motion, $\Sigma F_x = 0$ gives

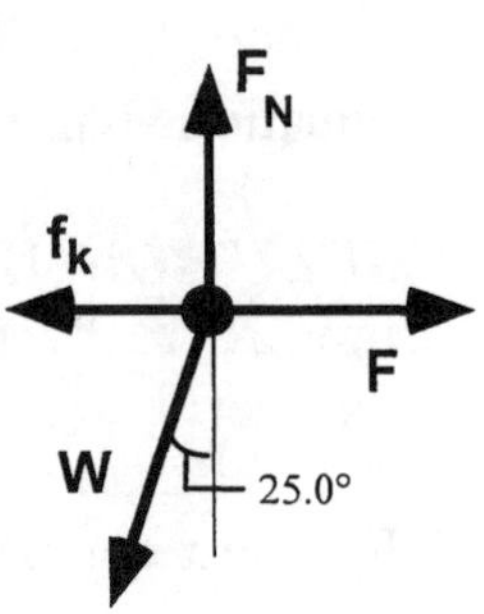

$$F - f_k - mg \sin\theta = 0$$

where

$$f_k = \mu_k F_N$$

Then

$$F - \mu_k F_N - mg \sin\theta = 0 \qquad (1)$$

Also, $\Sigma F_y = 0$ gives

$$F_N - mg \cos\theta = 0$$

so

$$F_N = mg \cos\theta \qquad (2)$$

Substituting Equation (2) into Equation (1) and solving for F yields

$$F = mg(\sin\theta + \mu_k \cos\theta)$$

$$F = (55.0\text{ kg})(9.80\text{ m/s}^2)[\sin 25.0° + (0.120)\cos 25.0°] = \boxed{286\text{ N}}$$

---

60. ***REASONING AND SOLUTION***

a. If the block is not to slide down the wall, then the vertical forces acting on the block must sum to zero.

$$F\cos 40.0° - mg + \mu_s F_N = 0$$

Additionally, the horizontal forces must sum to zero.

$$F\sin 40.0° - F_N = 0$$

Eliminating F_N gives

$$F = \frac{mg}{\cos 40.0° + \mu_s \sin 40.0°} = \boxed{79.0\text{ N}}$$

b. The above analysis applies to the case where the block is starting to slide up the wall except, that the frictional force will be in the opposite direction. Hence,

$$F = \frac{mg}{\cos 40.0° - \mu_s \sin 40.0°} = \boxed{219\text{ N}}$$

---

61. SSM ***REASONING*** When the bicycle is coasting straight down the hill, the forces that act on it are the normal force F_N exerted by the surface of the hill, the force of gravity mg,

and the force of air resistance R. When the bicycle climbs the hill, there is one additional force; it is the applied force that is required for the bicyclist to climb the hill at constant speed. We can use our knowledge of the motion of the bicycle down the hill to find R. Once R is known, we can analyze the motion of the bicycle as it climbs the hill.

SOLUTION The figure to the left below shows the free-body diagram for the forces during the downhill motion. The hill is inclined at an angle θ above the horizontal. The figure to the right shows these forces resolved into components parallel to and perpendicular to the line of motion.

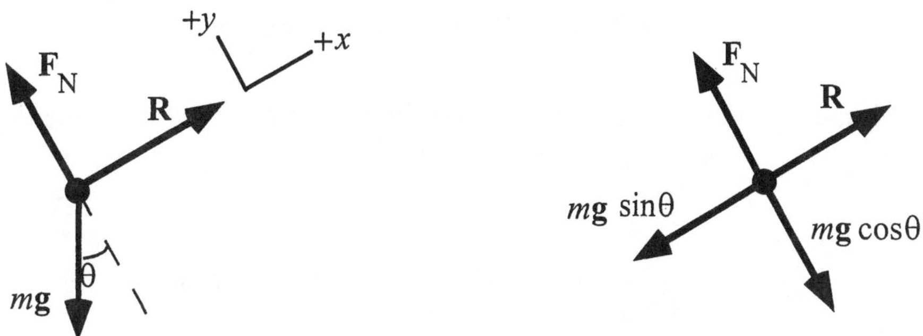

Since the bicyclist is traveling at a constant velocity, his acceleration is zero. Therefore, according to Newton's second law, we have $\Sigma F_x = 0$ and $\Sigma F_y = 0$. Taking the direction up the hill as positive, we have $\Sigma F_x = R - mg \sin\theta = 0$, or

$$R = mg \sin\theta \;=\; (80.0 \text{ kg})(9.80 \text{ m/s}^2) \sin 15.0° = 203 \text{ N}$$

When the bicyclist climbs the same hill at constant speed, an applied force P must push the system up the hill, and the force of air resistance will oppose the motion by pointing down the hill.

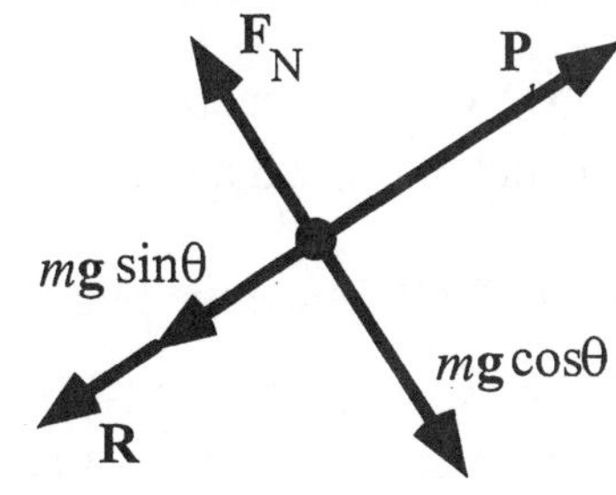

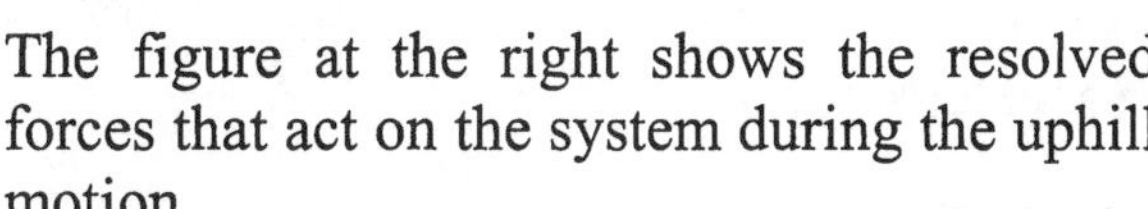
The figure at the right shows the resolved forces that act on the system during the uphill motion.

Using the same sign convention as above, we have $\Sigma F_x = P - mg \sin\theta - R = 0$, or

$$P = R + mg \sin\theta = 203 \text{ N} + 203 \text{ N} = \boxed{406 \text{ N}}$$

---

62. ***REASONING*** The weight of the part of the washcloth off the table is $m_{off}g$. At the instant just before the washcloth begins to slide, this weight is supported by a force that has magnitude equal to f_s^{MAX}, which is the static frictional force that the table surface applies to the part of the washcloth on the table. This force is transmitted "around the bend" in the washcloth hanging over the edge by the tension forces between the molecules of the

washcloth, in much the same way that a force applied to one end of a rope is transmitted along the rope as it passes around a pulley.

SOLUTION Since the static frictional supports the weight of the washcloth off the table, we have $f_s^{MAX} = m_{off}g$. The static frictional force is $f_s^{MAX} = \mu_s F_N$. The normal force F_N is applied by the table to the part of the washcloth on the table and has a magnitude equal to the weight of that part of the washcloth. This is so, because the table is assumed to be horizontal and the part of the washcloth on it does not accelerate in the vertical direction. Thus, we have

$$f_s^{MAX} = \mu_s F_N = \mu_s m_{on} g = m_{off} g$$

The magnitude g of the acceleration due to gravity can be eliminated algebraically from this result, giving $\mu_s m_{on} = m_{off}$. Dividing both sides by $m_{on} + m_{off}$ gives

$$\mu_s \left(\frac{m_{on}}{m_{on} + m_{off}} \right) = \frac{m_{off}}{m_{on} + m_{off}} \qquad \text{or} \qquad \mu_s f_{on} = f_{off}$$

where we have used f_{on} and f_{off} to denote the fractions of the washcloth on and off the table, respectively. Since $f_{on} + f_{off} = 1$, we can write the above equation on the left as

$$\mu_s \left(1 - f_{off} \right) = f_{off} \quad \text{or} \quad f_{off} = \frac{\mu_s}{1 + \mu_s} = \frac{0.40}{1 + 0.40} = \boxed{0.29}$$

---

63. ***REASONING*** According to Newton's second law, the acceleration has the same direction as the net force and a magnitude given by $a = \Sigma F/m$.

SOLUTION Since the two forces are perpendicular, the magnitude of the net force is given by the Pythagorean theorem as $\Sigma F = \sqrt{(40.0 \text{ N})^2 + (60.0 \text{ N})^2}$. Thus, according to Newton's second law, the magnitude of the acceleration is

$$a = \frac{\Sigma F}{m} = \frac{\sqrt{(40.0 \text{ N})^2 + (60.0 \text{ N})^2}}{4.00 \text{ kg}} = \boxed{18.0 \text{ m/s}^2}$$

The direction of the acceleration vector is given by

$$\theta = \tan^{-1} \left(\frac{60.0 \text{ N}}{40.0 \text{ N}} \right) = \boxed{56.3° \text{ above the } +x \text{ axis}}$$

---

64. ***REASONING*** Suppose the bobsled is moving along the $+x$ direction. There are two forces acting on it that are parallel to its motion; a force $+F_x$ propelling it forward and a force of

–450 N that is resisting its motion. The net force is the sum of these two forces. According to Newton's second law, Equation 4.2a, the net force is equal to the mass of the bobsled times its acceleration. Since the mass and acceleration are known, we can use the second law to determine the magnitude of the propelling force.

SOLUTION
a. Newton's second law states that

$$\underbrace{+F_x - 450\text{ N}}_{\Sigma F_x} = ma_x \tag{4.2a}$$

Solving this equation for F_x gives

$$F_x = ma_x + 450\text{ N} = (270\text{ kg})(2.4\text{ m/s}^2) + 450\text{ N} = \boxed{1100\text{ N}}$$

b. The magnitude of the net force that acts on the bobsled is

$$\Sigma F_x = ma_x = (270\text{ kg})(2.4\text{ m/s}^2) = \boxed{650\text{ N}} \tag{4.2a}$$

---

65. SSM ***REASONING*** If we assume that the acceleration is constant, we can use Equation 2.4 ($v = v_0 + at$) to find the acceleration of the car. Once the acceleration is known, Newton's second law ($\Sigma\mathbf{F} = m\mathbf{a}$) can be used to find the magnitude and direction of the net force that produces the deceleration of the car.

SOLUTION The average acceleration of the car is, according to Equation 2.4,

$$a = \frac{v - v_0}{t} = \frac{17.0\text{ m/s} - 27.0\text{ m/s}}{8.00\text{ s}} = -1.25\text{ m/s}^2$$

where the minus sign indicates that the direction of the acceleration is opposite to the direction of motion; therefore, the acceleration points due west.

According to Newton's Second law, the net force on the car is

$$\Sigma F = ma = (1380\text{ kg})(-1.25\text{ m/s}^2) = -1730\text{ N}$$

The magnitude of the net force is $\boxed{1730\text{ N}}$. From Newton's second law, we know that the direction of the force is the same as the direction of the acceleration, so the force also points $\boxed{\text{due west}}$.

---

66. ***REASONING*** The skydiver is falling along the $-y$ direction. There are two forces acting on him: the upward-acting force $+f_{air}$ of air resistance, and his weight, $-mg$. The net force is the sum of these forces. According to Newton's second law, Equation 4.2b, the net force is equal to the mass of the skydiver times his acceleration. We can use the second law to determine the acceleration.

SOLUTION
a. Newton's second law states that

$$\underbrace{+f_{\text{air}} - mg}_{\Sigma F_y} = ma_y \qquad (4.2b)$$

Solving this equation for a_y and noting that $f_{\text{air}} = \frac{1}{3}mg$, we have

$$a_y = \frac{-mg + \frac{1}{3}mg}{m} = -\tfrac{2}{3}g = -6.5 \text{ m/s}^2$$

The magnitude of the skydiver's acceleration is $\boxed{6.5 \text{ m/s}^2}$.

b. When the skydiver falls at a constant velocity, his acceleration is zero, and he is in equilibrium. According to Equation 4.9b, $\Sigma F_y = 0$, and the net force acting on him must be zero.

$$\underbrace{f_{\text{air}} - mg}_{\Sigma F_y} = 0$$

The force of air resistance is

$$f_{\text{air}} = mg = (110 \text{ kg})(9.8 \text{ m/s}^2) = \boxed{1100 \text{ N, upward}}$$

---

67. ***REASONING AND SOLUTION*** Newton's second law applied to object 1 (422 N) gives

$$T = m_1 a_1$$

Similarly, for object 2 (185 N)

$$T - m_2 g = m_2 a_2$$

If the string is not to break or go slack, both objects must have accelerations of the same magnitude.

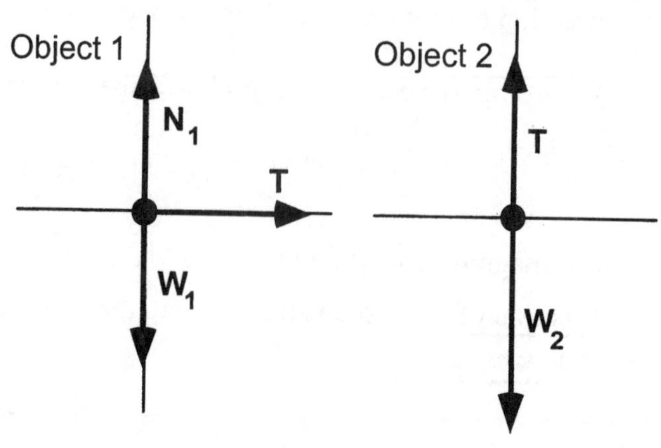

Then $a_1 = a$ and $a_2 = -a$. The above equations become

$$T = m_1 a \qquad (1)$$
$$T - m_2 g = -m_2 a \qquad (2)$$

a. Substituting Equation (1) into Equation (2) and solving for a yields

$$a = \frac{m_2 g}{m_1 + m_2} = \boxed{2.99 \text{ m/s}^2}$$

b. Using this value in Equation (1) gives

$$T = m_1 a = \boxed{129 \text{ N}}$$

---

68. ***REASONING*** The forces acting on the motorcycle are the normal force F_N, the 3150-N propulsion force, the 250-N force of air resistance, and the weight, which is $mg = (292 \text{ kg})(9.80 \text{ m/s}^2 = 2860 \text{ N}$. All of these forces must be considered when determining the net force for use with Newton's second law to determine the acceleration. In particular, we note that the motion occurs along the ramp and that both the propulsion force and air resistance are directed parallel to the ramp surface. In contrast, the normal force and the weight do not act parallel to the ramp. The normal force is perpendicular to the ramp surface, while the weight acts vertically downward. However, the weight does have a component along the ramp.

SOLUTION In drawing the free-body diagram for the motorcycle we choose the $+x$ axis to be parallel to the ramp surface and upward, the $+y$ direction being perpendicular to the ramp surface. The free-body diagram is as follows:

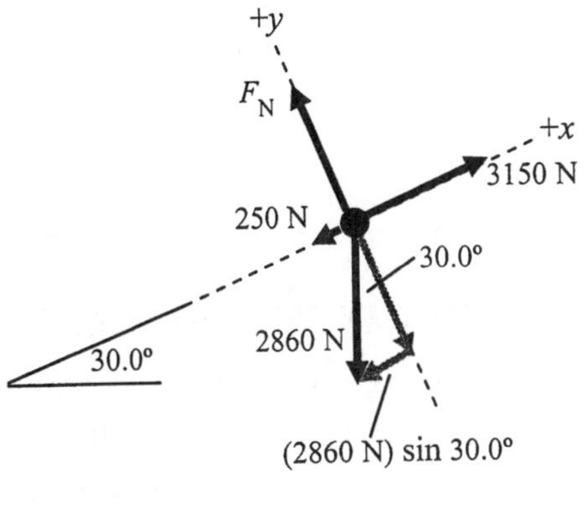

Since the motorcycle accelerates along the ramp and we seek only that acceleration, we can ignore the forces that point along the y axis (the normal force F_N and the y component of the weight). The x component of Newton's second law is

$$\Sigma F_x = 3150 \text{ N} - (2860 \text{ N})\sin 30.0° - 250 \text{ N} = ma_x$$

Solving for the acceleration gives

$$a_x = \frac{3150 \text{ N} - (2860 \text{ N})\sin 30.0° - 250 \text{ N}}{292 \text{ kg}} = \boxed{5.03 \text{ m/s}^2}$$

---

69. SSM WWW ***REASONING*** The speed of the skateboarder at the bottom of the ramp can be found by solving Equation 2.9 ($v^2 = v_0^2 + 2ax$, where x is the distance that the skater moves down the ramp) for v. The figure at the right shows the free-body diagram for the skateboarder. The net force ΣF, which accelerates the skateboarder down the ramp, is the component of the weight that is parallel to the incline: $\Sigma F = mg\sin\theta$. Therefore, we know from Newton's second law that the acceleration of the skateboarder down the ramp is

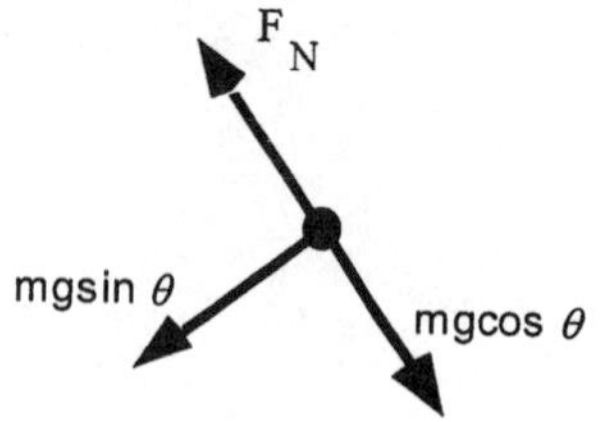

$$a = \frac{\Sigma F}{m} = \frac{mg\sin\theta}{m} = g\sin\theta$$

SOLUTION Thus, the speed of the skateboarder at the bottom of the ramp is

$$v = \sqrt{v_0^2 + 2ax} = \sqrt{v_0^2 + 2gx\sin\theta} = \sqrt{(2.6\text{ m/s})^2 + 2(9.80\text{ m/s}^2)(6.0\text{ m})\sin 18°} = \boxed{6.6\text{ m/s}}$$

---

70. ***REASONING AND SOLUTION***

a. In the vertical direction $\Sigma F_y = ma_y$ gives

$$T - mg = ma_y$$

so

$$T = ma_y + mg = mg(1 + a_y/g)$$

$$T = (822\text{ N})\left(1 + \frac{1.10\text{ m/s}^2}{9.80\text{ m/s}^2}\right) = \boxed{914\text{ N}}$$

b. The acceleration of the man is zero if his velocity is constant. From part a

$$T = mg = \boxed{822\text{ N}}$$

---

71. SSM ***REASONING AND SOLUTION***

a. Each cart has the same mass and acceleration; therefore, the net force acting on any one of the carts is, according to Newton's second law

$$\Sigma F = ma = (26\text{ kg})(0.050\text{ m/s}^2) = \boxed{1.3\text{ N}}$$

b. The fifth cart must essentially push the sixth, seventh, eight, ninth and tenth cart. In other words, it must exert on the sixth cart a total force of

$$\Sigma F = ma = 5(26\text{ kg})(0.050\text{ m/s}^2) = \boxed{6.5\text{ N}}$$

---

72. ***REASONING*** To determine the acceleration we will use Newton's second law $\Sigma \mathbf{F} = m\mathbf{a}$. Two forces act on the rocket, the thrust T and the rocket's weight W, which is $mg = (4.50 \times 10^5 \text{ kg})(9.80 \text{ m/s}^2) = 4.41 \times 10^6$ N. Both of these forces must be considered when determining the net force $\Sigma\mathbf{F}$. The direction of the acceleration is the same as the direction of the net force.

SOLUTION In constructing the free-body diagram for the rocket we choose upward and to the right as the positive directions. The free-body diagram is as follows:

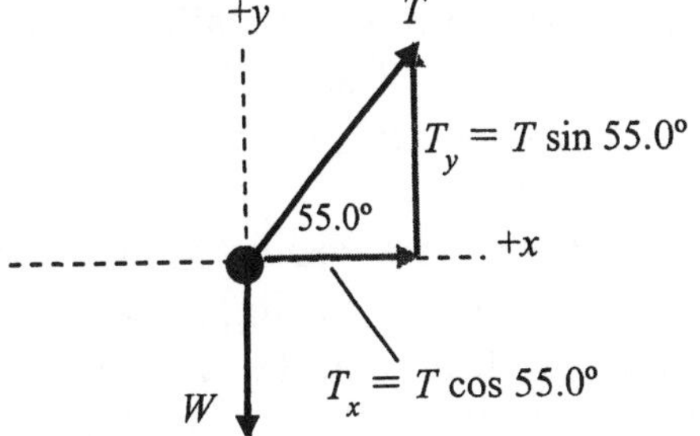

The x component of the net force is

$$\Sigma F_x = T\cos 55.0° = (7.50\times10^6 \text{ N})\cos 55.0° = 4.30\times10^6 \text{ N}$$

The y component of the net force is

$$\Sigma F_y = T\sin 55.0° - W = (7.50\times10^6 \text{ N})\sin 55.0° - 4.41\times10^6 \text{ N} = 1.73\times10^6 \text{ N}$$

The magnitudes of the net force and of the acceleration are

$$\Sigma F = \sqrt{(\Sigma F_x)^2 + (\Sigma F_y)^2}$$

$$a = \frac{\sqrt{(\Sigma F_x)^2 + (\Sigma F_y)^2}}{m} = \frac{\sqrt{(4.30\times10^6 \text{ N})^2 + (1.73\times10^6 \text{ N})^2}}{4.50\times10^5 \text{ kg}} = \boxed{10.3 \text{ m/s}^2}$$

The direction of the acceleration is the same as the direction of the net force. Thus, it is directed above the horizontal at an angle of

$$\theta = \tan^{-1}\left(\frac{\Sigma F_y}{\Sigma F_x}\right) = \tan^{-1}\left(\frac{1.73\times10^6 \text{ N}}{4.30\times10^6 \text{ N}}\right) = \boxed{21.9°}$$

---

73. ***REASONING AND SOLUTION*** The acceleration needed so that the craft touches down with zero velocity is

$$a = \frac{v^2 - v_0^2}{2s} = \frac{-(18.0 \text{ m/s})^2}{2(-165 \text{ m})} = 0.982 \text{ m/s}^2.$$

Newton's second law applied in the vertical direction gives

$$F - mg = ma$$

Then

$$F = m(a + g) = (1.14 \times 10^4 \text{ kg})(0.982 \text{ m/s}^2 + 1.60 \text{ m/s}^2) = \boxed{29\ 400 \text{ N}}$$

---

74. ***REASONING***

a. Just before the instant the crate begins to slide it is in equilibrium, so that the forces acting on the crate must balance, that is, the net force must be zero. There are three forces present, the static frictional force, the normal forçe from the inclined surface, and the weight mg of the crate. The static frictional force holding the crate in place has its maximum possible value. Using Newton's second law with the acceleration equal to zero, we will obtain the static friction coefficient.

b. Once the crate begins to slide, it accelerates under the influence of the net force that results from the following three forces, the kinetic frictional force, the normal force from the inclined surface, and the weight of the crate. Using the net force in Newton's second law, we will determine the acceleration.

SOLUTION

a. In drawing the free-body diagram for the crate we choose the $+x$ axis to be parallel to the ramp surface and downward, the $+y$ direction being perpendicular to the ramp surface. We also use f to symbolize the frictional force. The free-body diagram is as follows, and since the crate is in equilibrium, we find that

$$\Sigma F_x = mg\sin 38.0° - \mu_s F_N = 0$$
$$\Sigma F_y = F_N - mg\cos 38.0° = 0$$

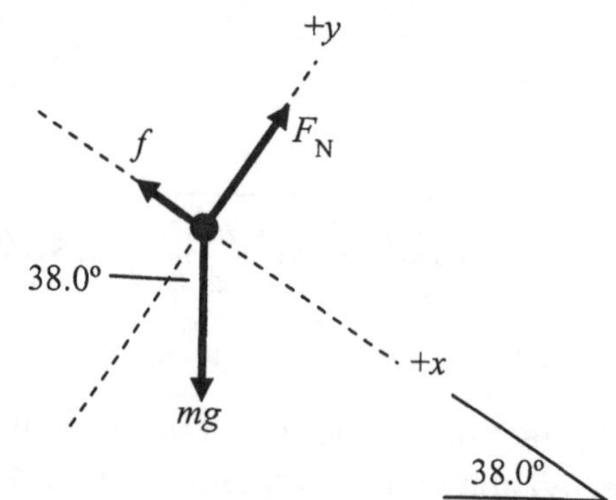

In the first of these expressions we have used Equation 4.7 for f to express the maximum static frictional force. Solving the second equation for the normal force F_N and substituting into the first equation gives

$$mg\sin 38.0° - \mu_s \underbrace{mg\cos 38.0°}_{F_N} = 0 \quad \text{or} \quad \mu_s = \frac{\sin 38.0°}{\cos 38.0°} = \tan 38.0° = \boxed{0.781}$$

b. We again apply Newton's second law, but now only the y component of the acceleration is zero, since the crate accelerates along the x axis. Therefore, x component of the net force equals the mass times the x component of the acceleration. It follows, then, that

$$\Sigma F_x = mg\sin 38.0° - \mu_k F_N = ma_x$$
$$\Sigma F_y = F_N - mg\cos 38.0° = 0$$

In the first of these expressions we have used Equation 4.8 to express the kinetic frictional force. Solving the second equation for F_N and substituting into the first shows that

$$mg\sin 38.0° - \mu_k \underbrace{mg\cos 38.0°}_{F_N} = ma_x$$

$$a_x = g\left(\sin 38.0° - \mu_k \cos 38.0°\right) = \left(9.80\text{ m/s}^2\right)\left[\sin 38.0° - (0.600)\cos 38.0°\right] = \boxed{1.40\text{ m/s}^2}$$

---

75. SSM ***REASONING AND SOLUTION*** Three forces act on the man. They are two upward forces of tension of equal magnitude T, and the force of gravity mg. Therefore, if we take up as the positive direction, Newton's second law gives

$$\Sigma F = 2T - mg = ma$$

Solving for the acceleration a, we find

$$a = \frac{2T - mg}{m} = \frac{2T}{m} - g = \frac{2(358\text{ N})}{72.0\text{ kg}} - 9.80\text{ m/s}^2 = \boxed{0.14\text{ m/s}^2}$$

---

76. ***REASONING AND SOLUTION*** If the +x axis is taken to be parallel to and up the ramp, then $\Sigma F_x = ma_x$ gives

$$T - f_k - mg\sin 30.0° = ma_x$$

where $f_k = \mu_k F_N$. Hence,

$$T = ma_x + \mu_k F_N + mg\sin 30.0° \qquad (1)$$

Also, $\Sigma F_y = ma_y$ gives

$$F_N - mg\cos 30.0° = 0$$

since no acceleration occurs in this direction. Then

$$F_N = mg\cos 30.0° \qquad (2)$$

Substitution of Equation (2) into Equation (1) yields

$$T = ma_x + \mu_k mg\cos 30.0° + mg\sin 30.0°$$

$$T = (205\text{ kg})(0.800\text{ m/s}^2) + (0.900)(205\text{ kg})(9.80\text{ m/s}^2)\cos 30.0° + (205\text{ kg})(9.80\text{ m/s}^2)\sin 30.0° = \boxed{2730\text{ N}}$$

---

77. ***REASONING*** The free-body diagrams for Robin (mass = m) and for the chandelier (mass = M) are given at the right. The tension T in the rope applies an upward force to both. Robin accelerates upward, while the chandelier accelerates downward, each acceleration having the same magnitude. Our solution is based on separate applications of Newton's second law to Robin and the chandelier.

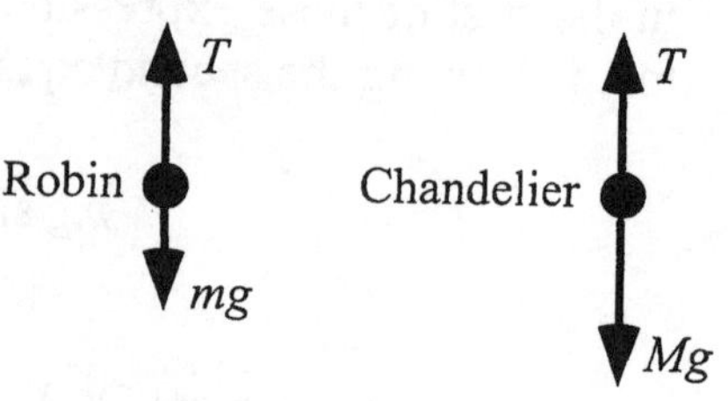

SOLUTION Applying Newton's second law, we find

$$\underbrace{T - mg = ma}_{\text{Robin Hood}} \qquad \text{and} \qquad \underbrace{T - Mg = -Ma}_{\text{Chandelier}}$$

In these applications we have taken upward as the positive direction, so that Robin's acceleration is a, while the chandelier's acceleration is $-a$. Solving the Robin-Hood equation for T gives

$$T = mg + ma$$

Substituting this expression for T into the Chandelier equation gives

$$mg + ma - Mg = -Ma \qquad \text{or} \qquad a = \left(\frac{M-m}{M+m}\right)g$$

a. Robin's acceleration is

$$a = \left(\frac{M-m}{M+m}\right)g = \left[\frac{(220\text{ kg}) - (82\text{ kg})}{(220\text{ kg}) + (82\text{ kg})}\right]\left(9.80\text{ m/s}^2\right) = \boxed{4.5\text{ m/s}^2}$$

b. Substituting the value of a into the expression for T gives

$$T = mg + ma = (82\text{ kg})\left(9.80\text{ m/s}^2 + 4.5\text{ m/s}^2\right) = \boxed{1200\text{ N}}$$

---

78. ***REASONING*** Newton's second law, Equation 4.2a, can be used to find the tension in the coupling between the cars, since the mass and acceleration are known. The tension in the coupling between the 30th and 31st cars is responsible for providing the acceleration for the 20 cars from the 31st to the 50th car. The tension in the coupling between the 49th and 50th cars is responsible only for pulling one car, the 50th.

SOLUTION

a. The tension T between the 30th and 31st cars is

$$T_x = (\text{Mass of 20 cars})a_x \tag{4.2a}$$

$$= (20\text{ cars})\left(6.8\times10^3\text{ kg/car}\right)\left(8.0\times10^{-2}\text{ m/s}^2\right) = \boxed{1.1\times10^4\text{ N}}$$

b. The tension T between the 49th and 50th cars is

$$T_x = (\text{Mass of 1 car})a_x \tag{4.2a}$$

$$= (1\text{ car})\left(6.8\times10^3\text{ kg/car}\right)\left(8.0\times10^{-2}\text{ m/s}^2\right) = \boxed{5.4\times10^2\text{ N}}$$

---

79. SSM ***REASONING*** The box comes to a halt because the kinetic frictional force and the component of its weight parallel to the incline oppose the motion and cause the box to slow down. The distance that the box travels up the incline can be can be found by solving Equation 2.9 $(v^2 = v_0^2 + 2ax)$ for x. Before we use this approach, however, we must first determine the acceleration of the box as it travels along the incline.

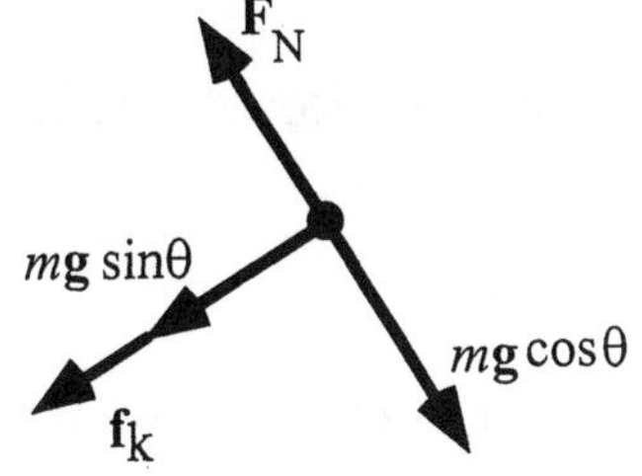

SOLUTION The figure above shows the free-body diagram for the box. It shows the resolved components of the forces that act on the box. If we take the direction up the incline as the positive x direction, then, Newton's second law gives

$$\Sigma F_x = -mg\sin\theta - f_k = ma_x \qquad \text{or} \qquad -mg\sin\theta - \mu_k F_N = ma_x$$

where we have used Equation 4.8, $f_k = \mu_k F_N$. In the y direction we have

$$\Sigma F_y = F_N - mg\cos\theta = 0 \qquad \text{or} \qquad F_N = mg\cos\theta$$

since there is no acceleration in the y direction. Therefore, the equation for the motion in the x direction becomes

$$-mg\sin\theta - \mu_k mg\cos\theta = ma_x \qquad \text{or} \qquad a_x = -g(\sin\theta + \mu_k\cos\theta)$$

According to Equation 2.9, with this value for the acceleration and the fact that $v = 0$ m/s, the distance that the box slides up the incline is

$$x = -\frac{v_0^2}{2a} = \frac{v_0^2}{2g(\sin\theta + \mu_k\cos\theta)} = \frac{(1.50\text{ m/s})^2}{2(9.80\text{ m/s}^2)[\sin 15.0° + (0.180)\cos 15.0°]} = \boxed{0.265\text{ m}}$$

---

80. ***REASONING AND SOLUTION*** If the x axis is taken parallel to the slope with $+x$ down the slope, then $\Sigma F_x = ma_x$ gives

$$F_\text{w} + mg \sin \theta - f_\text{k} = ma_x$$

where $f_\text{k} = \mu_\text{k} F_\text{N}$ and F_w is the force exerted by the wind on the person and sled. $\Sigma F_y = ma_y$ gives

$$F_\text{N} - mg \cos \theta = 0$$

since there is no acceleration of the sled in this direction. Hence,

$$F_\text{N} = mg \cos \theta$$

Substitution of this into the above result gives

$$a_x = F_\text{w}/m + g \sin \theta - \mu_\text{k} g \cos \theta$$

$$a_x = (105\text{ N})/(65.0\text{ kg}) + (9.80\text{ m/s}^2)\sin 30.0° - (0.150)(9.80\text{ m/s}^2)\cos 30.0° = 5.24\text{ m/s}^2$$

The time required for the sled to travel a distance, x, subject to this acceleration is found from

$$x = v_0 t + (1/2)a_x t^2$$

Therefore, and using the fact that $v_0 = 0$ m/s, the time is

$$t = \sqrt{\frac{2x}{a_x}} = \sqrt{\frac{2(175\text{ m})}{5.24\text{ m/s}^2}} = \boxed{8.17\text{ s}}$$

---

81. ***REASONING*** The tension in each coupling bar is responsible for accelerating the objects behind it. The masses of the cars are m_1, m_2, and m_3. We can use Newton's second law to express the tension in each coupling bar, since friction is negligible:

$$\underbrace{T_A = (m_1 + m_2 + m_3)a}_{\text{Coupling bar A}} \qquad \underbrace{T_B = (m_2 + m_3)a}_{\text{Coupling bar B}} \qquad \underbrace{T_C = m_3 a}_{\text{Coupling bar C}}$$

In these expressions $a = 0.12\text{ m/s}^2$ remains constant. Consequently, the tension in a given bar will change only if the total mass of the objects accelerated by that bar changes as a result of the luggage transfer. Using Δ (Greek capital delta) to denote a change in the usual fashion, we can express the changes in the above tensions as follows:

$$\underbrace{\Delta T_A = \left[\Delta\left(m_1 + m_2 + m_3\right)\right]a}_{\text{Coupling bar A}} \qquad \underbrace{\Delta T_B = \left[\Delta\left(m_2 + m_3\right)\right]a}_{\text{Coupling bar B}} \qquad \underbrace{\Delta T_C = \left(\Delta m_3\right)a}_{\text{Coupling bar C}}$$

SOLUTION

a. Moving luggage from car 2 to car 1 does not change the total mass $m_1 + m_2 + m_3$, so $\Delta(m_1 + m_2 + m_3) = 0$ kg and $\boxed{\Delta T_A = 0\text{ N}}$.

The transfer from car 2 to car 1 causes the total mass $m_2 + m_3$ to decrease by 39 kg, so $\Delta(m_2 + m_3) = -39$ kg and

$$\Delta T_B = \left[\Delta\left(m_2 + m_3\right)\right]a = (-39\text{ kg})\left(0.12\text{ m/s}^2\right) = \boxed{-4.7\text{ N}}$$

The transfer from car 2 to car 1 does not change the mass m_3, so $\Delta m_3 = 0$ kg and $\boxed{\Delta T_C = 0\text{ N}}$.

b. Moving luggage from car 2 to car 3 does not change the total mass $m_1 + m_2 + m_3$, so $\Delta(m_1 + m_2 + m_3) = 0$ kg and $\boxed{\Delta T_A = 0\text{ N}}$.

The transfer from car 2 to car 3 does not change the total mass $m_2 + m_3$, so $\Delta(m_2 + m_3) = 0$ kg and $\boxed{\Delta T_B = 0\text{ N}}$.

The transfer from car 2 to car 3 causes the mass m_3 to increase by 39 kg, so $\Delta m_3 = +39$ kg and

$$\Delta T_C = \left(\Delta m_3\right)a = (+39\text{ kg})\left(0.12\text{ m/s}^2\right) = \boxed{+4.7\text{ N}}$$

---

82. ***REASONING AND SOLUTION*** The distance required for the truck to stop is found from

$$x = \frac{v^2 - v_0^2}{2a} = \frac{(0\text{ m/s})^2 - v_0^2}{2a}$$

The acceleration of the truck is needed. The frictional force decelerates the crate. The maximum force that friction can supply is

$$f_s^{\text{MAX}} = \mu_s F_N = \mu_s mg$$

Newton's second law requires that

$$f_s^{MAX} = -\,ma \text{ so } a = -\,\mu_s g$$

Now the stopping distance is

$$x = \frac{v_0^2}{2\mu_s g} = \frac{(25 \text{ m/s})^2}{2(0.650)\left(9.80 \text{ m/s}^2\right)} = \boxed{49.1 \text{ m}}$$

---

83. **SSM** ***REASONING AND SOLUTION*** The penguin comes to a halt on the horizontal surface because the kinetic frictional force opposes the motion and causes it to slow down. The time required for the penguin to slide to a halt ($v = 0$ m/s) after entering the horizontal patch of ice is, according to Equation 2.4,

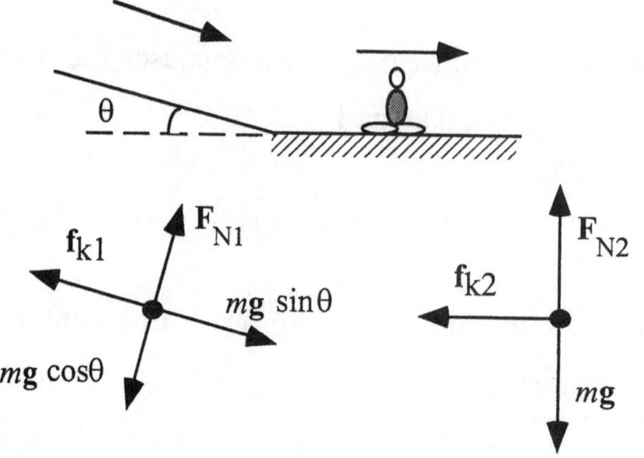

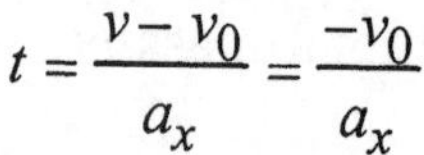

$$t = \frac{v - v_0}{a_x} = \frac{-v_0}{a_x}$$

Free-body diagram A   Free-body diagram B

We must, therefore, determine the acceleration of the penguin as it slides along the horizontal patch.

For the penguin sliding on the horizontal patch of ice, we find from free-body diagram B and Newton's second law in the x direction (motion to the right is taken as positive) that

$$\Sigma F_x = -f_{k2} = ma_x \qquad \text{or} \qquad a_x = \frac{-f_{k2}}{m} = \frac{-\mu_k F_{N2}}{m}$$

In the y direction in free-body diagram B, we have $\Sigma F_y = F_{N2} - mg = 0$, or $F_{N2} = mg$. Therefore, the acceleration of the penguin is

$$a_x = \frac{-\mu_k mg}{m} = -\mu_k g \qquad (1)$$

Equation (1) indicates that, in order to find the acceleration a_x, we must find the coefficient of kinetic friction.

We are told in the problem statement that the coefficient of kinetic friction between the penguin and the ice is the same for the incline as for the horizontal patch. Therefore, we can use the motion of the penguin on the incline to determine the coefficient of friction and use it in Equation (1).

For the penguin sliding down the incline, we find from free-body diagram A and Newton's second law (taking the direction of motion as positive) that

$$\Sigma F_x = mg\sin\theta - f_{k1} = ma_x = 0 \qquad \text{or} \qquad f_{k1} = mg\sin\theta \qquad (2)$$

Here, we have used the fact that the penguin slides down the incline with a constant velocity, so that it has zero acceleration. From Equation 4.8, we know that $f_{k1} = \mu_k F_{N1}$. Applying Newton's second law in the direction perpendicular to the incline, we have

$$\Sigma F_y = F_{N1} - mg\cos\theta = 0 \qquad \text{or} \qquad F_{N1} = mg\cos\theta$$

Therefore, $f_{k1} = \mu_k mg\cos\theta$, so that according to Equation (2), we find

$$f_{k1} = \mu_k mg\cos\theta = mg\sin\theta$$

Solving for the coefficient of kinetic friction, we have

$$\mu_k = \frac{\sin\theta}{\cos\theta} = \tan\theta$$

Finally, the time required for the penguin to slide to a halt after entering the horizontal patch of ice is

$$t = \frac{-v_0}{a_x} = \frac{-v_0}{-\mu_k g} = \frac{v_0}{g\tan\theta} = \frac{1.4 \text{ m/s}}{(9.80 \text{ m/s}^2)\tan 6.9°} = \boxed{1.2 \text{ s}}$$

---

84. ***REASONING AND SOLUTION***

a. The rope exerts a tension, T, acting upward on each block. Applying Newton's second law to the lighter block (block 1) gives

$$T - m_1 g = m_1 a$$

Similarly, for the heavier block (block 2)

$$T - m_2 g = -m_2 a$$

Subtracting the second equation from the first and rearranging yields

$$a = \left(\frac{m_2 - m_1}{m_2 + m_1}\right) g = \boxed{3.68 \text{ m/s}^2}$$

b. The tension in the rope is now 908 N since the tension is the reaction to the applied force exerted by the hand. Newton's second law applied to the block is

$$T - m_1 g = m_1 a$$

Solving for a gives

$$a = \frac{T}{m_1} - g = \frac{(908\text{ N})}{42.0\text{ kg}} - 9.80\text{ m/s}^2 = \boxed{11.8\text{ m/s}^2}$$

c. In the first case, the inertia of BOTH blocks affects the acceleration whereas, in the second case, only the lighter block's inertia remains.

---

85. ***REASONING AND SOLUTION***

a. Newton's second law for block 1 (10.0 kg) is

$$T = m_1 a \qquad (1)$$

Block 2 (3.00 kg) has two ropes attached each carrying a tension T. Also, block 2 only travels half the distance that block 1 travels in the same amount of time so its acceleration is only half of block 1's acceleration. Newton's second law for block 2 is then

$$2T - m_2 g = -(1/2)m_2 a \qquad (2)$$

Solving Equation (1) for a, substituting into Equation (2), and rearranging gives

$$T = \frac{\frac{1}{2}m_2 g}{1 + \frac{1}{4}(m_2/m_1)} = \boxed{13.7\text{ N}}$$

b. Using this result in Equation (1) yields

$$a = \frac{T}{m_1} = \frac{13.7\text{ N}}{10.0\text{ kg}} = \boxed{1.37\text{ m/s}^2}$$

---

86. ***REASONING AND SOLUTION***

a. The static frictional force is responsible for accelerating the top block so that it does not slip against the bottom one. The maximum force that can be supplied by friction is

$$f_s^{\text{MAX}} = \mu_s F_N = \mu_s m_1 g$$

Newton's second law requires that $f_s^{\text{MAX}} = m_1 a$, so

$$a = \mu_s g$$

The force necessary to cause BOTH blocks to have this acceleration is

$$F = (m_1 + m_2)a = (m_1 + m_2)\mu_s g$$

$$F = (5.00\text{ kg} + 12.0\text{ kg})(0.600)(9.80\text{ m/s}^2) = \boxed{1.00 \times 10^2\text{ N}}$$

b. The maximum acceleration that the two block combination can have before slipping occurs is

$$a = F/(17.0\text{ kg})$$

Newton's second law applied to the 5.00 kg block is

$$F - \mu_s m_1 g = m_1 a = (5.00\text{ kg})(F)/(17.0\text{ kg})$$

Hence

$$F = \boxed{41.6\text{ N}}$$

---

87. **SSM** ***REASONING AND SOLUTION***

a. Combining Equations 4.4 and 4.5, we see that the acceleration due to gravity on the surface of Saturn can be calculated as follows:

$$g_{\text{Saturn}} = G\frac{M_{\text{Saturn}}}{r^2_{\text{Saturn}}} = \left(6.67 \times 10^{-11}\text{ N}\cdot\text{m}^2/\text{kg}^2\right)\frac{\left(5.67 \times 10^{26}\text{ kg}\right)}{(6.00 \times 10^7\text{ m})^2} = \boxed{10.5\text{ m/s}^2}$$

b. The ratio of the person's weight on Saturn to that on earth is

$$\frac{W_{\text{Saturn}}}{W_{\text{earth}}} = \frac{mg_{\text{Saturn}}}{mg_{\text{earth}}} = \frac{g_{\text{Saturn}}}{g_{\text{earth}}} = \frac{10.5\text{ m/s}^2}{9.80\text{ m/s}^2} = \boxed{1.07}$$

---

88. ***REASONING*** We proceed in two steps. First we use Newton's second law to determine the total resistive force R that acts on the skier during the accelerated motion. Then, recognizing that motion at a constant velocity is an example of equilibrium, we apply Newton's second law a second time with the acceleration equal to zero and conclude that the net force acting on the skier must be zero. Since the net force in the horizontal direction is comprised of the pulling force and the resistive force, we conclude that the pulling force must balance the resistive force.

SOLUTION In constructing the free-body diagram we choose the horizontal direction to the right as the positive x direction. We also ignore forces in the vertical direction, since they do not play a role in the solution. The free-body diagram is as follows. Applying Newton's second law, we find that

$$\Sigma F_x = 520\text{ N} - R = ma_x$$

$$R = 520\text{ N} - ma_x = 520\text{ N} - (75\text{ kg})\left(2.4\text{ m/s}^2\right) = 340\text{ N}$$

R 520 N +x

When the skier is pulled at a constant velocity, the

pulling force is P. Since the acceleration is now zero, Newton's second law gives

$$\Sigma F_x = P - R = 0 \quad \text{or} \quad P = R = \boxed{340\text{ N}}$$

---

89. SSM ***REASONING*** The book is kept from falling as long as the total static frictional force balances the weight of the book. The forces that act on the book are shown in the following free-body diagram, where P is the pressing force applied by each hand.

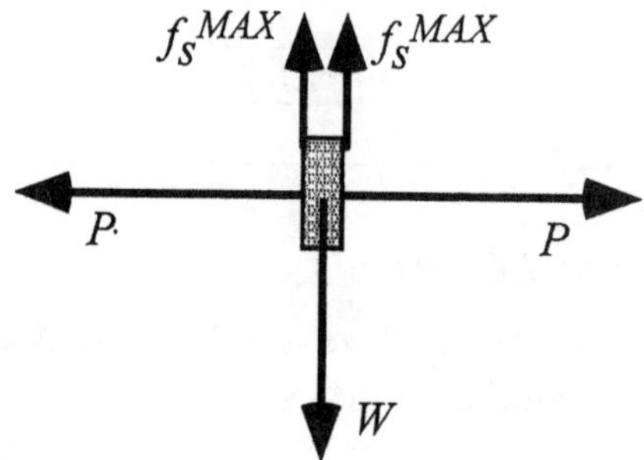

In this diagram, note that there are two pressing forces, one from each hand. Each hand also applies a static frictional force, and, therefore, two static frictional forces are shown. The maximum static frictional force is related in the usual way to a normal force F_N, but in this problem the normal force is provided by the pressing force, so that $F_N = P$.

SOLUTION Since the frictional forces balance the weight, we have

$$2f_s^{MAX} = 2(\mu_s F_N) = 2(\mu_s P) = W$$

Solving for P, we find that

$$P = \frac{W}{2\mu_s} = \frac{31\text{ N}}{2(0.40)} = \boxed{39\text{ N}}$$

---

90. ***REASONING*** The magnitudes of the initial ($v_0 = 0$ m/s) and final ($v = 805$ m/s) velocities are known. In addition, data is given for the mass and the thrust, so that Newton's second law can be used to determine the acceleration of the probe. Therefore, kinematics Equation 2.4 ($v = v_0 + at$) can be used to determine the time t.

SOLUTION Solving Equation 2.4 for the time gives

$$t = \frac{v - v_0}{a}$$

Newton's second law gives the acceleration as $a = (\Sigma F)/m$. Using this expression in Equation 2.4 gives

$$t = \frac{(v - v_0)}{(\Sigma F)/m} = \frac{m(v - v_0)}{\Sigma F} = \frac{(474\text{ kg})(805\text{ m/s} - 0\text{ m/s})}{56 \times 10^{-3}\text{ N}} = 6.8 \times 10^6\text{ s}$$

Since one day contains 8.64×10^4 s, the time is

$$t = \left(6.8 \times 10^6\ \cancel{\text{s}}\right)\frac{1\text{ day}}{8.64 \times 10^4\ \cancel{\text{s}}} = \boxed{79\text{ days}}$$

---

91. SSM ***REASONING*** In order to start the crate moving, an external agent must supply a force that is at least as large as the maximum value $f_s^{\text{MAX}} = \mu_s F_N$, where μ_s is the coefficient of static friction (see Equation 4.7). Once the crate is moving, the magnitude of the frictional force is very nearly constant at the value $f_k = \mu_k F_N$, where μ_k is the coefficient of kinetic friction (see Equation 4.8). In both cases described in the problem statement, there are only two vertical forces that act on the crate; they are the upward normal force F_N, and the downward pull of gravity (the weight) mg. Furthermore, the crate has no vertical acceleration in either case. Therefore, if we take upward as the positive direction, Newton's second law in the vertical direction gives $F_N - mg = 0$, and we see that, in both cases, the magnitude of the normal force is $F_N = mg$.

SOLUTION

a. Therefore, the applied force needed to start the crate moving is

$$f_s^{\text{MAX}} = \mu_s mg = (0.760)(60.0\text{ kg})(9.80\text{ m/s}^2) = \boxed{447\text{ N}}$$

b. When the crate moves in a straight line at constant speed, its velocity does not change, and it has zero acceleration. Thus, Newton's second law in the horizontal direction becomes $P - f_k = 0$, where P is the required pushing force. Thus, the applied force required to keep the crate sliding across the dock at a constant speed is

$$P = f_k = \mu_k mg = (0.410)(60.0\text{ kg})(9.80\text{ m/s}^2) = \boxed{241\text{ N}}$$

---

92. ***REASONING AND SOLUTION*** From Newton's second law and the equation: $v = v_0 + at$, we have

$$F = ma = m\frac{v - v_0}{t}$$

a. When the skier accelerates from rest (v_0 = 0 m/s) to a speed of 11 m/s in 8.0 s, the required net force is

$$F = m\frac{v - v_0}{t} = (73\text{ kg})\frac{(11\text{ m/s}) - 0\text{ m/s}}{8.0\ s} = \boxed{1.0 \times 10^2\text{ N}}$$

b. When the skier lets go of the tow rope and glides to a halt (v = 0 m/s) in 21 s, the net force acting on the skier is

$$F = m\frac{v - v_0}{t} = (73\text{ kg})\frac{0\text{ m/s} - (11\text{ m/s})}{21\text{ s}} = -38\text{ N}$$

The magnitude of the net force is $\boxed{38\text{ N}}$.

---

93. ***REASONING AND SOLUTION***

a. A free-body diagram for the crate gives

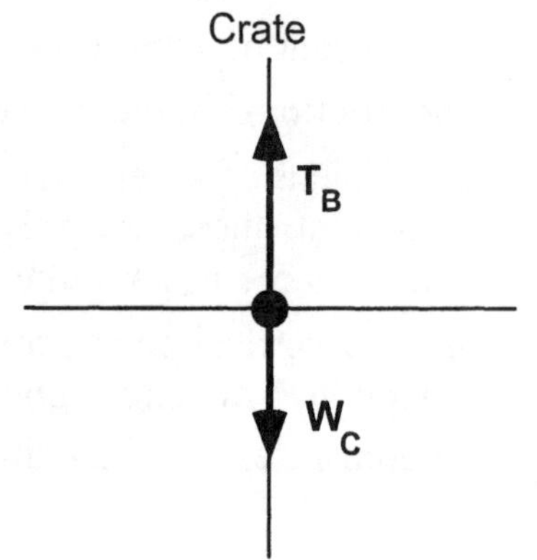

$$T_\text{B} - W_\text{C} = m_\text{C}a$$

$$T_\text{B} = W_\text{C} + m_\text{C}a$$

$$T_\text{B} = 1510\text{ N} + \left(\frac{1510\text{ N}}{9.80\text{ m/s}^2}\right)\left(0.620\text{ m/s}^2\right) = \boxed{1610\text{ N}}$$

b. An analysis of the free-body diagram for the platform yields

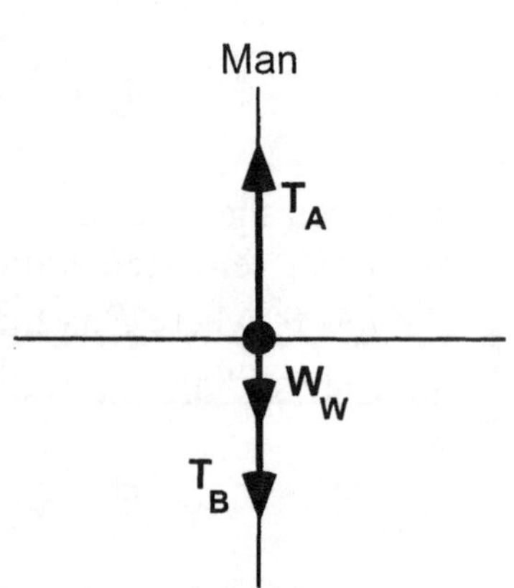

$$T_\text{A} - T_\text{B} - W_\text{W} = m_\text{W}a$$

$$T_\text{A} = T_\text{B} + W_\text{W} + m_\text{W}a$$

$$T_\text{A} = 1610\text{ N} + \left(\frac{965\text{ N}}{9.80\text{ m/s}^2}\right)\left(0.620\text{ m/s}^2\right) + 965\text{ N}$$

$$= \boxed{2640\text{ N}}$$

---

94. ***REASONING AND SOLUTION***
a. The apparent weight of the person is given by Equation 4.6 as

$$F_N = mg + ma$$
$$= (95.0 \text{ kg})(9.80 \text{ m/s}^2 + 1.80 \text{ m/s}^2) = \boxed{1.10 \times 10^3 \text{ N}}$$

b. $$F_N = (95.0 \text{ kg})(9.80 \text{ m/s}^2) = \boxed{931 \text{ N}}$$

c. $$F_N = (95.0 \text{ kg})(9.80 \text{ m/s}^2 - 1.30 \text{ m/s}^2) = \boxed{808 \text{ N}}$$

---

95. SSM ***REASONING*** We can use the appropriate equation of kinematics to find the acceleration of the bullet. Then Newton's second law can be used to find the average net force on the bullet.

SOLUTION According to Equation 2.4, the acceleration of the bullet is

$$a = \frac{v - v_0}{t} = \frac{715 \text{ m/s} - 0 \text{ m/s}}{2.50 \times 10^{-3} \text{ s}} = 2.86 \times 10^5 \text{ m/s}^2$$

Therefore, the net average force on the bullet is

$$\Sigma F = ma = (15 \times 10^{-3} \text{ kg})(2.86 \times 10^5 \text{ m/s}^2) = \boxed{4290 \text{ N}}$$

---

96. ***REASONING AND SOLUTION*** The acceleration is obtained from

$$x = v_0 t + \frac{1}{2} a t^2$$

where $v_0 = 0$ m/s. So

$$a = 2x/t^2$$

Newton's second law gives

$$\Sigma F = ma = m\left(\frac{2x}{t^2}\right) = (72 \text{ kg})\left[\frac{2(18 \text{ m})}{(0.95 \text{ s})^2}\right] = \boxed{2900 \text{ N}}$$

---

97. ***REASONING*** The magnitude of the gravitational force exerted on the satellite by the earth is given by Equation 4.3 as $F = Gm_{\text{satellite}} m_{\text{earth}} / r^2$, where r is the distance between the satellite and the center of the earth. This expression also gives the magnitude of the gravitational force exerted on the earth by the satellite. According to Newton's second law, the magnitude of the earth's acceleration is equal to the magnitude of the gravitational force

exerted on it divided by its mass. Similarly, the magnitude of the satellite's acceleration is equal to the magnitude of the gravitational force exerted on it divided by its mass.

SOLUTION

a. The magnitude of the gravitational force exerted on the satellite when it is a distance of two earth radii from the center of the earth is

$$F = \frac{Gm_{\text{satellite}}m_{\text{earth}}}{r^2} = \frac{\left(6.67\times10^{-11}\ \text{N}\cdot\text{m}^2/\text{kg}^2\right)\left(425\ \text{kg}\right)\left(5.98\times10^{24}\ \text{kg}\right)}{\left[(2)\left(6.38\times10^{6}\ \text{m}\right)\right]^2} = \boxed{1.04\times10^{3}\ \text{N}}$$

b. The magnitude of the gravitational force exerted on the earth when it is a distance of two earth radii from the center of the satellite is

$$F = \frac{Gm_{\text{satellite}}m_{\text{earth}}}{r^2} = \frac{\left(6.67\times10^{-11}\ \text{N}\cdot\text{m}^2/\text{kg}^2\right)\left(425\ \text{kg}\right)\left(5.98\times10^{24}\ \text{kg}\right)}{\left[(2)\left(6.38\times10^{6}\ \text{m}\right)\right]^2} = \boxed{1.04\times10^{3}\ \text{N}}$$

c. The acceleration of the satellite can be obtained from Newton's second law.

$$a_{\text{satellite}} = \frac{F}{m_{\text{satellite}}} = \frac{1.04\times10^{3}\ \text{N}}{425\ \text{kg}} = \boxed{2.45\ \text{m/s}^2}$$

d. The acceleration of the earth can also be obtained from Newton's second law.

$$a_{\text{earth}} = \frac{F}{m_{\text{earth}}} = \frac{1.04\times10^{3}\ \text{N}}{5.98\times10^{24}\ \text{kg}} = \boxed{1.74\times10^{-22}\ \text{m/s}^2}$$

---

98. ***REASONING*** Each particle experiences two gravitational forces, one due to each of the remaining particles. To get the net gravitational force, we must add the two contributions, taking into account the directions. The magnitude of the gravitational force that any one particle exerts on another is given by Newton's law of gravitation as $F = Gm_1m_2/r^2$. Thus, for particle A, we need to apply this law to its interaction with particle B and with particle C. For particle B, we need to apply the law to its interaction with particle A and with particle C. Lastly, for particle C, we must apply the law to its interaction with particle A and with particle B. In considering the directions, we remember that the gravitational force between two particles is always a force of attraction.

SOLUTION We begin by calculating the magnitude of the gravitational force for each pair of particles:

$$F_{AB} = \frac{Gm_A m_B}{r^2} = \frac{\left(6.67 \times 10^{-11}\ \text{N}\cdot\text{m}^2/\text{kg}^2\right)(363\ \text{kg})(517\ \text{kg})}{(0.500\ \text{m})^2} = 5.007 \times 10^{-5}\ \text{N}$$

$$F_{BC} = \frac{Gm_B m_C}{r^2} = \frac{\left(6.67 \times 10^{-11}\ \text{N}\cdot\text{m}^2/\text{kg}^2\right)(517\ \text{kg})(154\ \text{kg})}{(0.500\ \text{m})^2} = 8.497 \times 10^{-5}\ \text{N}$$

$$F_{AC} = \frac{Gm_A m_C}{r^2} = \frac{\left(6.67 \times 10^{-11}\ \text{N}\cdot\text{m}^2/\text{kg}^2\right)(363\ \text{kg})(154\ \text{kg})}{(0.500\ \text{m})^2} = 6.629 \times 10^{-6}\ \text{N}$$

In using these magnitudes we take the direction to the right as positive.

a. Both particles B and C attract particle A to the right, the net force being

$$F_A = F_{AB} + F_{AC} = 5.007 \times 10^{-5}\ \text{N} + 6.629 \times 10^{-6}\ \text{N} = \boxed{5.67 \times 10^{-5}\ \text{N, right}}$$

b. Particle C attracts particle B to the right, while particle A attracts particle B to the left, the net force being

$$F_B = F_{BC} - F_{AB} = 8.497 \times 10^{-5}\ \text{N} - 5.007 \times 10^{-5}\ \text{N} = \boxed{3.49 \times 10^{-5}\ \text{N, right}}$$

c. Both particles A and B attract particle C to the left, the net force being

$$F_C = F_{AC} + F_{BC} = 6.629 \times 10^{-6}\ \text{N} + 8.497 \times 10^{-5}\ \text{N} = \boxed{9.16 \times 10^{-5}\ \text{N, left}}$$

---

99. SSM ***REASONING AND SOLUTION***

The system is shown in the drawing. We will let $m_1 = 21.0$ kg, and $m_2 = 45.0$ kg. Then, m_1 will move upward, and m_2 will move downward. There are two forces that act on each object; they are the tension T in the cord and the weight mg of the object. The forces are shown in the free-body diagrams at the far right.

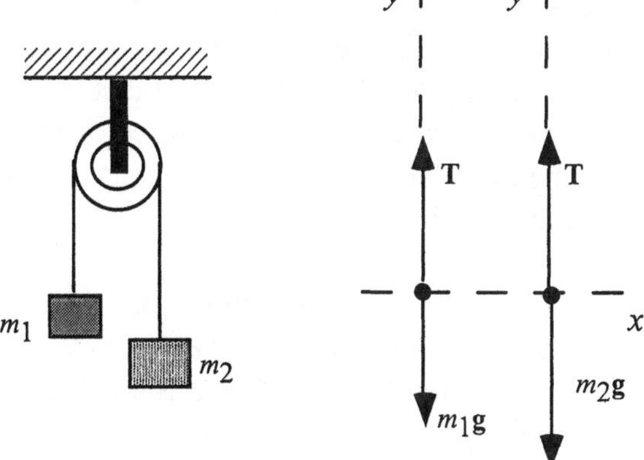

We will take up as the positive direction. If the acceleration of m_1 is a, then the acceleration of m_2 must be $-a$.

From Newton's second law, we have for m_1

$$\Sigma F_y = T - m_1 g = m_1 a \qquad (1)$$

and for m_2

$$\Sigma F_y = T - m_2 g = -m_2 a \qquad (2)$$

a. Eliminating T between these two equations, we obtain

$$a = \frac{m_2 - m_1}{m_2 + m_1} g = \left(\frac{45.0\text{ kg}-21.0\text{ kg}}{45.0\text{ kg}+21.0\text{ kg}} \right)(9.80\text{ m/s}^2) = \boxed{3.56\text{ m/s}^2}$$

b. Eliminating a between Equations (1) and (2), we find

$$T = \frac{2m_1 m_2}{m_1 + m_2} g = \left[\frac{2(21.0\text{ kg})(45.0\text{ kg})}{21.0\text{ kg}+45.0\text{ kg}} \right](9.80\text{ m/s}^2) = \boxed{281\text{ N}}$$

---

100. ***REASONING AND SOLUTION*** The deceleration produced by the frictional force is

$$a = -\frac{f_k}{m} = \frac{-\mu_k mg}{m} = -\mu_k g$$

The time it takes for the car to come to a halt is given by Equation 2.4 as

$$t = \frac{v - v_0}{a} = \frac{v - v_0}{-\mu_k g} = \frac{0\text{ m/s} - 16.1\text{ m/s}}{-(0.720)(9.80\text{ m/s}^2)} = \boxed{2.28\text{ s}}$$

---

101. ***REASONING*** The toboggan has a constant velocity, so it has no acceleration and is in equilibrium. Therefore, the forces acting on the toboggan must balance, that is, the net force acting on the toboggan must be zero. There are three forces present, the kinetic frictional force, the normal force from the inclined surface, and the weight mg of the toboggan. Using Newton's second law with the acceleration equal to zero, we will obtain the kinetic friction coefficient.

SOLUTION In drawing the free-body diagram for the toboggan we choose the $+x$ axis to be parallel to the hill surface and downward, the $+y$ direction being perpendicular to the hill surface. We also use f_k to symbolize the frictional force. Since the toboggan is in equilibrium, the zero net force components in the x and y directions are

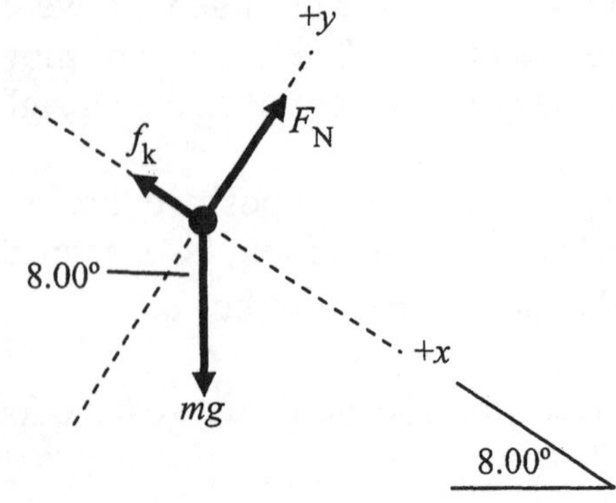

$$\Sigma F_x = mg \sin 8.00° - \mu_k F_N = 0$$

$$\Sigma F_y = F_N - mg \cos 8.00° = 0$$

In the first of these expressions we have used Equation 4.8 for f_k to express the kinetic frictional force. Solving the second equation for the normal force F_N and substituting into the first equation gives

$$mg \sin 8.00° - \mu_k \underbrace{mg \cos 8.00°}_{F_N} = 0 \quad \text{or} \quad \mu_k = \frac{\sin 8.00°}{\cos 8.00°} = \tan 8.00° = \boxed{0.141}$$

---

102. ***REASONING*** The drawing shows the bicycle (represented as a circle) moving down the hill. Since the bicycle is moving at a constant velocity, its acceleration is zero and it is in equilibrium. Choosing the x axis to be parallel to the hill, Equation 4.9a states that $\sum F_x = 0$, so the net force along the x axis is zero. This relation will allow us to find the value of the numerical constant c that appears in the expression for $\mathbf{f}_{\text{air}}$.

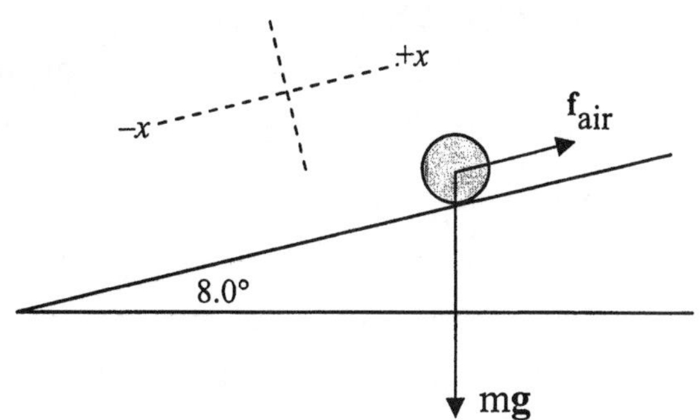

SOLUTION Taking up the hill as the $+x$ direction, the x-component of the weight is −mg sin 8.0°, and the force due to air resistance is $+f_{\text{air}} = +cv$, where the plus sign indicates that this force points opposite to the motion of the bicycle, or up the hill. Equation 4.9a can be written as

$$\underbrace{+cv - mg \sin 8.0°}_{\Sigma F_x} = 0$$

Solving this equation for the constant c gives

$$c = \frac{mg \sin 8.0°}{v} = \frac{(85 \text{ kg})(9.80 \text{ m/s}^2)\sin 8.0°}{8.9 \text{ m/s}} = \boxed{13 \text{ kg/s}}$$

---

103. SSM ***REASONING*** The shortest time to pull the person from the cave corresponds to the maximum acceleration, a_y, that the rope can withstand. We first determine this acceleration and then use kinematic Equation 3.5b ($y = v_{0y}t + \frac{1}{2} a_y t^2$) to find the time t.

SOLUTION As the person is being pulled from the cave, there are two forces that act on him; they are the tension T in the rope that points vertically upward, and the weight of the person mg that points vertically downward. Thus, if we take upward as the positive direction, Newton's second law gives $\Sigma F_y = T - mg = ma_y$. Solving for a_y, we have

$$a_y = \frac{T}{m} - g = \frac{T}{W/g} - g = \frac{569\text{ N}}{(5.20\times 10^2\text{ N})/(9.80\text{ m/s}^2)} - 9.80\text{ m/s}^2 = 0.92\text{ m/s}^2$$

Therefore, from Equation 3.5b with $v_{0y} = 0$ m/s, we have $y = \frac{1}{2}a_y t^2$. Solving for t, we find

$$t = \sqrt{\frac{2y}{a_y}} = \sqrt{\frac{2(35.1\text{ m})}{0.92\text{ m/s}^2}} = \boxed{8.7\text{ s}}$$

---

104. ***REASONING*** According to Newton's second law, the acceleration of the probe is $a = \Sigma F/m$. Using this value for the acceleration in Equation 2.8 and noting that the probe starts from rest ($v_0 = 0$ m/s), we can write the distance traveled by the probe as

$$x = v_0 t + \frac{1}{2}at^2 = \frac{1}{2}\left(\frac{\Sigma F}{m}\right)t^2$$

This equation is the basis for our solution.

SOLUTION Since each engine produces the same amount of force or thrust T, the net force is $\Sigma F = 2T$ when the engines apply their forces in the same direction and $\Sigma F = \sqrt{T^2 + T^2} = \sqrt{2}T$ when they apply their forces perpendicularly. Thus, we write the distances traveled in the two situations as follows:

$$\underbrace{x = \frac{1}{2}\left(\frac{2T}{m}\right)t^2}_{\text{Engines fired in same direction}} \quad \text{and} \quad \underbrace{x = \frac{1}{2}\left(\frac{\sqrt{2}T}{m}\right)t_\perp^2}_{\text{Engines fired perpendicularly}}$$

Since the distances are the same, we have

$$\frac{1}{2}\left(\frac{2T}{m}\right)t^2 = \frac{1}{2}\left(\frac{\sqrt{2}T}{m}\right)t_\perp^2 \quad \text{or} \quad \sqrt{2}\,t^2 = t_\perp^2$$

The firing time when the engines apply their forces perpendicularly is, then,

$$t_\perp = \left(\sqrt[4]{2}\right)t = \left(\sqrt[4]{2}\right)(28\text{ s}) = \boxed{33\text{ s}}$$

---

105. ***REASONING AND SOLUTION*** The figure to the left below shows the forces that act on the sports car as it accelerates up the hill. The figure to the right below shows these forces resolved into components parallel to and perpendicular to the line of motion. Forces pointing up the hill will be taken as positive.

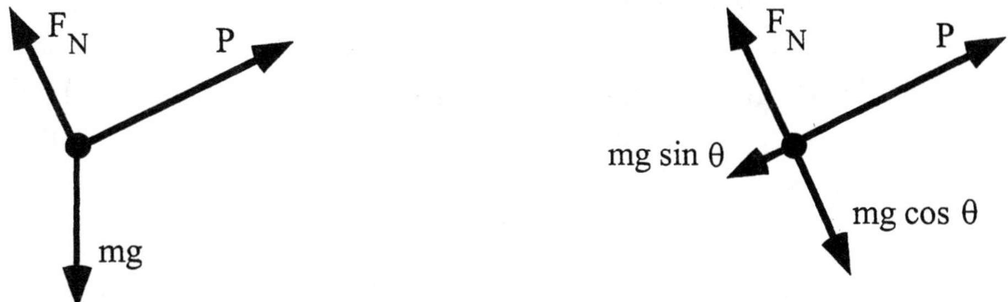

a. The acceleration will be a maximum when $P = f_s^{max}$. From the forces along the line of motion:

$$f_s^{max} - mg \sin \theta = ma$$

The force f_s^{max} is equal to $\mu_s F_N$. The normal force can be found from the forces perpendicular to the line of motion:

$$F_N = mg \cos \theta$$

Then

$$\mu_s (mg \cos \theta) - mg \sin \theta = ma$$

and

$$a = g(\mu_s \cos\theta - \sin\theta) = (9.80 \text{ m/s}^2)(0.88 \cos 18° - \sin 18°) = \boxed{5.2 \text{ m/s}^2}$$

b. When the car is being driven downhill, P (= f_s^{max}) now points down the hill in the same direction as (mg sin θ). Taking the *direction of motion* as positive, we have

$$f_s^{max} + mg \sin \theta = ma$$

Following the same steps as above we obtain

$$\mu_s (mg \cos \theta) + mg \sin \theta = ma$$

and

$$a = g(\mu_s \cos\theta + \sin\theta) = (9.80 \text{ m/s}^2)(0.88 \cos 18° + \sin 18°) = \boxed{11 \text{ m/s}^2}$$

106. ***REASONING AND SOLUTION*** According to Equations 4.4 and 4.5, the acceleration due to gravity at the surface of the neutron star is

$$a = \frac{Gm}{r^2} = \frac{(6.67\times10^{-11}\,\mathrm{N\cdot m^2/kg^2})(2.0\times10^{30}\,\mathrm{kg})}{(5.0\times10^{3}\,\mathrm{m})^{2}} = 5.3\times10^{12}\ \mathrm{m/s^2}$$

Since the gravitational force is assumed to be constant, the acceleration will be constant and the speed of the object can be calculated from $v^2 = v_0^2 + 2ay$, with $v_0 = 0\ \mathrm{m/s}$ since the object falls from rest. Solving for v yields

$$v = \sqrt{2ay} = \sqrt{2\left(5.3\times10^{12}\ \mathrm{m/s^2}\right)\left(0.010\ \mathrm{m}\right)} = \boxed{3.3\times10^{5}\ \mathrm{m/s}}$$

---

107. SSM ***REASONING*** There are four forces that act on the chandelier; they are the forces of tension T in each of the three wires, and the downward force of gravity *mg*. Under the influence of these forces, the chandelier is at rest and, therefore, in equilibrium. Consequently, the sum of the *x* components as well as the sum of the *y* components of the forces must each be zero. The figure below shows the free-body diagram for the chandelier and the force components for a suitable system of *x*, *y* axes. Note that the free-body diagram only shows one of the forces of tension; the second and third tension forces are not shown. The triangle at the right shows the geometry of one of the cords, where ℓ is the length of the cord, and *d* is the distance from the ceiling.

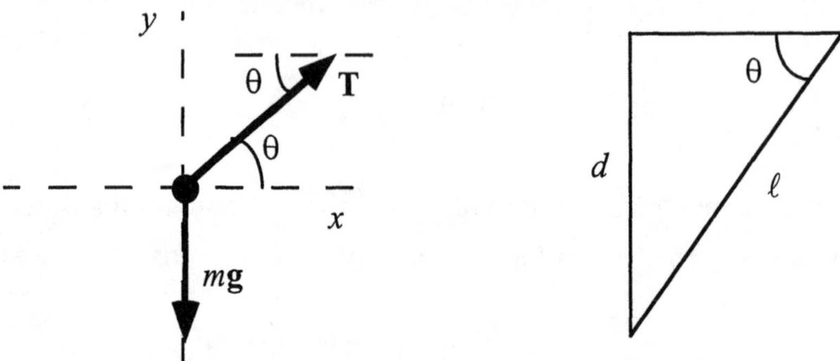

We can use the forces in the *y* direction to find the magnitude *T* of the tension in any one wire.

SOLUTION Remembering that there are three tension forces, we see from the free-body diagram that

$$3T\sin\theta = mg \qquad \text{or} \qquad T = \frac{mg}{3\sin\theta} = \frac{mg}{3(d/\ell)} = \frac{mg\ell}{3d}$$

Therefore, the magnitude of the tension in any one of the cords is

$$T = \frac{(44\ \mathrm{kg})(9.80\ \mathrm{m/s^2})(2.0\ \mathrm{m})}{3(1.5\ \mathrm{m})} = \boxed{1.9\times10^{2}\,\mathrm{N}}$$

---

108. ***REASONING AND SOLUTION***

a. The force acting on the sphere which accelerates it is the horizontal component of the tension in the string. Newton's second law for the horizontal motion of the sphere gives

$$T \sin \theta = ma$$

The vertical component of the tension in the string supports the weight of the sphere so

$$T \cos \theta = mg$$

Eliminating T from the above equations results in $\boxed{a = g \tan \theta}$.

b. $$a = g \tan \theta = \left(9.80 \text{ m/s}^2\right) \tan 10.0° = \boxed{1.73 \text{ m/s}^2}$$

c. Rearranging the result of part a and setting $a = 0 \text{ m/s}^2$ gives

$$\theta = \tan^{-1}(a/g) = \boxed{0°}$$

---

109. ***REASONING AND SOLUTION***

a. The left mass (mass 1) has a tension T_1 pulling it up. Newton's second law gives

$$T_1 - m_1 g = m_1 a \qquad (1)$$

The right mass (mass 3) has a different tension, T_3, trying to pull it up. Newton's second for it is

$$T_3 - m_3 g = -m_3 a \qquad (2)$$

The middle mass (mass 2) has both tensions acting on it along with friction. Newton's second law for its horizontal motion is

$$T_3 - T_1 - \mu_k m_2 g = m_2 a \qquad (3)$$

Solving Equation (1) and Equation (2) for T_1 and T_3, respectively, and substituting into Equation (3) gives

$$a = \frac{(m_3 - m_1 - \mu_k m_2)g}{m_1 + m_2 + m_3}.$$

Hence,

$$a = \boxed{0.60 \text{ m/s}^2}$$

b. From part a

$$T_1 = m_1(g + a) = \boxed{104\ \text{N}} \qquad \text{and} \qquad T_3 = m_3(g - a) = \boxed{230\ \text{N}}$$

---

110. ***REASONING*** The diagram at the right shows the two applied forces that act on the crate. These two forces, plus the kinetic frictional force $\mathbf{f}_k$ constitute the net force that acts on the crate. Once the net force has been determined, Newtons' second law, $\Sigma\mathbf{F} = m\mathbf{a}$ (Equation 4.1) can be used to find the acceleration of the crate.

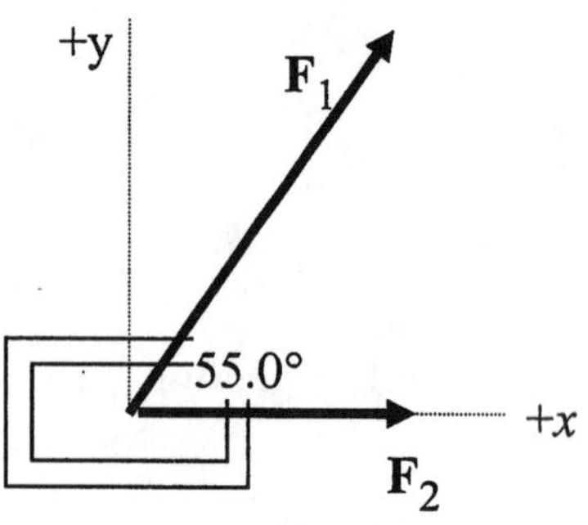

SOLUTION The sum of the two applied forces is $\mathbf{F} = \mathbf{F}_1 + \mathbf{F}_2$. The *x*-component of this sum is $F_x = F_1 \cos 55.0° + F_2$ $= (88.0\text{ N}) \cos 55.0° + 54.0\text{ N} = 104\text{ N}$. The *y*-component of **F** is $F_y = F_1 \sin 55.0° =$ $(88.0\text{ N}) \sin 55.0° = 72.1\text{ N}$. The magnitude of **F** is

$$F = \sqrt{F_x^2 + F_y^2} = \sqrt{(104\text{ N})^2 + (72.1\text{ N})^2} = 127\text{ N}$$

Since the crate starts from rest, it moves along the direction of **F**. The kinetic frictional force $\mathbf{f}_k$ opposes the motion, so it points opposite to **F**. The net force acting on the crate is the sum of **F** and $\mathbf{f}_k$. The magnitude *a* of the crate's acceleration is equal to the magnitude ΣF of the net force divided by the mass *m* of the crate

$$a = \frac{\Sigma F}{m} = \frac{-f_k + F}{m} \qquad (4.1)$$

According to Equation 4.8, the magnitude f_k of the kinetic frictional force is given by $f_k = \mu_k F_N$, where F_N is the magnitude of the normal force. In this situation, F_N is equal to the magnitude of the crate's weight, so $F_N = mg$. Thus, the *x*-component of the acceleration is

$$a = \frac{-\mu_k mg + F}{m} = \frac{-(0.350)(25.0\text{ kg})\left(9.80\text{ m/s}^2\right) + 127\text{ N}}{25.0\text{ kg}} = \boxed{1.65\text{ m/s}^2}$$

The crate moves along the direction of **F**, whose *x* and *y* components have been determined previously. Therefore, the acceleration is also along **F**. The angle ϕ that **F** makes with the *x*-axis can be found using the inverse tangent function:

$$\phi = \tan^{-1}\left(\frac{F_y}{F_x}\right) = \tan^{-1}\left(\frac{F_1 \sin 55.0°}{F_1 \cos 55.0° + F_2}\right)$$

$$= \tan^{-1}\left[\frac{(88.0\text{ N})\sin 55.0°}{(88.0\text{ N})\cos 55.0° + 54.0\text{ N}}\right] = \boxed{34.6° \text{ above the } x \text{ axis}}$$

---

111. SSM ***REASONING*** The following figure shows the crate on the incline and the free body diagram for the crate. The diagram at the far right shows all the forces resolved into components that are parallel and perpendicular to the surface of the incline. We can analyze the motion of the crate using Newton's second law. The coefficient of friction can be determined from the resulting equations.

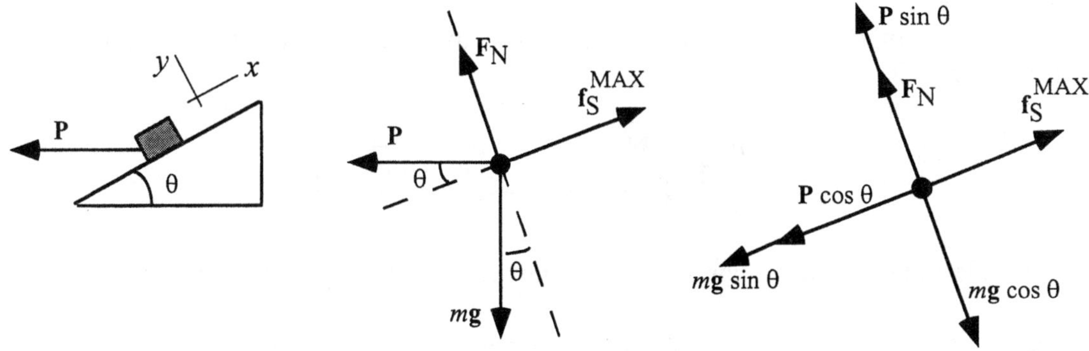

SOLUTION Since the crate is at rest, it is in equilibrium and its acceleration is zero in all directions. If we take the direction down the incline as positive, Newton's second law indicates that

$$\Sigma F_x = P\cos\theta + mg\sin\theta - f_s^{\text{MAX}} = 0$$

According to Equation 4.7, $f_s^{\text{MAX}} = \mu_s F_N$. Therefore, we have

$$P\cos\theta + mg\sin\theta - \mu_s F_N = 0 \qquad (1)$$

The expression for the normal force can be found from analyzing the forces that are perpendicular to the incline. Taking up to be positive, we have

$$\Sigma F_y = P\sin\theta + F_N - mg\cos\theta = 0 \qquad \text{or} \qquad F_N = mg\cos\theta - P\sin\theta$$

Equation (1) then becomes

$$P\cos\theta + mg\sin\theta - \mu_s(mg\cos\theta - P\sin\theta) = 0$$

Solving for the coefficient of static friction, we find that

$$\mu_s = \frac{P\cos\theta + mg\sin\theta}{mg\cos\theta - P\sin\theta} = \frac{(535\text{ N})\cos 20.0° + (225\text{ kg})(9.80\text{ m/s}^2)\sin 20.0°}{(225\text{ kg})(9.80\text{ m/s}^2)\cos 20.0° - (535\text{ N})\sin 20.0°} = \boxed{0.665}$$

---

112. ***CONCEPT QUESTIONS***

a. Yes. Since there is only one force acting on the man in the horizontal direction, it is the net force. According to Newton's second law, Equation 4.1, the man must accelerate under the action of this force. The factors that determine this acceleration are (1) the magnitude and (2) the direction of the force exerted on the man, and (3) the mass of the man.

b. Yes. When the woman exerts a force on the man, the man exerts a force of equal magnitude, but opposite direction, on the woman (Newton's third law).

SOLUTION

a. The acceleration of the man is, according to Equation 4.1, equal to the net force acting on him divided by his mass.

$$a_{\text{man}} = \frac{\Sigma F}{m} = \frac{45\text{ N}}{82\text{ kg}} = \boxed{0.55\text{ m/s}^2\ \text{(due east)}}$$

b. The acceleration of the woman is equal to the net force acting on her divided by her mass.

$$a_{\text{woman}} = \frac{\Sigma F}{m} = \frac{45\text{ N}}{48\text{ kg}} = \boxed{0.94\text{ m/s}^2\ \text{(due west)}}$$

---

113. ***CONCEPT QUESTIONS***

a. The other force is the static frictional force exerted on the refrigerator by the floor.

b. Since the refrigerator does not move, the static frictional force must be equal in magnitude, but opposite in direction, to the horizontal pushing force that the person exerts on the refrigerator.

c. The magnitude of the maximum static frictional force is given by Equation 4.7 as $f_s^{\text{MAX}} = \mu_s F_{\text{N}}$. This is also the largest possible force that the person can exert on the refrigerator before it begins to move. Thus, the factors that determine this force magnitude are the coefficient of static friction μ_s and the magnitude F_{N} of the normal force (which is equal to the weight of the refrigerator in this case).

SOLUTION

a. Since the refrigerator does not move, it is in equilibrium, and the magnitude of the static frictional force must be equal to the magnitude of the horizontal pushing force. Thus, the magnitude of the static frictional force is $\boxed{267\text{ N}}$. The direction of this force must be opposite to that of the pushing force, so the static frictional force is in the $\boxed{+x\text{ direction}}$.

b. The magnitude of the largest pushing force is given by Equation 4.7 as

$$f_s^{\text{MAX}} = \mu_s F_N = \mu_s mg = (0.65)(57\text{ kg})(9.80\text{ m/s}^2) = \boxed{360\text{ N}}$$

---

114. ***CONCEPT QUESTIONS***

a. The gravitational force exerted on the rock is greater than that on the pebble, because the rock has the greater mass.

b. The accelerations are equal, because all objects, regardless of their mass, fall to the earth with the same acceleration due to gravity.

SOLUTION

a. The magnitude of the gravitational force exerted on the rock by the earth is given by Equation 4.3 as

$$F_{\text{rock}} = \frac{Gm_{\text{earth}}m_{\text{rock}}}{r_{\text{earth}}^2}$$

$$= \frac{\left(6.67\times10^{-11}\text{ N}\cdot\text{m}^2/\text{kg}^2\right)\left(5.98\times10^{24}\text{ kg}\right)\left(5.0\text{ kg}\right)}{\left(6.38\times10^6\text{ m}\right)^2} = \boxed{49\text{ N}}$$

The magnitude of the gravitational force exerted on the pebble by the earth is

$$F_{\text{pebble}} = \frac{Gm_{\text{earth}}m_{\text{pebble}}}{r_{\text{earth}}^2}$$

$$= \frac{\left(6.67\times10^{-11}\text{ N}\cdot\text{m}^2/\text{kg}^2\right)\left(5.98\times10^{24}\text{ kg}\right)\left(3.0\times10^{-4}\text{ kg}\right)}{\left(6.38\times10^6\text{ m}\right)^2} = \boxed{2.9\times10^{-3}\text{ N}}$$

b. The acceleration of the rock is equal to the gravitational force exerted on the rock divided by its mass.

$$a_{\text{rock}} = \frac{F_{\text{rock}}}{m_{\text{rock}}} = \frac{Gm_{\text{earth}}}{r_{\text{earth}}^2}$$

$$= \frac{\left(6.67\times10^{-11}\text{ N}\cdot\text{m}^2/\text{kg}^2\right)\left(5.98\times10^{24}\text{ kg}\right)}{\left(6.38\times10^6\text{ m}\right)^2} = \boxed{9.80\text{ m/s}^2\text{, downward}}$$

The acceleration of the pebble is equal to the gravitational force exerted on the pebble divided by its mass.

$$a_{\text{pebble}} = \frac{F_{\text{pebble}}}{m_{\text{pebble}}} = \frac{Gm_{\text{earth}}}{r_{\text{earth}}^2}$$

$$= \frac{\left(6.67 \times 10^{-11} \text{ N} \cdot \text{m}^2/\text{kg}^2\right)\left(5.98 \times 10^{24} \text{ kg}\right)}{\left(6.38 \times 10^6 \text{ m}\right)^2} = \boxed{9.80 \text{ m/s}^2\text{, downward}}$$

---

115. ***CONCEPT QUESTION*** Yes, the raindrop exerts a gravitational force on the earth. This gravitational force is equal in magnitude to the gravitational force that the earth exerts on the raindrop. The forces that the raindrop and the earth exert on each other are Newton's third law (action–reaction) forces.

SOLUTION

a. The magnitude of the gravitational force exerted on the raindrop by the earth is given by Equation 4.3:

$$F_{\text{raindrop}} = \frac{Gm_{\text{earth}} m_{\text{raindrop}}}{r_{\text{earth}}^2}$$

$$= \frac{\left(6.67 \times 10^{-11} \text{ N} \cdot \text{m}^2/\text{kg}^2\right)\left(5.98 \times 10^{24} \text{ kg}\right)\left(5.2 \times 10^{-7} \text{ kg}\right)}{\left(6.38 \times 10^6 \text{ m}\right)^2} = \boxed{5.1 \times 10^{-6} \text{ N}}$$

b. The magnitude of the gravitational force exerted on the earth by the raindrop is

$$F_{\text{earth}} = \frac{Gm_{\text{earth}} m_{\text{raindrop}}}{r_{\text{earth}}^2}$$

$$= \frac{\left(6.67 \times 10^{-11} \text{ N} \cdot \text{m}^2/\text{kg}^2\right)\left(5.98 \times 10^{24} \text{ kg}\right)\left(5.2 \times 10^{-7} \text{ kg}\right)}{\left(6.38 \times 10^6 \text{ m}\right)^2} = \boxed{5.1 \times 10^{-6} \text{ N}}$$

---

116. ***CONCEPT QUESTIONS***

a. $\mathbf{F}_2$ can be positive and can have any magnitude. $\mathbf{F}_2$ can also be negative, provided that its magnitude is less than the magnitude of $\mathbf{F}_1$.

b. $\mathbf{F}_2$ must be negative and have a magnitude that is greater than that of $\mathbf{F}_1$.

c. $\mathbf{F}_2$ must be negative and have a magnitude that is equal to that of $\mathbf{F}_1$.

SOLUTION

a. We may use Newton's second law, $\Sigma F_x = ma_x$, to find the force F_2. Taking the positive x direction to be to the right, we have

$$\underbrace{F_1 + F_2}_{\Sigma F_x} = ma_x \qquad \text{so} \qquad F_2 = ma_x - F_1$$

$$F_2 = (3.0\text{ kg})(+5.0\text{ m/s}^2) - (+9.0\text{ N}) = \boxed{+6\text{ N}}$$

b. Applying Newton's second law again gives

$$F_2 = ma_x - F_1 = (3.0\text{ kg})(-5.0\text{ m/s}^2) - (+9.0\text{ N}) = \boxed{-24\text{ N}}$$

c. An application of Newton's second law gives

$$F_2 = ma_x - F_1 = (3.0\text{ kg})(0\text{ m/s}^2) - (+9.0\text{ N}) = \boxed{-9.0\text{ N}}$$

---

117. ***CONCEPT QUESTIONS***

a. No, the magnitude of the normal force is not equal to the weight of the car. As the drawing shows, the normal force F_N points perpendicular to the hill, while the weight W points vertically down. Since the car does not leave the surface of the hill, the magnitude of the perpendicular component of the weight $W\cos\theta$ must equal the magnitude of the normal force, so $F_N = W\cos\theta$. Thus, the magnitude of the normal force is less than the magnitude of the weight.

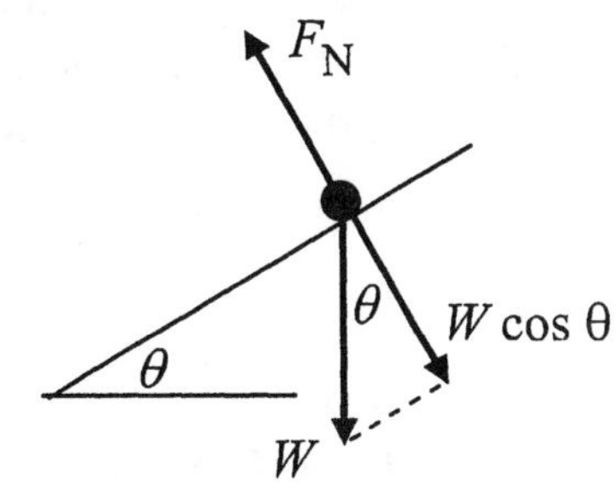

b. As the hill becomes steeper, θ increases, and $\cos\theta$ decreases. Consequently, the normal force decreases as the hill becomes steeper.

c. The magnitude of the normal force does not depend on whether the car is traveling up or down the hill.

SOLUTION

a. From part a of the Concept Questions, we have that $F_N = W\cos\theta$. The ratio of the magnitude of the normal force to the magnitude W of the weight is

$$\frac{F_N}{W} = \frac{W\cos\theta}{W} = \cos 15° = \boxed{0.97}$$

b. When the angle is 35°, the ratio is

$$\frac{F_N}{W} = \frac{W\cos\theta}{W} = \cos 35° = \boxed{0.82}$$

---

118. ***CONCEPT QUESTIONS***

a. Only one force contributes to the horizontal net force acting on block 1, as shown in the free-body diagram. This is the force **–P** with which block 2 pushes on block 1. The minus sign in the free-body diagram indicates the direction of the force is to the left. This force is part of the action-reaction pair of forces that is consistent with Newton's third law. Block 1 pushes forward and to the right against block 2, and block 2 pushes backward and to the left against block 1 with an oppositely directed force of equal magnitude.

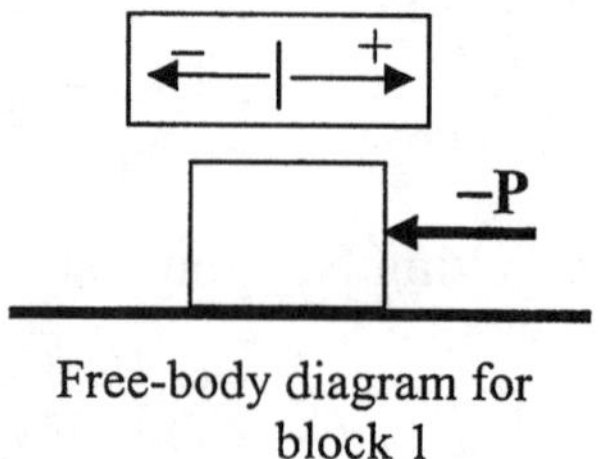

Free-body diagram for block 1

b. Two forces contribute to the horizontal net force acting on block 2, as shown in the free-body diagram. One is the force **P** with which block 1 pushes on block 2. According to Newton's third law, this force has the same magnitude but the opposite direction as the force with which block 2 pushes on block 1. The other force is the kinetic frictional force $\mathbf{f}_k$, which points to the left, in opposition to the relative motion between the block and the surface on which it slides.

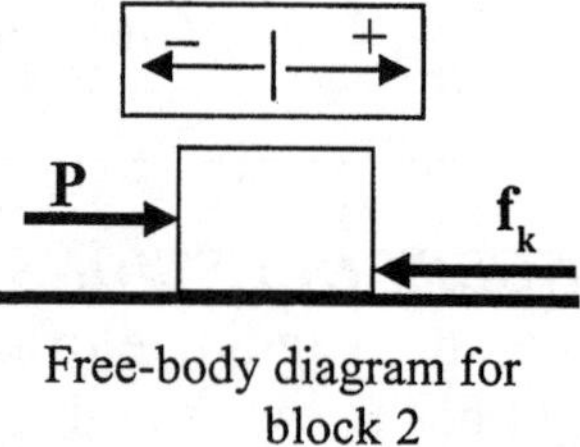

Free-body diagram for block 2

c. The two blocks push against each other with forces **P** and –**P** that are greater in Case B than in Case A. Think about why block 1 pushes on block 2. This occurs because the frictional force acting on block 2 opposes the motion. No frictional force acts on block 1, however, so block 1 runs into block 2, exerting a force on it. When it has a greater mass (all else remaining the same), as in Case B, block 1 has a greater inertia, or a greater tendency to keep moving, and, therefore, exerts a greater force on block 2. According to Newton's third law, block 2 pushes back on block 1 with an oppositely directed force of equal magnitude.

d. Both blocks decelerate, the magnitude of the deceleration being the same for each block. The have the same deceleration, because they are pressed together. Since the blocks are moving to the right in the drawing, the acceleration vector points to the left, for it reflects the slowing down of the motion. The magnitude of the deceleration is greater in Case A than in Case B. This follows directly from Newton's second law, which says that, for a given net force, the acceleration of an object is inversely proportional to the object's mass. The greater the mass, the smaller the acceleration. The two blocks (viewed as a single combined object) have less mass in Case A, and therefore, the deceleration has a greater magnitude in this case.

SOLUTION Referring to the free-body diagram for block 1, we write Newton's second law as follows:

$$\underbrace{-P}_{\substack{\text{Net force}\\ \text{on block 1}}} = m_1 a \tag{1}$$

Referring to the free-body diagram for block 2, we write Newton's second law as follows:

$$\underbrace{P - f_k}_{\substack{\text{Net force}\\ \text{on block 2}}} = m_2 a \tag{2}$$

Solving Equation (1) for a gives $a = -P/m_1$. Substituting this result into Equation (2) gives

$$P - f_k = m_2\left(\frac{-P}{m_1}\right) \quad \text{or} \quad P = \frac{m_1 f_k}{m_1 + m_2}$$

Substituting this result for P into $a = -P/m_1$ gives

$$a = \frac{-P}{m_1} = \frac{-m_1 f_k}{m_1\left(m_1 + m_2\right)} = \frac{-f_k}{m_1 + m_2}$$

We can now use these results to calculate P and a in both cases.

a. **Case A**

$$P = \frac{m_1 f_k}{m_1 + m_2} = \frac{(3.0\text{ kg})(5.8\text{ N})}{3.0\text{ kg} + 3.0\text{ kg}} = \boxed{2.9\text{ N}}$$

Case B

$$P = \frac{m_1 f_k}{m_1 + m_2} = \frac{(6.0\text{ kg})(5.8\text{ N})}{6.0\text{ kg} + 3.0\text{ kg}} = \boxed{3.9\text{ N}}$$

b. **Case A**

$$a = \frac{-f_k}{m_1 + m_2} = \frac{-5.8\text{ N}}{3.0\text{ kg} + 3.0\text{ kg}} = \boxed{-0.97\text{ m/s}^2}$$

Case B

$$a = \frac{-f_k}{m_1 + m_2} = \frac{-5.8\text{ N}}{6.0\text{ kg} + 3.0\text{ kg}} = \boxed{-0.64\text{ m/s}^2}$$

These results are consistent with our answers to the Concept Questions.

---

119. ***CONCEPT QUESTIONS***

a. The block is in equilibrium in each case. Since the block moves at a constant velocity in each case, it is not accelerating. A zero acceleration is the hallmark of equilibrium.

b. The direction of the kinetic frictional force is not the same in each case. The frictional force always opposes the relative motion between the surface of the block and the wall. Therefore, when the block slides upward, the frictional force points downward. When the block slides downward, the frictional force points upward. These directions are shown in the free-body diagrams (not to scale) for the two cases. In these diagrams **W** is the weight of the block and $\mathbf{f}_k$ is the kinetic frictional force.

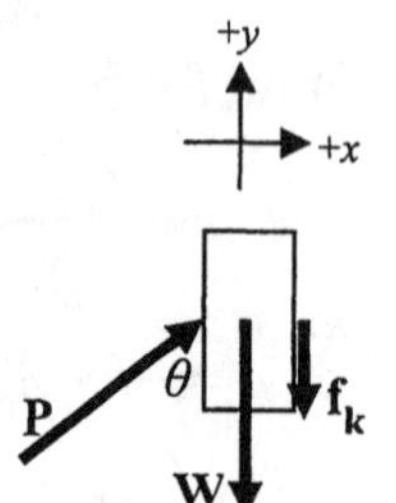

Free-body diagram for upward motion of the block

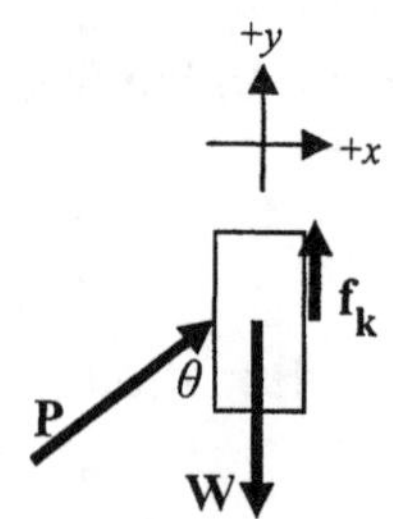

Free-body diagram for downward motion of the

c. In each case the magnitude of the frictional force is the same. It is given by Equation 4.8 as $f_k = \mu_k F_N$, where μ_k is the coefficient of kinetic friction and F_N is the magnitude of the normal force. The coefficient of kinetic friction does not depend on the direction of the motion. Furthermore, the magnitude of the normal force is the same in each case and is the component of the pushing force that is perpendicular to the wall, or $F_N = P \sin\theta$.

d. The magnitude of the pushing force is greater when the block slides up the wall. Since the block is in equilibrium in both cases, the upward vertical components of the forces must balance the downward components. When the block slides upward, the frictional force points downward, like the weight. Thus, the upward vertical component of the pushing force must balance the sum of the two. However, when the block slides downward, the frictional force points upward. In this case, the upward component of the pushing force must balance only part of the weight, since the frictional force also balances part of it. A greater upward component of the pushing force means that the magnitude of the pushing force is greater.

SOLUTION Using Equations 4.9a and 4.9b to describe the equilibrium of the block and referring to the free-body diagrams, we have

Upward motion $$\Sigma F_y = P\cos\theta - W - f_k = 0 \quad (4.9a)$$

Downward motion $$\Sigma F_y = P\cos\theta - W + f_k = 0 \quad (4.9b)$$

According to Equation 4.8, the magnitude of the kinetic frictional force is $f_k = \mu_k F_N$, where we have pointed out in the answer to Concept Question (c) that the magnitude of the normal force is $F_N = P \sin\theta$. Substituting into Equations (4.9a) and (4.9b), we obtain

Upward motion $\Sigma F_y = P\cos\theta - W - \mu_k P\sin\theta = 0$ (4.9a)

Downward motion $\Sigma F_y = P\cos\theta - W + \mu_k P\sin\theta = 0$ (4.9b)

Solving each case for P, we find that

Upward motion

$$P = \frac{W}{\cos\theta - \mu_k \sin\theta} = \frac{39.0\text{ N}}{\cos 30.0° - (0.250)\sin 30.0°} = \boxed{52.6\text{ N}}$$

Downward motion

$$P = \frac{W}{\cos\theta + \mu_k \sin\theta} = \frac{39.0\text{ N}}{\cos 30.0° + (0.250)\sin 30.0°} = \boxed{39.4\text{ N}}$$

These results are consistent with our answers to the Concept Questions.

CHAPTER 5 | *DYNAMICS OF UNIFORM CIRCULAR MOTION*

PROBLEMS

---

1. SSM ***REASONING*** The speed of the plane is given by Equation 5.1: $v = 2\pi r / T$, where T is the period or the time required for the plane to complete one revolution.

SOLUTION Solving Equation 5.1 for T we have

$$T = \frac{2\pi r}{v} = \frac{2\pi(2850\ \text{m})}{110\ \text{m/s}} = \boxed{160\ \text{s}}$$

---

2. ***REASONING*** The magnitude a_c of the car's centripetal acceleration is given by Equation 5.2 as $a_c = v^2 / r$, where v is the speed of the car and r is the radius of the track. The radius is $r = 2.6 \times 10^3$ m. The speed can be obtained from Equation 5.1 as the circumference ($2\pi r$) of the track divided by the period T of the motion. The period is the time for the car to go once around the track ($T = 360$ s).

SOLUTION Since $a_c = v^2 / r$ and $v = (2\pi r)/T$, the magnitude of the car's centripetal acceleration is

$$a_c = \frac{v^2}{r} = \frac{\left(\dfrac{2\pi r}{T}\right)^2}{r} = \frac{4\pi^2 r}{T^2} = \frac{4\pi^2\left(2.6\times10^3\ \text{m}\right)}{(360\ \text{s})^2} = \boxed{0.79\ \text{m/s}^2}$$

---

3. ***REASONING*** Since the tip of the blade moves on a circular path, it experiences a centripetal acceleration whose magnitude a_c is given by Equation 5.2 as, $a_c = v^2 / r$, where v is the speed of blade tip and r is the radius of the circular path. The radius is known, and the speed can be obtained by dividing the distance that the tip travels by the time t of travel. Since an angle of 90° corresponds to one fourth of the circumference of a circle, the distance is $\frac{1}{4}(2\pi r)$.

SOLUTION Since $a_c = v^2 / r$ and $v = \frac{1}{4}(2\pi r)/t = \pi r / (2t)$, the magnitude of the centripetal acceleration of the blade tip is

$$a_c = \frac{v^2}{r} = \frac{\left(\frac{\pi r}{2t}\right)^2}{r} = \frac{\pi^2 r}{4t^2} = \frac{\pi^2 (0.45\text{ m})}{4(0.40\text{ s})^2} = \boxed{6.9\text{ m/s}^2}$$

---

4. ***REASONING AND SOLUTION*** Since the speed of the object on and off the circle remains constant at the same value, the object always travels the same distance in equal time intervals, both on and off the circle. Furthermore since the object travels the distance *OA* in the same time it would have moved from *O* to *P* on the circle, we know that the distance *OA* is equal to the distance along the arc of the circle from *O* to *P*.

The circumference of the circle is $2\pi r = 2\pi(3.6\text{ m}) = 22.6\text{ m}$. The arc *OP* subtends an angle of $\theta = 25°$; therefore, since any circle contains 360°, the arc *OP* is 25/360 or 6.9 per cent of the circumference of the circle. Thus,

$$OP = (22.6\text{ m})(0.069) = 1.6\text{ m}$$

and, from the argument given above, we conclude that the distance *OA* is $\boxed{1.6\text{ m}}$.

---

5. **SSM** ***REASONING AND SOLUTION*** Since the magnitude of the centripetal acceleration is given by Equation 5.2, $a_c = v^2/r$, we can solve for *r* and find that

$$r = \frac{v^2}{a_c} = \frac{(98.8\text{ m/s})^2}{3.00(9.80\text{ m/s}^2)} = \boxed{332\text{ m}}$$

---

6. ***REASONING*** The centripetal acceleration is given by Equation 5.2 as $a_c = v^2/r$. The value of the radius *r* is given, so to determine a_c we need information about the speed *v*. But the speed is related to the period *T* by $v = (2\pi r)/T$, according to Equation 5.1. We can substitute this expression for the speed into Equation 5.2 and see that

$$a_c = \frac{v^2}{r} = \frac{(2\pi r/T)^2}{r} = \frac{4\pi^2 r}{T^2}$$

SOLUTION To use the expression obtained in the reasoning, we need a value for the period *T*. The period is the time for one revolution. Since the container is turning at 2.0 revolutions per second, the period is $T = (1\text{ s})/(2.0\text{ revolutions}) = 0.50\text{ s}$. Thus, we find that the centripetal acceleration is

$$a_c = \frac{4\pi^2 r}{T^2} = \frac{4\pi^2 (0.12 \text{ m})}{(0.50 \text{ s})^2} = \boxed{19 \text{ m/s}^2}$$

---

7. ***REASONING AND SOLUTION***

a. We know

$$a_c = v^2/r = (1.4 \text{ m/s})^2/(0.039 \text{ m}) = \boxed{5.0\times10^1 \text{ m/s}^2}$$

b. Between sprockets, the chain is straight and travels at a constant speed, so its acceleration is $\boxed{\text{zero}}$.

c. For the front sprocket

$$a_c = v^2/r = (1.4 \text{ m/s})^2/(0.10 \text{ m}) = \boxed{2.0\times10^1 \text{ m/s}^2}$$

---

8. ***REASONING AND SOLUTION*** The centripetal acceleration for any point on the blade a distance r from center of the circle, according to Equation 5.2, is $a_c = v^2/r$. From Equation 5.1, we know that $v = 2\pi r/T$ where T is the period of the motion. Combining these two equations, we obtain

$$a_c = \frac{(2\pi r/T)^2}{r} = \frac{4\pi^2 r}{T^2}$$

a. Since the turbine blades rotate at 617 rev/s, all points on the blades rotate with a period of $T = (1/617)\text{ s} = 1.62\times10^{-3}\text{ s}$. Therefore, for a point with $r = 0.020\text{ m}$, the magnitude of the centripetal acceleration is

$$a_c = \frac{4\pi^2 (0.020 \text{ m})}{(1.62\times10^{-3} \text{ s})^2} = \boxed{3.0\times10^5 \text{ m/s}^2}$$

b. Expressed as a multiple of g, this centripetal acceleration is

$$a_c = \left(3.0\times10^5 \text{ m/s}^2\right)\left(\frac{1.00\, g}{9.80 \text{ m/s}^2}\right) = \boxed{3.1\times10^4\ g}$$

---

9. **SSM** ***REASONING*** The magnitude of the centripetal acceleration of any point on the helicopter blade is given by Equation 5.2, $a_c = v^2/r$, where r is the radius of the circle on which that point moves. From Equation 5.1: $v = 2\pi r/T$. Combining these two expressions, we obtain

$$a_c = \frac{4\pi^2 r}{T^2}$$

All points on the blade move with the same period T.

SOLUTION The ratio of the centripetal acceleration at the end of the blade (point 1) to that which exists at a point located 3.0 m from the center of the circle (point 2) is

$$\frac{a_{C1}}{a_{C2}} = \frac{4\pi^2 r_1 / T^2}{4\pi^2 r_2 / T^2} = \frac{r_1}{r_2} = \frac{6.7 \text{ m}}{3.0 \text{ m}} = \boxed{2.2}$$

---

10. ***REASONING AND SOLUTION*** The sample makes one revolution in time T as given by $T = 2\pi r/v$. The speed is

$$v^2 = ra_c = (5.00 \times 10^{-2} \text{ m})(6.25 \times 10^3)(9.80 \text{ m/s}^2) \quad \text{so that} \quad v = 55.3 \text{ m/s}$$

The period is

$$T = 2\pi(5.00 \times 10^{-2} \text{ m})/(55.3 \text{ m/s}) = 5.68 \times 10^{-3} \text{ s} = 9.47 \times 10^{-5} \text{ min}$$

The number of revolutions per minute $= 1/T = \boxed{10\ 600 \text{ rev/min}}$.

---

11. SSM ***REASONING AND SOLUTION*** The magnitude of the centripetal force on the ball is given by Equation 5.3: $F_C = mv^2/r$. Solving for v, we have

$$v = \sqrt{\frac{F_C r}{m}} = \sqrt{\frac{(0.028 \text{ N})(0.25 \text{ m})}{0.015 \text{ kg}}} = \boxed{0.68 \text{ m/s}}$$

---

12. ***REASONING*** The magnitude F_c of the centripetal force that acts on the skater is given by Equation 5.3 as $F_c = mv^2/r$, where m and v are the mass and speed of the skater, and r is the distance of the skater from the pivot. Since all of these variables are known, we can find the magnitude of the centripetal force.

SOLUTION The magnitude of the centripetal force is

$$F_c = \frac{mv^2}{r} = \frac{(80.0 \text{ kg})(6.80 \text{ m/s})^2}{6.10 \text{ m}} = \boxed{606 \text{ N}}$$

---

13. ***REASONING AND SOLUTION***

a. In terms of the period of the motion, the centripetal force is written as

$$F_c = 4\pi^2 mr/T^2 = 4\pi^2\ (0.0120\ \text{kg})(0.100\ \text{m})/(0.500\ \text{s})^2 = \boxed{0.189\ \text{N}}$$

b. The centripetal force varies as the square of the speed. Thus, doubling the speed would increase the centripetal force by a factor of $\boxed{2^2 = 4}$.

---

14. ***REASONING*** The person feels the centripetal force acting on his back. This force is $F_c = mv^2/r$, according to Equation 5.3. This expression can be solved directly to determine the radius r of the chamber.

SOLUTION Solving Equation 5.3 for the radius r gives

$$r = \frac{mv^2}{F_c} = \frac{(83\ \text{kg})(3.2\ \text{m/s})^2}{560\ \text{N}} = \boxed{1.5\ \text{m}}$$

---

15. ***REASONING AND SOLUTION*** The centripetal force is provided by the maximum force of static friction, $f_s^{max} = mv^2/r$. The new frictional force, f_s', is one-third the original value so $f_s' = mv'^2/r = f_s^{max}/3 = mv^2/3r$. Solving for the new velocity, v', we obtain

$$v' = \frac{v}{\sqrt{3}} = \frac{21\ \text{m/s}}{\sqrt{3}} = \boxed{12\ \text{m/s}}$$

---

16. ***REASONING AND SOLUTION*** Initially, the stone executes uniform circular motion in a circle of radius r which is equal to the radius of the tire. At the instant that the stone flies out of the tire, the force of static friction just exceeds its maximum value $f_s^{MAX} = \mu_s F_N$ (see Equation 4.7). The force of static friction that acts on the stone from one side of the tread channel is, therefore,

$$f_s^{MAX} = 0.90\ (1.8\ \text{N}) = 1.6\ \text{N}$$

and the magnitude of the total frictional force that acts on the stone just before it flies out is $2 \times 1.6\ \text{N} = 3.2\ \text{N}$. If we assume that only static friction supplies the centripetal force, then, $F_c = 3.2\ \text{N}$. Solving Equation 5.3 ($F_c = mv^2/r$) for the radius r, we have

$$r = \frac{mv^2}{F_c} = \frac{6.0 \times 10^{-3}\ \text{kg}\ (13\ \text{m/s})^2}{3.2\ \text{N}} = \boxed{0.31\ \text{m}}$$

---

17. SSM WWW ***REASONING*** Let v_0 be the initial speed of the ball as it begins its projectile motion. Then, the centripetal force is given by Equation 5.3: $F_C = mv_0^2/r$. We are given the values for m and r; however, we must determine the value of v_0 from the details of the projectile motion after the ball is released.

In the absence of air resistance, the x component of the projectile motion has zero acceleration, while the y component of the motion is subject to the acceleration due to gravity. The horizontal distance traveled by the ball is given by Equation 3.5a (with $a_x = 0$):

$$x = v_{0x}t = (v_0 \cos\theta)t$$

with t equal to the flight time of the ball while it exhibits projectile motion. The time t can be found by considering the vertical motion. From Equation 3.3b,

$$v_y = v_{0y} + a_y t$$

After a time t, $v_y = -v_{0y}$. Assuming that up and to the right are the positive directions, we have

$$t = \frac{-2v_{0y}}{a_y} = \frac{-2v_0 \sin\theta}{a_y}$$

and

$$x = (v_0 \cos\theta)\left(\frac{-2v_0 \sin\theta}{a_y}\right)$$

Using the fact that $2\sin\theta\cos\theta = \sin 2\theta$, we have

$$x = -\frac{2v_0^2 \cos\theta \sin\theta}{a_y} = -\frac{v_0^2 \sin 2\theta}{a_y} \qquad (1)$$

Equation (1) (with upward and to the right chosen as the positive directions) can be used to determine the speed v_0 with which the ball begins its projectile motion. Then Equation 5.3 can be used to find the centripetal force.

SOLUTION Solving equation (1) for v_0, we have

$$v_0 = \sqrt{\frac{-x\,a_y}{\sin 2\theta}} = \sqrt{\frac{-(86.75\text{ m})(-9.80\text{ m/s}^2)}{\sin 2(41°)}} = 29.3\text{ m/s}$$

Then, from Equation 5.3,

$$F_C = \frac{mv_0^2}{r} = \frac{(7.3 \text{ kg})(29.3 \text{ m/s})^2}{1.8 \text{ m}} = \boxed{3500 \text{ N}}$$

---

18. ***REASONING AND SOLUTION*** The centripetal acceleration of the block is

$$a_C = v^2/r = (28 \text{ m/s})^2/(150 \text{ m}) = 5.2 \text{ m/s}^2$$

The angle θ can be obtained from

$$\theta = \tan^{-1}\left(\frac{a_C}{g}\right) = \tan^{-1}\left(\frac{5.2 \text{ m/s}^2}{9.80 \text{ m/s}^2}\right) = \boxed{28°}$$

---

19. ***REASONING***

a. The free body diagram shows the swing ride and the two forces that act on a chair: the tension **T** in the cable, and the weight ***mg*** of the chair and its occupant. We note that the chair does not accelerate vertically, so the net force $\sum F_y$ in the vertical direction must be zero, $\sum F_y = 0$. The net force consists of the upward vertical component of the tension and the downward weight of the chair. The fact that the net force is zero will allow us to determine the magnitude of the tension.

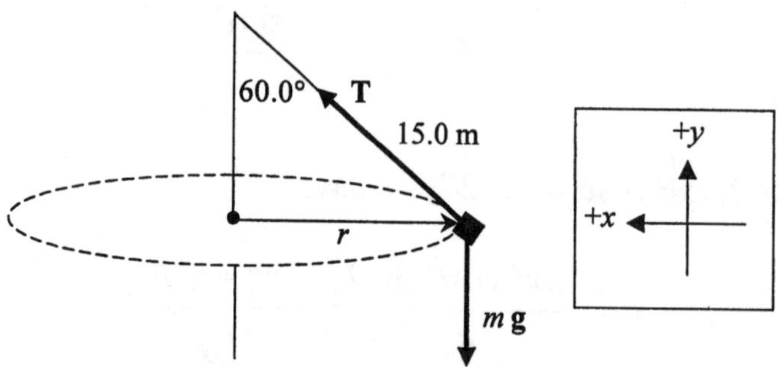

b. According to Newton's second law, the net force $\sum F_x$ in the horizontal direction is equal to the mass m of the chair and its occupant times the centripetal acceleration $(a_c = v^2/r)$, so that $\sum F_x = ma_c = mv^2/r$. There is only one force in the horizontal direction, the horizontal component of the tension, so it is the net force. We will use Newton's second law to find the speed v of the chair.

SOLUTION

a. The vertical component of the tension is $+T \cos 60.0°$, and the weight is $-mg$, where we have chosen "up" as the + direction. Since the chair and its occupant have no vertical acceleration, we have that $\sum F_y = 0$, so

$$\underbrace{+T\cos 60.0° - mg}_{\Sigma F_y} = 0 \qquad (1)$$

Solving for the magnitude T of the tension gives

$$T = \frac{mg}{\cos 60.0°} = \frac{(179\text{ kg})(9.80\text{ m/s}^2)}{\cos 60.0°} = \boxed{3510\text{ N}}$$

b. The horizontal component of the tension is $+T$ sin 60.0°, where we have chosen the direction to the left in the diagram as the + direction. Since the chair and its occupant have a centripetal acceleration in this direction, we have

$$\underbrace{T\sin 60.0°}_{\Sigma F_x} = ma_c = m\left(\frac{v^2}{r}\right) \qquad (2)$$

From the drawing we see that the radius r of the circular path is r = (15.0 m) sin 60.0° = 13.0 m. Solving Equation (2) for the speed v gives

$$v = \sqrt{\frac{rT\sin 60.0°}{m}} = \sqrt{\frac{(13.0\text{ m})(3510\text{ N})\sin 60.0°}{179\text{ kg}}} = \boxed{14.9\text{ m/s}}$$

---

20. ***REASONING AND SOLUTION*** Assuming that there is no friction between the tires and the ice, Equation 5.4 applies.

$$\tan\theta = v^2/rg = (25\text{ m/s})^2/(150\text{ m})(9.80\text{ m/s}^2) = 0.43$$

or

$$\boxed{\theta = 23°}$$

---

21. SSM ***REASONING AND SOLUTION*** Equation 5.4 gives the relationship between the speed v the angle of banking, and the radius of curvature. Solving for v, we obtain

$$v = \sqrt{r\,g\,\tan\theta} = \sqrt{(120\text{ m})(9.80\text{ m/s}^2)\tan 18°} = \boxed{2.0\times 10^1\text{ m/s}}$$

---

22. ***REASONING*** The angle θ at which a friction-free curve is banked depends on the radius r of the curve and the speed v with which the curve is to be negotiated, according to Equation 5.4: $\tan\theta = v^2/(rg)$. For known values of θ and r, the safe speed is

$$v = \sqrt{rg\,\tan\theta}$$

Before we can use this result, we must determine $\tan\theta$ for the banking of the track.

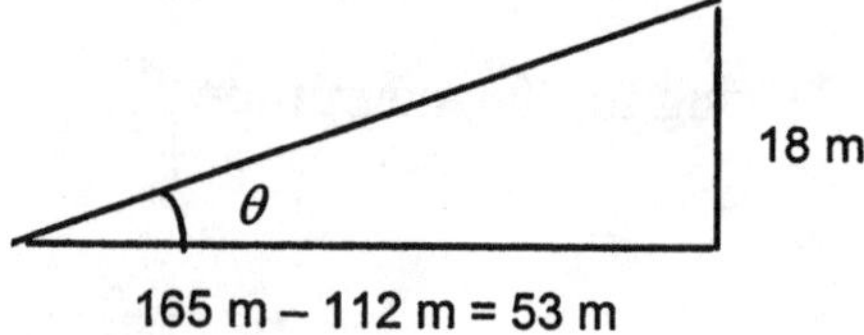

SOLUTION The drawing at the right shows a cross-section of the track. From the drawing we have

$$\tan\theta = \frac{18\text{ m}}{53\text{ m}} = 0.34$$

a. Therefore, the smallest speed at which cars can move on this track without relying on friction is

$$v_{\text{min}} = \sqrt{(112\text{ m})(9.80\text{ m/s}^2)(0.34)} = \boxed{19\text{ m/s}}$$

b. Similarly, the largest speed is

$$v_{\text{max}} = \sqrt{(165\text{ m})(9.80\text{ m/s}^2)(0.34)} = \boxed{23\text{ m/s}}$$

---

23. ***REASONING*** The distance d is related to the radius r of the circle on which the car travels by $d = r/\sin 50.0°$ (see the drawing).

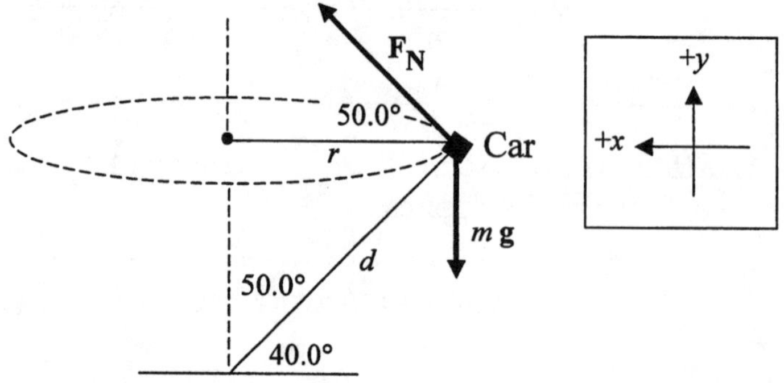

We can obtain the radius by noting that the car experiences a centripetal force that is directed toward the center of the circular path. This force is provided by the component, $F_N \cos 50.0°$, of the normal force that is parallel to the radius. Setting this force equal to the mass m of the car times the centripetal acceleration $(a_c = v^2/r)$ gives $F_N \cos 50.0° = ma_c = mv^2/r$. Solving for the radius r and substituting it into the relation $d = r/\sin 50.0°$ gives

$$d = \frac{r}{\sin 50.0°} = \frac{\dfrac{mv^2}{F_N \cos 50.0°}}{\sin 50.0°} = \frac{mv^2}{(F_N \cos 50.0°)(\sin 50.0°)} \quad (1)$$

The magnitude F_N of the normal force can be obtained by observing that the car has no vertical acceleration, so the net force in the vertical direction must be zero, $\sum F_y = 0$. The net force consists of the upward vertical component of the normal force and the downward weight of the car. The vertical component of the normal force is $+F_N \sin 50.0°$, and the weight is $-mg$, where we have chosen the "up" direction as the + direction. Thus, we have that

$$\underbrace{+F_N \sin 50.0° - mg}_{\Sigma F_y} = 0 \qquad (2)$$

Solving this equation for F_N and substituting it into the equation above will yield the distance *d*.

SOLUTION Solving Equation (2) for F_N and substituting the result into Equation (1) gives

$$d = \frac{mv^2}{(F_N \cos 50.0°)(\sin 50.0°)} = \frac{mv^2}{\left(\dfrac{mg}{\sin 50.0°}\right)(\cos 50.0°)(\sin 50.0°)}$$

$$= \frac{v^2}{g\cos 50.0°} = \frac{(34.0 \text{ m/s})^2}{(9.80 \text{ m/s}^2)\cos 50.0°} = \boxed{184 \text{ m}}$$

---

24. ***REASONING*** From the discussion on banked curves in Section 5.4, we know that a car can safely round a banked curve without the aid of static friction if the angle θ of the banked curve is given by $\tan\theta = v_0^2/(rg)$, where v_0 is the speed of the car and *r* is the radius of the curve (see Equation 5.4). The maximum speed that a car can have when rounding an unbanked curve is $v_0 = \sqrt{\mu_s g r}$ (see Example 7). By combining these two relations, we can find the angle θ.

SOLUTION The angle of the banked curve is $\theta = \tan^{-1}\left[v_0^2/(rg)\right]$. Substituting the expression $v_0 = \sqrt{\mu_s g r}$ into this equation gives

$$\theta = \tan^{-1}\left(\frac{v_0^2}{rg}\right) = \tan^{-1}\left(\frac{\mu_s g r}{rg}\right)$$

$$= \tan^{-1}(\mu_s) = \tan^{-1}(0.81) = \boxed{39°}$$

---

25. SSM WWW ***REASONING*** Refer to Figure 5.10 in the text. The horizontal component of the lift **L** is the centripetal force that holds the plane in the circle. Thus,

$$L\sin\theta = \frac{mv^2}{r} \qquad (1)$$

The vertical component of the lift supports the weight of the plane; therefore,

$$L\cos\theta = mg \qquad (2)$$

Dividing the first equation by the second gives

$$\tan\theta = \frac{v^2}{rg} \qquad (3)$$

Equation (3) can be used to determine the angle θ of banking. Once θ is known, then the magnitude of L can be found from either equation (1) or equation (2).

SOLUTION Solving equation (3) for θ gives

$$\theta = \tan^{-1}\left[\frac{(123\text{ m/s})^2}{(3810\text{ m})(9.80\text{ m/s}^2)}\right] = 22.1°$$

The lifting force is, from equation (2),

$$L = \frac{mg}{\cos\theta} = \frac{(2.00\times10^5\text{ kg})(9.80\text{ m/s}^2)}{\cos 22.1°} = \boxed{2.12\times10^6\text{ N}}$$

---

26. ***REASONING*** The centripetal force F_c required to keep an object of mass m that moves with speed v on a circle of radius r is $F_c = mv^2/r$ (Equation 5.3). From Equation 5.1, we know that $v = 2\pi r/T$, where T is the period or the time for the suitcase to go around once. Therefore, the centripetal force can be written as

$$F_c = \frac{m(2\pi r/T)^2}{r} = \frac{4m\pi^2 r}{T^2} \qquad (1)$$

This expression can be solved for T. However, we must first find the centripetal force that acts on the suitcase.

SOLUTION Three forces act on the suitcase. They are the weight $m\mathbf{g}$ of the suitcase, the force of static friction $\mathbf{f}_s^{MAX}$, and the normal force $\mathbf{F}_N$ exerted on the suitcase by the surface of the carousel. The following figure shows the free body diagram for the suitcase.

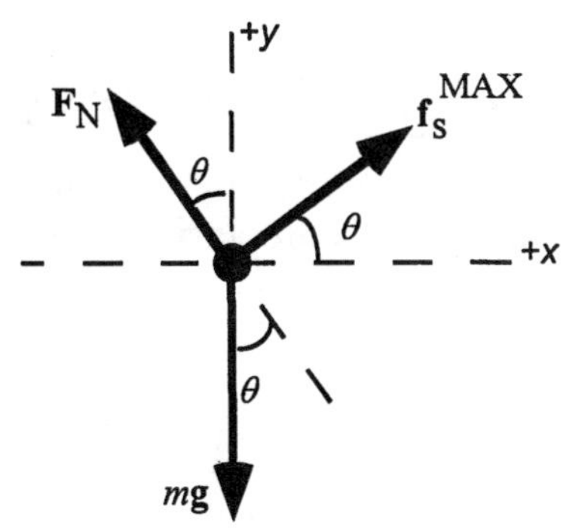

In this diagram, the y axis is oriented along the vertical direction. The force of gravity acts, then, in the $-y$ direction. The centripetal force that causes the suitcase to move on its circular path is provided by the net force in the $+x$ direction in the diagram. From the diagram, we can see that only the forces $\mathbf{F}_N$ and $\mathbf{f}_s^{MAX}$ have horizontal components. Thus, we have $F_c = f_s^{MAX}\cos\theta - F_N\sin\theta$, where the minus sign indicates that the x component of $\mathbf{F}_N$ points to the left in the diagram. Using Equation 4.7 for the maximum static frictional force, we can write this result as in equation (2).

$$F_c = \mu_s F_N \cos\theta - F_N \sin\theta = F_N(\mu_s \cos\theta - \sin\theta) \qquad (2)$$

If we apply Newton's second law in the y direction, we see from the diagram that

$$F_N \cos\theta + f_s^{MAX}\sin\theta - mg = ma_y = 0 \qquad \text{or} \qquad F_N\cos\theta + \mu_s F_N \sin\theta - mg = 0$$

where we again have used Equation 4.7 for the maximum static frictional force. Solving for the normal force, we find

$$F_N = \frac{mg}{\cos\theta + \mu_s \sin\theta}$$

Using this result in equation (2), we obtain the magnitude of the centripetal force that acts on the suitcase:

$$F_c = F_N(\mu_s\cos\theta - \sin\theta) = \frac{mg(\mu_s \cos\theta - \sin\theta)}{\cos\theta + \mu_s\sin\theta}$$

With this expression for the centripetal force, equation (1) becomes

$$\frac{mg(\mu_s\cos\theta - \sin\theta)}{\cos\theta + \mu_s \sin\theta} = \frac{4m\pi^2 r}{T^2}$$

Solving for the period T, we find

$$T = \sqrt{\frac{4\pi^2 r\left(\cos\theta + \mu_s \sin\theta\right)}{g(\mu_s\cos\theta - \sin\theta)}} = \sqrt{\frac{4\pi^2(11.0\text{ m})\left(\cos 36.0^\circ + 0.760\sin 36.0^\circ\right)}{\left(9.80\text{ m/s}^2\right)\left(0.760\cos 36.0^\circ - \sin 36.0^\circ\right)}} = \boxed{45\text{ s}}$$

---

27. SSM WWW ***REASONING*** Equation 5.5 gives the orbital speed for a satellite in a circular orbit around the earth. It can be modified to determine the orbital speed around any planet **P** by replacing the mass of the earth M_E by the mass of the planet M_P: $v = \sqrt{GM_P / r}$.

SOLUTION The ratio of the orbital speeds is, therefore,

$$\frac{v_2}{v_1} = \frac{\sqrt{GM_P / r_2}}{\sqrt{GM_P / r_1}} = \sqrt{\frac{r_1}{r_2}}$$

Solving for v_2 gives

$$v_2 = v_1 \sqrt{\frac{r_1}{r_2}} = (1.70 \times 10^4 \text{ m/s}) \sqrt{\frac{5.25 \times 10^6 \text{ m}}{8.60 \times 10^6 \text{ m}}} = \boxed{1.33 \times 10^4 \text{ m/s}}$$

---

28. ***REASONING AND SOLUTION*** The radius of a synchronous satellite is calculated in Example 11 as $r = 4.23 \times 10^7$ m. The speed is therefore,

$$v = \sqrt{\frac{GM_E}{r}} = \sqrt{\frac{(6.67 \times 10^{-11} \text{ N} \cdot \text{m}^2/\text{kg}^2)(5.98 \times 10^{24} \text{ kg})}{4.23 \times 10^7 \text{ m}}} = \boxed{3070 \text{ m/s}}$$

---

29. ***REASONING AND SOLUTION*** We have for Jupiter $v^2 = GM_J/r$, where

$$r = 6.00 \times 10^5 \text{ m} + 7.14 \times 10^7 \text{ m} = 7.20 \times 10^7 \text{ m}$$

Thus,

$$v = \sqrt{\frac{(6.67 \times 10^{-11} \text{ N} \cdot \text{m}^2/\text{kg}^2)(1.90 \times 10^{27} \text{ kg})}{7.20 \times 10^7 \text{ m}}} = \boxed{4.20 \times 10^4 \text{ m/s}}$$

---

30. ***REASONING AND SOLUTION*** The period of the moon's motion (approximately the length of a month) is given by

$$T = \sqrt{\frac{4\pi^2 r^3}{GM_E}} = \sqrt{\frac{4\pi^2 (3.85 \times 10^8 \text{ m})^3}{(6.67 \times 10^{-11} \text{ N} \cdot \text{m}^2/\text{kg}^2)(5.98 \times 10^{24} \text{ kg})}}$$

$$= 2.38 \times 10^6 \text{ s} = \boxed{27.5 \text{ days}}$$

---

31. ***REASONING*** In Section 5.5 it is shown that the period T of a satellite in a circular orbit about the earth is given by (see Equation 5.6)

$$T = \frac{2\pi r^{3/2}}{\sqrt{GM_E}}$$

where r is the radius of the orbit, G is the universal gravitational constant, and M_E is the mass of the earth. The ratio of the periods of satellites A and B is, then,

$$\frac{T_A}{T_B} = \frac{\dfrac{2\pi r_A^{3/2}}{\sqrt{GM_E}}}{\dfrac{2\pi r_B^{3/2}}{\sqrt{GM_E}}} = \frac{r_A^{3/2}}{r_B^{3/2}} \qquad (1)$$

We do not know the radii r_A and r_B. However we do know that the speed v of a satellite is equal to the circumference ($2\pi r$) of its orbit divided by the period T, so $v = 2\pi r / T$.

SOLUTION Solving the relation $v = 2\pi r / T$ for r gives $r = vT / 2\pi$. Substituting this value for r into Equation (1) yields

$$\frac{T_A}{T_B} = \frac{r_A^{3/2}}{r_B^{3/2}} = \frac{\left[v_A T_A / (2\pi)\right]^{3/2}}{\left[v_B T_B / (2\pi)\right]^{3/2}} = \frac{(v_A T_A)^{3/2}}{(v_B T_B)^{3/2}}$$

Squaring both sides of this equation, algebraically solving for the ratio T_A/T_B, and using the fact that $v_A = 3v_B$ gives

$$\frac{T_A}{T_B} = \frac{v_B^3}{v_A^3} = \frac{v_B^3}{(3v_B)^3} = \boxed{\frac{1}{27}}$$

---

32. ***REASONING*** Equation 5.2 for the centripetal acceleration applies to both the plane and the satellite, and the centripetal acceleration is the same for each. Thus, we have

$$a_c = \frac{v_{\text{plane}}^2}{r_{\text{plane}}} = \frac{v_{\text{satellite}}^2}{r_{\text{satellite}}} \quad \text{or} \quad v_{\text{plane}} = \left(\sqrt{\frac{r_{\text{plane}}}{r_{\text{satellite}}}}\right) v_{\text{satellite}}$$

The speed of the satellite can be obtained directly from Equation 5.5.

SOLUTION Using Equation 5.5, we can express the speed of the satellite as

$$v_{\text{satellite}} = \sqrt{\frac{Gm_E}{r_{\text{satellite}}}}$$

Substituting this expression into the expression obtained in the reasoning for the speed of the plane gives

$$v_{\text{plane}} = \left(\sqrt{\frac{r_{\text{plane}}}{r_{\text{satellite}}}}\right) v_{\text{satellite}} = \left(\sqrt{\frac{r_{\text{plane}}}{r_{\text{satellite}}}}\right)\sqrt{\frac{Gm_E}{r_{\text{satellite}}}} = \frac{\sqrt{r_{\text{plane}} Gm_E}}{r_{\text{satellite}}}$$

$$v_{\text{plane}} = \frac{\sqrt{(15\text{ m})(6.67\times10^{-11}\text{ N}\cdot\text{m}^2/\text{kg}^2)(5.98\times10^{24}\text{ kg})}}{6.7\times10^6\text{ m}} = \boxed{12\text{ m/s}}$$

---

33. **SSM** ***REASONING*** The true weight of the satellite when it is at rest on the planet's surface can be found from Equation 4.4: $W = (GM_P m)/r^2$ where M_P and m are the masses of the planet and the satellite, respectively, and r is the radius of the planet. However, before we can use Equation 4.4, we must determine the mass M_P of the planet.

The mass of the planet can be found by replacing M_E by M_P in Equation 5.6 and solving for M_P. When using Equation 5.6, we note that r corresponds to the radius of the circular orbit *relative to the center of the planet.*

SOLUTION The period of the satellite is $T = 2.00\text{ h} = 7.20\times10^3\text{ s}$. From Equation 5.6,

$$M_P = \frac{4\pi^2 r^3}{GT^2} = \frac{4\pi^2\left[(4.15\times10^6\text{ m})+(4.1\times10^5\text{ m})\right]^3}{(6.67\times10^{-11}\text{ N}\cdot\text{m}^2/\text{kg}^2)(7.20\times10^3\text{ s})^2} = 1.08\times10^{24}\text{ kg}$$

Using Equation 4.4, we have

$$W = \frac{GM_P m}{r^2} = \frac{(6.67\times10^{-11}\text{ N}\cdot\text{m}^2/\text{kg}^2)(1.08\times10^{24}\text{ kg})(5850\text{ kg})}{(4.15\times10^6\text{ m})^2} = \boxed{2.45\times10^4\text{ N}}$$

---

34. ***REASONING AND SOLUTION*** The period of rotation is given by $T^2 = 4\pi^2 r^3/GM$. Comparing the Earth and Venus yields

$$(T_V/T_E)^2 = (r_V/r_E)^3 \text{ so that } T_V/T_E = 0.611$$

The earth's orbital period is 365 days so

$$T_V = (0.611)(365\text{ days}) = \boxed{223\text{ days}}$$

---

35. ***REASONING AND SOLUTION***

a. The centripetal acceleration of a point on the rim of chamber A is the artificial acceleration due to gravity,

$$a_A = v_A^2/r_A = 10.0 \text{ m/s}^2$$

A point on the rim of chamber A moves with a speed $v_A = 2\pi r_A/T$ where T is the period of revolution, 60.0 s. Substituting the second equation into the first and rearranging yields

$$r_A = a_A T^2/(4\pi^2) = \boxed{912 \text{ m}}$$

b. Now

$$r_B = r_A/4.00 = \boxed{228 \text{ m}}$$

c. A point on the rim of chamber B has a centripetal acceleration $a_B = v_B^2/r_B$. The point moves with a speed $v_B = 2\pi r_B/T$. Substituting the second equation into the first yields

$$a_B = \frac{4\pi^2 r_B}{T^2} = \frac{4\pi^2 (228 \text{ m})}{(60.0 \text{ s})^2} = \boxed{2.50 \text{ m/s}^2}$$

---

36. ***REASONING*** According to Equation 5.3, the magnitude F_c of the centripetal force that acts on each passenger is $F_c = mv^2/r$, where m and v are the mass and speed of a passenger and r is the radius of the turn. From this relation we see that the speed is given by $v = \sqrt{F_c r/m}$. The centripetal force is the net force required to keep each passenger moving on the circular path and points toward the center of the circle. With the aid of a free-body diagram, we will evaluate the net force and, hence, determine the speed.

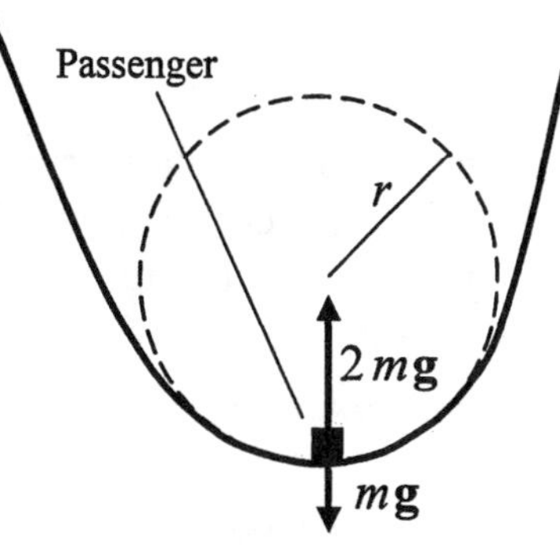

SOLUTION The free-body diagram shows a passenger at the bottom of the circular dip. There are two forces acting: her downward-acting weight mg and the upward-acting force $2mg$ that the seat exerts on her. The net force is $+2mg - mg = +mg$, where we have taken "up" as the positive direction. Thus, $F_c = mg$. The speed of the passenger can be found by using this result in the equation above.

Substituting $F_c = mg$ into the relation $v = \sqrt{F_c r/m}$ yields

$$v=\sqrt{\frac{F_c r}{m}}=\sqrt{\frac{(mg)r}{m}}=\sqrt{gr}=\sqrt{\left(9.80\text{ m/s}^2\right)(20.0\text{ m})}=\boxed{14.0\text{ m/s}}$$

---

37. SSM ***REASONING*** As the motorcycle passes over the top of the hill, it will experience a centripetal force, the magnitude of which is given by Equation 5.3: $F_C = mv^2/r$. The centripetal force is provided by the net force on the cycle + driver system. At that instant, the net force on the system is composed of the normal force, which points upward, and the weight, which points downward. Taking the direction toward the center of the circle (downward) as the positive direction, we have $F_C = mg - F_N$. This expression can be solved for F_N, the normal force.

SOLUTION
a. The magnitude of the centripetal force is

$$F_C=\frac{mv^2}{r}=\frac{(342\text{ kg})(25.0\text{ m/s})^2}{126\text{ m}}=\boxed{1.70\times10^3\text{ N}}$$

b. The magnitude of the normal force is

$$F_N=mg-F_C=(342\text{ kg})(9.80\text{ m/s}^2)-1.70\times10^3\text{ N}=\boxed{1.66\times10^3\text{ N}}$$

---

38. ***REASONING*** The centripetal force is the name given to the net force pointing toward the center of the circular path. At point 3 at the top the net force pointing toward the center of the circle consists of the normal force and the weight, both pointing toward the center. At point 1 at the bottom the net force consists of the normal force pointing upward toward the center and the weight pointing downward or away from the center. In either case the centripetal force is given by Equation 5.3 as $F_c = mv^2/r$.

SOLUTION At point 3 we have

$$F_c = F_N + mg = \frac{mv_3^2}{r}$$

At point 1 we have

$$F_c = F_N - mg = \frac{mv_1^2}{r}$$

Subtracting the second equation from the first gives

$$2mg=\frac{mv_3^2}{r}-\frac{mv_1^2}{r}$$

Rearranging gives

Thus, we find that

$$v_3^2 = 2gr + v_1^2$$

$$v_3 = \sqrt{2(9.80 \text{ m/s}^2)(3.0 \text{ m}) + (15 \text{ m/s})^2} = \boxed{17 \text{ m/s}}$$

---

39. SSM ***REASONING*** This situation is similar to the loop-the-loop trick discussed in Section 5.7 of the text. When the plane passes over the top of a vertical circle of radius *r* such that the passengers experience apparent weightlessness, the centripetal force is provided entirely by the true weight *mg*. Thus, $mg = mv^2/r$.

SOLUTION Solving the above expression for *r* gives

$$r = \frac{v^2}{g} = \frac{(215 \text{ m/s})^2}{9.80 \text{ m/s}^2} = \boxed{4.72 \times 10^3 \text{ m}}$$

---

40. ***REASONING*** As the motorcycle passes over the top of the hill, it experiences a centripetal force, the magnitude of which is given by Equation 5.3 as $F_c = mv^2/r$, where *m* and *v* are the mass and speed of the motorcycle, and *r* is the radius of the circular crest in the road. The speed of the motorcycle is then $v = \sqrt{F_c r/m}$. The centripetal force is the net force acting on the motorcycle and is directed toward the center of the circle. When the motorcycle crests the hill, there are two forces that act along the radial direction, the normal force $\mathbf{F_N}$ (upward) that the road exerts on the motorcycle and the weight *mg* (downward) of the motorcycle and rider. Taking the direction toward the center of the circle (downward) as the positive direction, we have that $F_c = +mg - F_N$. When the motorcycle just loses contact with the road, the normal force becomes zero. With this information, we can find the maximum speed that the cycle can have.

SOLUTION Substituting $F_c = +mg - F_N$ into the relation $v = \sqrt{F_c r/m}$ gives

$$v = \sqrt{\frac{(mg - F_N)r}{m}}$$

The maximum speed v_{max} occurs when the motorcycle just loses contact with the road. At this instant the normal force becomes zero. Setting $F_N = 0$ N, we have

$$v_{max} = \sqrt{gr} = \sqrt{(9.80 \text{ m/s}^2)(45.0 \text{ m})} = \boxed{21.0 \text{ m/s}}$$

---

41. ***REASONING*** When the stone is whirled in a horizontal circle, the centripetal force is provided by the tension T_h in the string and is given by Equation 5.3 as

$$\underbrace{T_h}_{\text{Centripetal force}} = \frac{mv^2}{r} \qquad (1)$$

where m and v are the mass and speed of the stone, and r is the radius of the circle. When the stone is whirled in a vertical circle, the maximum tension occurs when the stone is at the lowest point in its path. The free-body diagram shows the forces that act on the stone in this situation: the tension $\mathbf{T}_v$ in the string and the weight $m\mathbf{g}$ of the stone. The centripetal force is the net force that points toward the center of the circle. Setting the centripetal force equal to mv^2/r, as per Equation 5.3, we have

$$\underbrace{+T_v - mg}_{\text{Centripetal force}} = \frac{mv^2}{r} \qquad (2)$$

Here, we have assumed upward to be the positive direction. We are given that the maximum tension in the string in the case of vertical motion is 15.0% larger than that in the case of horizontal motion. We can use this fact, along with Equations 1 and 2, to find the speed of the stone.

Solution Since the maximum tension in the string in the case of vertical motion is 15.0% larger than that in the horizontal motion, $T_v = (1.000 + 0.150)\ T_h$. Substituting the values of T_h and T_v from Equations (1) and (2) into this relation gives

$$T_v = (1.000 + 0.150)T_h$$

$$\frac{mv^2}{r} + mg = (1.000 + 0.150)\left(\frac{mv^2}{r}\right)$$

Solving this equation for the speed v of the stone yields

$$v = \sqrt{\frac{gr}{0.150}} = \sqrt{\frac{(9.80\text{ m/s}^2)\ (1.10\text{ m})}{0.150}} = \boxed{8.48\text{ m/s}}$$

42. ***REASONING*** The drawing at the right shows the two forces that act on a piece of clothing *just before* it loses contact with the wall of the cylinder. At that instant the centripetal force is provided by the normal force $\mathbf{F}_N$ and the radial component of the weight. From the drawing, the radial component of the weight is given by

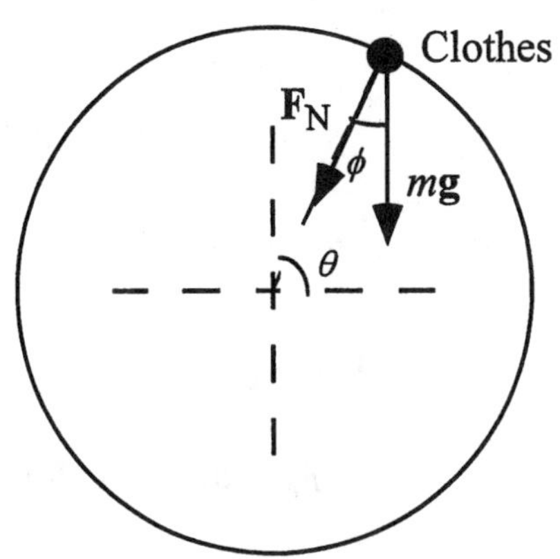

$$mg\cos\phi = mg\cos(90°-\theta) = mg\sin\theta$$

Therefore, with inward taken as the positive direction, Equation 5.3 ($F_c = mv^2/r$) gives

$$F_N + mg\sin\theta = \frac{mv^2}{r}$$

At the instant that a piece of clothing loses contact with the surface of the drum, $F_N = 0$, and the above expression becomes

$$mg\sin\theta = \frac{mv^2}{r}$$

According to Equation 5.1, $v = 2\pi r/T$, and with this substitution we obtain

$$g\sin\theta = \frac{(2\pi r/T)^2}{r} = \frac{4\pi^2 r}{T^2}$$

This expression can be solved for the period *T*. Since the period is the required time for one revolution, the number of revolutions per second can be found by calculating 1/*T*.

SOLUTION Solving for the period, we obtain

$$T = \sqrt{\frac{4\pi^2 r}{g\sin\theta}} = 2\pi\sqrt{\frac{r}{g\sin\theta}} = 2\pi\sqrt{\frac{0.32\text{ m}}{\left(9.80\text{ m/s}^2\right)\sin 70.0°}} = 1.17\text{ s}$$

Therefore, the number of revolutions per second that the cylinder should make is

$$\frac{1}{T} = \frac{1}{1.17\text{ s}} = \boxed{0.85\text{ rev/s}}$$

---

43. SSM ***REASONING*** In Example 3, it was shown that the magnitudes of the centripetal acceleration for the two cases are

$$[\text{Radius} = 33 \text{ m}] \qquad a_C = 35 \text{ m/s}^2$$
$$[\text{Radius} = 24 \text{ m}] \qquad a_C = 48 \text{ m/s}^2$$

According to Newton's second law, the centripetal force is $F_C = ma_C$ (see Equation 5.3).

SOLUTION a. Therefore, when the sled undergoes the turn of radius 33 m,

$$F_C = ma_C = (350 \text{ kg})(35 \text{ m/s}^2) = \boxed{1.2 \times 10^4 \text{ N}}$$

b. Similarly, when the radius of the turn is 24 m,

$$F_C = ma_C = (350 \text{ kg})(48 \text{ m/s}^2) = \boxed{1.7 \times 10^4 \text{ N}}$$

---

44. ***REASONING AND SOLUTION*** The normal force exerted by the wall on each astronaut is the centripetal force needed to keep him in the circular path, i.e., $F_C = mv^2/r$. Rearranging and letting $F_C = (1/2)mg$ yields

$$r = 2v^2/g = 2(35.8 \text{ m/s})^2/(9.80 \text{ m/s}^2) = \boxed{262 \text{ m}}$$

---

45. ***REASONING AND SOLUTION*** Let s represent the length of the path of the pebble after it is released. From Conceptual Example 2, we know that the pebble will fly off tangentially. Therefore, the path s is perpendicular to the radius r of the circle. Thus, the distances r, s, and d form a right triangle with hypotenuse d as shown in the figure at the right. From the figure we see that

Target
s
Pebble
d
α
r
35°
θ
C

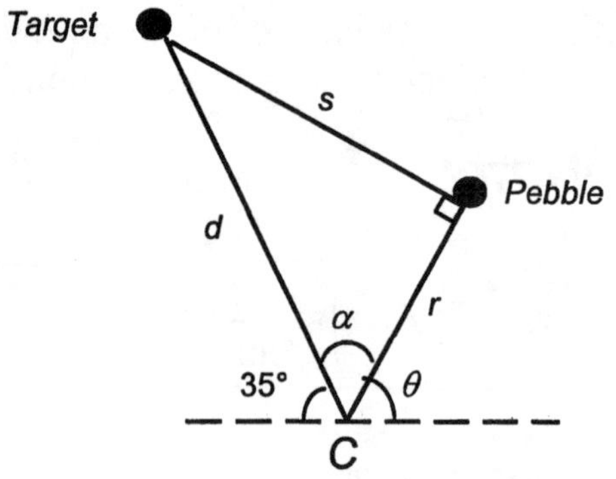

$$\cos\alpha = \frac{r}{d} = \frac{r}{10r} = \frac{1}{10}$$

or

$$\alpha = \cos^{-1}\left(\frac{1}{10}\right) = 84°$$

Furthermore, from the figure, we see that $\alpha + \theta + 35° = 180°$. Therefore,

$$\theta = 145° - \alpha = 145° - 84° = \boxed{61°}$$

---

46. ***REASONING AND SOLUTION*** The force **P** supplied by the man will be largest when the partner is at the lowest point in the swing. The diagram at the right shows the forces acting on the partner in this situation. The centripetal force necessary to keep the partner swinging along the arc of a circle is provided by the resultant of the force supplied by the man and the weight of the partner. From the figure

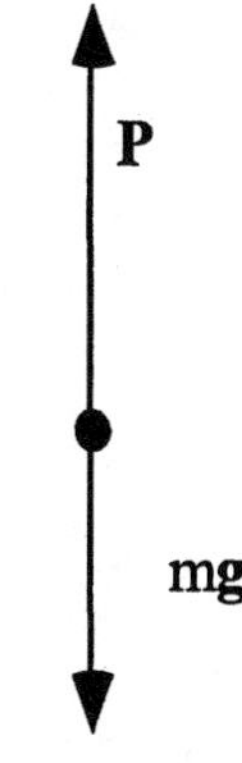

$$P - mg = \frac{mv^2}{r}$$

Therefore,

$$P = \frac{mv^2}{r} + mg$$

Since the weight of the partner, W, is equal to mg, it follows that $m = (W/g)$ and

$$P = \frac{(W/g)v^2}{r} + W = \frac{[(475\text{ N})/(9.80\text{ m/s}^2)]\,(4.00\text{ m/s})^2}{(6.50\text{ m})} + (475\text{ N}) = \boxed{594\text{ N}}$$

---

47. SSM ***REASONING AND SOLUTION*** In each case, the magnitude of the centripetal acceleration is given by Equation 5.2, $a_c = v^2/r$. Therefore,

$$\frac{a_{cA}}{a_{cB}} = \frac{v_A^2/r_A}{v_B^2/r_B}$$

Since each boat experiences the same centripetal acceleration, $a_{cA}/a_{cB} = 1$. Solving for the ratio of the speeds gives

$$\frac{v_A}{v_B} = \sqrt{\frac{r_A}{r_B}} = \sqrt{\frac{120\text{ m}}{240\text{ m}}} = \boxed{0.71}$$

---

48. ***REASONING*** The centripetal force is the name given to the net force pointing toward the center of the circular path. At the lowest point the net force consists of the tension in the arm pointing upward toward the center and the weight pointing downward or away from the center. In either case the centripetal force is given by Equation 5.3 as $F_c = mv^2/r$.

SOLUTION (a) The centripetal force is

$$F_c = \frac{mv^2}{r} = \frac{(9.5\text{ kg})(2.8\text{ m/s})^2}{0.85\text{ m}} = \boxed{88\text{ N}}$$

(b) Using T to denote the tension in the arm, at the bottom of the circle we have

$$F_c = T - mg = \frac{mv^2}{r}$$

$$T = mg + \frac{mv^2}{r} = (9.5\text{kg})(9.80\text{ m/s}^2) + \frac{(9.5\text{ kg})(2.8\text{ m/s})^2}{0.85\text{ m}} = \boxed{181\text{ N}}$$

---

49. SSM ***REASONING AND SOLUTION*** Since the tension serves the same purpose as the normal force at point 1 in Figure 5.21, we have, using the equation for the situation at point 1 with F_{N1} replaced by T,

$$\frac{mv^2}{r} = T - mg$$

Solving for T gives

$$T = \frac{mv^2}{r} + mg = m\left(\frac{v^2}{r} + g\right) = (2100\text{ kg})\left[\frac{(7.6\text{ m/s})^2}{15\text{ m}} + (9.80\text{ m/s}^2)\right] = \boxed{2.9\times10^4\text{ N}}$$

---

50. ***REASONING AND SOLUTION*** The period of a satellite is given by

$$T = \sqrt{\frac{4\pi^2 r^3}{GM_E}} = \sqrt{\frac{4\pi^2\left[2(6.38\times10^6\text{ m})\right]^3}{(6.67\times10^{-11}\text{ N}\cdot\text{m}^2/\text{kg}^2)(5.98\times10^{24}\text{ kg})}} = \boxed{1.43\times10^4\text{ s}}$$

---

51. ***REASONING*** The centripetal acceleration for any point that is a distance r from the center of the disc is, according to Equation 5.2, $a_c = v^2/r$. From Equation 5.1, we know that $v = 2\pi r/T$ where T is the period of the motion. Combining these two equations, we obtain

$$a_c = \frac{(2\pi r/T)^2}{r} = \frac{4\pi^2 r}{T^2}$$

SOLUTION Using the above expression for a_c, the ratio of the centripetal accelerations of the two points in question is

$$\frac{a_2}{a_1} = \frac{4\pi^2 r_2/T_2^2}{4\pi^2 r_1/T_1^2} = \frac{r_2/T_2^2}{r_1/T_1^2}$$

Since the disc is rigid, all points on the disc must move with the same period, so $T_1 = T_2$. Making this cancellation and solving for a_2, we obtain

$$a_2 = a_1 \frac{r_2}{r_1} = (120 \text{ m/s}^2)\left(\frac{0.050 \text{ m}}{0.030 \text{ m}}\right) = \boxed{2.0 \times 10^2 \text{ m/s}^2}$$

Note that even though $T_1 = T_2$, it is not true that $v_1 = v_2$. Thus, the simplest way to approach this problem is to express the centripetal acceleration in terms of the period T which cancels in the final step.

---

52. ***REASONING AND SOLUTION***
a. At the equator a person travels in a circle whose radius equals the radius of the earth, $r = R_e = 6.38 \times 10^6$ m, and whose period of rotation is $T = 1$ day $= 86\,400$ s. We have

$$v = 2\pi R_e / T = \boxed{464 \text{ m/s}}$$

The centripetal acceleration is

$$a_c = \frac{v^2}{r} = \frac{(464 \text{ m/s})^2}{6.38 \times 10^6 \text{ m}} = \boxed{3.37 \times 10^{-2} \text{ m/s}^2}$$

b. At 30.0° latitude a person travels in a circle of radius,

$$r = R_e \cos 30.0° = 5.53 \times 10^6 \text{ m}$$

Thus,

$$v = 2\pi r / T = \boxed{402 \text{ m/s}} \quad \text{and} \quad a_c = v^2 / r = \boxed{2.92 \times 10^{-2} \text{ m/s}^2}$$

---

53. ***REASONING AND SOLUTION*** If F is the force on mass 2, then $2F$ is the force on mass 1, and we have

$$\text{For mass 1:} \quad 2F = m_1 v_1^2 / r_1 \qquad\qquad \text{For mass 2:} \quad F = m_2 v_2^2 / r_2$$

Dividing the equations and rearranging gives

$$m_2/m_1 = (1/2)(r_2/r_1)(v_1/v_2)^2 = (v_1/v_2)^2 \text{ since } r_2 = 2r_1$$

The period of revolution is the same for both masses so $v_1 = 2\pi r_1/T$ and $v_2 = 2\pi r_2/T$. Dividing these gives $v_1/v_2 = r_1/r_2 = 1/2$. Now

$$m_2/m_1 = (1/2)^2 = \boxed{\frac{1}{4}}$$

---

54. ***REASONING AND SOLUTION***
a. The centripetal force is provided by $\boxed{\text{the normal force exerted on the rider by the wall}}$.

b. Newton's second law applied in the horizontal direction gives

$$F_N = mv^2/r = (55.0\text{ kg})(10.0\text{ m/s})^2/(3.30\text{ m}) = \boxed{1670\text{ N}}$$

c. Newton's second law applied in the vertical direction gives $\mu_s F_N - mg = 0$ or

$$\mu_s = (mg)/F_N = \boxed{0.323}$$

---

55. SSM WWW ***REASONING*** If the effects of gravity are not ignored in Example 5, the plane will make an angle θ with the vertical as shown in figure **A** below. The figure **B** shows the forces that act on the plane, and figure **C** shows the horizontal and vertical components of these forces.

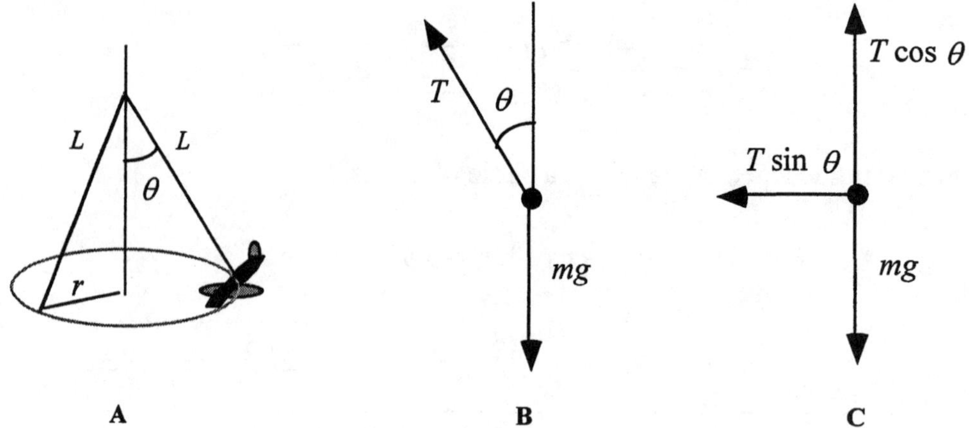

From figure **C** we see that the resultant force in the horizontal direction is the horizontal component of the tension in the guideline and provides the centripetal force. Therefore,

$$T\sin\theta = \frac{mv^2}{r}$$

From figure **A**, the radius r is related to the length L of the guideline by $r = L\sin\theta$; therefore,

$$T\sin\theta = \frac{mv^2}{L\sin\theta} \qquad (1)$$

The resultant force in the vertical direction is zero: $T\cos\theta - mg = 0$, so that

$$T\cos\theta = mg \qquad (2)$$

From equation (2) we have

$$T = \frac{mg}{\cos\theta} \qquad (3)$$

Equation (3) contains two unknown, T and θ. First we will solve equations (1) and (3) simultaneously to determine the value(s) of the angle θ. Once θ is known, we can calculate the tension using equation (3).

SOLUTION Substituting equation (3) into equation (1):

$$\left(\frac{mg}{\cos\,\theta}\right)\sin\,\theta = \frac{mv^2}{L\,\sin\,\theta}$$

Thus,

$$\frac{\sin^2\,\theta}{\cos\,\theta} = \frac{v^2}{gL} \qquad (4)$$

Using the fact that $\cos^2\,\theta + \sin^2\,\theta = 1$, equation (4) can be written

$$\frac{1-\cos^2\,\theta}{\cos\,\theta} = \frac{v^2}{gL}$$

or

$$\frac{1}{\cos\,\theta} - \cos\,\theta = \frac{v^2}{gL}$$

This can be put in the form of an equation that is quadratic in $\cos\,\theta$. Multiplying both sides by $\cos\,\theta$ and rearranging yields:

$$\cos^2\,\theta + \frac{v^2}{gL}\cos\,\theta - 1 = 0 \qquad (5)$$

Equation (5) is of the form

$$ax^2 + bx + c = 0 \qquad (6)$$

with $x = \cos\,\theta$, $a = 1$, $b = v^2/(gL)$, and $c = -1$. The solution to equation (6) is found from the quadratic formula:

$$x = \frac{-b \pm \sqrt{b^2 - 4ac}}{2a}$$

When $v = 19.0$ m/s, $b = 2.17$. The positive root from the quadratic formula gives $x = \cos\,\theta = 0.391$. Substitution into equation (3) yields

$$T = \frac{mg}{\cos\,\theta} = \frac{(0.900\text{ kg})(9.80\text{ m/s}^2)}{0.391} = \boxed{23\text{ N}}$$

When $v = 38.0$ m/s, $b = 8.67$. The positive root from the quadratic formula gives $x = \cos\,\theta = 0.114$. Substitution into equation (3) yields

$$T = \frac{mg}{\cos\theta} = \frac{(0.900\text{ kg})(9.80\text{ m/s}^2)}{0.114} = \boxed{77\text{ N}}$$

---

56. ***CONCEPT QUESTIONS*** a. The period for the second hand is the time it takes it to go once around the circle or $T_{\text{second}} = 60$ s. Similarly, for the minute hand the period is $T_{\text{minute}} = 1\text{ h} = 3600$ s.

b. The relationship between the centripetal acceleration and the period can be obtained by using Equations 5.2 and 5.1:

$$a_c = \frac{v^2}{r} \quad (5.2)$$

$$v = \frac{2\pi r}{T} \quad (5.1)$$

$$a_c = \frac{(2\pi r/T)^2}{r} = \frac{4\pi^2 r}{T^2}$$

SOLUTION Using the expression for the centripetal acceleration obtained in the answer to concept question b, we have

$$\frac{a_{\text{c, second}}}{a_{\text{c, minute}}} = \frac{4\pi^2 r/T_{\text{second}}^2}{4\pi^2 r/T_{\text{minute}}^2} = \frac{T_{\text{minute}}^2}{T_{\text{second}}^2}$$

$$= \frac{(3600\text{ s})^2}{(60\text{ s})^2} = \boxed{3600}$$

---

57. ***CONCEPT QUESTIONS*** a. Example 2 has the smallest centripetal acceleration. Uniform circular motion on a circle with an infinitely large radius is like motion along a straight line at a constant velocity. In such a case there is no centripetal acceleration.

b. Example 1 has the greatest centripetal acceleration. According to Equation 5.2, the acceleration is $a_c = v^2/r$. We note that the speed is in the numerator and the radius is in the denominator of this expression. Therefore, the greatest speed and the smallest radius, as is the case for Example 1, produce the greatest centripetal acceleration.

SOLUTION With Equation 5.2, we find the following values of the centripetal acceleration.

Example 1 $$a_c = \frac{v^2}{r} = \frac{(12\text{ m/s})^2}{0.50\text{ m}} = \boxed{290\text{ m/s}^2}$$

Example 2 $$a_c = \frac{v^2}{r} = \frac{(35\text{ m/s})^2}{\infty} = \boxed{0\text{ m/s}^2}$$

Example 3 $$a_c = \frac{v^2}{r} = \frac{(2.3\text{ m/s})^2}{1.8\text{ m}} = \boxed{2.9\text{ m/s}^2}$$

---

58. ***CONCEPT QUESTIONS*** a. The centripetal acceleration depends only on the speed v and the radius r of the curve, according to Equation 5.2 ($a_c = v^2/r$). The speeds of the cars are the same, and since they are negotiating the same curve, the radius is the same. Therefore, the cars have the same centripetal acceleration.

b. The centripetal force depends on the mass m, as well as the speed and the radius of the curve, according to Equation 5.3 ($F_c = mv^2/r$). Since the speed and the radius are the same for each car, the car with the greater mass, which is car B, experiences the greater centripetal acceleration.

SOLUTION Using Equations 5.2 and 5.3, we find the following values for the centripetal acceleration and force:

Car A $$a_c = \frac{v^2}{r} = \frac{(27\text{ m/s})^2}{120\text{ m}} = \boxed{6.1\text{ m/s}^2} \quad (5.2)$$

$$F_c = \frac{m_A v^2}{r} = \frac{(1100\text{ kg})(27\text{ m/s})^2}{120\text{ m}} = \boxed{6700\text{ N}} \quad (5.3)$$

Car B $$a_c = \frac{v^2}{r} = \frac{(27\text{ m/s})^2}{120\text{ m}} = \boxed{6.1\text{ m/s}^2} \quad (5.2)$$

$$F_c = \frac{m_B v^2}{r} = \frac{(1600\text{ kg})(27\text{ m/s})^2}{120\text{ m}} = \boxed{9700\text{ N}} \quad (5.3)$$

---

59. ***CONCEPT QUESTIONS*** a. Static, rather than kinetic, friction provides the centripetal force, because the penny is stationary and not sliding relative to the record.

b. The speed can be determined from the period T of the motion and the radius r, according to Equation 5.1 ($v = 2\pi r/T$). The period can be found from the fact that the record makes $33\frac{1}{3}$ revolutions in each minute or 60 s. Therefore,

$$T = \frac{1 \text{ min}}{33.3 \text{ rev}} \left(\frac{60 \text{ s}}{1 \text{ min}} \right) = 1.80 \text{ s}$$

SOLUTION Using Equation 5.3 for the centripetal force ($F_c = mv^2/r$) and Equation 5.1 for the speed in terms of the period ($v = 2\pi r/T$), we have

$$F_c = \frac{mv^2}{r} = \frac{m(2\pi r/T)^2}{r} = \frac{m4\pi^2 r}{T^2} \quad (1)$$

According to Equation 4.7, the maximum force of static friction is $f_s^{max} = \mu_s F_N$, where F_N is the normal force. Since the penny does not accelerate in the vertical direction, the upward normal force must be balanced by the downward-pointing weight, so that $F_N = mg$ and $f_s^{max} = \mu_s mg$. Using equation (1), we find

$$\underbrace{\mu_s mg}_{\text{Centripetal force}} = \frac{m4\pi^2 r}{T^2}$$

$$\mu_s = \frac{4\pi^2 r}{gT^2} = \frac{4\pi^2 (0.150 \text{ m})}{(9.80 \text{ m/s}^2)(1.80 \text{ s})^2} = \boxed{0.187}$$

---

60. ***CONCEPT QUESTIONS*** a. The satellite with the lower orbit has the greater speed, according to Equation 5.5 ($v = \sqrt{GM_E/r}$), where r is the radius of the orbit. Thus, satellite A has the greater speed.

b. In Equation 5.5 ($v = \sqrt{GM_E/r}$) the term r is the orbital radius, as measured from the center of the earth, not the surface of the earth. Therefore, you do not substitute the heights of 360×10^3 m and 720×10^3 m for the term r. Instead, these heights must be added to the radius of the earth (6.38×10^6 m) in order to get the radii.

SOLUTION First we add the orbital heights to the radius of the earth to obtain the orbital radii. Then we use Equation 5.5 to calculate the speeds.

Satellite A $r_A = 6.38 \times 10^6 \text{ m} + 360 \times 10^3 \text{ m} = 6.74 \times 10^6 \text{ m}$

$$v = \sqrt{\frac{GM_E}{r_A}} = \sqrt{\frac{(6.67 \times 10^{-11} \text{ N} \cdot \text{m}^2/\text{kg}^2)(5.98 \times 10^{24} \text{ kg})}{6.74 \times 10^6 \text{ m}}} = \boxed{7690 \text{ m/s}}$$

Satellite B $r_A = 6.38 \times 10^6 \text{ m} + 720 \times 10^3 \text{ m} = 7.10 \times 10^6 \text{ m}$

$$v = \sqrt{\frac{GM_E}{r_A}} = \sqrt{\frac{(6.67 \times 10^{-11} \text{ N} \cdot \text{m}^2/\text{kg}^2)(5.98 \times 10^{24} \text{ kg})}{7.10 \times 10^6 \text{ m}}} = \boxed{7500 \text{ m/s}}$$

---

61. ***CONCEPT QUESTIONS*** a. Since the speed and mass are constant and the radius is fixed, the centripetal force is the same at each point on the circle.

b. When the ball is at the three o'clock position, the force of gravity, acting downward, is perpendicular to the string and cannot contribute to the centripetal force. (See Figure 5.21, point 2 for a similar situation.) At this point, only the tension of $T = 16$ N contributes to the centripetal force. Considering that the centripetal force is the same everywhere, we can conclude that it has a value of 16 N everywhere.

c. At the twelve o'clock position the tension T and the force of gravity mg both act downward (the negative direction) toward the center of the circle, with the result that the centripetal force at this point is $-T - mg$. (See Figure 5.21, point 3.) The magnitude of the centripetal force here, then, is $T + mg$. At the six o'clock position the tension points upward toward the center of the circle, while the force of gravity points downward, with the result that the centripetal force at this point is $T - mg$. (See Figure 5.21, point 1.) The only way for centripetal force to have the same magnitude of 16 N at both of these places is for the tension at the six o'clock position to be greater. The greater tension compensates for the fact that the force of gravity points away from the center of the circle.

SOLUTION Assuming that upward is the positive direction, we find at the twelve and six o'clock positions that

Twelve o'clock

$$\underbrace{-T - mg}_{\text{Centripetal force}} = -16 \text{ N}$$

$$T = 16 \text{ N} - (0.20 \text{ kg})(9.80 \text{ m/s}^2) = \boxed{14 \text{ N}}$$

Six o'clock

$$\underbrace{T - mg}_{\text{Centripetal force}} = 16 \text{ N}$$

$$T = 16 \text{ N} + (0.20 \text{ kg})(9.80 \text{ m/s}^2) = \boxed{18 \text{ N}}$$

CHAPTER 6 | *WORK AND ENERGY*

PROBLEMS

---

1. SSM ***REASONING AND SOLUTION*** The work done by the retarding force is given by Equation 6.1: $W = (F \cos\theta)s$. Since the force is a retarding force, it must point opposite to the direction of the displacement, so that $\theta = 180°$. Thus, we have

$$W = (F\cos\theta)s = (3.0 \times 10^3 \text{ N})(\cos 180°)(850 \text{ m}) = \boxed{-2.6 \times 10^6 \text{ J}}$$

The work done by this force is $\boxed{\text{negative}}$, because the retarding force is directed opposite to the direction of the displacement of the truck.

---

2. ***REASONING AND SOLUTION*** Each locomotive does work

$$W = Ts\cos\theta = (5.00 \times 10^3 \text{ N})(2.00 \times 10^3 \text{ m}) \cos 20.0° = 9.40 \times 10^6 \text{ J}$$

The net work is then

$$W_T = 2W = \boxed{1.88 \times 10^7 \text{ J}}$$

---

3. ***REASONING AND SOLUTION*** According to Equation 6.1, $W = Fs \cos\theta$, the work is

a. $$W = (94.0 \text{ N})(35.0 \text{ m}) \cos 25.0° = \boxed{2980 \text{ J}}$$

b. $$W = (94.0 \text{ N})(35.0 \text{ m}) \cos 0° = \boxed{3290 \text{ J}}$$

---

4. ***REASONING***

a. The work done by the gravitational force is given by Equation 6.1 as $W = (F\cos\theta)\,s$. The gravitational force points downward, opposite to the upward vertical displacement of 4.60 m. Therefore, the angle θ is 180°.

b. The work done by the escalator is done by the upward normal force that the escalator exerts on the man. Since the man is moving at a constant velocity, he is in equilibrium, and the net force acting on him must be zero. This means that the normal force must balance the man's weight. Thus, the magnitude of the normal force is $F_N = mg$, and the work that the escalator does is also given by Equation 6.1. However, since the normal force and the upward vertical displacement point in the same direction, the angle θ is 0°.

SOLUTION

a. According to Equation 6.1, the work done by the gravitational force is

$$W = (F\cos\theta)s = (mg\cos\theta)s$$
$$= (75.0\text{ kg})(9.80\text{ m/s}^2)\cos 180°(4.60\text{ m}) = \boxed{-3.38\times10^3\text{ J}}$$

b. The work done by the escalator is

$$W = (F\cos\theta)s = (F_N\cos\theta)s$$
$$= (75.0\text{ kg})(9.80\text{ m/s}^2)\cos 0°(4.60\text{ m}) = \boxed{3.38\times10^3\text{ J}}$$

---

5. **SSM** ***REASONING AND SOLUTION*** Solving Equation 6.1 for the angle θ, we obtain

$$\theta = \cos^{-1}\left(\frac{W}{Fs}\right) = \cos^{-1}\left[\frac{1.10\times10^3\text{ J}}{(30.0\text{ N})(50.0\text{ m})}\right] = \boxed{42.8°}$$

---

6. ***REASONING AND SOLUTION***

a. In both cases, the lift force **L** is perpendicular to the displacement of the plane, and, therefore, does no work. As shown in the drawing in the text, when the plane is in the dive, there is a component of the weight **W** that points in the direction of the displacement of the plane. When the plane is climbing, there is a component of the weight that points opposite to the displacement of the plane. Thus, since the thrust **T** is the same for both cases, the net force in the direction of the displacement is greater for the case where the plane is diving. Since the displacement **s** is the same in both cases, $\boxed{\text{more net work is done during the dive}}$.

b. The work done during the dive is $W_{\text{dive}} = (T + W\cos 75°)\,s$, while the work done during the climb is $W_{\text{climb}} = (T + W\cos 115°)\,s$. Therefore, the difference between the net work done during the dive and the climb is

$$W_{\text{dive}} - W_{\text{climb}} = (T + W\cos 75°)\,s - (T + W\cos 115°)\,s = Ws\,(\cos 75° - \cos 115°)$$
$$= (5.9\times10^4\text{ N})(1.7\times10^3\text{ m})(\cos 75° - \cos 115°) = \boxed{6.8\times10^7\text{ J}}$$

---

7. ***REASONING*** The drawing shows three of the forces that act on the cart: **F** is the pushing force that the shopper exerts, **f** is the frictional force that opposes the motion of the cart, and *mg* is its weight. The displacement **s** of the cart is also shown. Since the cart moves at a constant velocity along the $+x$

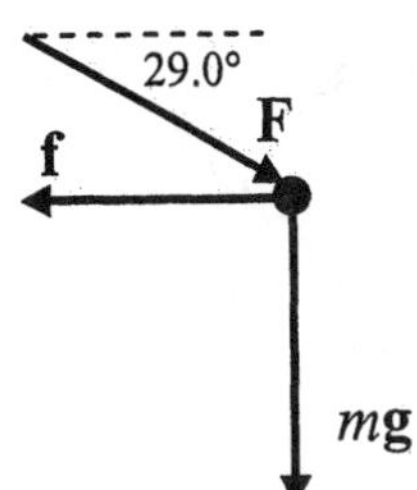

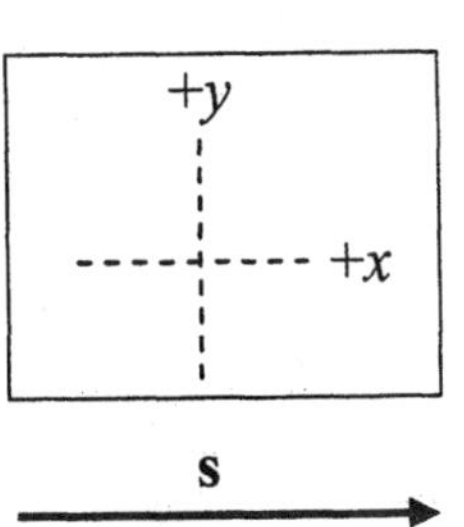

direction, it is in equilibrium. The net force acting on it in this direction is zero, $\Sigma F_x = 0$. This relation can be used to find the magnitude of the pushing force. The work done by a constant force is given by Equation 6.1 as $W = (F\cos\theta)s$, where F is the magnitude of the force, s is the magnitude of the displacement, and θ is the angle between the force and the displacement.

SOLUTION

a. The x-component of the net force is zero, $\Sigma F_x = 0$, so that

$$\underbrace{F\cos 29.0° - f}_{\Sigma F_x} = 0 \qquad (4.9a)$$

The magnitude of the force that the shopper exerts is $F = \dfrac{f}{\cos 29.0°} = \dfrac{48.0\text{ N}}{\cos 29.0°} = \boxed{54.9\text{ N}}$.

b. The work done by the pushing force **F** is

$$W = (F\cos\theta)s = (54.9\text{ N})(\cos 29.0°)(22.0\text{ m}) = \boxed{1060\text{ J}} \qquad (6.1)$$

c. The angle between the frictional force and the displacement is 180°, so the work done by the frictional force **f** is

$$W = (f\cos\theta)s = (48.0\text{ N})(\cos 180.0°)(22.0\text{ m}) = \boxed{-1060\text{ J}}$$

d. The angle between the weight of the cart and the displacement is 90°, so the work done by the weight $m\mathbf{g}$ is

$$W = (mg\cos\theta)s = (16.0\text{ kg})(9.80\text{ m/s}^2)(\cos 90°)(22.0\text{ m}) = \boxed{0\text{ J}}$$

---

8. ***REASONING AND SOLUTION*** The applied force does work

$$W_P = Ps\cos 0° = (150\text{ N})(7.0\text{ m}) = \boxed{1.0\times 10^3\text{ J}}$$

The frictional force does work

$$W_f = f_k s\cos 180° = -\mu_k F_N s$$

where $F_N = mg$, so

$$W_f = -(0.25)(55\text{ kg})(9.80\text{ m/s}^2)(7.0\text{ m}) = \boxed{-940\text{ J}}$$

The normal force and gravity do no work, since they both act at a 90° angle to the displacement.

---

9. SSM ***REASONING AND SOLUTION*** According to Equation 6.1, the work done by the husband and wife are, respectively,

$$\text{[Husband]} \qquad W_H = (F_H \cos\theta_H)s$$

$$\text{[Wife]} \qquad W_W = (F_W \cos\theta_W)s$$

Since both the husband and the wife do the same amount of work,

$$(F_H \cos\theta_H)s = (F_W \cos\theta_W)s$$

Since the displacement has the same magnitude s in both cases, the magnitude of the force exerted by the wife is

$$F_W = F_H \frac{\cos\theta_H}{\cos\theta_W} = (67\text{ N})\frac{\cos 58°}{\cos 38°} = \boxed{45\text{ N}}$$

---

10. ***REASONING*** The net work done by the pushing force and the frictional force is zero, and our solution is focused on this fact. Thus, we express this net work as $W_P + W_f = 0$, where W_P is the work done by the pushing force and W_f is the work done by the frictional force. We will substitute for each individual work using Equation 6.1 [$W = (F\cos\theta)\,s$] and solve the resulting equation for the magnitude P of the pushing force.

SOLUTION According to Equation 6.1, the work done by the pushing force is

$$W_P = (P\cos 30.0°)\,s = 0.866\,P\,s$$

The frictional force opposes the motion, so the angle between the force and the displacement is 180°. Thus, the work done by the frictional force is

$$W_f = (f_k \cos 180°)\,s = -f_k\,s$$

Equation 4.8 indicates that the magnitude of the kinetic frictional force is $f_k = \mu_k F_N$, where F_N is the magnitude of the normal force acting on the crate. The free-body diagram shows the forces acting on the crate. Since there is no acceleration in the vertical direction, the y component of the net force must be zero:

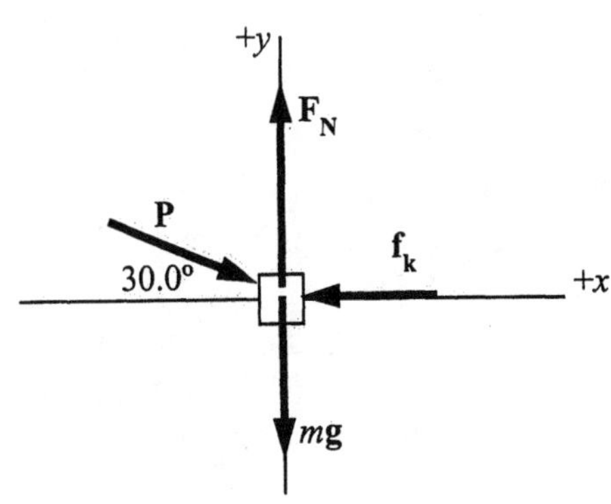

$$F_N - mg - P\sin 30.0° = 0$$

Therefore,

$$F_N = mg + P\sin 30.0°$$

It follows, then, that the magnitude of the frictional force is

$$f_k = \mu_k F_N = \mu_k (mg + P \sin 30.0°)$$

The work done by the frictional force is

$$W_f = -f_k s = -(0.200)[(1.00 \times 10^2 \text{ kg})(9.80 \text{ m/s}^2) + 0.500P]s = -(0.100P + 196)s$$

Since the net work is zero, we have

$$W_P + W_f = 0.866\, Ps - (0.100P + 196)s = 0$$

Eliminating s algebraically and solving for P gives $\boxed{P = 256 \text{ N}}$.

---

11. ***REASONING AND SOLUTION*** The net work done on the car is

$$W_T = W_F + W_f + W_g + W_N$$

$$W_T = Fs \cos 0.0° + fs \cos 180° - mgs \sin 5.0° + F_N s \cos 90°$$

Rearranging this result gives

$$F = \frac{W_T}{s} + f + mg \sin 5.0°$$

$$= \frac{150 \times 10^3 \text{ J}}{290 \text{ m}} + 524 \text{ N} + (1200 \text{ kg})(9.80 \text{ m/s}^2) \sin 5.0° = \boxed{2.07 \times 10^3 \text{ N}}$$

---

12. ***REASONING*** The work done by the catapult $W_{catapult}$ is one contribution to the work done by the net external force that changes the kinetic energy of the plane. The other contribution is the work done by the thrust force of the plane's engines W_{thrust}. According to the work-energy theorem (Equation 6.3), the work done by the net external force $W_{catapult}$ + W_{thrust} is equal to the change in the kinetic energy. The change in the kinetic energy is the given kinetic energy of 4.5×10^7 J at lift-off minus the initial kinetic energy, which is zero since the plane starts at rest. The work done by the thrust force can be determined from Equation 6.1 [$W = (F \cos \theta)\, s$], since the magnitude F of the thrust is 2.3×10^5 N and the magnitude s of the displacement is 87 m. We note that the angle θ between the thrust and the displacement is 0°, because they have the same direction. In summary, we will calculate $W_{catapult}$ from $W_{catapult} + W_{thrust} = KE_f - KE_0$.

SOLUTION According to the work-energy theorem, we have

$$W_{\text{catapult}} + W_{\text{thrust}} = \text{KE}_{\text{f}} - \text{KE}_0$$

Using Equation 6.1 and noting that $\text{KE}_0 = 0$ J, we can write the work energy theorem as follows:

$$W_{\text{catapult}} + \underbrace{(F\cos\theta)s}_{\text{Work done by thrust}} = \text{KE}_{\text{f}}$$

Solving for W_{catapult} gives

$$W_{\text{catapult}} = \text{KE}_{\text{f}} - \underbrace{(F\cos\theta)s}_{\text{Work done by thrust}}$$

$$= 4.5\times10^7\ \text{J} - (2.3\times10^5\ \text{N})\cos 0°(87\ \text{m}) = \boxed{2.5\times10^7\ \text{J}}$$

---

13. SSM ***REASONING*** The work done to launch either object can be found from Equation 6.3, the work-energy theorem, $W = \text{KE}_{\text{f}} - \text{KE}_0 = \frac{1}{2}mv_{\text{f}}^2 - \frac{1}{2}mv_0^2$.

SOLUTION

a. The work required to launch the hammer is

$$W = \tfrac{1}{2}mv_{\text{f}}^2 - \tfrac{1}{2}mv_0^2 = \tfrac{1}{2}m\left(v_{\text{f}}^2 - v_0^2\right) = \tfrac{1}{2}(7.3\ \text{kg})\left[(29\ \text{m/s})^2 - (0\ \text{m/s})^2\right] = \boxed{3.1\times10^3\ \text{J}}$$

b. Similarly, the work required to launch the bullet is

$$W = \tfrac{1}{2}m\left(v_{\text{f}}^2 - v_0^2\right) = \tfrac{1}{2}(0.0026\ \text{kg})\left[(410\ \text{m/s})^2 - (0\ \text{m/s})^2\right] = \boxed{2.2\times10^2\ \text{J}}$$

---

14. ***REASONING AND SOLUTION*** The work done on the arrow by the bow is given by

$$W = Fs\cos 0° = Fs$$

This work is converted into kinetic energy according to the work energy theorem.

$$W = \tfrac{1}{2}mv_{\text{f}}^2 - \tfrac{1}{2}mv_0^2$$

Solving for v_{f}, we find that

$$v_{\text{f}} = \sqrt{\frac{2W}{m} + v_0^2} = \sqrt{\frac{2(65\ \text{N})(0.90\ \text{m})}{75\times10^{-3}\ \text{kg}} + (0\ \text{m/s})^2} = \boxed{39\ \text{m/s}}$$

---

15. ***REASONING AND SOLUTION***

a. The work-energy theorem gives

$$W = (1/2)mv_f^2 - (1/2)mv_o^2 = (1/2)(45 \times 10^{-3}\text{ kg})(41\text{ m/s})^2 = \boxed{38\text{ J}}$$

b. From the definition of work

$$W = Fs\cos 0°$$

so

$$F = W/s = (38\text{ J})/(1.0 \times 10^{-2}\text{ m}) = \boxed{3.8 \times 10^3\text{ N}}$$

---

16. ***REASONING AND SOLUTION*** From the work-energy theorem, Equation 6.3,

$$W = \tfrac{1}{2}mv_f^2 - \tfrac{1}{2}mv_0^2 = \tfrac{1}{2}m\left[v_f^2 - v_0^2\right]$$

a. $W = \tfrac{1}{2}(7420\text{ kg})\left[(8450\text{ m/s})^2 - (2820\text{ m/s})^2\right] = \boxed{2.35\times 10^{11}\text{ J}}$

b. $\text{W} = \tfrac{1}{2}(7420\text{ kg})\left[(2820\text{ m/s})^2 - (8450\text{ m/s})^2\right] = \boxed{-2.35\times 10^{11}\text{ J}}$

---

17. SSM WWW ***REASONING AND SOLUTION*** The work required to bring each car up to speed is, from the work-energy theorem, $W = \text{KE}_f - \text{KE}_0 = \frac{1}{2}mv_f^2 - \frac{1}{2}mv_0^2$. Therefore,

$$W_B = \tfrac{1}{2}m\left(v_f^2 - v_0^2\right) = \tfrac{1}{2}(1.20\times 10^3\text{ kg})\left[(40.0\text{ m/s})^2 - \left(0\text{ m/s}\right)^2\right] = 9.60\times 10^5\text{ J}$$

$$W_B = \tfrac{1}{2}m\left(v_f^2 - v_0^2\right) = \tfrac{1}{2}(2.00\times 10^3\text{ kg})\left[(40.0\text{ m/s})^2 - \left(0\text{ m/s}\right)^2\right] = 1.60\times 10^6\text{ J}$$

The *additional* work required to bring car B up to speed is, therefore,

$$W_B - W_A = (1.6\times 10^6\text{ J}) - (9.60\times 10^5\text{ J}) = \boxed{6.4\times 10^5\text{ J}}$$

---

18. ***REASONING AND SOLUTION*** The work energy theorem, $W = \frac{1}{2}mv_f^2 - \frac{1}{2}mv_0^2$, gives

$$v_f = \sqrt{\frac{2W}{m} + v_0^2}$$

where

$$W = Fs \cos 180° = -(4.0 \times 10^5 \text{ N})(2500 \times 10^3 \text{ m}) = -1.0 \times 10^{12} \text{ J}$$

Now

$$v_f = \sqrt{\frac{2\left(-1.0 \times 10^{12} \text{ J}\right)}{5.0 \times 10^4 \text{ kg}} + \left(11\,000 \text{ m/s}\right)^2} = \boxed{9 \times 10^3 \text{ m/s}}$$

---

19. SSM ***REASONING*** According to the work-energy theorem, the kinetic energy of the sled increases in each case because work is done on the sled. The work-energy theorem is given by Equation 6.3: $W = \text{KE}_f - \text{KE}_0 = \frac{1}{2}mv_f^2 - \frac{1}{2}mv_0^2$. The work done on the sled is given by Equation 6.1: $W = (F\cos\theta)s$. The work done in each case can, therefore, be expressed as

$$W_1 = (F \cos 0°)s = \frac{1}{2}mv_f^2 - \frac{1}{2}mv_0^2 = \Delta\text{KE}_1$$

and

$$W_2 = (F \cos 62°)s = \frac{1}{2}mv_f^2 - \frac{1}{2}mv_0^2 = \Delta\text{KE}_2$$

The fractional increase in the kinetic energy of the sled when $\theta = 0°$ is

$$\frac{\Delta\text{KE}_1}{\text{KE}_0} = \frac{(F \cos 0°)s}{\text{KE}_0} = 0.38$$

Therefore,

$$Fs = (0.38)\ \text{KE}_0 \qquad (1)$$

The fractional increase in the kinetic energy of the sled when $\theta = 62°$ is

$$\frac{\Delta\text{KE}_2}{\text{KE}_0} = \frac{(F \cos 62°)s}{\text{KE}_0} = \frac{Fs}{\text{KE}_0}(\cos 62°) \qquad (2)$$

Equation (1) can be used to substitute in Equation (2) for Fs.

SOLUTION Combining Equations (1) and (2), we have

$$\frac{\Delta\text{KE}_2}{\text{KE}_0} = \frac{Fs}{\text{KE}_0}(\cos 62°) = \frac{(0.38)\ \text{KE}_0}{\text{KE}_0}(\cos 62°) = (0.38)(\cos 62°) = 0.18$$

Thus, the sled's kinetic energy would increase by $\boxed{18\ \%}$.

---

20. ***REASONING*** To find the coefficient of kinetic friction μ_k, we need to find the force of kinetic friction f_k and the normal force F_N (see Equation 4.8, $f_k = \mu_k F_N$). The normal force and the weight mg of the sled balance, since they are the only two forces acting vertically and the sled does not accelerate in the vertical direction. The force of kinetic friction can be obtained from the work W_f done by the frictional force, according to Equation 6.1 [$W_f = (f_k \cos\theta)\, s$], where s is the magnitude of the displacement. To find the work, we will employ the work-energy theorem, as given in Equation 6.3 ($W = \text{KE}_f - \text{KE}_0$). In this equation W is the work done by the net force, but the normal force and the weight balance, so the net force is that due to the pulling force P and the frictional force. As a result $\text{W} = \text{W}_{\text{pull}} + \text{W}_f$.

SOLUTION According to the work-energy theorem, we have

$$\text{W} = \text{W}_{\text{pull}} + \text{W}_f = \text{KE}_f - \text{KE}_0$$

Using Equation 6.1 [$W = (F\cos\theta)\, s$] to express each work contribution, writing the kinetic energy as $\frac{1}{2}mv^2$, and noting that the initial kinetic energy is zero (the sled starts from rest), we obtain

$$\underbrace{(P\cos 0°)s}_{W_{\text{pull}}} + \underbrace{(f_k \cos 180°)s}_{W_f} = \tfrac{1}{2}mv^2.$$

The angle θ between the force and the displacement is 0° for the pulling force (it points in the same direction as the displacement) and 180° for the frictional force (it points opposite to the displacement). Equation 4.8 indicates that the magnitude of the frictional force is $f_k = \mu_k F_N$, and we know that the magnitude of the normal force is $F_N = mg$. With these substitutions the work-energy theorem becomes

$$\underbrace{(P\cos 0°)s}_{W_{\text{pull}}} + \underbrace{(\mu_k mg \cos 180°)s}_{W_f} = \tfrac{1}{2}mv^2$$

Solving for the coefficient of kinetic friction gives

$$\mu_k = \frac{\frac{1}{2}mv^2 - (P\cos 0°)s}{(mg\cos 180°)s} = \frac{\frac{1}{2}(16\text{ kg})(2.0\text{ m/s})^2 - (24\text{ N})(8.0\text{ m})}{-(16\text{ kg})(9.80\text{ m/s}^2)(8.0\text{ m})} = \boxed{0.13}$$

---

21. **SSM** **WWW** ***REASONING*** When the satellite goes from the first to the second orbit, its kinetic energy changes. The net work that the external force must do to change the orbit can be found from the work-energy theorem: $W = \text{KE}_f - \text{KE}_0 = \frac{1}{2}mv_f^2 - \frac{1}{2}mv_0^2$. The speeds v_f and v_0 can be obtained from Equation 5.5 for the speed of a satellite in a circular orbit of radius r. Given the speeds, the work energy theorem can be used to obtain the work.

SOLUTION According to Equation 5.5, $v = \sqrt{GM_E/r}$. Substituting into the work-energy theorem, we have

$$W = \tfrac{1}{2}mv_f^2 - \tfrac{1}{2}mv_0^2 = \tfrac{1}{2}m\left(v_f^2 - v_0^2\right) = \tfrac{1}{2}m\left[\left(\sqrt{\frac{GM_E}{r_f}}\right)^2 - \left(\sqrt{\frac{GM_E}{r_0}}\right)^2\right] = \frac{GM_E m}{2}\left(\frac{1}{r_f} - \frac{1}{r_0}\right)$$

Therefore,

$$W = \frac{(6.67\times10^{-11}\ \text{N}\cdot\text{m}^2/\text{kg}^2)(5.98\times10^{24}\ \text{kg})(6200\ \text{kg})}{2}$$

$$\times\left(\frac{1}{7.0\times10^6\ \text{m}} - \frac{1}{3.3\times10^7\ \text{m}}\right) = \boxed{1.4\times10^{11}\ \text{J}}$$

---

22. ***REASONING*** Since the person has an upward acceleration, there must be a net force acting in the upward direction. The net force ΣF_y is related to the acceleration a_y by Newton's second law, $\Sigma F_y = ma_y$, where m is the mass of the person. This relation will allow us to determine the tension in the cable. The work done by the tension and the person's weight can be found directly from the definition of work, Equation 6.1.

SOLUTION

a. The free-body diagram at the right shows the two forces that act on the person. Applying Newton's second law, we have

T

s

+y

mg

$$\underbrace{T - mg}_{\Sigma F_y} = ma_y$$

Solving for the magnitude of the tension in the cable yields

$$T = m(a_y + g) = (79\ \text{kg})(0.70\ \text{m/s}^2 + 9.80\ \text{m/s}^2) = \boxed{8.3\times10^2\ \text{N}}$$

b. The work done by the tension in the cable is

$$W_T = (T\cos\theta)s = (8.3\times10^2\ \text{N})(\cos 0°)(11\ \text{m}) = \boxed{9.1\times10^3\ \text{J}} \quad (6.1)$$

c. The work done by the person's weight is

$$W_W = (mg\cos\theta)s = (79\ \text{kg})(9.8\ \text{m/s}^2)(\cos 180°)(11\ \text{m}) = \boxed{-8.5\times10^3\ \text{J}} \quad (6.1)$$

d. The work-energy theorem relates the work done by the two forces to the change in the kinetic energy of the person. The work done by the two forces is $W = W_{\mathrm{T}} + W_{\mathrm{W}}$:

$$\underbrace{\mathrm{W_T} + \mathrm{W_W}}_{W} = \tfrac{1}{2}mv_{\mathrm{f}}^2 - \tfrac{1}{2}mv_0^2 \qquad (6.3)$$

Solving this equation for the final speed of the person gives

$$v_{\mathrm{f}} = \sqrt{v_0^2 + \frac{2}{m}\left(W_{\mathrm{T}} + W_{\mathrm{W}}\right)}$$

$$= \sqrt{(0\ \mathrm{m/s})^2 + \frac{2}{79\ \mathrm{kg}}\left(9.1\times10^3\ \mathrm{J} - 8.5\times10^3\ \mathrm{J}\right)} = \boxed{4\ \mathrm{m/s}}$$

---

23. ***REASONING*** It is useful to divide this problem into two parts. The first part involves the skier moving on the snow. We can use the work-energy theorem to find her speed when she comes to the edge of the cliff. In the second part she leaves the snow and falls freely toward the ground. We can again employ the work-energy theorem to find her speed just before she lands.

SOLUTION The drawing at the right shows the three forces that act on the skier as she glides on the snow. The forces are: her weight mg, the normal force $\mathbf{F}_{\mathrm{N}}$, and the kinetic frictional force $\mathbf{f}_{\mathrm{k}}$. Her displacement is labeled as **s**. The work-energy theorem, Equation 6.3, is

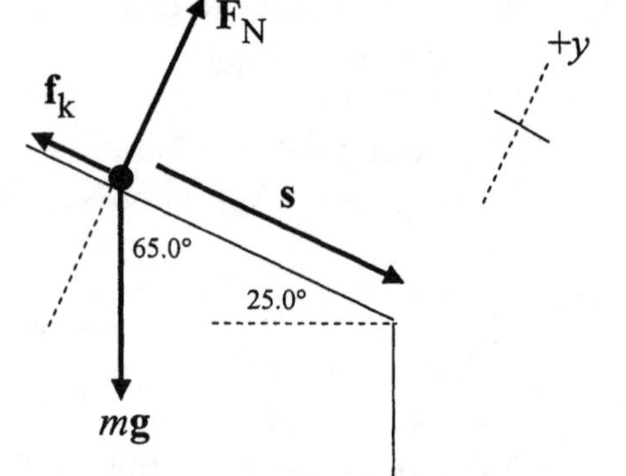

$$W = \tfrac{1}{2}mv_{\mathrm{f}}^2 - \tfrac{1}{2}mv_0^2$$

where W is the work done by the net external force that acts on the skier. The work done by each force is given by Equation 6.1, $W = (F\cos\theta)s$, so the work-energy theorem becomes

$$\underbrace{\left(mg\cos 65.0^\circ\right)s + \left(f_{\mathrm{k}}\cos 180^\circ\right)s + \left(F_{\mathrm{N}}\cos 90^\circ\right)s}_{W} = \tfrac{1}{2}mv_{\mathrm{f}}^2 - \tfrac{1}{2}mv_0^2$$

Since $\cos 90° = 0$, the third term on the left side can be eliminated. The magnitude f_{k} of the kinetic frictional force is given by Equation 4.8 as $f_{\mathrm{k}} = \mu_{\mathrm{k}}F_{\mathrm{N}}$. The magnitude F_{N} of the normal force can be determined by noting that the skier does not leave the surface of the slope, so $a_y = 0\ \mathrm{m/s^2}$. Thus, we have that $\Sigma F_y = 0$, so

$$\underbrace{F_N - mg\cos 25.0° = 0}_{\Sigma F_y} \quad \text{or} \quad F_N = mg\cos 25.0°$$

The magnitude of the kinetic frictional force becomes $f_k = \mu_k F_N = \mu_k mg\cos 25.0°$. Substituting this result into the work-energy theorem, we find that

$$\underbrace{(mg\cos 65.0°)s + (\mu_k mg\cos 25.0°)(\cos 180°)s}_{W} = \tfrac{1}{2}mv_f^2 - \tfrac{1}{2}mv_0^2$$

Algebraically eliminating the mass m of the skier from every term, setting cos 180° = –1 and $v_0 = 0$ m/s, and solving for the final speed v_f, gives

$$v_f = \sqrt{2gs(\cos 65.0° - \mu_k\cos 25.0°)}$$

$$= \sqrt{2(9.80\text{ m/s}^2)(10.4\text{ m})[\cos 65.0° - (0.200)\cos 25.0°]} = 7.01\text{ m/s}$$

The drawing at the right shows her displacement **s** during free fall. Note that the displacement is a vector that starts where she leaves the slope and ends where she touches the ground. The only force acting on her during the free fall is her weight $m\mathbf{g}$. The work-energy theorem, Equation 6.3, is

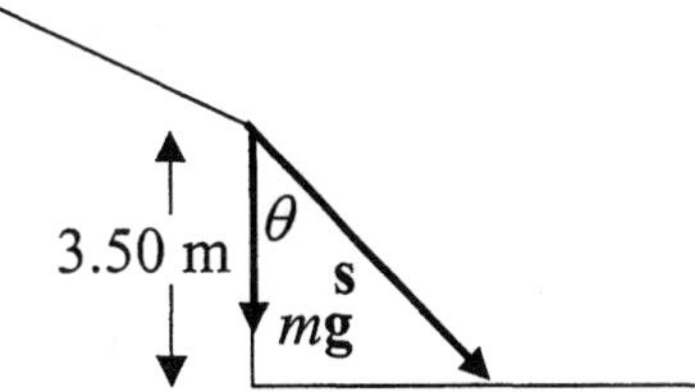

$$W = \tfrac{1}{2}mv_f^2 - \tfrac{1}{2}mv_0^2$$

The work W is that done by her weight, so the work-energy theorem becomes

$$\underbrace{(mg\cos\theta)s}_{W} = \tfrac{1}{2}mv_f^2 - \tfrac{1}{2}mv_0^2$$

In this expression θ is the angle between her weight (which points vertically downward) and her displacement. Note from the drawing that $s\cos\theta = 3.50$ m. Algebraically eliminating the mass m of the skier from every term in the equation above and solving for the final speed v_f gives

$$v_f = \sqrt{v_0^2 + 2g(s\cos\theta)}$$

$$= \sqrt{(7.01\text{ m/s})^2 + 2(9.80\text{ m/s}^2)(3.50\text{ m})} = \boxed{10.9\text{ m/s}}$$

---

24. ***REASONING AND SOLUTION*** The net work done on the plane can be found from the work-energy theorem:

$$W = \tfrac{1}{2}mv_f^2 - \tfrac{1}{2}mv_0^2 = \tfrac{1}{2}m\left[v_f^2 - v_0^2\right]$$

The tension in the guideline provides the centripetal force required to keep the plane moving in a circular path. Since the tension in the guideline becomes four times greater,

$$T_f = 4\,T_0$$

or

$$\frac{mv_f^2}{r_f} = 4\frac{mv_0^2}{r_0}$$

Solving for v_f^2 gives

$$v_f^2 = \frac{4v_0^2 r_f}{r_0}$$

Substitution of this expression into the work-energy theorem gives

$$W = \tfrac{1}{2}m\left[\frac{4v_0^2 r_f}{r_0} - v_0^2\right] = \tfrac{1}{2}mv_0^2\left[4\left(\frac{r_f}{r_0}\right) - 1\right]$$

Therefore,

$$W = \tfrac{1}{2}(0.90\text{ kg})(22\text{ m/s})^2\left[4\left(\frac{14\text{ m}}{16\text{ m}}\right) - 1\right] = \boxed{5.4\times10^2\text{ J}}$$

---

25. SSM ***REASONING*** During each portion of the trip, the work done by the resistive force is given by Equation 6.1, $W = (F\cos\theta)s$. Since the resistive force always points opposite to the displacement of the bicyclist, $\theta = 180°$; hence, on each part of the trip, $W = (F\cos 180°)s = -Fs$. The work done by the resistive force during the round trip is the algebraic sum of the work done during each portion of the trip.

SOLUTION

a. The work done during the round trip is, therefore,

$$W_{total} = W_1 + W_2 = -F_1 s_1 - F_2 s_2$$

$$= -(3.0\text{ N})(5.0\times10^3\text{ m}) - (3.0\text{ N})(5.0\times10^3\text{ m}) = \boxed{-3.0\times10^4\text{ J}}$$

b. Since the work done by the resistive force over the closed path is *not* zero, we can conclude that the resistive force is *not* a conservative force.

---

26. ***REASONING*** The work done by the weight of the basketball is given by Equation 6.1 as $W = (F \cos\theta)s$, where $F = mg$ is the magnitude of the weight, θ is the angle between the weight and the displacement, and s is the magnitude of the displacement. The drawing shows that the weight and displacement are parallel, so that $\theta = 0°$. The potential energy of the basketball is given by Equation 6.5 as PE = mgh, where h is the height of the ball above the ground.

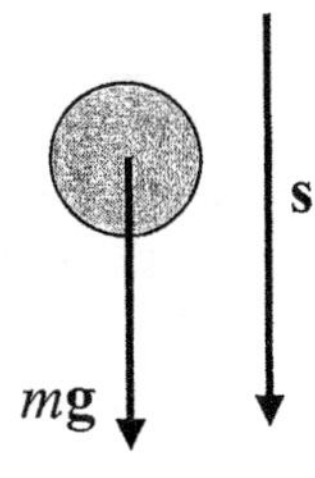

SOLUTION
a. The work done by the weight of the basketball is

$$W = (F \cos\theta)s = \text{mg}(\cos 0°)(h_0 - h_f) = (0.60 \text{ kg})(9.80 \text{ m/s}^2)(6.1 \text{ m} - 1.5 \text{ m}) = \boxed{27 \text{ J}}$$

b. The potential energy of the ball, relative to the ground, when it is released is

$$\text{PE}_0 = mgh_0 = (0.60 \text{ kg})(9.80 \text{ m/s}^2)(6.1 \text{ m}) = \boxed{36 \text{ J}} \quad (6.5)$$

c. The potential energy of the ball, relative to the ground, when it is caught is

$$\text{PE}_f = mgh_f = (0.60 \text{ kg})(9.80 \text{ m/s}^2)(1.5 \text{ m}) = \boxed{8.8 \text{ J}} \quad (6.5)$$

d. The change in the ball's gravitational potential energy is

$$\Delta\text{PE} = \text{PE}_f - \text{PE}_0 = 8.8 \text{ J} - 36 \text{ J} = -27 \text{ J}$$

We see that the change in the gravitational potential energy is equal to –27 J = $\boxed{-W}$, where W is the work done by the weight of the ball (see part a).

---

27. ***REASONING AND SOLUTION***

$$\text{PE} = \text{mgh} = (55.0 \text{ kg})(9.80 \text{ m/s}^2)(443\text{m}) = \boxed{2.39 \times 10^5 \text{ J}} \quad (6.5)$$

---

28. ***REASONING AND SOLUTION***
a. From the definition of work, $W = Fs \cos\theta$. For upward motion $\theta = 180°$ so

$$W = -mgs = -(71.1 \text{ N})(2.13 \text{ m} - 1.52 \text{ m}) = \boxed{-43 \text{ J}}$$

b. The change in potential energy is

$$\Delta\text{PE} = mgh_f - mgh_0 = (71.1 \text{ N})(2.13 \text{ m} - 1.52 \text{ m}) = \boxed{+43 \text{ J}}$$

---

29. ***REASONING AND SOLUTION***
The vertical height of the skier is $h = s \sin 14.6°$ so

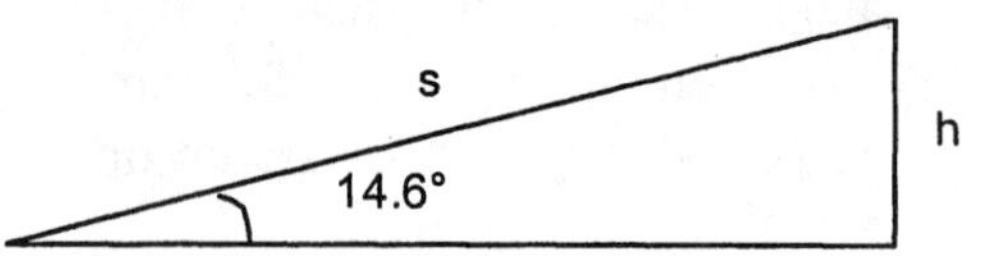

$$\Delta\text{PE} = mgs \sin 14.6°$$

$$= (75.0 \text{ kg})(9.80 \text{ m/s}^2)(2830 \text{ m}) \sin 14.6°$$

$$= \boxed{5.24 \times 10^5 \text{ J}}$$

---

30. ***REASONING*** The change in gravitational potential energy for both the adult and the child is $\text{PE} = mgh_f - mgh_0$, where we have used Equation 6.5. Therefore, $\text{PE} = mg(h_f - h_0)$. In this expression $h_f - h_0$ is the vertical height of the second floor above the first floor, and its value is not given. However, we know that it is the same for both staircases, a fact that will play the central role in our solution.

SOLUTION Solving $\text{PE} = mg(h_f - h_0)$ for $h_f - h_0$, we obtain

$$h_f - h_0 = \frac{(\Delta\text{PE})_{\text{Adult}}}{m_{\text{Adult}}g} \quad \text{and} \quad h_f - h_0 = \frac{(\Delta\text{PE})_{\text{Child}}}{m_{\text{Child}}g}$$

Since $h_f - h_0$ is the same for the adult and the child, we have

$$\frac{(\Delta\text{PE})_{\text{Adult}}}{m_{\text{Adult}}g} = \frac{(\Delta\text{PE})_{\text{Child}}}{m_{\text{Child}}g}$$

Solving this result for $(\Delta\text{PE})_{\text{Child}}$ gives

$$(\Delta\text{PE})_{\text{Child}} = \frac{(\Delta\text{PE})_{\text{Adult}}\, m_{\text{Child}}}{m_{\text{Adult}}} = \frac{(2.00 \times 10^3 \text{ J})(18.0 \text{ kg})}{81.0 \text{ kg}} = \boxed{444 \text{ J}}$$

---

31. SSM ***REASONING*** The only nonconservative force that acts on rocket man is the force generated by the propulsion unit. Thus, Equation 6.7b can be used to determine the net work done by this force.

SOLUTION From Equation 6.7b, we have

$$W_{\text{nc}} = \Delta\text{KE} + \Delta\text{PE} = \left(\tfrac{1}{2}mv_f^2 - \tfrac{1}{2}mv_0^2\right) + \left(mgh_f - mgh_0\right)$$

Since rocket man starts from rest, $v_0 = 0$ m/s. If we take $h_0 = 0$ m on the ground, we have

$$W_{\text{nc}} = \tfrac{1}{2}mv_{\text{f}}^2 + mgh_{\text{f}} = m\left(\tfrac{1}{2}v_{\text{f}}^2 + gh_{\text{f}}\right)$$

$$= (136\text{ kg})\left[\tfrac{1}{2}(5.0\text{ m/s})^2 + (9.80\text{ m/s}^2)(16\text{ m})\right] = \boxed{2.3\times 10^4\text{ J}}$$

---

32. ***REASONING*** The only two forces that act on the gymnast are his weight and the force exerted on his hands by the high bar. The latter is the (non-conservative) reaction force to the force exerted on the bar by the gymnast, as predicted by Newton's third law. This force, however, does no work because it points perpendicular to the circular path of motion. Thus, $W_{\text{nc}} = 0$ J, and we can apply the principle of conservation of mechanical energy.

SOLUTION The conservation principle gives

$$\underbrace{\tfrac{1}{2}mv_{\text{f}}^2 + mgh_{\text{f}}}_{E_{\text{f}}} = \underbrace{\tfrac{1}{2}mv_0^2 + mgh_0}_{E_0}$$

Since the gymnast's speed is momentarily zero at the top of the swing, $v_0 = 0$ m/s. If we take $h_{\text{f}} = 0$ m at the bottom of the swing, then $h_0 = 2r$, where r is the radius of the circular path followed by the gymnast's waist. Making these substitutions in the above expression and solving for v_{f}, we obtain

$$v_{\text{f}} = \sqrt{2gh_0} = \sqrt{2g(2r)} = \sqrt{2(9.80\text{ m/s}^2)(2\times 1.1\text{ m})} = \boxed{6.6\text{ m/s}}$$

---

33. ***REASONING*** We can find the landing speed v_{f} from the final kinetic energy KE_{f}, since $\text{KE}_{\text{f}} = \tfrac{1}{2}mv_{\text{f}}^2$. Furthermore, we are ignoring air resistance, so the conservation of mechanical energy applies. This principle states that the mechanical energy with which the pebble is initially launched is also the mechanical energy with which it finally strikes the ground. Mechanical energy is kinetic energy plus gravitational potential energy. With respect to potential energy we will use ground level as the zero level for measuring heights. The pebble initially has kinetic and potential energies. When it strikes the ground, however, the pebble has only kinetic energy, its potential energy having been converted into kinetic energy. It does not matter how the pebble is launched, because only the vertical height determines the gravitational potential energy. Thus, in each of the three parts of the problem the same amount of potential energy is converted into kinetic energy. As a result, the speed that we will calculate from the final kinetic energy will be the same in parts (a), (b), and (c).

SOLUTION

a. Applying the conservation of mechanical energy in the form of Equation 6.9b, we have

$$\underbrace{\tfrac{1}{2}mv_f^2 + mgh_f}_{\substack{\text{Final mechanical energy}\\ \text{at ground level}}} = \underbrace{\tfrac{1}{2}mv_0^2 + mgh_0}_{\substack{\text{Initial mechanical energy}\\ \text{at top of building}}}$$

Solving for the final speed gives

$$v_f = \sqrt{v_0^2 + 2g\left(h_0 - h_f\right)} = \sqrt{(14.0\text{ m/s})^2 + 2\left(9.80\text{ m/s}^2\right)\left[(31.0\text{ m}) - (0\text{ m})\right]} = \boxed{28.3\text{ m/s}}$$

b. The calculation and answer are the same as in part (a).

c. The calculation and answer are the same as in part (a).

---

34. ***REASONING AND SOLUTION*** The conservation of energy gives

$$\tfrac{1}{2}mv_f^2 + mgh_f = \tfrac{1}{2}mv_0^2 + mgh_0$$

Rearranging gives

$$h_f - h_0 = \frac{(14.0\text{ m/s})^2 - (13.0\text{ m/s})^2}{2\left(9.80\text{ m/s}^2\right)} = \boxed{1.4\text{ m}}$$

---

35. SSM ***REASONING AND SOLUTION*** Since we are ignoring friction and air resistance, the net work done by the nonconservative forces is zero, and the principle of conservation of mechanical energy holds. Let E_0 be the total mechanical energy of the vaulter at take-off and E_f be the total mechanical energy as he clears the bar. The principle of conservation of mechanical energy gives

$$\underbrace{\frac{1}{2}mv_f^2 + mgh_f}_{E_f} = \underbrace{\frac{1}{2}mv_0^2 + mgh_0}_{E_0}$$

For maximum height h_f, the vaulter just clears the bar with zero speed, $v_f = 0$ m/s. If we let $h_0 = 0$ m at ground level, we have

$$mgh_f = \frac{1}{2}mv_0^2$$

Solving for h_f, we find

$$h_f = \frac{v_0^2}{2g} = \frac{(9.00\text{ m/s})^2}{2(9.80\text{ m/s}^2)} = \boxed{4.13\text{ m}}$$

---

36. ***REASONING*** Since air resistance is being neglected, the only force that acts on the golf ball is the conservative gravitational force (its weight). Since the maximum height of the trajectory and the initial speed of the ball are known, the conservation of mechanical energy can be used to find the kinetic energy of the ball at the top of the highest point. The

conservation of mechanical energy can also be used to find the speed of the ball when it is 8.0 m below its highest point.

SOLUTION

a. The conservation of mechanical energy, Equation 6.9b, states that

$$\underbrace{\tfrac{1}{2}mv_f^2}_{\text{KE}_f} + mgh_f = \tfrac{1}{2}mv_0^2 + mgh_0$$

Solving this equation for the final kinetic energy, KE_f, yields

$$\text{KE}_f = \tfrac{1}{2}mv_0^2 + mg\left(h_0 - h_f\right)$$

$$= \tfrac{1}{2}(0.0470\text{ kg})(52.0\text{ m/s})^2 + (0.0470\text{ kg})\left(9.80\text{ m/s}^2\right)(0\text{ m} - 24.6\text{ m}) = \boxed{52.2\text{ J}}$$

b. The conservation of mechanical energy, Equation 6.9b, states that

$$\underbrace{\tfrac{1}{2}mv_f^2 + mgh_f}_{E_f} = \underbrace{\tfrac{1}{2}mv_0^2 + mgh_0}_{E_f}$$

The mass m can be eliminated algebraically from this equation, since it appears as a factor in every term. Solving for v_f and noting that the final height is $h_f = 24.6\text{ m} - 8.0\text{ m} = 16.6\text{ m}$, we have that

$$v_f = \sqrt{v_0^2 + 2g\left(h_0 - h_f\right)} = \sqrt{(52.0\text{ m/s})^2 + 2\left(9.80\text{ m/s}^2\right)(0\text{ m} - 16.6\text{ m})} = \boxed{48.8\text{ m/s}}$$

---

37. ***REASONING AND SOLUTION*** Mechanical energy is conserved if no friction acts. $E_f = E_0$ or

$$\tfrac{1}{2}mv_f^2 + mgh = \tfrac{1}{2}mv_0^2$$

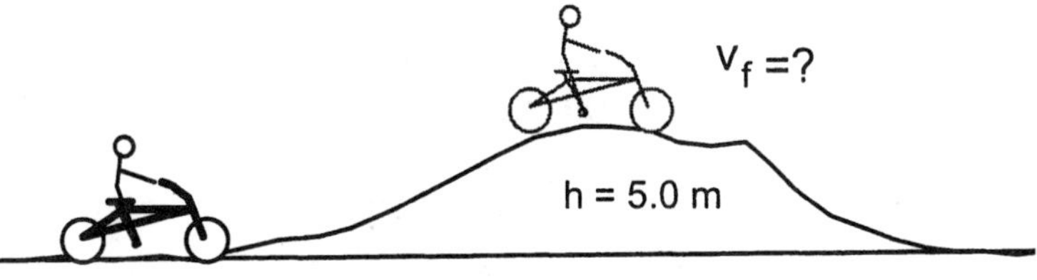

Taking $h_0 = 0$ m and rearranging yields

$$v_f = \sqrt{v_0^2 - 2gh} = \sqrt{(11.0\text{ m/s})^2 - 2\left(9.80\text{ m/s}^2\right)(5.0\text{ m})} = \boxed{4.8\text{ m/s}}$$

---

38. ***REASONING AND SOLUTION*** The conservation of energy applied between point A and the top of the trajectory gives

$$\text{KE}_{\text{A}} + mgh_{\text{A}} = mgh$$

where $h = 4.00$ m. Rearranging, we find

$$\text{KE}_{\text{A}} = mg(h - h_{\text{A}})$$

or

$$v_{\text{A}} = \sqrt{2g(h - h_{\text{A}})} = \sqrt{2(9.80 \text{ m/s}^2)(4.00 \text{ m} - 3.00 \text{ m})} = \boxed{4.43 \text{ m/s}}$$

---

39. ***REASONING AND SOLUTION*** Since friction and air resistance are negligible, mechanical energy is conserved. Thus, the ball will have the same speed at the bottom of its swing whether it is moving toward or away from the crane. As the drawing shows, when the cable swings to an angle of 20.0°, the ball will rise a distance of

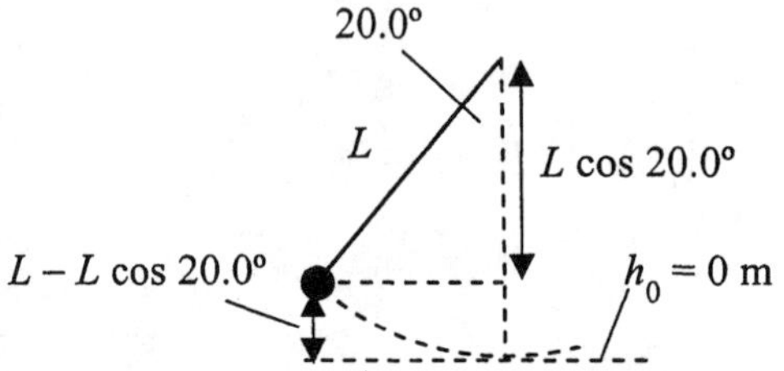

20.0°

L

$L \cos 20.0°$

$L - L \cos 20.0°$

$h_0 = 0$ m

$$h = L\,(1 - \cos 20.0°)$$

where L is the length of the cable. Applying the conservation of energy with $h_0 = 0$ m gives

$$\tfrac{1}{2}mv_0^2 = \tfrac{1}{2}mv_{\text{f}}^2 + mgh$$

Solving for v_{f} gives

$$v_{\text{f}} = \sqrt{v_0^2 - 2gL(1 - \cos 20.0°)}$$

$$= \sqrt{(5.00 \text{ m/s})^2 - 2(9.80 \text{ m/s}^2)(12.0 \text{ m})(1 - \cos 20.0°)} = \boxed{3.29 \text{ m/s}}$$

---

40. ***REASONING*** To find the maximum height H above the end of the track we will analyze the projectile motion of the skateboarder after she leaves the track. For this analysis we will use the principle of conservation of mechanical energy, which applies because friction and air resistance are being ignored. In applying this principle to the projectile motion, however, we will need to know the speed of the skateboarder when she leaves the track. Therefore, we will begin by determining this speed, also using the conservation principle in the process. Our approach, then, uses the conservation principle twice.

SOLUTION Applying the conservation of mechanical energy in the form of Equation 6.9b, we have

$$\underbrace{\tfrac{1}{2}mv_f^2 + mgh_f}_{\text{Final mechanical energy at end of track}} = \underbrace{\tfrac{1}{2}mv_0^2 + mgh_0}_{\text{Initial mechanical energy on flat part of track}}$$

We designate the flat portion of the track as having a height $h_0 = 0$ m and note from the drawing that its end is at a height of $h_f = 0.40$ m above the ground. Solving for the final speed at the end of the track gives

$$v_f = \sqrt{v_0^2 + 2g(h_0 - h_f)} = \sqrt{(5.4 \text{ m/s})^2 + 2(9.80 \text{ m/s}^2)[(0 \text{ m}) - (0.40 \text{ m})]} = 4.6 \text{ m/s}$$

This speed now becomes the initial speed $v_0 = 4.6$ m/s for the next application of the conservation principle. At the maximum height of her trajectory she is traveling horizontally with a speed v_f that equals the horizontal component of her launch velocity. Thus, for the next application of the conservation principle $v_f = (4.6 \text{ m/s}) \cos 48°$. Applying the conservation of mechanical energy again, we have

$$\underbrace{\tfrac{1}{2}mv_f^2 + mgh_f}_{\text{Final mechanical energy at maximum height of trajectory}} = \underbrace{\tfrac{1}{2}mv_0^2 + mgh_0}_{\text{Initial mechanical energy upon leaving the track}}$$

Recognizing that $h_0 = 0.40$ m and $h_f = 0.40 \text{ m} + H$ and solving for H give

$$\tfrac{1}{2}mv_f^2 + mg[(0.40 \text{ m}) + H] = \tfrac{1}{2}mv_0^2 + mg(0.40 \text{ m})$$

$$H = \frac{v_0^2 - v_f^2}{2g} = \frac{(4.6 \text{ m/s})^2 - [(4.6 \text{ m/s})\cos 48°]^2}{2(9.80 \text{ m/s}^2)} = \boxed{0.60 \text{ m}}$$

---

41. SSM ***REASONING*** Friction and air resistance are being ignored. The normal force from the slide is perpendicular to the motion, so it does no work. Thus, no net work is done by nonconservative forces, and the principle of conservation of mechanical energy applies.

SOLUTION Applying the principle of conservation of mechanical energy to the swimmer at the top and the bottom of the slide, we have

$$\underbrace{\frac{1}{2}mv_f^2 + mgh_f}_{E_f} = \underbrace{\frac{1}{2}mv_0^2 + mgh_0}_{E_0}$$

If we let h be the height of the bottom of the slide above the water, $h_f = h$, and $h_0 = H$. Since the swimmer starts from rest, $v_0 = 0$ m/s, and the above expression becomes

$$\frac{1}{2}v_f^2 + gh = gH$$

Solving for H, we obtain

$$H = h + \frac{v_f^2}{2g}$$

Before we can calculate H, we must find v_f and h. Since the velocity in the horizontal direction is constant,

$$v_f = \frac{\Delta x}{\Delta t} = \frac{5.00 \text{ m}}{0.500 \text{ s}} = 10.0 \text{ m/s}$$

The vertical displacement of the swimmer after leaving the slide is, from Equation 3.5b (with down being negative),

$$y = \frac{1}{2}a_y t^2 = \frac{1}{2}(-9.80 \text{ m/s}^2)(0.500 \text{ s})^2 = -1.23 \text{ m}$$

Therefore, h = 1.23 m. Using these values of v_f and h in the above expression for H, we find

$$H = h + \frac{v_f^2}{2g} = 1.23 \text{ m} + \frac{(10.0 \text{ m/s})^2}{2(9.80 \text{ m/s}^2)} = \boxed{6.33 \text{ m}}$$

---

42. ***REASONING AND SOLUTION*** If air resistance is ignored, the only nonconservative force that acts on the skier is the normal force exerted on the skier by the snow. Since this force is always perpendicular to the direction of the displacement, the work done by the normal force is zero. We can conclude, therefore, that mechanical energy is conserved.

$$\frac{1}{2}\text{m}v_0^2 + \text{mgh}_0 = \frac{1}{2}\text{m}v_f^2 + \text{mgh}_f$$

Since the skier starts from rest v_0 = 0 m/s. Let h_f define the zero level for heights, then the final gravitational potential energy is zero. This gives

$$\text{mgh}_0 = \frac{1}{2}\text{m}v_f^2 \qquad (1)$$

At the crest of the second hill, the two forces that act on the skier are the normal force and the weight of the skier. The resultant of these two forces provides the necessary centripetal force to keep the skier moving along the circular arc of the hill. When the skier *just loses contact* with the snow, the normal force is zero and the weight of the skier must provide the necessary centripetal force.

$$\text{mg} = \frac{\text{m}v_f^2}{\text{r}} \quad \text{so that} \quad v_f^2 = \text{gr} \qquad (2)$$

Substituting this expression for v_f^2 into Equation (1) gives

$$\mathrm{mgh_0} = \frac{1}{2}\mathrm{mgr}$$

Solving for h_0 gives

$$h_0 = \frac{r}{2} = \frac{36\ \mathrm{m}}{2} = \boxed{18\ \mathrm{m}}$$

---

43. ***REASONING AND SOLUTION*** At the bottom of the circular path of the swing, the centripetal force is provided by the tension in the rope less the weight of the swing and rider. That is,

$$\underbrace{\frac{mv^2}{r}}_{F_c} = T - mg$$

Solving for the mass yields

$$m = \frac{T}{\dfrac{v^2}{r} + g}$$

The energy of the swing is conserved if friction is ignored. The initial energy, E_0, when the swing is released is completely potential energy and is $E_0 = mgh_0$, where

$$h_0 = r\,(1 - \cos 60.0°) = \tfrac{1}{2}r$$

is the vertical height of the swing. At the bottom of the path the swing's energy is entirely kinetic and is

$$E_f = \tfrac{1}{2}mv^2$$

Setting $E_o = E_f$ and solving for v gives

$$v = \sqrt{2gh_0} = \sqrt{gr}$$

The expression for the mass now becomes

$$m = \frac{T}{2g} = \frac{8.00 \times 10^2\ \mathrm{N}}{2\left(9.80\ \mathrm{m/s^2}\right)} = \boxed{40.8\ \mathrm{kg}}$$

---

44. ***REASONING AND SOLUTION*** When the car is at the top of the track the centripetal force consists of the full weight of the car.

$$mv^2/r = mg$$

Applying the conservation of energy between the bottom and the top of the track gives

$$(1/2)mv^2 + mg(2r) = (1/2)mv_0{}^2$$

Using both of the above equations

$$v_0{}^2 = 5gr$$

so

$$r = v_0{}^2/(5g) = (4.00\text{ m/s})^2/(49.0\text{ m/s}^2) = \boxed{0.327\text{ m}}$$

---

45. SSM ***REASONING AND SOLUTION*** The work-energy theorem can be used to determine the change in kinetic energy of the car according to Equation 6.8:

$$W_{\text{nc}} = \underbrace{\left(\tfrac{1}{2}mv_{\text{f}}^2 + mgh_{\text{f}}\right)}_{E_{\text{f}}} - \underbrace{\left(\tfrac{1}{2}mv_0^2 + mgh_0\right)}_{E_0}$$

The nonconservative forces are the forces of friction and the force due to the chain mechanism. Thus, we can rewrite Equation 6.8 as

$$W_{\text{friction}} + W_{\text{chain}} = \left(\text{KE}_{\text{f}} + mgh_{\text{f}}\right) - \left(\text{KE}_0 + mgh_0\right)$$

We will measure the heights from ground level. Solving for the change in kinetic energy of the car, we have

$$\text{KE}_{\text{f}} - \text{KE}_0 = W_{\text{friction}} + W_{\text{chain}} - mgh_{\text{f}} + mgh_0$$

$$= -2.00\times10^4\text{ J} + 3.00\times10^4\text{ J} - (375\text{ kg})(9.80\text{ m/s}^2)(20.0\text{ m})$$
$$+ (375\text{ kg})(9.80\text{ m/s}^2)(5.00\text{ m}) = \boxed{-4.51\times10^4\text{ J}}$$

---

46. ***REASONING AND SOLUTION***

The work done by air resistance is equal to the energy lost by the ball.

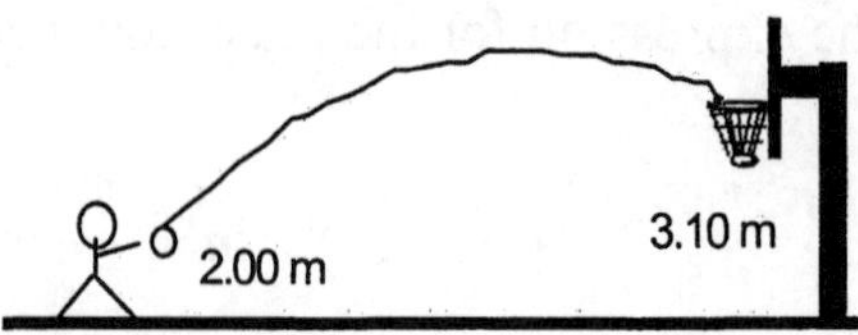

$$\text{W}_{\text{nc}} = \text{E}_{\text{f}} - \text{E}_0$$

$$W_{\text{nc}} = \left(\tfrac{1}{2}mv_{\text{f}}^2 + mgh_{\text{f}}\right) - \left(\tfrac{1}{2}mv_0^2 + mgh_0\right)$$

$$W_{nc} = \tfrac{1}{2}(0.600\text{ kg})(4.20\text{ m/s})^2 + (0.600\text{ kg})(9.80\text{ m/s}^2)(3.10\text{ m})$$
$$-\tfrac{1}{2}(0.600\text{ kg})(7.20\text{ m/s})^2 - (0.600\text{ kg})(9.80\text{ m/s}^2)(2.00\text{ m}) = \boxed{-3.8\text{ J}}$$

---

47. ***REASONING AND SOLUTION*** The work done by nonconservative forces on the balloon is

$$W_{nc} = \Delta\text{KE} + \Delta\text{PE} = +9.70 \times 10^4\text{ J}$$

The change in kinetic energy of the balloon is

$$\Delta\text{KE} = \tfrac{1}{2}mv_f^2 = \tfrac{1}{2}(5.00 \times 10^2\text{ kg})(8.00\text{ m/s})^2 = +1.60 \times 10^4\text{ J}$$

The change in the potential energy is

$$\Delta\text{PE} = mgh = (5.00 \times 10^2\text{ kg})(9.80\text{ m/s}^2)h = (4.90 \times 10^3\text{ kg·m/s}^2)h$$

Hence,

$$h = \frac{9.70 \times 10^4\text{ J} - 1.60 \times 10^4\text{ J}}{4.90 \times 10^3\text{ kg}\cdot\text{m/s}^2} = \boxed{16.5\text{ m}}$$

---

48. ***REASONING***

a. Since there is no air friction, the only force that acts on the projectile is the conservative gravitational force (its weight). The initial and final speeds of the ball are known, so the conservation of mechanical energy can be used to find the maximum height that the projectile attains.

b. When air resistance, a nonconservative force, is present, it does negative work on the projectile and slows it down. Consequently, the projectile does not rise as high as when there is no air resistance. The work-energy theorem, in the form of Equation 6.6, may be used to find the work done by air friction. Then, using the definition of work, Equation 6.1, the average force due to air resistance can be found.

SOLUTION

a. The conservation of mechanical energy, as expressed by Equation 6.9b, states that

$$\underbrace{\tfrac{1}{2}mv_f^2 + mgh_f}_{E_f} = \underbrace{\tfrac{1}{2}mv_0^2 + mgh_0}_{E_0}$$

The mass m can be eliminated algebraically from this equation since it appears as a factor in every term. Solving for the final height h_f gives

$$h_f = \frac{\frac{1}{2}\left(v_0^2 - v_f^2\right)}{g} + h_0$$

Setting $h_0 = 0$ m and $v_f = 0$ m/s, the final height, in the absence of air resistance, is

$$h_f = \frac{v_0^2 - v_f^2}{2g} = \frac{(18.0\text{ m/s})^2 - (0\text{ m/s})^2}{2\left(9.80\text{ m/s}^2\right)} = \boxed{16.5\text{ m}}$$

b. The work-energy theorem is

$$W_{nc} = \left(\tfrac{1}{2}mv_f^2 - \tfrac{1}{2}mv_0^2\right) + \left(mgh_f - mgh_0\right) \qquad (6.6)$$

where W_{nc} is the nonconservative work done by air resistance. According to Equation 6.1, the work can be written as $W_{nc} = \left(\bar{F}_R \cos 180°\right)s$, where $\bar{F}_R$ is the average force of air resistance. As the projectile moves upward, the force of air resistance is directed downward, so the angle between the two vectors is $\theta = 180°$ and $\cos\theta = -1$. The magnitude s of the displacement is the difference between the final and initial heights, $s = h_f - h_0 = 11.8$ m. With these substitutions, the work-energy theorem becomes

$$-\bar{F}_R\, s = \tfrac{1}{2}m\left(v_f^2 - v_0^2\right) + mg\left(h_f - h_0\right)$$

Solving for $\bar{F}_R$ gives

$$\bar{F}_R = \frac{\frac{1}{2}m\left(v_f^2 - v_0^2\right) + mg\left(h_f - h_0\right)}{-s}$$

$$= \frac{\frac{1}{2}(0.750\text{ kg})\left[(0\text{ m/s})^2 - (18.0\text{ m/s})^2\right] + (0.750\text{ kg})\left(9.80\text{ m/s}^2\right)(11.8\text{ m})}{-(11.8\text{ m})} = \boxed{2.9\text{ N}}$$

---

49. **SSM** ***REASONING*** The work-energy theorem can be used to determine the net work done on the car by the nonconservative forces of friction and air resistance. The work-energy theorem is, according to Equation 6.8,

$$W_{nc} = \left(\tfrac{1}{2}mv_f^2 + mgh_f\right) - \left(\tfrac{1}{2}mv_0^2 + mgh_0\right)$$

The nonconservative forces are the force of friction, the force of air resistance, and the force provided by the engine. Thus, we can rewrite Equation 6.8 as

$$W_{\text{friction}} + W_{\text{air}} + W_{\text{engine}} = \left(\tfrac{1}{2}mv_{\text{f}}^2 + mgh_{\text{f}}\right) - \left(\tfrac{1}{2}mv_0^2 + mgh_0\right)$$

This expression can be solved for $W_{\text{friction}} + W_{\text{air}}$.

SOLUTION We will measure the heights from sea level, where $h_0 = 0$ m. Since the car starts from rest, $v_0 = 0$ m/s. Thus, we have

$$W_{\text{friction}} + W_{\text{air}} = m\left(\tfrac{1}{2}v_{\text{f}}^2 + gh_{\text{f}}\right) - W_{\text{engine}}$$

$$= (1.50 \times 10^3 \text{ kg})\left[\tfrac{1}{2}(27.0 \text{ m/s})^2 + (9.80 \text{ m/s}^2)(2.00 \times 10^2 \text{ m})\right] - (4.70 \times 10^6 \text{ J})$$

$$= \boxed{-1.21 \times 10^6 \text{ J}}$$

---

50. ***REASONING AND SOLUTION*** The force exerted by the bat on the ball is the only non-conservative force acting. The work due to this force is

$$W_{\text{nc}} = \frac{1}{2}mv_{\text{f}}^2 - \frac{1}{2}mv_0^2 + mg\left(h_{\text{f}} - h_0\right)$$

Taking $h_0 = 0$ m at the level of the bat, $v_0 = 40.0$ m/s just before the bat strikes the ball and v_{f} to be the speed of the ball at $h_{\text{f}} = 25.0$ m, we have

$$v_{\text{f}} = \sqrt{\frac{2W_{\text{nc}}}{m} + v_0^2 - 2gh_{\text{f}}}$$

$$= \sqrt{\frac{2(70.0 \text{ J})}{0.140 \text{ kg}} + (40.0 \text{ m/s})^2 - 2(9.80 \text{ m/s}^2)(25.0 \text{ m})} = \boxed{45.9 \text{ m/s}}$$

---

51. ***REASONING AND SOLUTION***

a. The lost mechanical energy is

$$E_{\text{lost}} = E_0 - E_{\text{f}}$$

The ball is dropped from rest, so its initial energy is purely potential. The ball is momentarily at rest at the highest point in its rebound, so its final energy is also purely potential. Then

$$E_{\text{lost}} = mgh_0 - mgh_{\text{f}} = (0.60 \text{ kg})(9.80 \text{ m/s}^2)\left[(1.05 \text{ m}) - (0.57 \text{ m})\right] = \boxed{2.8 \text{ J}}$$

b. The work done by the player must compensate for this loss of energy.

$$E_{\text{lost}} = (F \cos\theta)\, s \quad \Rightarrow \quad F = \frac{E_{\text{lost}}}{(\cos\theta)\, s} = \frac{2.8\text{ J}}{(\cos 0°)(0.080\text{ m})} = \boxed{35\text{ N}}$$

---

52. ***REASONING*** When launched with the minimum initial speed v_0, the puck has just enough initial kinetic energy to reach the teammate with a final speed v_f of zero. All of the initial kinetic energy has served the purpose of compensating for the work done by friction in opposing the motion. Friction is a nonconservative force, so to determine v_0 we will use the work-energy theorem in the form of Equation 6.8: $W_{nc} = \left(\frac{1}{2}mv_f^2 + mgh_f\right) - \left(\frac{1}{2}mv_0^2 + mgh_0\right)$. W_{nc} is the work done by the net nonconservative force, which, in this case, is just the kinetic frictional force. This work can be expressed using Equation 6.1 as $W_{nc} = (f_k \cos\theta)s$, where f_k is the magnitude of the kinetic frictional force and s is the magnitude of the displacement, or the distance between the players. We are given neither the frictional force nor the distance between the players, but we are given the initial speed that enables the puck to travel half way. This information will be used to evaluate W_{nc}.

SOLUTION Noting that a hockey rink is flat ($h_f = h_0$), we write Equation 6.8 as follows

$$W_{nc} = \underbrace{\left(\tfrac{1}{2}mv_f^2 + mgh_f\right)}_{\text{Final mechanical energy}} - \underbrace{\left(\tfrac{1}{2}mv_0^2 + mgh_0\right)}_{\text{Initial mechanical energy}} = \tfrac{1}{2}mv_f^2 - \tfrac{1}{2}mv_0^2 = -\tfrac{1}{2}mv_0^2 \qquad (1)$$

where we have used the fact that the final speed is $v_f = 0$ m/s when the initial speed v_0 has its minimum value. According to Equation 6.1 the work done on the puck by the kinetic frictional force is

$$W_{nc} = (f_k \cos\theta)s = (f_k \cos 180°)s = -f_k s$$

In this expression the angle θ is 180°, because the frictional force points opposite to the displacement. With this substitution Equation (1) becomes

$$-f_k s = -\tfrac{1}{2}mv_0^2 \quad \text{or} \quad f_k s = \tfrac{1}{2}mv_0^2 \qquad (2)$$

To evaluate the term $f_k s$ we note that Equation (2) also applies to the failed attempt at the pass, so that

$$f_k\left(\tfrac{1}{2}s\right) = \tfrac{1}{2}m(1.7\text{ m/s})^2 \quad \text{or} \quad f_k s = m(1.7\text{ m/s})^2$$

Substituting this result into Equation (2) gives

$$m(1.7\text{ m/s})^2 = \tfrac{1}{2}mv_0^2 \quad \text{or} \quad v_0 = \sqrt{2(1.7\text{ m/s})^2} = \boxed{2.4\text{ m/s}}$$

---

53. SSM ***REASONING AND SOLUTION*** According to the work-energy theorem as given in Equation 6.8, we have

$$W_{nc} = \left(\tfrac{1}{2}mv_f^2 + mgh_f\right) - \left(\tfrac{1}{2}mv_0^2 + mgh_0\right)$$

The metal piece starts at rest and is at rest just as it barely strikes the bell, so that $v_f = v_0 = 0$ m/s. In addition, $h_f = h$ and $h_0 = 0$ m, while $W_{nc} = 0.25\left(\tfrac{1}{2}Mv^2\right)$, where M and v are the mass and speed of the hammer. Thus, the work-energy theorem becomes

$$0.25\left(\tfrac{1}{2}Mv^2\right) = mgh$$

Solving for the speed of the hammer, we find

$$v = \sqrt{\frac{2mgh}{0.25M}} = \sqrt{\frac{2(0.400\text{ kg})(9.80\text{ m/s}^2)(5.00\text{ m})}{0.25\,(9.00\text{ kg})}} = \boxed{4.17\text{ m/s}}$$

---

54. ***REASONING AND SOLUTION*** For the actual motion of the rocket

$$\Delta\text{KE} = W_{nc} - \Delta\text{PE} = W_{nc} - mgh \qquad (6.7\text{b})$$

$$\Delta\text{KE} = -8.00 \times 10^2\text{ J} - (3.00\text{ kg})(9.80\text{ m/s}^2)(1.00 \times 10^2\text{ m}) = -3740\text{ J}$$

Since $\Delta\text{KE} = \text{KE}_f - \text{KE}_0 = -\,\text{KE}_0$, the initial kinetic energy of the rocket is then

$$\text{KE}_0 = 3740\text{ J}$$

If the rocket were launched with this initial kinetic energy and no air resistance, the potential energy of the rocket at the top of its trajectory would be

$$\text{PE} = mgh = 3740\text{ J}$$

Hence,

$$h = \frac{3740\text{ J}}{(3.00\text{ kg})(9.80\text{ m/s}^2)} = \boxed{127\text{ m}}$$

---

55. ***REASONING*** The average power is given by Equation 6.10b ($\overline{P} = \frac{\text{Change in energy}}{\text{Time}}$). The time is given. Since the road is level, there is no change in the gravitational potential energy, and the change in energy refers only to the kinetic energy. According to Equation 6.2, the kinetic energy is $\tfrac{1}{2}mv^2$. The speed v is given, but the mass m is not. However, we can obtain the mass from the given weights, since the weight is mg.

SOLUTION

a. Using Equations 6.10b and 6.2, we find that the average power is

$$\overline{P} = \frac{\text{Change in energy}}{\text{Time}} = \frac{\text{KE}_\text{f} - \text{KE}_0}{\text{Time}} = \frac{\frac{1}{2}mv_\text{f}^2 - \frac{1}{2}mv_0^2}{\text{Time}}$$

Since the car starts from rest, $v_0 = 0$ m/s, and since the weight is $W = mg$, the mass is $m = W/g$. Therefore, the average power is

$$\overline{P} = \frac{\frac{1}{2}mv_\text{f}^2 - \frac{1}{2}mv_0^2}{\text{Time}} = \frac{Wv_\text{f}^2}{2g(\text{Time})} \tag{1}$$

Using the given values for the weight, final speed, and the time, we find that

$$\overline{P} = \frac{Wv_\text{f}^2}{2g(\text{Time})} = \frac{(9.0\times10^3\ \text{N})(20.0\ \text{m/s})^2}{2(9.80\ \text{m/s}^2)(5.6\ \text{s})} = \boxed{3.3\times10^4\ \text{W (44 hp)}}$$

b. From Equation (1) it follows that

$$\overline{P} = \frac{Wv_\text{f}^2}{2g(\text{Time})} = \frac{(1.4\times10^4\ \text{N})(20.0\ \text{m/s})^2}{2(9.80\ \text{m/s}^2)(5.6\ \text{s})} = \boxed{5.1\times10^4\ \text{W (68 hp)}}$$

---

56. ***REASONING AND SOLUTION*** The work done by the person in turning the crank one revolution is

$$W = F(2\pi r) = (22\ \text{N})(2\pi)(0.28\ \text{m}) = 39\ \text{J}$$

The power expended is then

$$\overline{P} = \frac{W}{t} = \frac{39\ \text{J}}{1.3\ \text{s}} = \boxed{3.0\times10^1\ \text{W}} \tag{6.10a}$$

---

57. SSM ***REASONING AND SOLUTION*** One kilowatt·hour is the amount of work or energy generated when one kilowatt of power is supplied for a time of one hour. From Equation 6.10a, we know that $W = \overline{P}t$. Using the fact that $1\ \text{kW} = 1.0\times10^3$ J/s and that $1\ \text{h} = 3600$ s, we have

$$1.0\ \text{kWh} = (1.0\times10^3\ \text{J/s})(1\ \text{h}) = (1.0\times10^3\ \text{J/s})(3600\ \text{s}) = \boxed{3.6\times10^6\ \text{J}}$$

---

58. ***REASONING AND SOLUTION*** The work done by the crane is the work done against gravity, $W = mgh$, so

$$t = \frac{mgh}{P} = \frac{(3.00\times10^2\text{ kg})(9.80\text{ m/s}^2)(10.0\text{ m})}{4.00\times10^2\text{ W}} = \boxed{73.5\text{ s}}$$

---

59. ***REASONING*** The average power generated by the tension in the cable is equal to the work done by the tension divided by the elapsed time (Equation 6.10a). The work can be related to the change in the lift's kinetic and potential energies via the work-energy theorem.

SOLUTION

a. The average power is

$$\overline{P} = \frac{W_{nc}}{t} \qquad (6.10a)$$

where W_{nc} is the work done by the nonconservative tension force. This work is related to the lift's kinetic and potential energies by Equation 6.6, $W_{nc} = \left(\frac{1}{2}mv_f^2 - \frac{1}{2}mv_0^2\right) + \left(mgh_f - mgh_0\right)$, so the average power is

$$\overline{P} = \frac{W_{nc}}{t} = \frac{\left(\frac{1}{2}mv_f^2 - \frac{1}{2}mv_0^2\right) + \left(mgh_f - mgh_0\right)}{t}$$

Since the lift moves at a constant speed, $v_0 = v_f$, the average power becomes

$$\overline{P} = \frac{mg\left(h_f - h_0\right)}{t} = \frac{(4\text{ skiers})\left(65\,\frac{\text{kg}}{\text{skier}}\right)\left(9.80\text{ m/s}^2\right)(140\text{ m})}{(2\text{ min})\left(\frac{60\text{ s}}{1\text{ min}}\right)} = \boxed{3.0\times10^3\text{ W}}$$

---

60. ***REASONING AND SOLUTION***

a. The power developed by the engine is

$$P = Fv = (2.00\times10^2\text{ N})(20.0\text{ m/s}) = \boxed{4.00\times10^3\text{ W}}$$

b. The force required of the engine in order to maintain a constant speed up the slope is

$$F = F_a + mg\sin 37.0°$$

The power developed by the engine is then

$$P = Fv = (F_a + mg \sin 37.0°)v \qquad (6.11)$$

$$P = [2.00 \times 10^2 \text{ N} + (2.50 \times 10^2 \text{ kg})(9.80 \text{ m/s}^2)\sin 37.0°](20.0 \text{ m/s}) = \boxed{3.35 \times 10^4 \text{ W}}$$

---

61. SSM ***REASONING AND SOLUTION*** In the drawings below, the positive direction is to the right. When the boat is not pulling a skier, the engine must provide a force F_1 to overcome the resistive force of the water F_R. Since the boat travels with constant speed, these forces must be equal in magnitude and opposite in direction.

$$-F_R + F_1 = ma = 0$$

or

$$F_R = F_1 \qquad (1)$$

F_R ← • → F_1

When the boat is pulling a skier, the engine must provide a force F_2 to balance the resistive force of the water F_R and the tension in the tow rope T.

$$-F_R + F_2 - T = ma = 0$$

or

$$F_2 = F_R + T \qquad (2)$$

T ← F_R ← • → F_2

The average power is given by $\overline{P} = F\overline{v}$, according to Equation 6.11 in the text. Since the boat moves with the same speed in both cases, $v_1 = v_2$, and we have

$$\frac{\overline{P}_1}{F_1} = \frac{\overline{P}_2}{F_2} \qquad \text{or} \qquad F_2 = F_1 \frac{\overline{P}_2}{\overline{P}_1}$$

Using Equations (1) and (2), this becomes

$$F_1 + T = F_1 \frac{\overline{P}_2}{\overline{P}_1}$$

Solving for T gives

$$T = F_1 \left(\frac{\overline{P}_2}{\overline{P}_1} - 1 \right)$$

The force F_1 can be determined from $\overline{P}_1 = F_1 v_1$, thereby giving

$$T = \frac{\overline{P}_1}{v_1} \left(\frac{\overline{P}_2}{\overline{P}_1} - 1 \right) = \frac{7.50 \times 10^4 \text{ W}}{12 \text{ m/s}} \left(\frac{8.30 \times 10^4 \text{ W}}{7.50 \times 10^4 \text{ W}} - 1 \right) = \boxed{6.7 \times 10^2 \text{ N}}$$

---

62. ***REASONING AND SOLUTION*** The following drawings show the free-body diagrams for the car in going both up and down the hill. The force $\mathbf{F_R}$ is the combined force of air resistance and friction, and the forces $\mathbf{F_U}$ and $\mathbf{F_D}$ are the forces supplied by the engine in going uphill and downhill respectively.

Going up the hill

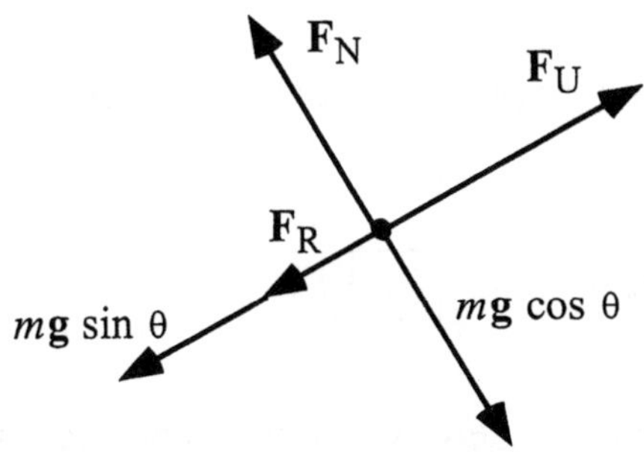

Going down the hill

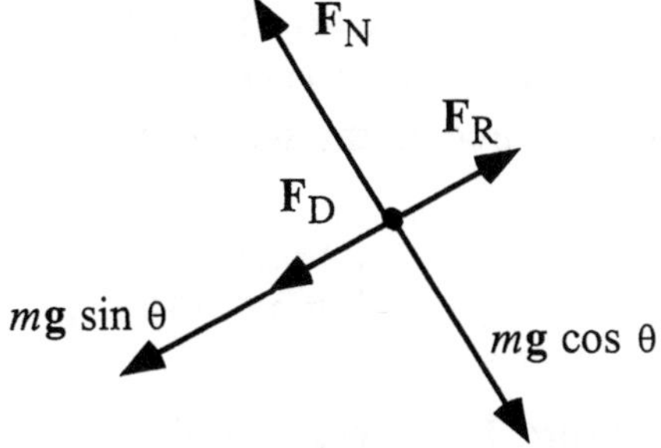

Writing Newton's second law in the direction of motion for the car as it goes uphill, taking uphill as the positive direction, we have

$$F_U - F_R - mg\,\sin\theta = ma = 0$$

Solving for F_U, we have

$$F_U = F_R + mg\,\sin\theta$$

Similarly, when the car is going downhill, Newton's second law in the direction of motion gives

$$F_R - F_D - mg\,\sin\theta = ma = 0$$

so that

$$F_D = F_R - mg\,\sin\theta$$

Since the car needs 47 hp more to sustain the constant uphill velocity than the constant downhill velocity, we can write

$$\overline{P}_U = \overline{P}_D + \Delta P$$

where $\overline{P}_U$ is the power needed to sustain the constant uphill velocity, $\overline{P}_D$ is the power needed to sustain the constant downhill velocity, and $\Delta P = 47\text{ hp}$. In terms of SI units,

$$\Delta P = (47\text{ hp})\left(\frac{746\text{ W}}{1\text{ hp}}\right) = 3.51\times10^4\text{ W}$$

Using Equation 6.11 ($\overline{P} = F\overline{v}$), the equation $\overline{P}_U = \overline{P}_D + \Delta P$ can be written as

$$F_U\overline{v} = F_D\overline{v} + \Delta P$$

Using the expressions for F_U and F_D, we have

$$(F_R + mg\sin\theta)\bar{v} = (F_R - mg\sin\theta)\bar{v} + \Delta P$$

Solving for θ, we find

$$\theta = \sin^{-1}\left(\frac{\Delta P}{2mg\bar{v}}\right) = \sin^{-1}\left(\frac{3.51\times10^4\ \text{W}}{2(1900\ \text{kg})(9.80\ \text{m/s}^2)(27\ \text{m/s})}\right) = \boxed{2.0°}$$

---

63. ***REASONING AND SOLUTION***

a. The work done is equal to the colored area under the graph. Therefore, the work done by the net force on the skater from 0 to 3.0 m is

$$\begin{bmatrix}\text{Area under graph}\\ \text{from 0 to 3.0 m}\end{bmatrix} = (31\ \text{N} - 0\ \text{N})(3.0\ \text{m} - 0\ \text{m}) = \boxed{93\ \text{J}}$$

b. From 3.0 m to 6.0 m, the net force component $F\cos\theta$ is zero, and there is no area under the curve. Therefore, $\boxed{\text{no work is done}}$.

c. The speed of the skater will increase from 1.5 m/s when $s = 0$ m to a final value v_f at $s = 3.0$ m. Since the force drops to zero at $s = 3.0$ m, the speed remains constant at the value v_f from $s = 3.0$ m to $s = 6.0$ m; therefore, the speed at $s = 6.0$ m is v_f. We can find v_f from the work-energy theorem (Equation 6.3). According to the work-energy theorem, the work done on the skater is given by

$$W = \tfrac{1}{2}mv_f^2 - \tfrac{1}{2}mv_0^2$$

Solving for v_f, we find

$$v_f = \sqrt{\frac{2W}{m} + v_0^2} = \sqrt{\frac{2(93\ \text{J})}{65\ \text{kg}} + (1.5\ \text{m/s})^2} = \boxed{2.3\ \text{m/s}}$$

---

64. ***REASONING*** When the force varies with the displacement, the work is the area beneath the graph of the force component $F\cos\theta$ along the displacement as a function of the magnitude s of the displacement. Here, the shape of this area is a triangle. The area of a triangle is one-half times the base times the height of the triangle.

SOLUTION The base of the triangle is 1.60 m, and the "height" is 62.0 N. Therefore, the area, which is the work, is

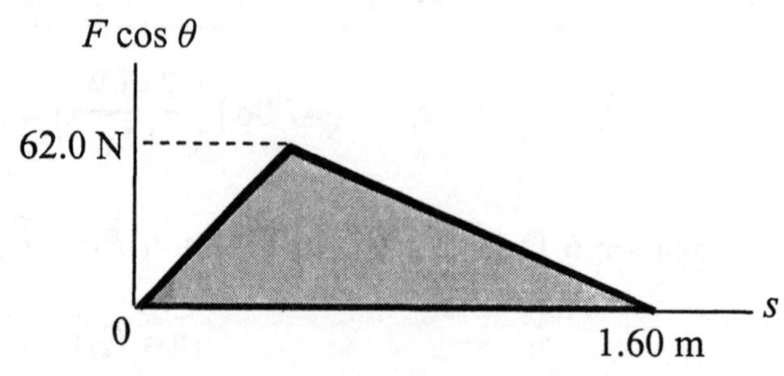

$$\begin{aligned} W &= \tfrac{1}{2}\text{Base}\times\text{Height} \\ &= \tfrac{1}{2}(1.60\ \text{m})(62.0\ \text{N}) \\ &= \boxed{49.6\ \text{J}} \end{aligned}$$

---

65. SSM ***REASONING*** The area under the force-versus-displacement graph over any displacement interval is equal to the work done over that displacement interval.

SOLUTION
a. Since the area under the force-versus-displacement graph from $s = 0$ to 0.50 m is greater for bow 1, it follows that $\boxed{\text{bow 1 requires more work to draw the bow fully}}$.

b. The work required to draw bow 1 is equal to the area of the triangular region under the force-versus-displacement curve for bow 1. Since the area of a triangle is equal one-half times the base of the triangle times the height of the triangle, we have

$$W_1 = \frac{1}{2}(0.50\text{m})(350\text{ N}) = 88\text{ J}$$

For bow 2, we note that each small square under the force-versus-displacement graph has an area of

$$(0.050\text{ m})(40.0\text{ N}) = 2.0\text{ J}$$

We estimate that there are approximately 31.3 squares under the force-versus-displacement graph for bow 2; therefore, the total work done is

$$(31.3\text{ squares})\left(\frac{2.0\text{ J}}{\text{square}}\right) = 63\text{ J}$$

Therefore, the additional work required to stretch bow 1 as compared to bow 2 is

$$88\text{ J} - 63\text{ J} = \boxed{25\text{ J}}$$

---

66. ***REASONING*** The area under the force-versus-displacement graph over any displacement interval is equal to the work done over that interval. From Example 16, we know that the total work done in drawing back the string of the bow is 60.5 J, corresponding to a total area under the curve of 242 squares (each square represents 0.250 J of work). The percentage of the total work done over any displacement interval is, therefore,

$$\frac{\text{Work done during interval}}{\text{Total work done during all intervals}} \times 100 = \frac{\text{Number of squares under curve during interval}}{\text{Total number of squares under entire curve}} \times 100$$

SOLUTION
a. We estimate that there are 130 squares under the curve from $s = 0$ to 0.306 m; therefore, the percentage of work done during the interval in question is

$$\frac{130\text{ squares}}{242\text{ squares}} \times 100 = \boxed{54\%}$$

b. Similarly, we estimate that there are 112 squares under the curve in the interval from $s = 0.306$ m to 0.500 m; therefore, the percentage of work done during the interval in question is

$$\frac{112 \text{ squares}}{242 \text{ squares}} \times 100 = \boxed{46\%}$$

---

67. ***REASONING AND SOLUTION***

a. The net work done is equal to the area under the curve. For the first 10 m, the area is triangular and the area under the curve is equal to (1/2) × height × base; therefore

$$W_{0,10} = \tfrac{1}{2}(10.0\text{ N})(10.0\text{ m} - 0\text{ m}) = 5.00 \times 10^1\text{ J}$$

From 10 to 20 m the region is rectangular and the area under the curve is

$$W_{10,20} = (10.0\text{ N})(20.0\text{ m} - 10.0\text{ m}) = 1.00 \times 10^2\text{ J}$$

Thus, the total work done by the force is

$$W_{0,10} + W_{10,20} = (5.00 \times 10^1\text{ J}) + (1.00 \times 10^2\text{ J}) = \boxed{1.50 \times 10^2\text{ J}}$$

b. From the work-energy theorem,

$$W = \tfrac{1}{2}mv_f^2 - \tfrac{1}{2}mv_0^2$$

Since the object is initially at rest, $v_0 = 0$ m/s. Solving for v_f gives

$$v_f = \sqrt{\frac{2W}{m}} = \sqrt{\frac{2(1.50 \times 10^2\text{ J})}{6.00\text{ kg}}} = \boxed{7.07\text{ m/s}}$$

---

68. ***REASONING*** The work done by the tension in the cable is given by Equation 6.1 as $W = (T\cos\theta)s$. Since the elevator is moving upward at a constant velocity, it is in equilibrium, and the magnitude T of the tension must be equal to the magnitude mg of the elevator's weight; $T = mg$.

SOLUTION

a. The tension and the displacement vectors point in the same direction (upward), so the angle between them is $\theta = 0°$. The work done by the tension is

$$W = (T\cos\theta)s = (mg\cos\theta)s \tag{6.1}$$
$$= (1200\ \text{kg})(9.80\ \text{m/s}^2)\cos 0°\,(35\ \text{m}) = \boxed{4.1\times10^5\ \text{J}}$$

b. The weight and the displacement vectors point in opposite directions, so the angle between them is $\theta = 180°$. The work done by the weight is

$$W = (mg\cos\theta)s = (1200\ \text{kg})(9.80\ \text{m/s}^2)\cos 180°\,(35\ \text{m}) = \boxed{-4.1\times10^5\ \text{J}} \tag{6.1}$$

---

69. SSM ***REASONING*** No forces other than gravity act on the rock since air resistance is being ignored. Thus, the net work done by nonconservative forces is zero, $W_{nc} = 0$ J. Consequently, the principle of conservation of mechanical energy holds, so the total mechanical energy remains constant as the rock falls.

If we take $h = 0$ m at ground level, the gravitational potential energy at any height h is, according to Equation 6.5, $\text{PE} = mgh$. The kinetic energy of the rock is given by Equation 6.2: $\text{KE} = \frac{1}{2}mv^2$. In order to use Equation 6.2, we must have a value for v^2 at each desired height h. The quantity v^2 can be found from Equation 2.9 with $v_0 = 0$ m/s, since the rock is released from rest. At a height h, the rock has fallen through a distance $(20.0\ \text{m}) - h$, and according to Equation 2.9, $v^2 = 2ay = 2a[(20.0\ \text{m}) - h]$. Therefore, the kinetic energy at any height h is given by $\text{KE} = ma[(20.0\ \text{m}) - h]$. The total energy of the rock at any height h is the sum of the kinetic energy and potential energy at the particular height.

SOLUTION The calculations are performed below for $h = 10.0$ m. The table that follows also shows the results for $h = 20.0$ m and $h = 0$ m.

$$\text{PE} = mgh = (2.00\ \text{kg})(9.80\ \text{m/s}^2)(10.0\ \text{m}) = \boxed{196\ \text{J}}$$

$$\text{KE} = ma[(20.0\ \text{m}) - h] = (2.00\ \text{kg})(9.80\ \text{m/s}^2)(20.0\ \text{m} - 10.0\ \text{m}) = \boxed{196\ \text{J}}$$

$$E = \text{KE} + \text{PE} = 196\ \text{J} + 196\ \text{J} = \boxed{392\ \text{J}}$$

***h* (m)**	**KE (J)**	**PE (J)**	**E (J)**
20.0	0	392	392
10.0	196	196	392
0	392	0	392

Note that in each case, the value of E is the same, because mechanical energy is conserved.

---

70. ***REASONING AND SOLUTION*** The work done by the pitcher is

$$W_{nc} = F(\pi r)$$

Assuming that only the pitcher does work on the ball

$$W_{nc} = \tfrac{1}{2}mv_f^2 - \tfrac{1}{2}mv_0^2 + mgh_f - mgh_0$$

Then,

$$v_f = \sqrt{\frac{2W_{nc}}{m} + v_0^2 - 2g(h_f - h_0)}$$

$$= \sqrt{\frac{2(28\text{ N})\pi(0.51\text{ m})}{0.25\text{ kg}} + (12\text{ m/s})^2 - 2(9.80\text{ m/s}^2)(-1.02\text{ m})} = \boxed{23\text{ m/s}}$$

---

71. SSM ***REASONING AND SOLUTION*** We will assume that the tug-of-war rope remains parallel to the ground, so that the force that moves team B is in the same direction as the displacement. According to Equation 6.1, the work done by team A is

$$W = (F\cos\theta)s = (1100\text{ N})(\cos 0°)(2.0\text{ m}) = \boxed{2.2 \times 10^3\text{ J}}$$

---

72. ***REASONING AND SOLUTION***

a. The work done by non-conservative forces is given by Equation 6.7b as

$$W_{nc} = \Delta\text{KE} + \Delta\text{PE} \qquad \text{so} \qquad \Delta\text{PE} = W_{nc} - \Delta\text{KE}$$

Now

$$\Delta\text{KE} = \tfrac{1}{2}mv_f^2 - \tfrac{1}{2}mv_0^2 = \tfrac{1}{2}(55.0\text{ kg})[(6.00\text{ m/s})^2 - (1.80\text{ m/s})^2] = 901\text{ J}$$

and

$$\Delta\text{PE} = 80.0\text{ J} - 265\text{ J} - 901\text{ J} = \boxed{-1086\text{ J}}$$

b. $\Delta\text{PE} = mg(h - h_0)$ so

$$h - h_0 = \frac{\Delta\text{PE}}{mg} = \frac{-1086\text{ J}}{(55.0\text{ kg})(9.80\text{ m/s}^2)} = -2.01\text{ m}$$

Thus, the skater's vertical position has changed by $\boxed{2.01\text{ m}}$, and the skater is $\boxed{\text{below the starting point}}$.

---

73. ***REASONING AND SOLUTION***

a. The work done by the applied force is

$$W = Fs\cos\theta = (2.40\times10^2\text{ N})(8.00\text{ m})\cos 20.0° = \boxed{1.80\times10^3\text{ J}}$$

b. The work done by the frictional force is $W_\text{f} = f_\text{k}s\cos\theta$, where

$$f_\text{k} = \mu_\text{s}(mg - F\sin\theta)$$

$$= (0.200)\left[(85.0\text{ kg})(9.80\text{ m/s}^2) - (2.40\times10^2\text{ N})(\sin 20.0°)\right] = 1.50\times10^2\text{ N}$$

Now

$$W_\text{f} = (1.50\times10^2\text{ N})(8.00\text{ m})\cos 180° = \boxed{-1.20\times10^3\text{ J}}$$

---

74. ***REASONING*** The average power developed by the cheetah is equal to the work done by the cheetah divided by the elapsed time (Equation 6.10a). The work, on the other hand, can be related to the change in the cheetah's kinetic energy by the work-energy theorem, Equation 6.3.

SOLUTION

a. The average power is

$$\bar{P} = \frac{W}{t} \tag{6.10a}$$

where W is the work done by the cheetah. This work is related to the change in the cheetah's kinetic energy by Equation 6.3, $W = \frac{1}{2}mv_\text{f}^2 - \frac{1}{2}mv_0^2$, so the average power is

$$\bar{P} = \frac{W}{t} = \frac{\frac{1}{2}mv_\text{f}^2 - \frac{1}{2}mv_0^2}{t}$$

$$= \frac{\frac{1}{2}(110\text{ kg})(27\text{ m/s})^2 - \frac{1}{2}(110\text{ kg})(0\text{ m/s})^2}{4.0\text{ s}} = \boxed{1.0\times10^4\text{ W}}$$

b. The power, in units of horsepower (hp), is

$$\bar{P} = \left(1.0\times10^4\text{ W}\right)\left(\frac{1\text{ hp}}{745.7\text{ W}}\right) = \boxed{13\text{ hp}}$$

---

75. ***REASONING*** We can determine the force exerted on the diver by the water if we first find the work done by the force. This is because the work can be expressed using Equation 6.1 as $W_\text{nc} = (F\cos\theta)s$, where F is the magnitude of the force from the water and s is the

magnitude of the displacement in the water. Since the force that the water exerts is nonconservative, we have included the subscript "nc" in labeling the work. To calculate the work we will use the work-energy theorem in the form of Equation 6.8: $W_{nc} = \left(\tfrac{1}{2}mv_f^2 + mgh_f\right) - \left(\tfrac{1}{2}mv_0^2 + mgh_0\right)$. In this theorem W_{nc} is the work done by the net nonconservative force, which, in this case, is just the force exerted by the water.

SOLUTION We write Equation 6.8 as follows:

$$W_{nc} = \underbrace{\left(\tfrac{1}{2}mv_f^2 + mgh_f\right)}_{\text{Final mechanical energy}} - \underbrace{\left(\tfrac{1}{2}mv_0^2 + mgh_0\right)}_{\text{Initial mechanical energy}} = mgh_f - mgh_0 \qquad (1)$$

where we have used the fact that the diver is at rest initially and finally, so $v_0 = v_f = 0$ m/s. According to Equation 6.1 the work done on the diver by the force of the water is

$$W_{nc} = (F \cos \theta)s = (F \cos 180°)s = -Fs$$

The angle θ between the force of the water and the displacement is 180°, because the force opposes the motion. Substituting this result into Equation (1) gives

$$-Fs = mgh_f - mgh_0 \qquad (2)$$

Identifying the final position under the water as $h_f = 0$ m und using upward as the positive direction, we know that the initial position on the tower must be $h_0 = 3.00 \text{ m} + 1.10 \text{ m} = 4.10$ m. Thus, solving Equation (2) for F, we find that

$$F = \frac{mg\left(h_0 - h_f\right)}{s} = \frac{(67.0 \text{ kg})(9.80 \text{ m/s}^2)\left[(4.10 \text{ m}) - (0 \text{ m})\right]}{1.10 \text{ m}} = \boxed{2450 \text{ N}}$$

---

76. ***REASONING***

a. Because the kinetic frictional force, a nonconservative force, is present, it does negative work on the skier. The work-energy theorem, in the form of Equation 6.6, may be used to find the work done by this force.

b. Once the work done by the kinetic frictional force is known, the magnitude of the kinetic frictional force can be determined by using the definition of work, Equation 6.1, since the magnitude of the skier's displacement is known.

SOLUTION

a. The work W_{nc} done by the kinetic frictional force is related to the object's kinetic and potential energies by Equation 6.6:

$$W_{nc} = \left(\tfrac{1}{2}mv_f^2 - \tfrac{1}{2}mv_0^2\right) + \left(mgh_f - mgh_0\right)$$

The initial height of the skier at the bottom of the hill is $h_0 = 0$ m, and the final height is $h_f = s \sin 25°$ (see the drawing). Thus, the work is

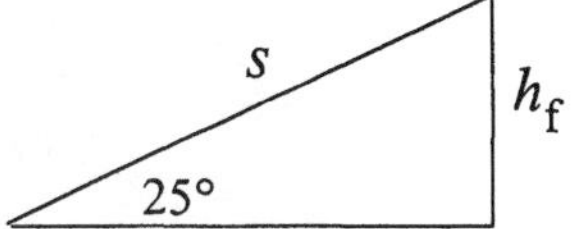

$$W_{\text{nc}} = \left(\tfrac{1}{2}mv_f^2 - \tfrac{1}{2}mv_0^2\right) + mg\left(s \sin 25° - h_0\right)$$

$$= \tfrac{1}{2}(63\text{ kg})(4.4\text{ m/s})^2 - \tfrac{1}{2}(63\text{ kg})(6.6\text{ m/s})^2$$

$$+ (63\text{ kg})\left(9.80\text{ m/s}^2\right)\left[(1.9\text{ m})\sin 25° - 0\text{ m}\right] = \boxed{-270\text{ J}}$$

b. The work done by the kinetic frictional force is, according to Equation 6.1, $W_{\text{nc}} = (f_k \cos 180°)s$, where f_k is the magnitude of the kinetic frictional force, and s is the magnitude of the skier's displacement. The displacement of the skier is up the hill and the kinetic frictional force is directed down the hill, so the angle between the two vectors is $\theta =$ 180° and $\cos\theta = -1$. Solving the equation above for f_k, we have

$$f_k = \frac{W_{\text{nc}}}{-s} = \frac{-270\text{ J}}{-1.9\text{ m}} = \boxed{140\text{ N}}$$

---

77. SSM WWW ***REASONING*** Gravity is the only force acting on the vaulters, since friction and air resistance are being ignored. Therefore, the net work done by the nonconservative forces is zero, and the principle of conservation of mechanical energy holds.

SOLUTION Let E_{2f} represent the total mechanical energy of the second vaulter at the ground, and E_{20} represent the total mechanical energy of the second vaulter at the bar. Then, the principle of mechanical energy is written for the second vaulter as

$$\underbrace{\frac{1}{2}mv_{2f}^2 + mgh_{2f}}_{E_{2f}} = \underbrace{\frac{1}{2}mv_{20}^2 + mgh_{20}}_{E_{20}}$$

Since the mass m of the vaulter appears in every term of the equation, m can be eliminated algebraically. The quantity $h_{20} = h$, where h is the height of the bar. Furthermore, when the vaulter is at ground level, $h_{2f} = 0$ m. Solving for v_{20} we have

$$v_{20} = \sqrt{v_{2f}^2 - 2gh} \qquad (1)$$

In order to use Equation (1), we must first determine the height h of the bar. The height h can be determined by applying the principle of conservation of mechanical energy to the first vaulter on the ground and at the bar. Using notation similar to that above, we have

$$\underbrace{\frac{1}{2}mv_{1f}^2 + mgh_{1f}}_{E_{1f}} = \underbrace{\frac{1}{2}mv_{10}^2 + mgh_{10}}_{E_{10}}$$

where E_{10} corresponds to the total mechanical energy of the first vaulter at the bar. The height of the bar is, therefore,

$$h = h_{10} = \frac{v_{1f}^2 - v_{10}^2}{2g} = \frac{(8.90 \text{ m/s})^2 - (1.00 \text{ m/s})^2}{2(9.80 \text{ m/s}^2)} = 3.99 \text{ m}$$

The speed at which the second vaulter clears the bar is, from Equation (1),

$$v_{20} = \sqrt{v_{2f}^2 - 2gh} = \sqrt{(9.00 \text{ m/s})^2 - 2(9.80 \text{ m/s}^2)(3.99 \text{ m})} = \boxed{1.7 \text{ m/s}}$$

---

78. ***REASONING AND SOLUTION*** If air resistance is ignored, the only nonconservative force that acts on the person is the normal force exerted on the person by the surface. Since this force is always perpendicular to the direction of the displacement, the work done by the normal force is zero. We can conclude, therefore, that mechanical energy is conserved.

$$\frac{1}{2}mv_0^2 + mgh_0 = \frac{1}{2}mv_f^2 + mgh_f \qquad (1)$$

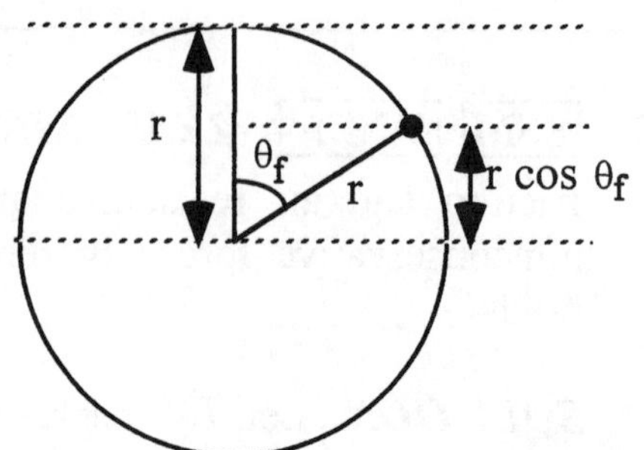

where the final state pertains to the position where the person leaves the surface. Since the person starts from rest $v_0 = 0$ m/s. Since the radius of the surface is r, $h_0 = r$, and $h_f = r \cos \theta_f$ where θ_f is the angle at which the person leaves the surface. Equation (1) becomes

$$mgr = \frac{1}{2}mv_f^2 + mg(r \cos \theta_f) \qquad (2)$$

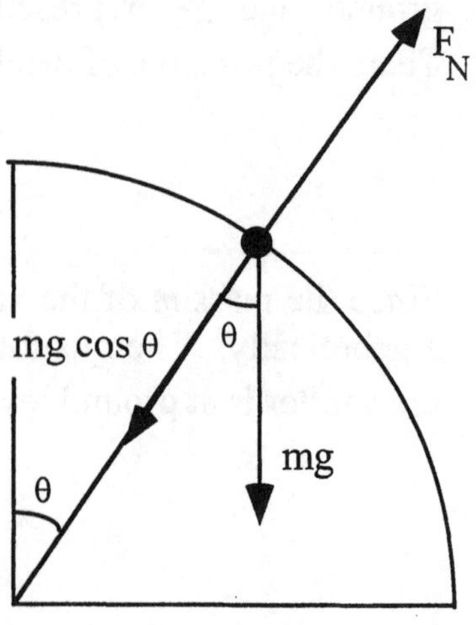

In general, as the person slides down the surface, the two forces that act on him are the normal force $\mathbf{F_N}$ and the weight mg. The centripetal force required to keep the person moving in the circular path is the resultant of $\mathbf{F_N}$ and the radial component of the weight, mg cos θ.

When the person leaves the surface, the normal force is zero, and the radial component of the weight provides the centripetal force.

$$mg \cos \theta_f = \frac{mv_f^2}{r} \qquad \Rightarrow \qquad v_f^2 = gr \cos \theta_f \qquad (3)$$

Substituting this expression for v_f^2 into Equation (2) gives

$$mgr = \tfrac{1}{2}mg(r\cos\theta_f) + mg(r\cos\theta_f)$$

Solving for θ_f gives

$$\theta_f = \cos^{-1}\left(\frac{2}{3}\right) = \boxed{48°}$$

---

79. SSM WWW ***REASONING*** After the wheels lock, the only nonconservative force acting on the truck is friction. The work done by this conservative force can be determined from the work-energy theorem. According to Equation 6.8,

$$W_{nc} = E_f - E_0 = \left(\tfrac{1}{2}mv_f^2 + mgh_f\right) - \left(\tfrac{1}{2}mv_0^2 + mgh_0\right) \qquad (1)$$

where W_{nc} is the work done by friction. According to Equation 6.1, $W = (F\cos\theta)s$; since the force of kinetic friction points opposite to the displacement, $\theta = 180°$. According to Equation 4.8, the kinetic frictional force has a magnitude of $f_k = \mu_k F_N$, where μ_k is the coefficient of kinetic friction and F_N is the magnitude of the normal force. Thus,

$$W_{nc} = (f_k\cos\theta)s = \mu_k F_N(\cos 180°)s = -\mu_k F_N s \qquad (2)$$

Since the truck is sliding down an incline, we refer to the free-body diagram in order to determine the magnitude of the normal force F_N. The free-body diagram at the right shows the three forces that act on the truck. They are the normal force $\mathbf{F}_N$, the force of kinetic friction $\mathbf{f}_k$, and the weight $m\mathbf{g}$. The weight has been resolved into its components, and these vectors are shown as dashed arrows. From the free body diagram and the fact that the truck does not accelerate perpendicular to the incline, we can see that the magnitude of the normal force is given by

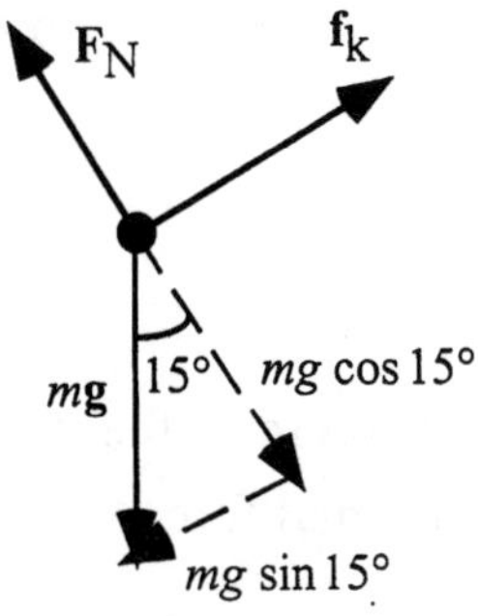

$$F_N = mg\cos 15°$$

Therefore, Equation (2) becomes

$$W_{nc} = -\mu_k(mg\cos 15°)s \qquad (3)$$

Combining Equations (1), (2), and (3), we have

$$-\mu_k(mg\cos 15°)s = \left(\tfrac{1}{2}mv_f^2 + mgh_f\right) - \left(\tfrac{1}{2}mv_0^2 + mgh_0\right) \qquad (4)$$

This expression may be solved for s.

SOLUTION When the car comes to a stop, $v_f = 0$ m/s. If we take $h_f = 0$ m at ground level, then, from the drawing at the right, we see that $h_0 = s \sin 15°$. Equation (4) then gives

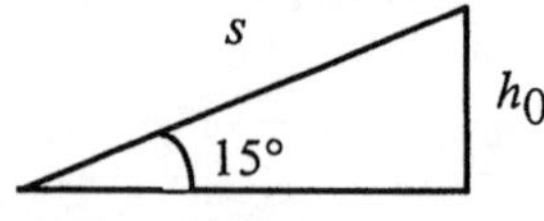

$$s = \frac{v_0^2}{2g(\mu_k \cos 15° - \sin 15°)} = \frac{(11.1 \text{ m/s})^2}{2(9.80 \text{ m/s}^2)[(0.750)(\cos 15°) - \sin 15°]} = \boxed{13.5 \text{ m}}$$

---

80. ***CONCEPT QUESTIONS***

a. Since the force exerted by the elevator and the displacement are in the same direction on the upward part of the trip, the angle between them is 0°, and so the work done by the force is positive.

b. Since the force exerted by the elevator and the displacement are in opposite directions on the downward part of the trip, the angle between them is 180°, and so the work done by the force is negative.

SOLUTION

a. The free-body diagram at the right shows the three forces that act on you: **W** is your weight, $\mathbf{W}_{\text{belongings}}$ is the weight of your belongings, and $\mathbf{F}_N$ is the normal force exerted on you by the floor of the elevator. Since you are moving upward at a constant velocity, you are in equilibrium and the net force in the y direction must be zero:

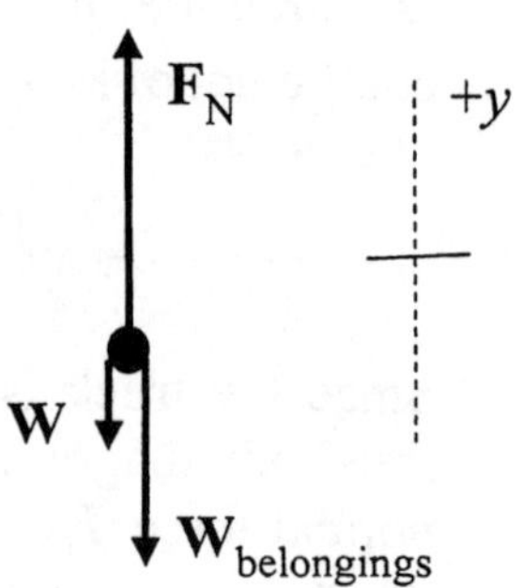

$$\underbrace{F_N - W - W_{\text{belongings}}}_{\Sigma F_y} = 0 \qquad (4.9b)$$

Therefore, the magnitude of the normal force is $F_N = W + W_{\text{belongings}}$. The work done by the normal force is

$$W = (F_N \cos\theta)s = (W + W_{\text{belongings}})(\cos\theta)s \qquad (6.1)$$

$$= (685 \text{ N} + 915 \text{ N})(\cos 0°)(15.2 \text{ m}) = \boxed{24\,300 \text{ J}}$$

b. During the downward trip, you are still in equilibrium since the elevator is moving with a constant velocity. The magnitude of the normal force is now $F_N = W$. The work done by the normal force is

$$W = (F_N \cos\theta)s = (W \cos\theta)s = (685 \text{ N})(\cos 180°)(15.2 \text{ m}) = \boxed{-10\,400 \text{ J}} \qquad (6.1)$$

---

81. ***CONCEPT QUESTIONS***

a. Since the asteroid is slowing down, it is decelerating. The force must point in a direction that is opposite to that of the displacement.

b. The direction of the force is opposite to the displacement of the asteroid, so the angle between them is 180°. Therefore, the force does negative work.

c. The asteroid is slowing down, so its speed is decreasing. Thus, its kinetic energy is changing.

d. The work done by the force is related to the change in the asteroid's kinetic energy by the work-energy theorem: $W = \frac{1}{2}mv_f^2 - \frac{1}{2}mv_0^2$, Equation 6.3.

SOLUTION

a. The work-energy theorem states that the work done by the force acting on the asteroid is equal to its final kinetic energy minus its initial kinetic energy:

$$W = \tfrac{1}{2}mv_f^2 - \tfrac{1}{2}mv_0^2 \qquad (6.3)$$

$$= \tfrac{1}{2}\left(4.5\times10^4 \text{ kg}\right)(5500 \text{ m/s})^2 - \tfrac{1}{2}\left(4.5\times10^4 \text{ kg}\right)(7100 \text{ m/s})^2 = \boxed{-4.5\times10^{11} \text{ J}}$$

b. The work done by the force acting on the asteroid is given by Equation 6.1 as $W = (F\cos\theta)s$. Since the force acts to slow down the asteroid, the force and the displacement vectors point in opposite directions, so the angle between them is $\theta = 180°$. The magnitude of the force is

$$F = \frac{W}{s\cos\theta} = \frac{-4.5\times10^{11} \text{ J}}{\left(1.8\times10^6 \text{ m}\right)\cos180°} = \boxed{2.5\times10^5 \text{ N}} \qquad (6.1)$$

---

82. ***CONCEPT QUESTIONS***

a. Since there is no friction, there are only two forces acting on each box, the gravitational force (its weight) and the normal force. The normal force, being perpendicular to the displacement of a box, does no work. The gravitational force does work. However, it is a conservative force, so the conservation of mechanical energy applies to the motion of each box. This principle states that the final speed depends only on the initial speed, and the initial and final heights. The final speed does not depend on the mass of the box or on the steepness of the slope. Thus, both boxes reach B with the same speed.

b. The kinetic energy is given by the expression $\text{KE} = \frac{1}{2}mv^2$. Both boxes have the same speed v when they reach B. However, the heavier box has the greater mass. Therefore, the heavier box has the greater kinetic energy.

SOLUTION
a. The conservation of mechanical energy states that

$$\tfrac{1}{2}mv_B^2 + mgh_B = \tfrac{1}{2}mv_A^2 + mgh_A \tag{6.9b}$$

Solving for the speed v_B at B gives

$$v_B = \sqrt{v_A^2 + 2g(h_A - h_B)} = \sqrt{(0\text{ m/s})^2 + 2(9.80\text{ m/s}^2)(4.5\text{ m} - 1.5\text{ m})} = \boxed{7.7\text{ m/s}}$$

b. The speed of the heavier box is the same as the lighter box: $\boxed{7.7\text{ m/s}}$.

c. The ratio of the kinetic energies is

$$\frac{(\text{KE}_B)_{\text{heavier box}}}{(\text{KE}_B)_{\text{lighter box}}} = \frac{\left(\frac{1}{2}mv_B^2\right)_{\text{heavier box}}}{\left(\frac{1}{2}mv_B^2\right)_{\text{lighter box}}} = \frac{m_{\text{heavier box}}}{m_{\text{lighter box}}} = \frac{44\text{ kg}}{11\text{ kg}} = \boxed{4.0}$$

---

83. ***CONCEPT QUESTIONS***
a. The kinetic frictional force does negative work, because the force is directed opposite to the displacement of the student.

b. Since the kinetic frictional force does negative work, the final total mechanical energy must be less than the initial total mechanical energy. Therefore, the total mechanical energy decreases.

c. The work done by the kinetic frictional force is related to the change in the student's total mechanical energy by the work-energy theorem, $W_{nc} = E_f - E_0$, Equation 6.8.

SOLUTION
The work-energy theorem states that

$$W_{nc} = \underbrace{\tfrac{1}{2}mv_f^2 + mgh_f}_{E_f} - \underbrace{\left(\tfrac{1}{2}mv_0^2 + mgh_0\right)}_{E_0} \tag{6.8}$$

Solving for the final speed gives

$$v_f = \sqrt{\frac{2W_{nc}}{m} + v_0^2 + 2g\left(h_0 - h_f\right)}$$

$$= \sqrt{\frac{2\left(-6.50 \times 10^3 \text{ J}\right)}{83.0 \text{ kg}} + (0 \text{ m/s})^2 + 2\left(9.80 \text{ m/s}^2\right)(11.8 \text{ m})} = \boxed{8.6 \text{ m/s}}$$

---

84. ***CONCEPT QUESTIONS***

a. The two types of energy that are changing are the kinetic energy and the gravitational potential energy. The kinetic energy is increasing, because the speed of the helicopter is increasing. The gravitational potential energy is increasing, because the height of the helicopter is increasing.

b. The work done by the lifting force is related to the kinetic and potential energies by the work-energy theorem, Equation 6.6.

c. The average power is defined as the work divided by the time, Equation 6.10a, so the time must be known in addition to the work.

SOLUTION

The average power generated by the lifting force is equal to the work done by the force divided by the elapsed time (Equation 6.10a). The work can be related to the change in the helicopter's total mechanical energy, which can be determined. The average power is

$$\overline{P} = \frac{W_{nc}}{t} \tag{6.10a}$$

where W_{nc} is the work done by the nonconservative lifting force. This work is related to the helicopter's kinetic and potential energies by Equation 6.6:

$$W_{nc} = \left(\tfrac{1}{2}mv_f^2 - \tfrac{1}{2}mv_0^2\right) + \left(mgh_f - mgh_0\right)$$

Thus, the average power is

$$\overline{P} = \frac{\text{W}_{nc}}{t} = \frac{\left(\frac{1}{2}mv_f^2 - \frac{1}{2}mv_0^2\right) + \left(mgh_f - mgh_0\right)}{t} = \frac{\frac{1}{2}m\left(v_f^2 - v_0^2\right) + mg\left(h_f - h_0\right)}{t}$$

$$\overline{P} = \frac{\frac{1}{2}(810 \text{ kg})\left[(7.0 \text{ m/s})^2 - (0 \text{ m/s})^2\right] + (810 \text{ kg})\left(9.80 \text{ m/s}^2\right)[8.2 \text{ m} - 0 \text{ m}]}{3.5 \text{ s}} = \boxed{2.4 \times 10^4 \text{ W}}$$

---

85. ***CONCEPT QUESTIONS***
a. The kinetic energy is changing because the skier's speed in changing.

b. The work is positive. Because the speed of the skier is increasing, the net external force must be in the same direction as the displacement of the skier.

c. The work done by the net external force is equal to the change in the kinetic energy of the skier.

SOLUTION The work done by the net external force acting on the skier can be found from the work-energy theorem in the form of Equation 6.3, because the final and initial kinetic energies of the skier can be determined:

$$W = \tfrac{1}{2}mv_f^2 - \tfrac{1}{2}mv_0^2$$

$$= \tfrac{1}{2}(70.3\text{ kg})(11.3\text{ m/s})^2 - \tfrac{1}{2}(70.3\text{ kg})(6.10\text{ m/s})^2 = \boxed{3.2 \times 10^3\text{ J}}$$

---

86. ***CONCEPT QUESTIONS***
a. During the coasting phase, the net force consists only of the horizontal force of kinetic friction $\mathbf{f}_k$ between the runners of the snowmobile and the snow, since the drive force is shut off. The downward-pointing weight of the snowmobile is balanced by the upward-pointing normal force that the snow exerts. Thus, W is the work done by the kinetic frictional force. According to Equation 6.1, this work is given by $W = (f_k \cos\theta)\, s$. Thus, W depends on f_k, which is the magnitude of the frictional force, on θ, which is the angle specifying the direction of the force relative to the displacement, and on s, which is the magnitude of the displacement of the snowmobile.

b. When the velocity is constant, the acceleration of the snowmobile is zero. Therefore, according to Newton's second law, the net force acting on the snowmobile is zero. This means that the downward-pointing weight of the snowmobile is balanced by the upward-pointing normal force that the snow exerts, as it is in the coasting phase. However, during the constant-velocity phase, the horizontal component of the net force must also be zero, so that the forward-pointing drive force must balance the opposing frictional force. Thus, the magnitude of the frictional force must be the same as the magnitude of the drive force.

c. By using only the work-energy theorem it is not possible to determine the time t during which the snowmobile coasts to a halt. To determine t it is necessary to use the equations of kinematics. We can use these equations, because the acceleration during the coasting phase is constant, being determined by the constant force of friction and the mass of the snowmobile, according to Newton's second law. For example, we can use Equation 2.7 $\left[s = \tfrac{1}{2}(v_0 + v_f)t\right]$ to determine t once we know the magnitude of the displacement s, which is the distance in which the snowmobile coasts to a halt.

SOLUTION

a. According to Equation 6.3, the work energy theorem, is

$$W = \tfrac{1}{2}mv_f^2 - \tfrac{1}{2}mv_0^2$$

The work W is the work done by the net force acting on the snowmobile. During the coasting phase, the net force acting on the snowmobile is the kinetic frictional force (magnitude $= f_k$), so the work W can be written as $W = (f_k \cos\theta)\, s$, as explained in the answer to Concept Question (a). Substituting this result into the work-energy theorem gives

$$(f_k \cos\theta)s = \tfrac{1}{2}mv_f^2 - \tfrac{1}{2}mv_0^2 \qquad (1)$$

The angle θ between the frictional force and the displacement is $\theta = 180°$, because the frictional force opposes the relative motion of the snowmobile and the snow. Furthermore, the final speed is $v_f = 0$ m/s, since the snowmobile coasts to a halt. Finally, as explained in the answer to Concept Question (b), the forward-pointing drive force balances the opposing frictional force during the constant-velocity phase of the motion, with the result that the magnitudes of these two forces are equal: $f_k = 205$ N. During the coasting phase, the frictional force has this same value. The distance traveled as the snowmobile coasts to a halt can now be obtained by solving Equation (1) for s, which is the magnitude of the displacement:

$$s = \frac{m(v_f^2 - v_0^2)}{2f_k \cos\theta} = \frac{(136\text{ kg})\left[(0\text{ m/s})^2 - (5.50\text{ m/s})^2\right]}{2(205\text{ N})\cos 180°} = \boxed{10.0\text{ m}}$$

b. As explained in the answer to Concept Question (c), we can use Equation 2.7 $\left[s = \tfrac{1}{2}(v_0 + v_f)t\right]$ to determine the time t during which the snowmobile coasts to a halt. Solving this equation for t gives

$$t = \frac{2s}{v_0 + v_f} = \frac{2(10.0\text{ m})}{5.50\text{ m/s} + 0\text{ m/s}} = \boxed{3.64\text{ s}}$$

---

87. ***CONCEPT QUESTIONS***

a. When the block on the longer track reaches its maximum height, its final total mechanical energy consists only of gravitational potential energy, because it comes to a momentary halt at this point. Thus, its final speed is zero, and, as a result, its kinetic energy is also zero.

b. When the block on the shorter track reaches the top of its trajectory after leaving the track, its final total mechanical energy consists of part kinetic and part potential energy. At the top of the trajectory the block is still moving horizontally, with a velocity that equals the horizontal velocity component that it had when it left the track (assuming that air resistance is negligible). Thus, at the top of the trajectory the block has some kinetic energy.

c. The height H is greater than the height $H_1 + H_2$. To see why, note that both blocks have the same initial kinetic energy, since each has the same initial speed. Moreover, the total mechanical energy is conserved, because the track is frictionless (assuming air resistance is negligible). Therefore, the initial kinetic energy is converted to potential energy as each block moves upward. On the longer track, the final total mechanical energy is all potential energy, whereas on the shorter track it is only part potential energy, as discussed in the answer to Concept Question (b). Since gravitational potential energy is proportional to height, the final height is greater for the block with the greater potential energy, or the one on the longer track.

SOLUTION The two inclines in the problem are as follows:

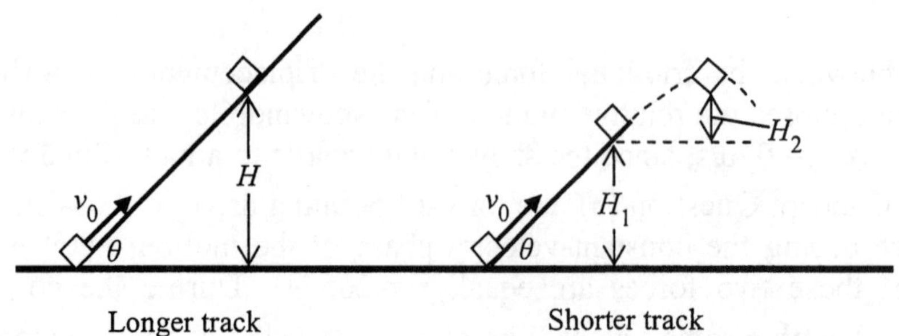

Longer track: Applying the principle of conservation of mechanical energy for the longer track gives

$$\underbrace{\tfrac{1}{2}mv_f^2 + mgh_f}_{\text{Final total mechanical energy}} = \underbrace{\tfrac{1}{2}mv_0^2 + mgh_0}_{\text{Initial total mechanical energy}}$$

Recognizing that $v_f = 0$ m/s, $h_f = H$, and $h_0 = 0$ m, we find that

$$\tfrac{1}{2}m(0\text{ m/s})^2 + mgH = \tfrac{1}{2}mv_0^2 + mg(0\text{ m})$$

$$H = \frac{v_0^2}{2g} = \frac{(7.00\text{ m/s})^2}{2(9.80\text{ m/s}^2)} = \boxed{2.50\text{ m}}$$

Shorter track: We apply the conservation principle to the motion of the block after it leaves the track, in order to calculate the final height at the top of the trajectory, where $h_f = H_1 + H_2$ and the final speed is v_f. For the initial position at the end of the track we have $h_0 = H_1$ and an initial speed of v_{0T}. Thus, it follows that

$$\underbrace{\tfrac{1}{2}mv_f^2 + mg(H_1 + H_2)}_{\text{Final total mechanical energy}} = \underbrace{\tfrac{1}{2}mv_{0T}^2 + mgH_1}_{\text{Initial total mechanical energy}}$$

Solving for $H_1 + H_2$ gives

$$H_1 + H_2 = \frac{v_{0T}^2 + 2gH_1 - v_f^2}{2g} \qquad (1)$$

At the top of its trajectory the block is moving horizontally with a velocity that equals the horizontal component of the velocity with which it left the track, that is, $v_f = v_{0T} \cos\theta$. Thus, a value for v_{0T} is needed and can be obtained by applying the conservation principle to the motion of the block that takes place on the track.

For the motion on the track the final speed is the speed v_{0T} at which the block leaves the track, and the final height is $h_f = H_1 = 1.25$ m. The initial speed is the given value of $v_0 = 7.00$ m/s, and the initial height is $h_0 = 0$ m. The conservation principle reveals that

$$\underbrace{\tfrac{1}{2}mv_{0T}^2 + mgH_1}_{\text{Final total mechanical energy}} = \underbrace{\tfrac{1}{2}mv_0^2 + mg(0\text{ m})}_{\text{Initial total mechanical energy}}$$

$$v_{0T} = \sqrt{v_0^2 - 2gH_1} = \sqrt{(7.00\text{ m/s})^2 - 2(9.80\text{ m/s}^2)(1.25\text{ m})} = 4.95\text{ m/s}$$

For use in Equation (1) we find, then, that the speed with which the block leaves the track is $v_{0T} = 4.95$ m/s and the final speed at the top of the trajectory is $v_f = v_{0T} \cos\theta = (4.95\text{ m/s}) \cos\theta$. In addition, $h_f = H_1 + H_2$ at the top of the trajectory, and $h_0 = H_1 = 1.25$ m for the initial position at the end of the track. Thus, Equation (1) reveals that

$$H_1 + H_2 = \frac{v_{0T}^2 + 2gH_1 - v_f^2}{2g}$$

$$= \frac{(4.95\text{ m/s})^2 + 2(9.80\text{ m/s}^2)(1.25\text{ m}) - [(4.95\text{ m/s})\cos 50.0°]^2}{2(9.80\text{ m/s}^2)} = \boxed{1.98\text{ m}}$$

As expected, $H_1 + H_2$ is less than H, and the block rises to a greater height on the longer track.

CHAPTER 7 | *IMPULSE AND MOMENTUM*

PROBLEMS

---

1. SSM ***REASONING*** The impulse that the volleyball player applies to the ball can be found from the impulse-momentum theorem, Equation 7.4. Two forces act on the volleyball while it's being spiked: an average force $\overline{\mathbf{F}}$ exerted by the player, and the weight of the ball. As in Example 1, we will assume that $\overline{\mathbf{F}}$ is much greater than the weight of the ball, so the weight can be neglected. Thus, the net average force $(\Sigma\overline{\mathbf{F}})$ is equal to $\overline{\mathbf{F}}$.

SOLUTION From Equation 7.4, the impulse that the player applies to the volleyball is

$$\underbrace{\overline{\mathbf{F}}\,\Delta t}_{\text{Impulse}} = \underbrace{m\mathbf{v}_f}_{\substack{\text{Final}\\\text{momentum}}} - \underbrace{m\mathbf{v}_0}_{\substack{\text{Initial}\\\text{momentum}}}$$

$$= m(\mathbf{v}_f - \mathbf{v}_0) = (0.35\text{ kg})\left[(-21\text{ m/s}) - (+4.0\text{ m/s})\right] = \boxed{-8.7\text{ kg}\cdot\text{m/s}}$$

The minus sign indicates that the direction of the impulse is the same as that of the final velocity of the ball.

---

2. ***REASONING AND SOLUTION*** According to the impulse-momentum theorem, Equation 7.4, $(\Sigma\overline{\mathbf{F}})\Delta t = m\mathbf{v}_f - m\mathbf{v}_0$, where $\Sigma\overline{\mathbf{F}}$ is the net average force acting on the person. Taking the direction of motion (downward) as the negative direction and solving for the net average force $\Sigma\overline{\mathbf{F}}$, we obtain

$$\Sigma\overline{\mathbf{F}} = \frac{m(\mathbf{v}_f - \mathbf{v}_0)}{\Delta t} = \frac{(62.0\text{ kg})\left[-1.10\text{ m/s} - (-5.50\text{ m/s})\right]}{1.65\text{ s}} = \boxed{+165\text{ N}}$$

The plus sign indicates that the force acts $\boxed{\text{upward}}$.

---

3. ***REASONING*** a. The change in momentum of the ball is the final momentum $m\mathbf{v}_f$ minus the initial momentum $m\mathbf{v}_f$, both of which can be determined.

b. According to the impulse-momentum theorem, $(\Sigma\overline{\mathbf{F}})\Delta t = m\mathbf{v}_f - m\mathbf{v}_0$, the net average force $\Sigma\overline{\mathbf{F}}$ applied to the ball is equal to the change $(m\mathbf{v}_f - m\mathbf{v}_0)$ in the ball's momentum, divided by the time Δt of impact. In this situation the tee upon which the ball is placed supports its weight, so the net average force is $\Sigma\overline{\mathbf{F}} = \overline{\mathbf{F}}$, the average force that the club applies to the ball.

SOLUTION a. The change $\Delta\mathbf{p}$ in the ball's momentum is

$$\Delta\mathbf{p} = m\mathbf{v}_f - m\mathbf{v}_0 = m(\mathbf{v}_f - \mathbf{v}_0)$$

$$= (0.045\text{ kg})(+38\text{ m/s} - 0\text{ m/s}) = \boxed{+1.7\text{ kg}\cdot\text{m/s}}$$

b. Solving the impulse-momentum theorem for the average force gives

$$\overline{\mathbf{F}} = \frac{m(\mathbf{v}_f - \mathbf{v}_0)}{\Delta t} = \frac{(0.045\text{ kg})(+38\text{ m/s} - 0\text{ m/s})}{3.0 \times 10^{-3}\text{ s}} = \boxed{+570\text{ N}}$$

---

4. ***REASONING*** During the collision, the bat exerts an impulse on the ball. The impulse is the product of the average force that the bat exerts and the time of contact. According to the impulse-momentum theorem, the impulse is also equal to the change in the momentum of the ball. We will use these two relations to determine the average force exerted by the bat on the ball.

SOLUTION The impulse $\mathbf{J}$ is given by Equation 7.1 as $\mathbf{J} = \overline{\mathbf{F}}\Delta t$, where $\overline{\mathbf{F}}$ is the average force that the bat exerts on the ball and Δt is the time of contact. According to the impulse-momentum theorem, Equation 7.4, the net average impulse $(\Sigma\overline{\mathbf{F}})\Delta t$ is equal to the change in the ball's momentum; $(\Sigma\overline{\mathbf{F}})\Delta t = m\mathbf{v}_f - m\mathbf{v}_0$. Since we are ignoring the weight of the ball, the bat's force is the net force, so $\Sigma\overline{\mathbf{F}} = \overline{\mathbf{F}}$. Substituting this value for the net average force into the impulse-momentum equation and solving for the average force gives

$$\overline{\mathbf{F}} = \frac{m\mathbf{v}_f - m\mathbf{v}_0}{\Delta t} = \frac{(0.149\text{ kg})(-45.6\text{ m/s}) - (0.149\text{ kg})(+40.2\text{ m/s})}{1.10\times10^{-3}\text{ s}} = \boxed{-11\ 600\text{ N}}$$

where the positive direction for the velocity has been chosen as the direction of the incoming ball.

---

5. $\boxed{\text{SSM}}$ ***REASONING AND SOLUTION*** The impulse $\mathbf{J}$ is given directly by Equation 7.1:

$$\mathbf{J} = \overline{\mathbf{F}}\,\Delta t = (+1400\text{ N})(7.9\times10^{-3}\text{ s}) = \boxed{+11\text{ N}\cdot\text{s}}$$

The plus sign indicates that the direction of the impulse is the same as that of $\overline{\mathbf{F}}$.

---

6. ***REASONING*** The impulse that the roof of the car applies to the hailstones can be found from the impulse-momentum theorem, Equation 7.4. Two forces act on the hailstones, the average force $\overline{\mathbf{F}}$ exerted by the roof, and the weight of the hailstones. Since it is assumed

that $\overline{\mathbf{F}}$ is much greater than the weight of the hailstones, the net average force $(\Sigma\overline{\mathbf{F}})$ is equal to $\overline{\mathbf{F}}$.

SOLUTION From Equation 7.4, the impulse that the roof applies to the hailstones is:

$$\underbrace{\overline{\mathbf{F}}\,\Delta t}_{\text{Impulse}} = \underbrace{m\mathbf{v}_\mathbf{f}}_{\substack{\text{Final}\\ \text{momentum}}} - \underbrace{m\mathbf{v}_0}_{\substack{\text{Initial}\\ \text{momentum}}} = m\left(\mathbf{v}_\mathbf{f} - \mathbf{v}_0\right)$$

Solving for $\overline{\mathbf{F}}$ (with *up* taken to be the positive direction) gives

$$\overline{\mathbf{F}} = \left(\frac{m}{\Delta t}\right)(\mathbf{v}_\mathbf{f} - \mathbf{v}_0) = (0.060\text{ kg/s})\left[(+15\text{ m/s}) - (-15\text{ m/s})\right] = +1.8\text{ N}$$

This is the average force exerted on the hailstones by the roof of the car. The positive sign indicates that this force points upward. From Newton's third law, the average force exerted by the hailstones on the roof is equal in magnitude and opposite in direction to this force. Therefore,

$$\textbf{Force on roof} = \boxed{-1.8\text{ N}}$$

The negative sign indicates that this force $\boxed{\text{points downward}}$.

---

7. ***REASONING*** The impulse that the wall exerts on the skater can be found from the impulse-momentum theorem, Equation 7.4. The average force $\overline{\mathbf{F}}$ exerted on the skater by the wall is the only force exerted on her in the horizontal direction, so it is the net force; $\Sigma\overline{\mathbf{F}} = \overline{\mathbf{F}}$.

SOLUTION From Equation 7.4, the average force exerted on the skater by the wall is

$$\overline{\mathbf{F}} = \frac{m\mathbf{v}_\mathbf{f} - m\mathbf{v}_0}{\Delta t} = \frac{(46\text{ kg})(-1.2\text{ m/s}) - (46\text{ kg})(0\text{ m/s})}{0.80\text{ s}} = -69\text{ N}$$

From Newton's third law, the average force exerted on the wall by the skater is equal in magnitude and opposite in direction to this force. Therefore,

$$\textbf{Force exerted on wall} = \boxed{+69\text{ N}}$$

The plus sign indicates that this force points $\boxed{\text{opposite to the velocity}}$ of the skater.

---

8. ***REASONING*** We will apply the impulse momentum theorem as given in Equation 7.4 to solve this problem. From this theorem we know that, for a given change in momentum,

greater forces are associated with shorter time intervals. Therefore, we expect that the force in the stiff-legged case will be greater than in the knees-bent case.

SOLUTION a. Assuming that upward is the positive direction, we find from the impulse-momentum theorem that

$$\Sigma\overline{\mathbf{F}} = \frac{m\mathbf{v}_f - m\mathbf{v}_0}{\Delta t} = \frac{(75\text{ kg})(0\text{ m/s}) - (75\text{ kg})(-6.4\text{ m/s})}{2.0\times10^{-3}\text{ s}} = \boxed{+2.4\times10^5\text{ N}}$$

b. Again using the impulse-momentum theorem, we find that

$$\Sigma\overline{\mathbf{F}} = \frac{m\mathbf{v}_f - m\mathbf{v}_0}{\Delta t} = \frac{(75\text{ kg})(0\text{ m/s}) - (75\text{ kg})(-6.4\text{ m/s})}{0.10\text{ s}} = \boxed{+4.8\times10^3\text{ N}}$$

c. The net average force acting on the man is $\Sigma\overline{\mathbf{F}} = \mathbf{F}_{\text{Ground}} + \mathbf{W}$, where F_{Ground} is the average upward force exerted on the man by the ground and **W** is the downward-acting weight of the man. It follows, then, that $\mathbf{F}_{\text{Ground}} = \Sigma\overline{\mathbf{F}} - \mathbf{W}$. Since the weight is $\mathbf{W} = -mg$, we have

Stiff – legged $\quad \mathbf{F}_{\text{Ground}} = \Sigma\overline{\mathbf{F}} - \mathbf{W}$

$$= +2.4\times10^5\text{ N} - \left[-(75\text{ kg})(9.80\text{ m/s}^2)\right] = \boxed{+2.4\times10^5\text{ N}}$$

Knees – bent $\quad \mathbf{F}_{\text{Ground}} = \Sigma\overline{\mathbf{F}} - \mathbf{W}$

$$= +4.8\times10^3\text{ N} - \left[-(75\text{ kg})(9.80\text{ m/s}^2)\right] = \boxed{+5.5\times10^3\text{ N}}$$

---

9. SSM WWW ***REASONING*** The impulse applied to the golf ball by the floor can be found from Equation 7.4, the impulse-momentum theorem: $(\Sigma\overline{\mathbf{F}})\Delta t = m\mathbf{v}_f - m\mathbf{v}_0$. Two forces act on the golf ball, the average force $\overline{\mathbf{F}}$ exerted by the floor, and the weight of the golf ball. Since $\overline{\mathbf{F}}$ is much greater than the weight of the golf ball, the net average force $(\Sigma\overline{\mathbf{F}})$ is equal to $\overline{\mathbf{F}}$.

Only the vertical component of the ball's momentum changes during impact with the floor. In order to use Equation 7.4 directly, we must first find the vertical components of the initial and final velocities. We begin, then, by finding these velocity components.

SOLUTION The figures below show the initial and final velocities of the golf ball.

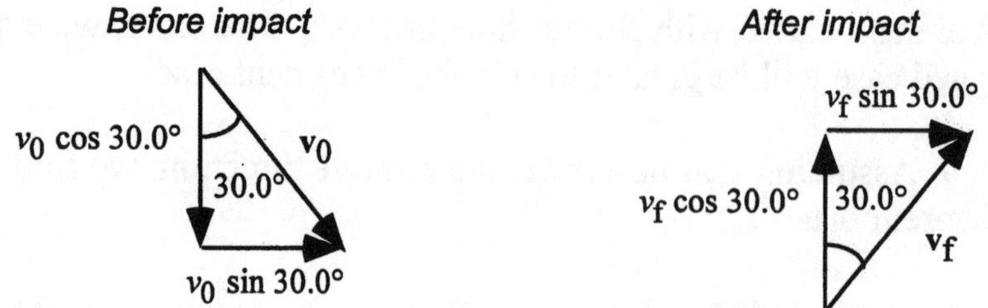

If we take up as the positive direction, then the vertical components of the initial and final velocities are, respectively, $\mathbf{v}_{0y} = -v_0 \cos 30.0°$ and $\mathbf{v}_{fy} = +v_f \cos 30.0°$. Then, from Equation 7.4 the impulse is

$$\overline{\mathbf{F}}\Delta t = m(\mathbf{v}_{fy} - \mathbf{v}_{0y}) = m\left[(+v_f \cos 30.0°) - (-v_0 \cos 30.0°)\right]$$

Since $v_0 = v_f = 45\ \text{m/s}$, the impulse applied to the golf ball by the floor is

$$\overline{\mathbf{F}}\Delta t = 2mv_0 \cos 30.0° = 2(0.047\ \text{kg})(45\ \text{m/s})(\cos 30.0°) = \boxed{3.7\ \text{N}\cdot\text{s}}$$

---

10. ***REASONING*** This solution can be divided into two parts. First, the student falls freely from rest and attains a certain velocity (which is unknown) just before hitting the ground. This impact velocity depends on the height H from which the student falls, and we will use an equation of kinematics from Chapter 3 to relate the height to the impact velocity. Second, the student then collides with the ground and comes to rest. We will employ the impulse-momentum theorem to determine the impact velocity from a knowledge of the average impact force and the time of impact, both of which are known.

SOLUTION Since the student falls freely from rest, we know that v_{0y} = 0 m/s, and a_y = –9.8 m/s^2. The acceleration is negative, because the downward direction is taken to be the negative direction. The displacement y of the student is $y = -H$, where H is the height; the minus sign indicates that the student falls downward. These variables are related by Equation 3.6b from the equations of kinematics as

$$v_y^2 = v_{0y}^2 + 2a_y y = (0\ \text{m/s})^2 + 2a_y(-H)$$

where v_y is the velocity of the student just before hitting the ground. Solving this expression for the height yields

$$H = \frac{-v_y^2}{2a_y} \tag{1}$$

At this point we do not know the impact velocity v_y. However, it can be determined by examining the collision of the student with the ground. Just before the collision, the velocity of the student is v_y. Since the student comes to rest, the final velocity is $v_f = 0$ m/s. The average force exerted on the student by the ground is $\overline{F} = +18\ 000$ N, and the time of collision is $\Delta t = 0.010$ s. The impulse-momentum theorem, Equation 7.4, relates these variables:

$$\overline{F}\Delta t = mv_f - mv_0 = m(0 \text{ m/s}) - mv_0 = -mv_0 \qquad (2)$$

Now v_0 is the impact velocity of the student, which was labeled as v_y in Equation (1); therefore, $v_0 = v_y$. Thus, Equation (2) becomes $\overline{F}\Delta t = -mv_y$. Solving this equation for v_y and substituting the result into Equation (1) yields

$$H = \frac{-\left(\dfrac{-\overline{F}\Delta t}{m}\right)^2}{2a_y} = \frac{-\overline{F}^2(\Delta t)^2}{2a_y m^2} = \frac{-(+18\ 000 \text{ N})^2(0.010 \text{ s})^2}{2(-9.80 \text{ m/s}^2)(63 \text{ kg})^2} = \boxed{0.42 \text{ m}}$$

---

11. ***REASONING AND SOLUTION*** According to the impulse-momentum theorem (Equation 7.4)

$$(\Sigma\overline{\mathbf{F}})\Delta t = m(\mathbf{v}_f - \mathbf{v}_0) \qquad (1)$$

Conservation of mechanical energy can be used to relate the velocities to the heights. If the floor is used to define the zero level for the heights, we have

$$mgh_0 = \tfrac{1}{2}mv_B^2$$

where h_0 is the height of the ball when it is dropped and v_B is the speed of the ball just before it strikes the ground. Solving for v_B gives

$$v_B = \sqrt{2gh_0} \qquad (2)$$

Similarly,

$$mgh_f = \tfrac{1}{2}mv_A^2$$

where h_f is the maximum height of the ball when it rebounds and v_A is the speed of the ball just after it rebounds from the ground. Solving for v_A gives

$$v_A = \sqrt{2gh_f} \qquad (3)$$

Substituting equations (2) and (3) into equation (1), where $\mathbf{v}_0 = -v_B$, and $\mathbf{v}_f = v_A$ gives (taking "upward" as the positive direction)

$$(\Sigma\overline{\mathbf{F}})\Delta t = m\sqrt{2g}\left[+\sqrt{h_f}-\left(-\sqrt{h_0}\right)\right]$$

$$= (0.500\text{ kg})\sqrt{2\left(9.80\text{ m/s}^2\right)}\left[+\sqrt{0.700\text{ m}}-\left(-\sqrt{1.20\text{ m}}\right)\right] = \boxed{+4.28\text{ N}\cdot\text{s}}$$

Since the impulse is positive, it is directed $\boxed{\text{upward}}$.

---

12. ***REASONING*** This is a problem in vector addition, and we will use the component method for vector addition. Using this method, we will add the components of the individual momenta in the direction due north to obtain the component of the vector sum in the direction due north. We will obtain the component of the vector sum in the direction due east in a similar fashion from the individual components in that direction. For each jogger the momentum is the mass times the velocity.

SOLUTION Assuming that the directions north and east are positive, the components of the joggers' momenta are as shown in the following table:

	Direction due east	Direction due north
85 kg jogger	$(85\text{ kg})(2.0\text{ m/s})$ $= 170\text{ kg}\cdot\text{m/s}$	0 kg·m/s
55 kg jogger	$(55\text{ kg})(3.0\text{ m/s})\cos 32°$ $= 140\text{ kg}\cdot\text{m/s}$	$(55\text{ kg})(3.0\text{ m/s})\sin 32°$ $= 87\text{ kg}\cdot\text{m/s}$
Total	310 kg·m /s	87 kg·m/s

Using the Pythagorean theorem, we find that the magnitude of the total momentum is

$$\sqrt{(310\text{ kg}\cdot\text{m/s})^2+(87\text{ kg}\cdot\text{m/s})^2} = \boxed{322\text{ kg}\cdot\text{m/s}}$$

The total momentum vector points north of east by an angle θ, which is given by

$$\theta = \tan^{-1}\left(\frac{87\text{ kg}\cdot\text{m/s}}{310\text{ kg}\cdot\text{m/s}}\right) = \boxed{16°}$$

---

13. ***REASONING*** During the time interval Δt, a mass m of water strikes the turbine blade. The incoming water has a momentum $m\mathbf{v}_0$ and that of the outgoing water is $m\mathbf{v}_f$. In order to change the momentum of the water, an impulse $(\Sigma\overline{\mathbf{F}})\Delta t$ is applied to it by the stationary turbine blade. Now $(\Sigma\overline{\mathbf{F}})\Delta t = \overline{\mathbf{F}}\Delta t$, since only the force of the blade is assumed to act on

the water in the horizontal direction. These variables are related by the impulse-momentum theorem, $\overline{\mathbf{F}}\Delta t = m\mathbf{v}_f - m\mathbf{v}_0$, which can be solved to find the average force $\overline{\mathbf{F}}$ exerted on the water by the blade.

SOLUTION Solving the impulse-momentum theorem for the average force gives

$$\overline{\mathbf{F}} = \frac{m\mathbf{v}_f - m\mathbf{v}_0}{\Delta t} = \frac{m}{\Delta t}(\mathbf{v}_f - \mathbf{v}_0)$$

The ratio $m/(\Delta t)$ is the mass of water per second that strikes the blade, or 30.0 kg/s, so the average force is

$$\overline{\mathbf{F}} = \frac{m}{\Delta t}(\mathbf{v}_f - \mathbf{v}_0) = (30.0\text{ kg/s})\left[(-16.0\text{ m/s}) - (+16.0\text{ m/s})\right] = -960\text{ N}$$

The magnitude of the average force is $\boxed{960\text{ N}}$.

---

14. ***REASONING AND SOLUTION*** The excess weight of the truck is due to the force exerted on the truck by the sand. Newton's third law requires that this force be equal in magnitude to the force exerted on the sand by the truck. In time *t*, a mass *m* of sand falls into the truck bed and comes to rest. The impulse is

$$\overline{\mathbf{F}}\Delta t = m(\mathbf{v}_f - \mathbf{v}_0) \qquad \text{so} \qquad \overline{\mathbf{F}} = \frac{m(\mathbf{v}_f - \mathbf{v}_0)}{\Delta t}$$

The sand gains a speed v_0 in falling a height *h* so

$$v_0 = \sqrt{2gh} = \sqrt{2(9.80\text{ m/s}^2)(2.00\text{ m})} = 6.26\text{ m/s}$$

The velocity of the sand just before it hits the truck is $\mathbf{v}_0 = -6.26$ m/s, where the downward direction is taken to be the negative direction. The final velocity of the sand is $\mathbf{v}_f = 0$ m/s. Thus, the average force exerted on the sand is

$$\overline{\mathbf{F}} = \left(\frac{m}{\Delta t}\right)(\mathbf{v}_f - \mathbf{v}_0) = (55.0\text{ kg/s})[(0\text{ m/s}) - (-6.26\text{ m/s})] = \boxed{+344\text{ N}}$$

---

15. ***REASONING AND SOLUTION*** The momentum is zero before the beat. Conservation of momentum requires that it is also zero after the beat; thus

$$0 = m_p v_p + m_b v_b$$

so that

$$v_p = -(m_b/m_p)v_b = -(0.050\text{ kg}/85\text{ kg})(0.25\text{ m/s}) = \boxed{-1.5\times 10^{-4}\text{ m/s}}$$

---

16. ***REASONING*** The sum of the external forces acting on the swimmer/raft system is zero, because the weight of the swimmer and raft is balanced by a corresponding normal force and friction is negligible. The swimmer and raft constitute an isolated system, so the principle of conservation of linear momentum applies. We will use this principle to find the recoil velocity of the raft.

SOLUTION As the swimmer runs off the raft, the total linear momentum of the swimmer/raft system is conserved:

$$\underbrace{m_s v_s + m_r v_r}_{\substack{\text{Total momentum}\\ \text{after swimmer}\\ \text{runs off raft}}} = \underbrace{0}_{\substack{\text{Total momentum}\\ \text{before swimmer}\\ \text{starts running}}}$$

where m_s and v_s are the mass and final velocity of the swimmer, and m_r and v_r are the mass and final velocity of the raft. Solving for v_r gives

$$v_r = -\frac{m_s v_s}{m_r} = -\frac{(55\text{ kg})(+4.6\text{ m/s})}{210\text{ kg}} = \boxed{-1.2\text{ m/s}}$$

---

17. SSM ***REASONING AND SOLUTION*** Bonzo and Ender constitute an isolated system. Therefore, the principle of conservation of linear momentum holds. Since both members of the system are initially stationary, the initial linear momentum is zero, and must remain zero during and after any interaction between Bonzo and Ender.

a. Since the total momentum of the system is conserved, and the total momentum of the system is zero, Bonzo and Ender must have equal but opposite linear momenta. **Bonzo** must have the larger mass, since **he flies off with the smaller recoil velocity**.

b. Conservation of linear momentum gives $0 = m_{\text{Bonzo}} v_{\text{Bonzo}} + m_{\text{Ender}} v_{\text{Ender}}$. Solving for the ratio of the masses, we have

$$\frac{m_{\text{Bonzo}}}{m_{\text{Ender}}} = -\frac{v_{\text{Ender}}}{v_{\text{Bonzo}}} = -\frac{-2.5\text{ m/s}}{1.5\text{ m/s}} = \boxed{1.7}$$

---

18. ***REASONING*** Let m be Al's mass, which means that Jo's mass is 168 kg − m. Since friction is negligible and since the downward-acting weight of each person is balanced by

the upward-acting normal force from the ice, the net external force acting on the two-person system is zero. Therefore, the system is isolated, and the conservation of linear momentum applies. The initial total momentum must be equal to the final total momentum.

SOLUTION Applying the principle of conservation of linear momentum and assuming that the direction in which Al moves is the positive direction, we find

$$\underbrace{m(0\text{ m/s})+(168\text{ kg}-m)(0\text{ m/s})}_{\text{Initial total momentum}}=\underbrace{m(0.90\text{ m/s})+(168\text{ kg}-m)(-1.2\text{ m/s})}_{\text{Final total momentum}}$$

Solving this equation for m and suppressing the units in the interests of clarity, we find

$$0=m(0.90)-(168)(1.2)+m(1.2)$$

$$m=\frac{(168)(1.2)}{0.90+1.2}=\boxed{96\text{ kg}}$$

---

19. ***REASONING*** During the breakup, the linear momentum of the system is conserved, since the force causing the breakup is an internal force. We will assume that the $+x$ axis is along the original line of motion (before the breakup), and the $+y$ axis is perpendicular to this line and points upward. We will apply the conservation of linear momentum twice, once for the momentum components along the x axis and again for the momentum components along the y axis.

SOLUTION The mass of each piece of the rocket after breakup is m, and so the mass of the rocket before breakup is $2m$. Applying the conservation of momentum theorem along the original line of motion (the x axis) gives

$$\underbrace{mv_1\cos 30.0°+mv_2\cos 60.0°}_{P_{f,x}}=\underbrace{2mv_0}_{P_{0,x}}\quad\text{or}\quad v_1\cos 30.0°+v_2\cos 60.0°=2v_0 \qquad (1)$$

Applying the conservation of momentum along the y axis gives

$$\underbrace{mv_1\sin 30.0°-mv_2\sin 60.0°}_{P_{f,y}}=\underbrace{0}_{P_{0,y}}\quad\text{or}\quad v_2=\frac{v_1\sin 30.0°}{\sin 60.0°} \qquad (2)$$

a. To find the speed v_1 of the first piece, we substitute the value for v_2 from Equation (2) into Equation (1). The result is

$$v_1\cos 30.0°+\left(\frac{v_1\sin 30.0°}{\sin 60.0°}\right)\cos 60.0°=2v_0$$

Solving for v_1 and setting $v_0 = 45.0$ m/s yields

$$v_1 = \frac{2v_0}{\cos 30.0° + \left(\frac{\sin 30.0°}{\sin 60.0°}\right)\cos 60.0°} = \frac{2(45.0 \text{ m/s})}{\cos 30.0° + \left(\frac{\sin 30.0°}{\sin 60.0°}\right)\cos 60.0°} = \boxed{77.9 \text{ m/s}}$$

b. The speed v_2 of the second piece can be found by substituting $v_1 = 77.9$ m/s into Equation (2):

$$v_2 = \frac{v_1 \sin 30.0°}{\sin 60.0°} = \frac{(77.9 \text{ m/s})\sin 30.0°}{\sin 60.0°} = \boxed{45.0 \text{ m/s}}$$

20\. ***REASONING*** During the time that the skaters are pushing against each other, the sum of the external forces acting on the two-skater system is zero, because the weight of each skater is balanced by a corresponding normal force and friction is negligible. The skaters constitute an isolated system, so the principle of conservation of linear momentum applies. We will use this principle to find an expression for the ratio of the skater's masses in terms of their recoil velocities. We will then obtain expressions for the recoil velocities by noting that each skater, after pushing off, comes to rest in a certain distance. The recoil velocity, acceleration, and distance are related by Equation 2.9 of the equations of kinematics.

SOLUTION While the skaters are pushing against each other, the total linear momentum of the two-skater system is conserved:

$$\underbrace{m_1 v_{f1} + m_2 v_{f2}}_{\substack{\text{Total momentum} \\ \text{after pushing}}} = \underbrace{0}_{\substack{\text{Total momentum} \\ \text{before pushing}}}$$

Solving this expression for the ratio of the masses gives

$$\frac{m_1}{m_2} = -\frac{v_{f2}}{v_{f1}} \quad (1)$$

For each skater the (initial) recoil velocity v_f, final velocity v, acceleration a, and displacement x are related by Equation 2.9 of the equations of kinematics: $v^2 = v_f^2 + 2ax$. Solving for the recoil velocity gives $v_f = \pm\sqrt{v^2 - 2ax}$. If we assume that skater 1 recoils in the positive direction and skater 2 recoils in the negative direction, the recoil velocities are

Skater 1 $\qquad v_{f1} = +\sqrt{v_1^2 - 2a_1 x_1}$

Skater 2 $\qquad v_{f2} = -\sqrt{v_2^2 - 2a_2 x_2}$

Substituting these expressions into Equation (1) gives

$$\frac{m_1}{m_2} = -\frac{-\sqrt{v_2^2-2a_2x_2}}{\sqrt{v_1^2-2a_1x_1}} = \frac{\sqrt{v_2^2-2a_2x_2}}{\sqrt{v_1^2-2a_1x_1}} \tag{2}$$

Since the skaters come to rest, their final velocities are zero, so $v_1 = v_2 = 0$ m/s. We also know that their accelerations have the same magnitudes. This means that $a_2 = -a_1$, where the minus sign denotes that the acceleration of skater 2 is opposite that of skater 1, since they are moving in opposite directions and are both slowing down. Finally, we are given that skater 1 glides twice as far as skater 2. Thus, the displacement of skater 1 is related to that of skater 2 by $x_1 = -2x_2$, where, the minus sign denotes that the skaters move in opposite directions. Substituting these values into Equation (2) yields

$$\frac{m_1}{m_2} = \frac{\sqrt{(0\text{ m/s})^2-2(-a_1)(x_2)}}{\sqrt{(0\text{ m/s})^2-2a_1(-2x_2)}} = \frac{1}{\sqrt{2}} = \boxed{0.707}$$

---

21. SSM ***REASONING*** No net external force acts on the plate parallel to the floor; therefore, the component of the momentum of the plate that is parallel to the floor is conserved as the plate breaks and flies apart. Initially, the total momentum parallel to the floor is zero. After the collision with the floor, the component of the total momentum parallel to the floor must remain zero. The drawing in the text shows the pieces in the plane parallel to the floor just after the collision. Clearly, the linear momentum in the plane parallel to the floor has two components; therefore the linear momentum of the plate must be conserved in each of these two mutually perpendicular directions. Using the drawing in the text, with the positive directions taken to be up and to the right, we have

$[x\text{ direction}]$ $$-m_1v_1\,(\sin 25.0°) + m_2v_2\,(\cos 45.0°) = 0$$

$[y\text{ direction}]$ $$m_1v_1\,(\cos 25.0°) + m_2v_2\,(\sin 45.0°) - m_3v_3 = 0$$

These equations can be solved simultaneously for the masses m_1 and m_2.

SOLUTION Using the values given in the drawing for the velocities after the plate breaks, we have, suppressing units,

$$-1.27m_1 + 1.27m_2 = 0 \tag{1}$$

$$2.72m_1 + 1.27m_2 - 3.99 = 0 \tag{2}$$

Subtracting (1) from (2) gives $\boxed{m_1 = 1.00\text{ kg}}$. Substituting this value into either (1) or (2) then yields $\boxed{m_2 = 1.00\text{ kg}}$.

---

22. ***REASONING*** For the system consisting of the female character, the gun and the bullet, the sum of the external forces is zero, because the weight of each object is balanced by a corresponding upward (normal) force, and we are ignoring friction. The female character, the gun and the bullet, then, constitute an isolated system, and the principle of conservation of linear momentum applies.

SOLUTION a. The total momentum of the system before the gun is fired is zero, since all parts of the system are at rest. Momentum conservation requires that the total momentum remains zero after the gun has been fired.

$$\underbrace{m_1 v_{f1} + m_2 v_{f2}}_{\text{Total momentum after gun is fired}} = \underbrace{0}_{\text{Total momentum before gun is fired}}$$

where the subscripts 1 and 2 refer to the woman (plus gun) and the bullet, respectively. Solving for v_{f1}, the recoil velocity of the woman (plus gun), gives

$$v_{f1} = -\frac{m_2 v_{f2}}{m_1} = \frac{-(0.010\text{ kg})(720\text{ m/s})}{51\text{ kg}} = \boxed{-0.14\text{ m/s}}$$

b. Repeating the calculation for the situation in which the woman shoots a blank cartridge, we have

$$v_{f1} = -\frac{m_2 v_{f2}}{m_1} = \frac{-(5.0\times10^{-4}\text{ kg})(720\text{ m/s})}{51\text{ kg}} = \boxed{-7.1\times10^{-3}\text{ m/s}}$$

In both cases, the minus sign means that the bullet and the woman move in opposite directions when the gun is fired. The total momentum of the system remains zero, because momentum is a vector quantity, and the momenta of the bullet and the woman have equal magnitudes, but opposite directions.

---

23. SSM WWW ***REASONING*** The cannon and the shell constitute the system. Since no external force hinders the motion of the system after the cannon is unbolted, conservation of linear momentum applies in that case. If we assume that the burning gun powder imparts the same kinetic energy to the system in each case, we have sufficient information to develop a mathematical description of this situation, and solve it for the velocity of the shell fired by the loose cannon.

SOLUTION For the case where the cannon is unbolted, momentum conservation gives

$$\underbrace{m_1 v_{f1} + m_2 v_{f2}}_{\text{Total momentum after shell is fired}} = \underbrace{0}_{\text{Initial momentum of system}} \qquad (1)$$

where the subscripts "1" and "2" refer to the cannon and shell, respectively. In both cases, the burning gun power imparts the same kinetic energy to the system. When the cannon is bolted to the ground, only the shell moves and the kinetic energy imparted to the system is

$$\text{KE} = \tfrac{1}{2} m_{\text{shell}} v_{\text{shell}}^2 = \tfrac{1}{2}(85.0 \text{ kg})(551 \text{ m/s})^2 = 1.29\times10^7 \text{ J}$$

The kinetic energy imparted to the system when the cannon is unbolted has the same value and can be written using the same notation as equation (1):

$$\text{KE} = \tfrac{1}{2} m_1 v_{f1}^2 + \tfrac{1}{2} m_2 v_{f2}^2 \qquad (2)$$

Solving equation (1) for v_{f1}, the velocity of the cannon after the shell is fired, and substituting the resulting expression into Equation (2) gives

$$\text{KE} = \frac{m_2^2 v_{f2}^2}{2\text{m}_1} + \tfrac{1}{2} m_2 v_{f2}^2 \qquad (3)$$

Solving equation (3) for v_{f2} gives

$$v_{f2} = \sqrt{\frac{2\text{KE}}{m_2\left(\frac{m_2}{m_1}+1\right)}} = \sqrt{\frac{2(1.29\times10^7 \text{ J})}{(85 \text{ kg})\left(\frac{85 \text{ kg}}{5.80\times10^3 \text{ kg}}+1\right)}} = \boxed{+547 \text{ m/s}}$$

---

24. ***REASONING AND SOLUTION*** Since no net external force acts in the horizontal direction, the total horizontal momentum of the system is conserved regardless of which direction the mass is thrown. The momentum of the system before the mass is thrown off the wagon is $m_W v_A$, where m_W and v_A are the mass and velocity of the wagon, respectively. When ten percent of the wagon's mass is thrown forward, the wagon is brought to a halt so its final momentum is zero. The momentum of the mass thrown forward is $0.1 m_W(v_A + v_M)$, where v_M is the velocity of the mass relative to the wagon and $(v_A + v_M)$ is the velocity of the mass relative to the ground. The conservation of momentum gives

$$\underbrace{0.1 m_W \left(v_A + v_M\right)}_{\text{Momentum after mass is thrown forward}} = \underbrace{m_W v_A}_{\text{Momentum before mass is thrown forward}}$$

Solving for v_M yields

$$v_M = 9v_A \tag{1}$$

When the direction in which the mass is thrown is reversed, the velocity of the mass relative to the ground is now $(v_A - v_M)$. The momentum of the mass and wagon is, therefore,

$$\underbrace{0.1m_W(v_A - v_M)}_{\text{Momentum of the mass}} + \underbrace{0.9m_W v_B}_{\text{Momentum of the wagon}}$$

where v_B is the velocity of the wagon. The conservation of momentum gives

$$\underbrace{0.1m_W(v_A - v_M) + 0.9m_W v_B}_{\substack{\text{Momentum after}\\ \text{mass is thrown off}}} = \underbrace{m_W v_A}_{\substack{\text{Momentum before}\\ \text{mass is thrown off}}} \tag{2}$$

Substituting equation (1) into equation (2) and solving for v_B/v_A gives $\boxed{\frac{v_B}{v_A} = 2}$.

---

25. ***REASONING AND SOLUTION*** The collision is an inelastic one, with the total linear momentum being conserved:

$$m_1 v_1 = (m_1 + m_2)V$$

The mass m_2 of the receiver is

$$m_2 = \frac{m_1 v_1}{V} - m_1 = \frac{(115 \text{ kg})(4.5 \text{ m/s})}{2.6 \text{ m/s}} - 115 \text{ kg} = \boxed{84 \text{ kg}}$$

---

26. ***REASONING*** Since the collision is an elastic collision, both the linear momentum and kinetic energy of the two-vehicle system are conserved. The final velocities of the car and van are given in terms of the initial velocity of the car by Equations 7.8a and 7.8b.

SOLUTION

a. The final velocity v_{f1} of the car is given by Equation 7.8a as

$$v_{f1} = \left(\frac{m_1 - m_2}{m_1 + m_2}\right) v_{01}$$

where m_1 and m_2 are, respectively, the masses of the car and van, and v_{01} is the initial velocity of the car. Thus,

$$v_{f1} = \left(\frac{715\text{ kg} - 1055\text{ kg}}{715\text{ kg} + 1055\text{ kg}} \right) (+2.25\text{ m/s}) = \boxed{-0.432\text{ m/s}}$$

b. The final velocity of the van is given by Equation 7.8b:

$$v_{f2} = \left(\frac{2m_1}{m_1 + m_2} \right) v_{01} = \left[\frac{2(715\text{ kg})}{715\text{ kg} + 1055\text{ kg}} \right] (+2.25\text{ m/s}) = \boxed{+1.82\text{ m/s}}$$

---

27. SSM ***REASONING*** Since all of the collisions are elastic, the total mechanical energy of the ball is conserved. However, since gravity affects its vertical motion, its linear momentum is *not* conserved. If h_f is the maximum height of the ball on its final bounce, conservation of energy gives

$$\underbrace{\frac{1}{2}mv_f^2 + mgh_f}_{E_f} = \underbrace{\frac{1}{2}mv_0^2 + mgh_0}_{E_0}$$

$$h_f = \frac{v_0^2 - v_f^2}{2g} + h_0$$

SOLUTION In order to use this expression, we must obtain the values for the velocities v_0 and v_f. The initial velocity has only a horizontal component, $v_0 = v_{0x}$. The final velocity also has only a horizontal component since the ball is at the top of its trajectory, $v_f = v_{fx}$. No forces act in the horizontal direction so the momentum of the ball in this direction is conserved, hence $v_0 = v_f$. Therefore,

$$h_f = h_0 = \boxed{3.00\text{ m}}$$

---

28. ***REASONING*** a. During the collision between the bullet and the wooden block, linear momentum is conserved, since no net external force acts on the bullet and the block. The weight of each is balanced by the tension in the suspension wire, and the forces that the bullet and block exert on each other are internal forces. This conservation law will allow us to find the speed of the bullet/block system immediately after the collision.

b. Just after the collision, the bullet/block rise up, ultimately reaching a final height h_f before coming to a momentary rest. During this phase, the tension in the wire (a nonconservative force) does no work, since it acts perpendicular to the motion. Thus, the work done by nonconservative forces is zero, and the total mechanical energy of the system is conserved. An application of this conservation law will enable us to determine the height h_f.

SOLUTION a. The principle of conservation of linear momentum states that the total momentum after the collision is equal to that before the collision.

$$\underbrace{(m_{\text{bullet}}+m_{\text{block}})v_{\text{f}}}_{\text{Momentum after collision}} = \underbrace{m_{\text{bullet}}v_{0,\text{bullet}}+m_{\text{block}}v_{0,\text{block}}}_{\text{Momentum before collision}}$$

Solving this equation for the speed v_{f} of the bullet/block system just after the collision gives

$$v_{\text{f}} = \frac{m_{\text{bullet}}v_{0,\text{bullet}}+m_{\text{block}}v_{0,\text{block}}}{m_{\text{bullet}}+m_{\text{block}}}$$

$$= \frac{(0.00250\text{ kg})(425\text{ m/s})+(0.215\text{ kg})(0\text{ m/s})}{0.00250\text{ kg}+0.215\text{ kg}} = \boxed{4.89\text{ m/s}}$$

b. Just after the collision, the total mechanical energy of the system is all kinetic energy, since we take the zero-level for the gravitational potential energy to be at the initial height of the block. As the bullet/block system rises, kinetic energy is converted into potential energy. At the highest point, the total mechanical energy is all gravitational potential energy. Since the total mechanical energy is conserved, we have

$$\underbrace{(m_{\text{bullet}}+m_{\text{block}})gh_{\text{f}}}_{\substack{\text{Total mechanical energy at}\\ \text{the top of the swing,}\\ \text{all potential}}} = \underbrace{\tfrac{1}{2}(m_{\text{bullet}}+m_{\text{block}})v_{\text{f}}^2}_{\substack{\text{Total mechanical energy at}\\ \text{the bottom of the swing,}\\ \text{all kinetic}}}$$

Solving this expression for the height h_{f} gives

$$h_{\text{f}} = \frac{\frac{1}{2}v_{\text{f}}^2}{g} = \frac{\frac{1}{2}(4.89\text{ m/s})^2}{9.80\text{ m/s}^2} = \boxed{1.22\text{ m}}$$

---

29. ***REASONING*** The velocity of the second ball just after the collision can be found from Equation 7.8b (see Example 7). In order to use Equation 7.8b, however, we must know the velocity of the first ball just before it strikes the second ball. Since we know the impulse delivered to the first ball by the pool stick, we can use the impulse-momentum theorem (Equation 7.4) to find the velocity of the first ball just before the collision.

SOLUTION According to the impulse-momentum theorem, $\overline{F}\Delta t = mv_{\text{f}} - mv_0$, and setting $v_0 = 0$ m/s and solving for v_{f}, we find that the velocity of the first ball after it is struck by the pool stick and just before it hits the second ball is

$$v_{\text{f}} = \frac{\overline{F}\Delta t}{m} = \frac{+1.50\text{ N}\cdot\text{s}}{0.165\text{ kg}} = 9.09\text{ m/s}$$

Substituting values into Equation 7.8b (with $v_{01} = 9.09$ m/s), we have

$$v_{f2} = \left(\frac{2m_1}{m_1 + m_2}\right) v_{01} = \left(\frac{2m}{m + m}\right) v_{01} = v_{01} = \boxed{+9.09\text{ m/s}}$$

---

30. ***REASONING*** The net external force acting on the two-puck system is zero (the weight of each ball is balanced by an upward normal force, and we are ignoring friction due to the layer of air on the hockey table). Therefore, the two pucks constitute an isolated system, and the principle of conservation of linear momentum applies.

SOLUTION Conservation of linear momentum requires that the total momentum is the same before and after the collision. Since linear momentum is a vector, the x and y components must be conserved separately. Using the drawing in the text, momentum conservation in the x direction yields

$$m_A v_{0A} = m_A v_{fA}(\cos\ 65°) + m_B v_{fB}(\cos\ 37°) \qquad (1)$$

while momentum conservation in the y direction yields

$$0 = m_A v_{fA}(\sin\ 65°) - m_B v_{fB}(\sin\ 37°) \qquad (2)$$

Solving equation (2) for v_{fB}, we find that

$$v_{fB} = \frac{m_A v_{fA}(\sin 65°)}{m_B(\sin 37°)} \qquad (3)$$

Substituting equation (3) into Equation (1) leads to

$$m_A v_{0A} = m_A v_{fA}(\cos\ 65°) + \left(\frac{m_A v_{fA}(\sin 65°)}{\sin 37°}\right)(\cos\ 37°)$$

a. Solving for v_{fA} gives

$$v_{fA} = \frac{v_{0A}}{\cos 65° + \left(\dfrac{\sin 65°}{\tan 37°}\right)} = \frac{+5.5\text{ m/s}}{\cos 65° + \left(\dfrac{\sin 65°}{\tan 37°}\right)} = \boxed{3.4\text{ m/s}}$$

b. From equation (3), we find that

$$v_{fB} = \frac{(0.025\text{ kg})\,(3.4\text{ m/s})\,(\sin 65°)}{(0.050\text{ kg})\,(\sin 37°)} = \boxed{2.6\text{ m/s}}$$

---

31. SSM WWW ***REASONING*** The system consists of the two balls. The total linear momentum of the two-ball system is conserved because the net external force acting on it is zero. The principle of conservation of linear momentum applies whether or not the collision is elastic.

$$\underbrace{m_1 v_{f1} + m_2 v_{f2}}_{\text{Total momentum after collision}} = \underbrace{m_1 v_{01} + 0}_{\text{Total momentum before collision}}$$

When the collision is elastic, the kinetic energy is also conserved during the collision

$$\underbrace{\tfrac{1}{2} m_1 v_{f1}^2 + \tfrac{1}{2} m_2 v_{f2}^2}_{\text{Total kinetic energy after collision}} = \underbrace{\tfrac{1}{2} m_1 v_{01}^2 + 0}_{\text{Total kinetic energy before collision}}$$

SOLUTION
a. The final velocities for an elastic collision are determined by simultaneously solving the above equations for the final velocities. The procedure is discussed in Example 7 in the text, and leads to Equations 7.8a and 7.8b. According to Equation 7.8:

$$v_{f1} = \left(\frac{m_1 - m_2}{m_1 + m_2} \right) v_{01} \qquad \text{and} \qquad v_{f2} = \left(\frac{2m_1}{m_1 + m_2} \right) v_{01}$$

Let the initial direction of motion of the 5.00-kg ball define the positive direction. Substituting the values given in the text, these equations give

$$[5.00\text{-kg ball}] \qquad v_{f1} = \left(\frac{5.00 \text{ kg} - 7.50 \text{ kg}}{5.00 \text{ kg} + 7.50 \text{ kg}} \right) (2.00 \text{ m/s}) = \boxed{-0.400 \text{ m/s}}$$

$$[7.50\text{-kg ball}] \qquad v_{f2} = \left(\frac{2(5.00 \text{ kg})}{5.00 \text{ kg} + 7.50 \text{ kg}} \right) (2.00 \text{ m/s}) = \boxed{+1.60 \text{ m/s}}$$

The signs indicate that, after the collision, the 5.00-kg ball reverses its direction of motion, while the 7.50-kg ball moves in the direction in which the 5.00-kg ball was initially moving.

b. When the collision is completely inelastic, the balls stick together, giving a composite body of mass $m_1 + m_2$ which moves with a velocity v_f. The statement of conservation of linear momentum then becomes

$$\underbrace{(m_1 + m_2) v_f}_{\text{Total momentum after collision}} = \underbrace{m_1 v_{01} + 0}_{\text{Total momentum before collision}}$$

The final velocity of the two balls after the collision is, therefore,

$$v_f = \frac{m_1 v_{01}}{m_1 + m_2} = \frac{(5.00 \text{ kg})(2.00 \text{ m/s})}{5.00 \text{ kg} + 7.50 \text{ kg}} = \boxed{+0.800 \text{ m/s}}$$

---

32. ***REASONING AND SOLUTION*** The conservation of momentum gives

$$m_b v_b + m_p v_p = m_p v_0$$

Therefore,

$$v_p = v_0 - \frac{m_b}{m_p} v_b = 715 \text{ m/s} - \frac{2.00 \text{ kg}}{0.150 \text{ kg}}(40.0 \text{ m/s}) = \boxed{182 \text{ m/s}}$$

---

33. ***REASONING AND SOLUTION*** The total linear momentum of the two-car system is conserved because no net external force acts on the system during the collision. We are ignoring friction *during the collision*, and the weights of the cars are balanced by the normal forces exerted by the ground. Momentum conservation gives

$$\underbrace{(m_1 + m_2)v_f}_{\text{Total momentum after collision}} = \underbrace{m_1 v_{01} + m_2 v_{02}}_{\text{Total momentum after collision}}$$

where $v_{02} = 0$ m/s since the 1900-kg car is stationary before the collision.

a. Solving for v_f, we find that the velocity of the two cars just after the collision is

$$v_f = \frac{m_1 v_{01} + m_2 v_{02}}{m_1 + m_2} = \frac{(2100 \text{ kg})(+17 \text{ m/s}) + (1900 \text{ kg})(0 \text{ m/s})}{2100 \text{ kg} + 1900 \text{ kg}} = \boxed{+8.9 \text{ m/s}}$$

The plus sign indicates that the velocity of the two cars just after the collision is in the same direction as the direction of the velocity of the 2100-kg car before the collision.

b. According to the impulse-momentum theorem, Equation 7.4, we have

$$\underbrace{\overline{F}\Delta t}_{\text{Impulse due to friction}} = \underbrace{(m_1 + m_2)v_{\text{final}}}_{\text{Final momentum when cars come to a halt}} - \underbrace{(m_1 + m_2)v_{\text{after}}}_{\text{Total momentum just after collision}}$$

where $v_{\text{final}} = 0$ m/s since the cars come to a halt, and $v_{\text{after}} = v_f = +8.9$ m/s. Therefore, we have

$$\overline{F}\Delta t = (2100 \text{ kg} + 1900 \text{ kg})(0 \text{ m/s}) - (2100 \text{ kg} + 1900 \text{ kg})(8.9 \text{ m/s}) = \boxed{-3.6 \times 10^4 \text{ N} \cdot \text{s}}$$

The minus sign indicates that the impulse due to friction acts opposite to the direction of motion of the locked, two-car system. This is reasonable since the velocity of the cars is decreasing in magnitude as the cars skid to a halt.

c. Using the same notation as in part (b) above, we have from the equations of kinematics (Equation 2.9) that

$$v_{\text{final}}^2 = v_{\text{after}}^2 + 2ax$$

where $v_{\text{final}} = 0$ m/s and $v_{\text{after}} = v_{\text{f}} = +8.9$ m/s. From Newton's second law we have that $a = -f_{\text{k}}/(m_1 + m_2)$, where f_{k} is the force of kinetic friction that acts on the cars as they skid to a halt. Therefore,

$$0 = v_{\text{f}}^2 + 2\left(\frac{-f_{\text{k}}}{m_1 + m_2}\right)x \quad \text{or} \quad x = \frac{(m_1 + m_2)v_{\text{f}}^2}{2f_{\text{k}}}$$

According to Equation 4.8, $f_{\text{k}} = \mu_{\text{k}} F_{\text{N}}$, where μ_{k} is the coefficient of kinetic friction and F_{N} is the magnitude of the normal force that acts on the two-car system. There are only two vertical forces that act on the system; they are the upward normal force $\mathbf{F}_{\text{N}}$ and the weight $(m_1 + m_2)g$ of the cars. Taking upward as the positive direction and applying Newton's second law in the vertical direction, we have $F_{\text{N}} - (m_1 + m_2)g = (m_1 + m_2)a_y = 0$, or $F_{\text{N}} = (m_1 + m_2)g$. Therefore, $f_{\text{k}} = \mu_{\text{k}}(m_1 + m_2)g$, and we have

$$x = \frac{(m_1 + m_2)v_{\text{f}}^2}{2\mu_{\text{k}}(m_1 + m_2)g} = \frac{v_{\text{f}}^2}{2\mu_{\text{k}}g} = \frac{(8.9\text{ m/s})^2}{2(0.68)(9.80\text{ m/s}^2)} = \boxed{5.9\text{ m}}$$

---

34. ***REASONING AND SOLUTION*** Momentum is conserved in the horizontal direction during the "collision." Let the coal be object 1 and the car be object 2. Then

$$(m_1 + m_2)v_f = m_1 v_1 \cos 25.0° + m_2 v_2$$

$$v_{\text{f}} = \frac{m_1 v_1 \cos 25.0° + m_2 v_2}{m_1 + m_2} = \frac{(150\text{ kg})(0.80\text{ m/s})\cos 25.0° + (440\text{ kg})(0.50\text{ m/s})}{150\text{ kg} + 440\text{ kg}} = \boxed{0.56\text{ m/s}}$$

The direction of the final velocity is $\boxed{\text{to the right}}$.

---

35. SSM ***REASONING*** The two skaters constitute the system. Since the net external force acting on the system is zero, the total linear momentum of the system is conserved. In the x direction (the east/west direction), conservation of linear momentum gives $P_{fx} = P_{0x}$, or

$$(m_1 + m_2)v_f \cos\theta = m_1 v_{01}$$

Note that since the skaters hold onto each other, they move away with a common velocity v_f. In the y direction, $P_{fy} = P_{0y}$, or

$$(m_1 + m_2)v_f \sin\theta = m_2 v_{02}$$

These equations can be solved simultaneously to obtain both the angle θ and the velocity v_f.

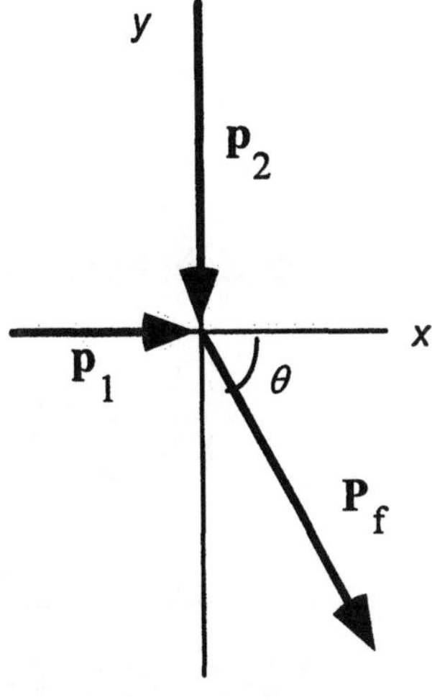

SOLUTION
a. Division of the equations above gives

$$\theta = \tan^{-1}\left(\frac{m_2 v_{02}}{m_1 v_{01}}\right) = \tan^{-1}\left[\frac{(70.0 \text{ kg})(7.00 \text{ m/s})}{(50.0 \text{ kg})(3.00 \text{ m/s})}\right] = \boxed{73.0°}$$

b. Solution of the first of the momentum equations gives

$$v_f = \frac{m_1 v_{01}}{(m_1 + m_2)\cos\theta} = \frac{(50.0 \text{ kg})(3.00 \text{ m/s})}{(50.0 \text{ kg} + 70.0 \text{ kg})(\cos 73.0°)} = \boxed{4.28 \text{ m/s}}$$

---

36. ***REASONING*** The ratio of the kinetic energy of the hydrogen atom after the collision to that of the electron before the collision is

$$\frac{\text{KE}_{\text{hydrogen, after collision}}}{\text{KE}_{\text{electron, before collison}}} = \frac{\frac{1}{2}m_H v_{f,H}^2}{\frac{1}{2}m_e v_{0,e}^2}$$

where $v_{f,H}$ is the final speed of the hydrogen atom, and $v_{0,e}$ is the initial speed of the electron. The ratio m_H/m_e of the masses is known. Since the electron and the stationary hydrogen atom experience an elastic head-on collision, we can employ Equation 7.8b to determine how $v_{f,H}$ is related to $v_{0,e}$.

SOLUTION According to Equation 7.8b, the final speed $v_{f,H}$ of the hydrogen atom after the collision is related to the initial speed $v_{0,e}$ of the electron by

$$v_{f,H} = \left(\frac{2m_e}{m_e + m_H}\right) v_{0,e}$$

Substituting this expression into the ratio of the kinetic energies gives

$$\frac{KE_{\text{hydrogen, after collision}}}{KE_{\text{electron, before collison}}} = \frac{\frac{1}{2}m_H v_{f,H}^2}{\frac{1}{2}m_e v_{0,e}^2} = \frac{m_H\left(\frac{2m_e}{m_e + m_H}\right)^2}{m_e}$$

The right hand side of this equation can be algebraically rearranged to give

$$\frac{KE_{\text{hydrogen, after collision}}}{KE_{\text{electron, before collison}}} = \left(\frac{m_H}{m_e}\right)\left(\frac{2}{1+\frac{m_H}{m_e}}\right)^2 = (1837)\left(\frac{2}{1+1837}\right)^2 = \boxed{2.175 \times 10^{-3}}$$

---

37. ***REASONING AND SOLUTION***
a. The total momentum of the sled and person is conserved since no external force acts in the horizontal direction. $(m_p + m_s)v_f = m_p v_{0p}$ gives

$$v_f = \frac{m_p v_{0p}}{m_p + m_s} = \frac{(60.0\text{ kg})(3.80\text{ m/s})}{60.0\text{ kg} + 12.0\text{ kg}} = \boxed{3.17\text{ m/s}}$$

The direction of v_f is the same as the direction of v_{0p}.

b. The frictional force acting on the sled is $f_k = -\mu_k F_N = -\mu_k mg$. Newton's second law then gives the acceleration to be

$$a = f_k/m = -\mu_k g$$

The acceleration is constant so $a = -v_f^2/(2x)$ from the equations of kinematics. Equating these two expressions and solving yields

$$\mu_k = \frac{v_f^2}{2gx} = \frac{(3.17\text{ m/s})^2}{2(9.80\text{ m/s}^2)(30.0\text{ m})} = \boxed{0.0171}$$

---

38. ***REASONING AND SOLUTION*** During the elastic collision both the total kinetic energy and the total momentum of the balls are conserved. The conservation of momentum gives

$$mv_{f1} + mv_{f2} = mv_{01} + mv_{02}$$

where ball 1 has $\mathbf{v}_{01} = +\,7.0$ m/s and ball 2 has $\mathbf{v}_{02} = -\,4.0$ m/s. Hence,

$$v_{f1} + v_{f2} = v_{01} + v_{02} = 3.0 \text{ m/s}$$

The conservation of kinetic energy gives

$$(1/2)mv_{f1}^2 + (1/2)mv_{f2}^2 = (1/2)mv_{01}^2 + (1/2)mv_{02}^2$$

or

$$v_{f1}^2 + v_{f2}^2 = v_{01}^2 + v_{02}^2 = 65 \text{ m}^2/\text{s}^2$$

Solving the first equation for v_{f1} and substituting into the second gives

$$2v_{f2}^2 - (6.0 \text{ m/s})v_{f2} - 56 \text{ m}^2/\text{s}^2 = 0$$

The quadratic formula yields two solutions, $v_{f2} = 7.0$ m/s and $v_{f2} = -\,4.0$ m/s.

The first equation now gives $v_{f1} = -\,4.0$ m/s and $v_{f1} = +\,7.0$ m/s. Hence, the 7.0-m/s-ball has a final velocity of $\boxed{-4.0 \text{ m/s}}$, opposite its original direction.

The 4.0-m/s-ball has a final velocity of $\boxed{+7.0 \text{ m/s}}$, opposite its original direction.

---

39. **SSM** ***REASONING*** The two balls constitute the system. The tension in the wire is the only nonconservative force that acts on the ball. The tension does no work since it is perpendicular to the displacement of the ball. Since $W_{nc} = 0$ J, the principle of conservation of mechanical energy holds and can be used to find the speed of the 1.50-kg ball just before the collision. Momentum is conserved during the collision, so the principle of conservation of momentum can be used to find the velocities of both balls just after the collision. Once the collision has occurred, energy conservation can be used to determine how high each ball rises.

SOLUTION
a. Applying the principle of energy conservation to the 1.50-kg ball, we have

$$\underbrace{\tfrac{1}{2}mv_f^2 + mgh_f}_{E_f} = \underbrace{\tfrac{1}{2}mv_0^2 + mgh_0}_{E_0}$$

If we measure the heights from the lowest point in the swing, $h_f = 0$ m, and the expression above simplifies to

$$\tfrac{1}{2}mv_f^2 = \tfrac{1}{2}mv_0^2 + mgh_0$$

Solving for v_f, we have

$$v_f = \sqrt{v_0^2 + 2gh_0} = \sqrt{(5.00\text{ m/s})^2 + 2(9.80\text{ m/s}^2)(0.300\text{ m})} = \boxed{5.56\text{ m/s}}$$

b. If we assume that the collision is elastic, then the velocities of both balls just after the collision can be obtained from Equations 7.8a and 7.8b:

$$v_{f1} = \left(\frac{m_1 - m_2}{m_1 + m_2}\right)v_{01} \qquad \text{and} \qquad v_{f2} = \left(\frac{2m_1}{m_1 + m_2}\right)v_{01}$$

Since v_{01} corresponds to the speed of the 1.50-kg ball just before the collision, it is equal to the quantity v_f calculated in part (a). With the given values of $m_1 = 1.50$ kg and $m_2 = 4.60$ kg, and the value of $v_{01} = 5.56$ m/s obtained in part (a), Equations 7.8a and 7.8b yield the following values:

$$\boxed{v_{f1} = -2.83\text{ m/s}} \qquad \text{and} \qquad \boxed{v_{f2} = +2.73\text{ m/s}}$$

The minus sign in v_{f1} indicates that the first ball reverses its direction as a result of the collision.

c. If we apply the conservation of mechanical energy to either ball after the collision we have

$$\underbrace{\tfrac{1}{2}mv_f^2 + mgh_f}_{E_f} = \underbrace{\tfrac{1}{2}mv_0^2 + mgh_0}_{E_0}$$

where v_0 is the speed of the ball just after the collision, and h_f is the final height to which the ball rises. For either ball, $h_0 = 0$ m, and when either ball has reached its maximum height, $v_f = 0$ m/s. Therefore, the expression of energy conservation reduces to

$$gh_f = \tfrac{1}{2}v_0^2 \qquad \text{or} \qquad h_f = \frac{v_0^2}{2g}$$

Thus, the heights to which each ball rises after the collision are

[1.50-kg ball] $$h_f = \frac{v_0^2}{2g} = \frac{(2.83\text{ m/s})^2}{2(9.80\text{ m/s}^2)} = \boxed{0.409\text{ m}}$$

[4.60 - kg ball] $$h_f = \frac{v_0^2}{2g} = \frac{(2.73 \text{ m/s})^2}{2(9.80 \text{ m/s}^2)} = \boxed{0.380 \text{ m}}$$

---

40. ***REASONING AND SOLUTION*** Initially, the ball has a total mechanical energy given by $E_0 = mgh_0$. After one bounce it reaches the top of its trajectory with an energy of

$$E_1 = 0.900\ E_0 = 0.900\ mgh_0$$

After two bounces it has energy

$$E_2 = (0.900)^2 mgh_0$$

After N bounces it has a remaining energy of

$$E_N = (0.900)^N mgh_0$$

In order to just reach the sill the ball must have $E_N = mgh$ where $h = 2.44$ m. Hence,

$$(0.900)^N mgh_0 = mgh \quad \text{or} \quad (0.900)^N = h/h_0$$

Taking the log of both sides gives

$$N \log(0.900) = \log(h/h_0)$$

Then

$$N = \frac{\log\left(\dfrac{2.44 \text{ m}}{6.10 \text{ m}}\right)}{\log (0.900)} = 8.7$$

The ball can make $\boxed{\text{8 bounces}}$ and still reach the sill.

---

41. $\boxed{\text{SSM}}$ ***REASONING AND SOLUTION*** The location of the center of mass for a two-body system is given by Equation 7.10:

$$x_{\text{cm}} = \frac{m_1 x_1 + m_2 x_2}{m_1 + m_2}$$

where the subscripts "1" and "2" refer to the earth and the moon, respectively. For convenience, we will let the center of the earth be coincident with the origin so that $x_1 = 0$ and $x_2 = d$, the center-to-center distance between the earth and the moon. Direct calculation then gives

$$x_{\text{cm}} = \frac{m_2 d}{m_1 + m_2} = \frac{(7.35\times10^{22}\text{ kg})(3.85\times10^{8}\text{ m})}{5.98\times10^{24}\text{ kg}+7.35\times10^{22}\text{ kg}} = \boxed{4.67\times10^{6}\text{ m}}$$

---

42. ***REASONING AND SOLUTION*** The velocity of the center of mass of a system is given by Equation 7.11. Using the data and the results obtained in Example 5, we obtain the following:

a. The velocity of the center of mass of the two-car system before the collision is

$$(v_{\text{cm}})_{\text{before}} = \frac{m_1 v_{01} + m_2 v_{02}}{m_1 + m_2}$$
$$= \frac{(65\times10^{3}\text{ kg})(+0.80\text{ m/s})+(92\times10^{3}\text{ kg})(+1.2\text{ m/s})}{65\times10^{3}\text{ kg}+92\times10^{3}\text{ kg}} = \boxed{+1.0\text{ m/s}}$$

b. The velocity of the center of mass of the two-car system after the collision is

$$(v_{\text{cm}})_{\text{after}} = \frac{m_1 v_{\text{f}} + m_2 v_{\text{f}}}{m_1 + m_2} = v_{\text{f}} = \boxed{+1.0\text{ m/s}}$$

c. The answer in part (b) $\boxed{\text{should be the same}}$ as the common velocity v_{f}. Since the cars are coupled together, every point of the two-car system, including the center of mass, must move with the same velocity.

---

43. ***REASONING AND SOLUTION*** Equation 7.10 gives the center of mass of this two-atom system as

$$x_{\text{cm}} = \frac{m_{\text{c}} x_{\text{c}} + m_{\text{o}} x_{\text{o}}}{m_{\text{c}} + m_{\text{o}}}$$

If we take the origin at the center of the carbon atom, then $x_{\text{c}} = 0\text{ m}$, and we have

$$x_{\text{cm}} = \frac{m_{\text{o}} x_{\text{o}}}{m_{\text{c}} + m_{\text{o}}} = \frac{x_{\text{o}}}{(m_{\text{c}}/m_{\text{o}})+1} = \frac{1.13\times10^{-10}\text{ m}}{(0.750 m_{\text{o}}/m_{\text{o}}) + 1} = \boxed{6.46\times10^{-11}\text{ m}}$$

---

44. ***REASONING AND SOLUTION*** The drawing below shows the relative positions and distances of the individual atoms that make up the nitric acid molecule.

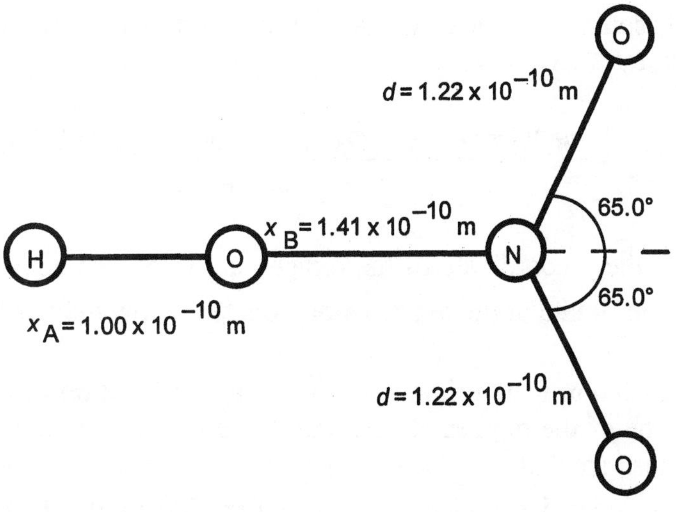

Since the molecules are distributed in a plane, we must specify both the *x* and *y* components of the center of mass. Generalizing Equation 7.10 to account for all five atoms, we find that the *x* coordinate of the center of mass is given by

$$x_{\text{cm}} = \frac{m_{\text{H}}x_{\text{H}} + m_{\text{O}}x_{\text{O1}} + m_{\text{N}}x_{\text{N}} + 2m_{\text{O}}x_{\text{O2}}}{m_{\text{H}} + m_{\text{N}} + 3m_{\text{O}}}$$

where the subscripts on the masses refer to the name of the atom. The distances x_{H}, x_{O1}, x_{N}, and x_{O2} refer to the *x* distances of the atoms from the coordinate origin. The distance x_{O1} represents the location of the oxygen atom that is collinear with the hydrogen atom and the nitrogen atom, while x_{O2} refers to the *x* component of the location of the oxygen atoms that are off the *x* axis. For convenience, we take the origin of coordinates at the location of the hydrogen atom, so $x_{\text{H}} = 0$, $x_{\text{O1}} = x_{\text{A}}$, $x_{\text{N}} = x_{\text{A}} + x_{\text{B}}$, and $x_{\text{O2}} = x_{\text{A}} + x_{\text{B}} + d\cos 65.0°$, where the distances x_{A}, x_{B}, and d are identified in the drawing above. Thus, we have

$$x_{\text{cm}} = \frac{m_{\text{H}}(0) + m_{\text{O}}x_{\text{A}} + m_{\text{N}}(x_{\text{A}} + x_{\text{B}}) + 2m_{\text{O}}(x_{\text{A}} + x_{\text{B}} + d\cos 65.0°)}{m_{\text{H}} + m_{\text{N}} + 3m_{\text{O}}}$$

From the drawing we have

$$d\cos 65.0° = \left(1.22\times10^{-10}\,\text{m}\right)\cos 65.0° = 5.16\times10^{-11}\,\text{m}$$

Substituting values, we obtain

$$x_{\text{cm}} = \frac{\left[\begin{array}{r}(26.6\times10^{-27}\,\text{kg})(1.00\times10^{-10}\,\text{m}) + (23.3\times10^{-27}\,\text{kg})(2.41\times10^{-10}\,\text{m}) \\ +\ 2(26.6\times10^{-27}\,\text{kg})(2.93\times10^{-10}\,\text{m})\end{array}\right]}{1.67\times10^{-27}\,\text{kg} + 23.3\times10^{-27}\,\text{kg} + 3(26.6\times10^{-27}\,\text{kg})} = 2.28\times10^{-10}\,\text{m}$$

Using similar reasoning, we deduce that the y coordinate of the center of mass of the molecule is given by

$$y_{\text{cm}} = \frac{m_{\text{H}} y_{\text{H}} + m_{\text{O}} y_{\text{O1}} + m_{\text{N}} y_{\text{N}} + m_{\text{O}} y_{\text{O2 above}} + m_{\text{O}} y_{\text{O2 below}}}{m_{\text{H}} + m_{\text{N}} + 3m_{\text{O}}}$$

where $y_{\text{O2 above}}$ is the y coordinate of the oxygen atom located above the H–O–N line and $y_{\text{O2 below}}$ is the y coordinate of the oxygen atom located below the H–O–N line.

However, since the coordinate origin was chosen at the location of the hydrogen atom, the hydrogen atom, one of the oxygen atoms, and the nitrogen atom are on the x axis and have zero y coordinates, so that $y_{\text{H}} = y_{\text{O1}} = y_{\text{N}} = 0$. Furthermore, since the remaining two oxygen atoms are located symmetrically on either side of the H–O–N line, we see that $y_{\text{O2 above}} = -y_{\text{O2 below}}$. Thus, the y coordinate of the center of mass of the nitric acid molecule is $y_{\text{cm}} = 0$. Therefore, the center of mass lies along the x axis at

$$x_{\text{cm}} = \boxed{2.28 \times 10^{-10}\ \text{m}}$$

---

45. **SSM** **WWW** ***REASONING*** The system consists of the lumberjack and the log. For this system, the sum of the external forces is zero. This is because the weight of the system is balanced by the corresponding normal force (provided by the buoyant force of the water) and the water is assumed to be frictionless. The lumberjack and the log, then, constitute an isolated system, and the principle of conservation of linear momentum holds.

SOLUTION

a. The total linear momentum of the system before the lumberjack begins to move is zero, since all parts of the system are at rest. Momentum conservation requires that the total momentum remains zero during the motion of the lumberjack.

$$\underbrace{m_1 v_{\text{f1}} + m_2 v_{\text{f2}}}_{\substack{\text{Total momentum}\\ \text{just before the jump}}} = \underbrace{0}_{\text{Initial momentum}}$$

Here the subscripts "1" and "2" refer to the first log and lumberjack, respectively. Let the direction of motion of the lumberjack be the positive direction. Then, solving for v_{1f} gives

$$v_{\text{f1}} = -\frac{m_2 v_{\text{f2}}}{m_1} = -\frac{(98\ \text{kg})(+3.6\ \text{m/s})}{230\ \text{kg}} = \boxed{-1.5\ \text{m/s}}$$

The minus sign indicates that the first log recoils as the lumberjack jumps off.

b. Now the system is composed of the lumberjack, just before he lands on the second log, and the second log. Gravity acts on the system, but for the short time under consideration while the lumberjack lands, the effects of gravity in changing the linear momentum of the system are negligible. Therefore, to a very good approximation, we can say that the linear momentum of the system is very nearly conserved. In this case, the initial momentum is not zero as it was in part (a); rather the initial momentum of the system is the momentum of the lumberjack just before he lands on the second log. Therefore,

$$\underbrace{m_1 v_{f1} + m_2 v_{f2}}_{\substack{\text{Total momentum} \\ \text{just after lumberjack lands}}} = \underbrace{m_1 v_{01} + m_2 v_{02}}_{\text{Initial momentum}}$$

In this expression, the subscripts "1" and "2" now represent the second log and lumberjack, respectively. Since the second log is initially at rest, $v_{01} = 0$. Furthermore, since the lumberjack and the second log move with a common velocity, $v_{f1} = v_{f2} = v_f$. The statement of momentum conservation then becomes

$$m_1 v_f + m_2 v_f = m_2 v_{02}$$

Solving for v_f, we have

$$v_f = \frac{m_2 v_{02}}{m_1 + m_2} = \frac{(98 \text{ kg})(+3.6 \text{ m/s})}{230 \text{ kg} + 98 \text{ kg}} = \boxed{+1.1 \text{ m/s}}$$

The positive sign indicates that the system moves in the same direction as the original direction of the lumberjack's motion.

---

46. ***REASONING AND SOLUTION*** The momentum of the spaceship is transferred to the rocket, so $(m_s + m_r)v_s = m_r v_r$, and the rocket's velocity is

$$v_r = \frac{(m_s + m_r)v_s}{m_r} = \frac{(4.0 \times 10^6 \text{ kg} + 1300 \text{ kg})(230 \text{ m/s})}{1300 \text{ kg}} = \boxed{7.1 \times 10^5 \text{ m/s}}$$

---

47. SSM ***REASONING*** Batman and the boat with the criminal constitute the system. Gravity acts on this system as an external force; however, gravity acts vertically, and we are concerned only with the horizontal motion of the system. If we neglect air resistance and friction, there are no external forces that act horizontally; therefore, the total linear momentum in the horizontal direction is conserved. When Batman collides with the boat, the horizontal component of his velocity is zero, so the statement of conservation of linear momentum in the horizontal direction can be written as

$$\underbrace{(m_1 + m_2)v_f}_{\substack{\text{Total horizontal momentum} \\ \text{after collision}}} = \underbrace{m_1 v_{01} + 0}_{\substack{\text{Total horizontal momentum} \\ \text{before collision}}}$$

Here, m_1 is the mass of the boat, and m_2 is the mass of Batman. This expression can be solved for v_f, the velocity of the boat after Batman lands in it.

SOLUTION Solving for v_f gives

$$v_f = \frac{m_1 v_{01}}{m_1 + m_2} = \frac{(510\text{ kg})(+11\text{ m/s})}{510\text{ kg} + 91\text{ kg}} = \boxed{+9.3\text{ m/s}}$$

The plus sign indicates that the boat continues to move in its initial direction of motion; it does not recoil.

---

48. ***REASONING AND SOLUTION*** According to the impulse-momentum theorem, Equation 7.4, the impulse of the net average force is equal to the change in the momentum of the car:

$$(\Sigma\overline{\mathbf{F}})\Delta t = m\mathbf{v}_f - m\mathbf{v}_0$$

Since the initial velocity of the car is $\mathbf{v}_0 = 0$ m/s, the final momentum $m\mathbf{v}_f$ of the car is

$$m\mathbf{v}_f = (\Sigma\overline{\mathbf{F}})\Delta t = (+680\text{ N})(7.2\text{ s}) = \boxed{+4900\text{ kg}\cdot\text{m/s}}$$

---

49. ***REASONING*** Since friction is negligible and since the downward-acting weight of each person is balanced by the upward-acting normal force from the sidewalk, the net external force acting on the two-person system is zero. Therefore, the system is isolated, and the conservation of linear momentum applies. The initial total momentum must be equal to the final total momentum.

SOLUTION Applying the principle of conservation of linear momentum and assuming that the direction in which Kevin is moving is the positive direction, we find

$$\underbrace{(87\text{ kg})v_{\text{Kevin}} + (22\text{ kg})(0\text{ m/s})}_{\text{Initial total momentum}} = \underbrace{(87\text{ kg} + 22\text{ kg})(2.4\text{ m/s})}_{\text{Final total momentum}}$$

$$v_{\text{Kevin}} = \frac{(87\text{ kg} + 22\text{ kg})(2.4\text{ m/s})}{87\text{ kg}} = \boxed{3.0\text{ m/s}}$$

---

50. ***REASONING AND SOLUTION***

a. According to Equation 7.4, the impulse-momentum theorem, $(\Sigma\overline{\mathbf{F}})\Delta t = m\mathbf{v}_f - m\mathbf{v}_0$. Since the only horizontal force exerted on the puck is the force $\overline{\mathbf{F}}$ exerted by the goalie,

$\Sigma\overline{\mathbf{F}} = \overline{\mathbf{F}}$. Since the goalie catches the puck, $\mathbf{v}_f = 0$ m/s. Solving for the average force exerted on the puck, we have

$$\overline{\mathbf{F}} = \frac{m(\mathbf{v}_f - \mathbf{v}_0)}{\Delta t} = \frac{(0.17\text{ kg})\left[(0\text{ m/s}) - (+65\text{ m/s})\right]}{5.0\times10^{-3}\text{ s}} = -2.2\times10^{3}\text{ N}$$

By Newton's third law, the force exerted on the goalie by the puck is equal in magnitude and opposite in direction to the force exerted on the puck by the goalie. Thus, the average force exerted on the goalie is $\boxed{+2.2\times10^{3}\text{N}}$.

b. If, instead of catching the puck, the goalie slaps it with his stick and returns the puck straight back to the player with a velocity of –65 m/s, then the average force exerted on the puck by the goalie is

$$\overline{\mathbf{F}} = \frac{m(\mathbf{v}_f - \mathbf{v}_0)}{\Delta t} = \frac{(0.17\text{ kg})[(-65\text{ m/s}) - (+65\text{ m/s})]}{5.0\times10^{-3}\text{ s}} = -4.4\times10^{3}\text{ N}$$

The average force exerted on the goalie by the puck is thus $\boxed{+4.4\times10^{3}\text{N}}$.

The answer in part (b) is twice that in part (a). This is consistent with the conclusion of Conceptual Example 3. The change in the momentum of the puck is greater when the puck rebounds from the stick. Thus, the puck exerts a greater impulse, and hence a greater force, on the goalie.

---

51. SSM ***REASONING*** The two-stage rocket constitutes the system. The forces that act to cause the separation during the explosion are, therefore, forces that are internal to the system. Since no external forces act on this system, it is isolated and the principle of conservation of linear momentum applies:

$$\underbrace{m_1 v_{f1} + m_2 v_{f2}}_{\substack{\text{Total momentum}\\ \text{after separation}}} = \underbrace{(m_1 + m_2)v_0}_{\substack{\text{Total momentum}\\ \text{before separation}}}$$

where the subscripts "1" and "2" refer to the lower and upper stages, respectively. This expression can be solved for v_{f1}.

SOLUTION Solving for v_{f1} gives

$$v_{f1} = \frac{(m_1 + m_2)v_0 - m_2 v_{f2}}{m_1}$$

$$= \frac{[2400\text{ kg} + 1200\text{ kg}](4900\text{ m/s}) - (1200\text{ kg})(5700\text{ m/s})}{2400\text{ kg}} = \boxed{+4500\text{ m/s}}$$

Since v_{f1} is positive $\boxed{\text{its direction is the same as the rocket before the explosion}}$.

---

52. ***REASONING AND SOLUTION*** The velocity of the center of mass is given by Equation 7.11,

$$v_{cm} = \frac{m_1 v_1 + m_2 v_2}{m_1 + m_2} = \frac{(m_1/m_2)v_1 + v_2}{(m_1/m_2) + 1}$$

a. When the masses are equal, $m_1/m_2 = 1$, and we have

$$v_{cm} = \frac{v_1 + v_2}{2} = \frac{9.70\text{ m/s} + (-11.8\text{ m/s})}{2} = \boxed{-1.05\text{ m/s}}$$

b. When the mass of the ball moving at 9.70 m/s is twice the mass of the other ball, we have $m_1/m_2 = 2$, and the velocity of the center of mass is

$$v_{cm} = \frac{2v_1 + v_2}{2+1} = \frac{2(9.70\text{ m/s}) + (-11.8\text{ m/s})}{3} = \boxed{+2.53\text{ m/s}}$$

---

53. SSM ***REASONING AND SOLUTION*** The comet piece and Jupiter constitute an isolated system, since no external forces act on them. Therefore, the head-on collision obeys the conservation of linear momentum:

$$\underbrace{(m_{\text{comet}} + m_{\text{Jupiter}})v_f}_{\text{Total momentum after collision}} = \underbrace{m_{\text{comet}}v_{\text{comet}} + m_{\text{Jupiter}}v_{\text{Jupiter}}}_{\text{Total momentum before collision}}$$

where v_f is the common velocity of the comet piece and Jupiter after the collision. We assume initially that Jupiter is moving in the $+x$ direction so $v_{\text{Jupiter}} = +1.3 \times 10^4$ m/s. The comet piece must be moving in the opposite direction so $v_{\text{comet}} = -6.0 \times 10^4$ m/s. The final velocity v_f of Jupiter and the comet piece can be written as $v_f = v_{\text{Jupiter}} + \Delta v$ where Δv is the change in velocity of Jupiter due to the collision. Substituting this expression into the conservation of momentum equation gives

$$(m_{\text{comet}} + m_{\text{Jupiter}})(v_{\text{Jupiter}} + \Delta v) = m_{\text{comet}}v_{\text{comet}} + m_{\text{Jupiter}}v_{\text{Jupiter}}$$

Multiplying out the left side of this equation, algebraically canceling the term $m_{\text{Jupiter}} v_{\text{Jupiter}}$ from both sides of the equation, and solving for Δv yields

$$\Delta v = \frac{m_{\text{comet}}(v_{\text{comet}} - v_{\text{Jupiter}})}{m_{\text{comet}} + m_{\text{Jupiter}}}$$

$$= \frac{(4.0\times10^{12}\text{ kg})(-6.0\times10^{4}\text{ m/s} - 1.3\times10^{4}\text{ m/s})}{4.0\times10^{12}\text{ kg} + 1.9\times10^{27}\text{ kg}} = -1.5\times10^{-10}\text{ m/s}$$

The change in Jupiter's speed is $\boxed{1.5\times10^{-10}\text{ m/s}}$.

---

54. ***REASONING AND SOLUTION*** The conservation of momentum law applied in the horizontal direction gives

$$m_{\text{c}}v_{\text{fc}} + m_{\text{s}}v_{\text{fs}}\cos 30.0° = 0$$

so that

$$v_{\text{fc}} = \frac{-m_{\text{s}}v_{\text{fs}}\cos 30.0°}{m_{\text{c}}} = \frac{-(5.00\text{ kg})(8.00\text{ m/s})\cos 30.0°}{105\text{ kg}} = -0.330\text{ m/s}$$

The magnitude of the velocity is $\boxed{0.330\text{ m/s}}$. The minus sign indicates that the direction is

$\boxed{\text{opposite the horizontal velocity component of the stone}}$.

---

55. ***REASONING*** We will divide the problem into two parts: (a) the motion of the freely falling block after it is dropped from the building and before it collides with the bullet, and (b) the collision of the block with the bullet.

During the falling phase we will use an equation of kinematics that describes the velocity of the block as a function of time (which is unknown). During the collision with the bullet, the external force of gravity acts on the system. This force changes the momentum of the system by a negligibly small amount since the collision occurs over an extremely short time interval. Thus, to a good approximation, the sum of the external forces acting on the system during the collision is negligible, so the linear momentum of the system is conserved. The principle of conservation of linear momentum can be used to provide a relation between the momenta of the system before and after the collision. This relation will enable us to find a value for the time it takes for the bullet/block to reach the top of the building.

SOLUTION Falling from rest ($v_{0,\text{ block}} = 0$ m/s), the block attains a final velocity v_{block} just before colliding with the bullet. This velocity is given by Equation 2.4 as

$$\underbrace{v_{\text{block}}}_{\substack{\text{Final velocity of}\\\text{block just before}\\\text{bullet hits it}}} = \underbrace{v_{0,\text{ block}}}_{\substack{\text{Initial velocity of}\\\text{block at top of}\\\text{building}}} + at$$

where a is the acceleration due to gravity ($a = -9.8\text{ m/s}^2$) and t is the time of fall. The upward direction is assumed to be positive. Therefore, the final velocity of the falling block is

$$v_{\text{block}} = at \qquad (1)$$

During the collision with the bullet, the total linear momentum of the bullet/block system is conserved, so we have that

$$\underbrace{\left(m_{\text{bullet}} + m_{\text{block}}\right)v_{\text{f}}}_{\substack{\text{Total linear momentum}\\\text{after collision}}} = \underbrace{m_{\text{bullet}}v_{\text{bullet}} + m_{\text{block}}v_{\text{block}}}_{\substack{\text{Total linear momentum}\\\text{before collision}}} \qquad (2)$$

Here v_{f} is the final velocity of the bullet/block system after the collision, and v_{bullet} and v_{block} are the initial velocities of the bullet and block just before the collision. We note that the bullet/block system reverses direction, rises, and comes to a momentary halt at the top of the building. This means that v_{f}, the final velocity of the bullet/block system after the collision must have the same magnitude as v_{block}, the velocity of the falling block just before the bullet hits it. Since the two velocities have opposite directions, it follows that $v_{\text{f}} = -v_{\text{block}}$. Substituting this relation and Equation (1) into Equation (2) gives

$$\left(m_{\text{bullet}} + m_{\text{block}}\right)\left(-at\right) = m_{\text{bullet}}v_{\text{bullet}} + m_{\text{block}}\left(at\right)$$

Solving for the time, we find that

$$t = \frac{-m_{\text{bullet}}v_{\text{bullet}}}{a\left(m_{\text{bullet}} + 2m_{\text{block}}\right)} = \frac{-(0.015\text{ kg})(+810\text{ m/s})}{\left(-9.80\text{ m/s}^2\right)\left[(0.015\text{ kg}) + 2(1.8\text{ kg})\right]} = \boxed{0.34\text{ s}}$$

---

56. ***REASONING*** The x- and y-coordinates of the center of mass can be found by applying Equation 7.10 to the three atoms in the sulfur dioxide molecule. The x-coordinate of the center of mass uses the x-coordinates of the centers of the atoms, and the y-coordinate of the center of mass uses the y-coordinates of the centers of the atoms.

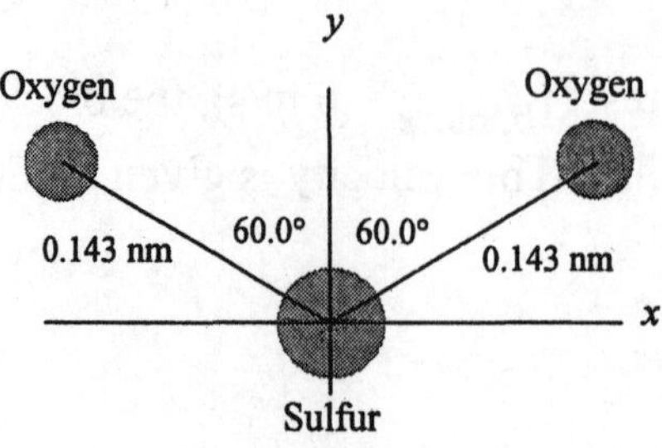

SOLUTION The *x*- and *y*-coordinates for the three atoms are listed in the table:

	x-coordinate	***y-coordinate***
Oxygen atom (on the left)	$x_1 = -(0.143 \text{ nm}) \sin 60.0° = -0.124 \text{ nm}$	$y_1 = +(0.143 \text{ nm}) \cos 60.0° = +0.0715 \text{ nm}$
Oxygen atom (on the right)	$x_2 = +(0.143 \text{ nm}) \sin 60.0° = +0.124 \text{ nm}$	$y_2 = +(0.143 \text{ nm}) \cos 60.0° = +0.0715 \text{ nm}$
Sulfur atom	$x_3 = 0 \text{ nm}$	$y_3 = 0 \text{ nm}$

a. Let m_1 and m_2 represent the masses of the oxygen atoms on the left and right, respectively. Likewise, let m_3 be the mass of the sulfur atom. The *x*-coordinate of the center of mass is

$$x_{\text{cm}} = \frac{m_1 x_1 + m_2 x_2 + m_3 x_3}{m_1 + m_2 + m_3} = \frac{m_1(-0.124 \text{ nm}) + m_2(+0.124 \text{ nm})x_2 + m_3(0 \text{ nm})}{m_1 + m_2 + m_3}$$

Since $m_1 = m_2$ (both are oxygen atoms), the equation above yields $x_{\text{cm}} = \boxed{0 \text{ m}}$.

b. The *y*-coordinate of the center of mass is

$$y_{\text{cm}} = \frac{m_1 y_1 + m_2 y_2 + m_3 y_3}{m_1 + m_2 + m_3} = \frac{m_1(+0.0715 \text{ nm}) + m_2(+0.0715 \text{ nm})x_2 + m_3(0 \text{ nm})}{m_1 + m_2 + m_3}$$

Substituting $m_1 = m_2 = m$ and $m_3 = 2m$ into the equation above yields

$$y_{\text{cm}} = \frac{m(+0.0715 \text{ nm}) + m(+0.0715 \text{ nm})x_2 + 2m(0 \text{ nm})}{m + m + 2m} = \boxed{0.0358 \text{ nm}}$$

---

57. SSM ***REASONING*** We will define the system to be the platform, the two people and the ball. Since the ball travels nearly horizontally, the effects of gravity are negligible. Momentum is conserved. Since the initial momentum of the system is zero, it must remain zero as the ball is thrown and caught. While the ball is in motion, the platform will recoil in such a way that the total momentum of the system remains zero. As the ball is caught, the system must come to rest so that the total momentum remains zero. The distance that the platform moves before coming to rest again can be determined by using the expressions for momentum conservation and the kinematic description for this situation.

SOLUTION While the ball is in motion, we have

$$MV + mv = 0 \tag{1}$$

where M is the combined mass of the platform and the two people, V is the recoil velocity of the platform, m is the mass of the ball, and v is the velocity of the ball.

The distance that the platform moves is given by

$$x = Vt \tag{2}$$

where t is the time that the ball is in the air. The time that the ball is in the air is given by

$$t = \frac{L}{v - V} \tag{3}$$

where L is the length of the platform, and the quantity $(v - V)$ is the velocity of the ball relative to the platform. Remember, both the ball and the platform are moving while the ball is in the air. Combining equations (2) and (3) gives

$$x = \left(\frac{V}{v - V}\right) L \tag{4}$$

From equation (1) the ratio of the velocities is $V/v = -(m/M)$. Equation (4) then gives

$$x = \frac{(V/v)L}{1 - (V/v)} = \frac{(-m/M)L}{1 + (m/M)} = -\frac{mL}{M + m} = -\frac{(6.0\text{ kg})(2.0\text{ m})}{118\text{ kg} + 6.0\text{ kg}} = -0.097\text{ m}$$

The minus sign indicates that displacement of the platform is in the opposite direction to the displacement of the ball. The distance moved by the platform is the magnitude of this displacement, or $\boxed{0.097\text{ m}}$.

---

58. ***CONCEPT QUESTIONS*** a. The change in velocity is the same for each car, and momentum is mass times velocity. Therefore, the car with the greatest mass has the greatest change in momentum, which is car B.

b. According to the impulse-momentum theorem, the impulse is equal to the change in momentum. Since car B has the greatest change in momentum, it experiences the greatest impulse.

c. Impulse is the average net force multiplied by the time during which the force acts. But the test time is the same for each car. Consequently, the car that experiences the greatest average net force is the car that experiences the greatest impulse, or car B.

SOLUTION Using the impulse-momentum theorem as given in Equation 7.4, we have

$$\underbrace{\overline{F}\Delta t}_{\text{Impulse}} = \underbrace{mv_{\text{f}}}_{\substack{\text{Final}\\\text{momentum}}} - \underbrace{mv_0}_{\substack{\text{Initial}\\\text{momentum}}}$$

Car A
$$\overline{F} = \frac{mv_{\text{f}} - mv_0}{\Delta t} = \frac{(1400\text{ kg})(27\text{ m/s}) - (1400\text{ kg})(0\text{ m/s})}{9.0\text{ s}} = \boxed{4200\text{ N}}$$

Car B
$$\overline{F} = \frac{mv_{\text{f}} - mv_0}{\Delta t} = \frac{(1900\text{ kg})(27\text{ m/s}) - (1900\text{ kg})(0\text{ m/s})}{9.0\text{ s}} = \boxed{5700\text{ N}}$$

---

59. ***CONCEPT QUESTIONS*** a. The spring exerts a force on disk 1 that points to the left and exerts a force on disk 2 that points to the right.

b. Disk 1 is moving to the right. The force applied by the expanding spring points to the left, however, and opposes the motion. Disk 1, therefore, slows down and its final speed is less than the speed v_0. In contrast, disk 2 is moving to the right, and the force applied to it by the expanding spring also points to the right. The speed of disk 2, therefore, increases to a value that is greater than the speed v_0.

SOLUTION Since friction is negligible and the air cushion balances the weight of the disks, we can apply the principle of conservation of linear momentum. Remembering that disk 1 comes to a halt after the spring is released, we have

$$\underbrace{(1.2\text{ kg})(0\text{ m/s}) + (2.4\text{ kg})v_{\text{f2}}}_{\substack{\text{Total momentum after}\\\text{spring is released}}} = \underbrace{(1.2\text{ kg} + 2.4\text{ kg})(5.0\text{ m/s})}_{\substack{\text{Total momentum before}\\\text{spring is released}}}$$

$$v_{\text{f2}} = \frac{(1.2\text{ kg} + 2.4\text{ kg})(5.0\text{ m/s})}{2.4\text{ kg}} = \boxed{7.5\text{ m/s}}$$

---

60. ***CONCEPT QUESTIONS*** a. Friction is negligible and the downward-pointing weight of the wagon and its contents is balanced by the upward-pointing normal force from the ground. Therefore, the net external force acting on the wagon and its contents is zero, and the principle of conservation of linear momentum applies. The total momentum of the system, then, must remain the same before and after the rock is thrown. When the rock is thrown forward, its forward momentum is increased. To keep the total momentum of the system constant, the forward momentum of the remaining wagon and its rider must decrease.

b. Our reasoning is similar to that in part (a), except that now the rock is given a momentum that points backward. Unless compensated for, this backward momentum would reduce the total momentum of the system. Thus, the forward momentum of the remaining wagon and its rider must increase.

c. After the rock is thrown, the wagon and its rider have a greater momentum when it is thrown backward. Momentum is mass times velocity, so the wagon acquires a greater speed when the rock is thrown backward.

SOLUTION In applying the momentum conservation principle, we assume that the forward direction is positive. Using v_f to denote the final speed of the wagon and its rider, we have

Rock thrown forward

$$\underbrace{(0.300\text{ kg})(16.0\text{ m/s})+(95.0\text{ kg}-0.300\text{ kg})v_f}_{\text{Total momentum after rock is thrown}} = \underbrace{(95.0\text{kg})(0.500\text{ m/s})}_{\text{Total momentum before rock is thrown}}$$

$$v_f = \frac{(95.0\text{kg})(0.500\text{ m/s})-(0.300\text{ kg})(16.0\text{ m/s})}{95.0\text{ kg}-0.300\text{ kg}} = \boxed{0.451\text{ m/s}}$$

Rock thrown backward

$$\underbrace{(0.300\text{ kg})(-16.0\text{ m/s})+(95.0\text{ kg}-0.300\text{ kg})v_f}_{\text{Total momentum after rock is thrown}} = \underbrace{(95.0\text{kg})(0.500\text{ m/s})}_{\text{Total momentum before rock is thrown}}$$

$$v_f = \frac{(95.0\text{kg})(0.500\text{ m/s})+(0.300\text{ kg})(16.0\text{ m/s})}{95.0\text{ kg}-0.300\text{ kg}} = \boxed{0.552\text{ m/s}}$$

---

61. ***CONCEPT QUESTIONS*** a. Since momentum is conserved, the total momentum of the two-object system after the collision must be the same as it was before the collision. Thus, to answer this question, we need to consider the initial total momentum. Momentum is mass times velocity. Since one of the objects is at rest initially, the total initial momentum comes only from the moving object. The greater the mass, the greater the momentum. Therefore, the momentum after the collision is greater when the large-mass object is moving initially.

b. Since momentum is mass times velocity, the speed of the system after the collision will be greater in the case when the system has the greater final momentum. Since the final total momentum is greater when the large-mass object is moving initially, the final speed is greater in that case also.

SOLUTION In applying the momentum conservation principle, we assume that the initially moving object is moving in the positive direction. In addition, we remember that one of the objects is initially at rest. Using v_f to denote the final speed of the two-object system, we have

Large - mass object moving initially

$$\underbrace{(3.0\text{ kg}+8.0\text{ kg})v_{\text{f}}}_{\text{Total momentum after collision}} = \underbrace{(3.0\text{ kg})(0\text{ m/s})+(8.0\text{ kg})(25\text{ m/s})}_{\text{Total momentum before collision}}$$

$$v_{\text{f}} = \frac{(3.0\text{ kg})(0\text{ m/s})+(8.0\text{ kg})(25\text{ m/s})}{3.0\text{ kg}+8.0\text{ kg}} = \boxed{18\text{ m/s}}$$

Small - mass object moving initially

$$\underbrace{(3.0\text{ kg}+8.0\text{ kg})v_{\text{f}}}_{\text{Total momentum after collision}} = \underbrace{(3.0\text{ kg})(25\text{ m/s})+(8.0\text{ kg})(0\text{ m/s})}_{\text{Total momentum before collision}}$$

$$v_{\text{f}} = \frac{(3.0\text{ kg})(25\text{ m/s})+(8.0\text{ kg})(0\text{ m/s})}{3.0\text{ kg}+8.0\text{ kg}} = \boxed{6.8\text{ m/s}}$$

---

62. ***CONCEPT QUESTIONS*** a. Momentum is mass times velocity and is a vector that has the same direction as the velocity. The initial total momentum is the sum of the two initial momentum vectors of the objects. One of the vectors points due east and one due north, so that they are perpendicular. Being perpendicular, these vectors can never add to give zero, because they have different directions. But according to the momentum-conservation principle, the final total momentum is the same as the initial total momentum. Therefore, the final total momentum cannot be zero after the collision.

b. Based on the conservation principle, the direction of the final total momentum must be the same as the direction of the initial total momentum. From the drawing, we can see that the initial total momentum has a component $\mathbf{p}_{0\text{A}}$ pointing due east and a component $\mathbf{p}_{0\text{B}}$ pointing due north. The final total momentum $\mathbf{P}_{\text{f}}$ has these same components and, therefore, must point north of east at an angle θ.

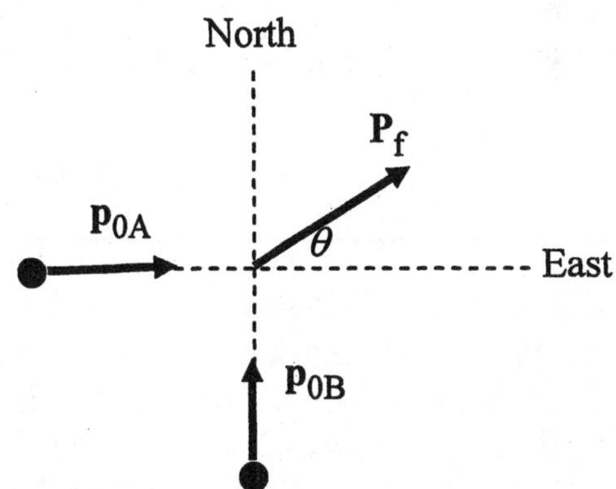

SOLUTION Based on the conservation principle, we know that the magnitude of the final total momentum is the same as the magnitude of the initial total momentum. Using the Pythagorean theorem with the initial momentum vectors of the two objects, we find that the magnitude of the final total momentum is

$$P_{\text{f}} = \sqrt{p_{0\text{A}}^2 + p_{0\text{B}}^2} = \sqrt{(m_{\text{A}}v_{0\text{A}})^2 + (m_{\text{B}}v_{0\text{B}})^2}$$

$$P_{\text{f}} = \sqrt{(17.0\text{ kg})^2(8.00\text{ m/s})^2 + (29.0\text{ kg})^2(5.00\text{ m/s})^2} = \boxed{199\text{ kg}\cdot\text{m/s}}$$

The direction is given by

$$\theta = \tan^{-1}\left(\frac{m_B v_{0B}}{m_A v_{0A}}\right) = \tan^{-1}\left[\frac{(29.0 \text{ kg})(5.00 \text{ m/s})}{(17.0 \text{ kg})(8.00 \text{ m/s})}\right] = \boxed{46.8° \text{ north of east}}$$

---

63. ***CONCEPT QUESTIONS*** a. The center of mass is always closer to the object with the largest mass, in this case John. Therefore, it is closer to the 9.0-m point.

b. The center of mass is again closer to John, so it is now closer to the 2.0-m point.

c. Considering our answers to (a) and (b), we conclude that the center of mass shifts toward the origin as a result of the switch.

SOLUTION Using Equation 7.10, we can calculate the location of the center of mass before and after the switch:

Before $$x_{cm} = \frac{m_N x_N + m_B x_B}{m_N + m_B} = \frac{(86 \text{ kg})(9.0 \text{ m}) + (55 \text{ kg})(2.0 \text{ m})}{86 \text{ kg} + 55 \text{ kg}} = 6.3 \text{ m}$$

After $$x_{cm} = \frac{m_N x_N + m_B x_B}{m_N + m_B} = \frac{(86 \text{ kg})(2.0 \text{ m}) + (55 \text{ kg})(9.0 \text{ m})}{86 \text{ kg} + 55 \text{ kg}} = 4.7 \text{ m}$$

The center of mass moves from the 6.3-m point to the 4.7-m point, which is toward the origin. The amount by which it moves is

$$6.3 \text{ m} - 4.7 \text{ m} = \boxed{1.6 \text{ m}}$$

---

64. ***CONCEPT QUESTIONS*** a. As the ball swings downward, its speed increases. Therefore, its momentum and kinetic energy, which depend on the ball's speed, also increase. Thus, these quantities are not conserved (they do not remain constant) during the motion. Since air resistance is negligible, however, the work done by nonconservative forces is zero, $W_{nc} = 0$ J, so the total mechanical energy, which is the sum of the kinetic and potential energies, is conserved (see Section 6.5).

b. During the collision, the *x* component of the total momentum of the ball/block system and the total kinetic energy of the system are conserved. The horizontal component of the total momentum is conserved, because the horizontal surface is frictionless and so the net average external force acting in the horizontal direction is zero. The total kinetic energy is conserved, since the collision is known to be elastic.

SOLUTION a. As the ball falls, the total mechanical energy is conserved. Thus, the total mechanical energy at the top of the swing is equal to that at the bottom:

$$\underbrace{m_{\text{ball}} g h_0}_{\substack{\text{Total mechanical}\\ \text{energy at top of}\\ \text{swing, all potential}}} = \underbrace{\tfrac{1}{2} m_{\text{ball}} v_{\text{f}}^2}_{\substack{\text{Total mechanical}\\ \text{energy at bottom of}\\ \text{swing, all kinetic}}}$$

In this expression h_0 is the initial height of the ball above its level at the bottom of the swing, which is where $h = 0$ m. Solving for the speed v_{f} of the ball just before the collision gives

$$v_{\text{f}} = \sqrt{2gh_0} = \sqrt{2\left(9.80 \text{ m/s}^2\right)\left(1.20 \text{ m}\right)} = 4.85 \text{ m/s}$$

At the bottom of the swing the ball is moving horizontally and to the right, which we take to be the $+x$ direction. Thus the velocity of the ball is $v_{\text{f}} = \boxed{+4.85 \text{ m/s}}$.

b. The conservation of linear momentum and the conservation of the total kinetic energy can be used to describe the behavior of the system during the collision. This situation is identical to that in Example 7 in Section 7.3, so Equation 7.8a applies:

$$v_{\text{f1}} = \left(\frac{m_1 - m_2}{m_1 + m_2}\right) v_{01}$$

where v_{f1} is the final velocity of the ball, m_1 and m_2 are, respectively, the masses of the ball and block, and v_{01} is the velocity of the ball just before the collision. Since $v_{01} = +4.85$ m/s, we find that

$$v_{\text{f1}} = \left(\frac{1.60 \text{ kg} - 2.40 \text{ kg}}{1.60 \text{ kg} + 2.40 \text{ kg}}\right)\left(+4.85 \text{ m/s}\right) = \boxed{-0.97 \text{ m/s}}$$

The minus sign indicates that the ball rebounds to the left after the collision.

---

65. ***CONCEPT QUESTIONS*** a. Yes, the conservation of linear momentum can be applied to this three-object system, even though the second collision occurs later than the first one. Since there is no friction between the blocks and the horizontal surface, air resistance is negligible, and the weight of each block is balanced by a normal force, the net average force is zero. Thus, the conservation of momentum principle applies.

b. The total kinetic energy of the three-body system is not conserved. Both collisions are inelastic, and the collision with block 2 is completely inelastic since the bullet comes to rest within the block. As with any inelastic collision, the total kinetic energy after the collisions is less than that before the collisions.

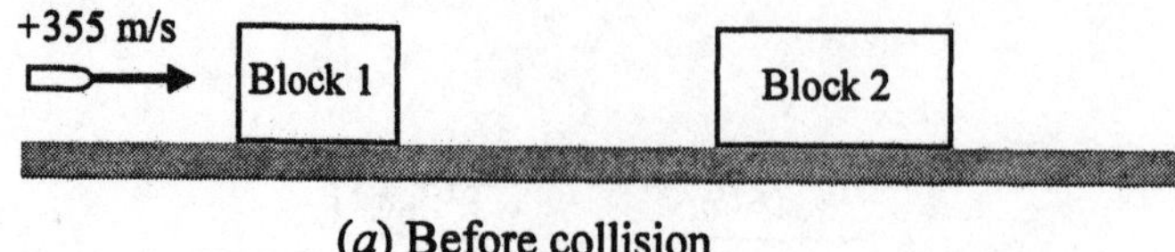

(a) Before collision

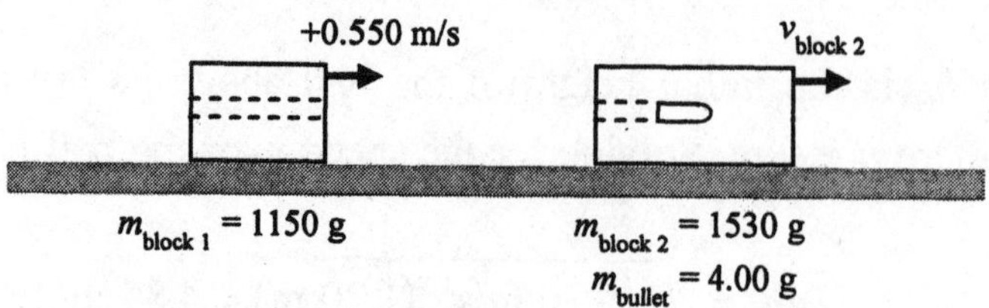

(b) After collision

SOLUTION a. The conservation of linear momentum states that the total momentum of the system after the collisions [see part (b) of the drawing] is equal to that before the collision [part (a) of the drawing]:

$$\underbrace{m_{block\,1}v_{block\,1} + \left(m_{block\,2} + m_{bullet}\right)v_{block\,2}}_{\text{Total momentum after collisions}} = \underbrace{m_{bullet}v_{bullet}}_{\substack{\text{Total momentum}\\ \text{before collisions}}}$$

Solving for the velocity $v_{block\,2}$ of block 2 after the collisions gives

$$v_{block\,2} = \frac{m_{bullet}v_{bullet} - m_{block\,1}v_{block\,1}}{m_{block\,2} + m_{bullet}}$$

$$= \frac{\left(4.00\times10^{-3}\text{ kg}\right)\left(+355\text{ m/s}\right) - \left(1.150\text{ kg}\right)\left(+0.550\text{ m/s}\right)}{1.530\times10^{-3}\text{ kg} + 4.00\times10^{-3}\text{ kg}} = \boxed{+0.513\text{ m/s}}$$

b. The ratio of the total kinetic energy (KE) after the collision to that before the collision is

$$\frac{\text{KE}_{after}}{\text{KE}_{before}} = \frac{\frac{1}{2}m_{block\,1}v_{block\,1}^2 + \frac{1}{2}\left(m_{block\,2} + m_{bullet}\right)v_{block\,2}^2}{\frac{1}{2}m_{bullet}v_{bullet}^2}$$

$$= \frac{\frac{1}{2}\left(1.150\text{ kg}\right)\left(0.550\text{ m/s}\right)^2 + \frac{1}{2}\left(1.530\text{ kg} + 4.00\times10^{-3}\text{ kg}\right)\left(0.513\text{ m/s}\right)^2}{\frac{1}{2}\left(4.00\times10^{-3}\text{ kg}\right)\left(355\text{ m/s}\right)^2} = \boxed{1.49\times10^{-3}}$$

CHAPTER 8 | *ROTATIONAL KINEMATICS*

PROBLEMS

---

1. SSM ***REASONING AND SOLUTION*** Equation 8.2 gives the desired result. Since the answer is to be expressed in rad/s, we must convert revolutions to radians.

$$\bar{\omega} = \frac{\Delta\theta}{\Delta t} = \left(\frac{3.5 \text{ rev}}{1.7 \text{ s}}\right)\left(\frac{2\pi \text{ rad}}{1 \text{ rev}}\right) = \boxed{13 \text{ rad/s}}$$

---

2. ***REASONING*** The average angular velocity is equal to the angular displacement divided by the elapsed time (Equation 8.2). Thus, the angular displacement of the baseball is equal to the product of the average angular velocity and the elapsed time. However, the problem gives the travel time in seconds and asks for the displacement in radians, while the angular velocity is given in revolutions per minute. Thus, we will begin by converting the angular velocity into radians per second.

SOLUTION Since 2π rad = 1 rev and 1 min = 60 s, the average angular velocity $\bar{\omega}$ (in rad/s) of the baseball is

$$\bar{\omega} = \left(\frac{330 \text{ rev}}{\text{min}}\right)\left(\frac{2\pi \text{ rad}}{1 \text{ rev}}\right)\left(\frac{1 \text{ min}}{60 \text{ s}}\right) = \boxed{35 \text{ rad/s}}$$

Since the average angular velocity of the baseball is equal to the angular displacement $\Delta\theta$ divided by the elapsed time Δt, the angular displacement is

$$\Delta\theta = \bar{\omega}\,\Delta t = (35 \text{ rad/s})(0.60 \text{ s}) = \boxed{21 \text{ rad}} \qquad (8.2)$$

---

3. SSM ***REASONING AND SOLUTION*** Since there are 2π radians per revolution, and it is stated in the problem that there are 100 grads in one-quarter of a circle, we find that the number of grads in one radian is

$$(1.00 \text{ rad})\left(\frac{1 \text{ rev}}{2\pi \text{ rad}}\right)\left(\frac{100 \text{ grad}}{0.250 \text{ rev}}\right) = \boxed{63.7 \text{ grad}}$$

---

4. ***REASONING AND SOLUTION*** In one revolution the pulsar turns through 2π radians. The average angular speed of the pulsar is, from Equation 8.2,

$$\bar{\omega} = \frac{\Delta\theta}{\Delta t} = \frac{2\pi \text{ rad}}{0.033 \text{ s}} = \boxed{1.9 \times 10^2 \text{ rad/s}}$$

---

5. ***REASONING AND SOLUTION*** Using Equation 8.4 and the appropriate conversion factors, the average angular acceleration of the CD in rad/s^2 is

$$\bar{\alpha} = \frac{\Delta\omega}{\Delta t} = \left(\frac{210 \text{ rev/min} - 480 \text{ rev/min}}{74 \text{ min}}\right)\left(\frac{2\pi \text{ rad}}{1 \text{ rev}}\right)\left(\frac{1 \text{ min}}{60 \text{ s}}\right)^2 = -6.4\times10^{-3} \text{ rad/s}^2$$

The magnitude of the average angular acceleration is $\boxed{6.4 \times 10^{-3} \text{ rad/s}^2}$.

---

6. ***REASONING*** Equation 8.4 $\left[\bar{\alpha} = (\omega - \omega_0)/t\right]$ indicates that the average angular acceleration is equal to the change in the angular velocity divided by the elapsed time. Since the wheel starts from rest, its initial angular velocity is $\omega_0 = 0$ rad/s. Its final angular velocity is given as $\omega = 0.24$ rad/s. Since the average angular acceleration is given as $\bar{\alpha} = 0.030 \text{ rad/s}^2$, Equation 8.4 can be solved to determine the elapsed time t.

SOLUTION Solving Equation 8.4 for the elapsed time gives

$$t = \frac{\omega - \omega_0}{\bar{\alpha}} = \frac{0.24 \text{ rad/s} - 0 \text{ rad/s}}{0.030 \text{ rad/s}^2} = \boxed{8.0 \text{ s}}$$

---

7. SSM ***REASONING AND SOLUTION*** Equation 8.4 gives the desired result. Assuming $t_0 = 0$ s, the final angular velocity is

$$\omega = \omega_0 + \alpha t = 0 \text{ rad/s} + (328 \text{ rad/s}^2)(1.50 \text{ s}) = \boxed{492 \text{ rad/s}}$$

---

8. ***REASONING*** The distance s traveled by a spot on the outer edge of a disk of radius r can be determined from the angular displacement $\Delta\theta$ (in radians) of the disk by using Equation 8.1 ($\Delta\theta = s/r$). The radius is given as $r = 0.15$ m. The angular displacement is not given. However, the angular velocity is given as $\omega = 1.4$ rev/s and the elapsed time as $\Delta t = 45$ s, so the displacement can be obtained from the definition of angular velocity stated in Equation 8.2 ($\omega = \Delta\theta/\Delta t$). We must remember that Equation 8.1 is only valid when $\Delta\theta$ is expressed in radians. It will, therefore, be necessary to convert the given angular velocity from rev/s into rad/s.

SOLUTION From Equation 8.1 we have

$$s = r\Delta\theta$$

Using Equation 8.2, we can write the displacement as $\Delta\theta = \omega\,\Delta t$. With this substitution Equation 8.1 becomes

$$s = r\omega\,\Delta t = (0.15\text{ m})\underbrace{(1.4\text{ rev/s})\left(\frac{2\pi\text{ rad}}{1\text{ rev}}\right)}_{\text{Conversion of rev/s into rad/s}}(45\text{ s}) = \boxed{59\text{ m}}$$

The radian, being a quantity without units, is dropped from the final result, leaving the answer in meters.

---

9. ***REASONING AND SOLUTION*** The angular displacements of the astronauts are equal.

For A $\qquad \theta = s_A / r_A \qquad$ (8.1)

For B $\qquad \theta = s_B / r_B$

Equating these two equations for θ and solving for s_B gives

$$s_B = (r_B/r_A)s_A = [(1.10 \times 10^3\text{ m})/(3.20 \times 10^2\text{ m})](2.40 \times 10^2\text{ m}) = \boxed{825\text{ m}}$$

---

10. ***REASONING*** The average angular velocity is defined as the angular displacement divided by the elapsed time (Equation 8.2). Therefore, the angular displacement is equal to the product of the average angular velocity and the elapsed time The elapsed time is given, so we need to determine the average angular velocity. We can do this by using the graph of angular velocity versus time that accompanies the problem.

SOLUTION The angular displacement $\Delta\theta$ is related to the average angular velocity $\bar{\omega}$ and the elapsed time Δt by Equation 8.2, $\Delta\theta = \bar{\omega}\,\Delta t$. The elapsed time is given as 8.0 s. To obtain the average angular velocity, we need to extend the graph that accompanies this problem from a time of 5.0 s to 8.0 s. It can be seen from the graph that the angular velocity increases by +3.0 rad/s during each second. Therefore, when the time increases from 5.0 to 8.0 s, the angular velocity increases from +6.0 rad/s to 6 rad/s + 3×(3.0 rad/s) = +15 rad/s. A graph of the angular velocity from 0 to 8.0 s is shown at the right. The average angular velocity during this time is equal to one half the sum of the initial and final angular velocities:

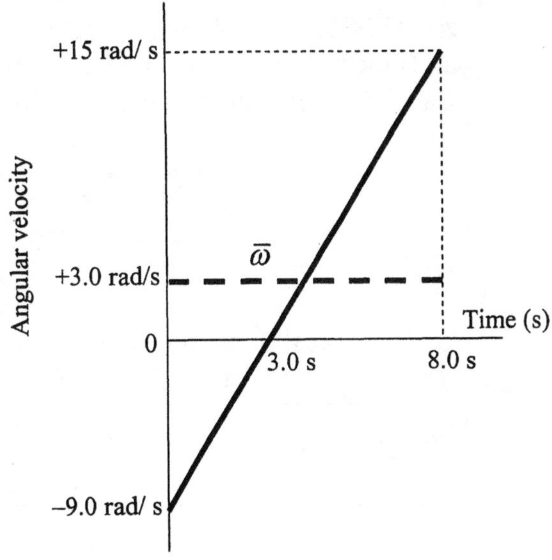

$$\bar{\omega} = \tfrac{1}{2}(\omega_0 + \omega) = \tfrac{1}{2}(-9.0 \text{ rad/s} + 15 \text{ rad/s}) = +3.0 \text{ rad/s}$$

The angular displacement of the wheel from 0 to 8.0 s is

$$\Delta\theta = \bar{\omega}\,\Delta t = (+3.0 \text{ rad/s})(8.0 \text{ s}) = \boxed{+24 \text{ rad}}$$

---

11. SSM ***REASONING*** The time required for the bullet to travel the distance d is equal to the time required for the discs to undergo an angular displacement of 0.240 rad. The time can be found from Equation 8.2; once the time is known, the speed of the bullet can be found using Equation 2.2.

SOLUTION From the definition of average angular velocity:

$$\bar{\omega} = \frac{\Delta\theta}{\Delta t}$$

the required time is

$$\Delta t = \frac{\Delta\theta}{\bar{\omega}} = \frac{0.240 \text{ rad}}{95.0 \text{ rad/s}} = 2.53 \times 10^{-3} \text{ s}$$

Note that $\bar{\omega} = \omega$ because the angular speed is constant. The (constant) speed of the bullet can then be determined from the definition of average speed:

$$\bar{v} = \frac{\Delta x}{\Delta t} = \frac{d}{\Delta t} = \frac{0.850 \text{ m}}{2.53 \times 10^{-3} \text{ s}} = \boxed{336 \text{ m/s}}$$

---

12. ***REASONING AND SOLUTION***

a. If the propeller is to appear stationary, each blade must move through an angle of 120° or $2\pi/3$ rad between flashes. The time required is

$$t = \frac{\theta}{\omega} = \frac{(2\pi/3) \text{ rad}}{(16.7 \text{ rev/s})\left(\dfrac{2\pi \text{ rad}}{1 \text{ rev}}\right)} = \boxed{2.00 \times 10^{-2} \text{ s}}$$

b. The next shortest time occurs when each blade moves through an angle of 240°, or $4\pi/3$ rad, between successive flashes. This time is twice that found in part a, or $\boxed{4.00 \times 10^{-2} \text{ s}}$.

---

13. ***REASONING AND SOLUTION*** The baton will make four revolutions in a time t given by

$$t = \frac{\theta}{\omega}$$

Half of this time is required for the baton to reach its highest point. The magnitude of the initial vertical velocity of the baton is then

$$v_0 = g\left(\tfrac{1}{2}t\right) = g\left(\frac{\theta}{2\omega}\right)$$

With this initial velocity the baton can reach a height of

$$h = \frac{v_0^2}{2g} = \frac{g\theta^2}{8\omega^2} = \frac{\left(9.80 \text{ m/s}^2\right)(8\pi \text{ rad})^2}{8\left[\left(1.80\,\frac{\text{rev}}{\text{s}}\right)\left(\frac{2\pi \text{ rad}}{1 \text{ rev}}\right)\right]^2} = \boxed{6.05 \text{ m}}$$

---

14. ***REASONING AND SOLUTION***
The figure at the right shows the relevant angles and dimensions for either one of the celestial bodies under consideration.

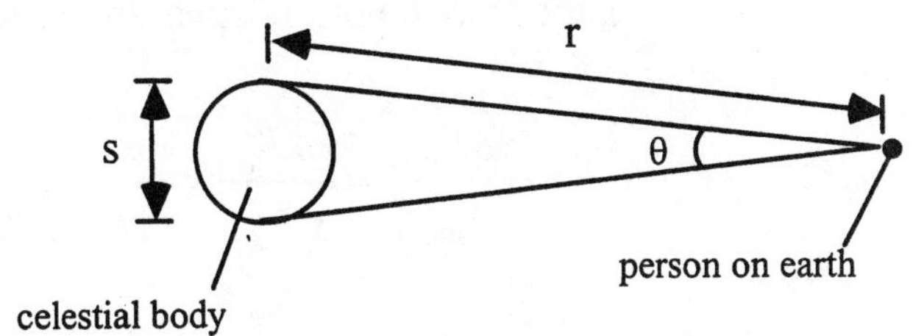

a. Using the figure above

$$\theta_{\text{moon}} = \frac{s_{\text{moon}}}{r_{\text{moon}}} = \frac{3.48 \times 10^6 \text{ m}}{3.85 \times 10^8 \text{ m}} = \boxed{9.04 \times 10^{-3} \text{ rad}}$$

$$\theta_{\text{sun}} = \frac{s_{\text{sun}}}{r_{\text{sun}}} = \frac{1.39 \times 10^9 \text{ m}}{1.50 \times 10^{11} \text{ m}} = \boxed{9.27 \times 10^{-3} \text{ rad}}$$

b. Since the sun subtends a slightly larger angle than the moon, as measured by a person standing on the earth, the sun cannot be completely blocked by the moon. Therefore, $\boxed{\text{a "total" eclipse of the sun is not really total}}$.

c. The relevant geometry is shown below.

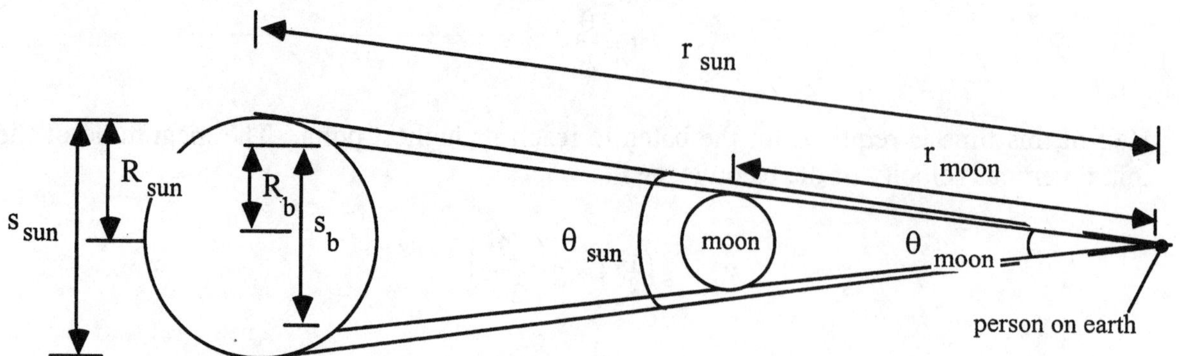

The apparent circular area of the sun as measured by a person standing on the earth is given by: $A_{\text{sun}} = \pi R_{\text{sun}}^2$, where R_{sun} is the radius of the sun. The apparent circular area of the sun that is blocked by the moon is $A_{\text{blocked}} = \pi R_{\text{b}}^2$, where R_{b} is shown in the figure above. Also from the figure above, it follows that

$$R_{\text{sun}} = (1/2)\, s_{\text{sun}} \quad \text{and} \quad R_{\text{b}} = (1/2)\, s_{\text{b}}$$

Therefore, the fraction of the apparent circular area of the sun that is blocked by the moon is

$$\frac{A_{\text{blocked}}}{A_{\text{sun}}} = \frac{\pi R_{\text{b}}^2}{\pi R_{\text{sun}}^2} = \frac{\pi (s_{\text{b}}/2)^2}{\pi (s_{\text{sun}}/2)^2} = \left(\frac{s_{\text{b}}}{s_{\text{sun}}}\right)^2 = \left(\frac{\theta_{\text{moon}}\, r_{\text{sun}}}{\theta_{\text{sun}}\, r_{\text{sun}}}\right)^2$$

$$= \left(\frac{\theta_{\text{moon}}}{\theta_{\text{sun}}}\right)^2 = \left(\frac{9.04\times 10^{-3}\ \text{rad}}{9.27\times 10^{-3}\ \text{rad}}\right)^2 = 0.951$$

The moon blocks out **95.1 percent** of the apparent circular area of the sun.

---

15. ***REASONING AND SOLUTION*** The ball has a time of flight given by the kinematics equation $0\ \text{m} = v_0 t - \frac{1}{2} a_y t^2$. Solving for the time gives

$$t = \frac{2v_0}{g} = \frac{2(19\ \text{m/s})(\sin 55°)}{9.80\ \text{m/s}^2} = 3.2\ \text{s}$$

The ball then makes (7.7 rev/s)(3.2 s) = **25 rev**

---

16. ***REASONING*** The angular displacement is given as $\theta = 0.500$ rev, while the initial angular velocity is given as $\omega_0 = 3.00$ rev/s and the final angular velocity as $\omega = 5.00$ rev/s. Since we seek the time t, we can use Equation 8.6 $\left[\theta = \frac{1}{2}(\omega_0 + \omega)t\right]$ from the equations of rotational kinematics to obtain it.

SOLUTION Solving Equation 8.6 for the time t, we find that

$$t = \frac{2\theta}{\omega_0 + \omega} = \frac{2(0.500 \text{ rev})}{3.00 \text{ rev/s} + 5.00 \text{ rev/s}} = \boxed{0.125 \text{ s}}$$

---

17. SSM ***REASONING AND SOLUTION*** Since the angular speed of the fan decreases, the sign of the angular acceleration must be opposite to the sign for the angular velocity. Taking the angular velocity to be positive, the angular acceleration, therefore, must be a negative quantity. Using Equation 8.4 we obtain

$$\omega_0 = \omega - \alpha t = 83.8 \text{ rad/s} - (-42.0 \text{ rad/s}^2)(1.75 \text{ s}) = \boxed{157.3 \text{ rad/s}}$$

---

18. ***REASONING AND SOLUTION*** From Equation 8.6, $\theta = \frac{1}{2}(\omega_0 + \omega)t$.

Solving for t gives

$$t = \frac{2\theta}{\omega_0 + \omega} = \frac{2(85.1 \text{ rad})}{18.5 \text{ rad/s} + 14.1 \text{ rad/s}} = \boxed{5.22 \text{ s}}$$

---

19. SSM WWW ***REASONING AND SOLUTION***

a. Since the flywheel comes to rest, its final angular velocity is zero. Furthermore, if the initial angular velocity ω_0 is assumed to be a positive number and the flywheel decelerates, the angular acceleration must be a negative number. Solving Equation 8.8 for θ , we obtain

$$\theta = \frac{\omega^2 - \omega_0^2}{2\alpha} = \frac{(0 \text{ rad/s})^2 - (220 \text{ rad/s})^2}{2(-2.0 \text{ rad/s}^2)} = \boxed{1.2 \times 10^4 \text{ rad}}$$

b. The time required for the flywheel to come to rest can be found from Equation 8.4. Solving for t, we obtain

$$t = \frac{\omega - \omega_0}{\alpha} = \frac{0 \text{ rad/s} - 220 \text{ rad/s}}{-2.0 \text{ rad/s}^2} = \boxed{1.1 \times 10^2 \text{ s}}$$

---

20. ***REASONING AND SOLUTION***

a. $$\theta = \tfrac{1}{2}(\omega + \omega_0)t = \tfrac{1}{2}(1420 \text{ rad/s} + 420 \text{ rad/s})(5.00 \text{ s}) = \boxed{4.60 \times 10^3 \text{ rad}}$$

b. $$\alpha = \frac{\omega - \omega_0}{t} = \frac{1420 \text{ rad/s} - 420 \text{ rad/s}}{5.00 \text{ s}} = \boxed{2.00 \times 10^2 \text{ rad/s}^2}$$

---

21. ***REASONING*** Equation 8.8 $\left(\omega^2 = \omega_0^2 + 2\alpha\theta\right)$ from the equations of rotational kinematics can be employed to find the final angular velocity ω. The initial angular velocity is $\omega_0 = 0$ rad/s since the top is initially at rest, and the angular acceleration is given as $\alpha = 12$ rad/s^2. The angle θ (in radians) through which the pulley rotates is not given, but it can be obtained from Equation 8.1 ($\theta = s/r$), where the arc length s is the 64-cm length of the string and r is the 2.0-cm radius of the top.

SOLUTION Solving Equation 8.8 for the final angular velocity gives

$$\omega = \pm\sqrt{\omega_0^2 + 2\alpha\theta}$$

We choose the positive root, because the angular acceleration is given as positive and the top is at rest initially. Substituting $\theta = s/r$ from Equation 8.1 gives

$$\omega = +\sqrt{\omega_0^2 + 2\alpha\left(\frac{s}{r}\right)} = +\sqrt{(0 \text{ rad/s})^2 + 2\left(12 \text{ rad/s}^2\right)\left(\frac{64 \text{ cm}}{2.0 \text{ cm}}\right)} = \boxed{28 \text{ rad/s}}$$

---

22. ***REASONING*** The equations of kinematics for rotational motion cannot be used directly to find the angular displacement, because the final angular velocity (not the initial angular velocity), the acceleration, and the time are known. We will combine two of the equations, Equations 8.4 and 8.6 to obtain an expression for the angular displacement that contains the three known variables.

SOLUTION The angular displacement of each wheel is equal to the average angular velocity multiplied by the time

$$\theta = \underbrace{\tfrac{1}{2}\left(\omega_0 + \omega\right)}_{\bar{\omega}} t \qquad (8.6)$$

The initial angular velocity ω_0 is not known, but it can be found in terms of the angular acceleration and time, which are known. The angular acceleration is defined as (with $t_0 = 0$ s)

$$\bar{\alpha} = \frac{\omega - \omega_0}{t} \quad \text{or} \quad \omega_0 = \omega - \alpha t \qquad (8.4)$$

Substituting this expression for ω_0 into Equation 8.6 gives

$$\theta = \tfrac{1}{2}\Big[\underbrace{(\omega - \alpha t)}_{\omega_0} + \omega\Big] t = \omega t - \tfrac{1}{2}\alpha t^2$$

$$= (+74.5 \text{ rad/s})(4.50 \text{ s}) - \tfrac{1}{2}(+6.70 \text{ rad/s}^2)(4.50 \text{ s})^2 = \boxed{+267 \text{ rad}}$$

---

23. SSM ***REASONING*** The time required for the change in the angular velocity to occur can be found by solving Equation 8.4 for t. In order to use Equation 8.4, however, we must know the initial angular velocity ω_0. Equation 8.6 can be used to find the initial angular velocity.

SOLUTION From Equation 8.6 we have

$$\theta = \frac{1}{2}(\omega_0 + \omega)t$$

Solving for ω_0 gives

$$\omega_0 = \frac{2\theta}{t} - \omega$$

Since the angular displacement θ is zero, $\omega_0 = -\omega$. Solving Equation 8.4 for t, and using the fact that $\omega_0 = -\omega$, gives

$$t = \frac{2\omega}{\alpha} = \frac{2(-25.0 \text{ rad/s})}{-4.00 \text{ rad/s}^2} = \boxed{12.5 \text{ s}}$$

---

24. ***REASONING AND SOLUTION*** The angular acceleration is found for the first circumstance.

$$\alpha = \frac{\omega^2 - \omega_0^2}{2\theta} = \frac{(3.14 \times 10^4 \text{ rad/s})^2 - (1.05 \times 10^4 \text{ rad/s})^2}{2(1.88 \times 10^4 \text{ rad})} = 2.33 \times 10^4 \text{ rad/s}^2$$

For the second circumstance

$$t = \frac{\omega - \omega_0}{\alpha} = \frac{7.85 \times 10^4 \text{ rad/s} - 0 \text{ rad/s}}{2.33 \times 10^4 \text{ rad/s}^2} = \boxed{3.37 \text{ s}}$$

---

25. ***REASONING*** There are three segments to the propeller's angular motion, and we will calculate the angular displacement for each separately. In these calculations we will remember that the final angular velocity for one segment is the initial velocity for the next segment. Then, we will add the separate displacements to obtain the total.

SOLUTION For the first segment the initial angular velocity is $\omega_0 = 0$ rad/s, since the propeller starts from rest. Its acceleration is $\alpha = 2.90 \times 10^{-3}$ rad/s^2 for a time $t = 2.10 \times 10^3$ s. Therefore, we can obtain the angular displacement θ_1 from Equation 8.7 of the equations of rotational kinematics as follows:

[First segment]

$$\theta_1 = \omega_0 t + \tfrac{1}{2}\alpha t^2 = (0\text{ rad/s})(2.10\times10^3\text{ s}) + \tfrac{1}{2}(2.90\times10^{-3}\text{ rad/s}^2)(2.10\times10^3\text{ s})^2$$
$$= 6.39\times10^3\text{ rad}$$

The initial angular velocity for the second segment is the final velocity for the first segment, and according to Equation 8.4, we have

$$\omega = \omega_0 + \alpha t = 0\text{ rad/s} + (2.90\times10^{-3}\text{ rad/s}^2)(2.10\times10^3\text{ s}) = 6.09\text{ rad/s}$$

Thus, during the second segment, the initial angular velocity is $\omega_0 = 6.09$ rad/s and remains constant at this value for a time of $t = 1.40 \times 10^3$ s. Since the velocity is constant, the angular acceleration is zero, and Equation 8.7 gives the angular displacement θ_2 as

[Second segment]

$$\theta_2 = \omega_0 t + \tfrac{1}{2}\alpha t^2 = (6.09\text{ rad/s})(1.40\times10^3\text{ s}) + \tfrac{1}{2}(0\text{ rad/s}^2)(1.40\times10^3\text{ s})^2 = 8.53\times10^3\text{ rad}$$

During the third segment, the initial angular velocity is $\omega_0 = 6.09$ rad/s, the final velocity is $\omega = 4.00$ rad/s, and the angular acceleration is $\alpha = -2.30 \times 10^{-3}$ rad/s^2. When the propeller picked up speed in segment one, we assigned positive values to the acceleration and subsequent velocity. Therefore, the deceleration or loss in speed here in segment three means that the acceleration has a negative value. Equation 8.8 $\left(\omega^2 = \omega_0^2 + 2\alpha\theta_3\right)$ can be used to find the angular displacement θ_3. Solving this equation for θ_3 gives

[Third segment]

$$\theta_3 = \frac{\omega^2 - \omega_0^2}{2\alpha} = \frac{(4.00\text{ rad/s})^2 - (6.09\text{ rad/s})^2}{2(-2.30\times10^{-3}\text{ rad/s}^2)} = 4.58\times10^3\text{ rad}$$

The total angular displacement, then, is

$$\theta_{\text{Total}} = \theta_1 + \theta_2 + \theta_3 = 6.39\times10^3\text{ rad} + 8.53\times10^3\text{ rad} + 4.58\times10^3\text{ rad} = \boxed{1.95\times10^4\text{ rad}}$$

---

26. ***REASONING*** According to Equation 3.5b, the time required for the diver to reach the water, assuming free-fall conditions, is $t=\sqrt{2y/a_y}$. If we assume that the "ball" formed by the diver is rotating at the instant that she begins falling vertically, we can use Equation 8.2 to calculate the number of revolutions made on the way down.

SOLUTION Taking upward as the positive direction, the time required for the diver to reach the water is

$$t=\sqrt{\frac{2(-8.3\text{ m})}{-9.80\text{ m/s}^2}}=1.3\text{ s}$$

Solving Equation 8.2 for $\Delta\theta$, we find

$$\Delta\theta=\bar{\omega}\,\Delta t=(1.6\text{ rev/s})(1.3\text{ s})=\boxed{2.1\text{ rev}}$$

---

27. **SSM** **WWW** ***REASONING*** The angular displacement of the child when he catches the horse is, from Equation 8.2, $\theta_c=\omega_c t$. In the same time, the angular displacement of the horse is, from Equation 8.7 with $\omega_0=0$ rad/s, $\theta_h=\frac{1}{2}\alpha t^2$. If the child is to catch the horse $\theta_c=\theta_h+(\pi/2)$.

SOLUTION Using the above conditions yields

$$\tfrac{1}{2}\alpha t^2-\omega_c t+\tfrac{\pi}{2}=0$$

or (suppressing the units)

$$\tfrac{1}{2}(0.0100)t^2-0.250t+\tfrac{\pi}{2}=0$$

The quadratic formula yields t = 7.37 s and 42.6 s; therefore, the shortest time needed to catch the horse is $\boxed{t=7.37\text{ s}}$.

---

28. ***REASONING AND SOLUTION***

a. According to Equation 8.9, the tangential speed of the sun is

$$v_T=r\omega=(2.2\times10^{20}\text{ m})(1.2\times10^{-15}\text{ rad/s})=\boxed{2.6\times10^5\text{ m/s}}$$

b. According to Equation 8.2, $\bar{\omega}=\Delta\theta/\Delta t$. Since the angular speed of the sun is constant, $\bar{\omega}=\omega$. Solving for Δt, we have

$$\Delta t = \frac{\Delta\theta}{\omega} = \left(\frac{2\pi \text{ rad}}{1.2\times10^{-15} \text{ rad/s}}\right)\left(\frac{1 \text{ h}}{3600 \text{ s}}\right)\left(\frac{1 \text{ day}}{24 \text{ h}}\right)\left(\frac{1 \text{ y}}{365.25 \text{ day}}\right) = \boxed{1.7\times10^{8} \text{ y}}$$

---

29. SSM ***REASONING AND SOLUTION*** Equation 8.9 gives the desired result.

$$v_{\text{T}} = r\omega = (2.00\times10^{-3} \text{ m})(7.85\times10^{4} \text{ rad/s}) = \boxed{157 \text{ m/s}}$$

---

30. ***REASONING*** The angular speed ω and tangential speed v_{T} are related by Equation 8.9 ($v_{\text{T}} = r\omega$), and this equation can be used to determine the radius r. However, we must remember that this relationship is only valid if we use radian measure. Therefore, it will be necessary to convert the given angular speed in rev/s into rad/s.

SOLUTION Solving Equation 8.9 for the radius gives

$$r = \frac{v_{\text{T}}}{\omega} = \frac{54 \text{ m/s}}{\underbrace{(47 \text{ rev/s})\left(\frac{2\pi \text{ rad}}{1 \text{ rev}}\right)}_{\text{Conversion from rev/s into rad/s}}} = \boxed{0.18 \text{ m}}$$

where we have used the fact that 1 rev corresponds to 2π rad to convert the given angular speed from rev/s into rad/s.

---

31. ***REASONING*** The angular speed ω of the sprocket can be calculated from the tangential speed v_{T} and the radius r using Equation 8.9 ($v_{\text{T}} = r\omega$). The radius is given as $r = 4.0\times10^{-2}$ m. The tangential speed is identical to the linear speed given for a chain link at point A, so that $v_{\text{T}} = 5.6$ m/s. We need to remember, however, that Equation 8.9 is only valid if radian measure is used. Thus, the value calculated for ω will be in rad/s, and we will have to convert to rev/s using the fact that 2π rad equals 1 rev.

SOLUTION Solving Equation 8.9 for the angular speed ω gives

$$\omega = \frac{v_{\text{T}}}{r} = \frac{5.6 \text{ m/s}}{4.0\times10^{-2} \text{ m}} = 140 \text{ rad/s}$$

Using the fact that 2π rad equals 1 rev, we can convert this result as follows:

$$\omega = (140 \text{ rad/s})\left(\frac{1 \text{ rev}}{2\pi \text{ rad}}\right) = \boxed{22 \text{ rev/s}}$$

---

32. ***REASONING AND SOLUTION***

a. In one lap, the car undergoes an angular displacement of 2π radians. Therefore, from the definition of average angular speed

$$\bar{\omega} = \frac{\Delta\theta}{\Delta t} = \frac{2\pi \text{ rad}}{18.9 \text{ s}} = \boxed{0.332 \text{ rad/s}}$$

b. Equation 8.9 relates the average angular speed $\bar{\omega}$ of an object moving in a circle to its average linear speed $\bar{v}_T$ tangent to the path of motion:

$$\bar{v}_T = r\bar{\omega}$$

Solving for r gives

$$r = \frac{\bar{v}_T}{\bar{\omega}} = \frac{42.6 \text{ m/s}}{0.332 \text{ rad/s}} = \boxed{128 \text{ m}}$$

Notice that the unit "rad," being dimensionless, does not appear in the final answer.

---

33. ***REASONING AND SOLUTION***

a. A person living in Ecuador makes one revolution (2π rad) every 23.9 hr (8.60×10^4 s). The angular speed of this person is $\omega = (2\pi \text{ rad})/(8.60 \times 10^4 \text{ s}) = 7.31 \times 10^{-5}$ rad/s. According to Equation 8.9, the tangential speed of the person is, therefore,

$$v_T = r\omega = \left(6.38 \times 10^6 \text{ m}\right)\left(7.31 \times 10^{-5} \text{ rad/s}\right) = \boxed{4.66 \times 10^2 \text{ m/s}}$$

b. The relevant geometry is shown in the drawing at the right. Since the tangential speed is one-third of that of a person living in Ecuador, we have,

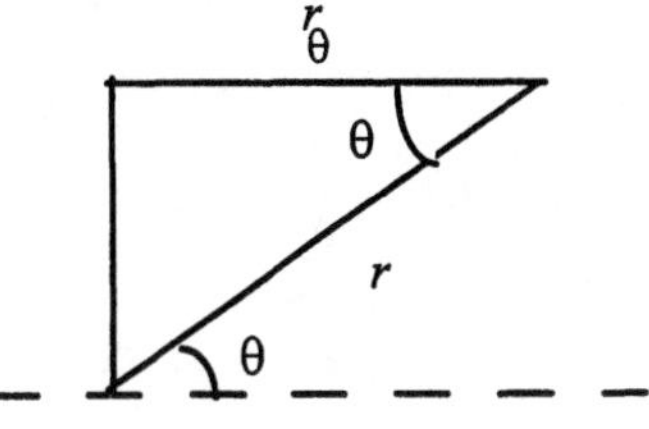

$$\frac{v_T}{3} = r_\theta \omega$$

or

$$r_\theta = \frac{v_T}{3\omega} = \frac{4.66 \times 10^2 \text{ m/s}}{3\left(7.31 \times 10^{-5} \text{ rad/s}\right)} = 2.12 \times 10^6 \text{ m}$$

The angle θ is, therefore,

$$\theta = \cos^{-1}\left(\frac{2.12 \times 10^6 \text{ m}}{6.38 \times 10^6 \text{ m}}\right) = \boxed{70.6°}$$

---

34. ***REASONING*** The tangential speed v_T of a point on the "equator" of the baseball is given by Equation 8.9 as $v_T = r\omega$, where r is the radius of the baseball and ω is its angular speed. The radius is given in the statement of the problem. The (constant) angular speed is related to that angle θ through which the ball rotates by Equation 8.2 as $\omega = \theta/t$, where we have assumed for convenience that $\theta_0 = 0$ rad when $t_0 = 0$ s. Thus, the tangential speed of the ball is

$$v_T = r\omega = r\left(\frac{\theta}{t}\right)$$

The time t that the ball is in the air is equal to the distance x it travels divided by its linear speed v, $t = x/v$, so the tangential speed can be written as

$$v_T = r\left(\frac{\theta}{t}\right) = r\left(\frac{\theta}{\frac{x}{v}}\right) = \frac{r\theta v}{x}$$

SOLUTION The tangential speed of a point on the equator of the baseball is

$$v_T = \frac{r\theta v}{x} = \frac{(3.67\times10^{-2}\text{ m})(49.0\text{ rad})(42.5\text{ m/s})}{16.5\text{ m}} = \boxed{4.63\text{ m/s}}$$

---

35. SSM ***REASONING AND SOLUTION***

a. From Equation 8.9, and the fact that 1 revolution = 2π radians, we obtain

$$v_T = r\omega = (0.0568\text{ m})\left(3.50\ \frac{\text{rev}}{\text{s}}\right)\left(\frac{2\pi\text{ rad}}{1\text{ rev}}\right) = \boxed{1.25\text{ m/s}}$$

b. Since the disk rotates at *constant* tangential speed,

$$v_{T1} = v_{T2} \qquad \text{or} \qquad \omega_1 r_1 = \omega_2 r_2$$

Solving for ω_2, we obtain

$$\omega_2 = \frac{\omega_1 r_1}{r_2} = \frac{(3.50\text{ rev/s})(0.0568\text{ m})}{0.0249\text{ m}} = \boxed{7.98\text{ rev/s}}$$

---

36. ***REASONING AND SOLUTION*** The figure below shows the initial and final states of the system.

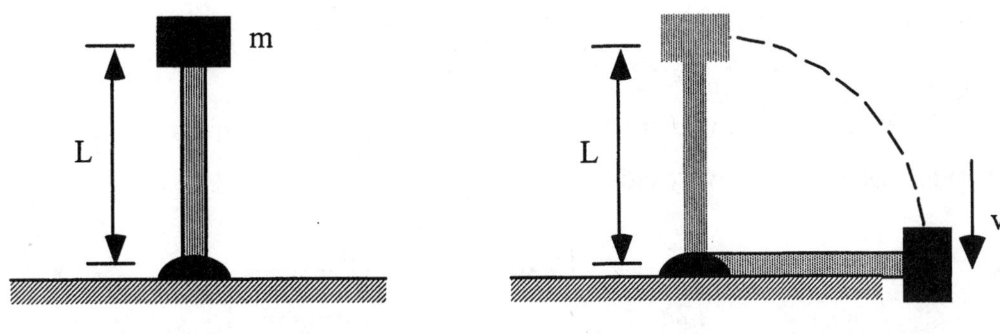

a. From the principle of conservation of mechanical energy:

$$E_0 = E_f$$

Initially the system has only gravitational potential energy. If the level of the hinge is chosen as the zero level for measuring heights, then: $E_0 = mgh_0 = mgL$. Just before the object hits the floor, the system has only kinetic energy.

Therefore

$$mgL = \frac{1}{2}mv^2$$

Solving for v gives

$$v = \sqrt{2gL}$$

From Equation 8.9, $v_T = r\omega$. Solving for ω gives $\omega = v_T/r$. As the object rotates downward, it travels in a circle of radius L. Its speed just before it strikes the floor is its tangential speed. Therefore,

$$\omega = \frac{v_T}{r} = \frac{v}{L} = \frac{\sqrt{2gL}}{L} = \sqrt{\frac{2g}{L}} = \sqrt{\frac{2(9.80\text{ m/s}^2)}{1.50\text{ m}}} = \boxed{3.61\text{ rad/s}}$$

b. From Equation 8.10:

$$a_T = r\alpha$$

Solving for α gives $\alpha = a_T/r$. Just before the object hits the floor, its tangential acceleration is the acceleration due to gravity. Thus,

$$\alpha = \frac{a_T}{r} = \frac{g}{L} = \frac{9.80\text{ m/s}^2}{1.50\text{ m}} = \boxed{6.53\text{ rad/s}^2}$$

37. ***REASONING AND SOLUTION*** The stone leaves the circular path with a horizontal speed

$$v_0 = v_T = r\omega$$

so $\omega = v_0/r$. We are given that $r = x/30$ so $\omega = 30v_0/x$. Kinematics gives $x = v_0 t$ and $h = (1/2)\,gt^2$. Using the above yields

$$\omega = 30\sqrt{\frac{g}{2h}} = 30\sqrt{\frac{9.80\text{ m/s}^2}{2(20.0\text{ m})}} = \boxed{14.8\text{ rad/s}}$$

---

38. ***REASONING AND SOLUTION*** The centripetal acceleration of any point on the fan blade a distance *r* from the rotation axis is given by Equation 8.11, $a_c = r\omega^2$.

Thus, for any given point a distance *r* from the rotation axis,

$$\frac{a_1}{a_2} = \frac{r\,\omega_1^2}{r\,\omega_2^2} = \left(\frac{\omega_1}{\omega_2}\right)^2 = \left(\frac{440\text{ rev/min}}{110\text{ rev/min}}\right)^2 = \boxed{16}$$

In general, Equation 8.11 is only valid if the angular displacement is expressed in radians. In this problem, however, we are dealing with the *ratio* of two values of ω, so it is not necessary to convert the angular speeds to rad/s. The conversions factors will cancel.

---

39. SSM ***REASONING*** Since the car is traveling with a constant speed, its tangential acceleration must be zero. The radial or centripetal acceleration of the car can be found from Equation 5.2. Since the tangential acceleration is zero, the total acceleration of the car is equal to its radial acceleration.

SOLUTION

a. Using Equation 5.2, we find that the car's radial acceleration, and therefore its total acceleration, is

$$a = a_R = \frac{v_T^2}{r} = \frac{(75.0\text{ m/s})^2}{625\text{ m}} = \boxed{9.00\text{ m/s}^2}$$

b. The direction of the car's total acceleration is the same as the direction of its radial acceleration. That is, the direction is $\boxed{\text{radially inward}}$.

---

40. ***REASONING***

a. According to Equation 8.2, the average angular speed is equal to the magnitude of the angular displacement divided by the elapsed time. The magnitude of the angular displacement is one revolution, or 2π rad. The elapsed time is one year, expressed in seconds.

b. The tangential speed of the earth in its orbit is equal to the product of its orbital radius and its orbital angular speed (Equation 8.9).

c. Since the earth is moving on a nearly circular orbit, it has a centripetal acceleration that is directed toward the center of the orbit. The magnitude a_c of the centripetal acceleration is given by Equation 8.11 as $a_c = r\omega^2$.

SOLUTION
a. The average angular speed is

$$\omega = \bar{\omega} = \frac{\Delta\theta}{\Delta t} = \frac{2\pi \text{ rad}}{3.16\times10^7 \text{ s}} = \boxed{1.99\times10^{-7} \text{ rad/s}} \qquad (8.2)$$

b. The tangential speed of the earth in its orbit is

$$v_T = r\omega = \left(1.50\times10^{11} \text{ m}\right)\left(1.99\times10^{-7} \text{ rad/s}\right) = \boxed{2.98\times10^4 \text{ m/s}} \qquad (8.9)$$

c. The centripetal acceleration of the earth due to its circular motion around the sun is

$$a_c = r\omega^2 = \left(1.50\times10^{11} \text{ m}\right)\left(1.99\times10^{-7} \text{ rad/s}\right)^2 = \boxed{5.94\times10^{-3} \text{ m/s}^2} \qquad (8.11)$$

$\boxed{\text{The acceleration is directed toward the center of the orbit.}}$

---

41. SSM ***REASONING*** The tangential acceleration $\mathbf{a}_T$ of the speedboat can be found by using Newton's second law, $\mathbf{F}_T = m\mathbf{a}_T$, where $\mathbf{F}_T$ is the net tangential force. Once the tangential acceleration of the boat is known, Equation 2.4 can be used to find the tangential speed of the boat 2.0 s into the turn. With the tangential speed and the radius of the turn known, Equation 5.2 can then be used to find the centripetal acceleration of the boat.

SOLUTION
a. From Newton's second law, we obtain

$$a_T = \frac{F_T}{m} = \frac{550 \text{ N}}{220 \text{ kg}} = \boxed{2.5 \text{ m/s}^2}$$

b. The tangential speed of the boat 2.0 s into the turn is, according to Equation 2.4,

$$v_T = v_{0T} + a_T t = 5.0 \text{ m/s} + (2.5 \text{ m/s}^2)(2.0 \text{ s}) = 1.0\times10^1 \text{ m/s}$$

The centripetal acceleration of the boat is then

$$a_c = \frac{v_T^2}{r} = \frac{(1.0 \times 10^1 \text{ m/s})^2}{32 \text{ m}} = \boxed{3.1 \text{ m/s}^2}$$

---

42. ***REASONING*** The drawing shows a top view of the race car as it travels around the circular turn. Its acceleration **a** has two perpendicular components: a centripetal acceleration $\mathbf{a}_c$ that arises because the car is moving on a circular path and a tangential acceleration $\mathbf{a}_T$ due to the fact that the car has an angular acceleration and its angular velocity is increasing. We can determine the magnitude of the centripetal acceleration from Equation 8.11 as $a_c = r\omega^2$, since both r and ω are given in the statement of the problem. As the drawing shows, we can use trigonometry to determine the magnitude a of the total acceleration, since the angle (35.0°) between **a** and $\mathbf{a}_c$ is given.

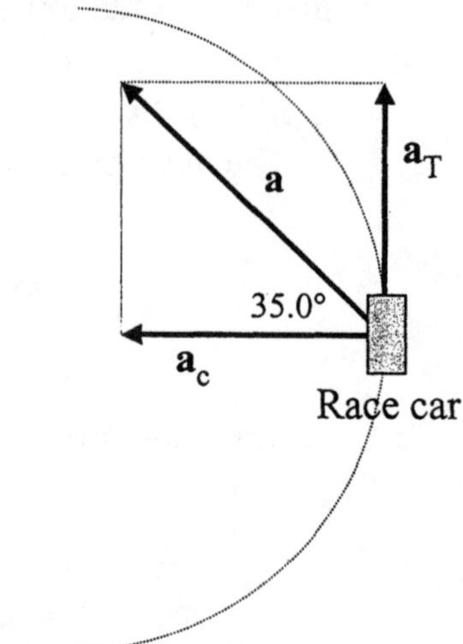

SOLUTION Since the vectors $\mathbf{a}_c$ and **a** are one side and the hypotenuse of a right triangle, we have that

$$a = \frac{a_c}{\cos 35.0°}$$

The magnitude of the centripetal acceleration is given by Equation 8.11 as $a_c = r\omega^2$, so the magnitude of the total acceleration is

$$a = \frac{a_c}{\cos 35.0°} = \frac{r\omega^2}{\cos 35.0°} = \frac{(23.5 \text{ m})(0.571 \text{ rad/s})^2}{\cos 35.0°} = \boxed{9.35 \text{ m/s}^2}$$

---

43. ***REASONING AND SOLUTION*** The tangential acceleration of point A is $a_{TA} = r_A\alpha$ and of point B is $a_{TB} = r_B\alpha$. Eliminating α gives

$$r_A/r_B = a_{TA}/a_{TB}$$

We are given that $a_{TA} = 2a_{TB}$, so $r_A/r_B = 2$.

From the figure in the text it is seen that

$$r_A = \sqrt{L_1^2 + L_2^2}$$

and

$$r_B = L_1$$

which give

$$(L_2/L_1)^2 = (r_A/r_B)^2 - 1$$

Then

$$\frac{L_1}{L_2} = \boxed{\frac{1}{\sqrt{3}}}$$

---

44. ***REASONING*** The centripetal acceleration experienced by the astronaut is given by Equation 8.11 $\left(a_c = r\omega^2\right)$, where the acceleration is $a_c = 4(9.80 \text{ m/s}^2)$ and the radius is $r = 3.0$ m. The angular displacement θ does not appear in Equation 8.11. However, the angular velocity ω does appear and can be related to θ by Equation 8.8 $\left(\omega^2 = \omega_0^2 + 2\alpha\theta\right)$, which is one of the equations of rotational kinematics. In this equation, the initial angular velocity is $\omega_0 = 0$ rad/s, since the centrifuge starts from rest, and the angular acceleration is $\alpha = 0.25 \text{ rad/s}^2$.

SOLUTION Substituting Equation 8.8 into Equation 8.11, we find that

$$a_c = r\omega^2 = r\left(\omega_0^2 + 2\alpha\theta\right)$$

Solving this result for θ gives

$$\theta = \frac{\left(a_c / r\right) - \omega_0^2}{2\alpha} = \frac{\left[4\left(9.80 \text{ m/s}^2\right)/\left(3.0 \text{ m}\right)\right] - \left(0 \text{ rad/s}\right)^2}{2\left(0.25 \text{ rad/s}^2\right)} = \boxed{26 \text{ rad}}$$

---

45. **SSM** ***REASONING*** The tangential acceleration and the centripetal acceleration of a point at a distance r from the rotation axis are given by Equations 8.10 and 8.11, respectively: $a_T = r\alpha$ and $a_c = r\omega^2$. After the drill has rotated through the angle in question, $a_c = 2a_T$, or

$$r\omega^2 = 2r\alpha$$

This expression can be used to find the angular acceleration α. Once the angular acceleration is known, Equation 8.8 can be used to find the desired angle.

SOLUTION Solving the expression obtained above for α gives

$$\alpha = \frac{\omega^2}{2}$$

Solving Equation 8.8 for θ (with $\omega_0 = 0$ rad/s since the drill starts from rest), and using the expression above for the angular acceleration α gives

$$\theta = \frac{\omega^2}{2\alpha} = \frac{\omega^2}{2(\omega^2/2)} = \left(\frac{\omega^2}{2}\right)\left(\frac{2}{\omega^2}\right) = \boxed{1.00 \text{ rad}}$$

Note that since both Equations 8.10 and 8.11 require that the angles be expressed in radians, the final result for θ is in radians.

---

46. ***REASONING AND SOLUTION***
a. If the wheel does not slip, a point on the rim rotates about the axle with a speed

$$v_T = v = 15.0 \text{ m/s}$$

For a point on the rim

$$\omega = v_T/r = (15.0 \text{ m/s})/(0.330 \text{ m}) = \boxed{45.5 \text{ rad/s}}$$

b.

$$v_T = r\omega = (0.175 \text{ m})(45.5 \text{ rad/s}) = \boxed{7.96 \text{ m/s}}$$

---

47. SSM WWW ***REASONING AND SOLUTION*** From Equation 2.4, the linear acceleration of the motorcycle is

$$a = \frac{v - v_0}{t} = \frac{22.0 \text{ m/s} - 0 \text{ m/s}}{9.00 \text{ s}} = 2.44 \text{ m/s}^2$$

Since the tire rolls without slipping, the linear acceleration equals the tangential acceleration of a point on the outer edge of the tire: $a = a_T$. Solving Equation 8.13 for α gives

$$\alpha = \frac{a_T}{r} = \frac{2.44 \text{ m/s}^2}{0.280 \text{ m}} = \boxed{8.71 \text{ rad/s}^2}$$

---

48. ***REASONING AND SOLUTION*** The bike would travel with the same speed as a point on the wheel $v = r\omega$. It would then travel a distance

$$x = vt = r\omega t = (0.45 \text{ m})(9.1 \text{ rad/s})(35 \text{ min})\left(\frac{60 \text{ s}}{1 \text{ min}}\right) = \boxed{8.6 \times 10^3 \text{ m}}$$

---

49. ***REASONING*** The angular displacement θ of each wheel is given by Equation 8.7 $\left(\theta = \omega_0 t + \frac{1}{2}\alpha t^2\right)$, which is one of the equations of rotational kinematics. In this expression ω_0 is the initial angular velocity, and α is the angular acceleration, neither of which is given directly. Instead the initial linear velocity v_0 and the linear acceleration a are given. However, we can relate these linear quantities to their analogous angular counterparts by means of the assumption that the wheels are rolling and not slipping. Then, according to Equation 8.12 ($v_0 = r\omega_0$), we know that $\omega_0 = v_0/r$, where r is the radius of the wheels. Likewise, according to Equation 8.13 ($a = r\alpha$), we know that $\alpha = a/r$. Both Equations 8.12 and 8.13 are only valid if used with radian measure. Therefore, when we substitute the expressions for ω_0 and α into Equation 8.7, the resulting value for the angular displacement θ will be in radians.

SOLUTION Substituting ω_0 from Equation 8.12 and α from Equation 8.13 into Equation 8.7, we find that

$$\theta = \omega_0 t + \tfrac{1}{2}\alpha t^2 = \left(\frac{v_0}{r}\right)t + \frac{1}{2}\left(\frac{a}{r}\right)t^2$$

$$= \left(\frac{20.0 \text{ m/s}}{0.300 \text{ m}}\right)(8.00 \text{ s}) + \tfrac{1}{2}\left(\frac{1.50 \text{ m/s}^2}{0.300 \text{ m}}\right)(8.00 \text{ s})^2 = \boxed{693 \text{ rad}}$$

---

50. ***REASONING*** The angle through which the tire rotates is equal to its average angular velocity $\bar{\omega}$ multiplied by the elapsed time t, $\theta = \bar{\omega} t$. According to Equation 8.6, this angle is related to the initial and final angular velocities of the tire by

$$\theta = \bar{\omega} t = \tfrac{1}{2}(\omega_0 + \omega)t$$

The tire is assumed to roll at a constant angular velocity, so that $\omega_0 = \omega$ and $\theta = \omega t$. Since the tire is rolling, its angular speed is related to its linear speed v by Equation 8.12, $v = r\omega$, where r is the radius of the tire. The angle of rotation then becomes

$$\theta = \omega t = \left(\frac{v}{r}\right)t$$

The time t that it takes for the tire to travel a distance x is equal to $t = x/v$, according to Equation 2.1. Thus, the angle that the tire rotates through is

$$\theta = \left(\frac{v}{r}\right)t = \left(\frac{v}{r}\right)\left(\frac{x}{v}\right) = \frac{x}{r}$$

SOLUTION Since 1 rev = 2π rad, the angle (in revolutions) is

$$\theta = \frac{x}{r} = \frac{96\,000 \times 10^3 \text{ m}}{0.31 \text{ m}} = 3.1 \times 10^8 \text{ rad}$$

$$\theta = \left(3.1 \times 10^8 \text{ rad}\right)\left(\frac{1 \text{ rev}}{2\pi \text{ rad}}\right) = \boxed{4.9 \times 10^7 \text{ rev}}$$

---

51. SSM WWW ***REASONING*** Assuming that the belt does not slip on the platter or the shaft pulley, the tangential speed of points on the platter and shaft pulley must be equal; therefore,

$$r_S \omega_S = r_P \omega_P$$

SOLUTION Solving the above expression for ω_S gives

$$\omega_S = \frac{r_P \omega_P}{r_S} = \frac{(3.49 \text{ rad/s})(0.102 \text{ m})}{1.27 \times 10^{-2} \text{ m}} = \boxed{28.0 \text{ rad/s}}$$

---

52. ***REASONING*** As a penny-farthing moves, both of its wheels roll without slipping. This means that the axle for each wheel moves through a linear distance (the distance through which the bicycle moves) that equals the circular arc length measured along the outer edge of the wheel. Since both axles move through the same linear distance, the circular arc length measured along the outer edge of the large front wheel must equal the circular arc length measured along the outer edge of the small rear wheel. In each case the arc length s is equal to the number n of revolutions times the circumference $2\pi r$ of the wheel (r = radius).

SOLUTION Since the circular arc length measured along the outer edge of the large front wheel must equal the circular arc length measured along the outer edge of the small rear wheel, we have

$$\underbrace{n_{\text{Rear}} 2\pi r_{\text{Rear}}}_{\text{Arc length for rear wheel}} = \underbrace{n_{\text{Front}} 2\pi r_{\text{Front}}}_{\text{Arc length for front wheel}}$$

Solving for n_{Rear} gives

$$n_{\text{Rear}} = \frac{n_{\text{Front}} r_{\text{Front}}}{r_{\text{Rear}}} = \frac{276(1.20 \text{ m})}{0.340 \text{ m}} = \boxed{974 \text{ rev}}$$

---

53. ***REASONING AND SOLUTION*** The angular speed of the ball can be found from Equation 8.12, $v = r\omega$. Solving for ω gives

$$\omega = \frac{v}{r} = \frac{3.60\ \text{m/s}}{0.200\ \text{m}} = 18.0\ \text{rad/s}$$

The time it takes for the ball to fall through the vertical distance y can be found from Equation 3.5b:

$$y = v_{0y}t + \frac{1}{2}a_y t^2$$

Solving for t (with $v_{0y} = 0$ m/s, and taking "up" to be positive) gives

$$t = \sqrt{\frac{2y}{a_y}} = \sqrt{\frac{2(-2.10\ \text{m})}{(-9.80\ \text{m/s}^2)}} = 0.655\ \text{s}$$

Since the angular speed of the ball remains constant while it is falling,

$$\theta = \omega t = (18.0\ \text{rad/s})(0.655\ \text{s}) = \boxed{11.8\ \text{rad}}$$

---

54. ***REASONING AND SOLUTION***
a. If the rope is not slipping on the cylinder, then the tangential speed of the teeth on the larger gear (gear 1) is 2.50 m/s. The angular speed of gear 1 is then

$$\omega_1 = v/r_1 = (2.50\ \text{m/s})/(0.300\ \text{m}) = \boxed{8.33\ \text{rad/s}}$$

The direction of the larger gear is $\boxed{\text{counterclockwise}}$.

b. The gears are in contact and do not slip. This requires that the teeth on both gears move with the same tangential speed.

$$v_{T1} = v_{T2}$$

or

$$\omega_1 r_1 = \omega_2 r_2$$

So

$$\omega_2 = \left(\frac{r_1}{r_2}\right)\omega_1 = \left(\frac{0.300\ \text{m}}{0.170\ \text{m}}\right)(8.33\ \text{rad/s}) = \boxed{14.7\ \text{rad/s}}$$

The direction of the smaller gear is $\boxed{\text{clockwise}}$.

---

55. ***REASONING AND SOLUTION*** The distance d traveled by the axle of the wheel is given by Equation 8.1 as $d = r\,\theta$, where r is the distance from the center of the circular hill to the axle (r = 9.00 m – 0.400 m = 8.60 m) and θ = 0.960 rad. Thus,

$$d = (8.60 \text{ m})(0.960 \text{ rad}) = 8.26 \text{ m}$$

According to the discussion in Section 8.6, the distance d traveled by the axle equals the circular arc length s along the outer edge of the rotating wheel; $d = s$. But $s = r_w \theta_w$, where r_w is the radius of the wheel and θ_w is the angle through which the wheel rotates; so, $d = s = r_w \theta_w$. Solving for θ_w,

$$\theta_w = \frac{d}{r_w} = \frac{8.26 \text{ m}}{0.400 \text{ m}} = \boxed{20.6 \text{ rad}}$$

---

56. ***REASONING AND SOLUTION***

a. $\omega_A = v/r = (0.381 \text{ m/s})/(0.0508 \text{ m}) = \boxed{7.50 \text{ rad/s}}$

b. $\omega_B = v/r = (0.381 \text{ m/s})(0.114 \text{ m}) = 3.34 \text{ rad/s}$

$$\bar{\alpha} = \frac{\omega_B - \omega_A}{t} = \frac{3.34 \text{ rad/s} - 7.50 \text{ rad/s}}{2.40 \times 10^3 \text{ s}} = \boxed{-1.73 \times 10^{-3} \text{ rad/s}^2}$$

The angular velocity is $\boxed{\text{decreasing}}$.

---

57. SSM ***REASONING*** The top of the racket has both tangential and centripetal acceleration components given by Equations 8.10 and 8.11, respectively: $a_T = r\alpha$ and $a_c = r\omega^2$. The total acceleration of the top of the racket is the resultant of these two components. Since these acceleration components are mutually perpendicular, their resultant can be found by using the Pythagorean theorem.

SOLUTION Employing the Pythagorean theorem, we obtain

$$a = \sqrt{a_T^2 + a_c^2} = \sqrt{(r\alpha)^2 + (r\omega^2)^2} = r\sqrt{\alpha^2 + \omega^4}$$

Therefore,

$$a = (1.5 \text{ m})\sqrt{(160 \text{ rad/s}^2)^2 + (14 \text{ rad/s})^4} = \boxed{380 \text{ m/s}^2}$$

---

58. ***REASONING AND SOLUTION***

a. $\omega = \omega_0 + \alpha t = 0 \text{ rad/s} + (3.00 \text{ rad/s}^2)(18.0 \text{ s}) = \boxed{54.0 \text{ rad/s}}$

b. $\theta = (1/2)(\omega_0 + \omega)t = (1/2)(0 \text{ rad/s} + 54.0 \text{ rad/s})(18.0 \text{ s}) = \boxed{486 \text{ rad}}$

---

59. SSM ***REASONING AND SOLUTION***

a. From Equation 8.7 we obtain

$$\theta = \omega_0 t + \tfrac{1}{2}\alpha t^2 = (5.00 \text{ rad/s})(4.00 \text{ s}) + \tfrac{1}{2}(2.50 \text{ rad/s}^2)(4.00 \text{ s})^2 = \boxed{4.00 \times 10^1 \text{ rad}}$$

b. From Equation 8.4, we obtain

$$\omega = \omega_0 + \alpha t = 5.00 \text{ rad/s} + (2.50 \text{ rad/s}^2)(4.00 \text{ s}) = \boxed{15.0 \text{ rad/s}}$$

---

60. ***REASONING AND SOLUTION***

a. The tangential acceleration of the train is given by Equation 8.10 as

$$a_T = r\alpha = (2.00 \times 10^2 \text{ m})(1.50 \times 10^{-3} \text{ rad/s}^2) = 0.300 \text{ m/s}^2$$

The centripetal acceleration of the train is given by Equation 8.11 as

$$a_c = r\omega^2 = (2.00 \times 10^2 \text{ m})(0.0500 \text{ rad/s})^2 = 0.500 \text{ m/s}^2$$

The magnitude of the total acceleration is found from the Pythagorean theorem to be

$$a = \sqrt{a_T^2 + a_c^2} = \boxed{0.583 \text{ m/s}^2}$$

b. The total acceleration vector makes an angle relative to the radial acceleration of

$$\theta = \tan^{-1}\left(\frac{a_T}{a_c}\right) = \tan^{-1}\left(\frac{0.300 \text{ m/s}^2}{0.500 \text{ m/s}^2}\right) = \boxed{31.0^\circ}$$

---

61. ***REASONING*** The length of tape that passes around the reel is just the average tangential speed of the tape times the time t. The average tangential speed $\bar{v}_T$ is given by Equation 8.9 $(\bar{v}_T = r\bar{\omega})$ as the radius r times the average angular speed $\bar{\omega}$ in rad/s.

SOLUTION The length L of tape that passes around the reel in t = 13 s is $L = \bar{v}_T t$. Using Equation 8.9 to express the tangential speed, we find

$$L = \bar{v}_T t = r\bar{\omega} t = (0.014 \text{ m})(3.4 \text{ rad/s})(13 \text{ s}) = \boxed{0.62 \text{ m}}$$

---

62. ***REASONING***

a. The tangential speed v_T of the sun as it orbits about the center of the Milky Way is related to the orbital radius r and angular speed ω by Equation 8.9, $v_T = r\omega$. Before we use this relation, however, we must first convert r to meters from light-years.

b. The centripetal force is the net force required to keep an object, such as the sun, moving on a circular path. According to Newton's second law of motion, the magnitude F_c of the centripetal force is equal to the product of the object's mass m and the magnitude a_c of its centripetal acceleration (see Section 5.3): $F_c = ma_c$. The magnitude of the centripetal acceleration is expressed by Equation 8.11 as $a_c = r\omega^2$, where r is the radius of the circular path and ω is the angular speed of the object.

SOLUTION

a. The radius of the sun's orbit about the center of the Milky Way is

$$r = \left(2.3\times10^{4}\text{ light-years}\right)\left(\frac{9.5\times10^{15}\text{ m}}{1\text{ light-year}}\right) = 2.2\times10^{20}\text{ m}$$

The tangential speed of the sun is

$$v_T = r\omega = \left(2.2\times10^{20}\text{ m}\right)\left(1.1\times10^{-15}\text{ rad/s}\right) = \boxed{2.4\times10^{5}\text{ m/s}} \qquad (8.9)$$

b. The magnitude of the centripetal force that acts on the sun is

$$\underbrace{F_c}_{\substack{\text{Centripetal}\\\text{force}}} = ma_c = mr\omega^2$$

$$= \left(1.99\times10^{30}\text{ kg}\right)\left(2.2\times10^{20}\text{ m}\right)\left(1.1\times10^{-15}\text{ rad/s}\right)^2 = \boxed{5.3\times10^{20}\text{ N}}$$

---

63. SSM ***REASONING AND SOLUTION*** According to Equation 8.9, the crank handle has an angular speed of

$$\omega = \frac{v_T}{r} = \frac{1.20\text{ m/s}}{\frac{1}{2}(0.400\text{ m})} = 6.00\text{ rad/s}$$

The crank barrel must have the same angular speed as the handle; therefore, the tangential speed of a point on the barrel is, again using Equation 8.9,

$$v_T = r\omega = (5.00\times10^{-2}\text{ m})(6.00\text{ rad/s}) = 0.300\text{ m/s}$$

If we assume that the rope does not slip, then all points on the rope must move with the same speed as the tangential speed of any point on the barrel. Therefore the linear speed with which the bucket moves down the well is $\boxed{v = 0.300\text{ m/s}}$.

---

64. ***REASONING AND SOLUTION*** The people meet at time t. At this time the magnitudes of their angular displacements must total 2π rad.

$$\theta_1 + \theta_2 = 2\pi \text{ rad}$$

Then

$$\omega_1 t + \omega_2 t = 2\pi \text{ rad}$$

$$t = \frac{2\pi \text{ rad}}{\omega_1 + \omega_2} = \frac{2\pi \text{ rad}}{1.7\times10^{-3}\text{ rad/s} + 3.4\times10^{-3}\text{ rad/s}} = \boxed{1200\text{ s}}$$

---

65. ***REASONING AND SOLUTION*** The time that it takes the fan to rotate from rest to the beginning of the 11 s period is

$$t_1 = \frac{\omega_1}{\alpha}$$

Kinematics applied to the 11.0-s period gives

$$\omega_1 = \frac{\theta}{t_2} - \tfrac{1}{2}\alpha t_2 = \frac{285\text{ rad}}{11.0\text{ s}} - \tfrac{1}{2}\left(2.00\text{ rad/s}^2\right)\left(11.0\text{ s}\right) = 14.9\text{ rad/s}$$

Then

$$t_1 = (14.9\text{ rad/s})/(2.00\text{ rad/s}^2) = \boxed{7.45\text{ s}}$$

---

66. ***REASONING AND SOLUTION***

a. The linear speed of all points on either sprocket, as well as the linear speed of all points on the chain, must be equal (otherwise the chain would bunch up or break). Therefore, according to Equation 8.9, the linear speed of the chain as it moves between the sprockets is

$$v_T = r\omega = (9.00\text{ cm})(9.40\text{ rad/s}) = \boxed{84.6\text{ cm/s}}$$

b. The centripetal acceleration of the chain as it passes around the rear sprocket is, according to Equation 8.11, $a_c = r_{\text{rear}}\omega^2$, where $\omega = v_T / r_{\text{rear}}$. Therefore,

$$a_c = r_{\text{rear}}\left(\frac{v_T}{r_{\text{rear}}}\right)^2 = \frac{v_T^2}{r_{\text{rear}}} = \frac{(84.6\text{ cm/s})^2}{5.10\text{ cm}} = \boxed{1.40\times10^3\text{ cm/s}^2}$$

---

67. SSM ***REASONING AND SOLUTION*** By inspection, the distance traveled by the "axle" or the center of the moving quarter is

$$d = 2\pi(2r) = 4\pi r$$

where r is the radius of the quarter. The distance d traveled by the "axle" of the moving quarter must be equal to the circular arc length s along the outer edge of the quarter. This arc length is $s = r\theta$, where θ is the angle through which the quarter rotates. Thus,

$$4\pi r = r\theta$$

so that $\theta = 4\pi$ rad. This is equivalent to

$$(4\pi \text{ rad})\left(\frac{1 \text{ rev}}{2\pi \text{ rad}}\right) = \boxed{2 \text{ revolutions}}$$

---

68. ***REASONING*** The golf ball must travel a distance equal to its diameter in a maximum time equal to the time required for one blade to move into the position of the previous blade.

SOLUTION The time required for the golf ball to pass through the opening between two blades is given by $\Delta t = \Delta\theta/\omega$, with $\omega = 1.25$ rad/s and $\Delta\theta = (2\pi \text{ rad})/16 = 0.393$ rad. Therefore, the ball must pass between two blades in a maximum time of

$$\Delta t = \frac{0.393 \text{ rad}}{1.25 \text{ rad/s}} = 0.314 \text{ s}$$

The minimum speed of the ball is

$$v = \frac{\Delta x}{\Delta t} = \frac{4.50 \times 10^{-2} \text{ m}}{0.314 \text{ s}} = \boxed{1.43 \times 10^{-1} \text{ m/s}}$$

---

69. ***CONCEPT QUESTION*** The average angular velocity $\bar{\omega}$ has the same direction as $\theta - \theta_0$, because $\bar{\omega} = \dfrac{\theta - \theta_0}{t - t_0}$ according to Equation 8.2. If θ is greater than θ_0, then $\bar{\omega}$ is positive. If θ is less than θ_0, then $\bar{\omega}$ is negative.

(a) $\bar{\omega}$ is positive, because 0.75 rad is greater than 0.45 rad.

(b) $\bar{\omega}$ is negative, because 0.54 rad is less than 0.94 rad.

(c) $\bar{\omega}$ is negative, because 4.2 rad is less than 5.4 rad.

(d) $\bar{\omega}$ is positive, because 3.8 rad is greater than 3.0 rad.

SOLUTION The average angular velocity is given by Equation 8.2 as $\bar{\omega} = \dfrac{\theta - \theta_0}{t - t_0}$, where $t - t_0 = 2.0$ s is the elapsed time:

(a) $$\bar{\omega} = \frac{\theta - \theta_0}{t - t_0} = \frac{0.75\text{ rad} - 0.45\text{ rad}}{2.0\text{ s}} = \boxed{+0.15\text{ rad/s}}$$

(b) $$\bar{\omega} = \frac{\theta - \theta_0}{t - t_0} = \frac{0.54\text{ rad} - 0.94\text{ rad}}{2.0\text{ s}} = \boxed{-0.20\text{ rad/s}}$$

(c) $$\bar{\omega} = \frac{\theta - \theta_0}{t - t_0} = \frac{4.2\text{ rad} - 5.4\text{ rad}}{2.0\text{ s}} = \boxed{-0.60\text{ rad/s}}$$

(d) $$\bar{\omega} = \frac{\theta - \theta_0}{t - t_0} = \frac{3.8\text{ rad} - 3.0\text{ rad}}{2.0\text{ s}} = \boxed{+0.4\text{ rad/s}}$$

---

70. ***CONCEPT QUESTION*** The average angular acceleration has the same direction as $\omega - \omega_0$, because $\bar{\alpha} = \dfrac{\omega - \omega_0}{t - t_0}$, according to Equation 8.4. If ω is greater than ω_0, $\bar{\alpha}$ is positive. If ω is less than ω_0, $\bar{\alpha}$ is negative.

(a) $\bar{\alpha}$ is positive, because 5.0 rad/s is greater than 2.0 rad/s.

(b) $\bar{\alpha}$ is negative, because 2.0 rad/s is less than 5.0 rad/s.

(c) $\bar{\alpha}$ is positive, because −3.0 rad/s is greater than −7.0 rad/s.

(d) $\bar{\alpha}$ is negative, because −4.0 rad/s is less than +4.0 rad/s.

SOLUTION The average angular acceleration is given by Equation 8.4 as $\bar{\alpha} = \dfrac{\omega - \omega_0}{t - t_0}$, where $t - t_0 = 4.0$ s is the elapsed time.

(a) $$\bar{\alpha} = \frac{\omega - \omega_0}{t - t_0} = \frac{+5.0\text{ rad/s} - 2.0\text{ rad/s}}{4.0\text{ s}} = \boxed{+0.75\text{ rad/s}^2}$$

(b) $$\bar{\alpha} = \frac{\omega - \omega_0}{t - t_0} = \frac{+2.0\text{ rad/s} - 5.0\text{ rad/s}}{4.0\text{ s}} = \boxed{-0.75\text{ rad/s}^2}$$

(c) $$\bar{\alpha} = \frac{\omega - \omega_0}{t - t_0} = \frac{-3.0\text{ rad/s} - (-7.0\text{ rad/s})}{4.0\text{ s}} = \boxed{+1.0\text{ rad/s}^2}$$

(d) $$\bar{\alpha} = \frac{\omega - \omega_0}{t - t_0} = \frac{-4.0\text{ rad/s} - (+4.0\text{ rad/s})}{4.0\text{ s}} = \boxed{-2.0\text{ rad/s}^2}$$

---

71. ***CONCEPT QUESTION*** The relation between the final angular velocity ω, the initial angular velocity ω_0, and the angular acceleration α is given by Equation 8.4 (with $t_0 = 0$ s) as

$$\omega = \omega_0 + \alpha t$$

If α has the same sign as ω_0, then the angular speed, which is the magnitude of the angular velocity ω, is increasing. On the other hand, If α and ω_0 have opposite signs, then the angular speed is decreasing.

(a) ω_0 and α have the same sign, so the angular speed is increasing.

(b) ω_0 and α have opposite signs, so the angular speed is decreasing.

(c) ω_0 and α have opposite signs, so the angular speed is decreasing.

(d) ω_0 and α have the same sign, so the angular speed is increasing.

SOLUTION According to Equation 8.4, we know that $\omega = \omega_0 + \alpha t$. Therefore, we find:

(a) $\omega = +12 \text{ rad/s} + \left(+3.0 \text{ rad/s}^2\right)(2.0 \text{ s}) = +18 \text{ rad/s}$. The angular speed is $\boxed{18 \text{ rad/s}}$.

(b) $\omega = +12 \text{ rad/s} + \left(-3.0 \text{ rad/s}^2\right)(2.0 \text{ s}) = +6 \text{ rad/s}$. The angular speed is $\boxed{6 \text{ rad/s}}$.

(c) $\omega = -12 \text{ rad/s} + \left(+3.0 \text{ rad/s}^2\right)(2.0 \text{ s}) = -6 \text{ rad/s}$. The angular speed is $\boxed{6 \text{ rad/s}}$.

(d) $\omega = -12 \text{ rad/s} + \left(-3.0 \text{ rad/s}^2\right)(2.0 \text{ s}) = -18 \text{ rad/s}$. The angular speed is $\boxed{18 \text{ rad/s}}$.

---

72. ***CONCEPT QUESTIONS***

a. No. The angular velocity does not change, so the angular acceleration α is zero. The tangential acceleration a_T is given by Equation 8.10 as $a_T = r\alpha$, so it is also zero.

b. Yes, because the nonzero angular acceleration α gives rise to a nonzero tangential acceleration a_T according to Equation 8.10, $a_T = r\alpha$.

SOLUTION Let r be the radial distance of the point from the axis of rotation. Then, according to Equation 8.10, we have

$$\underbrace{g}_{a_T} = r\alpha$$

Thus,

$$r = \frac{g}{\alpha} = \frac{9.80\ \text{m/s}^2}{12.0\ \text{rad/s}^2} = \boxed{0.817\ \text{m}}$$

---

73. ***CONCEPT QUESTIONS***

a. The angular displacement is greater than $\omega_0 t$. When the angular displacement θ is given by the expression $\theta = \omega_0 t$, it is assumed that the angular velocity remains constant at its initial (and smallest) value of ω_0 for the entire time. This expression does not account for the additional angular displacement that occurs because the angular velocity is increasing.

b. The angular displacement is less than ωt. When the angular displacement is given by the expression $\theta = \omega t$, it is assumed that the angular velocity remains constant at its final (and largest) value of ω for the entire time. This expression does not account for the fact that the wheel was rotating at a smaller angular speed during the time interval. Thus, this expression "overestimates" the angular displacement.

c. The angular displacement is given as the product of the average angular velocity and the time

$$\theta = \bar{\omega} t = \underbrace{\tfrac{1}{2}(\omega_0 + \omega)}_{\text{Average angular velocity}}\ t$$

SOLUTION

a. If the angular velocity is constant and equals the initial angular velocity ω_0, then $\bar{\omega} = \omega_0$ and the angular displacement is

$$\theta = \omega_0 t = (+220\ \text{rad/s})(10.0\ \text{s}) = \boxed{+2200\ \text{rad}}$$

b. If the angular velocity is constant and equals the final angular velocity ω, then $\bar{\omega} = \omega$ and the angular displacement is

$$\theta = \omega t = (+280\ \text{rad/s})(10.0\ \text{s}) = \boxed{+2800\ \text{rad}}$$

c. Using the definition of average angular velocity, we have

$$\theta = \tfrac{1}{2}(\omega_0 + \omega)t = \tfrac{1}{2}(+220\ \text{rad/s} + 280\ \text{rad/s})(10.0\ \text{s}) = \boxed{+2500\ \text{rad}} \qquad (8.6)$$

---

74. ***CONCEPT QUESTIONS***

a. The wheels on the winning dragster roll without slipping. The wheels on the losing dragster slip during part of the time. During the slippage, the wheels rotate, but the speed of the car does not increase. Thus, its speed at any given instant is less than that of the winning car.

b. For the wheels that roll without slipping, the relationship between their linear speed v and the angular speed ω is given by Equation 8.12 as $v = r\omega$, where r is the radius of a wheel.

c. For the wheels that roll without slipping, the relationship between the magnitude a of their linear acceleration and the magnitude α of the angular acceleration is given by Equation 8.13 as $a = r\alpha$, where r is the radius of a wheel.

SOLUTION

a. From Equation 8.12 we have that

$$v = r\omega = (0.320\text{ m})(288\text{ rad/s}) = \boxed{92.2\text{ m/s}}$$

b. The magnitude of the angular acceleration is given by Equation 8.13 as $\alpha = a/r$. The linear acceleration a is related to the initial and final linear speeds and the displacement x by Equation 2.9 from the equations of kinematics for linear motion; $a = \dfrac{v^2 - v_0^2}{2x}$. Thus, the magnitude of the angular acceleration is

$$\alpha = \frac{a}{r} = \frac{\dfrac{v^2 - v_0^2}{2x}}{r}$$

$$= \frac{v^2 - v_0^2}{2xr} = \frac{(92.2\text{ m/s})^2 - (0\text{ m/s})^2}{2(384\text{ m})(0.320\text{ m})} = \boxed{34.6\text{ rad/s}^2}$$

---

75. ***CONCEPT QUESTIONS***

a. It does not matter whether the arrow is aimed closer to or farther away from the axis. It may be thought that it is better to aim farther away from the axis, since the space between the edges of the blades is greater there, giving the arrow more time to pass through before the propeller strikes it. However, this is not true, because the blade edge sweeps through the open angular space as a rigid unit. This means that a point closer to the axis has a smaller distance to travel along the circular in order to bridge the angular opening and correspondingly has a smaller tangential speed. A point farther from the axis has a greater distance to travel along the circular arc but correspondingly has a greater tangential speed. These speeds have just the right values so that all points on the blade edge bridge the angular opening at the same instant.

b. The rotational speed of the blades must not be so fast that one blade rotates into the open angular space while part of the arrow is still there. A faster arrow speed means that the arrow spends less time in the open space. Thus, the blades can rotate more quickly into the open space without hitting the arrow, so the maximum value of the angular speed ω increases with increasing arrow speed v.

c. A longer arrow traveling at a given speed means that some part of the arrow is in the open space for a longer time. To avoid hitting the arrow, then, the blades must rotate more slowly. Thus, the maximum value of the angular speed ω decreases with increasing arrow length L.

SOLUTION The time during which some part of the arrow remains in the open angular space is the time it takes the arrow to travel at a speed v through a distance equal to its own length L. This time is $t_{\text{Arrow}} = L/v$. The time it takes for the edge to rotate at an angular speed ω through the angle θ between the blades is $t_{\text{Blade}} = \theta/\omega$. The maximum angular speed is the angular speed such that these two times are equal. Therefore, we have

$$\underbrace{\frac{L}{v}}_{\text{Arrow}} = \underbrace{\frac{\theta}{\omega}}_{\text{Blade}}$$

In this expression we note that the value of the angular opening is $\theta = 60.0°$, which is $\theta = \frac{1}{6}(2\pi)\ \text{rad} = \frac{1}{3}\pi\ \text{rad}$. Solving the expression for ω gives

$$\omega = \frac{\theta v}{L} = \frac{\pi v}{3L}$$

Substituting the given values for v and L into this result, we find that

a. $$\omega = \frac{\pi v}{3L} = \frac{\pi(75.0\ \text{m/s})}{3(0.71\ \text{m})} = \boxed{111\ \text{rad/s}}$$

b. $$\omega = \frac{\pi v}{3L} = \frac{\pi(91.0\ \text{m/s})}{3(0.71\ \text{m})} = \boxed{134\ \text{rad/s}}$$

c. $$\omega = \frac{\pi v}{3L} = \frac{\pi(91.0\ \text{m/s})}{3(0.81\ \text{m})} = \boxed{118\ \text{rad/s}}$$

These answers are consistent with the answers to the Concept Questions.

---

76. ***CONCEPT QUESTIONS***

a. In addition to knowing the initial angular velocity ω_0 and the acceleration α, we know that final angular velocity ω is 0 rev/s, because the wheel comes to a halt. With values available for these three variables, the unknown angular displacement θ can be calculated from Equation 8.8 $\left(\omega^2 = \omega_0^2 + 2\alpha\theta\right)$.

b. When using any of the equations of rotational kinematics, it is not necessary to use radian measure. Any self-consistent set of units may be used to measure the angular quantities, such as revolutions for θ, rev/s for ω_0 and ω, and rev/s^2 for α.

c. A greater initial angular velocity does not necessarily mean that the wheel will come to a halt on an angular section labeled with a greater number. It is certainly true that greater initial angular velocities lead to greater angular displacements for a given deceleration. However, remember that the angular displacement of the wheel in coming to a halt may consist of a number of complete revolutions plus a fraction of a revolution. In deciding on which number the wheel comes to a halt, the number of complete revolutions must be subtracted from the angular displacement, leaving only the fraction of a revolution remaining.

SOLUTION Solving Equation 8.8 for the angular displacement gives

$$\theta = \frac{\omega^2 - \omega_0^2}{2\alpha}$$

a. We know that $\omega_0 = 1.20$ rev/s, $\omega = 0$ rev/s, and $\alpha = -0.200$ rev/s^2, where ω_0 is positive since the rotation is counterclockwise and, therefore, α is negative because the wheel decelerates. The value obtained for the displacement is

$$\theta = \frac{\omega^2 - \omega_0^2}{2\alpha} = \frac{(0\text{ rev/s})^2 - (1.20\text{ rev/s})^2}{2\left(-0.200\text{ rev/s}^2\right)} = 3.60\text{ rev}$$

To decide where the wheel comes to a halt, we subtract the three complete revolutions from this result, leaving 0.60 rev. Converting this value into degrees and noting that each angular section is 30.0°, we find the following number n for the section where the wheel comes to a halt:

$$n = (0.60\text{ rev})\left(\frac{360°}{1\text{ rev}}\right)\left(\frac{1\text{ angular section}}{30.0°}\right) = 7.2$$

A value of $n = 7.2$ means that the wheel comes to a halt in the section following number 7. Thus, it comes to a halt in $\boxed{\text{section 8}}$.

b. Following the same procedure as in part a, we find that

$$\theta = \frac{\omega^2 - \omega_0^2}{2\alpha} = \frac{(0 \text{ rev/s})^2 - (1.47 \text{ rev/s})^2}{2(-0.200 \text{ rev/s}^2)} = 5.40 \text{ rev}$$

Subtracting the five complete revolutions from this result leaves 0.40 rev. Converting this value into degrees and noting that each angular section is 30.0°, we find the following number n for the section where the wheel comes to a halt:

$$n = (0.40 \text{ rev})\left(\frac{360°}{1 \text{ rev}}\right)\left(\frac{1 \text{ angular section}}{30.0°}\right) = 4.8$$

A value of $n = 4.8$ means that the wheel comes to a halt in the section following number 4. Thus, it comes to a halt in $\boxed{\text{section 5}}$. We see, then, that a greater initial velocity does not mean that the wheel comes to a halt in a section labeled with a greater number.

CHAPTER 9 | *ROTATIONAL DYNAMICS*

PROBLEMS

---

1. SSM WWW ***REASONING*** The maximum torque will occur when the force is applied perpendicular to the diagonal of the square as shown. The lever arm ℓ is half the length of the diagonal. From the Pythagorean theorem, the lever arm is, therefore,

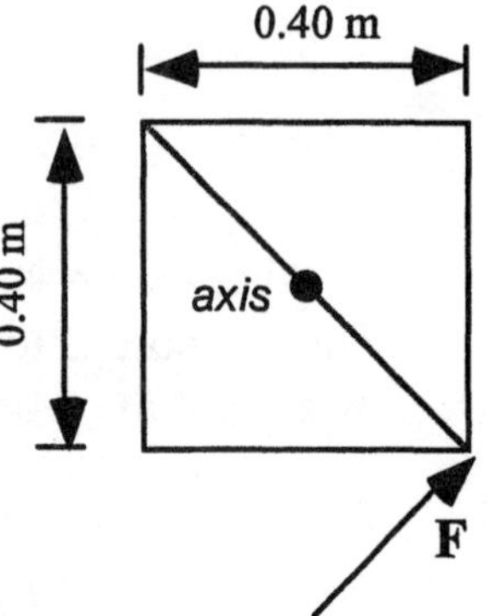

$$\ell = \tfrac{1}{2}\sqrt{(0.40\text{ m})^2 + (0.40\text{ m})^2} = 0.28\text{ m}$$

Since the lever arm is now known, we can use Equation 9.1 to obtain the desired result directly.

SOLUTION Equation 9.1 gives

$$\tau = F\ell = (15\text{ N})(0.28\text{ m}) = \boxed{4.2\text{ N}\cdot\text{m}}$$

---

2. ***REASONING*** The torque is given by Equation 9.1, $\tau = F\ell$, where F is the magnitude of the applied force and ℓ is the lever arm. From the figure in the text, the lever arm is given by $\ell = (0.28\text{ m})\sin 50.0°$. Since both τ and ℓ are known, Equation 9.1 can be solved for F.

SOLUTION Solving Equation 9.1 for F, we have

$$F = \frac{\tau}{\ell} = \frac{45\text{ N}\cdot\text{m}}{(0.28\text{ m})\sin 50.0°} = \boxed{2.1\times 10^2\text{ N}}$$

---

3. ***REASONING*** The drawing shows the wheel as it rolls to the right, so the torque applied by the engine is assumed to be clockwise about the axis of rotation. The force of static friction that the ground applies to the wheel is labeled as f_s. This force produces a counterclockwise torque τ about the axis of rotation, the magnitude of which is given by Equation 9.1 as $\tau = f_s\ell$, where ℓ is the lever arm. Using this relation we can find the magnitude f_s of the static frictional force.

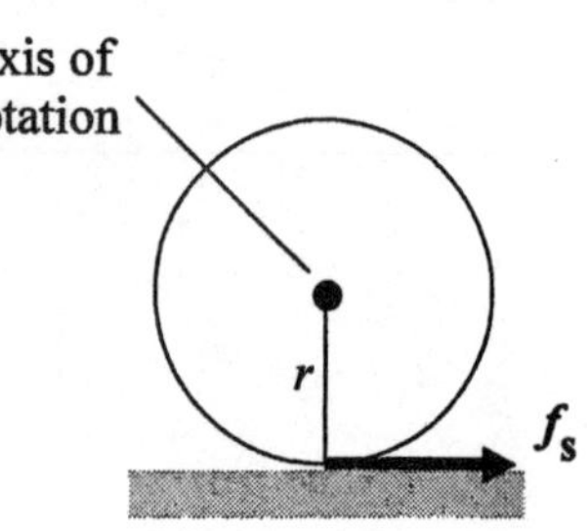

SOLUTION The countertorque is given as $\tau = f_s\ell$, where f_s is the magnitude of the static frictional force and ℓ is the lever arm. The lever arm is the distance between the line of action of the force and the axis of rotation; in this case the lever arm is just the radius r of the tire. Solving for f_s gives

$$f_s = \frac{\tau}{\ell} = \frac{295\ \text{N}\cdot\text{m}}{0.350\ \text{m}} = \boxed{843\ \text{N}}$$

---

4. ***REASONING AND SOLUTION*** The torque about the center of the cable car is $\tau = FL$, where L is the lever arm.

$$\tau = FL = 2(185\ \text{N})(4.60\ \text{m}) = \boxed{1.70\times10^3\ \text{N}\cdot\text{m}}$$

---

5. SSM ***REASONING*** The torque on either wheel is given by Equation 9.1, $\tau = F\ell$, where F is the magnitude of the force and ℓ is the lever arm. Regardless of how the force is applied, the lever arm will be proportional to the radius of the wheel.

SOLUTION The ratio of the torque produced by the force in the truck to the torque produced in the car is

$$\frac{\tau_{\text{truck}}}{\tau_{\text{car}}} = \frac{F\ell_{\text{truck}}}{F\ell_{\text{car}}} = \frac{Fr_{\text{truck}}}{Fr_{\text{car}}} = \frac{r_{\text{truck}}}{r_{\text{car}}} = \frac{0.25\ \text{m}}{0.19\ \text{m}} = \boxed{1.3}$$

---

6. ***REASONING*** To calculate the torques, we need to determine the lever arms for each of the forces. These lever arms are shown in the following drawings:

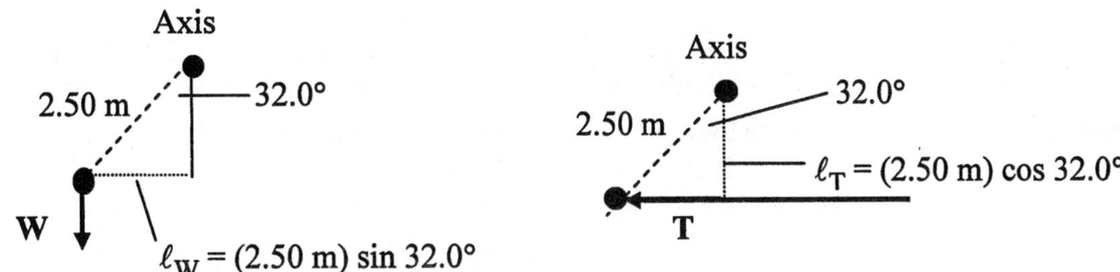

SOLUTION

a. Using Equation 9.1, we find that the magnitude of the torque due to the weight **W** is

$$\tau_{\text{W}} = W\ell_{\text{W}} = (10\,200\ \text{N})(2.5\ \text{m})\sin 32° = \boxed{13\,500\ \text{N}\cdot\text{m}}$$

b. Using Equation 9.1, we find that the magnitude of the torque due to the thrust **T** is

$$\tau_T = T\ell_T = (62\,300\text{ N})(2.5\text{ m})\cos 32° = \boxed{132\,000\text{ N}\cdot\text{m}}$$

---

7. ***REASONING AND SOLUTION*** The torque produced by a force of magnitude F is given by Equation 9.1, $\tau = F\ell$, where ℓ is the lever arm. In each case, the torque produced by the couple is equal to the sum of the individual torques produced by each member of the couple.

a. When the axis passes through point A, the torque due to the force at A is zero. The lever arm for the force at C is L. Therefore, taking counterclockwise as the positive direction, we have

$$\tau = \tau_A + \tau_C = 0 + \tau_C = \boxed{FL}$$

b. Each force produces a counterclockwise rotation. The magnitude of each force is F and each force has a lever arm of $L/2$. Taking counterclockwise as the positive direction, we have

$$\tau = \tau_A + \tau_C = F\left(\frac{L}{2}\right) + F\left(\frac{L}{2}\right) = \boxed{FL}$$

c. When the axis passes through point C, the torque due to the force at C is zero. The lever arm for the force at A is L. Therefore, taking counterclockwise as the positive direction, we have

$$\tau = \tau_A + \tau_C = \tau_A + 0 = \boxed{FL}$$

Note that the value of the torque produced by the couple is the same in all three cases; in other words, when the couple acts on the tire wrench, the couple produces a torque that does *not* depend on the location of the axis.

---

8. ***REASONING*** Each of the two forces produces a torque about the axis of rotation, one clockwise and the other counterclockwise. By setting the sum of the torques equal to zero $(\Sigma\tau = 0)$, we will be able to determine the distance x in the drawing

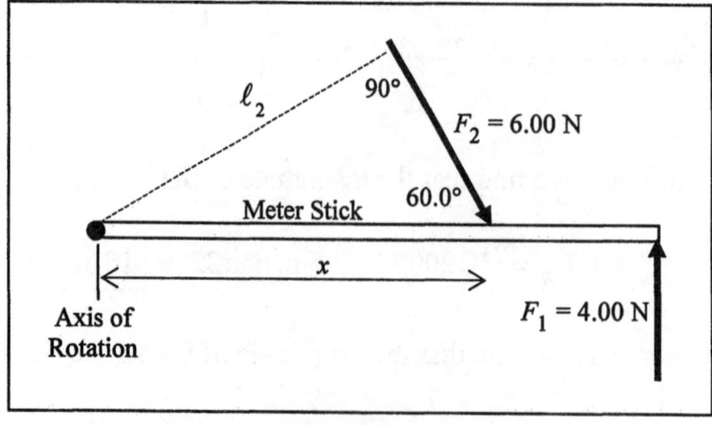

Table (Top view)

SOLUTION The torque τ_1 produced by the force F_1 is given by Equation 9.1 as $\tau_1 = +F_1\ell_1$, where the lever arm is $\ell_1 = 1.00$ m. It is a positive torque, since it tends to produce a counterclockwise rotation. The torque τ_2 produced by F_2 is $\tau_2 = -F_2\ell_2$, where $\ell_2 = x\sin 60.0°$. It is a negative torque, since it tends to produce a clockwise rotation. Setting the net torque equal to zero, we have

$$\underbrace{+F_1\ell_1 + (-F_2\ell_2)}_{\Sigma\tau} = 0 \qquad \text{or} \qquad +F_1\underbrace{(1.00\text{ m})}_{\ell_1} - F_2\underbrace{(x\sin 60.0°)}_{\ell_2} = 0$$

Solving for x gives

$$x = \frac{F_1(1.00\text{ m})}{F_2\sin 60.0°} = \frac{(4.00\text{ N})(1.00\text{ m})}{(6.00\text{ N})\sin 60.0°} = \boxed{0.770\text{ m}}$$

---

9. SSM ***REASONING*** Since the meter stick does not move, it is in equilibrium. The forces and the torques, therefore, each add to zero. We can determine the location of the 6.00 N force, by using the condition that the sum of the torques must vectorially add to zero.

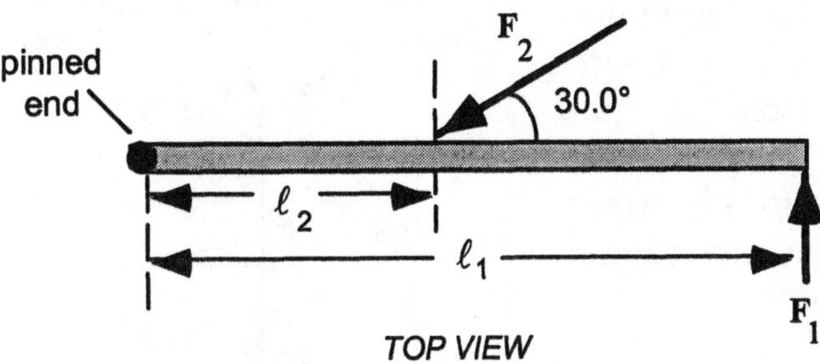

SOLUTION If we take counterclockwise torques as positive, then the torque of the first force about the pin is

$$\tau_1 = F_1\ell_1 = (2.00\text{ N})(1.00\text{ m}) = 2.00\text{ N}\cdot\text{m}$$

The torque due to the second force is

$$\tau_2 = -F_2(\sin 30.0°)\,\ell_2 = -(6.00\text{ N})(\sin 30.0°)\,\ell_2 = -(3.00\text{ N})\,\ell_2$$

The torque τ_2 is negative because the force F_2 tends to produce a clockwise rotation about the pinned end. Since the net torque is zero, we have

$$2.00\text{ N}\cdot\text{m} + [-(3.00\text{ N})\ell_2] = 0$$

Thus,

$$\ell_2 = \frac{2.00\ \text{N}\cdot\text{m}}{3.00\ \text{N}} = \boxed{0.667\ \text{m}}$$

---

10. ***REASONING AND SOLUTION*** The net torque about the axis in text drawing (a) is

$$\Sigma\tau = \tau_1 + \tau_2 = F_1 b - F_2 a = 0$$

Considering that $F_2 = 3F_1$, we have $b - 3a = 0$. The net torque in drawing (b) is then

$$\Sigma\tau = F_1(1.00\ \text{m} - a) - F_2 b = 0 \quad \text{or} \quad 1.00\ \text{m} - a - 3b = 0$$

Solving the first equation for b, substituting into the second equation and rearranging, gives

$$a = \boxed{0.100\ \text{m}} \quad \text{and} \quad b = \boxed{0.300\ \text{m}}$$

---

11. ***REASONING*** The drawing shows the forces acting on the person. It also shows the lever arms for a rotational axis perpendicular to the plane of the paper at the place where the person's toes touch the floor. Since the person is in equilibrium, the sum of the forces must be zero. Likewise, we know that the sum of the torques must be zero.

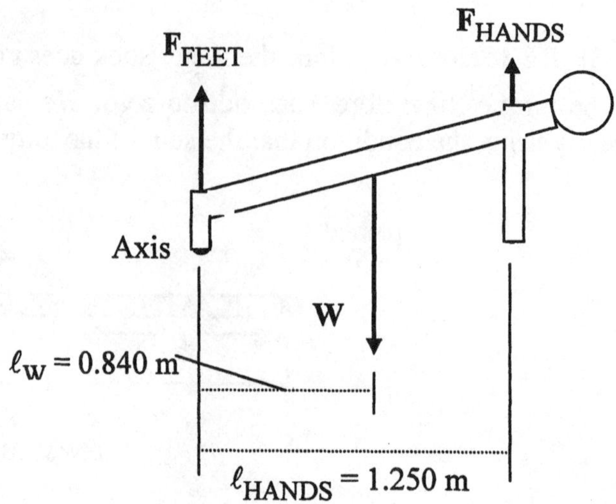

SOLUTION Taking upward to be the positive direction, we have

$$F_{\text{FEET}} + F_{\text{HANDS}} - W = 0$$

Remembering that counterclockwise torques are positive and using the axis and the lever arms shown in the drawing, we find

$$W\ell_{\text{W}} - F_{\text{HANDS}}\ell_{\text{HANDS}} = 0$$

$$F_{\text{HANDS}} = \frac{W\ell_{\text{W}}}{\ell_{\text{HANDS}}} = \frac{(584\ \text{N})(0.840\ \text{m})}{1.250\ \text{m}} = 392\ \text{N}$$

Substituting this value into the balance-of-forces equation, we find

$$F_{\text{FEET}} = W - F_{\text{HANDS}} = 584\text{ N} - 392\text{ N} = 192\text{ N}$$

The force on each hand is half the value calculated above, or $\boxed{196\text{ N}}$. Likewise, the force on each foot is half the value calculated above, or $\boxed{96\text{ N}}$.

---

12. ***REASONING*** When the board just begins to tip, three forces act on the board. They are the weight W of the board, the weight W_P of the person, and the force F exerted by the right support.

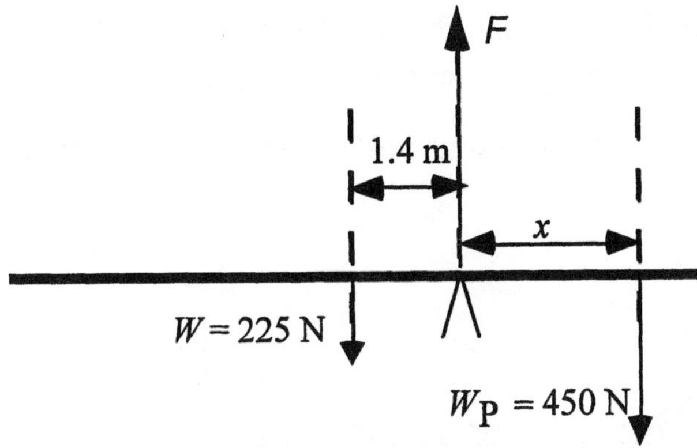

Since the board will rotate around the right support, the lever arm for this force is zero, and the torque exerted by the right support is zero. The lever arm for the weight of the board is equal to one-half the length of the board minus the overhang length: 2.5 m – 1.1 m = 1.4 m.

The lever arm for the weight of the person is x. Therefore, taking counterclockwise torques as positive, we have

$$-W_P x + W(1.4\text{ m}) = 0$$

This expression can be solved for x.

SOLUTION Solving the expression above for x, we obtain

$$x = \frac{W(1.4\text{ m})}{W_P} = \frac{(225\text{ N})(1.4\text{ m})}{450\text{ N}} = \boxed{0.70\text{ m}}$$

---

13. SSM ***REASONING*** The minimum value for the coefficient of static friction between the ladder and the ground, so that the ladder does not slip, is given by Equation 4.7:

$$f_s^{\text{MAX}} = \mu_s F_N$$

SOLUTION From Example 4, the magnitude of the force of static friction is $G_x = 727$ N. The magnitude of the normal force applied to the ladder by the ground is $G_y = 1230$ N. The minimum value for the coefficient of static friction between the ladder and the ground is

$$\mu_s = \frac{f_s^{\text{MAX}}}{F_N} = \frac{G_x}{G_y} = \frac{727\text{ N}}{1230\text{ N}} = \boxed{0.591}$$

---

14. ***REASONING*** The drawing shows the bridge and the four forces that act on it: the upward force $\mathbf{F}_1$ exerted on the left end by the support, the force due to the weight $\mathbf{W}_h$ of the hiker, the weight $\mathbf{W}_b$ of the bridge, and the upward force $\mathbf{F}_2$ exerted on the right side by the support. Since the bridge is in equilibrium, the sum of the torques about any axis of rotation must be zero $(\Sigma\tau = 0)$, and the sum of the forces in the vertical direction must be zero$\left(\Sigma F_y = 0\right)$. These two conditions will allow us to determine the magnitudes of $\mathbf{F}_1$ and $\mathbf{F}_2$.

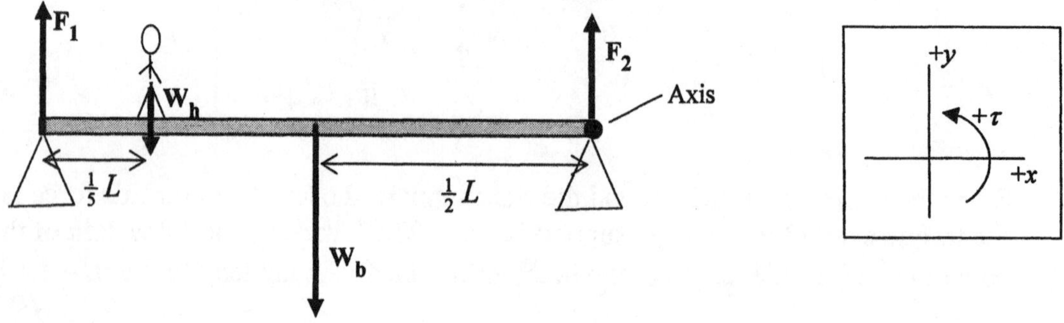

SOLUTION
a. We will begin by taking the axis of rotation about the right end of the bridge. The torque produced by $\mathbf{F}_2$ is zero, since its lever arm is zero. When we set the sum of the torques equal to zero, the resulting equation will have only one unknown, $\mathbf{F}_1$, in it. Setting the sum of the torques produced by the three forces equal to zero gives

$$\Sigma\tau = -F_1 L + W_h\left(\tfrac{4}{5}L\right) + W_b\left(\tfrac{1}{2}L\right) = 0$$

Algebraically eliminating the length L of the bridge from this equation and solving for F_1 gives

$$F_1 = \tfrac{4}{5}W_h + \tfrac{1}{2}W_b = \tfrac{4}{5}(985\text{ N}) + \tfrac{1}{2}(3610\text{ N}) = \boxed{2590\text{ N}}$$

b. Since the bridge is in equilibrium, the sum of the forces in the vertical direction must be zero:

$$\Sigma F_y = F_1 - W_h - W_b + F_2 = 0$$

Solving for F_2 gives

$$F_2 = -F_1 + W_h + W_b = -2590\text{ N} + 985\text{ N} + 3610\text{ N} = \boxed{2.01 \times 10^3\text{ N}}$$

---

15. SSM ***REASONING*** The figure at the right shows the door and the forces that act upon it. Since the door is uniform, the center of gravity, and, thus, the location of the weight **W**, is at the geometric center of the door.

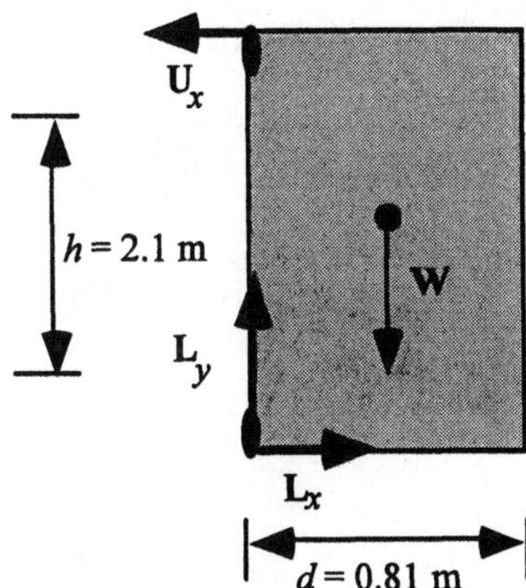

Let **U** represent the force applied to the door by the upper hinge, and **L** the force applied to the door by the lower hinge. Taking forces that point to the right and forces that point up as positive, we have

$$\Sigma F_x = L_x - U_x = 0 \quad \text{or} \quad L_x = U_x \qquad (1)$$

$$\Sigma F_y = L_y - W = 0 \quad \text{or} \quad L_y = W = 140\text{ N} \qquad (2)$$

Taking torques about an axis perpendicular to the plane of the door and through the lower hinge, with counterclockwise torques being positive, gives

$$\Sigma\tau = -W\left(\frac{d}{2}\right) + U_x h = 0 \qquad (3)$$

Equations (1), (2), and (3) can be used to find the force applied to the door by the hinges; Newton's third law can then be used to find the force applied by the door to the hinges.

SOLUTION

a. Solving Equation (3) for U_x gives

$$U_x = \frac{Wd}{2h} = \frac{(140\text{ N})(0.81\text{ m})}{2(2.1\text{ m})} = 27\text{ N}$$

From Equation (1), $L_x = 27$ N.

Therefore, the upper hinge exerts on the door a horizontal force of $\boxed{\text{27 N to the left}}$.

b. From Equation (1), L_x = 27 N, we conclude that the bottom hinge exerts on the door a horizontal force of $\boxed{\text{27 N to the right}}$.

c. From Newton's third law, the door applies a force of $\boxed{\text{27 N}}$ on the upper hinge. This force points horizontally to the right, according to Newton's third law.

d. Let **D** represent the force applied by the door to the lower hinge. Then, from Newton's third law, the force exerted on the lower hinge by the door has components $D_x = -27$ N and $D_y = -140$ N.

From the Pythagorean theorem,

$$D = \sqrt{D_x^2 + D_y^2} = \sqrt{(-27\text{ N})^2 + (-140\text{ N})^2} = \boxed{143\text{ N}}$$

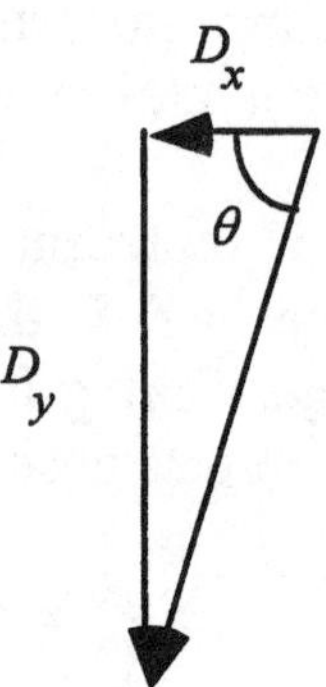

The angle θ is given by

$$\theta = \tan^{-1}\left(\frac{-140\text{ N}}{-27\text{ N}}\right) = 79°$$

The force is directed 79° below the horizontal.

---

16. ***REASONING AND SOLUTION*** The net torque about an axis through the contact point between the tray and the thumb is

$$\Sigma\tau = F(0.0400\text{ m}) - (0.250\text{ kg})(9.80\text{ m/s}^2)(0.320\text{ m}) - (1.00\text{ kg})(9.80\text{ m/s}^2)(0.180\text{ m}) - (0.200\text{ kg})(9.80\text{ m/s}^2)(0.140\text{ m}) = 0$$

$$F = \boxed{70.6\text{ N, up}}$$

Similarly, the net torque about an axis through the point of contact between the tray and the finger is

$$\Sigma\tau = T(0.0400\text{ m}) - (0.250\text{ kg})(9.80\text{ m/s}^2)(0.280\text{ m}) - (1.00\text{ kg})(9.80\text{ m/s}^2)(0.140\text{ m}) - (0.200\text{ kg})(9.80\text{ m/s}^2)(0.100\text{ m}) = 0$$

$$T = \boxed{56.4\text{ N, down}}$$

---

17. ***REASONING*** Since the forearm is in equilibrium, the sum of the torques about any axis of rotation must be zero $(\Sigma\tau = 0)$. For convenience, we will take the elbow joint to be the axis of rotation.

SOLUTION Let M and F be the magnitudes of the forces that the flexor muscle and test apparatus, respectively, exert on the forearm, and let ℓ_M and ℓ_F be the respective lever arms about the elbow joint. Setting the sum of the torques about the elbow joint equal to zero (with counterclockwise torques being taken as positive), we have

$$\Sigma\tau = M\ell_M - F\ell_F = 0$$

Solving for M yields

$$M = \frac{F\ell_F}{\ell_M} = \frac{(190\text{ N})(0.34\text{ m})}{0.054\text{ m}} = \boxed{1200\text{ N}}$$

The direction of the force is to the $\boxed{\text{left}}$.

---

18. ***REASONING AND SOLUTION*** The net torque about an axis through the elbow joint is

$$\Sigma\tau = (111\text{ N})(0.300\text{ m}) - (22.0\text{ N})(0.150\text{ m}) - (0.0250\text{ m})M = 0 \quad\text{or}\quad M = \boxed{1.20\times10^3\text{ N}}$$

---

19. ***REASONING*** The jet is in equilibrium, so the sum of the external forces is zero, and the sum of the external torques is zero. We can use these two conditions to evaluate the forces exerted on the wheels.

SOLUTION

a. Let F_f be the magnitude of the normal force that the ground exerts on the front wheel. Since the net torque acting on the plane is zero, we have (using an axis through the points of contact between the rear wheels and the ground)

$$\Sigma\tau = -W\ell_w + F_f\ell_f = 0$$

where W is the weight of the plane, and ℓ_w and ℓ_f are the lever arms for the forces W and F_f, respectively. Thus,

$$\Sigma\tau = -(1.00\times10^6\text{ N})(15.0\text{ m} - 12.6\text{ m}) + F_f(15.0\text{ m}) = 0$$

Solving for F_f gives $F_f = \boxed{1.60\times10^5\text{ N}}$.

b. Setting the sum of the vertical forces equal to zero yields

$$\Sigma F_y = F_f + 2F_r - W = 0$$

where the factor of 2 arises because there are two rear wheels. Substituting in the data,

$$\Sigma F_y = 1.60 \times 10^5\,\text{N} + 2F_\text{r} - 1.00 \times 10^6\,\text{N} = 0$$

$$F_\text{r} = \boxed{4.20\times10^5\ \text{N}}$$

---

20. ***REASONING*** When the wheel is resting on the ground it is in equilibrium, so the sum of the torques about any axis of rotation is zero$(\Sigma\tau = 0)$. This equilibrium condition will provide us with a relation between the magnitude of **F** and the normal force that the ground exerts on the wheel. When **F** is large enough, the wheel will rise up off the ground, and the normal force will become zero. From our relation, we can determine the magnitude of **F** when this happens.

SOLUTION The free body diagram shows the forces acting on the wheel: its weight **W**, the normal force $\mathbf{F}_\text{N}$, the horizontal force **F**, and the force $\mathbf{F}_\text{E}$ that the edge of the step exerts on the wheel. We select the axis of rotation to be at the edge of the step, so that the torque produced by $\mathbf{F}_\text{E}$ is zero. Letting ℓ_N, ℓ_W, and ℓ_F represent the lever arms for the forces $\mathbf{F}_\text{N}$, **W**, and **F**, the sum of the torques is

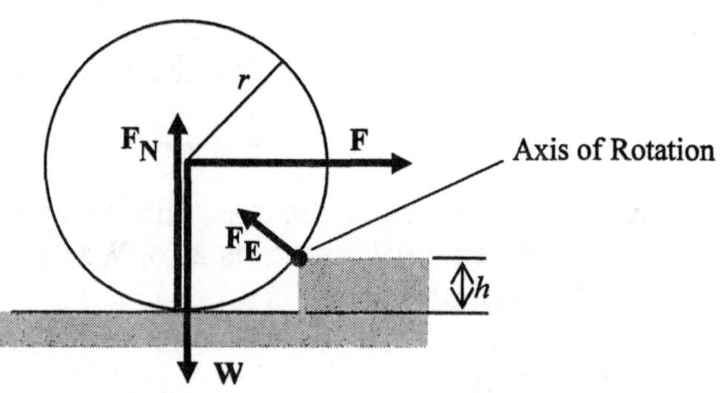

$$\Sigma\tau = -F_\text{N}\underbrace{\sqrt{r^2-(r-h)^2}}_{\ell_\text{N}} + W\underbrace{\sqrt{r^2-(r-h)^2}}_{\ell_\text{W}} - F\underbrace{(r-h)}_{\ell_\text{F}} = 0$$

Solving this equation for F gives

$$F = \frac{(W - F_\text{N})\sqrt{r^2-(r-h)^2}}{r-h}$$

When the bicycle wheel just begins to lift off the ground the normal force becomes zero $(F_\text{N} = 0\ \text{N})$. When this happens, the magnitude of **F** is

$$F = \frac{(W - F_\text{N})\sqrt{r^2-(r-h)^2}}{r-h} = \frac{(25.0\ \text{N} - 0\ \text{N})\sqrt{(0.340\ \text{m})^2 - (0.340\ \text{m} - 0.120\ \text{m})^2}}{0.340\ \text{m} - 0.120\ \text{m}} = \boxed{29\ \text{N}}$$

---

21. SSM ***REASONING*** The drawing shows the forces acting on the board, which has a length L. The ground exerts the vertical normal force **V** on the lower end of the board. The maximum force of static friction has a magnitude of $\mu_s V$ and acts horizontally on the lower end of the board. The weight **W** acts downward at the board's center. The vertical wall applies a force **P** to the upper end of the board, this force being perpendicular to the wall since the wall is smooth (i.e., there is no friction along the wall). We take upward and to the right as our positive directions. Then, since the horizontal forces balance to zero, we have

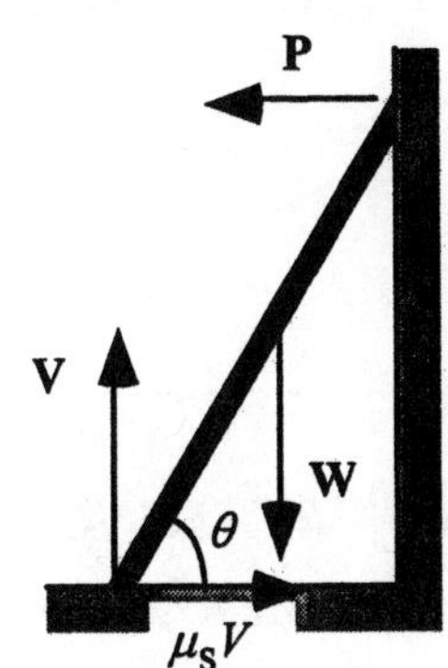

$$\mu_s V - P = 0 \qquad (1)$$

The vertical forces also balance to zero giving

$$V - W = 0 \qquad (2)$$

Using an axis through the lower end of the board, we express the fact that the torques balance to zero as

$$PL \sin \theta - W\left(\frac{L}{2}\right)\cos \theta = 0 \qquad (3)$$

Equations (1), (2), and (3) may then be combined to yield an expression for θ.

SOLUTION Rearranging Equation (3) gives

$$\tan \theta = \frac{W}{2P} \qquad (4)$$

But, $P = \mu_s V$ according to Equation (1), and $W = V$ according to Equation (2). Substituting these results into Equation (4) gives

$$\tan \theta = \frac{V}{2\mu_s V} = \frac{1}{2\mu_s}$$

Therefore,

$$\theta = \tan^{-1}\left(\frac{1}{2\mu_s}\right) = \tan^{-1}\left[\frac{1}{2(0.650)}\right] = \boxed{37.6°}$$

22. ***REASONING AND SOLUTION*** The net torque about an axis through the knee joint is

$$\Sigma\tau = M(\sin 25.0°)(0.100\text{ m}) - (44.5\text{ N})(\cos 30.0°)(0.250\text{ m}) = 0$$

$$M = \boxed{228\text{ N}}$$

---

23. SSM ***REASONING*** Since the man holds the ball motionless, the ball and the arm are in equilibrium. Therefore, the net force, as well as the net torque about any axis, must be zero.

SOLUTION
a. Using Equation 9.1, the net torque about an axis through the elbow joint is

$$\Sigma\tau = M(0.0510\text{ m}) - (22.0\text{ N})(0.140\text{ m}) - (178\text{ N})(0.330\text{ m}) = 0$$

Solving this expression for M gives $M = \boxed{1.21\times10^3\text{ N}}$.

b. The net torque about an axis through the center of gravity is

$$\Sigma\tau = -(1210\text{ N})(0.0890\text{ m}) + F(0.140\text{ m}) - (178\text{ N})(0.190\text{ m}) = 0$$

Solving this expression for F gives $F = \boxed{1.01\times10^3\text{ N}}$. Since the forces must add to give a net force of zero, we know that the direction of **F** is $\boxed{\text{downward}}$.

---

24. ***REASONING AND SOLUTION***
a. The net torque about an axis through the point of contact between the floor and her shoes is

$$\Sigma\tau = -(5.00\times10^2\text{ N})(1.10\text{ m})\sin 30.0° + F_N(\cos 30.0°)(1.50\text{ m}) = 0$$

$$F_N = \boxed{212\text{ N}}$$

b. Newton's second law applied in the horizontal direction gives $F_h - F_N = 0$, so $F_h = \boxed{212\text{ N}}$.

c. Newton's second law in the vertical direction gives $F_v - W = 0$, so $F_v = \boxed{5.00\times10^2\text{ N}}$.

---

25. SSM WWW ***REASONING AND SOLUTION*** Consider the left board, which has a length L and a weight of (356 N)/2 = 178 N. Let $\mathbf{F_V}$ be the upward normal force exerted by

the ground on the board. This force balances the weight, so $F_V = 178$ N. Let $\mathbf{f_s}$ be the force of static friction, which acts horizontally on the end of the board in contact with the ground. $\mathbf{f_s}$ points to the right. Since the board is in equilibrium, the net torque acting on the board through any axis must be zero. Measuring the torques with respect to an axis through the apex of the triangle formed by the boards, we have

$$+(178\text{ N})(\sin 30.0°)\left(\frac{L}{2}\right) + f_s(L\cos 30.0°) - F_V\,(L\sin 30.0°) = 0$$

or

$$44.5\text{ N} + f_s\cos 30.0° - F_V\,\sin 30.0° = 0$$

so that

$$f_s = \frac{(178\text{ N})(\sin 30.0°) - 44.5\text{ N}}{\cos 30.0°} = \boxed{51.4\text{ N}}$$

---

26. ***REASONING AND SOLUTION*** The weight W of the left side of the ladder, the normal force F_N of the floor on the left leg of the ladder, the tension T in the crossbar, and the reaction force R due to the right-hand side of the ladder, are shown in the figure below. In the vertical direction $-W + F_N = 0$, so that

$$F_N = W = mg = (10.0\text{ kg})(9.80\text{ m/s}^2) = 98.0\text{ N}$$

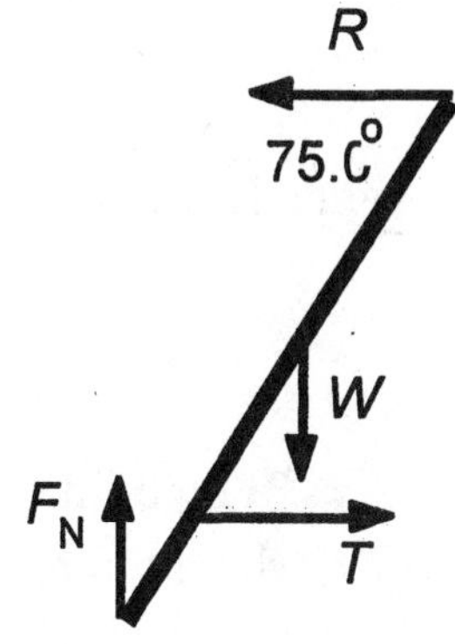

In the horizontal direction it is clear that $R = T$. The net torque about the base of the ladder is

$$\Sigma\tau = -T\,[(1.00\text{ m})\sin 75.0°] - W\,[(2.00\text{ m})\cos 75.0°] + R\,[(4.00\text{ m})\sin 75.0°] = 0$$

Substituting for W and using $R = T$, we obtain

$$T = \frac{(9.80\text{ N})(2.00\text{ m})\cos 75.0°}{(3.00\text{ m})\sin 75.0°} = \boxed{17.5\text{ N}}$$

---

27. ***REASONING*** The drawing shows the forces acting on the board, which has a length L. Wall 2 exerts a normal force P_2 on the lower end of the board. The maximum force of static friction that wall 2 can apply to the lower end of the board is $\mu_s P_2$ and is directed upward in the drawing. The weight W acts downward at the board's center. Wall 1 applies a normal force P_1 to the upper end of the board. We take upward and to the right as our positive directions.

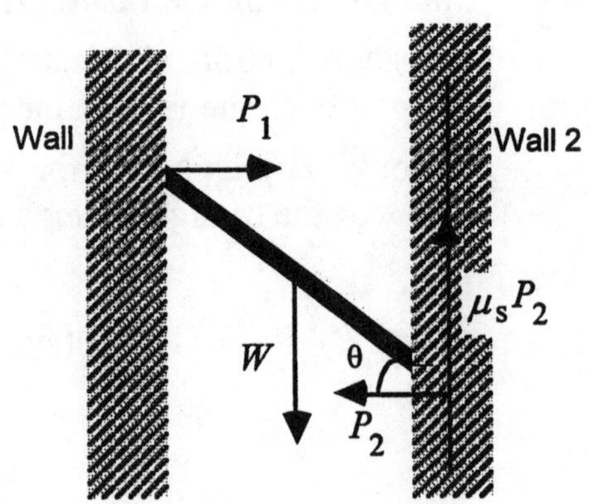

SOLUTION Then, since the horizontal forces balance to zero, we have

$$P_1 - P_2 = 0 \quad (1)$$

The vertical forces also balance to zero:

$$\mu_s P_2 - W = 0 \quad (2)$$

Using an axis through the lower end of the board, we now balance the torques to zero:

$$W\left(\frac{L}{2}\right)(\cos\theta) - P_1 L(\sin\theta) = 0 \quad (3)$$

Rearranging Equation (3) gives

$$\tan\theta = \frac{W}{2P_1} \quad (4)$$

But $W = \mu_s P_2$ according to Equation (2), and $P_2 = P_1$ according to Equation (1). Therefore, $W = \mu_s P_1$, which can be substituted in Equation (4) to show that

$$\tan\theta = \frac{\mu_s P_1}{2P_1} = \frac{\mu_s}{2} = \frac{0.98}{2}$$

or

$$\theta = \tan^{-1}(0.49) = 26°$$

From the drawing at the right,

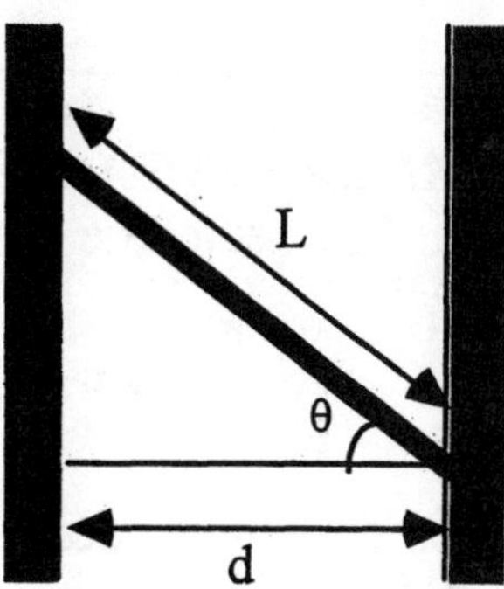

$$\cos\theta = \frac{d}{L}$$

Therefore, the longest board that can be propped between the two walls is

$$L = \frac{d}{\cos\theta} = \frac{1.5\ \text{m}}{\cos 26°} = \boxed{1.7\ \text{m}}$$

---

28. ***REASONING*** The moment of inertia of the stool is the sum of the individual moments of inertia of its parts. According to Table 9.1, a circular disk of radius R has a moment of inertia of $I_{\text{disk}} = \frac{1}{2} M_{\text{disk}} R^2$ with respect to an axis perpendicular to the disk center. Each thin rod is attached perpendicular to the disk at its outer edge. Therefore, each particle in a rod is located at a perpendicular distance from the axis that is equal to the radius of the disk. This means that each of the rods has a moment of inertia of $I_{\text{rod}} = M_{\text{rod}} R^2$.

SOLUTION Remembering that the stool has three legs, we find that the its moment of inertia is

$$I_{\text{stool}} = I_{\text{disk}} + 3I_{\text{rod}} = \tfrac{1}{2} M_{\text{disk}} R^2 + 3M_{\text{rod}} R^2$$

$$= \tfrac{1}{2}(1.2\ \text{kg})(0.16\ \text{m})^2 + 3(0.15\ \text{kg})(0.16\ \text{m})^2 = \boxed{0.027\ \text{kg}\cdot\text{m}^2}$$

---

29. SSM ***REASONING AND SOLUTION*** For a rigid body rotating about a fixed axis, Newton's second law for rotational motion is given by Equation 9.7, $\sum\tau = I\alpha$, where I is the moment of inertia of the body and α is the angular acceleration expressed in rad/s^2. Equation 9.7 gives

$$I = \frac{\Sigma\tau}{\alpha} = \frac{10.0\ \text{N}\cdot\text{m}}{8.00\ \text{rad/s}^2} = \boxed{1.25\ \text{kg}\cdot\text{m}^2}$$

---

30. ***REASONING*** According to Newton's second law for rotational motion, $\Sigma\tau = I\alpha$, the angular acceleration α of the blades is equal to the net torque $\Sigma\tau$ applied to the blades divided by their total moment of inertia I, both of which are known.

SOLUTION The angular acceleration of the fan blades is

$$\alpha = \frac{\Sigma\tau}{I} = \frac{1.8\ \text{N}\cdot\text{m}}{0.22\ \text{kg}\cdot\text{m}^2} = \boxed{8.2\ \text{rad/s}^2} \qquad (9.7)$$

---

31. ***REASONING AND SOLUTION***

a. The net torque on the disk about the axle is

$$\Sigma\tau = F_1R - F_2R = (0.314\ \text{m})(90.0\ \text{N} - 125\ \text{N}) = \boxed{-11\ \text{N}\cdot\text{m}}$$

b. The angular acceleration is given by $\alpha = \Sigma\tau/I$. From Table 9.1, the moment of inertia of the disk is

$$I = (1/2)\,MR^2 = (1/2)(24.3\ \text{kg})(0.314\ \text{m})^2 = 1.20\ \text{kg}\cdot\text{m}^2$$

$$\alpha = (-11\ \text{N}\cdot\text{m})/(1.20\ \text{kg}\cdot\text{m}^2) = \boxed{-9.2\ \text{rad/s}^2}$$

---

32. ***REASONING*** The rotational analog of Newton's second law is given by Equation 9.7, $\sum\tau = I\alpha$. Since the person pushes on the outer edge of one pane of the door with a force **F** that is directed perpendicular to the pane, the torque exerted on the door has a magnitude of *FL*, where the lever arm *L* is equal to the width of one pane of the door. Once the moment of inertia is known, Equation 9.7 can be solved for the angular acceleration α.

The moment of inertia of the door relative to the rotation axis is $I = 4I_P$, where I_P is the moment of inertia for one pane. According to Table 9.1, we find $I_P = \frac{1}{3}ML^2$, so that the rotational inertia of the door is $I = \frac{4}{3}ML^2$.

SOLUTION Solving Equation 9.7 for α, and using the expression for *I* determined above, we have

$$\alpha = \frac{FL}{\frac{4}{3}ML^2} = \frac{F}{\frac{4}{3}ML} = \frac{68\ \text{N}}{\frac{4}{3}(85\ \text{kg})(1.2\ \text{m})} = \boxed{0.50\ \text{rad/s}^2}$$

---

33. SSM ***REASONING AND SOLUTION***

a. The rim of the bicycle wheel can be treated as a hoop. Using the expression given in Table 9.1 in the text, we have

$$I_{\text{hoop}} = MR^2 = (1.20\ \text{kg})(0.330\ \text{m})^2 = \boxed{0.131\ \text{kg}\cdot\text{m}^2}$$

b. Any one of the spokes may be treated as a long, thin rod that can rotate about one end. The expression in Table 9.1 gives

$$I_{\text{rod}} = \tfrac{1}{3}ML^2 = \tfrac{1}{3}(0.010\text{ kg})(0.330\text{ m})^2 = \boxed{3.6\times10^{-4}\text{ kg}\cdot\text{m}^2}$$

c. The total moment of inertia of the bicycle wheel is the sum of the moments of inertia of each constituent part. Therefore, we have

$$I = I_{\text{hoop}} + 50I_{\text{rod}} = 0.131\text{ kg}\cdot\text{m}^2 + 50(3.6\times10^{-4}\text{ kg}\cdot\text{m}^2) = \boxed{0.149\text{ kg}\cdot\text{m}^2}$$

---

34. ***REASONING*** The net torque $\Sigma\tau$ acting on the CD is given by Newton's second law for rotational motion (Equation 9.7) as $\Sigma\tau = I\alpha$, where I is the moment of inertia of the CD and α is its angular acceleration. The moment of inertia can be obtained directly from Table 9.1, and the angular acceleration can be found from its definition (Equation 8.4) as the change in the CD's angular velocity divided by the elapsed time.

SOLUTION The net torque is $\Sigma\tau = I\alpha$. Assuming that the CD is a solid disk, its moment of inertia can be found from Table 9.1 as $I = \tfrac{1}{2}MR^2$, where M and R are the mass and radius of the CD. Thus, the net torque is

$$\Sigma\tau = I\alpha = \left(\tfrac{1}{2}MR^2\right)\alpha$$

The angular acceleration is given by Equation 8.4 as $\alpha = \left(\omega - \omega_0\right)/t$, where ω and ω_0 are the final and initial angular velocities, respectively, and t is the elapsed time. Substituting this expression for α into Newton's second law yields

$$\Sigma\tau = \left(\tfrac{1}{2}MR^2\right)\alpha = \left(\tfrac{1}{2}MR^2\right)\left(\frac{\omega-\omega_0}{t}\right)$$

$$= \left[\tfrac{1}{2}\left(17 \times 10^{-3}\text{ kg}\right)\left(6.0 \times 10^{-2}\text{ m}\right)^2\right]\left(\frac{21\text{ rad/s} - 0\text{ rad/s}}{0.80\text{ s}}\right) = \boxed{8.0 \times 10^{-4}\text{ N}\cdot\text{m}}$$

---

35. ***REASONING***

a. The angular acceleration α is defined as the change, $\omega - \omega_0$, in the angular velocity divided by the elapsed time t (see Equation 8.4). Since all these variables are known, we can determine the angular acceleration directly from this definition.

b. The magnitude τ of the torque is defined by Equation 9.1 as the product of the magnitude F of the force and the lever arm ℓ. The lever arm is the radius of the cylinder, which is known. Since there is only one torque acting on the cylinder, the magnitude of the force can be obtained by using Newton's second law for rotational motion, $\Sigma\tau = F\ell = I\alpha$.

SOLUTION

a. From Equation 8.4 we have that $\alpha = (\omega - \omega_0)/t$. We are given that $\omega_0 = 76.0$ rad/s, $\omega = \frac{1}{2}\omega_0 = 38.0$ rad/s, and $t = 6.40$ s, so

$$\alpha = \frac{\omega - \omega_0}{t} = \frac{38.0 \text{ rad/s} - 76.0 \text{ rad/s}}{6.40 \text{ s}} = -5.94 \text{ rad/s}^2$$

The magnitude of the angular acceleration is $\boxed{5.94 \text{ rad/s}^2}$.

b. Using Newton's second law for rotational motion, we have that $\Sigma\tau = F\ell = I\alpha$. Thus, the magnitude of the force is

$$F = \frac{I\alpha}{\ell} = \frac{\left(0.615 \text{ kg}\cdot\text{m}^2\right)\left(5.94 \text{ rad/s}^2\right)}{0.0830 \text{ m}} = \boxed{44.0 \text{ N}}$$

---

36. ***REASONING*** The drawing shows the two identical sheets and the axis of rotation for each.

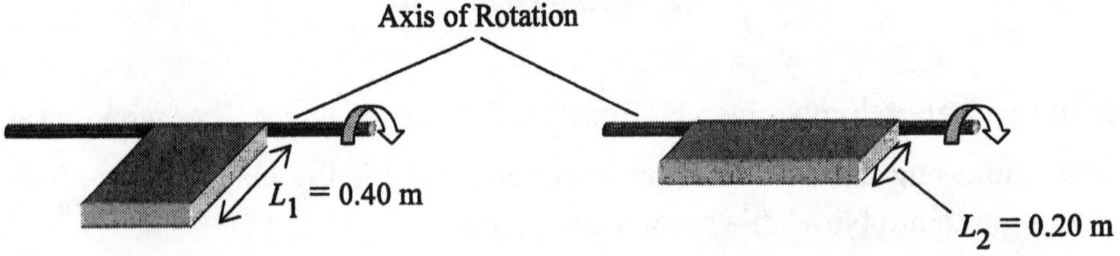

The time t it takes for each sheet to reach its final angular velocity depends on the angular acceleration α of the sheet. This relation is given by Equation 8.4 as $t = (\omega - \omega_0)/\alpha$, where ω and ω_0 are the final and initial angular velocities, respectively. We know that $\omega_0 = 0$ rad/s in each case and that the final angular velocities are the same. The angular acceleration can be determined by using Newton's second law for rotational motion, Equation 9.7, as $\alpha = \tau / I$, where τ is the torque applied to a sheet and I is its moment of inertia.

SOLUTION Substituting the relation $\alpha = \tau / I$ into $t = (\omega - \omega_0)/\alpha$ gives

$$t = \frac{(\omega - \omega_0)}{\alpha} = \frac{(\omega - \omega_0)}{\left(\frac{\tau}{I}\right)} = \frac{I(\omega - \omega_0)}{\tau}$$

The time it takes for each sheet to reach its final angular velocity is:

$$t_{\text{Left}} = \frac{I_{\text{Left}}(\omega - \omega_0)}{\tau} \qquad \text{and} \qquad t_{\text{Right}} = \frac{I_{\text{Right}}(\omega - \omega_0)}{\tau}$$

The moments of inertia I of the left and right sheets about the axes of rotation are given by the following relations, where M is the mass of each sheet (see Table 9.1 and the drawings above): $I_{\text{Left}} = \frac{1}{3}ML_1^2$ and $I_{\text{Right}} = \frac{1}{3}ML_2^2$. Note that the variables M, ω, ω_0, and τ are the same for both sheets. Dividing the time-expression for the right sheet by that for the left sheet gives

$$\frac{t_{\text{Right}}}{t_{\text{Left}}} = \frac{\dfrac{I_{\text{Right}}(\omega - \omega_0)}{\tau}}{\dfrac{I_{\text{Left}}(\omega - \omega_0)}{\tau}} = \frac{I_{\text{Right}}}{I_{\text{Left}}} = \frac{\frac{1}{3}ML_2^2}{\frac{1}{3}ML_1^2} = \frac{L_2^2}{L_1^2}$$

Solving this expression for t_{Right} yields

$$t_{\text{Right}} = t_{\text{Left}}\left(\frac{L_2^2}{L_1^2}\right) = (8.0\text{ s})\frac{(0.20\text{ m})^2}{(0.40\text{ m})^2} = \boxed{2.0\text{ s}}$$

---

37. SSM ***REASONING*** The angular acceleration of the bicycle wheel can be calculated from Equation 8.4. Once the angular acceleration is known, Equation 9.7 can be used to find the net torque caused by the brake pads. The normal force can be calculated from the torque using Equation 9.1.

SOLUTION The angular acceleration of the wheel is, according to Equation 8.4,

$$\alpha = \frac{\omega - \omega_0}{t} = \frac{3.7\text{ rad/s} - 13.1\text{ rad/s}}{3.0\text{ s}} = -3.1\text{ rad/s}^2$$

If we assume that all the mass of the wheel is concentrated in the rim, we may treat the wheel as a hollow cylinder. From Table 9.1, we know that the moment of inertia of a hollow cylinder of mass m and radius r about an axis through its center is $I = mr^2$. The net torque that acts on the wheel due to the brake pads is, therefore,

$$\sum \tau = I\alpha = (mr^2)\alpha \qquad (1)$$

From Equation 9.1, the net torque that acts on the wheel due to the action of the two brake pads is

$$\sum \tau = -2f_k\ell \qquad (2)$$

where f_k is the kinetic frictional force applied to the wheel by each brake pad, and $\ell = 0.33$ m is the lever arm between the axle of the wheel and the brake pad (see the drawing in the text). The factor of 2 accounts for the fact that there are two brake pads. The minus sign arises because the net torque must have the same sign as the angular acceleration. The kinetic frictional force can be written as (see Equation 4.8)

$$f_k = \mu_k F_N \tag{3}$$

where μ_k is the coefficient of kinetic friction and F_N is the magnitude of the normal force applied to the wheel by each brake pad.

Combining Equations (1), (2), and (3) gives

$$-2(\mu_k F_N)\ell = (mr^2)\alpha$$

$$F_N = \frac{-mr^2\alpha}{2\mu_k \ell} = \frac{-(1.3 \text{ kg})(0.33 \text{ m})^2(-3.1 \text{ rad/s}^2)}{2(0.85)(0.33 \text{ m})} = \boxed{0.78 \text{ N}}$$

---

38. ***REASONING AND SOLUTION*** The final angular speed of the arm is $\omega = v_T/r$, where $r = 0.28$ m. The angular acceleration needed to produce this angular speed is $\alpha = (\omega - \omega_0)/t$. The net torque required is $\Sigma\tau = I\alpha$. This torque is due solely to the force M, so that $\Sigma\tau = ML$. Thus,

$$M = \frac{\Sigma\tau}{L} = \frac{I\left(\dfrac{\omega - \omega_0}{t}\right)}{L}$$

Setting $\omega_0 = 0$ rad/s and $\omega = v_T/r$, the force becomes

$$M = \frac{I\left(\dfrac{v_T}{rt}\right)}{L} = \frac{Iv_T}{Lrt} = \frac{(0.065 \text{ kg}\cdot\text{m}^2)(5.0 \text{ m/s})}{(0.025 \text{ m})(0.28 \text{ m})(0.10 \text{ s})} = \boxed{460 \text{ N}}$$

---

39. SSM ***REASONING AND SOLUTION***

a. From Table 9.1 in the text, the moment of inertia of the rod relative to an axis that is perpendicular to the rod at one end is given by

$$I_{\text{rod}} = \tfrac{1}{3}ML^2 = \tfrac{1}{3}(2.00 \text{ kg})(2.00 \text{ m})^2 = \boxed{2.67 \text{ kg}\cdot\text{m}^2}$$

b. The moment of inertia of a point particle of mass m relative to a rotation axis located a perpendicular distance r from the particle is $I = mr^2$. Suppose that all the mass of the rod were located at a single point located a perpendicular distance R from the axis in part (a). If this point particle has the same moment of inertia as the rod in part (a), then the distance R, the radius of gyration, is given by

$$R = \sqrt{\frac{I}{m}} = \sqrt{\frac{2.67\ \text{kg}\cdot\text{m}^2}{2.00\ \text{kg}}} = \boxed{1.16\ \text{m}}$$

---

40. ***REASONING AND SOLUTION*** The moment of inertia of a solid cylinder about an axis coinciding with the cylinder axis which contains the center of mass is $I_{\text{cm}} = (1/2)\,MR^2$. The parallel axis theorem applied to an axis on the surface ($h = R$) gives $I = I_{\text{cm}} + MR^2$, so that

$$I = \boxed{\tfrac{3}{2}MR^2}$$

---

41. ***REASONING AND SOLUTION*** For door A, $\theta = (1/2)\,\alpha_A t_A^{\;2}$. For door B, $\theta = (1/2)\,\alpha_B t_B^{\;2}$. Equating gives

$$t_B = t_A\sqrt{\frac{\alpha_A}{\alpha_B}}$$

We need the angular accelerations. Applying Newton's second law for rotation to door A gives $FL = (1/3)ML^2\alpha_A$, and applying the second law to door B gives $F(1/2)L = (1/12)ML^2\alpha_B$.

Division of the previous two equations yields $\alpha_A/\alpha_B = 1/2$. Now

$$t_B = t_A/\sqrt{2} = (3.00\ \text{s})/(1.41) = \boxed{2.12\ \text{s}}$$

---

42. ***REASONING AND SOLUTION*** Newton's law applied to the 11.0-kg object gives

$$T_2 - (11.0\ \text{kg})(9.80\ \text{m/s}^2) = (11.0\ \text{kg})(4.90\ \text{m/s}^2) \quad \text{or} \quad T_2 = 162\ \text{N}$$

A similar treatment for the 44.0-kg object yields

$$T_1 - (44.0\ \text{kg})(9.80\ \text{m/s}^2) = (44.0\ \text{kg})(-4.90\ \text{m/s}^2) \quad \text{or} \quad T_1 = 216\ \text{N}$$

For an axis about the center of the pulley

$$T_2 r - T_1 r = I(-\alpha) = (1/2)\, Mr^2\, (-a/r)$$

Solving for the mass M we obtain

$$M = (-2/a)(T_2 - T_1) = [-2/(4.90\text{ m/s}^2)](162\text{ N} - 216\text{ N}) = \boxed{22.0\text{ kg}}$$

---

43. **SSM** ***REASONING AND SOLUTION***

a. The tangential speed of each object is given by Equation 8.9, $v_T = r\omega$. Therefore,

For object 1: $v_{T1} = (2.00\text{ m})(6.00\text{ rad/s}) = \boxed{12.0\text{ m/s}}$

For object 2: $v_{T2} = (1.50\text{ m})(6.00\text{ rad/s}) = \boxed{9.00\text{ m/s}}$

For object 3: $v_{T3} = (3.00\text{ m})(6.00\text{ rad/s}) = \boxed{18.0\text{ m/s}}$

b. The total kinetic energy of this system can be calculated by computing the sum of the kinetic energies of each object in the system. Therefore,

$$\text{KE} = \tfrac{1}{2}m_1 v_1^2 + \tfrac{1}{2}m_2 v_2^2 + \tfrac{1}{2}m_3 v_3^2$$

$$\text{KE} = \tfrac{1}{2}\left[(6.00\text{ kg})(12.0\text{ m/s})^2 + (4.00\text{ kg})(9.00\text{ m/s})^2 + (3.00\text{ kg})(18.0\text{ m/s})^2\right] = \boxed{1.08\times 10^3\text{ J}}$$

c. The total moment of inertia of this system can be calculated by computing the sum of the moments of inertia of each object in the system. Therefore,

$$I = \sum mr^2 = m_1 r_1^2 + m_2 r_2^2 + m_3 r_3^2$$

$$I = (6.00\text{ kg})(2.00\text{ m})^2 + (4.00\text{ kg})(1.50\text{ m})^2 + (3.00\text{ kg})(3.00\text{ m})^2 = \boxed{60.0\text{ kg}\cdot\text{m}^2}$$

d. The rotational kinetic energy of the system is, according to Equation 9.9,

$$\text{KE}_R = \tfrac{1}{2}I\omega^2 = \tfrac{1}{2}(60.0\text{ kg}\cdot\text{m}^2)(6.00\text{ rad/s})^2 = \boxed{1.08\times 10^3\text{ J}}$$

This agrees, as it should, with the result for part (b).

---

44. ***REASONING***

a. The kinetic energy is given by Equation 9.9 as $\text{KE}_\text{R} = \frac{1}{2}I\omega^2$. Assuming the earth to be a uniform solid sphere, we find from Table 9.1 that the moment of inertia is $I = \frac{2}{5}MR^2$. The mass and radius of the earth is $M = 5.98 \times 10^{24}$ kg and $R = 6.38 \times 10^6$ m (see the inside of the text's front cover). The angular speed ω must be expressed in rad/s, and we note that the earth turns once around its axis each day, which corresponds to 2π rad/day.

b. The kinetic energy for the earth's motion around the sun can be obtained from Equation 9.9 as $\text{KE}_\text{R} = \frac{1}{2}I\omega^2$. Since the earth's radius is small compared to the radius of the earth's orbit ($R_\text{orbit} = 1.50 \times 10^{11}$ m, see the inside of the text's front cover), the moment of inertia in this case is just $I = MR_\text{orbit}^2$. The angular speed ω of the earth as it goes around the sun can be obtained from the fact that it makes one revolution each year, which corresponds to 2π rad/year.

SOLUTION

a. According to Equation 9.9, we have

$$\text{KE}_\text{R} = \tfrac{1}{2}I\omega^2 = \tfrac{1}{2}\left(\tfrac{2}{5}MR^2\right)\omega^2$$

$$= \tfrac{1}{2}\left[\tfrac{2}{5}\left(5.98\times10^{24}\text{ kg}\right)\left(6.38\times10^{6}\text{ m}\right)^2\right]\left[\left(\frac{2\pi\text{ rad}}{1\text{ day}}\right)\left(\frac{1\text{ day}}{24\text{ h}}\right)\left(\frac{1\text{ h}}{3600\text{ s}}\right)\right]^2$$

$$= \boxed{2.57\times10^{29}\text{ J}}$$

b. According to Equation 9.9, we have

$$\text{KE}_\text{R} = \tfrac{1}{2}I\omega^2 = \tfrac{1}{2}\left(MR_\text{orbit}^2\right)\omega^2$$

$$= \tfrac{1}{2}\left(5.98\times10^{24}\text{ kg}\right)\left(1.50\times10^{11}\text{ m}\right)^2\left[\left(\frac{2\pi\text{ rad}}{1\text{ yr}}\right)\left(\frac{1\text{ yr}}{365\text{ day}}\right)\left(\frac{1\text{ day}}{24\text{ h}}\right)\left(\frac{1\text{ h}}{3600\text{ s}}\right)\right]^2$$

$$= \boxed{2.67\times10^{33}\text{ J}}$$

---

45. SSM ***REASONING*** The kinetic energy of the flywheel is given by Equation 9.9. The moment of inertia of the flywheel is the same as that of a solid disk, and, according to Table 9.1 in the text, is given by $I = \frac{1}{2}MR^2$. Once the moment of inertia of the flywheel is known, Equation 9.9 can be solved for the angular speed ω in rad/s. This quantity can then be converted to rev/min.

SOLUTION Solving Equation 9.9 for ω, we obtain,

$$\omega = \sqrt{\frac{2(\text{KE}_\text{R})}{I}} = \sqrt{\frac{2(\text{KE}_\text{R})}{\frac{1}{2}MR^2}} = \sqrt{\frac{4(1.2\times10^9\ \text{J})}{(13\,\text{kg})(0.30\ \text{m})^2}} = 6.4\times10^4\ \text{rad/s}$$

Converting this answer into rev/min, we find that

$$\omega = \left(6.4\times10^4\ \text{rad/s}\right)\left(\frac{1\ \text{rev}}{2\pi\,\text{rad}}\right)\left(\frac{60\ \text{s}}{1\ \text{min}}\right) = \boxed{6.1\times10^5\ \text{rev/min}}$$

---

46. ***REASONING*** Each blade can be approximated as a thin rod rotating about an axis perpendicular to the rod and passing through one end. The moment of inertia of a blade is given in Table 9.1 as $\frac{1}{3}ML^2$, where M is the mass of the blade and L is its length. The total moment of inertia I of the two blades is just twice that of a single blade. The rotational kinetic energy KE_R of the blades is given by Equation 9.9 as $\text{KE}_\text{R} = \frac{1}{2}I\omega^2$, where ω is the angular speed of the blades.

SOLUTION
a. The total moment of inertia of the two blades is

$$I = \tfrac{1}{3}ML^2 + \tfrac{1}{3}ML^2 = \tfrac{2}{3}ML^2 = \tfrac{2}{3}(240\ \text{kg})(6.7\ \text{m})^2 = \boxed{7200\ \text{kg}\cdot\text{m}^2}$$

b. The rotational kinetic energy is

$$\text{KE}_\text{R} = \tfrac{1}{2}I\omega^2 = \tfrac{1}{2}\left(7200\ \text{kg}\cdot\text{m}^2\right)\left(44\ \text{rad/s}\right)^2 = \boxed{7.0\times10^6\ \text{J}}$$

---

47. ***REASONING*** The rotational kinetic energy of a solid sphere is given by Equation 9.9 as $\text{KE}_\text{R} = \frac{1}{2}I\omega^2$, where I is its moment of inertia and ω its angular speed. The sphere has translational motion in addition to rotational motion, and its translational kinetic energy is $\text{KE}_\text{T} = \frac{1}{2}mv^2$ (Equation 6.2), where m is the mass of the sphere and v is the speed of its center of mass. The fraction of the sphere's total kinetic energy that is in the form of rotational kinetic energy is $\text{KE}_\text{R}/(\text{KE}_\text{R} + \text{KE}_\text{T})$.

SOLUTION The moment of inertia of a solid sphere about its center of mass is $I = \frac{2}{5}mR^2$, where R is the radius of the sphere (see Table 9.1). The fraction of the sphere's total kinetic energy that is in the form of rotational kinetic energy is

$$\frac{\text{KE}_\text{R}}{\text{KE}_\text{R}+\text{KE}_\text{T}}=\frac{\frac{1}{2}I\omega^2}{\frac{1}{2}I\omega^2+\frac{1}{2}mv^2}=\frac{\frac{1}{2}\left(\frac{2}{5}mR^2\right)\omega^2}{\frac{1}{2}\left(\frac{2}{5}mR^2\right)\omega^2+\frac{1}{2}mv^2}=\frac{\frac{2}{5}R^2\omega^2}{\frac{2}{5}R^2\omega^2+v^2}$$

Since the sphere is rolling without slipping on the surface, the translational speed v of the center of mass is related to the angular speed ω about the center of mass by $v = R\omega$ (see Equation 8.12). Substituting $v = R\omega$ into the equation above gives

$$\frac{\text{KE}_\text{R}}{\text{KE}_\text{R}+\text{KE}_\text{T}}=\frac{\frac{2}{5}R^2\omega^2}{\frac{2}{5}R^2\omega^2+v^2}=\frac{\frac{2}{5}R^2\omega^2}{\frac{2}{5}R^2\omega^2+(R\omega)^2}=\boxed{\frac{2}{7}}$$

---

48. ***REASONING AND SOLUTION*** The conservation of energy gives $(1/2)\,I\omega^2 = mg[(1/2)\,L]$. Now the tip of the rod has, as it hits the ground, a speed of $v = L\omega$, so $\omega = v/L$. The moment of inertia of the rod is given by $I = (1/3)\,mL^2$. Substituting the last two equations into the first equation and simplifying yields

$$v=\sqrt{3gL}=\sqrt{3\left(9.8\text{ m/s}^2\right)(2.00\text{ m})}=\boxed{7.67\text{ m/s}}$$

---

49. SSM ***REASONING AND SOLUTION*** The only force that does work on the cylinders as they move up the incline is the conservative force of gravity; hence, the total mechanical energy is conserved as the cylinders ascend the incline. We will let $h = 0$ on the horizontal plane at the bottom of the incline. Applying the principle of conservation of mechanical energy to the solid cylinder, we have

$$mgh_\text{s}=\tfrac{1}{2}mv_0^2+\tfrac{1}{2}I_\text{s}\omega_0^2 \qquad (1)$$

where, from Table 9.1, $I_\text{s}=\frac{1}{2}mr^2$. In this expression, v_0 and ω_0 are the initial translational and rotational speeds, and h_s is the final height attained by the solid cylinder. Since the cylinder rolls without slipping, the rotational speed ω_0 and the translational speed v_0 are related according to Equation 8.12, $\omega_0 = v_0/r$. Then, solving Equation (1) for h_s, we obtain

$$h_\text{s}=\frac{3v_0^2}{4g}$$

Repeating the above for the hollow cylinder and using $I_\text{h}=mr^2$ we have

$$h_\text{h}=\frac{v_0^2}{g}$$

The height h attained by each cylinder is related to the distance s traveled along the incline and the angle θ of the incline by

$$s_\text{s} = \frac{h_\text{s}}{\sin\theta} \qquad \text{and} \qquad s_\text{h} = \frac{h_\text{h}}{\sin\theta}$$

Dividing these gives

$$\frac{s_\text{s}}{s_\text{h}} = \boxed{3/4}$$

---

50. ***REASONING AND SOLUTION*** The conservation of energy gives

$$mgh + (1/2)\,mv^2 + (1/2)\,I\omega^2 = (1/2)\,mv_0{}^2 + (1/2)\,I\omega_0{}^2$$

If the ball rolls without slipping, $\omega = v/R$ and $\omega_0 = v_0/R$. We also know $I = (2/5)\,mR^2$. Substitution of the last two equations into the first and rearrangement gives

$$v = \sqrt{v_0^2 - \tfrac{10}{7}gh} = \sqrt{(3.50\text{ m/s})^2 - \tfrac{10}{7}(9.80\text{ m/s}^2)(0.760\text{ m})} = \boxed{1.3\text{ m/s}}$$

---

51. ***REASONING*** We first find the speed v_0 of the ball when it becomes airborne using the conservation of mechanical energy. Once v_0 is known, we can use the equations of kinematics to find its range x.

SOLUTION When the tennis ball starts from rest, its total mechanical energy is in the form of gravitational potential energy. The gravitational potential energy is equal to mgh if we take h = 0 m at the height where the ball becomes airborne. Just before the ball becomes airborne, its mechanical energy is in the form of rotational kinetic energy and translational kinetic energy. At this instant its total energy is $\frac{1}{2}mv_0^2 + \frac{1}{2}I\omega^2$. If we treat the tennis ball as a thin-walled spherical shell of mass m and radius r, and take into account that the ball rolls down the hill without slipping, its total kinetic energy can be written as

$$\frac{1}{2}mv_0^2 + \frac{1}{2}I\omega^2 = \frac{1}{2}mv_0^2 + \frac{1}{2}(\tfrac{2}{3}mr^2)\left(\frac{v_0}{r}\right)^2 = \frac{5}{6}mv_0^2$$

Therefore, from conservation of mechanical energy, he have

$$mgh = \frac{5}{6}mv_0^2 \qquad \text{or} \qquad v_0 = \sqrt{\frac{6gh}{5}}$$

The range of the tennis ball is given by $x = v_{0x}t = v_0(\cos\theta)\,t$, where t is the flight time of the ball. From Equation 3.3b, we find that the flight time t is given by

$$t = \frac{v - v_{0y}}{a_y} = \frac{(-v_{0y}) - v_{0y}}{a_y} = -\frac{2v_0\sin\theta}{a_y}$$

Therefore, the range of the tennis ball is

$$x = v_{0x}t = v_0(\cos\theta)\left(-\frac{2v_0\sin\theta}{a_y}\right)$$

If we take upward as the positive direction, then using the fact that $a_y = -g$ and the expression for v_0 given above, we find

$$x = \left(\frac{2\cos\theta\,\sin\theta}{g}\right)v_0^2 = \left(\frac{2\cos\theta\,\sin\theta}{g}\right)\left(\sqrt{\frac{6gh}{5}}\right)^2 = \frac{12}{5}h\cos\theta\,\sin\theta$$

$$= \frac{12}{5}(1.8\text{ m})(\cos 35°)(\sin 35°) = \boxed{2.0\text{ m}}$$

---

52. ***REASONING AND SOLUTION*** Angular momentum is conserved, $I\omega = I_0\omega_0$, so

$$\omega = \left(\frac{I_0}{I}\right)\omega_0 = \left(\frac{5.40\text{ kg}\cdot\text{m}^2}{3.80\text{ kg}\cdot\text{m}^2}\right)(5.00\text{ rad/s}) = \boxed{7.11\text{ rad/s}}$$

---

53. SSM WWW ***REASONING*** Let the two disks constitute the system. Since there are no external torques acting on the system, the principle of conservation of angular momentum applies. Therefore we have $L_{\text{initial}} = L_{\text{final}}$, or

$$I_A\omega_A + I_B\omega_B = (I_A + I_B)\omega_{\text{final}}$$

This expression can be solved for the moment of inertia of disk B.

SOLUTION Solving the above expression for I_B, we obtain

$$I_B = I_A\left(\frac{\omega_{\text{final}} - \omega_A}{\omega_B - \omega_{\text{final}}}\right) = (3.4\text{ kg}\cdot\text{m}^2)\left[\frac{-2.4\text{ rad/s} - 7.2\text{ rad/s}}{-9.8\text{ rad/s} - (-2.4\text{ rad/s})}\right] = \boxed{4.4\text{ kg}\cdot\text{m}^2}$$

54. ***REASONING*** Before any sand strikes the disk, only the disk is rotating. After the sand has landed on the disk, both the sand and the disk are rotating. If the sand and disk are taken to be the system of objects under consideration, we note that there are no external torques acting on the system. As the sand strikes the disk, each exerts a torque on the other. However, these torques are exerted by members of the system, and, as such, are internal torques. The conservation of angular momentum states that the total angular momentum of a system remains constant (is conserved) if the net average external torque acting on the system is zero. We will use this principle to find the final angular velocity of the system.

SOLUTION The angular momentum of the system (sand plus disk) is given by Equation 9.10 as the product of the system's moment of inertia I and angular velocity ω, or $L = I\omega$. The conservation of angular momentum can be written as

$$\underbrace{I\omega}_{\text{Final angular momentum}} = \underbrace{I_0\omega_0}_{\text{Initial angular momentum}}$$

where ω and ω_0 are the final and initial angular velocities, respectively, and I and I_0 are final and initial moments of inertia. The initial moment of inertia is given, while the final moment of inertia is the sum of the values for the rotating sand and disk, $I = I_{\text{sand}} + I_0$. We note that the sand forms a thin ring, so its moment of inertia is given by (see Table 9.1) $I_{\text{sand}} = M_{\text{sand}}R_{\text{sand}}^2$, where M_{sand} is the mass of the sand and R_{sand} is the radius of the ring. Thus, the final angular velocity of the system is, then,

$$\omega = \omega_0\left(\frac{I_0}{I}\right) = \omega_0\left(\frac{I_0}{I_{\text{sand}} + I_0}\right) = \omega_0\left(\frac{I_0}{M_{\text{sand}}R_{\text{sand}}^2 + I_0}\right)$$

$$= (0.067\text{ rad/s})\left[\frac{0.10\text{ kg}\cdot\text{m}^2}{(0.50\text{ kg})(0.40\text{ m})^2 + 0.10\text{ kg}\cdot\text{m}^2}\right] = \boxed{0.037\text{ rad/s}}$$

---

55. ***REASONING*** The rod and bug are taken to be the system of objects under consideration, and we note that there are no external torques acting on the system. As the bug crawls out to the end of the rod, each exerts a torque on the other. However, these torques are internal torques. The conservation of angular momentum states that the total angular momentum of a system remains constant (is conserved) if the net average external torque acting on the system is zero. We will use this principle to find the final angular velocity of the system.

SOLUTION The angular momentum L of the system (rod plus bug) is given by Equation 9.10 as the product of the system's moment of inertia I and angular velocity ω, or $L = I\omega$. The conservation of angular momentum can be written as

$$\underbrace{I\omega}_{\text{Final angular momentum}} = \underbrace{I_0\omega_0}_{\text{Initial angular momentum}}$$

where ω and ω_0 are the final and initial angular velocities, respectively, and I and I_0 are the final and initial moments of inertia. The initial moment of inertia is given. The initial moment of inertia of the bug is zero, because it is located at the axis of rotation. The final moment of inertia is the sum of the moment of inertia of the bug and that of the rod; $I = I_{\text{bug}} + I_0$. When the bug has reached the end of the rod, its moment of inertia is $I_{\text{bug}} = mL^2$, where m is its mass and L is the length of the rod. The final angular velocity of the system is, then,

$$\omega = \omega_0\left(\frac{I_0}{I}\right) = \omega_0\left(\frac{I_0}{I_{\text{bug}} + I_0}\right) = \omega_0\left(\frac{I_0}{mL^2 + I_0}\right)$$

$$= (0.32 \text{ rad/s})\left[\frac{1.1\times10^{-3} \text{ kg}\cdot\text{m}^2}{(4.2\times10^{-3} \text{ kg})(0.25 \text{ m})^2 + (1.1\times10^{-3} \text{ kg}\cdot\text{m}^2)}\right] = \boxed{0.26 \text{ rad/s}}$$

---

56. ***REASONING AND SOLUTION***
The conservation of angular momentum applies about the indicated axis

$$I_\text{d}\omega_\text{d} + M_\text{p}R_\text{p}^2\omega_\text{p} = 0 \quad \text{where} \quad \omega_\text{p} = v_\text{p}/R_\text{p}$$

Now $I_\text{d} = (1/2)\, M_\text{d}R_\text{d}^2$. Combining and solving yields

$$\omega_\text{d} = -2(M_\text{p}/M_\text{d})(R_\text{p}/R_\text{d}^2)v_\text{p} = -2\left(\frac{40.0 \text{ kg}}{1.00\times10^2 \text{ kg}}\right)\left[\frac{1.25 \text{ m}}{(2.00 \text{ m})^2}\right](2.00 \text{ m/s}) = -0.500 \text{ rad/s}$$

The magnitude of ω_d is $\boxed{0.500 \text{ rad/s}}$. The negative sign indicates that the disk rotates in a direction opposite to the motion of the person.

---

57. SSM ***REASONING*** Let the space station and the people within it constitute the system. Then as the people move radially from the outer surface of the cylinder toward the axis, any torques that occur are internal torques. Since there are no external torques acting on the system, the principle of conservation of angular momentum can be employed.

SOLUTION Since angular momentum is conserved,

$$I_{\text{final}}\,\omega_{\text{final}} = I_0\,\omega_0$$

Before the people move from the outer rim, the moment of inertia is

$$I_0 = I_{\text{station}} + 500 m_{\text{person}} r_{\text{person}}^2$$

or

$$I_0 = 3.00\times10^9\ \text{kg}\cdot\text{m}^2 + (500)(70.0\ \text{kg})(82.5\ \text{m})^2 = 3.24\times10^9\ \text{kg}\cdot\text{m}^2$$

If the people all move to the center of the space station, the total moment of inertia is

$$I_{\text{final}} = I_{\text{station}} = 3.00\times10^9\ \text{kg}\cdot\text{m}^2$$

Therefore,

$$\frac{\omega_{\text{final}}}{\omega_0} = \frac{I_0}{I_{\text{final}}} = \frac{3.24\times10^9\ \text{kg}\cdot\text{m}^2}{3.00\times10^9\ \text{kg}\cdot\text{m}^2} = 1.08$$

This fraction represents a percentage increase of $\boxed{\text{8 percent}}$.

---

58. ***REASONING AND SOLUTION*** Since the change occurs without the aid of external torques, the angular momentum of the system is conserved: $I_f\omega_f = I_0\omega_0$. Solving for ω_f gives

$$\omega_f = \omega_0\left(\frac{I_0}{I_f}\right)$$

where for a rod of mass M and length L, $I_0 = \frac{1}{12}ML^2$. To determine I_f, we will treat the arms of the "u" as point masses with mass $M/4$ a distance $L/4$ from the rotation axis. Thus,

$$I_f = \left[\frac{1}{12}\left(\frac{M}{2}\right)\left(\frac{L}{2}\right)^2\right] + \left[2\left(\frac{M}{4}\right)\left(\frac{L}{4}\right)^2\right] = \frac{1}{24}ML^2$$

and

$$\omega_f = \omega_0\left[\frac{\frac{1}{12}ML^2}{\frac{1}{24}ML^2}\right] = (7.0\ \text{rad/s})(2) = \boxed{14\ \text{rad/s}}$$

---

59. ***REASONING AND SOLUTION*** The block will just start to move when the centripetal force on the block just exceeds $f_s^{\max}$. Thus, if r_f is the smallest distance from the axis at

which the block stays at rest when the angular speed of the block is ω_f, then $\mu_s F_N = mr_f \omega_f^2$, or $\mu_s mg = mr_f \omega_f^2$. Thus,

$$\mu_s g = r_f \omega_f^2 \qquad (1)$$

Since there are no external torques acting on the system, angular momentum will be conserved.

$$I_0 \omega_0 = I_f \omega_f$$

where $I_0 = mr_0^2$, and $I_f = mr_f^2$. Making these substitutions yields

$$r_0^2 \omega_0 = r_f^2 \omega_f \qquad (2)$$

Solving Equation (2) for ω_f and substituting into Equation (1) yields:

$$\mu_s g = r_f \omega_0^2 \frac{r_0^4}{r_f^4}$$

Solving for r_f gives

$$r_f = \left(\frac{\omega_0^2 r_0^4}{\mu_s g} \right)^{1/3} = \left[\frac{(2.2 \text{ rad/s})^2 (0.30 \text{ m})^4}{(0.75)(9.80 \text{ m/s}^2)} \right]^{1/3} = \boxed{0.17 \text{ m}}$$

---

60. ***REASONING AND SOLUTION*** After the mass has moved inward to its final path the centripetal force acting on it is $T = 105$ N.

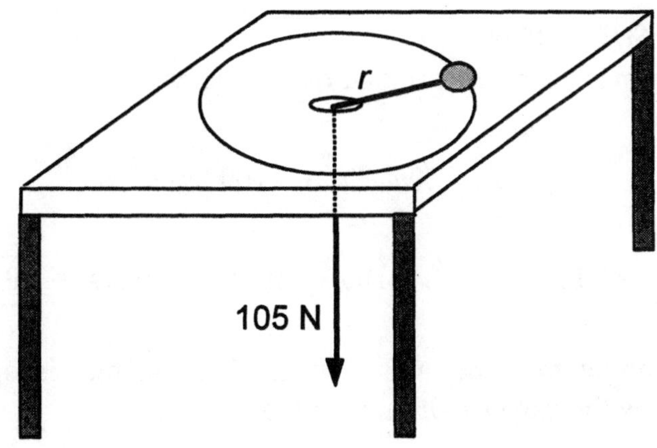

Its centripetal acceleration is

$$a_c = v^2/R = T/m$$

Now

$$v = \omega R \quad \text{so} \quad R = T/(\omega^2 m)$$

The centripetal force is parallel to the line of action (the string), so the force produces no torque on the object. Hence, angular momentum is conserved.

$$I\omega = I_0\omega_0 \quad \text{so that} \quad \omega = (I_0/I)\omega_0 = (R_0^{\ 2}/R^2)\omega_0$$

Substituting and simplifying

$$R^3 = (mR_0^{\ 4}\ \omega_0^{\ 2})/T, \text{ so that } R = \boxed{0.573 \text{ m}}$$

---

61. SSM ***REASONING AND SOLUTION*** Solving Equation 9.1 for the lever arm ℓ, we obtain

$$\ell = \frac{\tau}{F} = \frac{3.0 \text{ N}\cdot\text{m}}{12 \text{ N}} = \boxed{0.25 \text{ m}}$$

---

62. ***REASONING AND SOLUTION*** We know $\Sigma\tau = I\alpha$, where

$$I = (1/2)\ mr^2 = (1/2)(0.400 \text{ kg})(0.130 \text{ m})^2 = 3.38 \times 10^{-3} \text{ kg·m}^2$$

Also, $\omega = \omega_0 + \alpha t$, so that $\alpha = (85.0 \text{ rad/s} - 262 \text{ rad/s})/(18.0 \text{ s}) = -9.83 \text{ rad/s}^2$. Thus, the net torque is

$$\Sigma\tau = I\alpha = (3.38 \times 10^{-3} \text{ kg·m}^2)(-9.83 \text{ rad/s}^2) = \boxed{-3.32\times10^{-2} \text{ N}\cdot\text{m}}$$

---

63. ***REASONING AND SOLUTION***

a. Angular momentum is conserved, so that $I\omega = I_0\omega_0$, where

$$I = I_0 + 10m_b R_b^{\ 2} = 2100 \text{ kg·m}^2$$

$$\omega = (I_0/I)\omega_0 = [(1500 \text{ kg·m}^2)/(2100 \text{ kg·m}^2)](0.20 \text{ rad/s}) = \boxed{0.14 \text{ rad/s}}$$

b. A net external torque must be applied in a direction that is opposite to the angular deceleration caused by the baggage dropping onto the carousel.

---

64. ***REASONING*** Although this arrangement of body parts is vertical, we can apply Equation 9.3 to locate the overall center of gravity by simply replacing the horizontal position *x* by the vertical position *y*, as measured relative to the floor.

SOLUTION Using Equation 9.3, we have

$$y_{cg} = \frac{W_1 y_1 + W_2 y_2 + W_3 y_3}{W_1 + W_2 + W_3}$$

$$= \frac{(438\text{ N})(1.28\text{ m}) + (144\text{ N})(0.760\text{ m}) + (87\text{ N})(0.250\text{ m})}{438\text{ N} + 144\text{ N} + 87\text{ N}} = \boxed{1.03\text{ m}}$$

---

65. SSM ***REASONING*** The figure below shows eight particles, each one located at a different corner of an imaginary cube. As shown, if we consider an axis that lies along one edge of the cube, two of the particles lie on the axis, and for these particles $r = 0$. The next four particles closest to the axis have $r = \ell$, where ℓ is the length of one edge of the cube. The remaining two particles have $r = d$, where d is the length of the diagonal along any one of the faces. From the Pythagorean theorem, $d = \sqrt{\ell^2 + \ell^2} = \ell\sqrt{2}$.

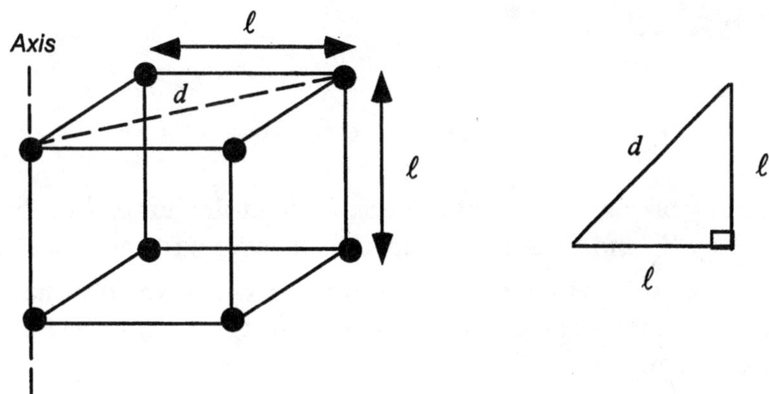

According to Equation 9.6, the moment of inertia of a system of particles is given by $I = \sum mr^2$.

SOLUTION Direct application of Equation 9.6 gives

$$I = \sum mr^2 = 4(m\ell^2) + 2(md^2) = 4(m\ell^2) + 2(2m\ell^2) = 8m\ell^2$$

or

$$I = 8(0.12\text{ kg})(0.25\text{ m})^2 = \boxed{0.060\text{ kg}\cdot\text{m}^2}$$

---

66. ***REASONING*** The drawing shows the beam and the five forces that act on it: the horizontal and vertical components $\mathbf{S}_x$ and $\mathbf{S}_y$ that the wall exerts on the left end of the beam, the weight $\mathbf{W}_b$ of the beam, the force due to the weight $\mathbf{W}_c$ of the crate, and the tension $\mathbf{T}$ in the cable. The beam is uniform, so its center of gravity is at the center of the beam, which is

where its weight can be assumed to act. Since the beam is in equilibrium, the sum of the torques about any axis of rotation must be zero $(\Sigma\tau = 0)$, and the sum of the forces in the horizontal and vertical directions must be zero $\left(\Sigma F_x = 0,\ \Sigma F_y = 0\right)$. These three conditions will allow us to determine the magnitudes of $\mathbf{S}_x$, $\mathbf{S}_y$, and **T**.

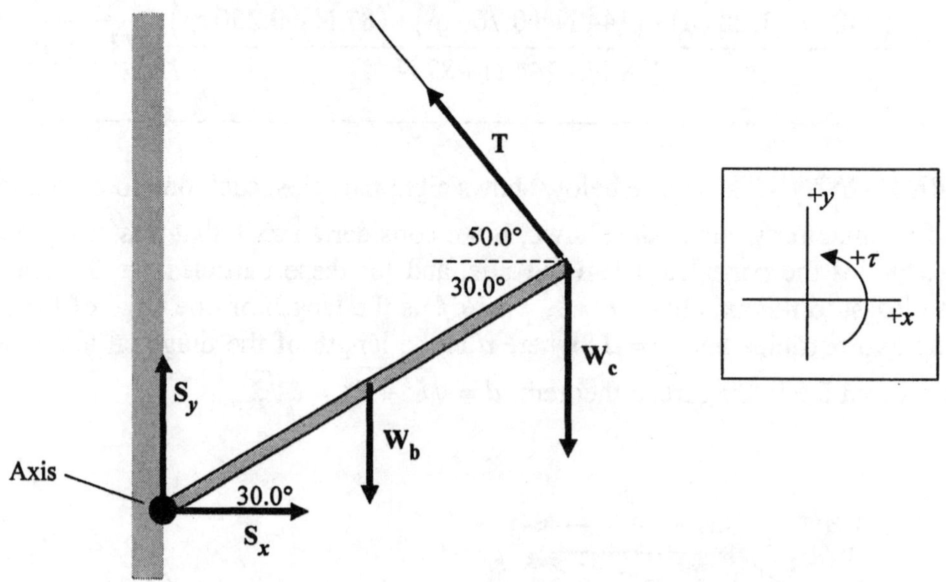

SOLUTION

a. We will begin by taking the axis of rotation to be at the left end of the beam. Then the torques produced by $\mathbf{S}_x$ and $\mathbf{S}_y$ are zero, since their lever arms are zero. When we set the sum of the torques equal to zero, the resulting equation will have only one unknown, *T*, in it. Setting the sum of the torques produced by the three forces equal to zero gives (with *L* equal to the length of the beam)

$$\Sigma\tau = -W_b\left(\tfrac{1}{2}L\cos 30.0°\right) - W_c\left(L\cos 30.0°\right) + T\left(L\sin 80.0°\right) = 0$$

Algebraically eliminating *L* from this equation and solving for *T* gives

$$T = \frac{W_b\left(\frac{1}{2}\cos 30.0°\right) + W_c\left(\cos 30.0°\right)}{\sin 80.0°}$$

$$= \frac{(1220\ \text{N})\left(\frac{1}{2}\cos 30.0°\right) + (1960\ \text{N})\left(\cos 30.0°\right)}{\sin 80.0°} = \boxed{2260\ \text{N}}$$

b. Since the beam is in equilibrium, the sum of the forces in the vertical direction must be zero:

$$\Sigma F_y = +S_y - W_b - W_c + T\sin 50.0° = 0$$

Solving for S_y gives

$$S_y = W_b + W_c - T\sin 50.0° = 1220\text{ N} + 1960\text{ N} - (2260\text{ N})\sin 50.0° = \boxed{1450\text{ N}}$$

The sum of the forces in the horizontal direction must also be zero:

$$\Sigma F_x = +S_x - T\cos 50.0° = 0$$

so that

$$S_x = T\cos 50.0° = (2260\text{ N})\cos 50.0° = \boxed{1450\text{ N}}$$

---

67. **SSM** **WWW** ***REASONING AND SOLUTION*** The figure below shows the massless board and the forces that act on the board due to the person and the scales.

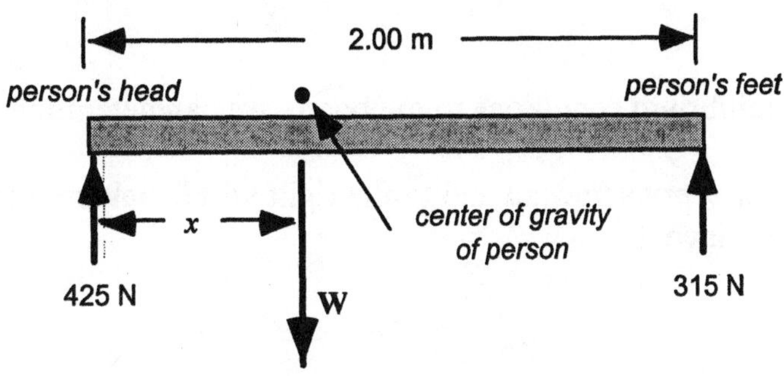

a. Applying Newton's second law to the vertical direction gives

$$315\text{ N} + 425\text{ N} - W = 0 \qquad \text{or} \qquad \mathbf{W} = \boxed{7.40\times 10^2\text{ N, downward}}$$

b. Let x be the position of the center of gravity relative to the scale at the person's head. Taking torques about an axis through the contact point between the scale at the person's head and the board gives

$$(315\text{ N})(2.00\text{ m}) - (7.40\times 10^2\text{ N})x = 0 \qquad \text{or} \qquad x = \boxed{0.851\text{ m}}$$

---

68. ***REASONING*** If we assume that the system is in equilibrium, we know that the vector sum of all the forces, as well as the vector sum of all the torques, that act on the system must be zero.

The figure below shows a free body diagram for the boom. Since the boom is assumed to be uniform, its weight $\mathbf{W}_B$ is located at its center of gravity, which coincides with its geometrical center. There is a tension **T** in the cable that acts at an angle θ to the horizontal, as shown. At the hinge pin P, there are two forces acting. The vertical force **V** that acts on the end of the boom prevents the boom from falling down. The horizontal force **H** that also acts at the hinge pin prevents the boom from sliding to the left. The weight $\mathbf{W}_L$ of the wrecking ball (the "load") acts at the end of the boom.

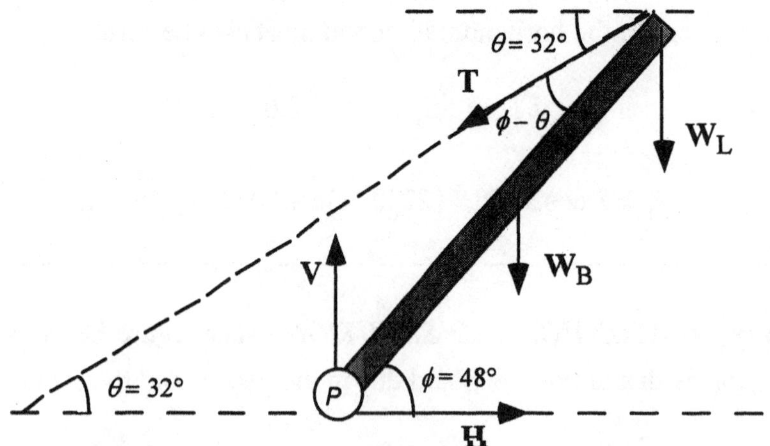

By applying the equilibrium conditions to the boom, we can determine the desired forces.

SOLUTION The directions upward and to the right will be taken as the positive directions. In the x direction we have

$$\sum F_x = H - T\cos\theta = 0 \tag{1}$$

while in the y direction we have

$$\sum F_y = V - T\sin\theta - W_L - W_B = 0 \tag{2}$$

Equations (1) and (2) give us two equations in three unknown. We must, therefore, find a third equation that can be used to determine one of the unknowns. We can get the third equation from the torque equation.

In order to write the torque equation, we must first pick an axis of rotation and determine the lever arms for the forces involved. Since both V and H are unknown, we can eliminate them from the torque equation by picking the rotation axis through the point P (then both V and H have zero lever arms). If we let the boom have a length L, then the lever arm for $\mathbf{W}_L$ is $L\cos\phi$, while the lever arm for $\mathbf{W}_B$ is $(L/2)\cos\phi$. From the figure, we see that the lever arm for $\mathbf{T}$ is $L\sin(\phi-\theta)$. If we take counterclockwise torques as positive, then the torque equation is

$$\sum \tau = -W_B\left(\frac{L\cos\phi}{2}\right) - W_L L\cos\phi + TL\sin\left(\phi-\theta\right) = 0$$

Solving for T, we have

$$T = \frac{\frac{1}{2}W_B + W_L}{\sin(\phi-\theta)}\cos\phi \tag{3}$$

a. From Equation (3) the tension in the support cable is

$$T = \frac{\frac{1}{2}(3600\text{ N}) + 4800\text{ N}}{\sin(48° - 32°)} \cos 48° = \boxed{1.6\times10^4\text{ N}}$$

b. The force exerted on the lower end of the hinge at the point P is the vector sum of the forces **H** and **V**. According to Equation (1),

$$H = T\cos\theta = \left(1.6\times10^4\text{ N}\right)\cos 32° = 1.4\times10^4\text{ N}$$

and, from Equation (2)

$$V = W_L + W_B + T\sin\theta = 4800\text{ N} + 3600\text{ N} + \left(1.6\times10^4\text{ N}\right)\sin 32° = 1.7\times10^4\text{ N}$$

Since the forces **H** and **V** are at right angles to each other, the magnitude of their vector sum can be found from the Pythagorean theorem:

$$F_P = \sqrt{H^2 + V^2} = \sqrt{(1.4\times10^4\text{ N})^2 + (1.7\times10^4\text{ N})^2} = \boxed{2.2\times10^4\text{ N}}$$

---

69. ***REASONING*** When the modules pull together, they do so by means of forces that are internal. These pulling forces, therefore, do not create a net external torque, and the angular momentum of the system is conserved. In other words, it remains constant. We will use the conservation of angular momentum to obtain a relationship between the initial and final angular speeds. Then, we will use Equation 8.9 ($v = r\omega$) to relate the angular speeds ω_0 and ω_f to the tangential speeds v_0 and v_f.

SOLUTION Let L be the initial length of the cable between the modules and ω_0 be the initial angular speed. Relative to the center-of-mass axis, the initial momentum of inertia of the two-module system is $I_0 = 2M(L/2)^2$, according to Equation 9.6. After the modules pull together, the length of the cable is $L/2$, the final angular speed is ω_f, and the momentum of inertia is $I_f = 2M(L/4)^2$. The conservation of angular momentum indicates that

$$\underbrace{I_f \omega_f}_{\text{Final angular momentum}} = \underbrace{I_0 \omega_0}_{\text{Initial angular momentum}}$$

$$\left[2M\left(\frac{L}{4}\right)^2\right]\omega_f = \left[2M\left(\frac{L}{2}\right)^2\right]\omega_0$$

$$\omega_f = 4\omega_0$$

According to Equation 8.9, $\omega_f = v_f/(L/4)$ and $\omega_0 = v_0/(L/2)$. With these substitutions, the result that $\omega_f = 4\omega_0$ becomes

$$\frac{v_f}{L/4} = 4\left(\frac{v_0}{L/2}\right) \quad \text{or} \quad v_f = 2v_0 = 2(17\text{ m/s}) = \boxed{34\text{ m/s}}$$

---

70. ***REASONING AND SOLUTION*** The conservation of energy gives for the cube $(1/2)\,mv_c^2 = mgh$, so $v_c = \sqrt{2gh}$. The conservation of energy gives for the marble $(1/2)\,mv_m^2 + (1/2)\,I\omega^2 = mgh$. Since the marble rolls without slipping: $\omega = v/R$. For a solid sphere we know $I = (2/5)\,mR^2$. Combining the last three equations gives

$$v_m = \sqrt{\frac{10gh}{7}}$$

Thus,

$$\frac{v_c}{v_m} = \sqrt{\frac{7}{5}} = \boxed{1.18}$$

---

71. ***REASONING AND SOLUTION*** Taking torques on the right leg about the apex of the "A" gives

$$\Sigma\tau = -(120\text{ N})\,L\cos 60.0° - FL\cos 30.0° + T(2L)\cos 60.0° = 0$$

where F is the horizontal component of the force exerted by the crossbar and T is the tension in the right string. Due to symmetry, the tension in the left string is also T. Newton's second law applied to the vertical gives

$$2T - 2W = 0 \quad \text{so} \quad T = 120\text{ N}$$

Substituting $T = 120$ N into the first equation yields $F = \boxed{69\text{ N}}$.

---

72. ***REASONING*** The drawing shows the drum, pulley, and the crate, as well as the tensions in the cord

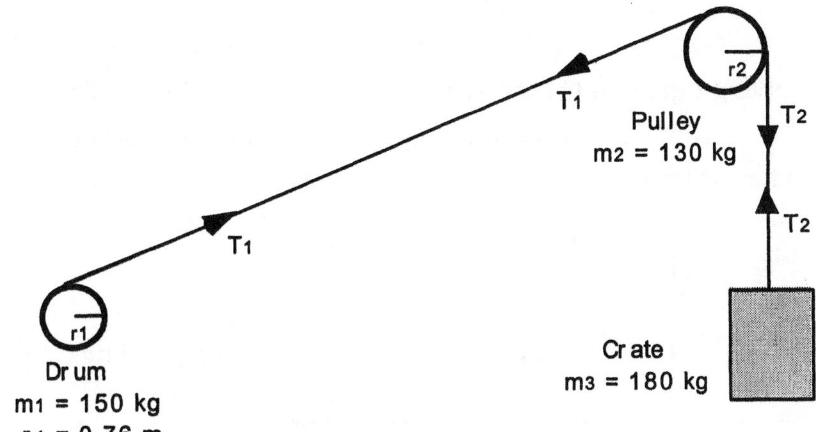

Let T_1 represent the magnitude of the tension in the cord between the drum and the pulley. Then, the net torque exerted on the drum must be, according to Equation 9.7, $\Sigma\tau = I_1\alpha_1$, where I_1 is the moment of inertia of the drum, and α_1 is its angular acceleration. If we assume that the cable does not slip, then Equation 9.7 can be written as

$$\underbrace{-T_1 r_1 + \tau}_{\Sigma\tau} = \underbrace{\left(m_1 r_1^2\right)}_{I_1}\underbrace{\left(\frac{a}{r_1}\right)}_{\alpha_1} \tag{1}$$

where τ is the counterclockwise torque provided by the motor, and a is the acceleration of the cord ($a = 1.2$ m/s^2). This equation cannot be solved for τ directly, because the tension T_1 is not known.

We next apply Newton's second law for rotational motion to the pulley in the drawing:

$$\underbrace{+T_1 r_2 - T_2 r_2}_{\Sigma\tau} = \underbrace{\left(\tfrac{1}{2} m_2 r_2^2\right)}_{I_2}\underbrace{\left(\frac{a}{r_2}\right)}_{\alpha_2} \tag{2}$$

where T_2 is the magnitude of the tension in the cord between the pulley and the crate, and I_2 is the moment of inertial of the pulley.

Finally, Newton's second law for translational motion ($\Sigma F_y = m\,a$) is applied to the crate, yielding

$$\underbrace{+T_2 - m_3 g}_{\Sigma F_y} = m_3 a \qquad (3)$$

SOLUTION Solving Equation (1) for T_1 and substituting the result into Equation (2), then solving Equation (2) for T_2 and substituting the result into Equation (3), results in the following value for the torque

$$\tau = r_1\left[a\left(m_1 + \tfrac{1}{2}m_2 + m_3\right) + m_3 g\right]$$

$$= (0.76\text{ m})\left[(1.2\text{ m/s}^2)\left(150\text{ kg} + \tfrac{1}{2}130\text{ kg} + 180\text{ kg}\right) + (180\text{ kg})(9.80\text{ m/s}^2)\right] = \boxed{1700\text{ N}\cdot\text{m}}$$

---

73. ***CONCEPT QUESTIONS*** a. The magnitude of a torque is the magnitude of the force times the lever arm of the force, according to Equation 9.1. The lever arm is the perpendicular distance between the line of action of the force and the axis. For the force at corner B, the lever arm is half the length of the short side of the rectangle. For the force at corner D, the lever arm is half the length of the long side of the rectangle. Therefore, the lever arm for the force at corner D is greater. Since the magnitudes of the two forces are the same, this means that the force at corner D creates the greater torque.

b. As we have seen, the torque at corner D (clockwise) has a greater magnitude than the torque at corner B (counterclockwise). Therefore, the net torque from these two contributions is clockwise. This means that the force at corner A must produce a counterclockwise torque, if the net torque produced by the three forces is to be zero. The force at corner A, then, must point toward corner B.

SOLUTION Let L be the length of the short side of the rectangle, so that the length of the long side is $2L$. We can use Equation 9.1 for the magnitude of each torque. Remembering that counterclockwise torques are positive and that the torque at corner A is counterclockwise, we can write the zero net torque as follows:

$$\Sigma\tau = F_A\ell_A + F_B\ell_B - F_D\ell_D = 0$$

$$F_A L + (12\text{ N})\left(\tfrac{1}{2}L\right) - (12\text{ N})L = 0$$

The length L can be eliminated algebraically from this result, which can then be solved for F_A:

$$F_A = -(12\text{ N})\left(\tfrac{1}{2}\right) + (12\text{ N}) = \boxed{6.0\text{ N (pointing toward corner B)}}$$

---

74. ***CONCEPT QUESTIONS*** a. In the left design, both the 60.0 and the 525-N forces create clockwise torques with respect to the rotational axis at the point where the tire contacts the ground. In the right design, only the 60.0-N force creates a torque, because the line of action

of the 525-N force passes through the axis. Thus, the total torque from the two forces is greater for the left design.

b. The man is supporting the wheelbarrow in equilibrium. Therefore, his force must create a torque that balances the total torque from the other two forces. To do this, his torque must be counterclockwise and have the same magnitude as the total torque from the other two forces. Since the two forces create more torque in the left design, the magnitude of the man's torque must be greater for that design.

c. The magnitude of a torque is the magnitude of the force times the lever arm of the force, according to Equation 9.1. The drawing shows that the lever arms of the man's forces in the two designs is the same (1.300 m). Thus, to create the greater torque necessitated by the left design, the man must apply a greater force for that design.

SOLUTION The lever arms for the forces can be obtained from the distances shown in the drawing for each design. We can use Equation 9.1 for the magnitude of each torque. Remembering that counterclockwise torques are positive, we can write an expression for the zero net torque for each design. These expressions can be solved for the man's force F in each case:

Left design

$$\Sigma\tau = -(525\text{ N})(0.400\text{ m}) - (60.0\text{ N})(0.600\text{ m}) + F(1.300\text{ m}) = 0$$

$$F = \frac{(525\text{ N})(0.400\text{ m}) + (60.0\text{ N})(0.600\text{ m})}{1.300\text{ m}} = \boxed{189\text{ N}}$$

Right design

$$\Sigma\tau = -(60.0\text{ N})(0.600\text{ m}) + F(1.300\text{ m}) = 0$$

$$F = \frac{(60.0\text{ N})(0.600\text{ m})}{1.300\text{ m}} = \boxed{27.7\text{ N}}$$

---

75. ***CONCEPT QUESTIONS*** a. The forces in part *b* of the drawing cannot keep the beam in equilibrium (no acceleration), because neither **V** nor **W**, both being vertical, can balance the horizontal component of **P**. This unbalanced component would accelerate the beam to the right.

b. The forces in part *c* of the drawing cannot keep the beam in equilibrium (no acceleration). To see why, consider a rotational axis perpendicular to the plane of the paper and passing through the pin. Neither **P** nor **H** create torques relative to this axis, because their lines of action pass directly through it. However, **W** does create a torque, which is unbalanced. This torque would cause the beam to have a counterclockwise angular acceleration.

c. The forces in part *d* are sufficient to eliminate the accelerations discussed above, so they can keep the beam in equilibrium.

SOLUTION Assuming that upward and to the right are the positive directions, we obtain the following expressions by setting the sum of the vertical and the sum of the horizontal forces equal to zero:

Horizontal forces $$P\cos\theta - H = 0 \qquad (1)$$

Vertical forces $$P\sin\theta + V - W = 0 \qquad (2)$$

Using a rotational axis perpendicular to the plane of the paper and passing through the pin and remembering that counterclockwise torques are positive, we also set the sum of the torques equal to zero. In doing so, we use L to denote the length of the beam and note that the lever arms for **W** and **V** are $L/2$ and L, respectively. The forces **P** and **H** create no torques relative to this axis, because their lines of action pass directly through it.

Torques $$W\left(\tfrac{1}{2}L\right) - VL = 0 \qquad (3)$$

Since L can be eliminated algebraically, Equation (3) may be solved immediately for V:

$$V = \tfrac{1}{2}W = \tfrac{1}{2}(340\text{ N}) = \boxed{170\text{ N}}$$

Substituting this result into Equation (2) gives

$$P\sin\theta + \tfrac{1}{2}W - W = 0$$

$$P = \frac{W}{2\sin\theta} = \frac{340\text{ N}}{2\sin 39°} = \boxed{270\text{ N}}$$

Substituting this result into Equation (1) gives

$$\left(\frac{W}{2\sin\theta}\right)\cos\theta - H = 0$$

$$H = \frac{W}{2\tan\theta} = \frac{340\text{ N}}{2\tan 39°} = \boxed{210\text{ N}}$$

---

76. ***CONCEPT QUESTIONS*** a. Equation 8.7 from the equations of rotational kinematics indicates that the angular displacement θ is $\theta = \omega_0 t + \tfrac{1}{2}\alpha t^2$, where ω_0 is the initial angular velocity, t is the time, and α is the angular acceleration. Since both wheels start from rest, ω_0 = 0 rad/s for each. Furthermore, each wheel makes the same number of revolutions in the same time, so θ and t are also the same for each. Therefore, the angular acceleration α must be the same for each.

b. For the hoop, all of the mass is located at a distance from the axis equal to the radius. For the disk, some of the mass is located closer to the axis. According to Equation 9.6, the moment of inertia is greater when the mass is located farther from the axis. Therefore, the moment of inertia of the hoop is greater. Table 9.1 indicates that the moment of inertia of a hoop is $I_{\text{hoop}} = MR^2$, while the moment of inertia of a disk is $I_{\text{disk}} = \frac{1}{2}MR^2$.

c. Newton's second law for rotational motion indicates that the net external torque is equal to the moment of inertia times the angular acceleration. Both disks have the same angular acceleration, but the moment of inertia of the hoop is greater. Thus, the net external torque must be greater for the hoop.

SOLUTION Using Equation 8.7, we can express the angular acceleration as follows:

$$\theta = \omega_0 t + \tfrac{1}{2}\alpha t^2 \quad \text{or} \quad \alpha = \frac{2(\theta - \omega_0 t)}{t^2}$$

This expression for the acceleration can now be used in Newton's second law for rotational motion. Accordingly, the net external torque is

$$\text{Hoop} \quad \Sigma\tau = I_{\text{hoop}}\alpha = MR^2\left[\frac{2(\theta - \omega_0 t)}{t^2}\right]$$

$$= (4.0\text{ kg})(0.35\text{ m})^2\left[\frac{2(13\text{ rad}) - 2(0\text{ rad/s})(8.0\text{ s})}{(8.0\text{ s})^2}\right] = \boxed{0.20\text{ N}\cdot\text{m}}$$

$$\text{Disk} \quad \Sigma\tau = I_{\text{disk}}\alpha = \tfrac{1}{2}MR^2\left[\frac{2(\theta - \omega_0 t)}{t^2}\right]$$

$$= \tfrac{1}{2}(4.0\text{ kg})(0.35\text{ m})^2\left[\frac{2(13\text{ rad}) - 2(0\text{ rad/s})(8.0\text{ s})}{(8.0\text{ s})^2}\right] = \boxed{0.10\text{ N}\cdot\text{m}}$$

---

77. ***CONCEPT QUESTIONS*** a. According to Equation 9.6, the moment of inertia for rod A is just that of the attached particle, since the rod itself is massless. For rod A with its attached particle, then, the moment of inertia is $I_A = ML^2$. According to Table 9.1, the moment of inertia for rod B is $I_B = \frac{1}{3}ML^2$. The moment of inertia for rod A with its attached particle is greater.

b. Since the moment of inertia for rod A is greater, its kinetic energy is also greater, according to Equation 9.9 $\text{KE}_R = \frac{1}{2}I\omega^2$.

SOLUTION Using Equation 9.9 to calculate the kinetic energy, we find

Rod A

$$\text{KE}_{\text{R}} = \tfrac{1}{2} I_{\text{A}} \omega^2 = \tfrac{1}{2}\left(ML^2 \right)\omega^2$$

$$= \tfrac{1}{2}(0.66 \text{ kg})(0.75 \text{ m})^2 (4.2 \text{ rad/s})^2 = \boxed{3.3 \text{ J}}$$

Rod B

$$\text{KE}_{\text{R}} = \tfrac{1}{2} I_{\text{B}} \omega^2 = \tfrac{1}{2}\left(\tfrac{1}{3} ML^2 \right)\omega^2$$

$$= \tfrac{1}{6}(0.66 \text{ kg})(0.75 \text{ m})^2 (4.2 \text{ rad/s})^2 = \boxed{1.1 \text{ J}}$$

---

78. ***CONCEPT QUESTIONS*** a. The force is applied to the person in the counterclockwise direction by the bar that he grabs when climbing aboard.

b. According to Newton's action-reaction law, the person must apply a force of equal magnitude and opposite direction to the bar and the carousel. This force acts on the carousel in the clockwise direction and, therefore, creates a clockwise torque.

c. Since the carousel is moving counterclockwise, the clockwise torque applied to it when the person climbs aboard decreases its angular speed.

SOLUTION We consider a system consisting of the person and the carousel. Since the carousel rotates on frictionless bearings, no net external torque acts on this system. Therefore, angular momentum is conserved, and we can set the total angular momentum of the system before the person hops on equal to the total angular momentum afterwards. Afterwards, the angular momentum includes a contribution from the person. According to Equation 9.6, his moment of inertia is $I_{\text{person}} = MR^2$, since he is at the outer edge of the carousel.

$$\underbrace{I_{\text{carousel}}\,\omega_{\text{f}} + I_{\text{person}}\,\omega_{\text{f}}}_{\text{Final total angular momentum}} = \underbrace{I_{\text{carousel}}\,\omega_0}_{\text{Initial total angular momentum}}$$

$$\omega_{\text{f}} = \frac{I_{\text{carousel}}\,\omega_0}{I_{\text{carousel}} + I_{\text{person}}} = \frac{I_{\text{carousel}}\,\omega_0}{I_{\text{carousel}} + MR^2}$$

$$= \frac{\left(125 \text{ kg}\cdot\text{m}^2\right)(3.14 \text{ rad/s})}{125 \text{ kg}\cdot\text{m}^2 + (40.0 \text{ kg})(1.50 \text{ m})^2} = \boxed{1.83 \text{ rad/s}}$$

---

79. ***CONCEPT QUESTIONS*** a. Two identical systems of objects may have different moments of inertia, depending on where the axis of rotation is located. The moment of inertia depends on the mass of each object and its distance from the axis of rotation. According to

Equation 9.6, the moment of inertia for the three-ball system is $I = m_1r_1^2 + m_2r_2^2 + m_3r_3^2$, where m is the mass of each ball and r is its distance from the axis. In system A, the ball whose mass is m_1 does not contribute to the moment of inertia, because the ball is located on the axis and $r_1 = 0$ m. In B, the ball whose mass is m_3 does not contribute to the moment of inertia, because it is located on the axis and $r_3 = 0$ m.

b. Even though the force acting on both systems is the same, the torque is not. The magnitude of the torque is equal to the magnitude F of the force times the lever arm ℓ (Equation 9.1). In system A the lever arm is $\ell = 3.00$ m. In B the lever arm is $\ell = 0$ m, since the line of action of the force passes through the axis of rotation.

c. The systems have different angular accelerations. According to Newton's second law for rotational motion, Equation 9.7, the angular acceleration α is given by $\alpha = (\Sigma\tau)/I$, where $\Sigma\tau$ is the net torque and I is the moment of inertia. Since the net torque and moment of inertia are different for the two systems, the angular accelerations are different. The angular velocity ω is given by Equation 8.4 as $\omega = \omega_0 + \alpha t$, where ω_0 is the initial angular velocity and t is the time. Since the angular accelerations are different for the two systems, the angular velocities will also be different at the same later time.

SOLUTION

a. The moment of inertia for each system is

System A $I = m_1r_1^2 + m_2r_2^2 + m_3r_3^2$

$$= (9.00\text{ kg})(0\text{ m})^2 + (6.00\text{ kg})(3.00\text{ m})^2 + (7.00\text{ kg})(5.00\text{ m})^2 = \boxed{229\text{ kg}\cdot\text{m}^2}$$

System B $I = m_1r_1^2 + m_2r_2^2 + m_3r_3^2$

$$= (9.00\text{ kg})(5.00\text{ m})^2 + (6.00\text{ kg})(4.00\text{ m})^2 + (7.00\text{ kg})(0\text{ m})^2 = \boxed{321\text{ kg}\cdot\text{m}^2}$$

As anticipated, the two systems have different moments of inertia.

b. The torque produced by the force has a magnitude that is equal to the product of the force magnitude and the lever arm:

System A $$\tau = -F\ell = -(424\text{ N})(3.00\text{ m}) = \boxed{-1270\text{ N}\cdot\text{m}}$$

The torque is negative because it produces a clockwise rotation about the axis.

System B $\tau = F\ell = (424\text{ N})(0\text{ m}) = \boxed{0\text{ N}\cdot\text{m}}$

As anticipated, the torques in the two cases are not the same.

c. The final angular velocity ω is related to the initial angular velocity ω_0, the angular acceleration α, and the time t by $\omega = \omega_0 + \alpha t$ (Equation 8.4). The angular acceleration is given by Newton's second law for rotational motion as $\alpha = (\Sigma\tau)/I$, where $\Sigma\tau$ is the net torque and I is the moment of inertia. Since there is only one torque acting on each system, it is the net torque, so $\Sigma\tau = \tau$. Substituting this expression for α into Equation 8.4 yields

$$\omega = \omega_0 + \alpha t = \omega_0 + \left(\frac{\tau}{I}\right)t$$

In both cases the initial angular velocity is ω_0 = 0 rad/s, since the systems start from rest. The final angular velocities after 5.00 s are:

System A
$$\omega = \omega_0 + \left(\frac{\tau}{I}\right)t = (0\text{ rad/s}) + \left(\frac{-1270\text{ N}\cdot\text{m}}{229\text{ kg}\cdot\text{m}^2}\right)(5.00\text{ s}) = \boxed{-27.7\text{ rad/s}}$$

System B
$$\omega = \omega_0 + \left(\frac{\tau}{I}\right)t = (0\text{ rad/s}) + \left(\frac{0\text{ N}\cdot\text{m}}{321\text{ kg}\cdot\text{m}^2}\right)(5.00\text{ s}) = \boxed{0\text{ rad/s}}$$

---

80. ***CONCEPT QUESTIONS***

a. The rolling wheel has the greater total kinetic energy. Its kinetic energy is the sum of its translational $\left(\frac{1}{2}mv^2\right)$ and rotational $\left(\frac{1}{2}I\omega^2\right)$ kinetic energies. The sliding wheel only has translational kinetic energy, since it does not rotate.

b. The rolling wheel has the greater potential energy when it comes to a halt. As the wheels move up the incline plane, the total mechanical energy is conserved, since only the conservative force of gravity does work on each wheel. Thus, the initial kinetic energy at the bottom of the incline is converted entirely into potential energy when the wheels come to a momentary halt. Since the rolling wheel has the greater kinetic energy to begin with, it also has the greater potential energy.

c. The rolling wheel rises to the greater height. The potential energy PE is given by Equation 6.5 as PE = *mgh*, where *h* is the height of the wheel above an arbitrary zero level. Since the rolling wheel has the greater potential energy when it comes to a halt, it also rises to a greater height.

d. The conclusion reached in question (c) would still be valid, because both the gravitational potential energy and the total kinetic energy are proportional to the mass. To find the height, we will set the two types of energy equal and, since each is proportional to the mass, the mass can be algebraically eliminated.

SOLUTION

a. The total kinetic energy KE of the rolling wheel is the sum of its translational and rotational kinetic energies

$$\text{KE} = \tfrac{1}{2}mv^2 + \tfrac{1}{2}I\omega^2$$

where ω is the angular speed of the wheel. Since the wheel is a disk, its moment of inertia is $I = \tfrac{1}{2}mR^2$ (see Table 9.1), where R is the radius of the disk. Furthermore, the angular speed ω of the rolling wheel is related to the linear speed v of its center of mass by Equation 8.12 as $\omega = v/R$. Thus, the total kinetic energy of the rolling wheel is

Rolling Wheel

$$\text{KE} = \tfrac{1}{2}mv^2 + \tfrac{1}{2}I\omega^2 = \tfrac{1}{2}mv^2 + \tfrac{1}{2}\left(\tfrac{1}{2}mR^2\right)\left(\frac{v}{R}\right)^2$$

$$= \tfrac{3}{4}mv^2 = \tfrac{3}{4}(2.0\text{ kg})(6.0\text{ m/s})^2 = \boxed{54\text{ J}}$$

The kinetic energy of the sliding wheel is

Sliding Wheel

$$\text{KE} = \tfrac{1}{2}mv^2 = \tfrac{1}{2}(2.0\text{ kg})(6.0\text{ m/s})^2 = \boxed{36\text{ J}}$$

As expected, the rolling wheel has the greater total kinetic energy.

b. As each wheel rolls up the incline, its total mechanical energy is conserved. The initial kinetic energy KE at the bottom of the incline is converted entirely into potential energy PE when the wheels come to a momentary halt. Thus, the potential energies of the wheels are:

Rolling Wheel

$$\text{PE} = \boxed{54\text{ J}}$$

Sliding Wheel

$$\text{PE} = \boxed{36\text{ J}}$$

As anticipated, the rolling wheel has the greater potential energy.

c. The potential energy of a wheel is given by Equation 6.5 as PE = mgh, where g is the acceleration due to gravity and h is the height relative to an arbitrary zero level. The height reached by each wheel is

Rolling Wheel

$$h = \frac{\text{PE}}{mg} = \frac{54\text{ J}}{(2.0\text{ kg})(9.80\text{ m/s}^2)} = \boxed{2.8\text{ m}}$$

Sliding Wheel

$$h = \frac{\text{PE}}{mg} = \frac{36\text{ J}}{(2.0\text{ kg})(9.80\text{ m/s}^2)} = \boxed{1.8\text{ m}}$$

d. The total kinetic energy of the rolling wheel is $\text{KE} = \frac{3}{4}mv^2$ (see the results of part a). The kinetic energy KE at the bottom of the incline is converted entirely into potential energy PE when the wheels come to a monetary halt. Since the potential energy is PE = mgh, we have that

$$\underbrace{\cancel{m}gh}_{\text{PE}} = \underbrace{\tfrac{3}{4}\cancel{m}v^2}_{\text{KE}} \quad \text{or} \quad h = \frac{3v^2}{4g}$$

The mass has been algebraically eliminated from this expression, so the final height reached by the wheel is independent of its mass. Thus, $h = \boxed{2.8\text{ m}}$, as expected.

CHAPTER 10 | SIMPLE HARMONIC MOTION AND ELASTICITY

PROBLEMS

---

1. **SSM** ***REASONING AND SOLUTION*** Using Equation 10.1, we first determine the spring constant:

$$k = \frac{F_{\text{Applied}}}{x} = \frac{89.0\text{ N}}{0.0191\text{ m}} = 4660\text{ N/m}$$

Again using Equation 10.1, we find that the force needed to compress the spring by 0.0508 m is

$$F_{\text{Applied}} = kx = (4660\text{ N/m})(0.0508\text{ m}) = \boxed{237\text{ N}}$$

---

2. ***REASONING AND SOLUTION*** Since $F = kx$,

$$x = \frac{F}{k} = \frac{9.6\text{ N}}{44\text{ N/m}} = \boxed{0.22\text{ m}}$$

---

3. ***REASONING AND SOLUTION*** $F_{\text{Applied}} = kx$ for either expansion or compression. Therefore,

a. $$F_{\text{Applied}} = kx = (248\text{ N/m})(0.0300\text{ m}) = \boxed{7.44\text{ N}} \qquad (10.1)$$

b. $$F_{\text{Applied}} = kx = (248\text{ N/m})(0.0300\text{ m}) = \boxed{7.44\text{ N}}$$

---

4. ***REASONING AND SOLUTION*** According to Equation 10.1 and the data from the graph, the effective spring constant is

$$k = \frac{F_{\text{Applied}}}{x} = \frac{160\text{ N}}{0.24\text{ m}} = \boxed{6.7\times 10^{2}\text{ N/m}}$$

---

5. **SSM** ***REASONING AND SOLUTION*** According to Newton's second law, the force required to accelerate the trailer is $F = ma$. The displacement of the spring is given by Equation 10.2. Solving Equation 10.2 for x and using $F = ma$, we obtain

$$x = -\frac{F}{k} = -\frac{ma}{k} = -\frac{(92 \text{ kg})(0.30 \text{ m/s}^2)}{2300 \text{ N/m}} = -0.012 \text{ m}$$

The amount that the spring stretches is $\boxed{0.012 \text{ m}}$.

---

6. ***REASONING*** The block is at equilibrium as it hangs on the spring. Therefore, the downward-directed weight of the block is balanced by the upward-directed force applied to the block by the spring. The weight of the block is mg, where m is the mass and g is the acceleration due to gravity. The force exerted by the spring is given by Equation 10.2 ($F = -kx$), where k is the spring constant and x is the displacement of the spring from its unstrained length. We will apply this reasoning twice, once to the single block hanging, and again to the two blocks. Although the spring constant is unknown, we will be able to eliminate it algebraically from the resulting two equations and determine the mass of the second block.

SOLUTION Let upward be the positive direction. Setting the weight mg equal to the force $-kx$ exerted by the spring in each case gives

$$\underbrace{m_1 g = -kx_1}_{\text{One hanging block}} \quad \text{and} \quad \underbrace{m_1 g + m_2 g = -kx_2}_{\text{Two hanging blocks}}$$

Dividing the equation on the right by the equation on the left, we can eliminate the unknown spring constant k and find that

$$\frac{m_1 g + m_2 g}{m_1 g} = \frac{-kx_2}{-kx_1} \quad \text{or} \quad 1 + \frac{m_2}{m_1} = \frac{x_2}{x_1}$$

It is given that $x_2/x_1 = 3.0$. Therefore, solving for the mass of the second block reveals that

$$m_2 = m_1\left(\frac{x_2}{x_1} - 1\right) = (0.70 \text{ kg})(3.0 - 1) = \boxed{1.4 \text{ kg}}$$

---

7. SSM ***REASONING*** When the ball is whirled in a horizontal circle of radius r at speed v, the centripetal force is provided by the restoring force of the spring. From Equation 5.3, the magnitude of the centripetal force is mv^2/r, while the magnitude of the restoring force is kx (see Equation 10.2). Thus,

$$\frac{mv^2}{r} = kx \qquad (1)$$

The radius of the circle is equal to $(L_0 + \Delta L)$, where L_0 is the unstretched length of the spring and ΔL is the amount that the spring stretches. Equation (1) becomes

$$\frac{mv^2}{L_0 + \Delta L} = k\Delta L \qquad (1')$$

If the spring were attached to the ceiling and the ball were allowed to hang straight down, motionless, the net force must be zero: $mg - kx = 0$, where $-kx$ is the restoring force of the spring. If we let Δy be the displacement of the spring in the vertical direction, then

$$mg = k\Delta y$$

Solving for Δy, we obtain

$$\Delta y = \frac{mg}{k} \qquad (2)$$

SOLUTION According to equation (1') above, the spring constant k is given by

$$k = \frac{mv^2}{\Delta L(L_0 + \Delta L)}$$

Substituting this expression for k into equation (2) gives

$$\Delta y = \frac{mg\Delta L(L_0 + \Delta L)}{mv^2} = \frac{g\Delta L(L_0 + \Delta L)}{v^2}$$

or

$$\Delta y = \frac{(9.80 \text{ m/s}^2)(0.010 \text{ m})(0.200 \text{ m} + 0.010 \text{ m})}{(3.00 \text{ m/s})^2} = \boxed{2.29 \times 10^{-3} \text{ m}}$$

---

8. ***REASONING AND SOLUTION*** The figure at the right shows the original situation before the spring is cut. The weight, W, of the object stretches the string by an amount x.

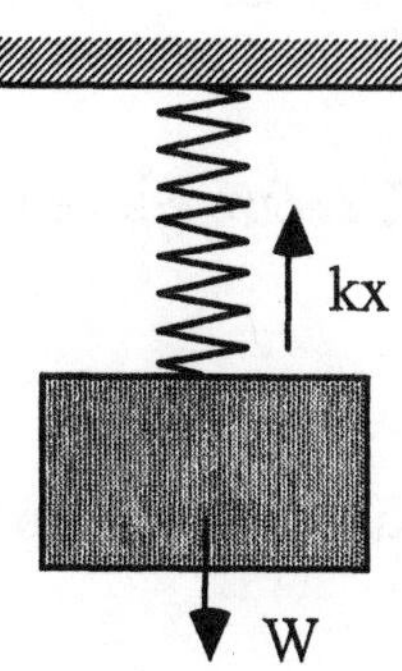

Applying F = kx to this situation gives

$$W = kx \qquad (1)$$

The figure at the right shows the situation after the spring is cut into two segments of equal length.

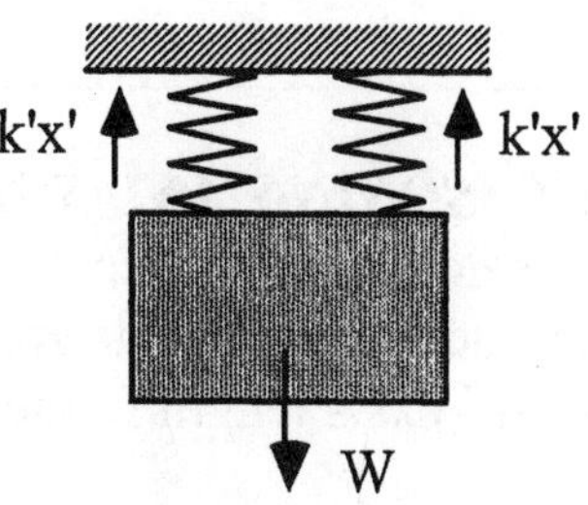

Let k' represent the spring constant of each half of the spring after it is cut. Now the weight, W, of the object stretches *each* segment by an amount x'.

Applying F = kx to this situation gives

$$W = k'x' + k'x' = 2k'x' \quad (2)$$

Combining Equations (1) and (2) yields

$$kx = 2k'x'$$

From Conceptual Example 2, we know that k' = 2k so that

$$kx = 2(2k)x'$$

Solving for x' gives

$$x' = \frac{x}{4} = \frac{0.160 \text{ m}}{4} = \boxed{0.040 \text{ m}}$$

---

9. ***REASONING*** The free-body diagram shows the magnitudes and directions of the forces acting on the block. The weight *mg* acts downward. The maximum force of static friction f_s^{max} acts upward just before the block begins to slip. The force from the spring $F = kx$ is directed to the right. The normal force F_N from the wall points to the left. The magnitude of the maximum force of static friction is related to the magnitude of the normal force according to Equation 4.7 $\left(f_s^{max} = \mu_s F_N\right)$, where μ_s is the coefficient of static friction. Since the block is at equilibrium just before it begins to slip, the forces in the *x* direction must balance and the forces in the *y* direction must balance. The balance of forces in the two directions will provide us with two equations, from which we will determine the coefficient of static friction.

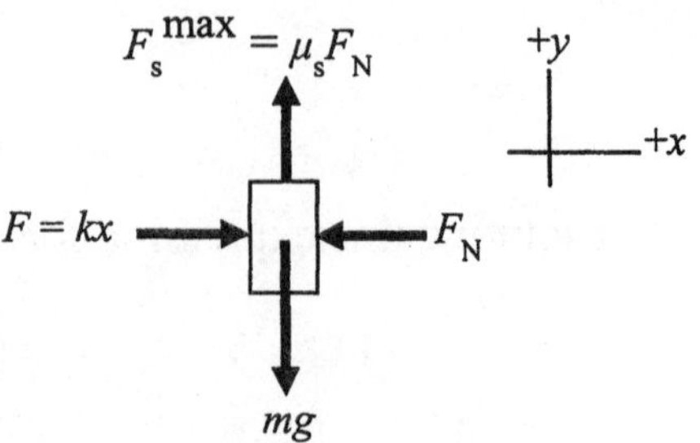

SOLUTION Since the forces in the *x* direction and in the *y* direction must balance, we have

$$mg = \mu_s F_N \quad \text{and} \quad F_N = kx$$

Substituting the second equation into the first equation gives

$$mg = \mu_s F_N = \mu_s(kx) \quad \text{or} \quad \mu_s = \frac{mg}{kx} = \frac{(1.6 \text{ kg})(9.80 \text{ m/s}^2)}{(510 \text{ N/m})(0.039 \text{ m})} = \boxed{0.79}$$

---

10. ***REASONING AND SOLUTION*** In order to find the amount the spring stretches we need to calculate the force, F = kx, acting on the spring. Since the board is in equilibrium, the net torque acting on it is zero. Taking the axis of rotation to be at the corner and assuming the board has a length L, the net torque is

$$\Sigma\tau = FL \sin 50.0° - mg(L/2) \cos 50.0° = 0$$

Solving for F gives

$$F = \frac{mg\cos 50.0°}{2\sin 50.0°} = \frac{mg}{2\tan 50.0°}$$

Therefore,

$$x = \frac{F}{k} = \frac{mg}{2k\tan 50.0°} = \frac{(10.1\text{ kg})(9.80\text{ m/s}^2)}{2(176\text{ N/m})\tan 50.0°} = \boxed{0.236\text{ m}}$$

---

11. **SSM** ***REASONING AND SOLUTION*** The force that acts on the block is given by Newton's Second law, $F = ma$. Since the block has a constant acceleration, the acceleration is given by Equation 2.8 with $v_0 = 0$ m/s; that is, $a = 2d/t^2$, where d is the distance through which the block is pulled. Therefore, the force that acts on the block is given by

$$F = ma = \frac{2md}{t^2}$$

The force acting on the block is the restoring force of the spring. Thus, according to Equation 10.2, $F = -kx$, where k is the spring constant and x is the displacement. Solving Equation 10.2 for x and using the expression above for F, we obtain

$$x = -\frac{F}{k} = -\frac{2md}{kt^2} = -\frac{2(7.00\text{ kg})(4.00\text{ m})}{(415\text{ N/m})(0.750\text{ s})^2} = -0.240\text{ m}$$

The amount that the spring stretches is $\boxed{0.240\text{ m}}$.

---

12. ***REASONING AND SOLUTION*** From the drawing given with the problem statement, we see that the kinetic frictional force on the bottom block (#1) is given by

$$F_{f1} = \mu_k(m_1 + m_2)g$$

and the maximum static frictional force on the top block (#2) is

$$F_{f2} = \mu_s m_2 g$$

Newton's second law applied to the bottom block gives

$$F - F_{f1} - kx = 0$$

and to the top block gives

$$F_{f2} - kx = 0$$

a. To find the compression, x, we have

$$x = F_{f2}/k = \mu_s m_2 g/k = (0.900)(15.0\text{ kg})(9.80\text{ m/s}^2)/(325\text{ N/m}) = \boxed{0.407\text{ m}}$$

b. $$F = kx + \mu_k(m_1 + m_2)g$$

$$F = (325\text{ N/m})(0.407\text{ m}) + (0.600)(45.0\text{ kg})(9.80\text{ m/s}^2) = \boxed{397\text{ N}}$$

---

13. ***REASONING AND SOLUTION*** In phase 1 of the block's motion (uniform acceleration) we find that the net force on the block is $F_1 - f_k = ma$ where the force of friction is $f_k = \mu_k mg$. Therefore $F_1 = m(a + \mu_k g)$, which is just the force exerted by the spring on the block, i.e., $F_1 = kx_1$. So we have

$$kx_1 = m(a + \mu_k g) \qquad (1)$$

We can find the acceleration using

$$a = \frac{v_f - v_0}{t} = \frac{5.00\text{ m/s} - 0\text{ m/s}}{0.500\text{ s}} = \boxed{10.0\text{ m/s}^2}$$

In phase 2 of the block's motion (constant speed) $a = 0\text{ m/s}^2$, so the force exerted by the spring is

$$k\,x_2 = m\mu_k g \qquad (2)$$

so that

$$\mu_k = \frac{k\,x_2}{m\,g}$$

Using this expression for μ_k in Equation (1) we obtain

$$k\,x_1 = m\left[a + \left(\frac{kx_2}{mg}\right)g\right]$$

a. Solving for k gives

$$k = \frac{ma}{x_1 - x_2} = \frac{(15.0\text{ kg})(10.0\text{ m/s}^2)}{0.200\text{ m} - 0.0500\text{ m}} = \boxed{1.00 \times 10^3\text{ N/m}}$$

b. Substituting this value for k into Equation (2) , we have

$$\mu_k = \frac{k\,x_2}{m\,g} = \frac{(1.00 \times 10^3\text{ N/m})(0.0500\text{ m})}{(15.0\text{ kg})(9.80\text{ m/s}^2)} = \boxed{0.340}$$

---

14. ***REASONING*** The number of times the diaphragm moves back and forth is the frequency f of the motion (in cycles/s or Hz) times the time interval t. The frequency f is related to the angular frequency ω (in rad/s) by Equation 10.6 ($\omega = 2\pi f$).

SOLUTION Solving Equation 10.6 ($\omega = 2\pi f$) for the frequency f gives

$$f = \frac{\omega}{2\pi}$$

The number of times the diaphragm moves back and forth in 2.5 s is

$$\text{Number of times} = ft = \left(\frac{\omega}{2\pi}\right)t = \left(\frac{7.54\times10^{4}\ \text{rad/s}}{2\pi}\right)(2.5\ \text{s}) = \boxed{3.00\times10^{4}}$$

---

15. SSM ***REASONING AND SOLUTION*** According to Equations 10.6 and 10.11, $2\pi f = \sqrt{k/m}$. According to the data in the problem, the frequency of vibration of either spring is

$$f = \frac{5.0\ \text{cycles}}{3.0\ \text{s}} = 1.67\ \text{Hz}$$

Squaring both sides of the equation $2\pi f = \sqrt{k/m}$ and solving for k, we obtain

$$k = 4\pi^2 f^2 m = 4\pi^2 (1.67\ \text{Hz})^2 (320\ \text{kg}) = \boxed{3.5\times10^{4}\ \text{N/m}}$$

---

16. ***REASONING AND SOLUTION*** From Conceptual Example 2, we know that when the spring is cut in half, the spring constant for each half is twice as large as that of the original spring. In this case, the spring is cut into four shorter springs. Thus, each of the four shorter springs with 25 coils has a spring constant of $4 \times 420\ \text{N/m} = 1680\ \text{N/m}$.

The angular frequency of simple harmonic motion is given by Equation 10.11:

$$\omega = \sqrt{\frac{k}{m}} = \sqrt{\frac{1680\ \text{N/m}}{46\ \text{kg}}} = \boxed{6.0\ \text{rad/s}}$$

---

17. ***REASONING AND SOLUTION***

a. Since the object oscillates between $\pm\ 0.080\ \text{m}$, the amplitude of the motion is $\boxed{0.080\ \text{m}}$.

b. From the graph, the period is $T = 4.0\ \text{s}$. Therefore, according to Equation 10.4,

$$\omega = \frac{2\pi}{T} = \frac{2\pi}{4.0\text{ s}} = \boxed{1.6\text{ rad/s}}$$

c. Equation 10.11 relates the angular frequency to the spring constant: $\omega = \sqrt{k/m}$. Solving for k we find

$$k = \omega^2 m = (1.6\text{ rad/s})^2\,(0.80\text{ kg}) = \boxed{2.0\text{ N/m}}$$

d. At $t = 1.0$ s, the graph shows that the spring has its maximum displacement. At this location, the object is momentarily at rest, so that its speed is $\boxed{v = 0\text{ m/s}}$.

e. The acceleration of the object at $t = 1.0$ s is a maximum, and its magnitude is

$$a_{max} = A\omega^2 = (0.080\text{ m})(1.6\text{ rad/s})^2 = \boxed{0.20\text{ m/s}^2}$$

---

18. ***REASONING AND SOLUTION*** The maximum acceleration is 25g, i.e., a_{max} = 25(9.80 m/s^2). We know that $a_{max} = A\omega^2$, where $\omega = 2\pi$ f, so

$$A = \frac{25\left(9.80\text{ m/s}^2\right)}{4\pi^2\,(9.5\text{ Hz})^2} = \boxed{0.069\text{ m}}$$

---

19. ***REASONING*** The maximum velocity v_{max} in simple harmonic motion has a magnitude given by Equation 10.8$\left(v_{max} = A\omega\right)$, where A is the amplitude and ω is the angular frequency. The angular frequency is related to the spring constant k and the mass m of the object by Equation 10.11 $\left(\omega = \sqrt{\frac{k}{m}}\right)$. These two expressions allow us to relate the magnitude of the maximum velocity to the amplitude, the spring constant, and the mass. Although the maximum velocity and the mass are not given, we can nonetheless apply the expressions to each spring/mass system. We will then be able to eliminate the unknown velocity and mass and obtain the desired spring constant.

SOLUTION Substituting Equation 10.11 for ω into Equation 10.8 for v_{max} gives

$$v_{max} = A\omega = A\sqrt{\frac{k}{m}}$$

Applying this result to each spring/mass system and recognizing that the maximum velocity and the mass are the same for each, we find that

$$v_{\text{max}} = A_1\sqrt{\frac{k_1}{m}} = A_2\sqrt{\frac{k_2}{m}} \quad \text{or} \quad A_1\sqrt{k_1} = A_2\sqrt{k_2}$$

Solving for k_2 shows that

$$k_2 = \left(\frac{A_1}{A_2}\right)^2 k_1 = \left(\frac{2A_2}{A_2}\right)^2 (174\text{ N/m}) = \boxed{696\text{ N/m}}$$

---

20. ***REASONING AND SOLUTION*** For mass 1 we have

$$f_1 = \frac{1}{2\pi}\sqrt{\frac{k}{m_1}} \qquad (10.11)$$

For mass 2 we have

$$f_2 = \frac{1}{2\pi}\sqrt{\frac{k}{m_1 + m_2}}$$

Therefore,

$$\frac{f_1}{f_2} = \sqrt{\frac{m_1 + m_2}{m_1}} = 3.00$$

Squaring and solving yields

$$\frac{m_1 + m_2}{m_1} = 9.00 \quad \text{or} \quad m_1 + m_2 = 9.00\ m_1$$

Solving for the ratio m_2/m_1 gives $m_2/m_1 = \boxed{8.00}$.

---

21. **SSM** ***REASONING*** The frequency of vibration of the spring is related to the added mass *m* by Equations 10.6 and 10.11:

$$f = \frac{1}{2\pi}\sqrt{\frac{k}{m}} \qquad (1)$$

The spring constant can be determined from Equation 10.1.

SOLUTION Since the spring stretches by 0.018 m when a 2.8-kg object is suspended from its end, the spring constant is, according to Equation 10.1,

$$k = \frac{F_{\text{Applied}}}{x} = \frac{mg}{x} = \frac{(2.8\text{ kg})(9.80\text{ m/s}^2)}{0.018\text{ m}} = 1.52\times10^3\text{ N/m}$$

Solving Equation (1) for *m*, we find that the mass required to make the spring vibrate at 3.0 Hz is

$$m = \frac{k}{4\pi^2 f^2} = \frac{1.52 \times 10^3 \text{ N/m}}{4\pi^2 (3.0 \text{ Hz})^2} = \boxed{4.3 \text{ kg}}$$

---

22. ***REASONING AND SOLUTION***

a. When the block passes through the position x = 0 m, its velocity is a maximum and can be found from Equation 10.8: $v_{max} = A\omega$.

We can find the angular frequency ω from the following reasoning: When the mass is given a displacement x, one spring is stretched by an amount x, while the other is compressed by an amount x. The total restoring force on the mass is, therefore,

$$F = -k_1x - k_2x = -(k_1 + k_2)x$$

Comparison with Equation 10.2 shows that the two spring system has an effective spring constant $k_{eff} = k_1 + k_2$. Thus, from Equation 10.11

$$\omega = \sqrt{\frac{k_{eff}}{m}} = \sqrt{\frac{k_1 + k_2}{m}}$$

Combining this with Equation 10.8 we obtain

$$v_{max} = A\sqrt{\frac{k_1 + k_2}{m}} = (0.070 \text{ m})\sqrt{\frac{650 \text{ N/m} + 450 \text{ N/m}}{3.0 \text{ kg}}} = \boxed{1.3 \text{ m/s}}$$

b. The angular frequency of the system is

$$\omega = \sqrt{\frac{k_1 + k_2}{m}} = \sqrt{\frac{650 \text{ N/m} + 450 \text{ N/m}}{3.0 \text{ kg}}} = \boxed{19 \text{ rad/s}}$$

---

23. ***REASONING AND SOLUTION*** The cup slips when the force of static friction is overcome. So $F = ma = \mu_s mg$, where the acceleration is the maximum value for the simple harmonic motion, i.e., so that

$$a_{max} = A\omega^2 = A(2\pi f)^2 \qquad (10.10)$$

$$\mu_s = A(2\pi f)^2/g = (0.0500 \text{ m})4\pi^2(2.00 \text{ Hz})^2/(9.80 \text{ m/s}^2) = \boxed{0.806}$$

---

24. ***REASONING AND SOLUTION***

a. $PE = (1/2)\,kx^2 = (1/2)\,(425 \text{ N/m})(0.470 \text{ m})^2 = \boxed{46.9 \text{ J}}$ (10.13)

b. Assuming that mechanical energy is conserved, KE = PE gives

$$v = \sqrt{\frac{2(\text{PE})}{m}} = \sqrt{\frac{2(46.9 \text{ J})}{0.0300 \text{ kg}}} = \boxed{55.9 \text{ m/s}}$$

---

25. SSM ***REASONING AND SOLUTION*** If we neglect air resistance, only the conservative forces of the spring and gravity act on the ball. Therefore, the principle of conservation of mechanical energy applies.

When the 2.00 kg object is hung on the end of the vertical spring, it stretches the spring by an amount x, where

$$x = \frac{F}{k} = \frac{mg}{k} = \frac{(2.00 \text{ kg})(9.80 \text{ m/s}^2)}{50.0 \text{ N/m}} = 0.392 \text{ m} \qquad (10.1)$$

This position represents the equilibrium position of the system with the 2.00-kg object suspended from the spring. The object is then pulled down another 0.200 m and released from rest ($v_0 = 0$ m/s). At this point the spring is stretched by an amount of $0.392 \text{ m} + 0.200 \text{m} = 0.592 \text{ m}$. This point represents the zero reference level ($h = 0$ m) for the gravitational potential energy.

***h* = 0 m:** The kinetic energy, the gravitational potential energy, and the elastic potential energy at the point of release are:

$$\text{KE} = \tfrac{1}{2}mv_0^2 = \tfrac{1}{2}m(0 \text{ m/s})^2 = \boxed{0 \text{ J}}$$

$$\text{PE}_{\text{gravity}} = mgh = mg(0 \text{ m}) = \boxed{0 \text{ J}}$$

$$\text{PE}_{\text{elastic}} = \frac{1}{2}kx_0^2 = \frac{1}{2}(50.0 \text{ N/m})(0.592 \text{ m})^2 = \boxed{8.76 \text{ J}}$$

The total mechanical energy E_0 at the point of release is the sum of the three energies above: $E_0 = \boxed{8.76 \text{ J}}$.

***h* = 0.200 m:** When the object has risen a distance of $h = 0.200$ m above the release point, the spring is stretched by an amount of $0.592 \text{ m} - 0.200 \text{ m} = 0.392 \text{ m}$. Since the total mechanical energy is conserved, its value at this point is still $E = \boxed{8.76 \text{ J}}$. The gravitational and elastic potential energies are:

$$\text{PE}_{\text{gravity}} = mgh = (2.00 \text{ kg})(9.80 \text{ m/s}^2)(0.200 \text{ m}) = \boxed{3.92 \text{ J}}$$

$$\mathrm{PE}_{\text{elastic}} = \frac{1}{2}kx^2 = \frac{1}{2}(50.0\text{ N/m})(0.392\text{ m})^2 = \boxed{3.84\text{ J}}$$

Since $\mathrm{KE} + \mathrm{PE}_{\text{gravity}} + \mathrm{PE}_{\text{elastic}} = E$,

$$\mathrm{KE} = \mathrm{E} - \mathrm{PE}_{\text{gravity}} - \mathrm{PE}_{\text{elastic}} = 8.76\text{ J} - 3.92\text{ J} - 3.84\text{ J} = \boxed{1.00\text{ J}}$$

***h* = 0.400 m:** When the object has risen a distance of $h = 0.400$ m above the release point, the spring is stretched by an amount of $0.592\text{ m} - 0.400\text{ m} = 0.192\text{ m}$. At this point, the total mechanical energy is still $E = \boxed{8.76\text{ J}}$. The gravitational and elastic potential energies are:

$$\mathrm{PE}_{\text{gravity}} = mgh = (2.00\text{ kg})(9.80\text{ m/s}^2)(0.400\text{ m}) = \boxed{7.84\text{ J}}$$

$$\mathrm{PE}_{\text{elastic}} = \frac{1}{2}kx^2 = \frac{1}{2}(50.0\text{ N/m})(0.192\text{ m})^2 = \boxed{0.92\text{ J}}$$

The kinetic energy is

$$\mathrm{KE} = \mathrm{E} - \mathrm{PE}_{\text{gravity}} - \mathrm{PE}_{\text{elastic}} = 8.76\text{ J} - 7.84\text{ J} - 0.92\text{ J} = \boxed{0\text{ J}}$$

The results are summarized in the table below:

h	**KE**	$\mathbf{PE}_{\text{grav}}$	$\mathbf{PE}_{\text{elastic}}$	E
0.000 m	0.00 J	0.00 J	8.76 J	8.76 J
0.200 m	1.00 J	3.92 J	3.84 J	8.76 J
0.400 m	0.00 J	7.84 J	0.92 J	8.76 J

---

26. ***REASONING AND SOLUTION*** Assuming that mechanical energy is conserved

$$(1/2)\,kx_0^2 = (1/2)\,mv_f^2 + mgh_f$$

or

$$v_f = \sqrt{\frac{kx_0^2}{m} - 2gh_f} = \sqrt{\frac{(675\text{ N/m})(0.0650\text{ m})^2}{0.0585\text{ kg}} - 2(9.80\text{ m/s}^2)(0.300\text{ m})} = \boxed{6.55\text{ m/s}}$$

---

27. ***REASONING*** As the block falls, only two forces act on it: its weight and the elastic force of the spring. Both of these forces are conservative forces, so the falling block obeys the principle of conservation of mechanical energy. We will use this conservation principle to determine the spring constant of the spring. Once the spring constant is known, Equation 10.11, $\omega = \sqrt{k/m}$, may be used to find the angular frequency of the block's vibrations.

SOLUTION
a. The conservation of mechanical energy states that the final total mechanical energy E_f is equal to the initial total mechanical energy E_0, or $E_f = E_0$ (Equation 6.9a). The expression for the total mechanical energy of an object oscillating on a spring is given by Equation 10.14. Thus, the conservation of total mechanical energy can be written as

$$\underbrace{\tfrac{1}{2}mv_f^2 + \tfrac{1}{2}I\omega_f^2 + mgh_f + \tfrac{1}{2}kx_f^2}_{E_f} = \underbrace{\tfrac{1}{2}mv_0^2 + \tfrac{1}{2}I\omega_0^2 + mgh_0 + \tfrac{1}{2}kx_0^2}_{E_0}$$

Before going any further, let's simplify this equation by noting which variables are zero. Since the block starts and ends at rest, $v_f = v_0 = 0$ m/s. The block does not rotate, so its angular speed is zero, $\omega_f = \omega_0 = 0$ rad/s. Initially, the spring is unstretched, so that $x_0 = 0$ m. Setting these terms equal to zero in the equation above gives

$$mgh_f + \tfrac{1}{2}kx_f^2 = mgh_0$$

Solving this equation for the spring constant *k*, we have that

$$k = \frac{mg\left(h_0 - h_f\right)}{\tfrac{1}{2}x_f^2} = \frac{(0.450\text{ kg})\left(9.80\text{ m/s}^2\right)(0.150\text{ m})}{\tfrac{1}{2}(0.150\text{ m})^2} = \boxed{58.8\text{ N/m}}$$

b. The angular frequency ω of the block's vibrations depends on the spring constant *k* and the mass *m* of the block:

$$\omega = \sqrt{\frac{k}{m}} = \sqrt{\frac{58.8\text{ N/m}}{0.450\text{ kg}}} = \boxed{11.4\text{ rad/s}} \qquad (10.11)$$

---

28. ***REASONING*** Since air resistance is negligible, we can apply the principle of conservation of mechanical energy, which indicates that the total mechanical energy of the block and the spring is the same at the instant it comes to a momentary halt on the spring and at the instant the block is dropped. Gravitational potential energy is one part of the total mechanical energy, and Equation 6.5 indicates that it is *mgh* for a block of mass *m* located at a height *h* relative to an arbitrary zero level. This dependence on *h* will allow us to determine the height at which the block was dropped.

SOLUTION The conservation of mechanical energy states that the final total mechanical energy E_f is equal to the initial total mechanical energy E_0. The expression for the total mechanical energy for an object on a spring is given by Equation 10.14, so that we have

$$\underbrace{\tfrac{1}{2}mv_f^2 + \tfrac{1}{2}I\omega_f^2 + mgh_f + \tfrac{1}{2}kx_f^2}_{E_f} = \underbrace{\tfrac{1}{2}mv_0^2 + \tfrac{1}{2}I\omega_0^2 + mgh_0 + \tfrac{1}{2}kx_0^2}_{E_0}$$

The block does not rotate, so the angular speeds ω_f and ω_0 are zero. Since the block comes to a momentary halt on the spring and is dropped from rest, the translational speeds v_f and v_0 are also zero. Because the spring is initially unstrained, the initial displacement x_0 of the spring is likewise zero. Thus, the above expression can be simplified as follows:

$$mgh_f + \tfrac{1}{2}kx_f^2 = mgh_0$$

The block was dropped at a height of $h_0 - h_f$ above the compressed spring. Solving the simplified energy-conservation expression for this quantity gives

$$h_0 - h_f = \frac{kx_f^2}{2mg} = \frac{(450\ \text{N/m})(0.025\ \text{m})^2}{2(0.30\ \text{kg})(9.80\ \text{m/s}^2)} = \boxed{0.048\ \text{m or } 4.8\ \text{cm}}$$

---

29. SSM ***REASONING*** The only force that acts on the block along the line of motion is the force due to the spring. Since the force due to the spring is a conservative force, the principle of conservation of mechanical energy applies. Initially, when the spring is unstrained, all of the mechanical energy is kinetic energy, $(1/2)mv_0^2$. When the spring is fully compressed, all of the mechanical energy is in the form of elastic potential energy, $(1/2)kx_{\text{max}}^2$, where x_{max}, the maximum compression of the spring, is the amplitude A. Therefore, the statement of energy conservation can be written as

$$\tfrac{1}{2}mv_0^2 = \tfrac{1}{2}kA^2$$

This expression may be solved for the amplitude A.

SOLUTION Solving for the amplitude A, we obtain

$$A = \sqrt{\frac{mv_0^2}{k}} = \sqrt{\frac{(1.00 \times 10^{-2}\ \text{kg})(8.00\ \text{m/s})^2}{124\ \text{N/m}}} = \boxed{7.18 \times 10^{-2}\ \text{m}}$$

---

30. ***REASONING*** The elastic potential energy of an ideal spring is given by Equation 10.13 $\left(\text{PE}_{\text{elastic}} = \tfrac{1}{2}kx^2\right)$, where k is the spring constant and x is the amount of stretch or compression from the spring's unstrained length. Thus, we need to determine x for an object of mass m hanging stationary from the spring. Since the object is stationary, it has no acceleration and is at equilibrium. The upward pull of the spring balances the downward-acting weight of the object. We can use this fact to determine x.

SOLUTION The magnitude of the pull of the spring is given by kx, according to Equation 10.2. The weight of the object is mg. Since these two forces must balance, we have

$kx = mg$ or $x = mg/k$. Substituting this result into Equation 10.13 for the elastic potential energy gives

$$\mathrm{PE}_{\text{elastic}} = \tfrac{1}{2}kx^2 = \tfrac{1}{2}k\left(\frac{mg}{k}\right)^2 = \frac{m^2g^2}{2k}$$

Applying this expression to the two spring/object systems, we find

$$\mathrm{PE}_1 = \frac{m_1^2g^2}{2k} \quad \text{and} \quad \mathrm{PE}_2 = \frac{m_2^2g^2}{2k}$$

Here, we have omitted the subscript "elastic" for convenience. Dividing the right-hand equation by the left-hand equation gives

$$\frac{\mathrm{PE}_2}{\mathrm{PE}_1} = \frac{\dfrac{m_2^2g^2}{2k}}{\dfrac{m_1^2g^2}{2k}} = \frac{m_2^2}{m_1^2} \quad \text{or} \quad \mathrm{PE}_2 = \mathrm{PE}_1\left(\frac{m_2^2}{m_1^2}\right) = (1.8\ \mathrm{J})\frac{(5.0\ \mathrm{kg})^2}{(3.2\ \mathrm{kg})^2} = \boxed{4.4\ \mathrm{J}}$$

---

31. ***REASONING*** Assuming that friction and air resistance are negligible, we can apply the principle of conservation of mechanical energy, which indicates that the total mechanical energy of the rod (or ram) and the spring is the same at the instant it contacts the staple and at the instant the spring is released. Kinetic energy $\frac{1}{2}mv^2$ is one part of the total mechanical energy and depends on the mass m and the speed v of the rod. The dependence on the speed will allow us to determine the speed of the ram at the instant of contact with the staple.

SOLUTION The conservation of mechanical energy states that the final total mechanical energy E_f is equal to the initial total mechanical energy E_0. The expression for the total mechanical energy for a spring/mass system is given by Equation 10.14, so that we have

$$\underbrace{\tfrac{1}{2}mv_f^2 + \tfrac{1}{2}I\omega_f^2 + mgh_f + \tfrac{1}{2}kx_f^2}_{E_f} = \underbrace{\tfrac{1}{2}mv_0^2 + \tfrac{1}{2}I\omega_0^2 + mgh_0 + \tfrac{1}{2}kx_0^2}_{E_0}$$

Since the ram does not rotate, the angular speeds ω_f and ω_0 are zero. Since the ram is initially at rest, the initial translational speed v_0 is also zero. Thus, the above expression can be simplified as follows:

$$\tfrac{1}{2}mv_f^2 + mgh_f + \tfrac{1}{2}kx_f^2 = mgh_0 + \tfrac{1}{2}kx_0^2$$

In falling from its initial height of h_0 to its final height of h_f, the ram falls through a distance of $h_0 - h_f = 0.022$ m, since the spring is compressed 0.030 m from its unstrained length to begin with and is still compressed 0.008 m when the ram makes contact with the staple. Solving the simplified energy-conservation expression for the final speed v_f gives

$$v_f = \sqrt{\frac{k\left(x_0^2 - x_f^2\right)}{m} + 2g\left(h_0 - h_f\right)}$$

$$= \sqrt{\frac{(32\,000 \text{ N/m})\left[(0.030 \text{ m})^2 - (0.008 \text{ m})^2\right]}{0.140 \text{ kg}} + 2\left(9.80 \text{ m/s}^2\right)(0.022 \text{ m})} = \boxed{14 \text{ m/s}}$$

---

32. ***REASONING AND SOLUTION***

a. Now look at conservation of energy before and after the split

Before $(1/2)\ mv_{max}^2 = (1/2)\ kA^2$

Solving for the amplitude A gives

$$A = v_{max}\sqrt{\frac{m}{k}}$$

After $(1/2)\ (m/2)v'^2 = (1/2)\ (m/2)(v_{max})^2 = (1/2)\ kA'^2$

Solving for the amplitude A' gives

$$A' = v_{max}\sqrt{\frac{m}{2k}}$$

Therefore, we find that

$$A' = A/\sqrt{2} = (5.08 \times 10^{-2} \text{ m})/\sqrt{2} = \boxed{3.59 \times 10^{-2} \text{ m}}$$

Similarly, for the frequency, we can show that

$$f' = f\sqrt{2} = (3.00 \text{ Hz})\sqrt{2} = \boxed{4.24 \text{ Hz}}$$

b. If the block splits at one of the extreme positions, the amplitude of the SHM would not change, so it would remain as $\boxed{5.08 \times 10^{-2} \text{ m}}$

The frequency would be

$$f' = f\sqrt{2} = (3.00 \text{ Hz})\sqrt{2} = \boxed{4.24 \text{ Hz}}$$

---

33. SSM ***REASONING AND SOLUTION*** The amount by which the spring stretches due to the weight of the 1.1 - kg object can be calculated using Equation 10.1, where the force *F* is equal to the weight of the object. The position of the object when the spring is stretched is the equilibrium position for the vertical harmonic motion of the object-spring system.

a. Solving Equation 10.1 for x with F equal to the weight of the object gives

$$x = \frac{F}{k} = \frac{mg}{k} = \frac{(1.1\text{ kg})(9.80\text{ m/s}^2)}{120\text{ N/m}} = \boxed{9.0\times10^{-2}\text{ m}}$$

b. The object is then pulled down another 0.20 m and released from rest ($v_0 = 0$ m/s). At this point the spring is stretched by an amount of 0.090 m + 0.20 m = 0.29 m. We will let this point be the zero reference level ($h = 0$ m) for the gravitational potential energy.

The kinetic energy, the gravitational potential energy, and the elastic potential energy at the point of release are:

$$\text{KE} = \tfrac{1}{2}mv_0^2 = \tfrac{1}{2}m(0\text{ m/s})^2 = 0\text{ J}$$

$$\text{PE}_{\text{gravity}} = mgh = mg(0\text{ m}) = 0\text{ J}$$

$$\text{PE}_{\text{elastic}} = \frac{1}{2}kx_0^2 = \frac{1}{2}(120\text{ N/m})(0.29\text{ m})^2 = 5.0\text{ J}$$

The total mechanical energy E_0 is the sum of these three energies, so $E_0 = 5.0$ J. When the object has risen a distance of $h = 0.20$ m above the release point, the spring is stretched by an amount of $0.29\text{ m} - 0.20\text{ m} = 0.090\text{ m}$. Since the total mechanical energy is conserved, its value at this point is still 5.0 J. Thus,

$$E = \text{KE} + \text{PE}_{\text{gravity}} + \text{PE}_{\text{elastic}}$$

$$E = \frac{1}{2}mv^2 + mgh + \frac{1}{2}kx^2$$

$$5.0\text{ J} = \frac{1}{2}(1.1\text{ kg})v^2 + (1.1\text{ kg})(9.80\text{ m/s}^2)(0.20\text{ m}) + \frac{1}{2}(120\text{ N/m})(0.090\text{ m})^2$$

Solving for v yields $v = \boxed{2.1\text{ m/s}}$.

---

34. ***REASONING*** As the climber falls, only two forces act on him: his weight and the elastic force of the nylon rope. Both of these forces are conservative forces, so the falling climber obeys the conservation of mechanical energy. We will use this conservation law to determine how much the rope is stretched when it breaks his fall and momentarily brings him to rest.

SOLUTION

The conservation of mechanical energy states that the final total mechanical energy E_f is equal to the initial total mechanical energy E_0, or $E_f = E_0$ (Equation 6.9a). The expression for the total mechanical energy of an object oscillating on a spring is given by Equation 10.14. Thus, the conservation of total mechanical energy can be written as

$$\underbrace{\tfrac{1}{2}mv_f^2 + \tfrac{1}{2}I\omega_f^2 + mgh_f + \tfrac{1}{2}kx_f^2}_{E_f} = \underbrace{\tfrac{1}{2}mv_0^2 + \tfrac{1}{2}I\omega_0^2 + mgh_0 + \tfrac{1}{2}kx_0^2}_{E_0}$$

Before going any further, let's simplify this equation by noting which variables are zero. Since the climber starts and ends at rest, $v_f = v_0 = 0$ m/s. The climber does not rotate, so his angular speed is zero, $\omega_f = \omega_0 = 0$ rad/s. Initially, the spring is unstretched, so that $x_0 = 0$ m. Setting these terms to zero in the equation above gives

$$mgh_f + \tfrac{1}{2}kx_f^2 = mgh_0$$

Rearranging the terms in this equation, we have

$$\tfrac{1}{2}kx_f^2 + mg\underbrace{(h_f - h_0)}_{x_f - 0.750\text{ m}} = 0$$

Note that the climber falls a distance of 0.750 m before the rope starts to stretch, and x_f is the displacement of the stretched rope. Since x_f points downward, it is considered to be negative. Thus, x_f – 0.750 m is the total downward displacement of the falling climber, which is also equal to $h_f - h_0$. With this substitution, the equation above becomes

$$\tfrac{1}{2}kx_f^2 + mgx_f - mg(0.750\text{ m}) = 0$$

This is a quadratic equation in the variable x_f. Using the quadratic formula gives

$$x_f = \frac{-mg \pm \sqrt{(mg)^2 - 4\left(\tfrac{1}{2}k\right)\left[-mg(0.750\text{ m})\right]}}{2\left(\tfrac{1}{2}k\right)}$$

$$x_f = \frac{-(86.0\text{ kg})(9.80\text{ m/s}^2)}{1.20\times10^3\text{ N/m}}$$

$$\pm\frac{\sqrt{\left[(86.0\text{ kg})(9.80\text{ m/s}^2)\right]^2 - 4\left(\tfrac{1}{2}\right)(1.20\times10^3\text{ N/m})\left[-(86.0\text{ kg})(9.80\text{ m/s}^2)(0.750\text{ m})\right]}}{1.20\times10^3\text{ N/m}}$$

There are two answers, x_f = +0.54 m and –1.95 m. Since the rope is stretched in the downward direction, which we have taken to be the negative direction, the displacement is –1.95 m. Thus, the amount that the rope is stretched is $x_f = \boxed{1.95\text{ m}}$.

35. ***REASONING*** The two blocks and the spring between them constitute the system in this problem. Since the surface is frictionless and the weights of the blocks are balanced by the normal forces from the surface, no net external force acts on the system. Thus, the system's total mechanical energy and total linear momentum are each conserved. The conservation of these two quantities will give us two equations containing the two unknown speeds with which the blocks move away. Using these equations, we will be able to obtain the speeds.

SOLUTION The conservation of mechanical energy states that the final total mechanical energy E_f is equal to the initial total mechanical energy E_0. Only the translational kinetic energies of the blocks and the elastic potential energy of the spring are of interest here. There is no rotational kinetic energy since there is no rotation. Gravitational potential energy plays no role, because the surface is horizontal and the vertical height does not change. Thus, the expression for the conservation of the total mechanical energy is

$$\underbrace{\tfrac{1}{2}m_1v_{f1}^2 + \tfrac{1}{2}m_2v_{f2}^2 + \tfrac{1}{2}kx_f^2}_{E_f} = \underbrace{\tfrac{1}{2}m_1v_{01}^2 + \tfrac{1}{2}m_2v_{02}^2 + \tfrac{1}{2}kx_0^2}_{E_0}$$

Since the blocks are initially at rest, the initial translational speeds v_{01} and v_{02} are zero. In addition, the spring is neither compressed nor stretched after it is released, so that $x_f = 0$ m. Thus, the above expression can be simplified as follows:

$$\tfrac{1}{2}m_1v_{f1}^2 + \tfrac{1}{2}m_2v_{f2}^2 = \tfrac{1}{2}kx_0^2 \qquad (1)$$

Remembering that linear momentum is mass times velocity, we can express the conservation of the total linear momentum of the system as follows:

$$\underbrace{m_1v_{f1} + m_2v_{f2}}_{\text{Final total momentum}} = \underbrace{m_1v_{01} + m_2v_{02}}_{\text{Initial total momentum}}$$

Since the blocks are initially at rest, v_{01} and v_{02} are zero, so that the expression for the conservation of linear momentum becomes

$$m_1v_{f1} + m_2v_{f2} = 0 \quad \text{or} \quad v_{f2} = -\frac{m_1v_{f1}}{m_2} \qquad (2)$$

Substituting this result into Equation (1) gives

$$\tfrac{1}{2}m_1v_{f1}^2 + \tfrac{1}{2}m_2\left(-\frac{m_1v_{f1}}{m_2}\right)^2 = \tfrac{1}{2}kx_0^2$$

Solving for v_{f1}, which we define to be the final speed of the 11.2-kg block, shows that

$$v_{f1} = \sqrt{\frac{m_2 k x_0^2}{m_1(m_2 + m_1)}} = \sqrt{\frac{(21.7\text{ kg})(1330\text{ N/m})(0.141\text{ m})^2}{(11.2\text{ kg})(21.7\text{ kg} + 11.2\text{ kg})}} = \boxed{1.25\text{ m/s}}$$

Substituting this value into Equation (2) gives

$$v_{f2} = -\frac{m_1 v_{f1}}{m_2} = -\frac{(11.2\text{ kg})(1.25\text{ m/s})}{21.7\text{ kg}} = -0.645\text{ m/s}$$

The speed of the 21.7-kg block is the magnitude of this result or $\boxed{0.645\text{ m/s}}$.

---

36. ***REASONING AND SOLUTION*** The bullet (mass m) moves with speed v, strikes the block (mass M) in an inelastic collision and the two move together with a final speed V. We first need to employ the conservation of linear momentum to the collision to obtain an expression for the final speed:

$$mv = (m + M)V \qquad \text{or} \qquad V = \frac{mv}{m + M}$$

The block/bullet system now compresses the spring by an amount x. During the compression the total mechanical energy is conserved so that

$$\tfrac{1}{2}(m + M)V^2 = \tfrac{1}{2}kx^2$$

Substituting the expression for V into this equation, we obtain

$$\tfrac{1}{2}(m + M)\left(\frac{mv}{m + M}\right)^2 = \tfrac{1}{2}kx^2$$

Solving this expression for v gives

$$v = \sqrt{\frac{kx^2(M + m)}{m^2}} = \sqrt{\frac{(845\text{ N/m})(0.200\text{ m})^2(2.51\text{ kg})^2}{(0.0100\text{ kg})^2}} = \boxed{921\text{ m/s}}$$

---

37. **SSM** ***REASONING*** Using the principle of conservation of mechanical energy, the initial elastic potential energy stored in the elastic bands must be equal to the sum of the kinetic energy and the gravitational potential energy of the performer at the point of ejection:

$$\frac{1}{2}kx^2 = \frac{1}{2}mv_0^2 + mgh$$

where v_0 is the speed of the performer at the point of ejection and, from the figure at the right, $h = x \sin\theta$.

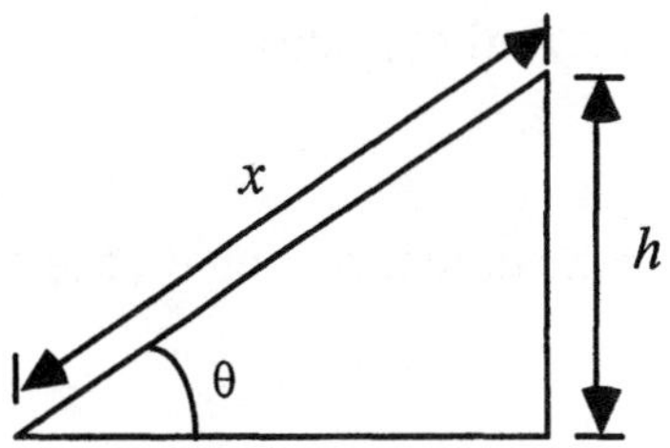

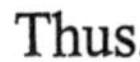

Thus,

$$\frac{1}{2}kx^2 = \frac{1}{2}mv_0^2 + mgx\sin\theta \qquad (1)$$

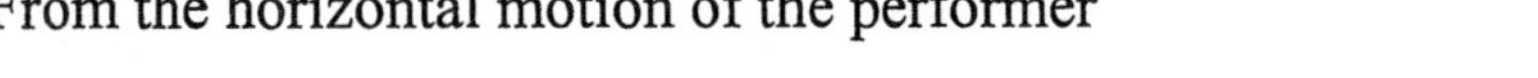

From the horizontal motion of the performer

$$v_{0x} = v_0 \cos\theta \qquad (2)$$

where

$$v_{0x} = \frac{s_x}{t} \qquad (3)$$

and $s_x = 26.8$ m. Combining equations (2) and (3) gives

$$v_0 = \frac{s_x}{t(\cos\theta)}$$

Equation (1) becomes:

$$\frac{1}{2}kx^2 = \frac{1}{2}m\frac{s_x^2}{t^2\cos^2\theta} + mgx\sin\theta$$

This expression can be solved for k, the spring constant of the firing mechanism.

SOLUTION Solving for k yields:

$$k = m\left(\frac{s_x}{xt\cos\theta}\right)^2 + \frac{2mg(\sin\theta)}{x}$$

$$k = (70.0\text{ kg})\left[\frac{26.8\text{ m}}{(3.00\text{ m})(2.14\text{ s})(\cos 40.0°)}\right]^2$$

$$+\frac{2(70.0\text{ kg})(9.80\text{ m/s}^2)(\sin 40.0°)}{3.00\text{ m}} = \boxed{2.37\times 10^3\text{ N/m}}$$

---

38. ***REASONING AND SOLUTION*** Using

$$f = 1/T = 1/(0.250\text{ s}) = 4.00\text{ Hz}$$

and also

$$f = \frac{1}{2\pi}\sqrt{\frac{k}{m}}$$

we can find the ratio

$$k/m = 4\pi^2 f^2 = 632 \text{ N/(kg·m)}$$

With the object resting on the spring, F = kx = mg so that,

$$x = \frac{g}{\frac{k}{m}} = 0.0155 \text{ m.}$$

When the mass leaves the spring, potential energy of the spring has been converted to gravitational energy, i.e.,

$$(1/2)\, kx'^2 = mgh$$

Where

$$x' = 0.0500 \text{ m} + 0.0155 \text{ m} = 0.0655 \text{ m}$$

Solving for h we get

$$h = \left(\frac{k}{m}\right)\left(\frac{x'^2}{2g}\right) = [632 \text{ N/(kg}\cdot\text{m)}]\left[\frac{(0.0655 \text{ m})^2}{2(9.80 \text{ m/s}^2)}\right] = \boxed{0.138 \text{ m}}$$

---

39. ***REASONING AND SOLUTION*** The period T of a simple pendulum can be obtained from Equations 10.5 and 10.16 as

$$T = 2\pi\sqrt{\frac{L}{g}}$$

Solving for L gives

$$L = \frac{T^2 g}{4\pi^2} = \frac{(2.0 \text{ s})^2 (9.80 \text{ m/s}^2)}{4\pi^2} = \boxed{0.99 \text{ m}}$$

---

40. ***REASONING*** As the ball swings down, it reaches it greatest speed at the lowest point in the motion. One complete cycle of the pendulum has four parts: the downward motion in which the ball attains its greatest speed at the lowest point, the subsequent upward motion in which the ball slows down and then momentarily comes to rest. The ball then retraces its motion, finally ending up where it originally began. The time it takes to reach the lowest point is one-quarter of the period of the pendulum, or $t = (1/4)T$. The period is related to the angular frequency ω of the pendulum by Equation 10.4, $T = 2\pi/\omega$. Thus, the time for the ball to reach its lowest point is

$$t = \tfrac{1}{4}T = \frac{1}{4}\left(\frac{2\pi}{\omega}\right)$$

The angular frequency ω of the pendulum depends on its length L and the acceleration g due to gravity through the relation $\omega = \sqrt{g/L}$ (Equation 10.16). Thus, the time is

$$t = \frac{1}{4}\left(\frac{2\pi}{\omega}\right) = \frac{1}{4}\left(\frac{2\pi}{\sqrt{\frac{g}{L}}}\right) = \frac{\pi}{2}\sqrt{\frac{L}{g}}$$

SOLUTION After the ball is released, the time that has elapsed before it attains its greatest speed is

$$t = \frac{\pi}{2}\sqrt{\frac{L}{g}} = \frac{\pi}{2}\sqrt{\frac{0.65\text{ m}}{9.80\text{ m/s}^2}} = \boxed{0.40\text{ s}}$$

---

41. **SSM** ***REASONING AND SOLUTION*** Recall that the relationship between frequency f and period T is $f = 1/T$. Then, according to Equations 10.6 and 10.16, the period of the simple pendulum is given by

$$T = 2\pi\sqrt{\frac{L}{g}}$$

where L is the length of the pendulum. Solving for g and noting that the period is $T = (280\text{ s})/100 = 2.8\text{ s}$, we obtain

$$g = \frac{4\pi^2 L}{T^2} = \frac{4\pi^2(1.2\text{ m})}{(2.8\text{ s})^2} = \boxed{6.0\text{ m/s}^2}$$

---

42. ***REASONING AND SOLUTION*** The period of a simple pendulum is given by

$$\text{T} = 2\pi\sqrt{\frac{\text{L}}{\text{g}}}$$

The gravitational acceleration changes, but we want the period to remain the same, therefore we can write

$$2\pi\sqrt{\frac{\text{L}}{\text{g}}} = 2\pi\sqrt{\frac{\text{L}'}{\text{g}'}}$$

Squaring each side of the equation and solving for L', the new length, we obtain

$$\text{L}' = \text{L}(\text{g}'/\text{g}) = (1.00\text{ m})(9.78\text{ m/s}^2)/(9.83\text{ m/s}^2) = \boxed{0.995\text{ m}}$$

---

43. SSM ***REASONING*** For small-angle displacements, the frequency of simple harmonic motion for a physical pendulum is determined by $2\pi f = \sqrt{mgL/I}$, where L is the distance between the axis of rotation and the center of gravity of the rigid body of moment of inertia I. Since the frequency f and the period T are related by $f = 1/T$, the period of pendulum A is given by

$$T_A = 2\pi\sqrt{\frac{I}{mgL}}$$

Since the pendulum is made from a thin, rigid, uniform rod, its moment of inertia is given by $I = (1/3)md^2$, where d is the length of the rod. Since the rod is uniform, its center of gravity lies at its geometric center, and $L = d/2$ Therefore, the period of pendulum A is given by

$$T_A = 2\pi\sqrt{\frac{2d}{3g}}$$

For the simple pendulum we have

$$T_B = 2\pi\sqrt{\frac{d}{g}}$$

SOLUTION The ratio of the periods is, therefore,

$$\frac{T_A}{T_B} = \frac{2\pi\sqrt{2d/(3g)}}{2\pi\sqrt{d/g}} = \sqrt{\frac{2}{3}} = \boxed{0.816}$$

---

44. ***REASONING*** The relation between the period T and angular frequency ω is $T = \frac{2\pi}{\omega}$ (Equation 10.6). The angular frequency of a physical pendulum is given by $\omega = \sqrt{\frac{mgL}{I}}$ (Equation 10.15), where m is the mass of the pendulum, g is the acceleration due to gravity, L is the distance between the axis of rotation at the pivot point and the center of gravity of the rod, and I is the moment of inertia of the rod. According to Table 9.1, the moment of inertia of a thin uniform rod of length D is $I = \frac{1}{3}mD^2$. Combining these three equations algebraically will give us an expression for the period that we seek. However, the length D of the rod is not given. Instead, the period of the simple pendulum is given. We will be able to use this information to eliminate the need for the missing length data.

SOLUTION Substituting Equation 10.15 for ω into Equation 10.6, shows that the period of the physical pendulum is

$$T = \frac{2\pi}{\omega} = \frac{2\pi}{\sqrt{\frac{mgL}{I}}} = 2\pi\sqrt{\frac{I}{mgL}}$$

Now we can use the expression $I = \frac{1}{3}mD^2$ for the moment of inertia of the rod. In addition, we recognize that the center of gravity of the uniform rod lies at the center of the rod, so that $L = \frac{1}{2}D$. With these two substitutions the expression for the period becomes

$$T = 2\pi\sqrt{\frac{I}{mgL}} = 2\pi\sqrt{\frac{\frac{1}{3}mD^2}{mg\left(\frac{1}{2}D\right)}} = 2\pi\sqrt{\frac{2D}{3g}} \qquad (1)$$

At this point, we must deal with the unknown length D of the rod. To this end, we note that the period of the simple pendulum is given by Equations 10.6 and 10.16 as

$$\omega_{\text{Simple}} = \frac{2\pi}{T_{\text{Simple}}} = \sqrt{\frac{g}{D}} \quad \text{or} \quad T_{\text{Simple}} = 2\pi\sqrt{\frac{D}{g}}$$

Solving this expression for D/g and substituting the result into Equation (1) gives

$$T = \left(\sqrt{\frac{2}{3}}\right)T_{\text{Simple}} = \left(\sqrt{\frac{2}{3}}\right)(0.66\text{ s}) = \boxed{0.54\text{ s}}$$

---

45. ***REASONING AND SOLUTION*** The period of the sphere (a physical pendulum) is

$$T_s = 2\pi\sqrt{\frac{I}{mgL}}$$

where for a solid sphere $I = (7/5)\,MR^2$ (see Table 9.1) and $L = R$. Therefore,

$$T_s = 2\pi\sqrt{\frac{7R}{5g}}$$

For the simple pendulum we know that

$$T_o = 2\pi\sqrt{\frac{L}{g}}$$

We want $T_s = T_o$ which leads to

$$7R/(5g) = L/g \quad \text{or} \quad \boxed{L = (7/5)R}$$

---

46. ***REASONING*** When the tow truck pulls the car out of the ditch, the cable stretches and a tension exists in it. This tension is the force that acts on the car. The amount ΔL that the cable stretches depends on the tension F, the length L_0 and cross-sectional area A of the

cable, as well as Young's modulus Y for steel. All of these quantities are given in the statement of the problem, except for Young's modulus, which can be found by consulting Table 10.1.

SOLUTION Solving Equation 10.17, $F = Y\left(\frac{\Delta L}{L_0}\right)A$, for the change in length, we have

$$\Delta L = \frac{FL_0}{AY} = \frac{(890\text{ N})(9.1\text{ m})}{\pi\left(0.50\times10^{-2}\text{ m}\right)^2\left(2.0\times10^{11}\text{ N/m}^2\right)} = \boxed{5.2\times10^{-4}\text{ m}}$$

---

47. **SSM** **WWW** ***REASONING AND SOLUTION*** The amount of compression can be obtained from Equation 10.17,

$$F = Y\left(\frac{\Delta L}{L_0}\right)A$$

where F is the magnitude of the force on the stand due to the weight of the statue. Solving for ΔL gives

$$\Delta L = \frac{FL_0}{YA} = \frac{mgL_0}{YA} = \frac{(3500\text{ kg})(9.80\text{ m/s}^2)(1.8\text{ m})}{(2.3\times10^{10}\text{ N/m}^2)(7.3\times10^{-2}\text{ m}^2)} = \boxed{3.7\times10^{-5}\text{ m}}$$

---

48. ***REASONING AND SOLUTION*** $F = S(\Delta X/L_0)A$ for the shearing force. The shear modulus S for copper is given in Table 10.2. From the figure we also see that $\tan\theta = (\Delta X/L_0)$ so that

$$\theta = \tan^{-1}\left(\frac{F}{SA}\right) = \tan^{-1}\left[\frac{6.0\times10^6\text{ N}}{\left(4.2\times10^{10}\text{ N/m}^2\right)\left(0.090\text{ m}^2\right)}\right] = \boxed{0.091°}$$

---

49. **SSM** ***REASONING AND SOLUTION*** The shearing stress is equal to the force per unit area applied to the rivet. Thus, when a shearing stress of 5.0×10^8 Pa is applied to each rivet, the force experienced by each rivet is

$$F = (\text{Stress})A = (\text{Stress})(\pi r^2) = (5.0\times10^8\text{ Pa})\left[\pi(5.0\times10^{-3}\text{ m})^2\right] = 3.9\times10^4\text{ N}$$

Therefore, the maximum tension T that can be applied to each beam, assuming that each rivet carries one-fourth of the total load, is $\boxed{4F = 1.6\times10^5\text{ N}}$.

---

50. ***REASONING AND SOLUTION*** Y = Stress/Strain where Stress = F/A and Strain = $\Delta L/L_0$.

a. $$\text{Stress} = \frac{(1800\text{ kg})(9.80\text{ m/s}^2)}{\pi\left(6.0\times10^{-3}\text{ m}\right)^2} = \boxed{1.6\times10^8\text{ N/m}^2}$$

b. Strain = $(8.0 \times 10^{-3}\text{ m})/(15\text{ m}) = \boxed{5.3\times10^{-4}}$

c. Y = Stress/Strain = $(1.6 \times 10^8\text{ N/m}^2)/(5.3 \times 10^{-4}) = \boxed{3.0\times10^{11}\text{ N/m}^2}$

---

51. ***REASONING AND SOLUTION*** We know that

$$\Delta P = -B(\Delta V/V_0) = -(2.6 \times 10^{10}\text{ N/m}^2)(-1.0 \times 10^{-10}\text{ m}^3)/(1.0 \times 10^{-6}\text{ m}^3) \qquad (10.20)$$

$$\Delta P = 2.6 \times 10^6\text{ N/m}^2$$

Since the pressure increases by $1.0 \times 10^4\text{ N/m}^2$ per meter of depth, the depth is

$$\frac{2.6\times10^6\text{ N/m}^2}{1.0\times10^4\,\dfrac{\text{N/m}^2}{\text{m}}} = \boxed{260\text{ m}}$$

---

52. ***REASONING***

a. According to the discussion in Section 10.8, the stress is the magnitude of the force per unit area required to cause an elastic deformation. We can determine the maximum stress that will fracture the femur by dividing the magnitude of the compressional force by the cross-sectional area of the femur.

b. The strain is defined in Section 10.8 as the change in length of the femur divided by its original length. Equation 10.17 shows how the strain $\Delta L/L_0$ is related to the stress F/A and Young's modulus Y ($Y = 9.4 \times 10^9\text{ N/m}^2$ for bone compression, according to Table 10.1).

SOLUTION

a. The maximum stress is equal to the maximum compressional force divided by the cross-sectional area of the femur:

$$\text{Maximum stress} = \frac{F}{A} = \frac{6.8\times10^4\text{ N}}{4.0\times10^{-4}\text{ m}^2} = \boxed{1.7\times10^8\text{ N/m}^2}$$

b. The strain $\Delta L/L$ can be found by rearranging Equation 10.17:

$$\underbrace{\frac{\Delta L}{L_0}}_{\text{Strain}} = \frac{1}{Y}\underbrace{\left(\frac{F}{A}\right)}_{\text{Stress}} = \left(\frac{1}{9.4\times10^9\ \text{N/m}^2}\right)\left(1.7\times10^8\ \text{N/m}^2\right) = \boxed{1.8\times10^{-2}}$$

---

53. **SSM** ***REASONING AND SOLUTION*** Equation 10.20 gives the desired result. Solving for $\Delta V / V_0$ and taking the value for the bulk modulus B of aluminum from Table 10.3, we obtain

$$\frac{\Delta V}{V_0} = -\frac{\Delta P}{B} = -\frac{-1.01\times10^5\ \text{Pa}}{7.1\times10^{10}\ \text{N/m}^2} = \boxed{1.4\times10^{-6}}$$

---

54. ***REASONING*** The amount ΔL by which the bone changes length when a compression force or a tension force acts on it is specified by $\Delta L = \frac{FL_0}{YA}$ (Equation 10.17), where F denotes the magnitude of either type of force, L_0 is the initial length of the bone, Y is the appropriate Young's modulus, and A is the cross-sectional area of the bone. The values of Young's modulus are given in Table 10.1 ($Y_{\text{Compression}} = 9.4\times10^9\ \text{N/m}^2$ and $Y_{\text{Tension}} = 1.6\times10^{10}\ \text{N/m}^2$). The values for F, L_0, and A are not given, but it is important to recognize that these variables have the ***same*** values for both types of forces. We will apply Equation 10.17 twice, once for the compression force and once for the tension force. Since F, L_0, and A have the same values in both of the resulting equations, we will be able to eliminate them algebraically and determine the amount $\Delta L_{\text{Tension}}$ by which the bone stretches.

SOLUTION Applying Equation 10.17 for both types of forces gives

$$\Delta L_{\text{Tension}} = \frac{FL_0}{Y_{\text{Tension}}A} \quad \text{and} \quad \Delta L_{\text{Compression}} = \frac{FL_0}{Y_{\text{Compression}}A}$$

Dividing the left-hand equation by the right-hand equation and eliminating the common variables algebraically shows that

$$\frac{\Delta L_{\text{Tension}}}{\Delta L_{\text{Compression}}} = \frac{\dfrac{FL_0}{Y_{\text{Tension}}A}}{\dfrac{FL_0}{Y_{\text{Compression}}A}} = \frac{Y_{\text{Compression}}}{Y_{\text{Tension}}}$$

Solving for $\Delta L_{\text{Tension}}$, we find that

$$\Delta L_{\text{Tension}} = \left(\frac{Y_{\text{Compression}}}{Y_{\text{Tension}}}\right)\Delta L_{\text{Compression}} = \left(\frac{9.4\times10^9\ \text{N/m}^2}{1.6\times10^{10}\ \text{N/m}^2}\right)\left(2.7\times10^{-5}\ \text{m}\right) = \boxed{1.6\times10^{-5}\ \text{m}}$$

---

55. ***REASONING AND SOLUTION*** $\Delta P = -B(\Delta V/V_0)$ where $\Delta V/V_0$ is the volume strain.

The change in pressure is $\Delta P = (6.5 \times 10^5 \text{ Pa}) - (1.8 \times 10^5 \text{ Pa}) = 4.7 \times 10^5$ Pa. Taking the value for the bulk modulus B of oil from Table 10.3, we find that

$$\Delta V/V_0 = -\,\Delta P/B = -\,(4.7 \times 10^5 \text{ Pa})/(1.7 \times 10^9 \text{ Pa}) = \boxed{-2.8 \times 10^{-4}} \qquad (10.20)$$

---

56. ***REASONING*** Both cylinders experience the same force **F**. The magnitude of this force is related to the change in length of each cylinder according to Equation 10.17: $F = Y(\Delta L / L_0)A$. See Table 10.1 for values of Young's modulus Y. Each cylinder decreases in length; the total decrease being the sum of the decreases for each cylinder.

SOLUTION The length of the copper cylinder decreases by

$$\Delta L_{\text{copper}} = \frac{FL_0}{YA} = \frac{FL_0}{Y(\pi r^2)} = \frac{(6500 \text{ N})(3.0 \times 10^{-2} \text{ m})}{(1.1 \times 10^{11} \text{ N/m}^2)\pi(0.25 \times 10^{-2} \text{ m})^2} = 9.0 \times 10^{-5} \text{ m}$$

Similarly, the length of the brass decreases by

$$\Delta L_{\text{brass}} = \frac{(6500 \text{ N})(5.0 \times 10^{-2} \text{ m})}{(9.0 \times 10^{10} \text{ N/m}^2)\pi(0.25 \times 10^{-2} \text{ m})^2} = 1.8 \times 10^{-4} \text{ m}$$

Therefore, the amount by which the length of the stack decreases is $\boxed{2.7 \times 10^{-4} \text{ m}}$.

---

57. ***REASONING AND SOLUTION*** From the drawing we have $\Delta x = 3.0$ mm and

$$A = 2\,\pi r \Delta x = 2\pi(1.00 \times 10^{-2} \text{ m})(3.0 \times 10^{-3} \text{ m})$$

We now have Stress $= F/A$. Therefore,

$$F = (\text{Stress})A = (3.5 \times 10^8 \text{ Pa})[2\pi(1.00 \times 10^{-2} \text{ m})(3.0 \times 10^{-3} \text{ m})] = \boxed{6.6 \times 10^4 \text{ N}}$$

---

58. ***REASONING AND SOLUTION***

a. Strain $= \Delta L/L_0 = F/(YA)$. In this case the area subjected to the compression is given by

$$A = \pi(r_{\text{out}}^2 - r_{\text{in}}^2) = 2.64 \times 10^{-4} \text{ m}^2$$

and the force is $F = mg$. Taking the value for Young's modulus Y for bone compression from Table 10.1, we find that

$$\text{Strain} = \frac{(63\text{ kg})(9.80\text{ m/s}^2)}{(9.4\times10^9\text{ Pa})(2.64\times10^{-4}\text{ m}^2)} = \boxed{2.5\times10^{-4}}$$

b. $\Delta L = \text{Strain} \times L_0 = (2.5\times10^{-4})(0.30\text{ m}) = \boxed{7.5\times10^{-5}\text{ m}}$

---

59. **SSM** ***REASONING AND SOLUTION***

a. When the jeep is suspended motionless, the tension T in the cable must be equal in magnitude and opposite in direction to the weight mg of the jeep. Therefore, from Equation 10.17,

$$mg = Y\left(\frac{\Delta L}{L_0}\right)A$$

Solving for ΔL and taking Y for steel from Table 10.1, we obtain

$$\Delta L = \frac{mgL_0}{YA} = \frac{mgL_0}{Y(\pi r^2)} = \frac{(2100\text{ kg})(9.80\text{ m/s}^2)(48\text{ m})}{(2.0\times10^{11}\text{ Pa})\pi(5.0\times10^{-3}\text{ m})^2} = \boxed{6.3\times10^{-2}\text{ m}}$$

b. When the jeep is hoisted upward with an acceleration a, Newton's second law gives

$$T - mg = ma \qquad \text{or} \qquad T = m(a+g)$$

Therefore, Equation 10.17 becomes

$$m(a+g) = Y\left(\frac{\Delta L}{L_0}\right)A$$

and solving for ΔL gives

$$\Delta L = \frac{m(a+g)L_0}{Y(\pi r^2)} = \frac{(2100\text{ kg})(1.5\text{ m/s}^2 + 9.80\text{ m/s}^2)(48\text{ m})}{(2.0\times10^{11}\text{ Pa})\pi(5.0\times10^{-3}\text{ m})^2} = \boxed{7.3\times10^{-2}\text{ m}}$$

---

60. ***REASONING AND SOLUTION*** The applied force, F_{applied}, may be resolved into two components, one which is parallel to the area of the top surface (a shearing force) and one which is perpendicular to the top surface (a tensile force).

a. The change in the height of the block is caused by the component of the applied force that is perpendicular to the top surface. Since this component is a tensile force, the change in the height of the block can be found from Equation 10.17 with $F = F_{\text{applied}}(\sin\theta)$, $L_0 = H_0$, and $\Delta L = \Delta H$:

$$F_{\text{applied}}(\sin\theta) = Y\left(\frac{\Delta H}{H_0}\right)A$$

From Table 10.1, the Young's modulus of copper is 1.1×10^{11} N/m^2. The area, A of the top surface is $A = (5.0 \times 10^{-2}$ m$)(3.0 \times 10^{-2}$ m$) = 1.5 \times 10^{-3}$ m^2. Solving for ΔH gives:

$$\Delta H = \frac{F_{\text{applied}}(\sin\theta)H_0}{YA} = \frac{(1800\text{ N})(\sin 25°)(0.040\text{ m})}{(1.1\times10^{11}\text{ N/m}^2)(1.5\times10^{-3}\text{ m}^2)} = \boxed{1.8\times10^{-7}\text{m}}$$

b. The shear deformation of the block is caused by the component of the applied force that is tangent to the top surface, and can be determined from Equation 10.18 with $F = F_{\text{applied}}(\cos\theta)$, $L_0 = 0.040$ m, and $A = 1.5 \times 10^{-3}$ m^2.

$$F_{\text{applied}}(\cos\theta) = S\left(\frac{\Delta X}{L_0}\right)A$$

From Table 10.2, the shear modulus of copper is 4.2×10^{10} N/m^2. Solving for ΔX gives

$$\Delta X = \frac{F_{\text{applied}}(\cos\theta)L_0}{SA} = \frac{(1800\text{ N})(\cos 25°)(0.040\text{ m})}{(4.2\times10^{10}\text{ N/m}^2)(1.5\times10^{-3}\text{ m}^2)} = \boxed{1.0\times10^{-6}\text{ m}}$$

---

61. **SSM** **WWW** ***REASONING*** The strain in the wire is given by $\Delta L / L_0$. From Equation 10.17, the strain is therefore given by

$$\frac{\Delta L}{L_0} = \frac{F}{YA} \qquad (1)$$

where F must be equal to the magnitude of the centripetal force that keeps the stone moving in the circular path of radius R. Table 10.1 gives the value of Y for steel.

SOLUTION Combining Equation (1) with Equation 5.3 for the magnitude of the centripetal force, we obtain

$$\frac{\Delta L}{L_0} = \frac{F}{Y(\pi r^2)} = \frac{(mv^2/R)}{Y(\pi r^2)} = \frac{(8.0\text{ kg})(12\text{ m/s})^2/(4.0\text{ m})}{(2.0\times10^{11}\text{ Pa})\pi(1.0\times10^{-3}\text{ m})^2} = \boxed{4.6\times10^{-4}}$$

---

62. ***REASONING*** Our approach is straightforward. We will begin by writing Equation 10.17 $\left[F = Y\left(\frac{\Delta L}{L_0}\right)A\right]$ as it applies to the composite rod. In so doing, we will use subscripts for only those variables that have different values for the composite rod and the aluminum and tungsten sections. Thus, we note that the force applied to the end of the composite rod (see Figure 10.29) is also applied to each section of the rod, with the result that the magnitude F of the force has no subscript. Similarly, the cross-sectional area A is the same for the composite rod and for the aluminum and tungsten sections. Next, we will express the

change $\Delta L_{\text{Composite}}$ in the length of the composite rod as the sum of the changes in lengths of the aluminum and tungsten sections. Lastly, we will take into account that the initial length of the composite rod is twice the initial length of either of its two sections and thereby simply our equation algebraically to the point that we can solve it for the effective value of Young's modulus that applies to the composite rod.

SOLUTION Applying Equation 10.17 to the composite rod, we obtain

$$F = Y_{\text{Composite}}\left(\frac{\Delta L_{\text{Composite}}}{L_{0,\text{ Composite}}}\right)A \qquad (1)$$

Since the change $\Delta L_{\text{Composite}}$ in the length of the composite rod is the sum of the changes in lengths of the aluminum and tungsten sections, we have $\Delta L_{\text{Composite}} = \Delta L_{\text{Aluminum}} + \Delta L_{\text{Tungsten}}$. Furthermore, the changes in length of each section can be expressed using Equation 10.17 $\left(\Delta L = \frac{FL_0}{YA}\right)$, so that

$$\Delta L_{\text{Composite}} = \Delta L_{\text{Aluminum}} + \Delta L_{\text{Tungsten}} = \frac{FL_{0,\text{ Aluminum}}}{Y_{\text{Aluminum}}A} + \frac{FL_{0,\text{ Tungsten}}}{Y_{\text{Tungsten}}A}$$

Substituting this result into Equation (1) gives

$$F = \left(\frac{Y_{\text{Composite}}A}{L_{0,\text{ Composite}}}\right)\Delta L_{\text{Composite}} = \left(\frac{Y_{\text{Composite}}A}{L_{0,\text{ Composite}}}\right)\left(\frac{FL_{0,\text{ Aluminum}}}{Y_{\text{Aluminum}}A} + \frac{FL_{0,\text{ Tungsten}}}{Y_{\text{Tungsten}}A}\right)$$

$$1 = Y_{\text{Composite}}\left(\frac{L_{0,\text{ Aluminum}}}{L_{0,\text{ Composite}}Y_{\text{Aluminum}}} + \frac{L_{0,\text{ Tungsten}}}{L_{0,\text{ Composite}}Y_{\text{Tungsten}}}\right)$$

In this result we now use the fact that $L_{0,\text{ Aluminum}}/L_{0,\text{ Composite}} = L_{0,\text{ Tungsten}}/L_{0,\text{ Composite}} = 1/2$ and obtain

$$1 = Y_{\text{Composite}}\left(\frac{1}{2Y_{\text{Aluminum}}} + \frac{1}{2Y_{\text{Tungsten}}}\right)$$

Solving for $Y_{\text{Composite}}$ shows that

$$Y_{\text{Composite}} = \frac{2Y_{\text{Tungsten}}Y_{\text{Aluminum}}}{Y_{\text{Tungsten}} + Y_{\text{Aluminum}}} = \frac{2\left(3.6\times10^{11}\text{ N/m}^2\right)\left(6.9\times10^{10}\text{ N/m}^2\right)}{\left(3.6\times10^{11}\text{ N/m}^2\right)+\left(6.9\times10^{10}\text{ N/m}^2\right)} = \boxed{1.2\times10^{11}\text{ N/m}^2}$$

The values for Y_{Tungsten} and Y_{Aluminum} have been taken from Table 10.1.

---

63. ***REASONING AND SOLUTION*** Strain = $\Delta L/L_0 = F/(YA)$ where F = mg and A = πr^2. Setting the strain for the spider web equal to the strain for the wire

$$F/(YA) = F'/(Y'A') \quad \text{so that} \quad F/(Yr^2) = F'/(Y'r'^2)$$

Thus,

$$r'^2 = F'Yr^2/(FY')$$

Taking the value for Young's modulus Y' from Table 10.1, we find that

$$r' = \sqrt{\frac{(95\text{ kg})(9.80\text{ m/s}^2)(4.5\times10^9\text{ Pa})(13\times10^{-6}\text{ m})^2}{(1.0\times10^{-3}\text{ kg})(9.80\text{ m/s}^2)(6.9\times10^{10}\text{ Pa})}} = \boxed{1.0\times10^{-3}\text{ m}}$$

---

64. ***REASONING AND SOLUTION*** We know $F = S(\Delta X/L_0)A$ where the force is due to friction, i.e.,

$$F = f_s^{\text{max}} = \mu_s mg = (0.90)(7.2\times10^{-2}\text{ kg})(9.80\text{ m/s}^2) = 0.64\text{ N}$$

Therefore,

a. Stress = $F/A = (0.64\text{ N})/(0.030\text{ m})^2 = \boxed{710\text{ N/m}^2}$

b. Strain = $\Delta X/L_0 = F/(AS) = (710\text{ N/m}^2)/(2.0\times10^{10}\text{ N/m}^2) = \boxed{3.5\times10^{-8}}$

c. ΔX = (Strain) $L_0 = (3.5\times10^{-8})(1.0\times10^{-2}\text{ m}) = \boxed{3.5\times10^{-10}\text{ m}}$

---

65. SSM WWW ***REASONING*** Equation 10.20 can be used to find the fractional change in volume of the brass sphere when it is exposed to the Venusian atmosphere. Once the fractional change in volume is known, it can be used to calculate the fractional change in radius.

SOLUTION According to Equation 10.20, the fractional change in volume is

$$\frac{\Delta V}{V_0} = -\frac{\Delta P}{B} = -\frac{8.9\times10^6\text{ Pa}}{6.7\times10^{10}\text{ Pa}} = -1.33\times10^{-4}$$

where we have taken the value for the bulk modulus B of brass from Table 10.3. The initial volume of the sphere is $V_0 = (4/3)\pi r^3$. If we assume that the change in the radius of the sphere is very small relative to the initial radius, we can think of the sphere's change in volume as the addition or subtraction of a spherical shell of volume ΔV, whose radius is r and whose thickness is Δr. Then, the change in volume of the sphere is equal to the volume of the shell and is given by $\Delta V = (4\pi r^2)\Delta r$. Combining the expressions for V_0 and ΔV, and solving for $\Delta r / r$, we have

$$\frac{\Delta r}{r} = \frac{1}{3}\frac{\Delta V}{V_0}$$

Therefore,

$$\frac{\Delta r}{r} = \frac{1}{3}(-1.33\times10^{-4}) = \boxed{-4.4\times10^{-5}}$$

---

66. ***REASONING*** If we compare Equation 10.17, which governs the stretching and compression of a solid cylinder, with Hooke's law (Equation 10.2), we find that x is analogous to ΔL and k is analogous to the term YA/L_0:

$$F = \underbrace{\left(\frac{YA}{L_0}\right)}_{k}\underbrace{\Delta L}_{x}$$

SOLUTION

a. Solving for k we have

$$k = \frac{YA}{L_0} = \frac{Y(\pi r^2)}{L_0} = \frac{(3.1\times10^6 \text{ N/m}^2)\pi(0.091\times10^{-2}\text{ m})^2}{2.5\times10^{-2}\text{ m}} = \boxed{3.2\times10^2 \text{ N/m}}$$

b. The work done by the variable force is equal to the area under the F-versus-x curve. The amount x of stretch is

$$x = \frac{F}{k} = \frac{3.0\times10^{-2}\text{ N}}{3.2\times10^2\text{ N/m}} = 9.4\times10^{-5}\text{ m}$$

The work done is

$$W = \tfrac{1}{2}Fx = \tfrac{1}{2}(3.0\times10^{-2}\text{ N})(9.4\times10^{-5}\text{ m}) = \boxed{1.4\times10^{-6}\text{ J}}$$

---

67. SSM WWW ***REASONING AND SOLUTION*** According to Equation 10.6, $\omega = 2\pi f$. The maximum speed and maximum acceleration of the atoms may be calculated from Equations 10.8 and 10.10, respectively.

a. Combining Equations 10.6 and 10.8, we obtain

$$v_{max} = \omega A = (2\pi f)A = 2\pi(2.0\times10^{12}\text{ Hz})(1.1\times10^{-11}\text{ m}) = \boxed{140 \text{ m/s}}$$

b. Combining Equations 10.6 and 10.10, we obtain

$$a_{max} = \omega^2 A = (4\pi^2 f^2)A = 4\pi^2(2.0\times10^{12}\ \text{Hz})^2(1.1\times10^{-11}\ \text{m}) = \boxed{1.7\times10^{15}\ \text{m/s}^2}$$

---

68. ***REASONING*** The force F that the spring exerts on the block just before it is released is equal to $-kx$, according to Equation 10.2. Here k is the spring constant and x is the displacement of the spring from its equilibrium position. Once the block has been released, it oscillates back and forth with an angular frequency given by Equation 10.11 as $\omega = \sqrt{k/m}$, where m is the mass of the block. The maximum speed that the block attains during the oscillatory motion is $v_{max} = A\omega$ (Equation 10.8). The magnitude of the maximum acceleration that the block attains is $a_{max} = A\omega^2$ (Equation 10.10).

SOLUTION
a. The force F exerted on the block by the spring is

$$F = -kx = -(82.0\ \text{N/m})(0.120\ \text{m}) = \boxed{-9.84\ \text{N}} \qquad (10.2)$$

b. The angular frequency ω of the resulting oscillatory motion is

$$\omega = \sqrt{\frac{k}{m}} = \sqrt{\frac{82.0\ \text{N/m}}{0.750\ \text{kg}}} = \boxed{10.5\ \text{rad/s}} \qquad (10.11)$$

c. The maximum speed v_{max} is the product of the amplitude and the angular frequency:

$$v_{max} = A\omega = (0.120\ \text{m})(10.5\ \text{rad/s}) = \boxed{1.26\ \text{m/s}} \qquad (10.8)$$

d. The magnitude a_{max} of the maximum acceleration is

$$a_{max} = A\omega^2 = (0.120\ \text{m})(10.5\ \text{rad/s})^2 = \boxed{13.2\ \text{m/s}^2} \qquad (10.10)$$

---

69. ***REASONING AND SOLUTION*** Applying Equation 10.16 and recalling that frequency and period are related by f = 1/T,

$$2\pi f = \frac{2\pi}{T} = \sqrt{\frac{g}{L}}$$

where L is the length of the pendulum. Thus,

$$T = 2\pi\sqrt{\frac{L}{g}}$$

Solving for L gives

$$L = g\left(\frac{T}{2\pi}\right)^2 = \left(9.80\ \text{m/s}^2\right)\left(\frac{9.2\ \text{s}}{2\pi}\right)^2 = \boxed{21\ \text{m}}$$

---

70. ***REASONING*** The amplitude of simple harmonic motion is the distance from the equilibrium position to the point of maximum height. The angular frequency ω is related to the period T of the motion by Equation 10.6. The maximum speed attained by the person is the product of the amplitude and the angular speed (Equation 10.8).

SOLUTION
a. Since the distance from the equilibrium position to the point of maximum height is the amplitude A of the motion, we have that $A = 45.0\ \text{cm} = \boxed{0.450\ \text{m}}$.

b. The angular frequency is inversely proportional to the period of the motion:

$$\omega = \frac{2\pi}{T} = \frac{2\pi}{1.90\ \text{s}} = \boxed{3.31\ \text{rad/s}} \qquad (10.6)$$

c. The maximum speed v_{max} attained by the person on the trampoline depends on the amplitude A and the angular frequency ω of the motion:

$$v_{max} = A\omega = (0.450\ \text{m})(3.31\ \text{rad/s}) = \boxed{1.49\ \text{m/s}} \qquad (10.8)$$

---

71. SSM ***REASONING AND SOLUTION*** If we assume that each block is subjected to one-fourth of the applied force F, then, according to Equation 10.18, the force on each block is given by

$$\frac{F}{4} = S\left(\frac{\Delta X}{L_0}\right)A$$

Solving for ΔX, we obtain

$$\Delta X = \frac{FL_0}{4SA} = \frac{(32\ \text{N})(0.030\ \text{m})}{4(2.6\times10^6\ \text{N/m}^2)(1.2\times10^{-3}\ \text{m}^2)} = \boxed{7.7\times10^{-5}\ \text{m}}$$

---

72. ***REASONING*** The weight of the person causes the spring in the scale to compress. The amount x of compression, according to Equation 10.1, depends on the magnitude F_{Applied} of the applied force and the spring constant k.

SOLUTION
a. Since the applied force is equal to the person's weight, the spring constant is

$$k = \frac{F_{\text{Applied}}}{x} = \frac{670 \text{ N}}{0.79 \times 10^{-2} \text{ m}} = \boxed{8.5 \times 10^{4} \text{ N/m}} \qquad (10.1)$$

b. When another person steps on the scale, it compresses by 0.34 cm. The weight (or applied force) that this person exerts on the scale is

$$F_{\text{Applied}} = kx = \left(8.5 \times 10^{4} \text{ N/m}\right)\left(0.34 \times 10^{-2} \text{ m}\right) = \boxed{290 \text{ N}} \qquad (10.1)$$

---

73. ***REASONING*** The shear stress is equal to the magnitude of the shearing force exerted on the bar divided by the cross sectional area of the bar. The vertical deflection ΔY of the right end of the bar is given by Equation 10.18 $[F = S(\Delta Y / L_0)A]$.

SOLUTION
a. The stress is

$$\frac{F}{A} = \frac{mg}{A} = \frac{(160 \text{ kg})(9.80 \text{ m/s}^2)}{3.2 \times 10^{-4} \text{ m}^2} = \boxed{4.9 \times 10^{6} \text{ N/m}^2}$$

b. Taking the value for the shear modulus S of steel from Table 10.2, we find that the vertical deflection ΔY of the right end of the bar is

$$\Delta Y = \left(\frac{F}{A}\right)\frac{L_0}{S} = (4.9 \times 10^{6} \text{ N/m}^2)\,\frac{0.10 \text{ m}}{8.1 \times 10^{10} \text{ N/m}^2} = \boxed{6.0 \times 10^{-6} \text{ m}}$$

---

74. ***REASONING AND SOLUTION***
a. $x = A \cos \omega t$ where $\omega = 2\pi f = 2\pi(2.00 \text{ Hz}) = 4\pi \text{ rad/s}$

$$x = (0.500 \text{ m}) \cos [(4\pi \text{ rad/s})(0.0500 \text{ s})] = \boxed{0.405 \text{ m}} \qquad (10.3)$$

b. $v = -A\omega \sin \omega t$

$$v = -(0.500 \text{ m})\,(4\pi \text{ rad/s}) \sin [(4\pi \text{ rad/s})(0.0500 \text{ s})] = -3.69 \text{ m/s} \qquad (10.7)$$

The magnitude of v is $\boxed{3.69 \text{ m/s}}$

c. $a = -A\omega^2 \cos \omega t$

$$a = -(0.500 \text{ m})\,(4\pi \text{ rad/s})^2 \cos [(4\pi \text{ rad/s})(0.0500 \text{ s})] = -63.9 \text{ m/s}^2 \qquad (10.9)$$

The magnitude of a is $\boxed{63.9 \text{ m/s}^2}$

---

75. ***REASONING***

a. The angular frequency ω (in rad/s) is given by Equation 10.11 $\left(\omega = \sqrt{\frac{k}{m}}\right)$, where k is the spring constant and m is the mass of the object. The frequency f (in Hz) can be obtained from the angular frequency by using Equation 10.6 ($\omega = 2\pi f$).

b. The block loses contact with the spring when the amplitude of the oscillation is sufficiently large. To understand why, consider the block at the very top of its oscillation cycle. There it is accelerating downward, with the maximum acceleration a_{max} of simple harmonic motion. Contact is maintained with the spring, as long as the magnitude of this acceleration is less than the magnitude g of the acceleration due to gravity. If a_{max} is greater than g, the end of the spring falls away from under the block. a_{max} is given by Equation 10.10 $\left(a_{max} = A\omega^2\right)$, from which we can obtain the amplitude A when $a_{max} = g$.

SOLUTION
a. From Equations 10.6 and 10.11 we have

$$\omega = 2\pi f = \sqrt{\frac{k}{m}} \quad \text{or} \quad f = \frac{1}{2\pi}\sqrt{\frac{k}{m}} = \frac{1}{2\pi}\sqrt{\frac{112 \text{ N/m}}{0.400 \text{ kg}}} = \boxed{2.66 \text{ Hz}}$$

b. Using Equation 10.10 with $a_{max} = g$ gives

$$a_{max} = g = A\omega^2 \quad \text{or} \quad A = \frac{g}{\omega^2}$$

Substituting Equation 10.11 $\left(\omega = \sqrt{\frac{k}{m}}\right)$ into this result gives

$$A = \frac{g}{\omega^2} = \frac{g}{\left(\sqrt{k/m}\right)^2} = \frac{gm}{k} = \frac{(9.80 \text{ m/s}^2)(0.400 \text{ kg})}{112 \text{ N/m}} = \boxed{0.0350 \text{ m}}$$

---

76. ***REASONING*** Since the surface is frictionless, we can apply the principle of conservation of mechanical energy, which indicates that the total mechanical energy of the spring/mass system is the same at the

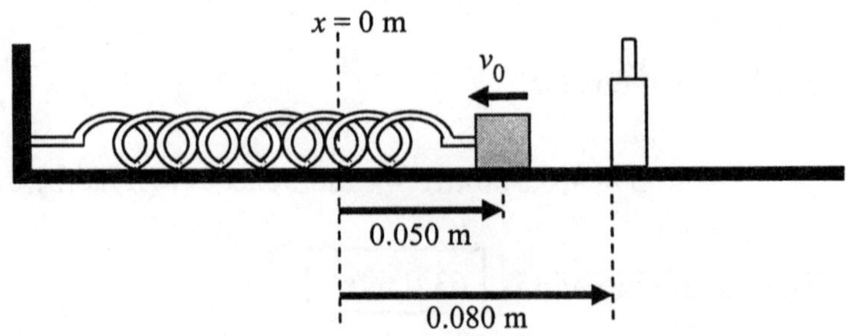

instant the block contacts the bottle (the final state of the system) and at the instant shown in the drawing (the initial state). Kinetic energy $\frac{1}{2}mv^2$ is one part of the total mechanical energy, and depends on the mass m and the speed v of the block. The dependence of the kinetic energy on speed is critical to our solution. In order for the block to knock over the bottle, it must at least reach the bottle. When launched with the minimum speed v_0 shown in the drawing, the block will reach the bottle with a final speed of $v_f = 0$ m/s. We will obtain the desired initial speed v_0 by solving the energy-conservation equation for this variable.

SOLUTION The conservation of mechanical energy states that the final total mechanical energy E_f is equal to the initial total mechanical energy E_0. The expression for the total mechanical energy for a spring/mass system is given by Equation 10.14, so that we have

$$\underbrace{\tfrac{1}{2}mv_f^2 + \tfrac{1}{2}I\omega_f^2 + mgh_f + \tfrac{1}{2}kx_f^2}_{E_f} = \underbrace{\tfrac{1}{2}mv_0^2 + \tfrac{1}{2}I\omega_0^2 + mgh_0 + \tfrac{1}{2}kx_0^2}_{E_0}$$

Since the block does not rotate, the angular speeds ω_f and ω_0 are zero. Moreover, the block reaches the bottle with a final speed of $v_f = 0$ m/s when the block is launched with the minimum initial speed v_0. In addition, the surface is horizontal, so that the final and initial heights, h_f and h_0, are the same. Thus, the above expression can be simplified as follows:

$$\tfrac{1}{2}kx_f^2 = \tfrac{1}{2}kx_0^2 + \tfrac{1}{2}mv_0^2$$

In this result, we are given no values for the spring constant k and the mass m. However, we are given a value for the angular frequency ω. This frequency is given by Equation 10.11 $\left(\omega = \sqrt{\frac{k}{m}}\right)$, which involves only the ratio k/m. Therefore, in solving the simplified energy-conservation expression for the speed v_0, we will divide both sides by m, so that the ratio k/m can be expressed using Equation 10.11.

$$\frac{\tfrac{1}{2}kx_f^2}{m} = \frac{\tfrac{1}{2}kx_0^2 + \tfrac{1}{2}mv_0^2}{m} \quad \text{or} \quad v_0 = \sqrt{\left(\frac{k}{m}\right)\left(x_f^2 - x_0^2\right)}$$

Substituting $\omega = \sqrt{\frac{k}{m}}$ from Equation 10.11, we find

$$v_0 = \omega\sqrt{\left(x_f^2 - x_0^2\right)} = (7.0 \text{ rad/s})\sqrt{(0.080 \text{ m})^2 - (0.050 \text{ m})^2} = \boxed{0.44 \text{ m/s}}$$

---

77. ***REASONING*** The angular frequency ω (in rad/s) is given by $\omega = \sqrt{\frac{k}{m}}$ (Equation 10.11), where k is the spring constant and m is the mass of the object. However, we are given neither k nor m. Instead, we are given information about how much the spring is

compressed and the launch speed of the object. Once launched, the object has kinetic energy, which is related to its speed. Before launching, the spring/object system has elastic potential energy, which is related to the amount by which the spring is compressed. This suggests that we apply the principle of conservation of mechanical energy in order to use the given information. This principle indicates that the total mechanical energy of the system is the same after the object is launched as it is before the launch. The resulting equation will provide us with the value of k/m that we need in order to determine the angular frequency from $\omega = \sqrt{\dfrac{k}{m}}$.

SOLUTION The conservation of mechanical energy states that the final total mechanical energy E_f is equal to the initial total mechanical energy E_0. The expression for the total mechanical energy for a spring/mass system is given by Equation 10.14, so that we have

$$\underbrace{\tfrac{1}{2}mv_\text{f}^2 + \tfrac{1}{2}I\omega_\text{f}^2 + mgh_\text{f} + \tfrac{1}{2}kx_\text{f}^2}_{E_\text{f}} = \underbrace{\tfrac{1}{2}mv_0^2 + \tfrac{1}{2}I\omega_0^2 + mgh_0 + \tfrac{1}{2}kx_0^2}_{E_0}$$

Since the object does not rotate, the angular speeds ω_f and ω_0 are zero. Since the object is initially at rest, the initial translational speed v_0 is also zero. Moreover, the motion takes place horizontally, so that the final height h_f is the same as the initial height h_0. Lastly, the spring is unstrained after the launch, so that x_f is zero. Thus, the above expression can be simplified as follows:

$$\tfrac{1}{2}mv_\text{f}^2 = \tfrac{1}{2}kx_0^2 \quad \text{or} \quad \frac{k}{m} = \frac{v_\text{f}^2}{x_0^2}$$

Substituting this result into Equation 10.11 shows that

$$\omega = \sqrt{\frac{k}{m}} = \sqrt{\frac{v_\text{f}^2}{x_0^2}} = \frac{v_\text{f}}{x_0} = \frac{1.50 \text{ m/s}}{0.0620 \text{ m}} = \boxed{24.2 \text{ rad/s}}$$

---

78. ***REASONING AND SOLUTION*** In this problem we want the strain = $\Delta L/L_0 = 0.01$. $F = Y(\Delta L/L_0)A$, where $F = mg$. The area of one piece of mohair is πr^2, so the total area of N identical pieces is $N\pi r^2$. Taking the value for Young's modulus Y for mohair from Table 10.1, substituting for the force and area in the equation above, and solving for N, we find that

$$N = \frac{mg}{Y\left(\dfrac{\Delta L}{L_0}\right)\pi r^2} = \frac{(75 \text{ kg})(9.80 \text{ m/s}^2)}{(2.9 \times 10^9 \text{ N/m}^2)(0.010)\pi(31 \times 10^{-6} \text{ m})^2} = \boxed{8400 \text{ pieces}}$$

---

79. ***REASONING AND SOLUTION*** The natural frequency of the suspension system is given by Equation 10.11:

$$\omega = \sqrt{\frac{k}{m}} = \sqrt{\frac{1.50 \times 10^6 \text{ N/m}}{215 \text{ kg}}} = 83.5 \text{ rad/s}$$

Thus, the wheel will resonate when its angular speed is 83.5 rad/s. This corresponds to a linear speed of

$$v = r\omega = (0.400 \text{ m})(83.5 \text{ rad/s}) = \boxed{33.4 \text{ m/s}}$$

---

80. ***REASONING AND SOLUTION*** Use conservation of energy to find the speed of point A (take the pivot to have zero gravitational PE).

$$E_{up} = mgh = E_{down} = (1/2)\, I\omega^2 + (1/2)\, kx^2$$

where the moment of inertia of the bar is $I = (1/3)mL^2$, L = bar length, and $\omega = v/L$. Substituting these into the energy equation, noting from the drawing accompanying the problem statement that $x = \sqrt{(0.100 \text{ m})^2 + (0.200 \text{ m})^2} - (0.100 \text{ m}) = 0.124 \text{ m}$ and that $h = L/2$, and solving for v, we find that

$$v = \sqrt{\frac{3\left(mgL - kx^2\right)}{m}} = \sqrt{\frac{3\left[(0.750 \text{ kg})(9.80 \text{ m/s}^2)(0.200 \text{ m}) - (25.0 \text{ N/m})(0.124 \text{ m})^2\right]}{0.750 \text{ kg}}} = \boxed{2.08 \text{ m/s}}$$

---

81. **SSM** ***REASONING AND SOLUTION*** If we assume that the wire has a circular cross-section, then Equation 10.17 becomes

$$F = Y\left(\frac{\Delta L}{L_0}\right)\left(\pi r^2\right)$$

where *F* is the tension in the wire and *r* is the radius of the wire. Solving for *r* gives

$$r = \sqrt{\frac{F L_0}{\pi Y \Delta L}} \qquad (1)$$

Young's modulus for steel is given in Table 10.1 as $Y = 2.0 \times 10^{11}$ N/m^2. The change in length ΔL can be determined from the figure at the right. If L_0 represents the length of the wire when it is horizontal, and L its length when stretched, then the figure shows that

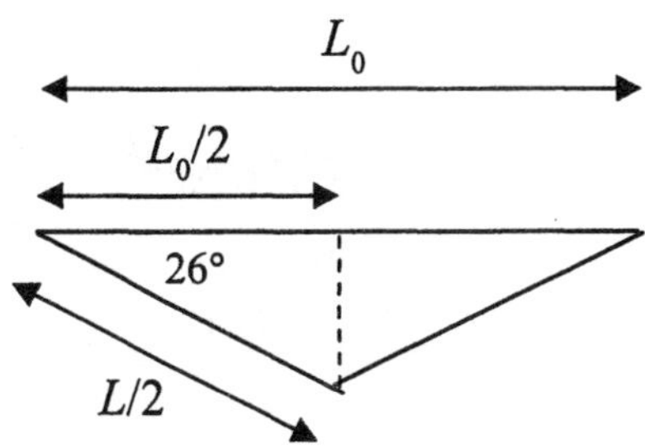

$$\cos 26° = \frac{L_0/2}{L/2} \quad \text{or} \quad L = \frac{L_0}{\cos 26°} = 1.11L_0 \qquad (2)$$

Therefore,

$$\Delta L = L - L_0 = 0.11L_0$$

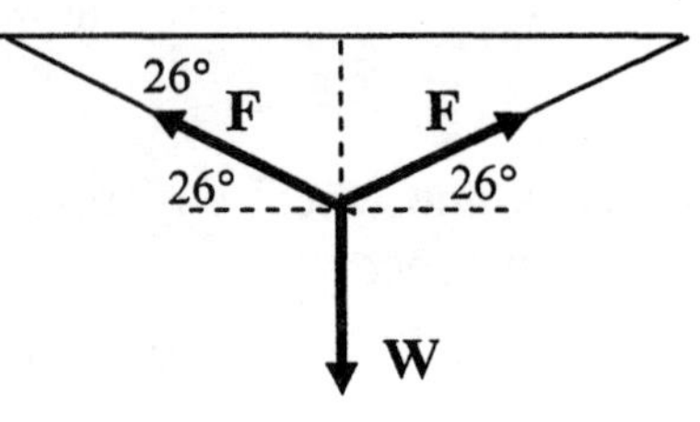

The tension in the wire can be determined from the free-body diagram at the right. Applying Newton's second law to the forces in the vertical direction yields

$$2F \sin 26° - W = 0 \quad \text{or} \quad F = \frac{W}{2 \sin 26°} \qquad (3)$$

Substituting Equations (2) and (3) into Equation (1) yields

$$r = \sqrt{\frac{W L_0}{2\pi(\sin 26°)Y(0.11L_0)}} = \sqrt{\frac{(96\text{ N})L_0}{2\pi(\sin 26°)(2.0\times10^{11}\text{ N/m}^2)(0.11L_0)}} = \boxed{4.0\times10^{-5}\text{ m}}$$

---

82. ***REASONING AND SOLUTION*** The frequency f of the simple harmonic motion is given by Equations 10.6 and 10.11 as $f = (1/2\pi)\sqrt{k/m}$. If we compare Equation 10.15, which governs the stretching and compression of a solid rod with Equation 10.1, we find that x is analogous to ΔL and k is analogous to the term YA/L_0:

$$F = \underbrace{\frac{YA}{L_0}}_{k}\;\underbrace{\Delta L}_{x}$$

The value of Young's modulus for copper is given in Table 10.1. Assuming that the rod has a circular cross-section, its area A is equal to πr^2, and we have

$$f = \frac{1}{2\pi}\sqrt{\frac{k}{m}} = \frac{1}{2\pi}\sqrt{\frac{YA}{L_0 m}} = \frac{1}{2\pi}\sqrt{\frac{Y(\pi r^2)}{L_0 m}}$$

$$= \frac{1}{2\pi}\sqrt{\frac{(1.1\times10^{11}\text{ N/m}^2)\pi(3.0\times10^{-3}\text{ m})^2}{(2.0\text{ m})(9.0\text{ kg})}} = \boxed{66\text{ Hz}}$$

---

83. SSM ***REASONING*** The angular frequency for simple harmonic motion is given by Equation 10.11 as $\omega = \sqrt{k/m}$. Since the frequency f is related to the angular frequency ω by $f = \omega/(2\pi)$ and f is related to the period T by $f = 1/T$, the period of the motion is given by

$$T = \frac{2\pi}{\omega} = 2\pi\sqrt{\frac{k}{m}}$$

SOLUTION
a. When $m_1 = m_2 = 3.0$ kg, we have that

$$T_1 = T_2 = 2\pi\sqrt{\frac{3.0 \text{ kg}}{120 \text{ N/m}}} = 0.99 \text{ s}$$

Both particles will pass through the position $x = 0$ m for the first time one-quarter of the way through one cycle, or

$$\Delta t = \tfrac{1}{4}T_1 = \tfrac{1}{4}T_2 = \frac{0.99 \text{ s}}{4} = \boxed{0.25 \text{ s}}$$

b. $T_1 = 0.99$ s, as in part (a) above, while

$$T_2 = 2\pi\sqrt{\frac{27.0 \text{ kg}}{120 \text{ N/m}}} = 3.0 \text{ s}$$

Each particle will pass through the position $x = 0$ m every odd-quarter of a cycle, $\tfrac{1}{4}T, \tfrac{3}{4}T, \tfrac{5}{4}T, \ldots$ Thus, the two particles will pass through $x = 0$ m when

3.0-kg particle $\quad t = \tfrac{1}{4}T_1, \tfrac{3}{4}T_1, \tfrac{5}{4}T_1, \ldots$

27.0-kg particle $\quad t = \tfrac{1}{4}T_2, \tfrac{3}{4}T_2, \tfrac{5}{4}T_2, \ldots$

Since $T_2 = 3T_1$, we see that both particles will be at $x = 0$ m simultaneously when $t = \tfrac{3}{4}T_1$, or $t = \tfrac{1}{4}T_2 = \tfrac{3}{4}T_1$. Thus,

$$t = \tfrac{3}{4}T_1 = \tfrac{3}{4}(0.99 \text{ s}) = \boxed{0.75 \text{ s}}$$

---

84. ***REASONING*** The change in length of the wire is, According to Equation 10.17, $\Delta L = FL_0 / YA$, where the force F is equal to the tension T in the wire. The tension in the wire can be found by applying Newton's second law to the two crates.

SOLUTION The drawing shows the free-body diagrams for the two crates. Taking up as the positive direction, Newton's second law for each of the two crates gives

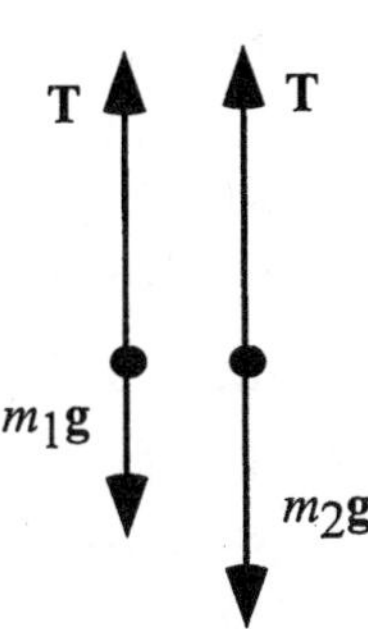

$$T - m_1 g = m_1 a \qquad (1)$$

$$T - m_2 g = -m_2 a \qquad (2)$$

Solving Equation (2) for *a*, we find $a = -\left(\frac{T - m_2 g}{m_2}\right)$. Substituting into Equation (1) gives

$$T - 2m_1 g + \frac{m_1}{m_2} T = 0$$

Solving for *T* we find

$$T = \frac{2m_1 m_2 g}{m_1 + m_2} = \frac{2(3.0\text{ kg})(5.0\text{ kg})(9.80\text{ m/s}^2)}{3.0\text{ kg} + 5.0\text{ kg}} = 37\text{ N}$$

Using the value given in Table 10.1 for Young's modulus *Y* of steel, we find, therefore, that the change in length of the wire is given by Equation 10.17 as

$$\Delta L = \frac{(37\text{ N})(1.5\text{ m})}{(2.0 \times 10^{11}\text{ N/m}^2)(1.3 \times 10^{-5}\text{ m}^2)} = \boxed{2.1 \times 10^{-5}\text{ m}}$$

---

85. ***CONCEPT QUESTIONS***

a. The restoring force of the spring and the static frictional force point in opposite directions. Since the block is in equilibrium, the net force in the horizontal direction is zero. Thus, these two forces must point in opposite directions.

b. The two forces are equal in magnitude. Since the block is in equilibrium, the net force in the horizontal direction is zero. Thus, the two forces must point in opposite directions and have equal magnitudes.

c. The maximum static frictional force is determined by the magnitude of the normal force exerted on the box and the coefficient of static friction (see Equation 4.7).

SOLUTION The drawing at the right shows the four forces that act on the block: its weight $m\mathbf{g}$, the normal force $\mathbf{F}_N$, the restoring force **F** exerted by the spring, and the maximum static frictional force $\mathbf{f}_s^{MAX}$. Since the box is not moving, it is in equilibrium. Let the *x* axis be parallel to the table top. According to Equation 4.9a, the net force ΣF_x in the *x* direction must be zero, $\Sigma F_x = 0$.

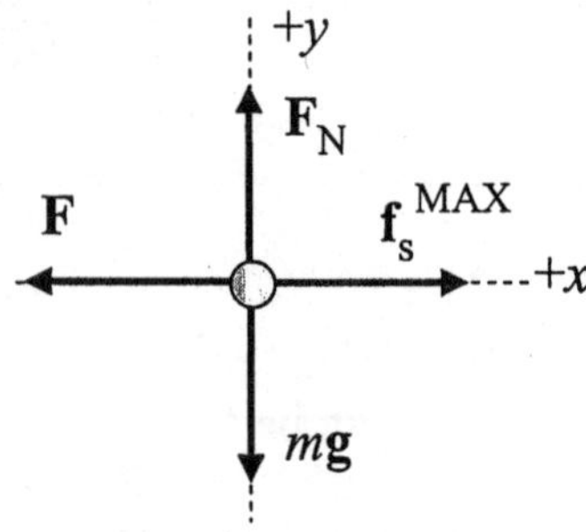

The restoring force is given by Equation 10.2 as $F = -kx$, where *k* is the spring constant and *x* is the displacement of the spring (assumed to be in the +*x* direction). The magnitude of the maximum static frictional force is given by Equation 4.7 as $f_s^{MAX} = \mu_s F_N$, where μ_s is the coefficient of static friction and F_N is the magnitude of the normal force. Thus, the condition for equilibrium can be written as

$$\underbrace{-kx + \mu_s F_N}_{\Sigma F_x} = 0 \quad \text{or} \quad x = \frac{\mu_s F_N}{k}$$

We can determine F_N by noting that the box is also in vertical equilibrium, so that

$$\underbrace{-mg + F_N}_{\Sigma F_y} = 0 \quad \text{or} \quad F_N = mg$$

The distance that the spring is stretched from its unstrained position is

$$x = \frac{\mu_s F_N}{k} = \frac{\mu_s mg}{k} = \frac{(0.74)(0.80\text{ kg})\left(9.80\text{ m/s}^2\right)}{59\text{ N/m}} = \boxed{9.8 \times 10^{-2}\text{ m}}$$

---

86. ***CONCEPT QUESTONS***

a. The restoring force points to the left. The displacement is to the right. The work is negative because the directions of the force and the displacement are opposite.

b. The work from $x_0 = -3.00$ to 0 m is positive, because the restoring force points to the right and the displacement is to the right. The work from 0 m to $x_f = +1.00$ m is negative, because the restoring force points to the left and the displacement is to the right. The magnitude of the positive work is greater than that of the negative work, so the total work is positive.

c. The work from $x_0 = -3.00$ to 0 m is positive, because the restoring force points to the right and the displacement is to the right. The work from 0 m to $x_0 = +3.00$ m is negative, because the restoring force points to the left and the displacement is to the right. The work from each segment has the same magnitude, so the total work is zero.

SOLUTION The work done in stretching or compressing a spring is given by Equation 10.12 as $W = \frac{1}{2}k\left(x_0^2 - x_f^2\right)$, where k is the spring constant and x_0 and x_f are, respectively, the initial and final displacements of the spring from its equilibrium position.

a. The work done in stretching the spring from +1.00 to +3.00 m is

$$W = \tfrac{1}{2}k\left(x_0^2 - x_f^2\right) = \tfrac{1}{2}(46.0\text{ N/m})\left[(1.00\text{ m})^2 - (3.00\text{ m})^2\right] = \boxed{-1.84 \times 10^2\text{ J}}$$

b. The work done in stretching the spring from –3.00 m to +1.00 m is

$$W = \tfrac{1}{2}k\left(x_0^2 - x_f^2\right) = \tfrac{1}{2}(46.0\text{ N/m})\left[(-3.00\text{ m})^2 - (1.00\text{ m})^2\right] = \boxed{+1.84 \times 10^2\text{ J}}$$

c. The work done in stretching the spring from –3.00 to +3.00 m is

$$W = \tfrac{1}{2}k\left(x_0^2 - x_f^2\right) = \tfrac{1}{2}(46.0\text{ N/m})\left[(-3.00\text{ m})^2 - (3.00\text{ m})^2\right] = \boxed{0\text{ J}}$$

---

87. ***CONCEPT QUESTIONS***

a. The acceleration of the box is zero when the net force acting on it is zero, in accord with Newton's second law of motion. The net force includes the box's weight (directed downward) and the restoring force of the spring (directed upward).

b. The speed of the box is zero when the spring is fully compressed.

c. The acceleration of the box is not zero when the spring is fully compressed. If the acceleration were zero, the box would be in equilibrium, and the net force on it would be zero. However, the box accelerates upward because the spring is exerting an upward force that is greater than the downward force due to the weight of the box.

SOLUTION

a. The drawing at the right shows the two forces acting on the box: its weight **mg** and the restoring force **F** exerted by the spring. At the instant the acceleration of the box is zero, it is in equilibrium. According to Equation 4.9b, the net force ΣF_y in the y direction must be zero, $\Sigma F_y = 0$.

The restoring force is given by Equation 10.2 as $F = -kx$, where k is the spring constant and x is the displacement of the spring (assumed to be in the downward, or negative, direction). Thus, the condition for equilibrium can be written as

$$\underbrace{-kx + mg}_{\Sigma F_y} = 0 \quad \text{or} \quad x = -\frac{mg}{k}$$

Solving for the *magnitude* of the spring's displacement gives

$$x = \frac{mg}{k} = \frac{(1.5\text{ kg})\left(9.80\text{ m/s}^2\right)}{450\text{ N/m}} = \boxed{3.3\times10^{-2}\text{ m}}$$

b. The conservation of mechanical energy states that the final total mechanical energy E_f is equal to the initial total mechanical energy E_0, or $E_f = E_0$ (Equation 6.9a). The expression for the total mechanical energy of an object is given by Equation 10.14. Thus, the conservation of total mechanical energy can be written as

$$\underbrace{\tfrac{1}{2}mv_f^2 + \tfrac{1}{2}I\omega_f^2 + mgh_f + \tfrac{1}{2}ky_f^2}_{E_f} = \underbrace{\tfrac{1}{2}mv_0^2 + \tfrac{1}{2}I\omega_0^2 + mgh_0 + \tfrac{1}{2}ky_0^2}_{E_0}$$

We can simplify this equation by noting which variables are zero. Since the box comes to a momentary halt, $v_f = 0$ m/s. The box does not rotate, so its angular speed is zero, $\omega_f = \omega_0 = 0$ rad/s. Initially, the spring is unstretched, so that $x_0 = 0$ m. Setting these terms equal to zero in the equation above gives

$$mgh_f + \tfrac{1}{2}kx_f^2 = \tfrac{1}{2}mv_0^2 + mgh_0$$

The vertical displacement $h_f - h_0$ through which the box falls is equal to the displacement x_f of the spring, so $x_f = h_f - h_0$. Note that x_f is negative, because h_f is less than h_0. The downward-moving box compresses the spring in the downward direction, which, as usual, we take to be the negative direction. Substituting this expression for x_f into the equation above and rearranging terms, we find that

$$\tfrac{1}{2}mv_0^2 - mg\underbrace{\left(h_f - h_0\right)}_{x_f} - \tfrac{1}{2}kx_f^2 = 0$$

or

$$\tfrac{1}{2}kx_f^2 + mgx_f - \tfrac{1}{2}mv_0^2 = 0$$

This is a quadratic equation in the variable x_f. The solution is

$$x_f = \frac{-mg \pm \sqrt{(mg)^2 - 4\left(\tfrac{1}{2}k\right)\left(-\tfrac{1}{2}mv_0^2\right)}}{2\left(\tfrac{1}{2}k\right)} = \frac{-mg}{k} \pm \sqrt{\left(\frac{mg}{k}\right)^2 + \left(\frac{m}{k}\right)v_0^2}$$

Substituting in the numbers, we find that

$$x_f = \frac{-(1.5\text{ kg})(9.80\text{ m/s}^2)}{450\text{ N/m}} \pm \sqrt{\left[\frac{(1.5\text{ kg})(9.80\text{ m/s}^2)}{450\text{ N/m}}\right]^2 + \left(\frac{1.5\text{ kg}}{450\text{ N/m}}\right)(0.49\text{ m/s})^2}$$

$$= 1.1\times10^{-2}\text{ m} \quad \text{or} \quad -7.6\times10^{-2}\text{ m}$$

The positive answer is discarded because the spring is compressed downward by the falling box, so the displacement of the spring is negative.

Therefore, the magnitude of the spring's displacement is $\boxed{7.6\times10^{-2}\text{ m}}$.

---

88. ***CONCEPT QUESTION*** She possesses gravitational potential energy with respect to the water and elastic potential energy. She has no kinetic energy since she comes to a momentary halt and has zero speed at the lowest point in the fall.

SOLUTION The conservation of mechanical energy states that the final total mechanical energy E_f is equal to the initial total mechanical energy E_0, or $E_f = E_0$ (Equation 6.9a). The expression for the total mechanical energy of an object is given by Equation 10.14. Thus, the conservation of total mechanical energy can be written as

$$\underbrace{\tfrac{1}{2}mv_f^2 + \tfrac{1}{2}I\omega_f^2 + mgh_f + \tfrac{1}{2}kx_f^2}_{E_f} = \underbrace{\tfrac{1}{2}mv_0^2 + \tfrac{1}{2}I\omega_0^2 + mgh_0 + \tfrac{1}{2}kx_0^2}_{E_0}$$

We can simplify this equation by noting which variables are zero. The jumper starts from rest and momentarily comes to a halt at the bottom of the jump; thus, $v_0 = v_f = 0$ m/s. She does not rotate, so her angular speed is zero; $\omega_f = \omega_0 = 0$ rad/s. Initially, the bungee cord is unstretched, so that $x_0 = 0$ m. Setting these terms to zero in the equation above gives

$$mgh_f + \tfrac{1}{2}kx_f^2 = mgh_0$$

We note from Figure 10.36 that $x_f = h_f - h_A$, where x_f is negative because the bungee cord is stretched downward, which is taken to be the negative direction. Substituting this expression for x_f into the equation above and rearranging terms, we find that

$$h_f^2 + \underbrace{\left(\frac{2mg}{k} - 2h_A\right)}_{b} h_f + \underbrace{h_A^2 - \frac{2mgh_0}{k}}_{c} = 0$$

$$\text{where } b = \frac{2(68.0\text{ kg})(9.80\text{ m/s}^2)}{66.0\text{ N/m}} - 2(37.0\text{ m}) = -53.8\text{ m}$$

$$\text{and} \quad c = (37.0\text{ m})^2 - \frac{2(68.0\text{ kg})(9.80\text{ m/s}^2)(46.0\text{ m})}{66.0\text{ N/m}} = 440.1\text{ m}^2$$

Thus, we have that

$$h_f^2 - (53.8\text{ m})h_f + (440.1\text{ m}^2) = 0$$

This is a quadratic equation in the variable h_f, and its solution is

$$h_f = \frac{-(-53.8\text{ m}) \pm \sqrt{(-53.8\text{ m})^2 - 4(440.1\text{ m}^2)}}{2} = 45.2\text{ m} \quad \text{or} \quad 10.1\text{ m}$$

The 45.2-m answer is discarded, because it implies that the jumper comes to a halt at a distance of only 46.0 m – 45.2 m = 0.81 m below the platform, which is above the point where the bungee cord is stretched. Thus, her height above the water when she reaches the lowest point in the fall is $\boxed{10.1\text{ m}}$.

---

89. ***CONCEPT QUESTIONS***

a. According to Equation 10.15 $\left(\omega = \sqrt{mgL/I}\right)$, the angular frequency of a physical pendulum depends on the ratio of the mass m to the moment of inertia I. Since the moment of inertia is directly proportional to the mass (see Equation 9.6), the mass algebraically cancels. Thus, the angular frequency is independent of the mass of the physical pendulum.

b. According to Equation 10.4, the period is $T = 2\pi/\omega$. Since the angular frequency ω is independent of the mass, so is the period. The period is the same for both the wood and metal pendulums.

SOLUTION The period T of a pendulum is given by Equation 10.4 as $T = 2\pi/\omega$, where ω is its angular frequency. The angular frequency of a physical pendulum is given by Equation 10.15 as $\omega = \sqrt{mgL/I}$, where m is its mass, L is the distance from the pivot to the center of mass, and I is the moment of inertia about the pivot. Combining these two relations yields

$$T = \frac{2\pi}{\omega} = \frac{2\pi}{\sqrt{\dfrac{mgL}{I}}} = 2\pi\sqrt{\frac{I}{mgL}}$$

The moment of inertia of a meter stick (a thin rod) that is oscillating about an axis at one end is given in Table 9.1 as $I = \frac{1}{3}mL_0^2$, where L_0 is the length of the stick. Since the meter stick is uniform, the distance L from one end to its center of mass is $L = \frac{1}{2}L_0$. Therefore, the period of oscillation becomes,

$$T = 2\pi\sqrt{\frac{I}{mgL}} = 2\pi\sqrt{\frac{\frac{1}{3}mL_0^2}{mg\left(\frac{1}{2}L_0\right)}} = 2\pi\sqrt{\frac{2L_0}{3g}}$$

$$= 2\pi\sqrt{\frac{2(1.00\ \text{m})}{3\left(9.80\ \text{m/s}^2\right)}} = \boxed{1.64\ \text{s}}$$

The period is the same for both pendulums, since the masses are eliminated algebraically.

---

90. ***CONCEPT QUESTIONS***

a. For each turn, the change in length of the wire is equal to the circumference of the tuning peg, $\Delta L = 2\pi r_\text{p}$, where r_p is the radius of the tuning peg.

b. It takes force to stretch the wire. This force arises because the tuning peg at one end of the wire pulls on the fixed support at the other end. In accord with Newton's action-reaction law, the fixed support pulls back. As a result of the oppositely-directed pulling forces at either end of the wire, the wire experiences an increased tension.

SOLUTION The tension is the force F that appears in Equation 10.17:

$$F = Y\left(\frac{\Delta L}{L_0}\right)A$$

where the A is the cross-sectional area of the wire. Assuming that the wire has a circular cross-section, $A = \pi r_{\text{w}}^2$, where r_{w} is the radius of the wire. When the tuning peg is turned through two revolutions, the length of the wire will increase by an amount equal to twice the circumference of the peg. Thus, $\Delta L = 2(2\pi r_{\text{p}})$, where r_{p} is the radius of the tuning peg. With these substitutions, Equation 10.17 becomes:

$$F = Y\left(\frac{4\pi r_{\text{p}}}{L_0}\right)\pi r_{\text{w}}^2 = \frac{4Yr_{\text{p}}}{L_0}\left(\pi r_{\text{w}}\right)^2$$

$$F = \frac{4(2.0\times10^{11}\ \text{N/m}^2)(1.8\times10^{-3}\ \text{m})}{0.76\ \text{m}}\left[\pi(0.80\times10^{-3}\ \text{m})\right]^2 = \boxed{1.2\times10^4\ \text{N}}$$

---

91. ***CONCEPT QUESTIONS***

a. As the block rests stationary in its equilibrium position, it has no acceleration. According to Newton's second law, the net force acting on the block is, therefore, zero. This means that the downward-directed weight of the block must be balanced by an upward-directed force. This upward force is the restoring force of the spring and is produced because the spring is compressed. The amount of compression is determined by the weight of the block. The amount must be enough for the spring to exert on the block a restoring force that has a magnitude equal to the block's weight.

b. As the block falls downward after being released, its speed is changing in the manner characteristic of simple harmonic motion. The block is not in equilibrium, and the forces acting on it do not balance to zero. Instead of thinking about forces, we may think about mechanical energy and its conservation. When the block is released from rest, the energy of the spring/block system is all in the form of gravitational potential energy, since we are taking this point to be the zero level for the gravitational potential energy. Being at rest, the block has no initial kinetic energy. It also has no initial elastic potential energy, since the spring is unstrained initially. When the block comes to a momentary halt at the lowest point in its fall, the energy is all in the form of elastic potential energy. Since the block is again at rest, it again has no kinetic energy. The spring has been compressed, and gravitational potential energy has been converted entirely into elastic potential energy. The amount by which the spring is compressed is determined by the amount of gravitational potential energy that must be converted into elastic potential energy. The amount must be enough that the elastic potential energy equals the gravitational potential energy.

c. The compression of the spring is greater in the non-equilibrium case than in the equilibrium case. The reason is that in the non-equilibrium case, the block has been allowed to move, and its inertia carries it beyond its stationary equilibrium position on the spring. The compression of the spring must increase beyond that corresponding to the stationary equilibrium position in order to produce the force that is needed to decelerate the block to a momentary halt.

SOLUTION

a. As the block rests stationary on the spring, the downward-directed weight balances the upward-directed restoring force from the spring. The magnitude of the weight is *mg*, and the magnitude of the restoring force is given by Equation 10.2 without the minus sign as *kx*. Thus, we have

$$\underbrace{mg}_{\substack{\text{Magnitude of}\\ \text{the weight}}} = \underbrace{kx}_{\substack{\text{Magnitude of}\\ \text{the spring force}}} \quad \text{or} \quad x = \frac{mg}{k} = \frac{(0.64\text{ kg})(9.80\text{ m/s}^2)}{170\text{ N/m}} = \boxed{0.037\text{ m}}$$

b. The conservation of mechanical energy states that the final total mechanical energy E_f is equal to the initial total mechanical energy E_0. The expression for the total mechanical energy for an object on a spring is given by Equation 10.14, so that we have

$$\underbrace{\tfrac{1}{2}mv_f^2 + \tfrac{1}{2}I\omega_f^2 + mgh_f + \tfrac{1}{2}kx_f^2}_{E_f} = \underbrace{\tfrac{1}{2}mv_0^2 + \tfrac{1}{2}I\omega_0^2 + mgh_0 + \tfrac{1}{2}kx_0^2}_{E_0}$$

The block does not rotate, so the angular speeds ω_f and ω_0 are zero. Since the block comes to a momentary halt on the spring and is released from rest, the translational speeds v_f and v_0 are also zero. Because the spring is initially unstrained, the initial displacement x_0 of the spring is likewise zero. Thus, the above expression can be simplified as follows:

$$mgh_f + \tfrac{1}{2}kx_f^2 = mgh_0 \quad \text{or} \quad \tfrac{1}{2}kx_f^2 = mg(h_0 - h_f)$$

The term $h_0 - h_f$ is the amount by which the spring has compressed, or $h_0 - h_f = x_f$. Making this substitution into the simplified energy-conservation equation gives

$$\tfrac{1}{2}kx_f^2 = mg(h_0 - h_f) = mgx_f \quad \text{or} \quad \tfrac{1}{2}kx_f = mg$$

Solving for x_f, we find

$$x = \frac{2mg}{k} = \frac{2(0.64\text{ kg})(9.80\text{ m/s}^2)}{170\text{ N/m}} = \boxed{0.074\text{ m}}$$

As expected, the spring compresses more in the non-equilibrium situation.

---

92. ***CONCEPT QUESTIONS***

a. According to Equation 10.1 the applied force needed to change the length of an ideal spring is $F_{\text{Applied}} = kx$, where k is the spring constant and x is the displacement. To change the length of a bone, the necessary applied force is given by Equation 10.17 as follows:

$$F_{\text{Applied}} = Y\left(\frac{\Delta L}{L_0}\right)A = \underbrace{(YA/L_0)}_{\substack{\text{The effective}\\ \text{spring constant}\\ k_{\text{Effective}}}} \underbrace{\Delta L}_{\substack{\text{The change in}\\ \text{length or the}\\ \text{displacement } x}}$$

In this expression Y is Young's modulus, A is the effective cross-sectional area of the bone, L_0 is the initial length of the bone, and ΔL is the change in length. Associating ΔL with x, we see that the effective spring constant of the bone is given by

$$k_{\text{Effective}} = \frac{YA}{L_0} \tag{1}$$

b. According to Equation 10.13, the elastic potential energy of an ideal spring is $\text{PE}_{\text{Elastic}} = \frac{1}{2}k_{\text{Effective}}x^2$. Equation 10.1 gives the applied force as $F_{\text{Applied}} = k_{\text{Effective}}x$. Solving Equation 10.1 for x and substituting the result into Equation 10.13 gives

$$\text{PE}_{\text{Elastic}} = \tfrac{1}{2}k_{\text{Effective}}x^2 = \tfrac{1}{2}k_{\text{Effective}}\left(\frac{F_{\text{Applied}}}{k_{\text{Effective}}}\right)^2 = \frac{F_{\text{Applied}}^2}{2k_{\text{Effective}}} \tag{2}$$

c. The person falls from rest and does not rotate, so initially he has only gravitational potential energy. Ignoring air resistance and friction, we may apply the conservation of mechanical energy. Since the person strikes the ground stiff-legged and comes to a halt without rotating, all of the energy he had to begin with must be absorbed by his legs as elastic potential energy. The height through which he falls determines the amount of his gravitational potential energy and, hence, the amount of energy his legs must absorb.

SOLUTION The conservation of mechanical energy states that the final total mechanical energy E_{f} is equal to the initial total mechanical energy E_0. The expression for the total mechanical energy for a spring/object system is given by Equation 10.14, so we have

$$\underbrace{\tfrac{1}{2}mv_{\text{f}}^2 + \tfrac{1}{2}I\omega_{\text{f}}^2 + mgh_{\text{f}} + \tfrac{1}{2}kx_{\text{f}}^2}_{E_{\text{f}}} = \underbrace{\tfrac{1}{2}mv_0^2 + \tfrac{1}{2}I\omega_0^2 + mgh_0 + \tfrac{1}{2}kx_0^2}_{E_0}$$

Since the person does not rotate, the angular speeds ω_{f} and ω_0 are zero. The person is at rest both initially and finally, so the initial and final translational speeds v_0 and v_{f} are also zero. Moreover, the thighbone is initially unstrained, with the result that x_0 is zero. Thus, the above expression can be simplified to give

$$mgh_f + \tfrac{1}{2}kx_f^2 = mgh_0$$

Using Equation (2) to express the final elastic potential energy of the thighbone, we can write the simplified energy-conservation equation as follows:

$$mgh_f + \frac{F_{\text{Applied}}^2}{2k_{\text{Effective}}} = mgh_0 \quad \text{or} \quad h_0 - h_f = \frac{F_{\text{Applied}}^2}{2mgk_{\text{Effective}}}$$

As the man falls, his center of gravity moves from its initial height of h_0 to its final height of h_f, which is a distance of $h_0 - h_f$. Using Equation (1) for the effective spring constant of the bone, we find

$$h_0 - h_f = \frac{F_{\text{Applied}}^2}{2mgk_{\text{Effective}}} = \frac{F_{\text{Applied}}^2 L_0}{2mgYA}$$

$$= \frac{\left(7.0\times10^4\ \text{N}\right)^2 (0.55\ \text{m})}{2(65\ \text{kg})\left(9.80\ \text{m/s}^2\right)\left(9.4\times10^9\ \text{N/m}^2\right)\left(4.0\times10^{-4}\ \text{m}^2\right)} = \boxed{0.56\ \text{m}}$$

The value used for Y is that for bone compression, as given in Table 10.1.

CHAPTER 11 | *FLUIDS*

PROBLEMS

---

1. SSM ***REASONING*** Equation 11.1 can be used to find the volume occupied by 1.00 kg of silver. Once the volume is known, the area of a sheet of silver of thickness d can be found from the fact that the volume is equal to the area of the sheet times its thickness.

SOLUTION Solving Equation 11.1 for V, the volume of 1.00 kg of silver is

$$V = \frac{m}{\rho} = \frac{1.00 \text{ kg}}{10\,500 \text{ kg/m}^3} = 9.52\times10^{-5} \text{ m}^3$$

The area of the silver, is, therefore,

$$A = \frac{V}{d} = \frac{9.52\times10^{-5} \text{ m}^3}{3.00\times10^{-7} \text{ m}} = \boxed{317 \text{ m}^2}$$

---

2. ***REASONING AND SOLUTION*** 14.0 karat gold is (14.0)/(24.0) gold or 58.3%. The weight of the gold in the necklace is then (1.27 N)(0.583) = 0.740 N. This corresponds to a volume given by $V = M/\rho = W/(\rho g)$. Thus,

$$V = \frac{0.740 \text{ N}}{\left(19\,300 \text{ kg/m}^3\right)\left(9.80 \text{ m/s}^2\right)} = \boxed{3.91\times10^{-6} \text{ m}^3}$$

---

3. ***REASONING AND SOLUTION*** The weight of the gold is $W = Mg = \rho Vg$. Therefore,

$$W = (19\,300 \text{ kg/m}^3)(0.30 \text{ m})(0.30 \text{ m})(0.20 \text{ m})(9.80 \text{ m/s}^2) = \boxed{3400 \text{ N}}$$

Since 1 N = 0.225 lb, W = (3400 N)[(0.225 lb)/(1 N)] = 760 lb. In other words, the movie pirate would have to be capable of carrying a 760-lb chest.

---

4. ***REASONING*** The weight W of the water bed is equal to the mass m of water times the acceleration g due to gravity; $W = mg$ (Equation 4.5). The mass, on the other hand, is equal to the density ρ of the water times its volume V, or $m = \rho V$ (Equation 11.1).

SOLUTION Substituting $m = \rho V$ into the relation $W = mg$ gives

$$W = mg = (\rho V)g$$

$$= (1.00 \times 10^3 \text{ kg/m}^3)(1.83 \text{ m} \times 2.13 \text{ m} \times 0.229 \text{ m})(9.80 \text{ m/s}^2) = \boxed{8750 \text{ N}}$$

We have taken the density of water from Table 11.1. Since the weight of the water bed is greater than the additional weight that the floor can tolerate, the bed should not be purchased.

---

5. SSM ***REASONING AND SOLUTION*** Equation 11.1 can be solved for the mass m. In order to use Equation 11.1, however, we must know the volume of the ice. Since the lake is circular, the volume of the ice is equal to the circular area of the lake $(A = \pi r^2)$ multiplied by the thickness d of the ice.

$$m = \rho V = \rho(\pi r^2)d = (917 \text{ kg/m}^3)\pi(480 \text{ m})^2(0.010 \text{ m}) = \boxed{6.6 \times 10^6 \text{ kg}}$$

---

6. ***REASONING AND SOLUTION*** If the concrete were solid, it would have a mass of

$$M = \rho V = (2200 \text{ kg/m}^3)(0.025 \text{ m}^3) = 55 \text{ kg}$$

The mass of concrete removed to make the hole is then 55 kg – 33 kg = 22 kg. This corresponds to a volume $V = (22 \text{ kg})/(2200 \text{ kg/m}^3) = 0.010 \text{ m}^3$. Since the hole is spherical $V = (4/3)\pi r^3$ so

$$r = \sqrt[3]{\frac{3V}{4\pi}} = \sqrt[3]{\frac{3(0.010 \text{ m}^3)}{4\pi}} = \boxed{0.13 \text{ m}}$$

---

7. ***REASONING*** According to the definition of density ρ given in Equation 11.1, the mass m of a substance is $m = \rho V$, where V is the volume. We will use this equation and the fact that the mass of the water and the gold are equal to find our answer. To convert from a volume in cubic meters to a volume in gallons, we refer to the inside of the front cover of the text to find that 1 gal = 3.785×10^{-3} m^3.

REASONING Using Equation 11.1, we find that

$$\rho_{\text{Water}} V_{\text{Water}} = \rho_{\text{Gold}} V_{\text{Gold}} \quad \text{or} \quad V_{\text{Water}} = \frac{\rho_{\text{Gold}} V_{\text{Gold}}}{\rho_{\text{Water}}}$$

Using the fact that 1 gal = 3.785×10^{-3} m^3 and densities for gold and water from Table 11.1, we find

$$V_{\text{Water}} = \frac{\rho_{\text{Gold}} V_{\text{Gold}}}{\rho_{\text{Water}}}$$

$$= \frac{(19\,300\text{ kg/m}^3)(0.15\text{ m})(0.050\text{ m})(0.050\text{ m})}{(1000\text{ kg/m}^3)}\left(\frac{1\text{ gal}}{3.785\times10^{-3}\text{ m}^3}\right) = \boxed{1.9\text{ gal}}$$

---

8. ***REASONING*** The period T of a satellite is the time for it to make one complete revolution around the planet. The period is the circumference of the circular orbit ($2\pi R$) divided by the speed v of the satellite, so that $T = (2\pi R)/v$ (see Equation 5.1). In Section 5.5 we saw that the centripetal force required to keep a satellite moving in a circular orbit is provided by the gravitational force. This relationship tells us that the speed of the satellite must be $v = \sqrt{GM/R}$ (Equation 5.5), where G is the universal gravitational constant and M is the mass of the planet. By combining this expression for the speed with that for the period, and using the definition of density, we can obtain the period of the satellite.

SOLUTION The period of the satellite is

$$T = \frac{2\pi R}{v} = \frac{2\pi R}{\sqrt{\dfrac{GM}{R}}} = 2\pi\sqrt{\frac{R^3}{GM}}$$

According to Equation 11.1, the mass of the planet is equal to its density ρ times its volume V. Since the planet is spherical, $V = \frac{4}{3}\pi R^3$. Thus, $M = \rho V = \rho\left(\frac{4}{3}\pi R^3\right)$. Substituting this expression for M into that for the period T gives

$$T = 2\pi\sqrt{\frac{R^3}{GM}} = 2\pi\sqrt{\frac{R^3}{G\rho\left(\frac{4}{3}\pi R^3\right)}} = \sqrt{\frac{3\pi}{G\rho}}$$

The density of iron is $\rho = 7860$ kg/m^3 (see Table 11.1), so the period of the satellite is

$$T = \sqrt{\frac{3\pi}{G\rho}} = \sqrt{\frac{3\pi}{\left(6.67\times10^{-11}\text{ N}\cdot\text{m}^2/\text{kg}^2\right)\left(7860\text{ kg/m}^3\right)}} = \boxed{4240\text{ s}}$$

---

9. $\boxed{\text{SSM}}$ $\boxed{\text{WWW}}$ ***REASONING*** The total mass of the solution is the sum of the masses of its constituents. Therefore,

$$\rho_s V_s = \rho_w V_w + \rho_g V_g \qquad (1)$$

where the subscripts s, w, and g refer to the solution, the water, and the ethylene glycol, respectively. The volume of the water can be written as $V_w = V_s - V_g$. Making this substitution for V_w, Equation (1) above can be rearranged to give

$$\frac{V_g}{V_s} = \frac{\rho_s - \rho_w}{\rho_g - \rho_w} \qquad (2)$$

Equation (2) can be used to calculate the relative volume of ethylene glycol in the solution.

SOLUTION The density of ethylene glycol is given in the problem. The density of water is given in Table 11.1 as 1.000×10^3 kg/m^3. The specific gravity of the solution is given as 1.0730. Therefore, the density of the solution is

$$\rho_s = (\text{specific gravity of solution}) \times \rho_w$$

$$= (1.0730)(1.000 \times 10^3 \text{ kg/m}^3) = 1.0730 \times 10^3 \text{ kg/m}^3$$

Substituting the values for the densities into Equation (2), we obtain

$$\frac{V_g}{V_s} = \frac{\rho_s - \rho_w}{\rho_g - \rho_w} = \frac{1.0730 \times 10^3 \text{ kg/m}^3 - 1.000 \times 10^3 \text{ kg/m}^3}{1116 \text{ kg/m}^3 - 1.000 \times 10^3 \text{ kg/m}^3} = 0.63$$

Therefore, the volume percentage of ethylene glycol is $\boxed{63\%}$.

---

10. ***REASONING*** The cap is in equilibrium, so the sum of all the forces acting on it must be zero. There are three forces in the vertical direction: the force $\mathbf{F}_{\text{inside}}$ due to the gas pressure inside the bottle, the force $\mathbf{F}_{\text{outside}}$ due to atmospheric pressure outside the bottle, and the force $\mathbf{F}_{\text{thread}}$ that the screw thread exerts on the cap. By setting the sum of these forces to zero, and using the relation $F = PA$, where P is the pressure and A is the area of the cap, we can determine the magnitude of the force that the screw threads exert on the cap.

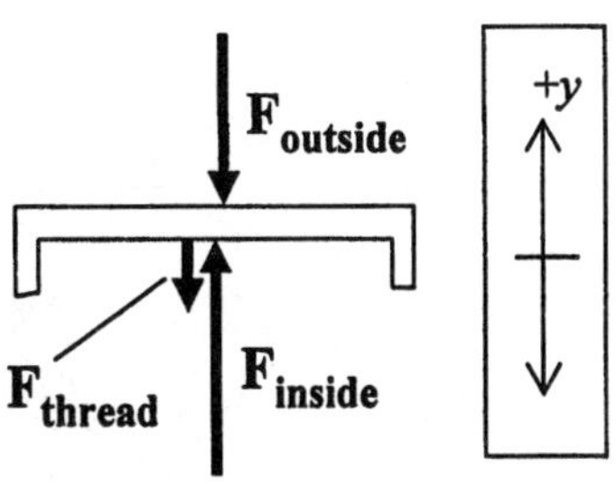

SOLUTION The drawing shows the free-body diagram of the cap and the three vertical forces that act on it. Since the cap is in equilibrium, the net force in the vertical direction must be zero.

$$\Sigma F_y = -F_{\text{thread}} + F_{\text{inside}} - F_{\text{outside}} = 0 \quad (4.9b)$$

Solving this equation for F_{thread}, and using the fact that force equals pressure times area, $F = PA$ (Equation 11.3), we have

$$F_{\text{thread}} = F_{\text{inside}} - F_{\text{outside}} = P_{\text{inside}}A - P_{\text{outside}}A$$

$$= \left(P_{\text{inside}} - P_{\text{outside}}\right)A = \left(1.80\times10^5\ \text{Pa} - 1.01\times10^5\ \text{Pa}\right)\left(4.10\times10^{-4}\ \text{m}^2\right) = \boxed{32\ \text{N}}$$

---

11. ***REASONING*** Since the inside of the box is completely evacuated, there is no air to exert an upward force on the lid from the inside. Furthermore, since the weight of the lid is negligible, there is only one force that acts on the lid; the downward force caused by the air pressure on the outside of the lid. In order to pull the lid off the box, one must supply a force that is at least equal in magnitude and opposite in direction to the force exerted on the lid by the outside air.

SOLUTION According to Equation 11.3, pressure is defined as $P = F/A$; therefore, the magnitude of the force on the lid due to the air pressure is

$$F = (0.85\times10^5\ \text{N/m}^2)(1.3\times10^{-2}\ \text{m}^2) = \boxed{1.1\times10^3\ \text{N}}$$

---

12. ***REASONING*** Pressure is the magnitude of the force applied perpendicularly to a surface divided by the area of the surface, according to Equation 11.3. The force magnitude, therefore, is equal to the pressure times the area.

SOLUTION According to Equation 11.3, we have

$$F = PA = \left(8.0\times10^4\ \text{lb/in.}^2\right)\left[(6.1\ \text{in.})(2.6\ \text{in.})\right] = \boxed{1.3\times10^6\ \text{lb}}$$

---

13. SSM ***REASONING AND SOLUTION*** Equation 11.3 gives the desired result, where F is equal to the weight of the woman. Assuming that the heel has a circular cross section, its area is $A = \pi r^2$, where r is the radius.

$$P = \frac{F}{A} = \frac{mg}{\pi r^2} = \frac{(50.0\ \text{kg})(9.80\ \text{m/s}^2)}{\pi\left(6.00\times10^{-3}\ \text{m}\right)^2} = \boxed{4.33\times10^6\ \text{Pa}}$$

---

14. ***REASONING*** Since the weight is distributed uniformly, each tire exerts one-half of the weight of the rider and bike on the ground. According to the definition of pressure, Equation 11.3, the force that each tire exerts on the ground is equal to the pressure P inside the tire times the area A of contact between the tire and the ground. From this relation, the area of contact can be found.

    SOLUTION The area of contact that *each tire* makes with the ground is

$$A = \frac{F}{P} = \frac{\frac{1}{2}\left(W_{\text{person}} + W_{\text{bike}}\right)}{P} = \frac{\frac{1}{2}(625\text{ N} + 98\text{ N})}{7.60\times10^{5}\text{ Pa}} = \boxed{4.76\times10^{-4}\text{ m}^2} \qquad (11.3)$$

---

15. ***REASONING*** According to Equation 11.3, the pressure P exerted on the ground by the stack of blocks is equal to the force F exerted by the blocks (their combined weight) divided by the area A of the block's surface in contact with the ground, or $P = F/A$. Since the pressure is largest when the area is smallest, the least number of blocks is used when the surface area in contact with the ground is the smallest. This area is 0.200 m × 0.100 m.

    SOLUTION The pressure exerted by N blocks stacked on top of one another is

$$P = \frac{F}{A} = \frac{NW_{\text{one block}}}{A} \qquad (11.3)$$

where $W_{\text{one block}}$ is the weight of one block. The least number of whole blocks required to produce a pressure of two atmospheres (2.02×10^5 Pa) is

$$N = \frac{PA}{W_{\text{one block}}} = \frac{\left(2.02\times10^{5}\text{ Pa}\right)\left(0.200\text{ m}\times0.100\text{ m}\right)}{169\text{ N}} = \boxed{24}$$

---

16. ***REASONING AND SOLUTION***

    a. The force exerted on the piston (surface area $A = \pi r^2 = 1.96 \times 10^{-3}\text{ m}^2$) is

$$F = PA = (1.013 \times 10^5\text{ N/m}^2)(1.96 \times 10^{-3}\text{ m}^2) = 199\text{ N}$$

The spring is therefore compressed by an amount x, where

$$x = F/k = (199\text{ N})/(3600\text{ N/m}) = \boxed{0.055\text{ m}}$$

    b. The work done by the spring is

$$W = (1/2)\,kx^2 = (1/2)(3600\text{ N/m})(0.055\text{ m})^2 = \boxed{5.4\text{ J}}$$

---

17. SSM ***REASONING AND SOLUTION*** Both the cylinder and the hemisphere have circular cross sections. According to Equation 11.3, the pressure exerted on the ground by the hemisphere is

$$P = \frac{W_\text{h}}{A_\text{h}} = \frac{W_\text{h}}{\pi r_\text{h}^2}$$

where W_h and r_h are the weight and radius of the hemisphere. Similarly, the pressure exerted on the ground by the cylinder is

$$P = \frac{W_\text{c}}{A_\text{c}} = \frac{W_\text{c}}{\pi r_\text{c}^2}$$

where W_c and r_c are the weight and radius of the cylinder. Since each object exerts the same pressure on the ground, we can equate the right-hand sides of the expressions above to obtain

$$\frac{W_\text{h}}{\pi r_\text{h}^2} = \frac{W_\text{c}}{\pi r_\text{c}^2}$$

Solving for r_h^2, we obtain

$$r_\text{h}^2 = r_\text{c}^2\,\frac{W_\text{h}}{W_\text{c}} \qquad (1)$$

The weight of the hemisphere is

$$W_\text{h} = \rho g V_\text{h} = \rho g\left[\tfrac{1}{2}\left(\tfrac{4}{3}\pi r_\text{h}^3\right)\right] = \tfrac{2}{3}\rho g \pi r_\text{h}^3$$

where ρ and V_h are the density and volume of the hemisphere, respectively. The weight of the cylinder is

$$W_\text{c} = \rho g V_\text{c} = \rho g \pi r_\text{c}^2 h$$

where ρ and V_c are the density and volume of the cylinder, respectively, and h is the height of the cylinder. Substituting these expressions for the weights into Equation (1) gives

$$r_\text{h}^2 = r_\text{c}^2\,\frac{W_\text{h}}{W_\text{c}} = r_\text{c}^2\,\frac{\tfrac{2}{3}\rho g \pi r_\text{h}^3}{\rho g \pi r_\text{c}^2 h}$$

Solving for r_h gives

$$r_h = \tfrac{3}{2}h = \tfrac{3}{2}(0.500 \text{ m}) = \boxed{0.750 \text{ m}}$$

---

18. ***REASONING AND SOLUTION*** The force exerted by a pressure is $F = PA$ and is perpendicular to the surface. The force on the outside of each half of the roof is

$$F = (10.0 \text{ mm Hg})\left(\frac{133 \text{ N/m}^2}{1 \text{ mm Hg}}\right)(14.5 \text{ m} \times 4.21 \text{ m}) = 8.11 \times 10^4 \text{N}$$

Adding the forces vectorially gives for the vertical force

$$F_V = 2F \cos 30.0° = 1.41 \times 10^5 \text{ N}$$

and for the horizontal force $F_h = 0$ N. The net force is then $\boxed{1.41 \times 10^5 \text{ N}}$.

The direction of the force is $\boxed{\text{downward}}$, since the horizontal components of the forces on the halves cancel.

---

19. $\boxed{\text{SSM}}$ ***REASONING AND SOLUTION*** Using Equation 11.4, we have

$$P_{heart} - P_{brain} = \rho gh = (1.000 \times 10^3 \text{ kg/m}^3)(9.80 \text{ m/s}^2)(12 \text{ m}) = \boxed{1.2 \times 10^5 \text{ Pa}}$$

---

20. ***REASONING*** The pressure P_2 at a lower point in a static fluid is related to the pressure P_1 at a higher point by Equation 11.4, $P_2 = P_1 + \rho gh$, where ρ is the density of the fluid, g is the magnitude of the acceleration due to gravity, and h is the difference in heights between the two points. This relation can be used directly to find the pressure in the artery in the brain.

SOLUTION Solving Equation 11.4 for pressure P_1 in the brain (the higher point), gives

$$P_1 = P_2 - \rho gh = 1.6 \times 10^4 \text{ Pa} - \left(1060 \text{ kg/m}^3\right)\left(9.80 \text{ m/s}^2\right)\left(0.45 \text{ m}\right) = \boxed{1.1 \times 10^4 \text{ Pa}}$$

---

21. ***REASONING*** The magnitude of the force that would be exerted on the window is given by Equation 11.3, $F = PA$, where the pressure can be found from Equation 11.4: $P_2 = P_1 + \rho gh$. Since P_1 represents the pressure at the surface of the water, it is equal to atmospheric pressure, P_{atm}. Therefore, the magnitude of the force is given by

$$F = (P_{atm} + \rho gh)A$$

where, if we assume that the window is circular with radius r, its area A is given by $A = \pi r^2$.

SOLUTION
a. Thus, the magnitude of the force is

$$F = \left[1.013\times10^5 \text{ Pa} + (1025 \text{ kg/m}^3)(9.80 \text{ m/s}^2)(11\ 000 \text{ m})\right]\pi(0.10 \text{ m})^2 = \boxed{3.5\times10^6 \text{ N}}$$

b. The weight of a jetliner whose mass is 1.2×10^5 kg is

$$W = mg = (1.2\times10^5 \text{ kg})(9.80 \text{ m/s}^2) = \boxed{1.2\times10^6 \text{ N}}$$

Therefore, the force exerted on the window at a depth of 11 000 m is about three times greater than the weight of a jetliner!

---

22. ***REASONING*** As the depth h increases, the pressure increases according to Equation 11.4 ($P_2 = P_1 + \rho gh$). In this equation, P_1 is the pressure at the shallow end, P_2 is the pressure at the deep end, and ρ is the density of water (1.00×10^3 kg/m^3, see Table 11.1). We seek a value for the pressure at the deep end minus the pressure at the shallow end.

SOLUTION Using Equation 11.4, we find

$$P_{\text{Deep}} = P_{\text{Shallow}} + \rho gh \quad \text{or} \quad P_{\text{Deep}} - P_{\text{Shallow}} = \rho gh$$

The drawing at the right shows that a value for h can be obtained from the 15-m length of the pool by using the tangent of the 11° angle:

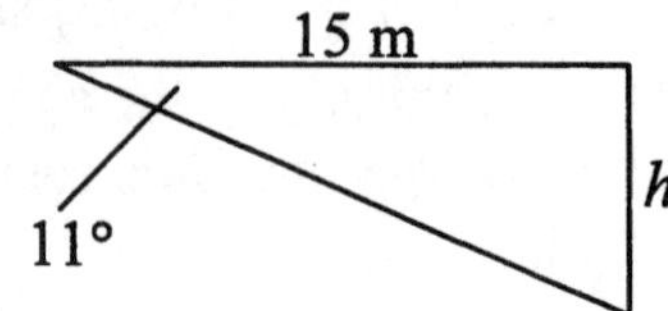

$$\tan 11° = \frac{h}{15 \text{ m}} \quad \text{or} \quad h = (15 \text{ m})\tan 11°$$

$$P_{\text{Deep}} - P_{\text{Shallow}} = \rho g(15 \text{ m})\tan 11°$$

$$= (1.00\times10^3 \text{ kg/m}^3)(9.80 \text{ m/s}^2)(15 \text{ m})\tan 11° = \boxed{2.9\times10^4 \text{ Pa}}$$

---

23. SSM ***REASONING AND SOLUTION***

a. The pressure at the level of house A is given by Equation 11.4 as $P = P_{\text{atm}} + \rho gh$. Now the height h consists of the 15.0 m and the diameter d of the tank. We first calculate the radius of the tank, from which we can infer d. Since the tank is spherical, its full mass is given by $M = \rho V = \rho[(4/3)\pi r^3]$. Therefore,

$$r^3 = \frac{3M}{4\pi\rho} \quad \text{or} \quad r = \left(\frac{3M}{4\pi\rho}\right)^{1/3} = \left[\frac{3(5.25\times10^5\,\text{kg})}{4\pi(1.000\times10^3\,\text{kg/m}^3)}\right]^{1/3} = 5.00\text{ m}$$

Therefore, the diameter of the tank is 10.0 m, and the height h is given by

$$h = 10.0\text{ m} + 15.0\text{ m} = 25.0\text{ m}$$

According to Equation 11.4, the gauge pressure in house A is, therefore,

$$P - P_{\text{atm}} = \rho gh = (1.000\times10^3\text{ kg/m}^3)(9.80\text{ m/s}^2)(25.0\text{ m}) = \boxed{2.45\times10^5\text{ Pa}}$$

b. The pressure at house B is $P = P_{\text{atm}} + \rho gh$, where

$$h = 15.0\text{ m} + 10.0\text{ m} - 7.30\text{ m} = 17.7\text{ m}$$

According to Equation 11.4, the gauge pressure in house B is

$$P - P_{\text{atm}} = \rho gh = (1.000\times10^3\text{ kg/m}^3)(9.80\text{ m/s}^2)(17.7\text{ m}) = \boxed{1.73\times10^5\text{ Pa}}$$

---

24. ***REASONING*** Since the diver uses a snorkel, the pressure in her lungs is atmospheric pressure. If she is swimming at a depth h below the surface, the pressure outside her lungs is atmospheric pressure plus that due to the water. The water pressure P_2 at the depth h is related to the pressure P_1 at the surface by Equation 11.4, $P_2 = P_1 + \rho gh$, where ρ is the density of the fluid, g is the magnitude of the acceleration due to gravity, and h is the depth. This relation can be used directly to find the depth.

SOLUTION We are given that the maximum difference in pressure between the outside and inside of the lungs is one-twentieth of an atmosphere, or $P_2 - P_1 = \frac{1}{20}\left(1.01\times10^5\text{ Pa}\right)$. Solving Equation 11.4 for the depth h gives

$$h = \frac{P_2 - P_1}{\rho g} = \frac{\frac{1}{20}\left(1.01\times10^5\text{ Pa}\right)}{\left(1025\text{ kg/m}^3\right)\left(9.80\text{ m/s}^2\right)} = \boxed{0.50\text{ m}}$$

---

25. ***REASONING AND SOLUTION*** The pump must generate an upward force to counteract the weight of the column of water above it. Therefore, $F = mg = (\rho hA)g$. The required pressure is then

$$P = F/A = \rho gh = (1.00 \times 10^3\ \text{kg/m}^3)(9.80\ \text{m/s}^2)(71\ \text{m}) = \boxed{7.0\times10^5\ \text{Pa}}$$

---

26. ***REASONING AND SOLUTION*** The difference in pressure between the ground and the roof is $\Delta P = \rho gh$, so

$$h = \frac{(13.0\ \text{mm Hg})\left(\dfrac{133\ \text{Pa}}{1\ \text{mm Hg}}\right)}{(1.29\ \text{kg/m}^3)(9.80\ \text{m/s}^2)} = \boxed{137\ \text{m}}$$

Note that we have used the fact that 133 Pa = 1 mm Hg as a conversion factor.

---

27. SSM WWW ***REASONING*** According to Equation 11.4, the pressure P_{mercury} at a point 7.10 cm below the ethyl alcohol-mercury interface is

$$P_{\text{mercury}} = P_{\text{interface}} + \rho_{\text{mercury}} g h_{\text{mercury}} \qquad (1)$$

where $P_{\text{interface}}$ is the pressure at the alcohol-mercury interface, and $h_{\text{mercury}} = 0.0710\ \text{m}$. The pressure at the interface is

$$P_{\text{interface}} = P_{\text{atm}} + \rho_{\text{ethyl}} g h_{\text{ethyl}} \qquad (2)$$

Equation (2) can be used to find the pressure at the interface. This value can then be used in Equation (1) to determine the pressure 7.10 cm below the interface.

SOLUTION Direct substitution of the numerical data into Equation (2) yields

$$P_{\text{interface}} = 1.01\times10^5\ \text{Pa} + (806\ \text{kg/m}^3)(9.80\ \text{m/s}^2)(1.10\ \text{m}) = 1.10\times10^5\ \text{Pa}$$

Therefore, the pressure 7.10 cm below the ethyl alcohol-mercury interface is

$$P_{\text{mercury}} = 1.10\times10^5\ \text{Pa} + (13\,600\ \text{kg/m}^3)(9.80\ \text{m/s}^2)(0.0710\ \text{m}) = \boxed{1.19\times10^5\ \text{Pa}}$$

---

28. ***REASONING*** In each case the pressure at point A in Figure 11.12 is atmospheric pressure and the pressure in the tube above the top of the liquid column is P. In Equation 11.4 ($P_2 = P_1 + \rho gh$), this means that $P_2 = P_{\text{atmosphere}}$ and $P_1 = P$. With this identification of the pressures and a value of 13 600 kg/m^3 for the density of mercury (see Table 11.1), Equation 11.4 provides a solution to the problem.

SOLUTION Rearranging Equation 11.4, we have $P_2 - P_1 = \rho gh$. Applying this expression to each setup gives

$$\left(P_2 - P_1\right)_{\text{Mercury}} = P_{\text{Atmosphere}} - P = \left(\rho gh\right)_{\text{Mercury}}$$

$$\left(P_2 - P_1\right)_{\text{Unknown}} = P_{\text{Unknown}} - P = \left(\rho gh\right)_{\text{Unknown}}$$

Since the left side of each of these equations is the same, we have

$$\rho_{\text{Mercury}} g h_{\text{Mercury}} = \rho_{\text{Unknown}} g h_{\text{Unknown}}$$

$$\rho_{\text{Unknown}} = \rho_{\text{Mercury}}\left(\frac{h_{\text{Mercury}}}{h_{\text{Unknown}}}\right) = \left(13\,600\text{ kg/m}^3\right)\left(\frac{h_{\text{Mercury}}}{16h_{\text{Mercury}}}\right) = \boxed{850\text{ kg/m}^3}$$

---

29. ***REASONING AND SOLUTION*** The pressure at the bottom of the container is

$$P = P_{\text{atm}} + \rho_{\text{w}} g h_{\text{w}} + \rho_{\text{m}} g h_{\text{m}}$$

We want $P = 2P_{\text{atm}}$, and we know $h = h_{\text{w}} + h_{\text{m}} = 1.00$ m. Using the above and rearranging gives

$$h_{\text{m}} = \frac{P_{\text{atm}} - \rho_{\text{w}} gh}{\left(\rho_{\text{m}} - \rho_{\text{w}}\right)g} = \frac{1.01\times10^5\text{ Pa} - \left(1.00\times10^3\text{ kg/m}^3\right)\left(9.80\text{ m/s}^2\right)\left(1.00\text{ m}\right)}{\left(13.6\times10^3\text{ kg/m}^3 - 1.00\times10^3\text{ kg/m}^3\right)\left(9.80\text{ m/s}^2\right)} = \boxed{0.74\text{ m}}$$

---

30. ***REASONING AND SOLUTION*** The force exerted on the top surface is

$$F_2 = P_2 A_2 = P_{\text{atm}} \pi R_2^{\,2}$$

The force exerted on the bottom surface is $F_1 = P_1 A_1 = (P_{\text{atm}} + \rho gh)\pi R_1^{\,2}$. Equating and rearranging yields

$$R_2^{\,2} = R_1^{\,2}(1 + \rho gh/P_{\text{atm}})$$

or

$$R_2^{\ 2} = 1.485\ R_1^{\ 2} \qquad (1)$$

Consider a right triangle formed by drawing a vertical line from a point on the circumference of the bottom circle to the plane of the top circle so that two sides are equal to $R_2 - R_1$ and h. Then $\tan 30.0° = (R_2 - R_1)/h$.

a. Now,

$$R_1 = R_2 - 2.887\ \text{m} \qquad (2)$$

Substituting (2) into (1) and rearranging yields

$$0.485\ R_2^{\ 2} - 8.57\ R_2 + 12.38 = 0$$

which has two roots, namely, R_2 = 16.1 m and 1.59 m. The value $R_2 = 1.59$ m leads to a negative value for R_1. Clearly, a radius cannot be negative, so we can eliminate the root R_2 = 1.59 m, and we conclude that $R_2 = \boxed{16.1\ \text{m}}$.

b. Now that R_2 is known, Equation (2) gives $R_1 = \boxed{13.2\ \text{m}}$.

---

31. SSM ***REASONING*** According to Equation 11.4, the initial pressure at the bottom of the pool is $P_0 = \left(P_{\text{atm}}\right)_0 + \rho gh$, while the final pressure is $P_\text{f} = \left(P_{\text{atm}}\right)_\text{f} + \rho gh$. Therefore, the change in pressure at the bottom of the pool is

$$\Delta P = P_\text{f} - P_0 = \left[\left(P_{\text{atm}}\right)_\text{f} + \rho gh\right] - \left[\left(P_{\text{atm}}\right)_0 + \rho gh\right] = \left(P_{\text{atm}}\right)_\text{f} - \left(P_{\text{atm}}\right)_0$$

According to Equation 11.3, $F = PA$, the change in the force at the bottom of the pool is

$$\Delta F = (\Delta P)\,A = \left[\left(P_{\text{atm}}\right)_\text{f} - \left(P_{\text{atm}}\right)_0\right]A$$

SOLUTION Direct substitution of the data given in the problem into the expression above yields

$$\Delta F = \left(765\ \text{mm Hg} - 755\ \text{mm Hg}\right)(12\ \text{m})(24\ \text{m})\left(\frac{133\ \text{Pa}}{1.0\ \text{mm Hg}}\right) = \boxed{3.8\times 10^5\ \text{N}}$$

Note that the conversion factor 133 Pa = 1.0 mm Hg is used to convert mm Hg to Pa.

---

32. ***REASONING AND SOLUTION*** We know that $F_2 = (A_2/A_1)F_1$. Since the pistons are circular in cross-section

$$F_2 = \frac{R_2^2}{R_1^2} F_1 = \left(\frac{5.1 \times 10^{-2} \text{ m}}{6.4 \times 10^{-3} \text{ m}} \right)^2 (330 \text{ N}) = \boxed{2.1 \times 10^4 \text{ N}}$$

---

33. ***REASONING*** We label the input piston as "2" and the output plunger as "1." When the bottom surfaces of the input piston and output plunger are at the same level, Equation 11.5, $F_2 = F_1 (A_2 / A_1)$, applies. However, this equation is not applicable when the bottom surface of the output plunger is $h = 1.50$ m above the input piston. In this case we must use Equation 11.4, $P_2 = P_1 + \rho gh$, to account for the difference in heights. In either case, we will see that the input force is less than the combined weight of the output plunger and car.

SOLUTION

a. Using $A = \pi r^2$ for the circular areas of the piston and plunger, the input force required to support the 24 500-N weight is

$$F_2 = F_1 \left(\frac{A_2}{A_1} \right) = (24\ 500 \text{ N}) \left[\frac{\pi (7.70 \times 10^{-3} \text{m})^2}{\pi (0.125 \text{ m})^2} \right] = \boxed{93.0 \text{ N}} \quad (11.5)$$

b. The pressure P_2 at the input piston is related to the pressure P_1 at the bottom of the output plunger by Equation 11.4, $P_2 = P_1 + \rho gh$, where h is the difference in heights. Setting $P_2 = F_2 / A_2 = F_2 / (\pi r_2^2)$, $P_1 = F_1 / (\pi r_1^2)$, and solving for F_2, we have

$$F_2 = F_1 \left(\frac{\pi r_2^2}{\pi r_1^2} \right) + \rho gh (\pi r_2^2) \quad (11.4)$$

$$= (24\ 500 \text{ N}) \left[\frac{\pi (7.70 \times 10^{-3} \text{m})^2}{\pi (0.125 \text{ m})^2} \right]$$

$$+ (8.30 \times 10^2 \text{ kg/m}^3)(9.80 \text{ m/s}^2)(1.30 \text{ m}) \pi (7.70 \times 10^{-3} \text{m})^2 = \boxed{94.9 \text{ N}}$$

---

34. ***REASONING*** Equation 11.5 gives the force F_2 of the output plunger in terms of the force F_1 applied to the input piston as $F_2 = F_1(A_2/A_1)$, where A_2 and A_1 are the corresponding areas. In this problem the chair begins to rise when the output force just barely exceeds the weight, so $F_2 = 2100$ N. We are given the input force as 55 N. We seek the ratio of the

radii, so we will express the area of each circular cross section as πr^2 when we apply Equation 11.5.

SOLUTION According to Equation 11.5, we have

$$\frac{A_2}{A_1} = \frac{F_2}{F_1} \quad \text{or} \quad \frac{\pi r_2^2}{\pi r_1^2} = \frac{F_2}{F_1}$$

Solving for the ratio of the radii yields

$$\frac{r_2}{r_1} = \sqrt{\frac{F_2}{F_1}} = \sqrt{\frac{2100 \text{ N}}{55 \text{ N}}} = \boxed{6.2}$$

---

35. SSM ***REASONING*** The pressure P' exerted on the bed of the truck by the plunger is $P' = P - P_{\text{atm}}$. According to Equation 11.3, $F = P'A$, so the force exerted on the bed of the truck can be expressed as $F = (P - P_{\text{atm}})A$. If we assume that the plunger remains perpendicular to the floor of the load bed, the torque that the plunger creates about the axis shown in the figure in the text is

$$\tau = F\ell = (P - P_{\text{atm}})A\ell = (P - P_{\text{atm}})(\pi r^2)\ell$$

SOLUTION Direct substitution of the numerical data into the expression above gives

$$\tau = (3.54 \times 10^6 \text{ Pa} - 1.01 \times 10^5 \text{ Pa})\pi(0.150 \text{ m})^2(3.50 \text{ m}) = \boxed{8.50 \times 10^5 \text{ N} \cdot \text{m}}$$

---

36. ***REASONING AND SOLUTION*** From Pascal's principle, the pressure in the brake fluid at the master cylinder is equal to the pressure in the brake fluid at the plungers: $P_C = P_P$, or

$$\frac{F_C}{A_C} = \frac{F_P}{A_P} \quad \text{or} \quad F_P = F_C \frac{A_P}{A_C} = F_C \frac{\pi r_P^2}{\pi r_C^2} = F_C \left(\frac{r_P}{r_C}\right)^2$$

The torque on the pedal is equal to the torque that is applied to the master cylinder so that

$$F\ell = F_C \ell_C \quad \text{or} \quad F_C = F \frac{\ell}{\ell_C}$$

Combining the expression for F_C with the expression for F_P above, we have

$$F_{\text{P}} = F\frac{\ell}{\ell_{\text{C}}}\left(\frac{r_{\text{P}}}{r_{\text{C}}}\right)^2 = (9.00\text{ N})\left(\frac{0.150\text{ m}}{0.0500\text{ m}}\right)\left(\frac{1.90\times10^{-2}\text{ m}}{9.50\times10^{-3}\text{ m}}\right)^2 = \boxed{108\text{ N}}$$

---

37. ***REASONING*** Since the piston and the plunger are at the same height, Equation 11.5, $F_2 = F_1(A_2 / A_1)$, applies, and we can find an expression for the force exerted on the spring. Then Equation 10.1, $F = kx$, can be used to determine the amount of compression of the spring.

SOLUTION Substituting the right hand side of Equation 11.5 into Equation 10.1, we find that

$$F_1\left(\frac{A_2}{A_1}\right) = kx$$

From the drawing in the text, we see that the force on the right piston must be equal in magnitude to the weight of the rock, or $F_1 = mg$. Therefore,

$$mg\left(\frac{A_2}{A_1}\right) = kx$$

Solving for x, we obtain

$$x = \frac{mg}{k}\left(\frac{A_2}{A_1}\right) = \frac{(40.0\text{ kg})(9.80\text{ m/s}^2)}{1600\text{ N/m}}\left(\frac{15\text{ cm}^2}{65\text{ cm}^2}\right) = \boxed{5.7\times10^{-2}\text{ m}}$$

---

38. ***REASONING*** Since the duck is in equilibrium, its downward-acting weight is balanced by the upward-acting buoyant force. According to Archimedes' principle, the magnitude of the buoyant force is equal to the weight of the water displaced by the duck. Setting the weight of the duck equal to the magnitude of the buoyant force will allow us to find the average density of the duck.

SOLUTION Since the weight W_{duck} of the duck is balanced by the magnitude F_{B} of the buoyant force, we have that $W_{\text{duck}} = F_{\text{B}}$. The duck's weight is $W_{\text{duck}} = mg = (\rho_{\text{duck}}V_{\text{duck}})g$, where ρ_{duck} is the average density of the duck and V_{duck} is its volume. The magnitude of the buoyant force, on the other hand, equals the weight of the water displaced by the duck, or $F_{\text{B}} = m_{\text{water}}g$, where m_{water} is the mass of the displaced water. But $m_{\text{water}} = \rho_{\text{water}}\left(\frac{1}{4}V_{\text{duck}}\right)$, since one-quarter of the duck's volume is beneath the water. Thus,

$$\underbrace{\rho_{\text{duck}} V_{\text{duck}} g}_{\text{Weight of duck}} = \underbrace{\rho_{\text{water}} \left(\tfrac{1}{4} V_{\text{duck}}\right) g}_{\substack{\text{Magnitude of} \\ \text{buoyant force}}}$$

Solving this equation for the average density of the duck (and taking the density of water from Table 11.1) gives

$$\rho_{\text{duck}} = \tfrac{1}{4}\rho_{\text{water}} = \tfrac{1}{4}\left(1.00\times10^{3}\ \text{kg/m}^{3}\right) = \boxed{250\ \text{kg/m}^{3}}$$

---

39. SSM ***REASONING*** The buoyant force exerted on the balloon by the air must be equal in magnitude to the weight of the balloon and its contents (load and hydrogen). The magnitude of the buoyant force is given by $\rho_{\text{air}} Vg$. Therefore,

$$\rho_{\text{air}} Vg = W_{\text{load}} + \rho_{\text{hydrogen}} Vg$$

where, since the balloon is spherical, $V = (4/3)\pi r^{3}$. Making this substitution for V and solving for r, we obtain

$$r = \left[\frac{3W_{\text{load}}}{4\pi g(\rho_{\text{air}} - \rho_{\text{hydrogen}})}\right]^{1/3}$$

SOLUTION Direct substitution of the data given in the problem yields

$$r = \left[\frac{3(5750\ \text{N})}{4\pi(9.80\ \text{m/s}^{2})\,(1.29\ \text{kg/m}^{3} - 0.0899\ \text{kg/m}^{3})}\right]^{1/3} = \boxed{4.89\ \text{m}}$$

---

40. ***REASONING AND SOLUTION*** The upward buoyant force of the water on the iceberg must equal the weight of the iceberg if it floats, $F_{\text{B}} = W$, so that $\rho_{\text{W}} gV = \rho_{\text{i}} gV_{\text{i}}$. Now

$$V/V_{\text{i}} = \rho_{\text{i}}/\rho_{\text{W}} = (917\ \text{kg/m}^{3})/(1025\ \text{kg/m}^{3}) = 0.895$$

The percent of the volume of the iceberg which is submerged is $\boxed{89.5\%}$.

---

41. ***REASONING*** The paperweight weighs less in water than in air, because of the buoyant force F_{B} of the water. The buoyant force points upward, while the weight points downward, leading to an effective weight in water of $W_{\text{In water}} = W - F_{\text{B}}$. There is also a buoyant force when the paperweight is weighed in air, but it is negligibly small. Thus, from the given

weights, we can obtain the buoyant force, which is the weight of the displaced water, according to Archimedes' principle. From the weight of the displaced water and the density of water, we can obtain the volume of the water, which is also the volume of the completely immersed paperweight.

SOLUTION We have

$$W_{\text{In water}} = W - F_B \quad \text{or} \quad F_B = W - W_{\text{In water}}$$

According to Archimedes' principle, the buoyant force is the weight of the displaced water, which is *mg*, where *m* is the mass of the displaced water. Using Equation 11.1, we can write the mass as the density times the volume or $m = \rho V$. Thus, for the buoyant force, we have

$$F_B = W - W_{\text{In water}} = \rho V g$$

Solving for the volume and using $\rho = 1.00 \times 10^3$ kg/m^3 for the density of water (see Table 11.1), we find

$$V = \frac{W - W_{\text{In water}}}{\rho g} = \frac{6.9\text{ N} - 4.3\text{ N}}{(1.00 \times 10^3\text{ kg/m}^3)(9.80\text{ m/s}^2)} = \boxed{2.7 \times 10^{-4}\text{ m}^3}$$

---

42. ***REASONING AND SOLUTION*** The upward buoyant force on the raft must equal the weight of the raft plus the weight of the swimmers, $F_B = W_r + W_s$. Thus,

$$\rho_w g V_r = \rho_p g V_r + M_s g$$

$$M_s = (\rho_w - \rho_p)V_r = (1.00 \times 10^3\text{ kg/m}^3 - 550\text{ kg/m}^3)(4.0\text{ m})(4.0\text{ m})(0.30\text{ m})$$

$$M_s = \boxed{2.2 \times 10^3\text{ kg}}$$

---

43. SSM WWW ***REASONING AND SOLUTION*** Under water, the weight of the person with empty lungs is $W_{\text{empty}} = W - \rho_{\text{water}} g V_{\text{empty}}$, where W is the weight of the person in air and V_{empty} is the volume of the empty lungs. Similarly, when the person's lungs are partially full under water, the weight of the person is $W_{\text{full}} = W - \rho_{\text{water}} g V_{\text{full}}$. Subtracting the second equation from the first equation and rearranging gives

$$V_{\text{full}} - V_{\text{empty}} = \frac{W_{\text{empty}} - W_{\text{full}}}{\rho_{\text{water}} g} = \frac{40.0\text{ N} - 20.0\text{ N}}{(1.00 \times 10^3\text{ kg/m}^3)(9.80\text{ m/s}^2)} = \boxed{2.04 \times 10^{-3}\text{ m}^3}$$

---

44. ***REASONING AND SOLUTION*** The buoyant force exerted by the water must at least equal the weight of the logs plus the weight of the people,

$$F_B = W_L + W_P$$

$$\rho_W g V = \rho_L g V + W_P$$

Now the volume of logs needed is

$$V = \frac{M_P}{\rho_W - \rho_L} = \frac{320 \text{ kg}}{1.00 \times 10^3 \text{ kg/m}^3 - 725 \text{ kg/m}^3} = 1.16 \text{ m}^3$$

The volume of one log is

$$V_L = \pi(8.00 \times 10^{-2} \text{ m})^2(3.00 \text{ m}) = 6.03 \times 10^{-2} \text{ m}^3$$

The number of logs needed is

$$N = V/V_L = (1.16)/(6.03 \times 10^{-2}) = 19.2$$

Therefore, $\boxed{\text{at least 20 logs are needed}}$.

---

45. ***REASONING AND SOLUTION*** The figure at the right shows the two forces that initially act on the box. Since the box is not accelerated, the two forces must have zero resultant: $F_{B0} - W_{box} = 0$. Therefore,

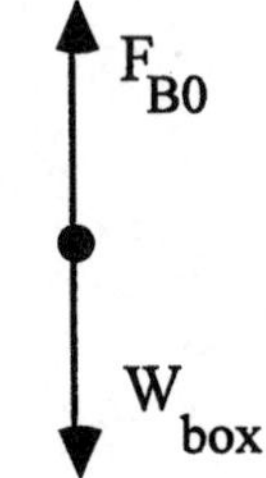

$$F_{B0} = W_{box} \tag{1}$$

From Archimedes' principle, the buoyant force on the box is equal to the weight of the water that is displaced by the box:

$$F_{B0} = W_{disp} \tag{2}$$

Combining (1) and (2) we have $W_{box} = W_{disp}$, or $m_{box}\, g = \rho_{water} g V_{disp}$. Therefore,

$$m_{box} = \rho_{water} V_{disp}$$

Since the box floats with one-third of its height beneath the water, $V_{disp} = (1/3)V_{box}$, or $V_{disp} = (1/3)L^3$. Therefore,

$$m_{\text{box}} = \frac{\rho_{\text{water}} L^3}{3} \tag{3}$$

The figure at the right shows the three forces that act on the box after water is poured into the box. The box begins to sink when

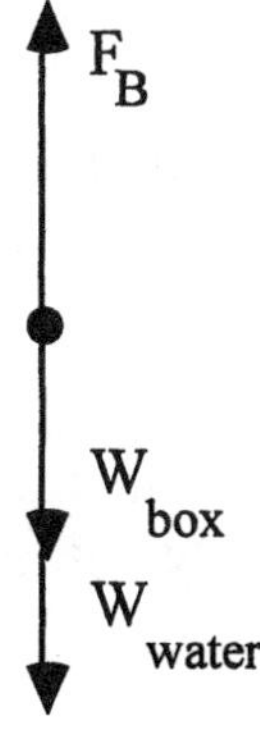

$$W_{\text{box}} + W_{\text{water}} \geq F_B \tag{4}$$

The box just begins to sink when the equality is satisfied. From Archimedes' principle, the buoyant force on the system is equal to the weight of the water that is displaced by the system: $F_B = W_{\text{displaced}}$.

The equality in (4) can be written as

$$m_{\text{box}}g + m_{\text{water}}g = m_{\text{displaced}}g \tag{5}$$

When the box begins to sink, the volume of the water displaced is equal to the volume of the box; Equation (5) then becomes

$$m_{\text{box}} + \rho_{\text{water}} V_{\text{water}} = \rho_{\text{water}} V_{\text{box}}.$$

The volume of water in the box at this instant is $V_{\text{water}} = L^2 h$, where h is the depth of the water in the box. Thus, the equation above becomes

$$m_{\text{box}} + \rho_{\text{water}} L^2 h = \rho_{\text{water}} L^3$$

Using Equation (3) for the mass of the box, we obtain

$$\frac{\rho_{\text{water}} L^3}{3} + \rho_{\text{water}} L^2 h = \rho_{\text{water}} L^3$$

Solving for h gives

$$h = \tfrac{2}{3} L = \tfrac{2}{3}(0.30 \text{ m}) = \boxed{0.20 \text{ m}}$$

---

46. ***REASONING*** When an object is completely submerged within a fluid, its apparent weight in the fluid is equal to its true weight *mg* minus the upward-acting buoyant force. According to Archimedes' principle, the magnitude of the buoyant force is equal to the weight of the fluid displaced by the object. The weight of the displaced fluid depends on the volume of the object. We will apply this principle twice, once for the object submerged in each fluid, to find the volume of the object.

SOLUTION The apparent weights of the object in ethyl alcohol and in water are:

Ethyl alcohol

$$\underbrace{15.2\text{ N}}_{\substack{\text{Weight in}\\ \text{alcohol}}} = \underbrace{mg}_{\substack{\text{True}\\ \text{weight}}} - \underbrace{\rho_{\text{alcohol}} gV}_{\substack{\text{Magnitude of}\\ \text{buoyant force}}} \qquad (1)$$

Water

$$\underbrace{13.7\text{ N}}_{\substack{\text{Weight}\\ \text{in water}}} = \underbrace{mg}_{\substack{\text{True}\\ \text{weight}}} - \underbrace{\rho_{\text{water}} gV}_{\substack{\text{Magnitude of}\\ \text{buoyant force}}} \qquad (2)$$

These equations contain two unknowns, the volume V of the object and its mass m. By subtracting Equation (2) from Equation (1), we can eliminate the mass algebraically. The result is

$$15.2\text{ N} - 13.7\text{ N} = gV\left(\rho_{\text{water}} - \rho_{\text{alcohol}}\right)$$

Solving this equation for the volume, and using the densities from Table 11.1, we have

$$V = \frac{15.2\text{ N} - 13.7\text{ N}}{g\left(\rho_{\text{water}} - \rho_{\text{alcohol}}\right)} = \frac{1.5\text{ N}}{\left(9.80\text{ m/s}^2\right)\left(1.00\times10^3\text{ kg/m}^3 - 806\text{ kg/m}^3\right)} = \boxed{7.9\times10^{-4}\text{ m}^3}$$

---

47. **SSM** ***REASONING*** The height of the cylinder that is in the oil is given by $h_{\text{oil}} = V_{\text{oil}}/(\pi r^2)$, where V_{oil} is the volume of oil displaced by the cylinder and r is the radius of the cylinder. We must, therefore, find the volume of oil displaced by the cylinder. After the oil is poured in, the buoyant force that acts on the cylinder is equal to the sum of the weight of the water displaced by the cylinder and the weight of the oil displaced by the cylinder. Therefore, the magnitude of the buoyant force is given by $F = \rho_{\text{water}} gV_{\text{water}} + \rho_{\text{oil}} gV_{\text{oil}}$. Since the cylinder floats in the fluid, the net force that acts on the cylinder must be zero. Therefore, the buoyant force that supports the cylinder must be equal to the weight of the cylinder, or

$$\rho_{\text{water}} gV_{\text{water}} + \rho_{\text{oil}} gV_{\text{oil}} = mg$$

where m is the mass of the cylinder. Substituting values into the expression above leads to

$$V_{\text{water}} + (0.725)V_{\text{oil}} = 7.00\times10^{-3}\text{ m}^3 \qquad (1)$$

From the figure in the text, $V_{\text{cylinder}} = V_{\text{water}} + V_{\text{oil}}$. Substituting values into the expression for V_{cylinder} gives

$$V_{\text{water}} + V_{\text{oil}} = 8.48\times10^{-3}\text{ m}^3 \qquad (2)$$

Subtracting Equation (1) from Equation (2) yields $V_{\text{oil}} = 5.38 \times 10^{-3}\ \text{m}^3$.

SOLUTION The height of the cylinder that is in the oil is, therefore,

$$h_{\text{oil}} = \frac{V_{\text{oil}}}{\pi r^2} = \frac{5.38 \times 10^{-3}\ \text{m}^3}{\pi(0.150\ \text{m})^2} = \boxed{7.6 \times 10^{-2}\ \text{m}}$$

---

48. ***REASONING AND SOLUTION*** Only the weight of the block compresses the spring. Applying Hooke's law gives $W = kx$. The spring is stretched by the buoyant force acting on the block minus the weight of the block. Hooke's law again gives $F_B - W = 2kx$. Eliminating kx gives $F_B = 3W$. Now $F_B = \rho_W gV$, so that the volume of the block is

$$V = 3M/\rho_W = 3(8.00\ \text{kg})/(1.00 \times 10^3\ \text{kg/m}^3) = 2.40 \times 10^{-2}\ \text{m}^3$$

The volume of wood in the block is

$$V_W = M/\rho_b = (8.00\ \text{kg})/(840\ \text{kg/m}^3) = 9.52 \times 10^{-3}\ \text{m}^3$$

The volume of the block that is hollow is $V - V_W = 1.45 \times 10^{-2}\ \text{m}^3$. The percentage of the block that is hollow is then

$$100(1.45 \times 10^{-2})/(2.40 \times 10^{-2}) = \boxed{60.3\ \%}.$$

---

49. ***REASONING AND SOLUTION*** The upward buoyant force must equal the weight of the shell if it is floating, $\rho_W gV = W$. The submerged volume of the shell is $V = (4/3)\pi R_2^3$, where R_2 is its outer radius. Now $R_2^3 = (3/4)\text{m}/(\rho_W \pi)$ gives

$$R_2 = \sqrt[3]{\frac{3m}{4\pi \rho_W}} = \sqrt[3]{\frac{3(1.00\ \text{kg})}{4\pi\left(1.00 \times 10^3\ \text{kg/m}^3\right)}} = \boxed{6.20 \times 10^{-2}\text{m}}$$

The weight of the shell is $W = \rho_g g(V_2 - V_1)$ so $R_1^3 = R_2^3 - (3/4)m/(\pi \rho_g)$, and

$$R_1 = \sqrt[3]{\frac{3m}{4\pi}\left(\frac{1}{\rho_w} - \frac{1}{\rho_g}\right)} = \sqrt[3]{\frac{3(1.00\text{ kg})}{4\pi}\left(\frac{1}{1.00\times10^3\text{ kg/m}^3} - \frac{1}{2.60\times10^3\text{ kg/m}^3}\right)}$$

$$= \boxed{5.28\times10^{-2}\text{ m}}$$

---

50. ***REASONING AND SOLUTION*** The volume flow rate is given by

$$Q = Av = \pi r^2 v = \pi(0.305\text{ m})^2(1.22\text{ m/s}) = 0.356\text{ m}^3/\text{s}$$

The number of gallons that flows in one day (8.64×10^4 s) is

$$(0.356\text{ m}^3/\text{s})\left(\frac{1.0\text{ gal}}{3.79\times10^{-3}\text{ m}^3}\right)(8.64\times10^4\text{ s}) = \boxed{8.12\times10^6\text{ gal}}$$

---

51. **SSM** ***REASONING AND SOLUTION*** Using Equation 11.10 for the volume flow rate, we have $Q = Av$, where A is the cross-sectional area of the pipe and v is the speed of the fluid. Since the pipe has a circular cross section, $A = \pi r^2$. Therefore, solving Equation 11.10 for v, we have

$$v = \frac{Q}{A} = \frac{Q}{\pi r^2} = \frac{1.50\text{ m}^3/\text{s}}{\pi(0.500\text{ m})^2} = \boxed{1.91\text{ m/s}}$$

---

52. ***REASONING***

a. According to Equation 11.10, the volume flow rate Q is equal to the product of the cross-sectional area A of the artery and the speed v of the blood, $Q = Av$. Since Q and A are known, we can determine v.

b. Since the volume flow rate Q_2 through the constriction is the same as the volume flow rate Q_1 in the normal part of the artery, $Q_2 = Q_1$. We can use this relation to find the blood speed in the constricted region.

SOLUTION

a. Since the artery is assumed to have a circular cross-section, its cross-sectional area is $A_1 = \pi r_1^2$, where r_1 is the radius. Thus, the speed of the blood is

$$v_1 = \frac{Q_1}{A_1} = \frac{Q_1}{\pi r_1^2} = \frac{3.6\times10^{-6}\text{m}^3/\text{s}}{\pi\left(5.2\times10^{-3}\text{m}\right)^2} = \boxed{4.2\times10^{-2}\text{m/s}} \qquad (11.10)$$

b. The volume flow rate is the same in the normal and constricted parts of the artery, so $Q_2 = Q_1$. Since $Q_2 = A_2 v_2$, the blood speed is $v_2 = Q_2/A_2 = Q_1/A_2$. We are given that the radius of the constricted part of the artery is one-third that of the normal artery, so $r_2 = \frac{1}{3}r_1$. Thus, the speed of the blood at the constriction is

$$v_2 = \frac{Q_1}{A_2} = \frac{Q_1}{\pi r_2^2} = \frac{Q_1}{\pi\left(\frac{1}{3}r_1\right)^2} = \frac{3.6\times10^{-6}\,\text{m}^3/\text{s}}{\pi\left[\frac{1}{3}\left(5.2\times10^{-3}\,\text{m}\right)\right]^2} = \boxed{0.38\ \text{m/s}}$$

---

53. ***REASONING*** The length L of the side of the square can be obtained, if we can find a value for the cross-sectional area A of the ducts. The area is related to the volume flow rate Q and the air speed v by Equation 11.10 ($Q = Av$). The volume flow rate can be obtained from the volume V of the room and the replacement time t as $Q = V/t$.

SOLUTION For a square cross section with sides of length L, we have $A = L^2$. And we know that the volume flow rate is $Q = V/t$. Therefore, using Equation 11.10 gives

$$Q = Av \quad \text{or} \quad \frac{V}{t} = L^2 v$$

Solving for L shows that

(a) ***Air speed = 3.0 m/s*** $\quad L = \sqrt{\frac{V}{tv}} = \sqrt{\frac{120\ \text{m}^3}{(1200\ \text{s})(3.0\ \text{m/s})}} = \boxed{0.18\ \text{m}}$

(b) ***Air speed = 5.0 m/s*** $\quad L = \sqrt{\frac{V}{tv}} = \sqrt{\frac{120\ \text{m}^3}{(1200\ \text{s})(5.0\ \text{m/s})}} = \boxed{0.14\ \text{m}}$

---

54. ***REASONING AND SOLUTION***

a. The mass flow rate is equal to $\rho A v$:

$$\text{Mass flow rate} = (1.00\times10^3\ \text{kg/m}^3)\pi(0.080\ \text{m})^2(3.0\ \text{m/s})\left(\frac{3600\ \text{s}}{1\ \text{h}}\right) = 2.2\times10^5\ \text{kg/hr}$$

The mass that flows in one hour is $(2.2\times10^5\ \text{kg/hr})(1.0\ \text{hr}) = \boxed{2.2\times10^5\ \text{kg}}$.

b. Using $A_1 v_1 = A_2 v_2$, we have $\pi r_1^2 v_1 = 3\pi r_2^2 v_2$ (since there are 3 hoses). Solving for v_2, we obtain

$$v_2 = (1/3)(r_1/r_2)^2 v_1 = (1/3)[(0.080\text{ m})/(0.020\text{ m})]^2(3.0\text{ m/s}) = \boxed{16\text{ m/s}}$$

---

55. SSM ***REASONING AND SOLUTION***

a. The volume flow rate is given by Equation 11.10. Assuming that the line has a circular cross section, $A = \pi r^2$, we have

$$Q = Av = (\pi r^2)v = \pi\ (0.0065\text{ m})^2(1.2\text{ m/s}) = \boxed{1.6\times10^{-4}\text{ m}^3/\text{s}}$$

b. The volume flow rate calculated in part (a) above is the flow rate for all twelve holes. Therefore, the volume flow rate through one of the twelve holes is

$$Q_{\text{hole}} = \frac{Q}{12} = (\pi r_{\text{hole}}^2)v_{\text{hole}}$$

Solving for v_{hole} we have

$$v_{\text{hole}} = \frac{Q}{12\pi r_{\text{hole}}^2} = \frac{1.6\times10^{-4}\text{ m}^3/\text{s}}{12\pi(4.6\times10^{-4}\text{ m})^2} = \boxed{2.0\times10^1\text{ m/s}}$$

---

56. ***REASONING*** We apply Bernoulli's equation as follows:

$$\underbrace{P_\text{S} + \tfrac{1}{2}\rho v_\text{S}^2 + \rho g y_\text{S}}_{\text{At surface of vaccine in reservoir}} = \underbrace{P_\text{O} + \tfrac{1}{2}\rho v_\text{O}^2 + \rho g y_\text{O}}_{\text{At opening}}$$

SOLUTION The vaccine's surface in the reservoir is stationary during the inoculation, so that $v_\text{S} = 0$ m/s. The vertical height between the vaccine's surface in reservoir and the opening can be ignored, so $y_\text{S} = y_\text{O}$. With these simplifications Bernoulli's equation becomes

$$P_\text{S} = P_\text{O} + \tfrac{1}{2}\rho v_\text{O}^2$$

Solving for the speed at the opening gives

$$v_\text{O} = \sqrt{\frac{2(P_\text{S} - P_\text{O})}{\rho}} = \sqrt{\frac{2(4.1\times10^6\text{ Pa})}{1100\text{ kg/m}^3}} = \boxed{86\text{ m/s}}$$

---

57. SSM ***REASONING AND SOLUTION***

a. Using Equation 11.12, the form of Bernoulli's equation with $y_1 = y_2$, we have

$$P_1 - P_2 = \tfrac{1}{2}\rho\left(v_2^2 - v_1^2\right) = \frac{1.29\text{ kg/m}^3}{2}\left[(15\text{ m/s})^2 - (0\text{ m/s})^2\right] = \boxed{150\text{ Pa}}$$

b. The pressure inside the roof is greater than the pressure on the outside. Therefore, there is a net outward force on the roof. If the wind speed is sufficiently high, some roofs are "blown outward."

---

58. ***REASONING*** We assume that region 1 contains the constriction and region 2 is the normal region. The difference in blood pressures between the two points in the horizontal artery is given by Bernoulli's equation (Equation 11.12) as $P_2 - P_1 = \frac{1}{2}\rho v_1^2 - \frac{1}{2}\rho v_2^2$, where v_1 and v_2 are the speeds at the two points. Since the volume flow rate is the same at the two points, the speed at 1 is related to the speed at 2 by Equation 11.9, the equation of continuity: $A_1 v_1 = A_2 v_2$, where A_1 and A_2 are the cross-sectional areas of the artery. By combining these two relations, we will be able to determine the pressure difference.

SOLUTION Solving the equation of continuity for the blood speed in region 1 gives $v_1 = v_2 A_2 / A_1$. Substituting this result into Bernoulli's equation yields

$$P_2 - P_1 = \tfrac{1}{2}\rho v_1^2 - \tfrac{1}{2}\rho v_2^2 = \tfrac{1}{2}\rho\left(\frac{v_2 A_2}{A_1}\right)^2 - \tfrac{1}{2}\rho v_2^2$$

Since $A_1 = \frac{1}{4}A_2$, the pressure difference is

$$P_2 - P_1 = \tfrac{1}{2}\rho\left(\frac{v_2 A_2}{\frac{1}{4}A_2}\right)^2 - \tfrac{1}{2}\rho v_2^2 = \tfrac{1}{2}\rho v_2^2(16 - 1)$$

$$= \tfrac{1}{2}\left(1060\text{ kg/m}^3\right)(0.11\text{ m/s})^2(15) = \boxed{96\text{ Pa}}$$

We have taken the density ρ of blood from Table 11.1.

---

59. SSM ***REASONING AND SOLUTION*** Let the speed of the air below and above the wing be given by v_1 and v_2, respectively. According to Equation 11.12, the form of Bernoulli's equation with $y_1 = y_2$, we have

$$P_1 - P_2 = \tfrac{1}{2}\rho\left(v_2^2 - v_1^2\right) = \frac{1.29\text{ kg/m}^3}{2}\left[(251\text{ m/s})^2 - (225\text{ m/s})^2\right] = 7.98 \times 10^3\text{ Pa}$$

From Equation 11.3, the lifting force is, therefore,

$$F=(P_1-P_2)A=(7.98\times10^3\ \text{Pa})(24.0\ \text{m}^2)=\boxed{1.92\times10^5\ \text{N}}$$

---

60. ***REASONING***

a. The drawing shows two points, labeled 1 and 2, in the fluid. Point 1 is at the top of the water, and point 2 is where it flows out of the dam at the bottom. Bernoulli's equation, Equation 11.11, can be used to determine the speed v_2 of the water exiting the dam.

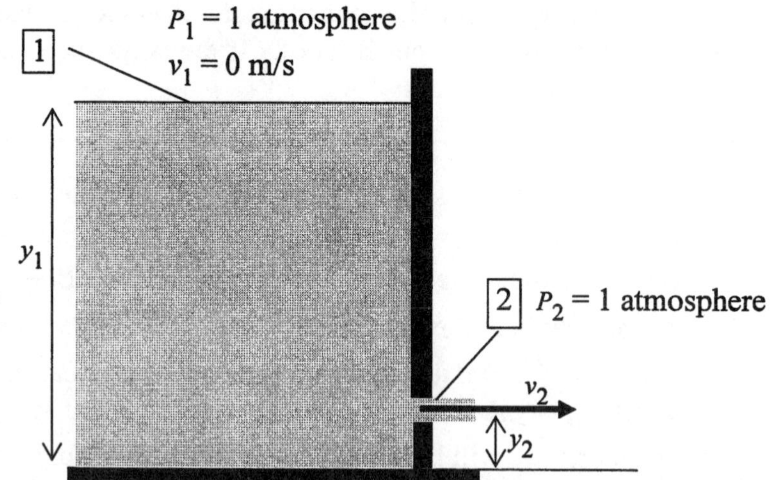

b. The number of cubic meters per second of water that leaves the dam is the volume flow rate Q. According to Equation 11.10, the volume flow rate is the product of the cross-sectional area A_2 of the crack and the speed v_2 of the water; $Q = A_2v_2$.

SOLUTION

a. According to Bernoulli's equation, as given in Equation 11.11, we have

$$P_1+\tfrac{1}{2}\rho v_1^2+\rho g y_1=P_2+\tfrac{1}{2}\rho v_2^2+\rho g y_2$$

Setting $P_1 = P_2$, $v_1 = 0$ m/s, and solving for v_2, we obtain

$$v_2=\sqrt{2g(y_1-y_2)}=\sqrt{2(9.80\ \text{m/s}^2)(15.0\ \text{m})}=\boxed{17.1\ \text{m/s}}$$

b. The volume flow rate of the water leaving the dam is

$$Q=A_2v_2=(1.30\times10^{-3}\text{m}^2)(17.1\ \text{m/s})=\boxed{2.22\times10^{-2}\ \text{m}^3/\text{s}} \qquad (11.10)$$

---

61. ***REASONING AND SOLUTION***

a. Using Bernoulli's equation we have

$$P_1=P_2+\rho g h$$
$$=(1.01\times10^5\ \text{Pa})+(1.00\times10^3\ \text{kg/m}^3)(9.80\ \text{m/s}^2)(15.0\ \text{m})=\boxed{2.48\times10^5\ \text{Pa}}$$

b. The pressure becomes equal to atmospheric pressure when the valve is opened, so $P_1 = \boxed{1.01 \times 10^5 \text{ Pa}}$.

c. The volume flow rate is given by $Q = Av$. We can find the speed v by using Bernoulli's equation, $P_2 + \rho gh = P_1 + (1/2)\rho v^2$, with $P_2 = P_1$. The result is

$$v = \sqrt{2gh} = \sqrt{2\left(9.80 \text{ m/s}^2\right)\left(15.0 \text{ m}\right)} = 17.1 \text{ m/s}$$

The volume flow rate is $Q = Av = (2.00 \times 10^{-2} \text{ m}^2)(17.1 \text{ m/s}) = \boxed{0.342 \text{ m}^3\text{/s}}$.

---

62. ***REASONING AND SOLUTION***
a. Taking the nozzle as position 1 and the top of the tank as position 2 we have, using Bernoulli's equation, $P_1 + (1/2)\rho v_1^{\;2} = P_2 + \rho gh$, we can solve for v_1 so that

$$v_1^{\;2} = 2[(P_2 - P_1)/\rho + gh]$$

$$v_1 = \sqrt{2[(5.00 \times 10^5 \text{ Pa})/(1.00 \times 10^3 \text{ kg/m}^3) + (9.80 \text{ m/s}^2)(4.00 \text{ m})]} = \boxed{32.8 \text{ m/s}}$$

b. To find the height of the water use $(1/2)\rho v_1^{\;2} = \rho gh$ so that

$$h = (1/2)v_1^{\;2}/g = (1/2)(32.8 \text{ m/s})^2/(9.80 \text{ m/s}^2) = \boxed{54.9 \text{ m}}$$

---

63. ***REASONING*** The top and bottom surfaces of the roof are at the same height, so we can use Bernoulli's equation in the form of Equation 11.12, $P_1 + \frac{1}{2}\rho v_1^2 = P_2 + \frac{1}{2}\rho v_2^2$, to determine the wind speed. We take point 1 to be inside the roof and point 2 to be outside the roof. Since the air inside the roof is not moving, $v_1 = 0$ m/s. The net outward force ΣF acting on the roof is the difference in pressure $P_1 - P_2$ times the area A of the roof, so $\Sigma F = (P_1 - P_2)A$.

SOLUTION Setting $v_1 = 0$ m/s in Bernoulli's equation and solving it for the speed v_2 of the wind, we obtain

$$v_2 = \sqrt{\frac{2\left(P_1 - P_2\right)}{\rho}}$$

Since the pressure difference is equal to the net outward force divided by the area of the roof, $(P_1 - P_2) = \Sigma F/A$, the speed of the wind is

$$v_2 = \sqrt{\frac{2(\Sigma F)}{\rho A}} = \sqrt{\frac{2(22\,000\text{ N})}{(1.29\text{ kg/m}^3)(5.0\text{ m} \times 6.3\text{ m})}} = \boxed{33\text{ m/s}}$$

---

64. ***REASONING AND SOLUTION*** We have

$$P_1 + (1/2)\rho v_1{}^2 + \rho g y_1 = P_2 + (1/2)\rho v_2{}^2 + \rho g y_2$$

Taking $v_1 = 0$ m/s, $y_2 = 0$ m, $P_1 = 1.01 \times 10^5$ Pa, and noting that v_2 is a maximum when $P_2 = 0$ Pa, we can solve for v_2. We thus obtain,

$$v_2 = \sqrt{\frac{2P_1}{\rho} + 2gy_1} = \sqrt{\frac{2\left(1.01\times10^5\text{ Pa}\right)}{1.00\times10^3\text{ kg/m}^3} + 2\left(9.80\text{ m/s}^2\right)\left(12\text{ m}\right)} = \boxed{21\text{ m/s}}$$

---

65. SSM ***REASONING*** Since the pressure difference is known, Bernoulli's equation can be used to find the speed v_2 of the gas in the pipe. Bernoulli's equation also contains the unknown speed v_1 of the gas in the Venturi meter; therefore, we must first express v_1 in terms of v_2. This can be done by using Equation 11.9, the equation of continuity.

SOLUTION

a. From the equation of continuity (Equation 11.9) it follows that $v_1 = \left(A_2 / A_1\right)v_2$. Therefore,

$$v_1 = \frac{0.0700\text{ m}^2}{0.0500\text{ m}^2}v_2 = (1.40)v_2$$

Substituting this expression into Bernoulli's equation (Equation 11.12), we have

$$P_1 + \tfrac{1}{2}\rho(1.40\,v_2)^2 = P_2 + \tfrac{1}{2}\rho v_2^2$$

Solving for v_2, we obtain

$$v_2 = \sqrt{\frac{2(P_2 - P_1)}{\rho\left[(1.40)^2 - 1\right]}} = \sqrt{\frac{2(120\text{ Pa})}{(1.30\text{ kg/m}^3)\left[(1.40)^2 - 1\right]}} = \boxed{14\text{ m/s}}$$

b. According to Equation 11.10, the volume flow rate is

$$Q = A_2 v_2 = (0.0700 \text{ m}^2)(14 \text{ m/s}) = \boxed{0.98 \text{ m}^3/\text{s}}$$

---

66. ***REASONING AND SOLUTION*** As seen in the figure, the lower pipe is at the level of zero potential energy.

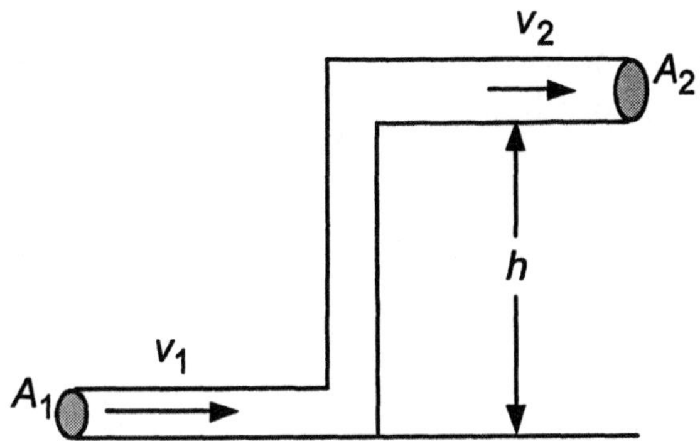

If the flow rate is uniform in both pipes, we have $(1/2)\rho v_1^2 = (1/2)\rho v_2^2 + \rho g h$ (since $P_1 = P_2$) and $A_1 v_1 = A_2 v_2$. We can solve for v_1, i.e., $v_1 = v_2(A_2/A_1)$, and plug into the previous expression to find v_2, so that

$$v_2 = \sqrt{\frac{2gh}{\left(\dfrac{A_2}{A_1}\right)^2 - 1}} = \sqrt{\frac{2\left(9.80 \text{ m/s}^2\right)\left(10.0 \text{ m}\right)}{\left[\dfrac{\pi\left(0.0400 \text{ m}\right)^2}{\pi\left(0.0200 \text{ m}\right)^2}\right]^2 - 1}} = 3.61 \text{ m/s}$$

The volume flow rate is then given by

$$Q = A_2 v_2 = \pi r_2^2 v_2 = \pi(0.0400 \text{ m})^2(3.61 \text{ m/s}) = \boxed{1.81\times10^{-2} \text{ m}^3/\text{s}}$$

---

67. ***REASONING*** In level flight the lift force must balance the plane's weight W, so its magnitude is also W. The lift force arises because the pressure P_B beneath the wings is greater than the pressure P_T on top of the wings. The lift force, then, is the pressure difference times the effective wing surface area A, so that $W = (P_B - P_T)A$. The area is given, and we can determine the pressure difference by using Bernoulli's equation.

SOLUTION According to Bernoulli's equation, we have

$$P_B + \tfrac{1}{2}\rho v_B^2 + \rho g y_B = P_T + \tfrac{1}{2}\rho v_T^2 + \rho g y_T$$

Since the flight is level, the height is constant and $y_B = y_T$, where we assume that the wing thickness may be ignored. Then, Bernoulli's equation simplifies and may be rearranged as follows:

$$P_B + \tfrac{1}{2}\rho v_B^2 = P_T + \tfrac{1}{2}\rho v_T^2 \quad \text{or} \quad P_B - P_T = \tfrac{1}{2}\rho v_T^2 - \tfrac{1}{2}\rho v_B^2$$

Recognizing that $W = (P_B - P_T)A$, we can substitute for the pressure difference from Bernoulli's equation to show that

$$W = \left(P_B - P_T\right)A = \tfrac{1}{2}\rho\left(v_T^2 - v_B^2\right)A$$

$$= \tfrac{1}{2}\left(1.29\text{ kg/m}^3\right)\left[\left(62.0\text{ m/s}\right)^2 - \left(54.0\text{ m/s}\right)^2\right]\left(16\text{ m}^2\right) = \boxed{9600\text{ N}}$$

We have used a value of 1.29 kg/m^3 from Table 11.1 for the density of air. This is an approximation, since the density of air decreases with increasing altitude above sea level.

---

68. ***REASONING AND SOLUTION***

a. At the surface of the water (position 1) and at the exit of the hose (position 2) the pressures are equal ($P_1 = P_2$) to the atmospheric pressure. If the hose exit defines $y = 0$ m, we have from Bernoulli's equation $(1/2)\rho v_1^2 + \rho g y = (1/2)\rho v_2^2$. If we take the speed at the surface of the water to be zero (i.e., $v_1 = 0$ m/s) we find that

$$v_2 = \boxed{\sqrt{2gy}}$$

b. The siphon will stop working when $v_2 = 0$ m/s, or $y = \boxed{0\text{ m}}$, i.e., when the end of the hose is at the water level in the tank.

c. At point A we have

$$P_A + \rho g(h + y) + (1/2)\rho v_A^2 = P_0 + (1/2)\rho v_2^2$$

But $v_A = v_2$ so that

$$\boxed{P_A = P_0 - \rho g(y + h)}$$

---

69. SSM ***REASONING AND SOLUTION*** Bernoulli's equation (Equation 11.12) is

$$P_1 + \tfrac{1}{2}\rho v_1^2 = P_2 + \tfrac{1}{2}\rho v_2^2$$

If we let the right hand side refer to the air above the plate, and the left hand side refer to the air below the plate, then $v_1 = 0$, since the air below the plate is stationary. We wish to find v_2 for the situation illustrated in part *b* of the figure shown in the text. Solving the equation above for v_2 (with $v_2 = v_{2b}$ and $v_1 = 0$) gives

$$v_{2b} = \sqrt{\frac{2(P_1 - P_2)}{\rho}} \quad (1)$$

In Equation (1), P_1 is atmospheric pressure and P_2 must be determined. We must first consider the situation in part *a* of the text figure.

The figure at the right shows the forces that act on the rectangular plate in part *a* of the text drawing. F_1 is the force exerted on the plate from the air below the plate, and F_2 is the force exerted on the plate from the air above the plate. Applying Newton's second law, we have (taking "up" to be the positive direction),

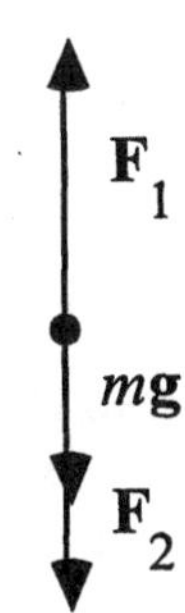

$$F_1 - F_2 - mg = 0$$

$$F_1 - F_2 = mg$$

Thus, the difference in pressures exerted by the air on the plate in part *a* of the drawing is

$$P_1 - P_2 = \frac{F_1 - F_2}{A} = \frac{mg}{A} \quad (2)$$

where *A* is the area of the plate. From Bernoulli's equation (Equation 11.12) we have, with $v_2 = v_{2a}$ and $v_{1a} = 0$,

$$P_1 - P_2 = \tfrac{1}{2}\rho v_{2a}^2 \quad (3)$$

where v_{2a} is the speed of the air along the top of the plate in part *a* of the text drawing. Combining Equations (2) and (3) we have

$$\frac{mg}{A} = \tfrac{1}{2}\rho v_{2a}^2 \quad (4)$$

The figure below, on the left, shows the forces that act on the plate in part *b* of the text drawing. The notation is the same as that used when the plate was horizontal (part *a* of the text figure). The figure at the right below shows the same forces resolved into components along the plate and perpendicular to the plate.

Applying Newton's second law we have $F_1 - F_2 - mg\sin\theta = 0$, or

$$F_1 - F_2 = mg\sin\theta$$

Thus, the difference in pressures exerted by the air on the plate in part *b* of the text figure is

$$P_1 - P_2 = \frac{F_1 - F_2}{A} = \frac{mg\sin\theta}{A}$$

Using Equation (4) above,

$$P_1 - P_2 = \tfrac{1}{2}\rho v_{2a}^2 \sin\theta$$

Thus, Equation (1) becomes

$$v_{2b} = \sqrt{\frac{2\left(\frac{1}{2}\rho v_{2a}^2 \sin\theta\right)}{\rho}}$$

Therefore,

$$v_{2b} = \sqrt{v_{2a}^2 \sin\,\theta} = \sqrt{(11.0\text{ m/s})^2 \sin 30.0^\circ} = \boxed{7.78\text{ m/s}}$$

---

70. ***REASONING AND SOLUTION*** We have $Q = \pi R^4(P_2 - P_1)/(8\eta L)$, so the difference in pressure $\Delta P = (P_2 - P_1)$ is $\Delta P = 8\eta LQ/(\pi R^4)$. Therefore,

$$\Delta P = \frac{8\left(4\times10^{-3}\text{ Pa}\cdot\text{s}\right)\left(0.10\text{ m}\right)\left(1.0\times10^{-7}\text{ m}^3/\text{s}\right)}{\pi\left(1.5\times10^{-3}\text{ m}\right)^4} = \boxed{20\text{ Pa}}$$

---

71. SSM WWW ***REASONING AND SOLUTION***

a. If the water behaves as an ideal fluid, and since the pipe is horizontal and has the same radius throughout, the speed and pressure of the water are the same at all points in the pipe. Since the right end of the pipe is open to the atmosphere, the pressure at the right end is atmospheric pressure; therefore, the pressure at the left end is also atmospheric pressure, or $\boxed{1.01\times 10^5 \text{ Pa}}$.

b. If the water is treated as a viscous fluid, the volume flow rate Q is described by Poiseuille's law (Equation 11.14):

$$Q = \frac{\pi R^4 (P_2 - P_1)}{8\eta L}$$

Let P_1 represent the pressure at the right end of the pipe, and let P_2 represent the pressure at the left end of the pipe. Solving for P_2 (with P_1 equal to atmospheric pressure), we obtain

$$P_2 = \frac{8\eta LQ}{\pi R^4} + P_1$$

Therefore,

$$P_2 = \frac{8(1.00\times 10^{-3} \text{ Pa}\cdot\text{s})(1.3 \text{ m})(9.0\times 10^{-3} \text{ m}^3/\text{s})}{\pi (6.4\times 10^{-3} \text{ m})^4} + 1.013\times 10^5 \text{ Pa} = \boxed{1.19\times 10^5 \text{ Pa}}$$

---

72. ***REASONING*** In this problem, we are treating air as a viscous fluid. According to Poiseuille's law, a fluid with viscosity η flowing through a pipe of radius R and length L has a volume flow rate Q given by Equation 11.14: $Q = \pi R^4 (P_2 - P_1)/(8\eta L)$. This expression can be solved for the quantity $P_2 - P_1$, the difference in pressure between the ends of the air duct. First, however, we must determine the volume flow rate Q of the air.

SOLUTION Since the fan forces air through the duct such that 280 m^3 of air is replenished every ten minutes, the volume flow rate is

$$Q = \left(\frac{280 \text{ m}^3}{10.0 \text{ min}}\right)\left(\frac{1.0 \text{ min}}{60 \text{ s}}\right) = 0.467 \text{ m}^3/\text{s}$$

The difference in pressure between the ends of the air duct is, according to Poiseuille's law,

$$P_2 - P_1 = \frac{8\eta LQ}{\pi R^4} = \frac{8(1.8\times10^{-5}\ \text{Pa}\cdot\text{s})(5.5\ \text{m})(0.467\ \text{m}^3/\text{s})}{\pi(7.2\times10^{-2}\ \text{m})^4} = \boxed{4.4\ \text{Pa}}$$

---

73. ***REASONING*** The volume flow rate Q of a viscous fluid flowing through a pipe of radius R is given by Equation 11.14 as $Q = \frac{\pi R^4 (P_2 - P_1)}{8\eta L}$, where $P_2 - P_1$ is the pressure difference between the ends of the pipe, L is the length of the pipe, and η is the viscosity of the fluid. Since all the variables are known except L, we can use this relation to find it.

SOLUTION Solving Equation 11.14 for the pipe length, we have

$$L = \frac{\pi R^4 (P_2 - P_1)}{8\eta Q} = \frac{\pi\left(5.1\times10^{-3}\ \text{m}\right)^4\left(1.8\times10^3\ \text{Pa}\right)}{8\left(1.0\times10^{-3}\ \text{Pa}\cdot\text{s}\right)\left(2.8\times10^{-4}\ \text{m}^3/\text{s}\right)} = \boxed{1.7\ \text{m}}$$

---

74. ***REASONING AND SOLUTION*** The Reynold's number, Re, can be written as $\text{Re} = 2\bar{v}\rho R/\eta$. To find the average speed $\bar{v}$,

$$\bar{v} = \frac{(\text{Re})\eta}{2\rho R} = \frac{(2000)\left(4\times10^{-3}\text{Pa}\cdot\text{s}\right)}{2\left(1060\ \text{kg/m}^3\right)\left(8.0\times10^{-3}\text{m}\right)} = \boxed{0.5\ \text{m/s}}$$

---

75. SSM ***REASONING AND SOLUTION***

a. Using Stoke's law, the viscous force is

$$F = 6\pi\eta Rv = 6\pi(1.00\times10^{-3}\ \text{Pa}\cdot\text{s})(5.0\times10^{-4}\ \text{m})(3.0\ \text{m/s}) = \boxed{2.8\times10^{-5}\ \text{N}}$$

b. When the sphere reaches its terminal speed, the net force on the sphere is zero, so that the magnitude of **F** must be equal to the magnitude of mg, or $F = mg$. Therefore, $6\pi\eta Rv_T = mg$, where v_T is the terminal speed of the sphere. Solving for v_T, we have

$$v_T = \frac{mg}{6\pi\eta R} = \frac{(1.0\times10^{-5}\ \text{kg})(9.80\ \text{m/s}^2)}{6\pi(1.00\times10^{-3}\ \text{Pa}\cdot\text{s})(5.0\times10^{-4}\ \text{m})} = \boxed{1.0\times10^1\ \text{m/s}}$$

---

76. ***REASONING AND SOLUTION*** Since water is a viscous fluid, its behavior is described by Poiseuille's law (Equation 11.14):

$$Q = \frac{\pi R^4 (P_2 - P_1)}{8\eta L}$$

Solving for R gives:

$$R = \left[\frac{8\eta LQ}{\pi(P_2 - P_1)} \right]^{1/4}$$

In this expression P_2 is the pressure just below the left vertical tube, P_1 is the pressure just below the right vertical tube, and L is the horizontal distance between the centers of the vertical tubes. The difference in pressures is given by

$$P_2 - P_1 = \rho g(y_2 - y_1) = \rho g \Delta y$$

Therefore,

$$R = \left[\frac{8\eta LQ}{\pi(P_2 - P_1)} \right]^{1/4}$$

$$= \left[\frac{8(1.00 \times 10^{-3} \text{ Pa} \cdot \text{s})(0.70 \text{ m})(0.014 \text{ m}^3/\text{s})}{\pi(1.000 \times 10^3 \text{ kg/m}^3)(9.80 \text{ m/s}^2)(0.045 \text{ m})} \right]^{1/4} = \boxed{1.5 \times 10^{-2} \text{ m}}$$

---

77. ***REASONING*** The drawing at the right shows the situation. As discussed in Conceptual Example 6, the job of the pump is to draw air out of the pipe that dips down into the water. The atmospheric pressure in the well then pushes the water upward into the pipe. In the drawing, the best the pump can do is to remove all of the air, in which case, the pressure P_1 at the top of the water in the pipe is zero. The pressure P_2 at the bottom of the pipe at point A is the same as that at the point B, namely, it is equal to atmospheric pressure (1.013×10^5 Pa), because the two points are at the same elevation, and point B is open to the atmosphere. Equation 11.4, $P_2 = P_1 + \rho g h$ can be applied to obtain the maximum depth h of the well.

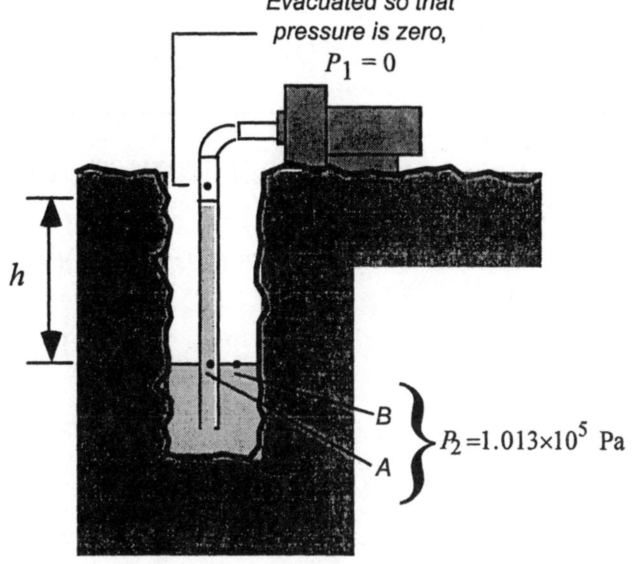

SOLUTION Setting $P_1 = 0$ Pa, and solving Equation 11.4 for h, we have

$$h = \frac{P_1}{\rho g} = \frac{1.013 \times 10^5 \text{ Pa}}{(1.000 \times 10^3 \text{ kg/m}^3)(9.80 \text{ m/s}^2)} = \boxed{10.3 \text{ m}}$$

---

78. ***REASONING AND SOLUTION*** Using Bernoulli's equation

$$\Delta P = P_1 - P_2 = (1/2)\rho v_2^2 - (1/2)\rho v_1^2 = (1/2)(1.29 \text{ kg/m}^3)[(8.5 \text{ m/s})^2 - (1.1 \text{ m/s})^2]$$

$$\Delta P = \boxed{46 \text{ Pa}}$$

The air flows from high pressure to low pressure (from lower to higher velocity), so it $\boxed{\text{enters at } B \text{ and exits at } A}$.

---

79. SSM ***REASONING*** According to Archimedes principle, the buoyant force that acts on the block is equal to the weight of the water that is displaced by the block. The block displaces an amount of water V, where V is the volume of the block. Therefore, the weight of the water displaced by the block is $W = mg = (\rho_{\text{water}} V)g$.

SOLUTION The buoyant force that acts on the block is, therefore,

$$F = \rho_{\text{water}} V g = (1.00 \times 10^3 \text{ kg/m}^3)(0.10 \text{ m} \times 0.20 \text{ m} \times 0.30 \text{ m})(9.80 \text{ m/s}^2) = \boxed{59 \text{ N}}$$

---

80. ***REASONING AND SOLUTION*** The gauge pressure of the solution at the location of the vein is

$$P = \rho g h = (1030 \text{ kg/m}^3)(9.80 \text{ m/s}^2)(0.610 \text{ m}) = 6.16 \times 10^3 \text{ Pa}$$

Now

$$1.013 \times 10^5 \text{ Pa} = 760 \text{ mm Hg} \quad \text{so} \quad 1 \text{ Pa} = 7.50 \times 10^{-3} \text{ mm Hg}$$

Then

$$P = \left(6.16 \times 10^3 \text{ Pa}\right)\left(\frac{7.50 \times 10^{-3} \text{ mm Hg}}{1 \text{ Pa}}\right) = \boxed{46.2 \text{ mm Hg}}$$

---

81. ***REASONING*** The density of the brass ball is given by Equation 11.1: $\rho = m/V$. Since the ball is spherical, its volume is given by the expression $V = (4/3)\pi r^3$, so that the density may be written as

$$\rho = \frac{m}{V} = \frac{3m}{4\pi r^3}$$

This expression can be solved for the radius r; but first, we must eliminate the mass m from the equation since its value is not specified explicitly in the problem. We can determine an expression for the mass m of the brass ball by analyzing the forces on the ball.

The only two forces that act on the ball are the upward tension **T** in the wire and the downward weight mg of the ball. If we take up as the positive direction, and we apply Newton's second law, we find that $T - mg = 0$, or $m = T/g$. Therefore, the density can be written as

$$\rho = \frac{3m}{4\pi r^3} = \frac{3T}{4\pi r^3 g}$$

This expression can now be solved for r.

SOLUTION We find that

$$r^3 = \frac{3T}{4\pi \rho g}$$

or, taking the cube root of both sides,

$$r = \sqrt[3]{\frac{3T}{4\pi \rho g}} = \sqrt[3]{\frac{3(120\text{ N})}{4\pi(8470\text{ kg/m}^3)(9.80\text{ m/s}^2)}} = \boxed{7.0\times10^{-2}\text{ m}}$$

---

82. ***REASONING AND SOLUTION***

a. We will treat the neutron star as spherical in shape, so that its volume is given by the familiar formula, $V = \frac{4}{3}\pi r^3$. Then, according to Equation 11.1, the density of the neutron star described in the problem statement is

$$\rho = \frac{m}{V} = \frac{m}{\frac{4}{3}\pi r^3} = \frac{3m}{4\pi r^3} = \frac{3(2.7\times10^{28}\text{ kg})}{4\pi(1.2\times10^3\text{ m})^3} = \boxed{3.7\times10^{18}\text{ kg/m}^3}$$

b. If a dime of volume $2.0\times10^{-7}\text{ m}^3$ were made of this material, it would weigh

$$W = mg = \rho V g = (3.7\times10^{18}\text{ kg/m}^3)(2.0\times10^{-7}\text{ m}^3)(9.80\text{ m/s}^2) = 7.3\times10^{12}\text{ N}$$

This weight corresponds to

$$7.3\times10^{12}\text{ N}\left(\frac{1\text{ lb}}{4.448\text{ N}}\right) = \boxed{1.6\times10^{12}\text{ lb}}$$

---

83. SSM ***REASONING AND SOLUTION*** The mass flow rate Q_{mass} is the amount of fluid mass that flows per unit time. Therefore,

$$Q_{\text{mass}} = \frac{m}{t} = \frac{\rho V}{t} = \frac{(1030\ \text{kg/m}^3)(9.5\times10^{-4}\ \text{m}^3)}{6.0\ \text{h}}\left(\frac{1.0\ \text{h}}{3600\ \text{s}}\right) = \boxed{4.5\times10^{-5}\ \text{kg/s}}$$

---

84. ***REASONING*** This is an application of Poiseuille's law, as given in Equation 11.14. According to this law, a pressure difference $P_2 - P_1$ is needed to make the viscous blood flow through the needle.

SOLUTION The pressure at the input of the needle is $P_2 = P_{\text{Atmosphere}} + \rho gh$, as given by Equation 11.4, whereas the pressure at the output of the needle is $P_1 = P_{\text{Atmosphere}}$. Therefore, we have

$$P_2 - P_1 = P_{\text{Atmosphere}} + \rho gh - P_{\text{Atmosphere}} = \rho gh$$

Using Poiseuille's law, we find

$$Q = \frac{\pi R^4 (P_2 - P_1)}{8\eta L} = \frac{\pi R^4 \rho gh}{8\eta L}$$

Solving for h gives

$$h = \frac{Q8\eta L}{\pi R^4 \rho g} = \frac{(4.5\times10^{-8}\ \text{m}^3/\text{s})8(4.0\times10^{-3}\ \text{Pa}\cdot\text{s})(0.030\ \text{m})}{\pi(2.5\times10^{-4}\ \text{m})^4(1060\ \text{kg/m}^3)(9.80\ \text{m/s}^2)} = \boxed{0.34\ \text{m}}$$

---

85. SSM ***REASONING*** Since the faucet is closed, the water in the pipe may be treated as a static fluid. The gauge pressure P_2 at the faucet on the first floor is related to the gauge pressure P_1 at the faucet on the second floor by Equation 11.4, $P_2 = P_1 + \rho gh$.

SOLUTION

a. Solving Equation 11.4 for P_1, we find the gauge pressure at the second-floor faucet is

$$P_1 = P_2 - \rho gh = 1.90\times10^5\ \text{Pa} - (1.00\times10^3\ \text{kg/m}^3)(9.80\ \text{m/s}^2)(6.50\ \text{m}) = \boxed{1.26\times10^5\ \text{Pa}}$$

b. If the second faucet were placed at a height h above the first-floor faucet so that the gauge pressure P_1 at the second faucet were zero, then no water would flow from the second faucet, even if it were open. Solving Equation 11.4 for h when P_1 equals zero, we obtain

$$h=\frac{P_2-P_1}{\rho g}=\frac{1.90\times10^5\ \text{Pa}-0}{(1.00\times10^3\ \text{kg/m}^3)(9.80\ \text{m/s}^2)}=\boxed{19.4\ \text{m}}$$

---

86. ***REASONING*** The density ρ of an object is equal to its mass m divided by its volume V, or $\rho = m/V$ (Equation 11.1). The volume of a sphere is $V=\frac{4}{3}\pi r^3$, where r is the radius. According to the discussion of Archimedes' principle in Section 11.6, any object that is solid throughout will float in a liquid if the density of the object is less than or equal to the density of the liquid. If not, the object will sink.

SOLUTION

a. The average density of the sun is

$$\rho=\frac{m}{V}=\frac{m}{\frac{4}{3}\pi r^3}=\frac{1.99\times10^{30}\ \text{kg}}{\frac{4}{3}\pi\left(6.96\times10^8\ \text{m}\right)^3}=\boxed{1.41\times10^3\ \text{kg/m}^3}$$

b. Since the average density of the solid object (1.41×10^3 kg/m^3) is greater than that of water (1.00×10^3 kg/m^3, see Table 11.1), the object will $\boxed{\text{sink}}$.

c. The average density of Saturn is

$$\rho=\frac{m}{V}=\frac{m}{\frac{4}{3}\pi r^3}=\frac{5.7\times10^{26}\ \text{kg}}{\frac{4}{3}\pi\left(6.0\times10^7\ \text{m}\right)^3}=0.63\times10^3\ \text{kg/m}^3$$

The average density of this solid object (0.63×10^3 kg/m^3) is less than that of water (1.00×10^3 kg/m^3), so the object will $\boxed{\text{float}}$.

---

87. ***REASONING AND SOLUTION*** Let r_h represent the inside radius of the hose, and r_p the radius of the plug, as suggested by the figure below.

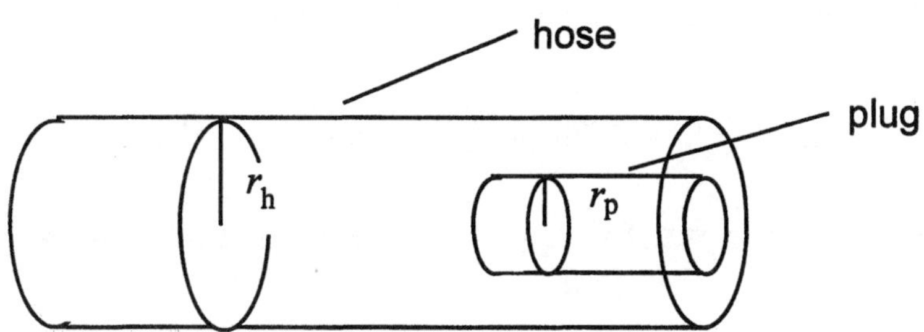

Then, from Equation 11.9, $A_1 v_1 = A_2 v_2$, we have $\pi r_h^2 v_1 = (\pi r_h^2 - \pi r_p^2) v_2$, or

$$\frac{v_1}{v_2} = \frac{\left(r_h^2 - r_p^2\right)}{r_h^2} = 1 - \left(\frac{r_p}{r_h}\right)^2 \quad \text{or} \quad \frac{r_p}{r_h} = \sqrt{1 - \frac{v_1}{v_2}}$$

According to the problem statement, $v_2 = 3v_1$, or

$$\frac{r_p}{r_h} = \sqrt{1 - \frac{v_1}{3v_1}} = \sqrt{\frac{2}{3}} = \boxed{0.816}$$

---

88. ***REASONING*** Pressure is the magnitude of the force applied perpendicularly to a surface divided by the area of the surface, according to Equation 11.3. To find the force that the suitcase applies to the floor, we first find the upward normal force F_N that the floor applies to the suitcase. By Newton's action-reaction law, the force that the suitcase applies to the floor has the same magnitude as F_N, but points downward. In determining F_N, we will use Newton's second law to account for the fact that the suitcase is accelerating.

SOLUTION Assuming that upward is the positive direction, we apply Newton's second law as follows:

$$\underbrace{\Sigma F}_{\text{Net force}} = \underbrace{F_N}_{\substack{\text{Normal}\\\text{force}}} - \underbrace{mg}_{\text{Weight}} = ma \quad \text{or} \quad F_N = mg + ma$$

According to Equation 11.3, the pressure, then, is

$$P = \frac{F_N}{A} = \frac{m(g+a)}{A} = \frac{(16\text{ kg})(9.80\text{ m/s}^2 + 1.5\text{ m/s}^2)}{(0.50\text{ m})(0.15\text{ m})} = \boxed{2.4 \times 10^3\text{ Pa}}$$

---

89. SSM ***REASONING AND SOLUTION*** The weight of the coin in air is equal to the sum of the weights of the silver and copper that make up the coin:

$$W_{\text{air}} = m_{\text{coin}} g = \left(m_{\text{silver}} + m_{\text{copper}}\right) g = \left(\rho_{\text{silver}} V_{\text{silver}} + \rho_{\text{copper}} V_{\text{copper}}\right) g \tag{1}$$

The weight of the coin in water is equal to its weight in air minus the buoyant force exerted on the coin by the water. Therefore,

$$W_{\text{water}} = W_{\text{air}} - \rho_{\text{water}} (V_{\text{silver}} + V_{\text{copper}})\, g \tag{2}$$

Solving Equation (2) for the sum of the volumes gives

$$V_{\text{silver}} + V_{\text{copper}} = \frac{W_{\text{air}} - W_{\text{water}}}{\rho_{\text{water}}\, g} = \frac{m_{\text{coin}}\, g - W_{\text{water}}}{\rho_{\text{water}}\, g}$$

or

$$V_{\text{silver}} + V_{\text{copper}} = \frac{(1.150 \times 10^{-2}\ \text{kg})(9.80\ \text{m/s}^2) - 0.1011\ \text{N}}{(1.000 \times 10^3\ \text{kg/m}^3)\,(9.80\ \text{m/s}^2)} = 1.184 \times 10^{-6}\ \text{m}^3$$

From Equation (1) we have

$$V_{\text{silver}} = \frac{m_{\text{coin}} - \rho_{\text{copper}} V_{\text{copper}}}{\rho_{\text{silver}}} = \frac{1.150 \times 10^{-2}\ \text{kg} - (8890\ \text{kg/m}^3)\text{V}_{\text{copper}}}{10\ 500\ \text{kg/m}^3}$$

or

$$V_{\text{silver}} = 1.095 \times 10^{-6}\ \text{m}^3 - (0.8467) V_{\text{copper}} \qquad (3)$$

Substitution of Equation (3) into Equation (2) gives

$$W_{\text{water}} = W_{\text{air}} - \rho_{\text{water}} \left[1.095 \times 10^{-6}\ \text{m}^3 - (0.8467) V_{\text{copper}} + V_{\text{copper}}\right] g$$

Solving for V_{copper} gives $V_{\text{copper}} = 5.806 \times 10^{-7}\ \text{m}^3$. Substituting into Equation (3) gives

$$V_{\text{silver}} = 1.095 \times 10^{-6}\ \text{m}^3 - (0.8467)(5.806 \times 10^{-7}\ \text{m}^3) = 6.034 \times 10^{-7}\ \text{m}^3$$

From Equation 11.1, the mass of the silver is

$$m_{\text{silver}} = \rho_{\text{silver}} V_{\text{silver}} = (10\ 500\ \text{kg/m}^3)(6.034 \times 10^{-7}\ \text{m}^3) = \boxed{6.3 \times 10^{-3}\ \text{kg}}$$

---

90. ***REASONING AND SOLUTION*** The mercury, being more dense, will flow from the right container into the left container until the pressure is equalized. Then the pressure at the bottom of the left container will be $P = \rho_W g h_W + \rho_m g h_{mL}$ and the pressure at the bottom of the right container will be $P = \rho_m g h_{mR}$. Equating gives

$$\rho_W g h_W + \rho_m g (h_{mL} - h_{mR}) = 0 \qquad (1)$$

Both liquids are incompressible and immiscible so

$$h_W = 1.00\ \text{m} \quad \text{and} \quad h_{mL} + h_{mR} = 1.00\ \text{m}$$

Using these in (1) and solving for h_{mL} gives, $h_{mL} = (1/2)(1.00 - \rho_w/\rho_m) = 0.46$ m.

So the fluid level in the left container is 1.00 m + 0.46 m = $\boxed{1.46 \text{ m}}$ from the bottom.

---

91. SSM ***REASONING*** Let the length of the tube be denoted by L, and let the length of the liquid be denoted by ℓ. When the tube is whirled in a circle at a constant angular speed about an axis through one end, the liquid collects at the other end and experiences a centripetal force given by (see Equation 8.11, and use $F = ma$) $F = mr\omega^2 = mL\omega^2$.

Since there is no air in the tube, this is the only radial force experienced by the liquid, and it results in a pressure of

$$P = \frac{F}{A} = \frac{mL\omega^2}{A}$$

where A is the cross-sectional area of the tube. The mass of the liquid can be expressed in terms of its density ρ and volume V: $m = \rho V = \rho A\ell$. The pressure may then be written as

$$P = \frac{\rho A\ell L\omega^2}{A} = \rho\ell L\omega^2 \tag{1}$$

If the tube were completely filled with liquid and allowed to hang vertically, the pressure at the bottom of the tube (that is, where $h = L$) would be given by

$$P = \rho g L \tag{2}$$

SOLUTION According to the statement of the problem, the quantities calculated by Equations (1) and (2) are equal, so that $\rho\ell L\omega^2 = \rho g L$. Solving for ω gives

$$\omega = \sqrt{\frac{g}{\ell}} = \sqrt{\frac{9.80 \text{ m/s}^2}{0.0100 \text{ m}}} = \boxed{31.3 \text{ rad/s}}$$

---

92. ***REASONING AND SOLUTION*** The mass of aluminum required to make the can is

$$m_{Al} = m_{total} - m_{soda}$$

where the mass of the soda is

$$m_{soda} = \rho_{soda}V_{soda} = (1.000 \times 10^3 \text{ kg/m}^3)(3.54 \times 10^{-4} \text{ m}^3) = 0.354 \text{ kg}$$

Therefore, the mass of the aluminum used to make the can is

$$m_{\text{Al}} = 0.416 \text{ kg} - 0.354 \text{ kg} = 0.062 \text{ kg}$$

Using the definition of density, $\rho = m/V$, we have

$$V_{\text{Al}} = \frac{m_{\text{Al}}}{\rho_{\text{Al}}} = \frac{0.062 \text{ kg}}{2700 \text{ kg/m}^3} = \boxed{2.3\times10^{-5} \text{ m}^3}$$

---

93. ***REASONING AND SOLUTION*** First find the velocity at the lower point using the equation of continuity, $A_1v_1 = A_2v_2$.

$$v_2 = v_1(A_1/A_2) = v_1(r_1/r_2)^2 = v_1(2r_2/r_2)^2 = 4v_1 = 4(0.50 \text{ m/s}) = 2.0 \text{ m/s}$$

Now use Bernoulli's equation, $(1/2)\rho v_1^{\,2} + \rho gy = (1/2)\rho v_2^{\,2}$, so

$$y = \tfrac{1}{2}\left(v_2^2 - v_1^2\right)/g = \tfrac{1}{2}\left[(2.0 \text{ m/s})^2 - (0.50 \text{ m/s})^2\right]/\left(9.80 \text{ m/s}^2\right) = \boxed{0.19 \text{ m}}$$

---

94. ***REASONING AND SOLUTION*** The figure at the right shows the forces that act on the balloon as it holds the passengers and the ballast stationary above the earth. W_0 is the combined weight of the balloon, the load of passengers, and the ballast. The quantity F_B is the buoyant force provided by the air outside the balloon and is given by

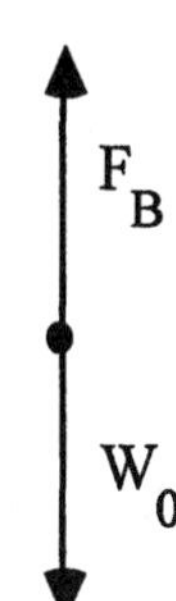

$$F_B = \rho_{\text{air}} g V_{\text{balloon}} \qquad (1)$$

Since the balloon is stationary, it follows that $F_B - W_0 = 0$, or

$$F_B = W_0 \qquad (2)$$

Then the ballast is dropped overboard, the balloon accelerates upward through a distance y in a time t with acceleration a_y, where (from kinematics) $a_y = \dfrac{2y}{t^2}$.

The figure at the right shows the forces that act on the balloon while it accelerates upward.

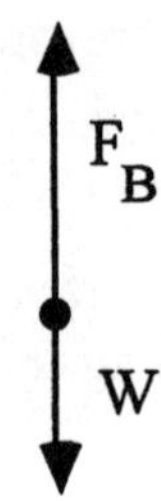

Applying Newton's second law, we have $F_B - W = ma_y$, where W is the weight of the balloon and the load of passengers. Replacing m by (W/g) we have

$$F_B - W = \frac{W}{g} a_y$$

Thus,

$$F_B = W + \frac{W}{g}\left(\frac{2y}{t^2}\right) = W\left(1 + \frac{2y}{gt^2}\right)$$

Solving for W gives

$$W = \frac{F_B}{\left(1 + \frac{2y}{gt^2}\right)} = F_B\left(\frac{gt^2}{gt^2 + 2y}\right)$$

The amount of ballast that must be thrown overboard is therefore [using Equations (1) and (2)]

$$\Delta W = W_0 - W = F_B - F_B\left(\frac{gt^2}{gt^2 + 2y}\right) = \left(1 - \frac{gt^2}{gt^2 + 2y}\right) \rho_{air} g V_{balloon}$$

$$\Delta W = r_{air} g\left(\tfrac{4}{3} \rho r_{balloon}^3\right)\left(1 - \frac{gt^2}{gt^2 + 2y}\right)$$

$$\Delta W = (1.29\ \text{kg/m}^3)(9.80\ \text{m/s}^2)\left[\tfrac{4}{3}\pi(6.25\ \text{m})^3\right]\left[1 - \frac{(9.80\ \text{m/s}^2)(15.0\ \text{s})^2}{(9.80\ \text{m/s}^2)(15.0\ \text{s})^2 + 2(105\ \text{m})}\right]$$

$$= \boxed{1120\ \text{N}}$$

95. ***REASONING AND SOLUTION*** The figures at the right show all the forces that act on each mass. T is the tension in the rope, F_B is the upward force exerted on mass 1 due to the pressure beneath the piston, and F_0 is the downward force exerted on mass 1 by the atmosphere.

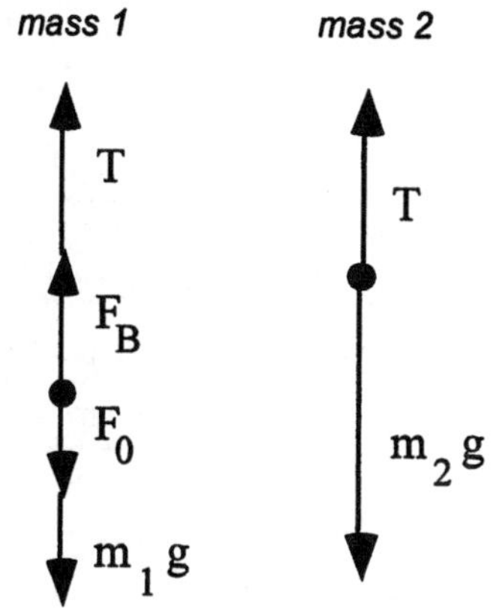

mass 1 mass 2

T T F_B F_0 $m_1 g$ $m_2 g$

Applying Newton's second law to mass 1, and taking "up" as the positive direction, we have

$$F_B + T - m_1 g - F_0 = m_1 a \qquad (1)$$

Similarly for mass 2 we have

$$T - m_2 g = -m_2 a \qquad (2)$$

Subtracting Equation (2) from Equation (1) gives

$$F_B + m_2 g - m_1 g - F_0 = (m_1 + m_2)\,a \qquad (3)$$

Solving Equation (3) for F_B gives

$$F_B = m_1(g + a) + m_2(a - g) + F_0 \qquad (4)$$

We must now find the acceleration a and the force F_0. From kinematics, we recall that $y = v_{0y}t + \frac{1}{2}a_y t^2$. Solving for a_y and setting $v_{oy} = 0$, we have

$$a_y = \frac{2y}{t^2} = \frac{2(1.25 \text{ m})}{(3.30 \text{ s})^2} = 0.230 \text{ m/s}^2$$

From the definition of pressure ($P = F/A$), the force F_0 is

$$F_0 = P_0 A = P_0(\pi r^2) = \pi(1.013 \times 10^5 \text{ N/m}^2)(2.50 \times 10^{-2} \text{ m})^2 = 1.99 \times 10^2 \text{ N}$$

Substituting these values into Equation (4) gives:

$$F_B = (0.500 \text{ kg})(9.80 \text{ m/s}^2 + 0.230 \text{ m/s}^2)$$

$$+ (9.50 \text{ kg})(0.230 \text{ m/s}^2 - 9.80 \text{ m/s}^2) + 1.99 \times 10^2 \text{ N} = 113 \text{ N}$$

Thus, the pressure beneath the piston is

$$P_{\text{B}} = \frac{F_{\text{B}}}{A} = \frac{F_{\text{B}}}{\pi r^2} = \frac{113 \text{ N}}{\pi (2.50 \times 10^{-2} \text{ m})^2} = \boxed{5.75 \times 10^4 \text{ Pa}}$$

---

96. ***REASONING AND SOLUTION***

a. Since the volume flow rate, $Q = Av$, is the same at each point, and since v is greater at the lower point, $\boxed{\text{the upper hole must have the larger area}}$.

b. Call the upper hole number 1 and the lower hole number 2 (the surface of the water is position 0). Take the zero level of potential energy at the bottom hole and then write Bernoulli's equation as

$$P_1 + (1/2)\rho v_1^{\ 2} + \rho gh = P_2 + (1/2)\rho v_2^{\ 2} = P_0 + \rho g(2h)$$

in which $P_1 = P_2 = P_0$ and from which we obtain

$$v_1 = \sqrt{2gh} \quad \text{and} \quad v_2 = \sqrt{4gh} \quad \text{or} \quad v_2 / v_1 = \sqrt{2}$$

Using the fact that $Q_1 = A_1 v_1 = Q_2 = A_2 v_2$, we have $v_2/v_1 = A_1/A_2$. But since the ratio of the areas is $A_1/A_2 = r_1^{\ 2}/r_2^{\ 2}$, we can write that

$$\frac{r_1}{r_2} = \sqrt{\frac{A_1}{A_2}} = \sqrt{\frac{v_2}{v_1}} = \sqrt[4]{2} = \boxed{1.19}$$

---

97. ***CONCEPT QUESTIONS***

a. The total mass is the sum of the mass of the gold and the mass of the quartz: $m_{\text{T}} = m_{\text{G}} + m_{\text{Q}}$.

b. The total volume is the sum of the volume of the gold and volume of the quartz: $V_{\text{T}} = V_{\text{G}} + V_{\text{Q}}$.

c. For any pure substance, the volume is related to the mass and the density according to the definition of density given in Equation 11.1 ($\rho = m/V$). Thus, the volume is $V = m/\rho$.

SOLUTION The total mass of the rock is

$$m_{\text{T}} = m_{\text{G}} + m_{\text{Q}} \tag{1}$$

The total volume of the rock is

$$V_{\text{T}} = V_{\text{G}} + V_{\text{Q}} \tag{2}$$

Using Equation 11.1 to express the individual volumes in terms of the individual masses and densities, we have from Equation (2) that

$$V_{\mathrm{T}} = \frac{m_{\mathrm{G}}}{\rho_{\mathrm{G}}} + \frac{m_{\mathrm{Q}}}{\rho_{\mathrm{Q}}} \tag{3}$$

Solving Equation (1) for m_{Q} and substituting into Equation (3) gives

$$V_{\mathrm{T}} = \frac{m_{\mathrm{G}}}{\rho_{\mathrm{G}}} + \frac{m_{\mathrm{T}} - m_{\mathrm{G}}}{\rho_{\mathrm{Q}}} = m_{\mathrm{G}}\left(\frac{1}{\rho_{\mathrm{G}}} - \frac{1}{\rho_{\mathrm{Q}}}\right) + \frac{m_{\mathrm{T}}}{\rho_{\mathrm{Q}}}$$

Solving this result for m_{G}, we find

$$m_{\mathrm{G}} = \frac{V_{\mathrm{T}} - \dfrac{m_{\mathrm{T}}}{\rho_{\mathrm{Q}}}}{\left(\dfrac{1}{\rho_{\mathrm{G}}} - \dfrac{1}{\rho_{\mathrm{Q}}}\right)} = \frac{\left(4.00 \times 10^{-3}\ \mathrm{m}^3\right) - \dfrac{12.0\ \mathrm{kg}}{2660\ \mathrm{kg/m^3}}}{\dfrac{1}{19\,300\ \mathrm{kg/m^3}} - \dfrac{1}{2660\ \mathrm{kg/m^3}}} = \boxed{1.6\ \mathrm{kg}}$$

The densities for gold and quartz have been taken from Table 11.1.

---

98. ***CONCEPT QUESTIONS***

a. The pressure outside the tube at the surface of the basting sauce is the atmospheric pressure of 1.013×10^5 Pa. The pressure in the bulb must be less than this value. If it were greater than atmospheric pressure, there would be a greater force on the top of the liquid in the tube than at the bottom, and the liquid could not have risen in the tube.

b. The atmospheric pressure outside the tube pushes the sauce up the tube, to the extent that the smaller pressure in the bulb allows it. The smaller the pressure in the bulb, the higher the sauce will rise. Conversely, the greater the pressure in the bulb, the less the sauce will rise. Thus, in the second trial the smaller value for h means that the pressure in the bulb is greater than in the first trial shown in the drawing.

SOLUTION According to Equation 11.4, we have

$$P_{\text{Atmospheric}} = P_{\text{Bulb}} + \rho g h$$

a. Solving for the pressure in the bulb for the first trial shown in the drawing, we find

$$P_{\text{Bulb}} = P_{\text{Atmospheric}} - \rho gh$$

$$= 1.013 \times 10^5 \text{ Pa} - \left(1200 \text{ kg/m}^3\right)\left(9.80 \text{ m/s}^2\right)\left(0.15 \text{ m}\right) = \boxed{9.95 \times 10^4 \text{ Pa}}$$

b. Solving for the pressure in the bulb for the second trial, we find

$$P_{\text{Bulb}} = P_{\text{Atmospheric}} - \rho gh$$

$$= 1.013 \times 10^5 \text{ Pa} - \left(1200 \text{ kg/m}^3\right)\left(9.80 \text{ m/s}^2\right)\left(0.10 \text{ m}\right) = \boxed{1.001 \times 10^5 \text{ Pa}}$$

---

99. ***CONCEPT QUESTIONS***

a. When the tube is floating, it is in equilibrium and there can be no net force acting on it. Therefore, the upward-directed buoyant force must have a magnitude that equals the magnitude of the tube's weight, which acts downward. This is true in either battery acid or antifreeze.

b. According to Archimedes' principle, the buoyant force has a magnitude that equals the weight of the displaced fluid. Since the antifreeze has the smaller density (smaller mass per unit volume), a greater volume of it is required to yield the displaced mass of fluid that has the necessary weight.

c. The farther up from the bottom of the tube the mark is, the greater is the volume of displaced fluid to which it corresponds. Since a greater volume of antifreeze must be displaced to float the tube, the antifreeze mark should be placed higher up the tube.

SOLUTION Since the magnitude of the buoyant force F_B equals the weight W of the tube, we have $F_B = W$. According to Archimedes' principle, the magnitude of the buoyant force equals the weight of the displaced fluid, which is the mass m of the displaced fluid times the acceleration due to gravity, or mg. But according to the definition of the density ρ, the mass is the density of the fluid times the displaced volume V, or $m = \rho V$. The result is that the weight of the displaced fluid is ρVg. Therefore, $F_B = W$ becomes

$$\rho Vg = W$$

The volume V of the displaced fluid equals the cross-sectional area A of the tube times the height h beneath the fluid surface, or $V = Ah$. With this substitution, our previous result becomes

$$\rho Ahg = W$$

a. Solving for h_{acid}, we find that

$$h_{\text{acid}} = \frac{W}{\rho_{\text{acid}} A g} = \frac{5.88 \times 10^{-2} \text{ N}}{(1280 \text{ kg/m}^3)(7.85 \times 10^{-5} \text{ m}^2)(9.80 \text{ m/s}^2)} = \boxed{5.97 \times 10^{-2} \text{ m}}$$

b. Solving for $h_{\text{antifreeze}}$, we find that

$$h_{\text{antifreeze}} = \frac{W}{\rho_{\text{antifreeze}} A g}$$

$$= \frac{5.88 \times 10^{-2} \text{ N}}{(1073 \text{ kg/m}^3)(7.85 \times 10^{-5} \text{ m}^2)(9.80 \text{ m/s}^2)} = \boxed{7.12 \times 10^{-2} \text{ m}}$$

---

100. ***Concept Questions***

a. Since the effects of air resistance and viscosity are being ignored, the water can be treated as a freely-falling object, as Chapter 2 discusses. It accelerates with the acceleration due to gravity. Therefore, it has a greater speed at the lower point than it did upon leaving the faucet.

b. The volume flow rate in cubic meters per second is the same as it was when the water left the faucet. This is because no water is added to or taken out of the stream after the water leaves the faucet.

c. The cross-sectional area of the water stream is less than it was when the water left the faucet. With the volume flow rate unchanging, the equation of continuity applies in the form of Equation 11.9. The volume flow rate is the cross-sectional area of the stream times the speed of the water. When the speed increases, as it does when the water falls, the cross-sectional area decreases.

SOLUTION Using the equation of continuity as stated in Equation 11.9, we have

$$\underbrace{A_1 v_1}_{\text{Below faucet}} = \underbrace{A_2 v_2}_{\text{At faucet}} \quad \text{or} \quad A_1 = \frac{A_2 v_2}{v_1}$$

To find the cross-sectional area A_1, we must find the speed v_1. To do this, we use Equation 2.9 from the equations of kinematics:

$$v_1^2 = v_2^2 + 2ay \quad \text{or} \quad v_1 = \sqrt{v_2^2 + 2ay}$$

In using this result, we choose upward as the positive direction, as usual. Substituting into the equation of continuity gives

$$A_1 = \frac{A_2 v_2}{\sqrt{v_2^2 + 2ay}} = \frac{(1.8\times10^{-4}\ \text{m}^2)(0.85\ \text{m/s})}{\sqrt{(0.85\ \text{m/s})^2 + 2(-9.80\ \text{m/s}^2)(-0.10\ \text{m})}} = \boxed{9.3\times10^{-5}\ \text{m}^2}$$

---

101. ***Concept Questions***

a. The fact that the pipe is horizontal means that the elevation of the water in the pipe and the water exiting the nozzle are the same.

b. The speed at which the water flows in the pipe is smaller than the speed of the water emerging from the nozzle. We can resort to the equation of continuity, which relates the cross-sectional areas A and the speeds v at two different points in the stream according to Equation 11.9 ($A_1v_1 = A_2v_2$). With its larger radius, the pipe has a greater cross-sectional area than that of the nozzle. According to Equation 11.9, the speed is smaller where the area is greater.

c. Knowing that the flow speed in the pipe is smaller than the flow speed in the nozzle, we can conclude that the absolute pressure of the water in the pipe is greater than the atmospheric pressure at the nozzle opening. This follows from Bernoulli's equation, which indicates for a horizontal pipe that the pressure is greater when the flow speed is smaller.

SOLUTION According to Bernoulli's equation as given in Equation 11.11, we have

$$\underbrace{P_1 + \tfrac{1}{2}\rho v_1^2 + \rho g y_1}_{\text{Pipe}} = \underbrace{P_2 + \tfrac{1}{2}\rho v_2^2 + \rho g y_2}_{\text{Nozzle}}$$

The pipe and nozzle are horizontal, so that $y_1 = y_2$ and Bernoulli's equation simplifies to

$$P_1 + \tfrac{1}{2}\rho v_1^2 = P_2 + \tfrac{1}{2}\rho v_2^2$$

We have values for the pressure P_2 at the nozzle opening and the speed v_1 in the pipe. However, to solve this expression for the pressure P_1 in the pipe, we also need a value for the speed v_2 at the nozzle opening. We obtain this value using the equation of continuity, as given in Equation 11.9:

$$A_1v_1 = A_2v_2 \quad \text{or} \quad \pi r_1^2 v_1 = \pi r_2^2 v_2 \quad \text{or} \quad v_2 = \frac{r_1^2 v_1}{r_2^2}$$

Here, we have also used that fact that the area of a circle is πr^2. Substituting this result for v_2 into Bernoulli's equation, we find that

$$P_1 + \tfrac{1}{2}\rho v_1^2 = P_2 + \tfrac{1}{2}\rho\left(\frac{r_1^2 v_1}{r_2^2}\right)^2$$

$$P_1 = P_2 + \tfrac{1}{2}\rho\left(\frac{r_1^4}{r_2^4} - 1\right)v_1^2$$

$$= 1.01\times10^5 \text{ Pa} + \tfrac{1}{2}\left(1.00\times10^3 \text{ kg/m}^3\right)\left[\frac{\left(1.9\times10^{-2}\text{ m}\right)^4}{\left(4.8\times10^{-3}\text{ m}\right)^4} - 1\right]\left(0.62 \text{ m/s}\right)^2$$

$$= \boxed{1.48\times10^5 \text{ Pa}}$$

The density of water is taken from Table 11.1 in the text.

---

102. ***CONCEPT QUESTIONS***

a. The amount of water per second leaking into the hold is the volume flow rate Q. According to Equation 11.10, the volume flow rate is the product of the area A of the hole and the speed v of the water entering the hold, $Q = Av$.

b. The speed of the water at the surface of the lake is approximately 0 m/s. Since the amount of water in the lake is large, the water level at the surface drops very, very slowly as water enters the hold of the ship. Thus, to a very good approximation, the speed of the water at the surface is zero.

c. As water moves downward from the surface to the hole, it is accelerated by the gravitational force, just like a ball that is dropped from rest at the top of a building.

SOLUTION

If we can find a value for the speed v_1 of the water entering the hold, the volume flow rate Q can be obtained from Equation 11.10 ($Q = Av_1$), since the area A of the hole is known. The place where the water enters the hold is just like point 1 in Figure 11.36*a*, whereas the lake surface, 2.0 m above, is just like point 2 in the figure. Bernoulli's equation, Equation 11.11 relates the pressure P, water speed v, and elevation y of these two points:

$$P_1 + \tfrac{1}{2}\rho v_1^2 + \rho g y_1 = P_2 + \tfrac{1}{2}\rho v_2^2 + \rho g y_2$$

Setting $P_1 = P_2$ (since the hold is open to the atmosphere), $v_2 = 0$ m/s, and solving for v_1, we obtain $v_1 = \sqrt{2g\left(y_2 - y_1\right)}$. The volume flow rate of the water leaving the hole is

$$Q = Av_1 = A\sqrt{2g\left(y_2 - y_1\right)} = \left(8.0\times10^{-3}\text{m}^2\right)\sqrt{2\left(9.80 \text{ m/s}^2\right)\left(2.0 \text{ m}\right)} = \boxed{5.0\times10^{-2} \text{ m}^3\text{/s}}$$

103. ***CONCEPT QUESTIONS***

a. The weight $W_{\text{hot air}}$ of the hot air is equal to its mass $m_{\text{hot air}}$ times the magnitude g of the acceleration due to gravity, or $W_{\text{hot air}} = m_{\text{hot air}}g$. According to Equation 11.1, the mass is equal to the product of the density $\rho_{\text{hot air}}$ and the volume V that the mass occupies (i.e., the volume of the balloon), so $m_{\text{hot air}} = \rho_{\text{hot air}}V$. Combining these two relations, the weight of the hot air can be expressed as $W_{\text{hot air}} = \rho_{\text{hot air}}Vg$.

b. According to Archimedes' principle, the magnitude of the buoyant force is equal to the weight of the fluid displaced by the balloon. In this case the fluid is the surrounding cool air. Thus, the buoyant force depends on the density of the cool air outside the balloon.

c. The free-body diagram shows the two forces acting on the balloon, its weight **W** and the buoyant force $\mathbf{F_B}$. Newton's second law, Equation 4.2b, states that the net force ΣF_y in the y-direction is equal to the mass m of the balloon times its acceleration a_y in that direction:

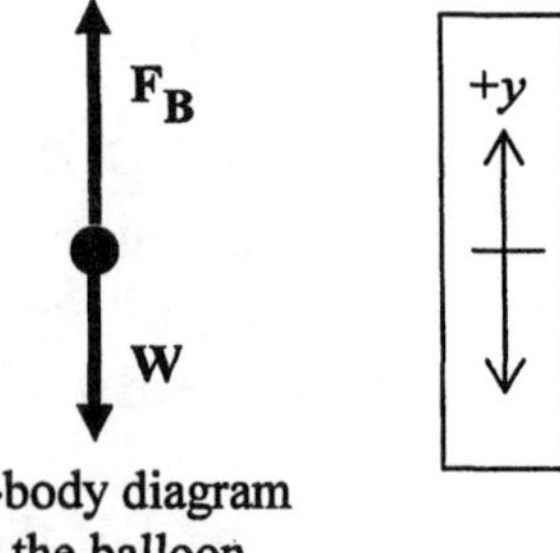

Free-body diagram for the balloon

$$\underbrace{F_B - W}_{\Sigma F_y} = ma_y$$

SOLUTION Solving Newton's second law for the acceleration of the balloon gives $a_y = \dfrac{F_B - W}{m}$. Since we are neglecting the weight of the balloon fabric and the basket, the only weight is that of the hot air inside the balloon. Thus, $m = m_{\text{hot air}} = \rho_{\text{hot air}}V$ and $W = m_{\text{hot air}}g = (\rho_{\text{hot air}}V)g$. The magnitude F_B of the buoyant force is equal to the weight of the cool air that the balloon displaces, so $F_B = m_{\text{cool air}}g = (\rho_{\text{cool air}}V)g$. Substituting these expressions for m, W, and F_B into Newton's second law gives

$$a_y = \frac{F_B - W}{m} = \frac{\rho_{\text{cool air}}Vg - \rho_{\text{hot air}}Vg}{\rho_{\text{hot air}}V} = \frac{\left(\rho_{\text{cool air}} - \rho_{\text{hot air}}\right)g}{\rho_{\text{hot air}}}$$

$$= \frac{\left(1.29\ \text{kg/m}^3 - 0.93\ \text{kg/m}^3\right)\left(9.80\ \text{m/s}^2\right)}{0.93\ \text{kg/m}^3} = \boxed{3.8\ \text{m/s}^2}$$

104. ***CONCEPT QUESTIONS***

a. In Poiseuille's law, the upstream pressure P_2 is the same for both hoses, because it is the pressure at the outlet. The downstream pressure P_1 is also the same for both hoses, since each opens to the atmosphere at the exit end. Therefore, the term $P_2 - P_1$ is the same for both hoses.

b. The volume flow rate Q is related to the radius R of a hose and the speed v of the water in the hose according to Equation 11.10: $Q = Av = \pi R^2 v$. Here, we have used the fact that the circular cross-sectional area of a hose is πR^2.

SOLUTION Using Poiseuille's law as given in Equation 11.14, we have

$$Q = \frac{\pi R^4 (P_2 - P_1)}{8\eta L}$$

Then, from Equation 11.10 and $A = \pi R^2$ for the area of a circle, we have

$$Q = Av = \pi R^2 v$$

Substituting this expression for Q into Poiseuille's law gives

$$\pi R^2 v = \frac{\pi R^4 (P_2 - P_1)}{8\eta L} \quad \text{or} \quad v = \frac{R^2 (P_2 - P_1)}{8\eta L}$$

To find the ratio v_B/v_A of the speeds, we apply this result to each hose, recognizing that the pressure difference $P_2 - P_1$, the length L, and the viscosity η are the same for both hoses:

$$\frac{v_B}{v_A} = \frac{\dfrac{R_B^2 (P_2 - P_1)}{8\eta L}}{\dfrac{R_A^2 (P_2 - P_1)}{8\eta L}} = \frac{R_B^2}{R_A^2} = (1.50)^2 = \boxed{2.25}$$

CHAPTER 12 | *TEMPERATURE AND HEAT*

PROBLEMS

---

1. **SSM** ***REASONING AND SOLUTION*** The difference between these two averages, expressed in Fahrenheit degrees, is

$$98.6\ °\text{F} - 98.2\ °\text{F} = 0.4\ \text{F}°$$

Since 1 C° is equal to $\frac{9}{5}$ F°, we can make the following conversion

$$(0.4\ \text{F}°)\left(\frac{1\ \text{C}°}{(9/5)\ \text{F}°}\right) = \boxed{0.2\ \text{C}°}$$

---

2. ***REASONING***

a. The relationship between the Kelvin temperature T and the Celsius temperature T_C is given by $T = T_C + 273.15$ (Equation 12.1).

b. The relationship between the Kelvin temperature T and the Fahrenheit temperature T_F can be obtained by following the procedure outlined in Examples 1 and 2 in the text. On the Kelvin scale the ice point is 273.15 K. Therefore, a Kelvin temperature T is $T - 273.15$ kelvins above the ice point. The size of the kelvin is larger than the size of a Fahrenheit degree by a factor of $\frac{9}{5}$. As a result a temperature that is $T - 273.15$ kelvins above the ice point on the Kelvin scale is $\frac{9}{5}(T - 273.15)$ F° above the ice point on the Fahrenheit scale. This amount must be added to the ice point of 32.0 °F. The relationship between the Kelvin and Fahrenheit temperatures, then, is given by

$$T_F = \tfrac{9}{5}(T - 273.15) + 32.0$$

SOLUTION

a. Solving Equation 12.1 for T_C, we find that

Day $$T_C = T - 273.15 = 375 - 273.15 = \boxed{102\ °\text{C}}$$

Night $$T_C = T - 273.15 = 1.00 \times 10^2 - 273.15 = \boxed{-173\ °\text{C}}$$

b. Using the equation developed in the reasoning, we find

Day $$T_F = \tfrac{9}{5}(T - 273.15) + 32.0 = \tfrac{9}{5}(375 - 273.15) + 32.0 = \boxed{215\ °\text{F}}$$

Night $T_F = \frac{9}{5}(T - 273.15) + 32.0 = \frac{9}{5}(1.00\times10^2 - 273.15) + 32.0 = \boxed{-2.80\times10^2\ °F}$

---

3. ***REASONING AND SOLUTION***

a. A temperature of 50.0 °F is 18.0 Fahrenheit degrees above the ice point of 32.0 °F. Since $1\ C° = \frac{9}{5}\ F°$, the difference of 18.0 F° is equivalent to

$$18.0\ F°\left(\frac{1\ C°}{\frac{9}{5}\ F°}\right) = 10.0\ C°$$

Adding 10.0 Celsius degrees to the ice point of 0 °C on the Celsius scale gives a Celsius temperature of $\boxed{10.0\ °C}$. Similar reasoning shows that a temperature of 104 °F is equivalent to a Celsius temperature of $\boxed{40.0\ °C}$.

b. The Kelvin temperature and the temperature on the Celsius scale are related by Equation 12.1: $T = T_c + 273.15$, where T is the Kelvin temperature and T_c is the Celsius temperature. Therefore, a temperature of 10.0 °C is equivalent to a Kelvin temperature of $\boxed{283.2\ K}$, and a temperature of 40.0 °C is equivalent to $\boxed{313.2\ K}$.

---

4. ***REASONING AND SOLUTION***

a. The Kelvin temperature and the temperature on the Celsius scale are related by Equation 12.1: $T = T_c + 273.15$, where T is the Kelvin temperature and T_c is the Celsius temperature. Therefore, a temperature of 77 K on the Celsius scale is

$$T_c = T - 273.15 = 77\ K\ -\ 273.15\ K\ =\ \boxed{-196\ °C}$$

b. The temperature of –196 °C is 196 Celsius degrees *below* the ice point of 0 °C. Since $1\ C° = \frac{9}{5}\ F°$, this number of Celsius degrees corresponds to

$$196\ C°\left(\frac{\frac{9}{5}\ F°}{1\ C°}\right) = 353\ F°$$

Subtracting 353 Fahrenheit degrees from the ice point of 32.0 °F on the Fahrenheit scale gives a Fahrenheit temperature of $\boxed{-321\ °F}$.

---

5. SSM **_REASONING AND SOLUTION_** The temperature of –273.15 °C is 273.15 Celsius degrees *below* the ice point of 0 °C. This number of Celsius degrees corresponds to

$$(273.15\text{ C°})\left(\frac{(9/5)\text{ F°}}{1\text{ C°}}\right) = 491.67\text{ F°}$$

Subtracting 491.67 F° from the ice point of 32.00 °F on the Fahrenheit scale gives a Fahrenheit temperature of $\boxed{-459.67\text{ °F}}$

---

6. **_REASONING AND SOLUTION_** If the voltage is proportional to the temperature difference between the junctions, then

$$\frac{V_1}{\Delta T_1} = \frac{V_2}{\Delta T_2} \quad \text{or} \quad \Delta T_2 = \frac{V_2}{V_1}\Delta T_1$$

Thus,

$$T_2 - 0.0\text{ °C} = \frac{1.90\times 10^{-3}\text{ V}}{4.75\times 10^{-3}\text{ V}}(110.0\text{ °C} - 0.0\text{ °C})$$

Solving for T_2 yields $\boxed{T_2 = 44.0\text{ °C}}$.

---

7. SSM **_REASONING_** The temperature at any pressure can be determined from the equation of the graph in Figure 12.4. Since the gas pressure is 5.00×10^3 Pa when the temperature is 0.00 °C, and the pressure is zero when the temperature is –273.15 °C, the slope of the line is given by

$$\frac{\Delta P}{\Delta T} = \frac{5.00\times 10^3\text{ Pa} - 0.00\text{ Pa}}{0.00\text{ °C} - (-273.15\text{ °C})} = 18.3\text{ Pa/C°}$$

Therefore, the equation of the line is $\Delta P = (18.3\text{ Pa/C°})\Delta T$, or

$$P - P_0 = (18.3\text{ Pa/C°})(T - T_0)$$

where $P_0 = 5.00\times 10^3$ Pa when $T_0 = 0.00$ °C.

SOLUTION Solving the equation above for *T*, we obtain

$$T = \frac{P - P_0}{18.3\text{ Pa/C°}} + T_0 = \frac{2.00\times 10^3\text{ Pa} - 5.00\times 10^3\text{ Pa}}{18.3\text{ Pa/C°}} + 0.00\text{ °C} = \boxed{-164\text{ °C}}$$

---

8. ***REASONING AND SOLUTION*** The space invaders temperature of 58 °I is 58 °I – 25 °I or 33 I° above the Space invaders ice point. Additionally, the space invaders degree is

$$\frac{100\text{ °C} - 0\text{ °C}}{156\text{ °I} - 25\text{ °I}} = 0.76\text{ C°/I°}$$

times smaller than the Celsius degree. Now

$$58\text{ °I} = (33\text{ °I})(0.76\text{ C°/I°}) = \boxed{25\text{ °C}}$$

---

9. ***REASONING AND SOLUTION*** Using the value for the coefficient of thermal expansion of steel given in Table 12.1, we find that the linear expansion of the aircraft carrier is

$$\Delta L = \alpha L_0\,\Delta T = (12 \times 10^{-6}\text{ C°}^{-1})(370\text{ m})(21\text{ °C} - 2.0\text{ °C}) = \boxed{0.084\text{ m}} \qquad (12.2)$$

---

10. ***REASONING AND SOLUTION*** The Concorde expands by an amount ΔL, where

$$\Delta L = \alpha L_0 \Delta T = (2.0 \times 10^{-5}\text{ C°}^{-1})(62\text{ m})(105\text{ °C} - 23\text{ °C}) = \boxed{0.10\text{ m}} \qquad (12.2)$$

---

11. **SSM** **WWW** ***REASONING AND SOLUTION*** The steel in the bridge expands according to Equation 12.2, $\Delta L = \alpha L_0 \Delta T$. Solving for L_0 and using the value for the coefficient of thermal expansion of steel given in Table 12.1, we find that the approximate length of the Golden Gate bridge is

$$L_0 = \frac{\Delta L}{\alpha\,\Delta T} = \frac{0.53\text{ m}}{\left[12 \times 10^{-6}\ (\text{C°})^{-1}\right](32\text{ °C} - 2\text{ °C})} = \boxed{1500\text{ m}}$$

---

12. ***REASONING AND SOLUTION***

a. The radius of the hole will be $\boxed{\text{larger}}$ when the plate is heated, because the hole expands as if it were made of copper.

b. The expansion of the radius is $\Delta r = \alpha r_0 \Delta T$. Using the value for the coefficient of thermal expansion of copper given in Table 12.1, we find that the fractional change in the radius is

$$\Delta r / r_0 = \alpha \Delta T = (17 \times 10^{-6}\text{ C°}^{-1})(110\text{ °C} - 11\text{ °C}) = \boxed{0.0017}$$

---

13. ***REASONING AND SOLUTION*** The value for the coefficient of thermal expansion of steel is given in Table 12.1. The relation, $\Delta L = \alpha L_0 \Delta T$, written in terms of the diameter d of the rod, is

$$\Delta T = \frac{\Delta d}{\alpha d_0} = \frac{0.0026 \text{ cm}}{\left(12 \times 10^{-6} \text{ C}^{\circ -1}\right)\left(2.0026 \text{ cm}\right)} = \boxed{110 \text{ C}^\circ} \qquad (12.2)$$

---

14. ***REASONING*** To determine the fractional decrease in length $\dfrac{\Delta L}{L_{0,\text{Silver}} + L_{0,\text{ Gold}}}$, we need the decrease ΔL in the rod's length. It is the sum of the decreases in the silver part and the gold part of the rod, or $\Delta L = \Delta L_{\text{Silver}} + \Delta L_{\text{Gold}}$. Each of the decreases can be expressed in terms of the coefficient of linear expansion α, the initial length L_0, and the change in temperature ΔT, according to Equation 12.2.

SOLUTION Using Equation 12.2 to express the decrease in length of each part of the rod, we find the total decrease in the rod's length to be

$$\Delta L = \underbrace{\alpha_{\text{Silver}} L_{0,\text{ Silver}} \Delta T}_{\Delta L_{\text{Silver}}} + \underbrace{\alpha_{\text{Gold}} L_{0,\text{ Gold}} \Delta T}_{\Delta L_{\text{Gold}}}$$

The fractional decrease in the rod's length is, then,

$$\frac{\Delta L}{L_{0,\text{Silver}} + L_{0,\text{ Gold}}} = \frac{\alpha_{\text{Silver}} L_{0,\text{ Silver}} \Delta T + \alpha_{\text{Gold}} L_{0,\text{ Gold}} \Delta T}{L_{0,\text{Silver}} + L_{0,\text{ Gold}}}$$

$$= \alpha_{\text{Silver}} \underbrace{\left(\frac{L_{0,\text{ Silver}}}{L_{0,\text{Silver}} + L_{0,\text{ Gold}}}\right)}_{\text{Silver fraction} = \frac{1}{3}} \Delta T + \alpha_{\text{Gold}} \underbrace{\left(\frac{L_{0,\text{ Gold}}}{L_{0,\text{Silver}} + L_{0,\text{ Gold}}}\right)}_{\text{Gold fraction} = \frac{2}{3}} \Delta T$$

Recognizing that one third of the rod is silver and two thirds is gold and taking values for the coefficients of linear expansion for silver and gold from Table 12.1, we have

$$\frac{\Delta L}{L_{0,\text{Silver}} + L_{0,\text{ Gold}}} = \alpha_{\text{Silver}} \left(\frac{1}{3}\right) \Delta T + \alpha_{\text{Gold}} \left(\frac{2}{3}\right) \Delta T = \left[\alpha_{\text{Silver}} \left(\frac{1}{3}\right) + \alpha_{\text{Gold}} \left(\frac{2}{3}\right)\right] \Delta T$$

$$= \left\{\left[19 \times 10^{-6} \left(\text{C}^\circ\right)^{-1}\right]\left(\frac{1}{3}\right) + \left[14 \times 10^{-6} \left(\text{C}^\circ\right)^{-1}\right]\left(\frac{2}{3}\right)\right\}\left(26 \text{ C}^\circ\right) = \boxed{4.1 \times 10^{-4}}$$

---

15. **SSM** ***REASONING AND SOLUTION*** Assuming that the rod expands linearly with heat, we first calculate the quantity $\Delta L / \Delta T$ using the data given in the problem.

$$\frac{\Delta L}{\Delta T} = \frac{8.47 \times 10^{-4} \text{ m}}{100.0 \text{ °C} - 25.0 \text{ °C}} = 1.13 \times 10^{-5} \text{ m/C°}$$

Therefore, when the rod is cooled from 25.0 °C, it will shrink by

$$\begin{aligned} \Delta L &= (1.13 \times 10^{-5} \text{ m/C°})\ \Delta T \\ &= (1.13 \times 10^{-5} \text{ m/C°})\ (0.00 \text{ °C} - 25.0 \text{ °C}) = \boxed{-2.82 \times 10^{-4} \text{ m}} \end{aligned}$$

---

16. ***REASONING*** The length of either heated strip is $L_0 + \Delta L$, where L_0 is the initial length and ΔL is the amount by which it expands. The expansion ΔL can be expressed in terms of the coefficient of linear expansion α, the initial length L_0, and the change in temperature ΔT, according to Equation 12.2. To find the change in temperature, we will set the length of the heated steel strip equal to the length of the heated aluminum strip and solve the resulting equation for ΔT.

SOLUTION According to Equation 12.2, the expansion is $\Delta L = \alpha\ L_0 \Delta T$. Using this equation we have

$$L_{0,\text{ Steel}} + \underbrace{\alpha_{\text{Steel}} L_{0,\text{ Steel}} \Delta T}_{\Delta L_{\text{Steel}}} = L_{0,\text{ Aluminum}} + \underbrace{\alpha_{\text{Aluminum}} L_{0,\text{ Aluminum}} \Delta T}_{\Delta L_{\text{Aluminum}}}$$

$$L_{0,\text{ Steel}} \left(1 + \alpha_{\text{Steel}} \Delta T\right) = L_{0,\text{ Aluminum}} \left(1 + \alpha_{\text{Aluminum}} \Delta T\right)$$

We know that the steel strip is 0.10 % longer than the aluminum strip, so that $L_{0,\text{ Steel}} = (1.0010) L_{0,\text{ Aluminum}}$. Substituting this result into the equation above, solving for ΔT, and taking values for the coefficients of linear expansion for aluminum and steel from Table 12.1give

$$(1.0010) L_{0,\text{ Aluminum}} \left(1 + \alpha_{\text{Steel}} \Delta T\right) = L_{0,\text{ Aluminum}} \left(1 + \alpha_{\text{Aluminum}} \Delta T\right)$$

$$0.0010 + (1.0010) \alpha_{\text{Steel}} \Delta T = \alpha_{\text{Aluminum}} \Delta T$$

$$\Delta T = \frac{0.0010}{\alpha_{\text{Aluminum}} - (1.0010) \alpha_{\text{Steel}}} = \frac{0.0010}{23 \times 10^{-6} \left(\text{C°}\right)^{-1} - (1.0010) \left[12 \times 10^{-6} \left(\text{C°}\right)^{-1}\right]} = \boxed{91 \text{ C°}}$$

---

17. ***REASONING AND SOLUTION*** $\Delta L = \alpha L_0 \Delta T$ gives for the expansion of the aluminum

$$\Delta L_A = \alpha_A L_A \Delta T \qquad (1)$$

and for the expansion of the brass

$$\Delta L_B = \alpha_B L_B \Delta T \qquad (2)$$

Taking the coefficients of thermal expansion for aluminum and brass from Table 10.1, adding Equations (1) and (2), and solving for ΔT give

$$\Delta T = \frac{\Delta L_A - \Delta L_B}{\alpha_A L_A + \alpha_B L_B} = \frac{1.3 \times 10^{-3}\text{ m}}{\left(23 \times 10^{-6}\ \text{C°}^{-1}\right)(1.0\text{ m}) + \left(19 \times 10^{-6}\ \text{C°}^{-1}\right)(2.0\text{ m})} = 21\,\text{C°}$$

The desired temperature is then

$$T = 28\text{ °C} + 21\text{ C°} = \boxed{49\text{ °C}}$$

---

18. ***REASONING AND SOLUTION*** The initial diameter of the sphere, d_s, is

$$d_s = (5.0 \times 10^{-4})d_r + d_r \qquad (1)$$

where d_r is the initial diameter of the ring. Applying $\Delta L = \alpha L_0 \Delta T$ to the diameter of the sphere gives

$$\Delta d_s = \alpha_s d_s \Delta T \qquad (2)$$

and to the ring gives

$$\Delta d_r = \alpha_r d_r \Delta T \qquad (3)$$

If the sphere is just to fit inside the ring, we must have

$$d_s + \Delta d_s = d_r + \Delta d_r$$

Using Equations (2) and (3) in this expression and solving for ΔT give

$$\Delta T = \frac{d_r - d_s}{\alpha_s d_s - \alpha_r d_r}$$

Substituting Equation (1) in this result and taking values for the coefficients of thermal expansion of steel and lead from Table 10.1 yield

$$\Delta T = \frac{-5.0 \times 10^{-4}}{\left(29 \times 10^{-6}\ \text{C}^{\circ -1}\right)\left(5.0 \times 10^{-4} + 1\right) - 12 \times 10^{-6}\ \text{C}^{\circ -1}} = \boxed{-29\ \text{C}^{\circ}}$$

The final temperature is

$$T_f = 70.0\ ^{\circ}\text{C} - 29\ \text{C}^{\circ} = \boxed{41\ ^{\circ}\text{C}}$$

---

19. **SSM** **WWW** ***REASONING AND SOLUTION*** Recall that $\omega = 2\pi / T$, Equation 10.6, where ω is the angular frequency of the pendulum and T is the period. Using this fact and Equation 10.16, we know that the period of the pendulum before the temperature rise is given by $T_1 = 2\pi\sqrt{L_0/g}$, where L_0 is the length of the pendulum. After the temperature has risen, the period becomes (using Equation 12.2), $T_2 = 2\pi\sqrt{\left[L_0 + \alpha L_0 \Delta T\right]/g}$. Dividing these expressions, solving for T_2, and taking the coefficient of thermal expansion of brass from Table 12.1, we find that

$$T_2 = T_1\sqrt{1 + \alpha \Delta T} = (2.0000\ \text{s})\sqrt{1 + (19 \times 10^{-6}/\text{C}^{\circ})\,(140\ \text{C}^{\circ})} = \boxed{2.0027\ \text{s}}$$

---

20. ***REASONING*** Each section of concrete expands as the temperature increases by an amount ΔT. The amount of the expansion ΔL is proportional to the initial length of the section, as indicated by Equation 12.2. Thus, to find the total expansion of the three sections, we can apply this expression to the total length of concrete, which is $L_0 = 3(2.4\ \text{m})$. Since the two gaps in the drawing are identical, each must have a minimum width that is one half the total expansion.

SOLUTION Using Equation 12.2 and taking the value for the coefficient of thermal expansion for concrete from Table 12.1, we find

$$\Delta L = \alpha L_0 \Delta T = \left[12 \times 10^{-6}\,(\text{C}^{\circ})^{-1}\right]\left[3(2.4\ \text{m})\right](32\ \text{C}^{\circ})$$

The minimum necessary gap width is one half this value or

$$\tfrac{1}{2}\left[12 \times 10^{-6}\,(\text{C}^{\circ})^{-1}\right]\left[3(2.4\ \text{m})\right](32\ \text{C}^{\circ}) = \boxed{1.4 \times 10^{-3}\ \text{m}}$$

---

21. ***REASONING AND SOLUTION***

a. The ruler will try to shrink as the temperature is lowered so a $\boxed{\text{tension}}$ is needed to keep it from shrinking.

b. The change in length of the ruler is given by $\Delta L = \alpha L_0 \Delta T$, Equation 12.2. The stress needed to stretch the ruler this amount is given by Stress = $F/A = Y(\Delta L/L_0)$, Equation 10.17. Substituting for ΔL and taking values for Young's modulus Y of steel from Table 10.1 and the coefficient of thermal expansion of steel from Table 12.1, we find that

$$\text{Stress} = Y\alpha\Delta T = (2.0 \times 10^{11}\ \text{N/m}^2)(12 \times 10^{-6}\ \text{C}^{\circ -1})[25\ ^\circ\text{C} - (-15\ ^\circ\text{C})] = \boxed{9.6 \times 10^7\ \text{N/m}^2}$$

---

22. ***REASONING AND SOLUTION*** Let $L_0 = 0.50$ m and L be the true length of the line at 40.0 °C. The ruler has expanded an amount

$$\Delta L_\text{r} = L - L_0 = \alpha_\text{r} L_0 \Delta T_\text{r} \qquad (1)$$

The copper plate must shrink by an amount

$$\Delta L_\text{p} = L_0 - L = \alpha_\text{p} L \Delta T_\text{p} \qquad (2)$$

Eliminating L from Equations (1) and (2), solving for ΔT_p, and using values of the coefficients of thermal expansion for copper and steel from Table 12.1, we find that

$$\Delta T_\text{p} = \frac{-\alpha_\text{r}\Delta T_\text{r}}{\alpha_\text{p}\left(1+\alpha_\text{r}\Delta T_\text{r}\right)}$$

$$= \frac{-\left[12\times10^{-6}\ (\text{C}^\circ)^{-1}\right](40.0\ ^\circ\text{C}-20.0\ ^\circ\text{C})}{\left[17\times10^{-6}\ (\text{C}^\circ)^{-1}\right]\left\{1+\left[12\times10^{-6}\ (\text{C}^\circ)^{-1}\right](40.0\ ^\circ\text{C}-20.0\ ^\circ\text{C})\right\}} = -14\ ^\circ\text{C}$$

Therefore, $T_\text{p} = 40.0\ ^\circ\text{C} - 14\ \text{C}^\circ = \boxed{26\ ^\circ\text{C}}$

---

23. **SSM** ***REASONING*** The change in length of the wire is the sum of the change in length of each of the two segments: $\Delta L = \Delta L_\text{al} + \Delta L_\text{st}$. Using Equation 12.2 to express the changes in length, we have

$$\alpha L_0 \Delta T = \alpha_\text{al} L_\text{0al} \Delta T + \alpha_\text{st} L_\text{0st} \Delta T$$

Dividing both sides by L_0 and algebraically canceling ΔT gives

$$\alpha = \alpha_\text{al}\left(\frac{L_\text{0al}}{L_0}\right) + \alpha_\text{st}\left(\frac{L_\text{0st}}{L_0}\right)$$

The length of the steel segment of the wire is given by $L_{0st} = L_0 - L_{0al}$. Making this substitution leads to

$$\alpha = \alpha_{al}\left(\frac{L_{0al}}{L_0}\right) + \alpha_{st}\left(\frac{L_0 - L_{0al}}{L_0}\right)$$

$$= \alpha_{al}\left(\frac{L_{0al}}{L_0}\right) + \alpha_{st}\left(\frac{L_0}{L_0}\right) - \alpha_{st}\left(\frac{L_{0al}}{L_0}\right)$$

This expression can be solved for the desired quantity, L_{0al}/L_0.

SOLUTION Solving for the ratio (L_{0al}/L_0) and taking values for the coefficients of thermal expansion for aluminum and steel from Table 12.1 gives

$$\frac{L_{0al}}{L_0} = \frac{\alpha - \alpha_{st}}{\alpha_{al} - \alpha_{st}} = \frac{19\times 10^{-6}(\text{C}^\circ)^{-1} - 12\times 10^{-6}(\text{C}^\circ)^{-1}}{23\times 10^{-6}(\text{C}^\circ)^{-1} - 12\times 10^{-6}(\text{C}^\circ)^{-1}} = \boxed{0.6}$$

---

24. ***REASONING AND SOLUTION*** The figure below (at the left) shows the forces that act on the middle of the aluminum wire for any value of the angle θ. The figure below (at the right) shows the same forces after they have been resolved into x and y components.

Applying Newton's second law to the vertical forces in figure on the right gives $2T(\sin\theta) - W = 0$. Solving for T gives:

$$T = \frac{W}{2(\sin\theta)} \qquad (1)$$

The following figures show how the angle is related to the initial length L_0 of the wire and to the final length L after the temperature drops.

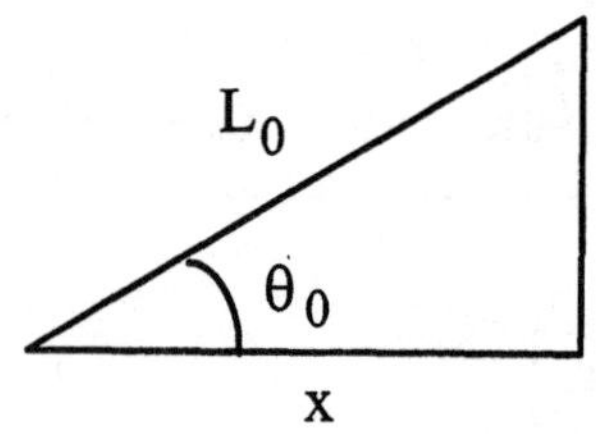

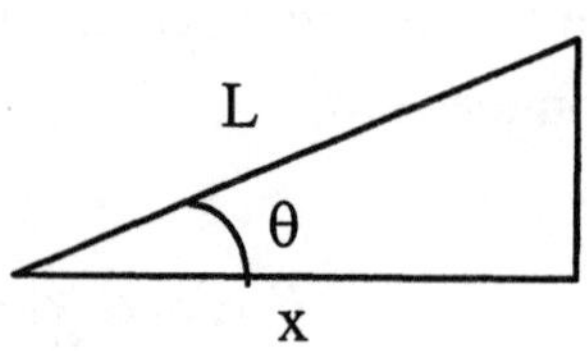

If the distance between the supports does not change, then the distance x is the same in both figures. Thus, at the original temperature,

$$x = L_0 (\cos \theta_0) \tag{2}$$

while at the lower temperature

$$x = L (\cos \theta) \tag{3}$$

Equating the right hand sides of Equations (2) and (3) leads to

$$\cos \theta = \frac{L_0 \cos \theta_0}{L} \tag{4}$$

Now, $L = L_0 + \Delta L$, and from Equation 12.2, it follows that

$$\frac{\Delta L}{L_0} = \alpha \Delta T$$

Thus, Equation (4) becomes

$$\cos \theta = \frac{L_0 \cos \theta_0}{L_0 + \Delta L} = \frac{\cos \theta_0}{1 + (\Delta L / L_0)} = \frac{\cos \theta_0}{1 + \alpha \Delta T}$$

From the figure in the text $\theta_0 = 3.00°$. Noting that the temperature of the wire drops by 20.0 C° ($\Delta T = -20.0$ C°) and taking the coefficient of thermal expansion of aluminum from Table 12.1, we find that the wire makes an angle θ with the horizontal that is

$$\theta = \cos^{-1}\left(\frac{\cos \theta_0}{1 + \alpha \Delta T}\right) = \cos^{-1}\left(\frac{\cos 3.00°}{1 + \left(23 \times 10^{-6}\ \text{C°}^{-1}\right)(-20.0\ \text{C°})}\right) = 2.446°$$

Using this value for θ in Equation (1) gives

$$T = \frac{W}{2 \sin \theta} = \frac{85.0\ \text{N}}{2 \sin 2.446°} = \boxed{996\ \text{N}}$$

---

25. ***REASONING AND SOLUTION*** The volume V_0 of an object changes by an amount ΔV when its temperature changes by an amount ΔT; the mathematical relationship is given by Equation 12.3: $\Delta V = \beta V_0 \Delta T$. Thus, the volume of the kettle at 24 °C can be found by solving Equation 12.3 for V_0. According to Table 12.1, the coefficient of volumetric expansion for copper is 51×10^{-6} $(\text{C°})^{-1}$. Solving Equation 12.3 for V_0, we have

$$V_0 = \frac{\Delta V}{\beta \Delta T} = \frac{1.2 \times 10^{-5} \text{ m}^3}{[51 \times 10^{-6} \, (\text{C°})^{-1}](100 \text{ °C} - 24 \text{ °C})} = \boxed{3.1 \times 10^{-3} \text{ m}^3}$$

---

26. ***REASONING AND SOLUTION*** Taking the coefficient of volumetric expansion β for water from Table 12.1, we find that the change in the volume of the water is

$$\Delta V = \beta V_0 \Delta T = (207 \times 10^{-6} \text{ C°}^{-1})(110 \text{ m}^3)(27 \text{ °C} - 17 \text{ °C}) = \boxed{0.23 \text{ m}^3} \qquad (12.3)$$

---

27. ***REASONING*** The change ΔV in the interior volume of the shell is given by Equation 12.3 as $\Delta V = \beta V_0 \Delta T$, where β is the coefficient of volume expansion, V_0 is the initial volume, and ΔT is the increase in temperature. The interior volume behaves as if it were filled with the surrounding silver. The interior volume is spherical, and the volume of a sphere is $\frac{4}{3}\pi r^3$, where r is the radius of the sphere.

SOLUTION In applying Equation 12.3, we note that the initial spherical volume of the interior space is $V_0 = \frac{4}{3}\pi r^3$, so that we have

$$\Delta V = \beta V_0 \Delta T = \beta \left(\tfrac{4}{3}\pi r^3 \right) \Delta T$$

$$= \left[57 \times 10^{-6} \left(\text{C°} \right)^{-1} \right] \tfrac{4}{3}\pi \left(2.0 \times 10^{-2} \text{ m} \right)^3 \left(147 \text{ °C} - 18 \text{ °C} \right) = \boxed{2.5 \times 10^{-7} \text{ m}^3}$$

We have taken the coefficient of volume expansion for silver from Table 12.1.

---

28. ***REASONING*** The increase ΔV in volume is given by Equation 12.3 as $\Delta V = \beta V_0 \Delta T$, where β is the coefficient of volume expansion, V_0 is the initial volume, and ΔT is the increase in temperature. The lead and quartz objects experience the same change in volume. Therefore, we can use Equation 12.3 to express the two volume changes and set them equal. We will solve the resulting equation for ΔT_{Quartz}.

SOLUTION Recognizing that the lead and quartz objects experience the same change in volume and expressing that change with Equation 12.3, we have

$$\underbrace{\beta_{\text{Lead}} V_0 \Delta T_{\text{Lead}}}_{\Delta V_{\text{Lead}}} = \underbrace{\beta_{\text{Quartz}} V_0 \Delta T_{\text{Quartz}}}_{\Delta V_{\text{Quartz}}}$$

In this result V_0 is the initial volume of each object. Solving for ΔT_{Quartz} and taking values for the coefficients of volume expansion for lead and quartz from Table 12.1 gives

$$\Delta T_{\text{Quartz}} = \frac{\beta_{\text{Lead}} \Delta T_{\text{Lead}}}{\beta_{\text{Quartz}}} = \frac{\left[87\times10^{-6}\,(\text{C}°)^{-1}\right](4.0\ \text{C}°)}{1.5\times10^{-6}\,(\text{C}°)^{-1}} = \boxed{230\ \text{C}°}$$

---

29. SSM ***REASONING AND SOLUTION*** According to Equation 12.3, $\Delta V = \beta V_0 \Delta T$. Taking the coefficient of volumetric expansion β for water from Table 12.1, we find that

$$\frac{\Delta V_{\text{air}}}{\Delta V_{\text{water}}} = \frac{\beta_{\text{air}} V_0 \Delta T}{\beta_{\text{water}} V_0 \Delta T} = \frac{\beta_{\text{air}}}{\beta_{\text{water}}} = \frac{3.7\times10^{-3}(\text{C}°)^{-1}}{207\times10^{-6}(\text{C}°)^{-1}} = \boxed{18}$$

---

30. ***REASONING AND SOLUTION*** Both the water and pipe expand as the temperature increases. For the expansion of the water

$$\Delta V_{\text{W}} = \beta_{\text{W}}\, V_0\, \Delta T$$

and for the expansion of the pipe

$$\Delta V_{\text{P}} = \beta_{\text{C}}\, V_0\, \Delta T$$

The initial volume of the pipe and water is $V_0 = \pi r^2 L$ The reservoir then needs a capacity of

$$\Delta V = \Delta V_{\text{W}} - \Delta V_{\text{P}} = (\beta_{\text{W}} - \beta_{\text{C}})\, V_0\, \Delta T$$

Taking the coefficients of volumetric expansion β_{W} and β_{C} for water and copper from Table 12.1, we find

$$\Delta V = (207\times10^{-6}\ \text{C}°^{-1} - 51\times10^{-6}\ \text{C}°^{-1})\pi(9.5\times10^{-3}\ \text{m})^2(76\ \text{m})(54\ \text{C}°) = \boxed{1.8\times10^{-4}\ \text{m}^3}$$

---

31. SSM ***REASONING AND SOLUTION*** Both the gasoline and the tank expand as the temperature increases. The coefficients of volumetric expansion β_g and β_s for gasoline and steel are available in Table 12.1. According to Equation 12.3, the volume expansion of the gasoline is

$$\Delta V_g = \beta_g V_0 \Delta T = \left[950\times10^{-6}\ (\text{C}°)^{-1}\right](20.0\ \text{gal})(18\ \text{C}°) = 0.34\ \text{gal}$$

while the volume of the steel tank expands by an amount

$$\Delta V_s = \beta_s V_0 \Delta T = \left[36\times10^{-6}\,(\text{C}°)^{-1}\right](20.0\ \text{gal})(18\ \text{C}°) = 0.013\ \text{gal}$$

The amount of gasoline which spills out is

$$\Delta V_g - \Delta V_s = \boxed{0.33\ \text{gal}}$$

---

32. ***REASONING AND SOLUTION*** Both the coffee and beaker expand as the temperature increases. For the expansion of the coffee

$$\Delta V_c = \beta_w V_0\, \Delta T$$

and for the expansion of the beaker

$$\Delta V_b = \beta_b\, V_0\, \Delta T$$

The excess expansion of the coffee, hence the amount which spills, is

$$\Delta V = \Delta V_c - \Delta V_b = (\beta_w - \beta_b)\, V_0\, \Delta T$$

Taking the coefficients of volumetric expansion β_w and β_c for coffee (water) and glass (Pyrex) from Table 12.1, we find

$$\Delta V = [207\times10^{-6}\ (\text{C}°)^{-1} - 9.9\times10^{-6}\ (\text{C}°)^{-1}](0.50\times10^{-3}\ \text{m}^3)(92\ °\text{C} - 18\ °\text{C})$$

$$= \boxed{7.3\times10^{-6}\ \text{m}^3}$$

---

33. ***REASONING*** The change in volume of the cylindrical mercury column is proportional to the coefficient of volume expansion β for mercury, its original volume V_0, and the change ΔT in temperature. The volume of a cylinder of radius r and height ΔL is $\Delta V = \pi r^2 \Delta L$. The coefficient of volume expansion for mercury can be found in Table 12.1.

SOLUTION The change in the volume of the mercury is

$$\underbrace{\pi r^2 \Delta L}_{\Delta V} = \beta V_0 \Delta T \qquad (12.3)$$

Solving for ΔL gives

$$\Delta L = \frac{\beta V_0 \Delta T}{\pi r^2} = \frac{\left[1.82 \times 10^{-4} \left(\text{C}°\right)^{-1}\right]\left(45 \text{ mm}^3\right)\left(1.0 \text{ C}°\right)}{\pi\left(1.7 \times 10^{-2} \text{ mm}\right)^2} = \boxed{9.0 \text{ mm}}$$

---

34. ***REASONING*** The change ΔV in volume is given by Equation 12.3 as $\Delta V = \beta V_0 \Delta T$, where β is the coefficient of volume expansion, V_0 is the initial volume, and ΔT is the increase in temperature. The increase in volume of the mercury is given directly by this equation, with V_0 being the initial volume of the interior space of the brass shell minus the initial volume of the steel ball. If the space occupied by the mercury did not change with temperature, the spillage would simply be the increase in volume of the mercury. However, the space occupied by the mercury does change with temperature. Both the brass shell and the steel ball expand. The interior volume of the brass shell expands as if it were solid brass, and this expansion provides more space for the mercury to occupy, thereby reducing the amount of spillage. The expansion of the steel ball, in contrast, takes up space that would otherwise be occupied by mercury, thereby increasing the amount of spillage. The total spillage, therefore, is $\Delta V_{\text{Mercury}} - \Delta V_{\text{Brass}} + \Delta V_{\text{Steel}}$.

SOLUTION Table 12.1 gives the coefficients of volume expansion for mercury, brass, and steel. Applying Equation 12.3 to the mercury, the brass cavity, and the steel ball, we have

$$\text{Spillage} = \Delta V_{\text{Mercury}} - \Delta V_{\text{Brass}} + \Delta V_{\text{Steel}}$$

$$= \beta_{\text{Mercury}} V_{0,\,\text{Mercury}} \Delta T - \beta_{\text{Brass}} V_{0,\,\text{Brass}} \Delta T + \beta_{\text{Steel}} V_{0,\,\text{Steel}} \Delta T$$

$$= \underbrace{\left[182 \times 10^{-6} \left(\text{C}°\right)^{-1}\right]\left[\left(1.60 \times 10^{-3} \text{ m}^3\right) - \left(0.70 \times 10^{-3} \text{ m}^3\right)\right]\left(12 \text{ C}°\right)}_{\text{Mercury}}$$

$$- \underbrace{\left[57 \times 10^{-6} \left(\text{C}°\right)^{-1}\right]\left(1.60 \times 10^{-3} \text{ m}^3\right)\left(12 \text{ C}°\right)}_{\text{Brass}}$$

$$+ \underbrace{\left[36 \times 10^{-6} \left(\text{C}°\right)^{-1}\right]\left(0.70 \times 10^{-3} \text{ m}^3\right)\left(12 \text{ C}°\right)}_{\text{Steel}} = \boxed{1.2 \times 10^{-6} \text{ m}^3}$$

---

35. SSM ***REASONING*** In order to keep the water from expanding as its temperature increases from 15 to 25 °C, the atmospheric pressure must be increased to compress the water as it tries to expand. The magnitude of the pressure change ΔP needed to compress a substance by an amount ΔV is, according to Equation 10.20, $\Delta P = B(\Delta V / V_0)$. The ratio $\Delta V / V_0$ is, according to Equation 12.3, $\Delta V / V_0 = \beta \Delta T$. Combining these two equations yields

$$\Delta P = B \beta \Delta T$$

SOLUTION Taking the value for the coefficient of volumetric expansion β for water from Table 12.1, we find that the change in atmospheric pressure that is required to keep the water from expanding is

$$\Delta P = (2.2 \times 10^9 \text{ N/m}^2)\left[207 \times 10^{-6} \, (\text{ C°})^{-1}\right](25 \text{ °C} - 15 \text{ °C})$$

$$= \left(4.6 \times 10^6 \text{ Pa}\right)\left(\frac{1 \text{ atm}}{1.01 \times 10^5 \text{ Pa}}\right) = \boxed{45 \text{ atm}}$$

---

36. ***REASONING AND SOLUTION***

a. The apparent weight of the sphere will be larger $\boxed{\text{after}}$ it cools. This is because the sphere shrinks while cooling, displacing less water, and hence decreasing the buoyant force acting on it.

b. The weight of the submerged sphere before cooling is

$$W_0 = \rho g V_0 - \rho_W g V_0$$

The weight of the submerged sphere after cooling is

$$W = \rho g V_0 - \rho_W g V$$

The weight difference is

$$\Delta W = W - W_0 = -\rho_W g \Delta V$$

where $\Delta V = V - V_0$. The volume change is $\Delta V = \beta V_0 \Delta T$. Now, $\Delta W = -\rho_W g \beta V_0 \Delta T$. The sphere's original volume is

$$V_0 = (4/3)\pi r^3 = (4/3)(\pi)(0.50 \text{ m})^3 = 0.52 \text{ m}^3$$

The coefficient of volumetric expansion β for aluminum is given in Table 12.1, so we have

$$\Delta W = -(1.0 \times 10^3 \text{ kg/m}^3)(9.80 \text{ m/s}^2)(69 \times 10^{-6} \text{ C°}^{-1})(0.52 \text{ m}^3)(-5.0 \times 10^1 \text{ C°}) = \boxed{18 \text{ N}}$$

---

37. **SSM** **WWW** ***REASONING*** The cavity that contains the liquid in either Pyrex thermometer expands according to Equation 12.3, $\Delta V_g = \beta_g V_0 \Delta T$. On the other hand, the volume of mercury expands by an amount $\Delta V_m = \beta_m V_0 \Delta T$, while the volume of alcohol expands by an amount $\Delta V_a = \beta_a V_0 \Delta T$. Therefore, the net change in volume for the mercury thermometer is

$$\Delta V_m - \Delta V_g = (\beta_m - \beta_g) V_0 \Delta T$$

while the net change in volume for the alcohol thermometer is

$$\Delta V_a - \Delta V_g = (\beta_a - \beta_g) V_0 \Delta T$$

In each case, this volume change is related to a movement of the liquid into a cylindrical region of the thermometer with volume $\pi r^2 h$, where r is the radius of the region and h is the height of the region. For the mercury thermometer, therefore,

$$h_m = \frac{(\beta_m - \beta_g) V_0 \Delta T}{\pi r^2}$$

Similarly, for the alcohol thermometer

$$h_a = \frac{(\beta_a - \beta_g) V_0 \Delta T}{\pi r^2}$$

These two expressions can be combined to give the ratio of the heights, h_a/h_m.

SOLUTION Taking the values for the coefficients of volumetric expansion for methyl alcohol, Pyrex glass, and mercury from Table 12.1, we divide the two expressions for the heights of the liquids in the thermometers and find that

$$\frac{h_a}{h_m} = \frac{\beta_a - \beta_g}{\beta_m - \beta_g} = \frac{1200 \times 10^{-6}\ (\text{C}^\circ)^{-1} - 9.9 \times 10^{-6}\ (\text{C}^\circ)^{-1}}{182 \times 10^{-6}\ (\text{C}^\circ)^{-1} - 9.9 \times 10^{-6}\ (\text{C}^\circ)^{-1}} = 6.9$$

Therefore, the degree marks are $\boxed{\text{6.9 times further apart}}$ on the alcohol thermometer than on the mercury thermometer.

---

38. ***REASONING AND SOLUTION*** When the temperature is 0.0 °C, $P = \rho_0 g h_0$, and when the temperature is 38.0 °C, $P = \rho g h$. Equating and solving for h gives $h = (\rho_0/\rho)h_0$. Now $\rho_0/\rho = V/V_0$, since the mass of mercury in the tube remains constant. Then $h = (V/V_0)h_0$. Now,

$$\Delta V = V - V_0 = \beta V_0 \Delta T \quad \text{or} \quad V/V_0 = 1 + \beta \Delta T$$

Therefore,

$$h = (1 + \beta\Delta T)h_0 = [1 + (182 \times 10^{-6}\ \text{C}^{\circ -1})(38.0\ °\text{C} - 0.0\ °\text{C})](0.760\ \text{m}) = \boxed{0.765\ \text{m}}$$

where we have taken the value for the coefficient of volumetric expansion β for mercury from Table 12.1.

---

39. ***REASONING AND SOLUTION*** The heat released by the blood is given by $Q = cm\Delta T$, in which the specific heat capacity c of the blood (water) is given in Table 12.2. Then

$$\Delta T = \frac{Q}{cm} = \frac{2000\ \text{J}}{[4186\ \text{J/(kg}\cdot\text{C}^\circ)](0.6\ \text{kg})} = 0.8\ \text{C}^\circ$$

Therefore,

$$T_f = T_i - \Delta T = \boxed{36.2\ °\text{C}}$$

---

40. ***REASONING*** Since the container of the glass and the liquid is being ignored and since we are assuming negligible heat exchange with the environment, the principle of conservation of energy applies. In reaching equilibrium the cooler liquid gains heat, and the hotter glass loses heat. We will apply this principle by equating the heat gained to the heat lost. The heat Q that must be supplied or removed to change the temperature of a substance of mass m by an amount ΔT is given by Equation 12.4 as $Q = cm\Delta T$, where c is the specific heat capacity. In using this equation as we apply the energy-conservation principle, we must remember to express the change in temperature ΔT as the higher minus the lower temperature.

SOLUTION Applying the energy-conservation principle and using Equation 12.4 give

$$\underbrace{c_{\text{Liquid}} m \Delta T_{\text{Liquid}}}_{\text{Heat gained by liquid}} = \underbrace{c_{\text{Glass}} m \Delta T_{\text{Glass}}}_{\text{Heat lost by glass}}$$

Since it is the same for both the glass and the liquid, the mass m can be eliminated algebraically from this equation. Solving for c_{Liquid} and taking the specific heat capacity for glass from Table 12.2, we find

$$c_{\text{Liquid}} = \frac{c_{\text{Glass}} \Delta T_{\text{Glass}}}{\Delta T_{\text{Liquid}}} = \frac{[840\ \text{J/(kg}\cdot\text{C}^\circ)](83.0\ \text{C}^\circ - 53.0\ \text{C}^\circ)}{(53.0\ \text{C}^\circ - 43.0\ \text{C}^\circ)} = \boxed{2500\ \text{J/(kg}\cdot\text{C}^\circ)}$$

---

41. **SSM** ***REASONING*** According to Equation 12.4, the heat required to warm the pool can be calculated from $Q = cm\Delta T$. The specific heat capacity c of water is given in Table 12.2. In order to use Equation 12.4, we must first determine the mass of the water in the pool. Equation 11.1 indicates that the mass can be calculated from $m = \rho V$, where ρ is the density of water and V is the volume of water in the pool.

SOLUTION Combining these two expressions, we have $Q = c\rho V\Delta T$, or

$$Q = \left[4186\ \text{J/(kg}\cdot\text{C}^\circ)\right](1.00\times10^{3}\ \text{kg/m}^{3})(12.0\ \text{m}\times 9.00\ \text{m}\times 1.5\ \text{m})\ (27\ ^\circ\text{C} - 15\ ^\circ\text{C}) = 8.14\times10^{9}\ \text{J}$$

Using the fact that 1 kWh = 3.6×10^{6} J, the cost of using electrical energy to heat the water in the pool at a cost of \$0.10 per kWh is

$$(8.14\times10^{9}\ \text{J})\left(\frac{\$0.10}{3.6\times10^{6}\ \text{J}}\right) = \boxed{\$230.}$$

---

42. ***REASONING*** Since there is no heat lost or gained by the system, the heat lost by the water in cooling down must be equal to the heat gained by the thermometer in warming up. The heat Q lost or gained by a substance is given by Equation 12.4 as $Q = cm\Delta T$, where c is the specific heat capacity, m is the mass, and ΔT is the change in temperature. Thus, we have that

$$\underbrace{c_{H_2O}\, m_{H_2O}\, \Delta T_{H_2O}}_{\text{Heat lost by water}} = \underbrace{c_{\text{therm}}\, m_{\text{therm}}\, \Delta T_{\text{therm}}}_{\text{Heat gained by thermometer}}$$

We can use this equation to find the temperature of the water before the insertion of the thermometer.

SOLUTION Solving the equation above for ΔT_{H_2O}, and using the value of c_{H_2O} from Table 12.2, we have

$$\Delta T_{H_2O} = \frac{c_{\text{therm}}\, m_{\text{therm}}\, \Delta T_{\text{therm}}}{c_{H_2O}\, m_{H_2O}}$$

$$= \frac{\left[815\ \text{J/}(\text{kg}\cdot\text{C}^\circ)\right](31.0\ \text{g})(41.5\ ^\circ\text{C} - 12.0\ ^\circ\text{C})}{\left[4186\ \text{J/}(\text{kg}\cdot\text{C}^\circ)\right](119\ \text{g})} = 1.50\ \text{C}^\circ$$

The temperature of the water before the insertion of the thermometer was

$$T = 41.5\ ^\circ\text{C} + 1.50\ \text{C}^\circ = \boxed{43.0\ ^\circ\text{C}}$$

---

43. ***REASONING*** The metabolic processes occurring in the person's body produce the heat that is added to the water. As a result, the temperature of the water increases. The heat Q that must be supplied to increase the temperature of a substance of mass m by an amount ΔT is given by Equation 12.4 as $Q = cm\Delta T$, where c is the specific heat capacity. The increase ΔT in temperature is the final higher temperature T_f minus the initial lower temperature T_0. Hence, we will solve Equation 12.4 for the desired final temperature.

SOLUTION From Equation 12.4, we have

$$Q = cm\Delta T = cm\left(T_f - T_0\right)$$

Solving for the final temperature, noting that the heat is $Q = (3.0 \times 10^5 \text{ J/h})(0.50 \text{ h})$ and taking the specific heat capacity of water from Table 12.2, we obtain

$$T_f = T_0 + \frac{Q}{cm} = 21.00\ ^\circ\text{C} + \frac{\left(3.0\times10^5 \text{ J/h}\right)\left(0.50 \text{ h}\right)}{\left[4186 \text{ J/}\left(\text{kg}\cdot\text{C}^\circ\right)\right]\left(1.2\times10^3 \text{ kg}\right)} = \boxed{21.03\ ^\circ\text{C}}$$

---

44. ***REASONING AND SOLUTION*** We require that the heat gained by the cold water equals the heat lost by the hot water, i.e.,

$$Q_{cw} = Q_{hw}$$

Therefore,

$$c_{cw}m_{cw}\Delta T_{cw} = c_{hw}m_{hw}\Delta T_{hw}$$

Since the specific heat capacity is that of water in each case, $c_{cw} = c_{hw}$, then

$$m_{cw}(36.0\ ^\circ\text{C} - 13.0\ ^\circ\text{C}) = m_{hw}(49.0\ ^\circ\text{C} - 36.0\ ^\circ\text{C})$$

Then

$$m_{cw} = (0.57)m_{hw}$$

We also know that $m_{cw} + m_{hw} = 191$ kg. Substituting for m_{cw}, and solving for m_{hw}, we have

$$m_{hw} = \frac{191 \text{ kg}}{1.57} = \boxed{121 \text{ kg}}$$

---

45. **SSM** ***REASONING*** Let the system be comprised only of the metal forging and the oil. Then, according to the principle of energy conservation, the heat lost by the forging equals the heat gained by the oil, or $Q_{metal} = Q_{oil}$. According to Equation 12.4, the heat lost by the forging is $Q_{metal} = c_{metal}m_{metal}(T_{0metal} - T_{eq})$, where T_{eq} is the final temperature of the

system at thermal equilibrium. Similarly, the heat gained by the oil is given by $Q_{\text{oil}} = c_{\text{oil}} m_{\text{oil}} (T_{\text{eq}} - T_{0\text{oil}})$.

SOLUTION

$$Q_{\text{metal}} = Q_{\text{oil}}$$

$$c_{\text{metal}} m_{\text{metal}} (T_{0\text{metal}} - T_{\text{eq}}) = c_{\text{oil}} m_{\text{oil}} (T_{\text{eq}} - T_{0\text{oil}})$$

Solving for $T_{0\text{metal}}$, we have

$$T_{0\text{metal}} = \frac{c_{\text{oil}} m_{\text{oil}} (T_{\text{eq}} - T_{0\text{oil}})}{c_{\text{metal}} m_{\text{metal}}} + T_{\text{eq}}$$

or

$$T_{0\text{metal}} = \frac{[2700\text{ J/(kg}\cdot\text{C}^\circ)](710\text{ kg})(47\text{ °C} - 32\text{ °C})}{[430\text{ J/(kg}\cdot\text{C}^\circ)](75\text{ kg})} + 47\text{ °C} = \boxed{940\text{ °C}}$$

---

46. ***REASONING AND SOLUTION*** We wish to convert 2.0% of the heat Q into gravitational potential energy, i.e., (0.020)Q = mgh. Thus,

$$\text{mg} = \frac{(0.020)\text{Q}}{\text{h}} = \frac{(0.020)(110\text{ Calories})\left(\dfrac{4186\text{ J}}{1\text{ Calorie}}\right)}{2.1\text{ m}} = \boxed{4.4 \times 10^3\text{ N}}$$

---

47. ***REASONING*** The heat Q that must be added to the water is given by Equation 12.4 as $Q = cm\Delta T$, where c is the specific heat capacity (see Table 12.2), m is the mass, and ΔT is the change in temperature. The mass of the water is related to its density ρ and volume V by $m = \rho V$ (Equation 11.1). Thus, the heat that must be added can be written as

$$Q = cm\Delta T = c(\rho V)\Delta T \qquad (1)$$

SOLUTION Solving Equation (1) for the volume V of water and converting cubic meters to barrels, we find that

$$V = \frac{Q}{c\rho\Delta T}$$

$$= \left(\frac{2.4 \times 10^8\text{ J}}{[4186\text{ J/(kg}\cdot\text{C}^\circ)](1.0 \times 10^3\text{ kg/m}^3)(17.0\text{ C}^\circ)}\right)\left(\frac{1\text{ barrel}}{0.16\text{ m}^3}\right) = \boxed{21\text{ barrels}}$$

---

48. ***REASONING AND SOLUTION*** The change ΔV in volume of the water in the swimming pool is given by $\Delta V = \beta V_0 \Delta T$, Equation 12.3, where β is the coefficient of volume expansion for water (see Table 12.1), V_0 is the original volume of the water, and ΔT is the change in its temperature. The change in temperature depends on the heat Q added to the water by $\Delta T = Q/cm$, Equation 12.4, where c is the specific heat capacity (see Table 12.2) and m is the mass. Substituting this expression for ΔT into the equation for ΔT gives

$$\Delta V = \beta V_0 \left(\frac{Q}{cm} \right) = \frac{\beta Q}{c\left(\dfrac{m}{V_0} \right)}$$

But m/V_0 is equal to the density ρ of water (Equation 11.1), so the change in volume of the water is

$$\Delta V = \frac{\beta Q}{c\rho}$$

$$= \frac{\left[2.07 \times 10^{-4}\ (\text{C}^\circ)^{-1}\right]\left(2.00 \times 10^{9}\ \text{J}\right)}{\left[4186\ \text{J}/(\text{kg}\cdot\text{C}^\circ)\right]\left(1.00 \times 10^{3}\ \text{kg}/\text{m}^3\right)} = \boxed{9.89 \times 10^{-2}\ \text{m}^3}$$

---

49. **SSM** ***REASONING AND SOLUTION*** As the rock falls through a distance h, its initial potential energy $m_{\text{rock}}gh$ is converted into kinetic energy. This kinetic energy is then converted into heat when the rock is brought to rest in the pail. If we ignore the heat absorbed by the pail, the principle of conservation of energy indicates that

$$m_{\text{rock}}gh = c_{\text{rock}}m_{\text{rock}}\Delta T + c_{\text{water}}m_{\text{water}}\Delta T$$

where we have used Equation 12.4 to express the heat absorbed by the rock and the water. Table 12.2 gives the specific heat capacity of the water. Solving for ΔT yields

$$\Delta T = \frac{m_{\text{rock}}gh}{c_{\text{rock}}m_{\text{rock}} + c_{\text{water}}m_{\text{water}}}$$

Substituting values yields

$$\Delta T = \frac{(0.20\ \text{kg})(9.80\ \text{m/s}^2)(15\ \text{m})}{[1840\ \text{J}/(\text{kg}\cdot\text{C}^\circ)](0.20\ \text{kg}) + [4186\ \text{J}/(\text{kg}\cdot\text{C}^\circ)](0.35\ \text{kg})} = \boxed{0.016\ \text{C}^\circ}$$

---

50. ***REASONING AND SOLUTION*** The heat lost by the steel rod is $Q = cm\Delta T = c\rho V_0 \Delta T$. Table 12.2 gives the specific heat capacity c of steel. The rod contracts according to the equation: $\Delta L = \alpha L_0 \Delta T$. Table 12.1 gives the coefficient of thermal expansion of steel. We also know that $F = AY(\Delta L/ L_0)$, Equation 10.17, so that $F = AY\alpha\Delta T$. Table 10.1 gives Young's modulus Y for steel.

Combining expressions yields,

$$F = \frac{\alpha QY}{c\rho L_0} = \frac{(12 \times 10^{-6}\ \text{C}^{\circ -1})(3300\ \text{J})(2.0 \times 10^{11}\ \text{Pa})}{[452\ \text{J/(kg} \cdot \text{C}^\circ)(7860\ \text{kg/m}^3)(2.0\ \text{m})} = \boxed{1.1 \times 10^3\ \text{N}}$$

---

51. **SSM** ***REASONING*** Heat Q_1 must be added to raise the temperature of the aluminum in its solid phase from 130 °C to its melting point at 660 °C. According to Equation 12.4, $Q_1 = cm\Delta T$. The specific heat c of aluminum is given in Table 12.2. Once the solid aluminum is at its melting point, additional heat Q_2 must be supplied to change its phase from solid to liquid. The additional heat required to melt or liquefy the aluminum is $Q_2 = mL_\text{f}$, where L_f is the latent heat of fusion of aluminum. Therefore, the total amount of heat which must be added to the aluminum in its solid phase to liquefy it is

$$Q_\text{total} = Q_1 + Q_2 = m(c\Delta T + L_\text{f})$$

SOLUTION Substituting values, we obtain

$$Q_\text{total} = (0.45\ \text{kg})\left\{[9.00 \times 10^2\ \text{J/(kg} \cdot \text{C}^\circ)](660\ ^\circ\text{C} - 130\ ^\circ\text{C}) + 4.0 \times 10^5\ \text{J/kg}\right\} = \boxed{3.9 \times 10^5\ \text{J}}$$

---

52. ***REASONING AND SOLUTION***

a. The latent heat of vaporization L_v of water is given in Table 12.3. To change water at 100.0 °C to steam we have from Equation 12.5 that

$$Q = mL_\text{v} = (2.00\ \text{kg})(22.6 \times 10^5\ \text{J/kg}) = \boxed{4.52 \times 10^6\ \text{J}}$$

b. For liquid water at 0.0 °C we must include the heat needed to raise the temperature to the boiling point. This heat is depends on the specific heat c of water, which is given in Table 12.2. Using Equations 12.4 and 12.5, we have $Q = (cm\Delta T)_\text{water} + mL_\text{v}$

$$Q = [4186\ \text{J/(kg} \cdot \text{C}^\circ)](2.00\ \text{kg})(100.0\ \text{C}^\circ) + 4.52 \times 10^6\ \text{J} = \boxed{5.36 \times 10^6\ \text{J}}$$

---

53. ***REASONING AND SOLUTION*** We want the same amount of heat removed from the water as from the ethyl alcohol, i.e.,

$$Q_{\text{water}} = Q_{\text{alcohol}}$$

or

$$(mL_f)_{\text{water}} = (mL_f)_{\text{alcohol}}$$

Then, taking values of the latent heats of fusion for water and ethyl alcohol from Table 12.3, we find

$$m_{\text{alcohol}} = m_{\text{water}} \frac{(L_f)_{\text{water}}}{(L_f)_{\text{alcohol}}} = (3.0 \text{ kg})\frac{33.5 \times 10^4 \text{ J/kg}}{10.8 \times 10^4 \text{ J/kg}} = \boxed{9.3 \text{ kg}}$$

---

54. ***REASONING*** As the body perspires, heat Q must be added to change the water from the liquid to the gaseous state. The amount of heat depends on the mass m of the water and the latent heat of vaporization L_v, according to $Q = mL_v$ (Equation 12.5).

SOLUTION The mass of water lost to perspiration is

$$m = \frac{Q}{L_v} = \frac{(240 \text{ Calories})\left(\frac{4186 \text{ J}}{1 \text{ Calorie}}\right)}{2.42 \times 10^6 \text{ J/kg}} = \boxed{0.42 \text{ kg}}$$

---

55. **SSM** ***REASONING*** From the conservation of energy, the heat lost by the mercury is equal to the heat gained by the water. As the mercury loses heat, its temperature decreases; as the water gains heat, its temperature rises to its boiling point. Any remaining heat gained by the water will then be used to vaporize the water.

According to Equation 12.4, the heat lost by the mercury is $Q_{\text{mercury}} = (cm\Delta T)_{\text{mercury}}$. The heat required to vaporize the water is, from Equation 12.5, $Q_{\text{vap}} = (m_{\text{vap}}L_v)_{\text{water}}$. Thus, the total amount of heat gained by the water is $Q_{\text{water}} = (cm\Delta T)_{\text{water}} + (m_{\text{vap}}L_v)_{\text{water}}$.

SOLUTION

$$Q_{\substack{\text{lost by}\\ \text{mercury}}} = Q_{\substack{\text{gained by}\\ \text{water}}}$$

$$(cm\Delta T)_{\text{mercury}} = (cm\Delta T)_{\text{water}} + (m_{\text{vap}}L_v)_{\text{water}}$$

where $\Delta T_{\text{mercury}} = (205\ ^\circ\text{C} - 100.0\ ^\circ\text{C})$ and $\Delta T_{\text{water}} = (100.0\ ^\circ\text{C} - 80.0\ ^\circ\text{C})$. The specific heats of mercury and water are given in Table 12.2, and the latent heat of vaporization of water is given in Table 12.3. Solving for the mass of the water that vaporizes gives

$$m_{\text{vap}} = \frac{c_{\text{mercury}} m_{\text{mercury}} \Delta T_{\text{mercury}} - c_{\text{water}} m_{\text{water}} \Delta T_{\text{water}}}{(L_{\text{v}})_{\text{water}}}$$

$$= \frac{[139 \text{ J/(kg}\cdot\text{C°)}](2.10 \text{ kg})(105 \text{ C°}) - [4186 \text{ J/(kg}\cdot\text{C°)}](0.110 \text{ kg})(20.0 \text{ C°})}{22.6 \times 10^5 \text{ J/kg}}$$

$$= \boxed{9.49 \times 10^{-3} \text{ kg}}$$

---

56. ***REASONING AND SOLUTION*** The heat required to evaporate the water is $Q = mL_{\text{V}}$, and to lower the temperature of the jogger we have $Q = m_{\text{j}}c\Delta T$. Equating these two expressions and solving for the mass of the water, m, we have

$$m = m_{\text{j}}c\Delta T/L_{\text{V}}$$

$$m = (75 \text{ kg})[3500 \text{ J/(kg·C°)}](1.\ 5 \text{ C°})/(2.42 \times 10^6 \text{ J/kg}) = \boxed{0.16 \text{ kg}}$$

---

57. ***REASONING AND SOLUTION*** The heat required is $Q = mL_{\text{f}} + cm\Delta T$, where $m = \rho V$. See Table 12.2 for the specific heat c and Table 12.3 for the latent heat L_{f}. Thus,

$$Q = \rho VL_{\text{f}} + c\rho V\Delta T$$

$$Q = (917 \text{ kg/m}^3)(4.50 \times 10^{-4} \text{ m})(1.25 \text{ m}^2)\{33.5 \times 10^4 \text{ J/kg}$$

$$+ [2.00 \times 10^3 \text{ J/(kg·C°)}](12.0 \text{ C°})\} = \boxed{1.85 \times 10^5 \text{ J}}$$

---

58. ***REASONING*** Since the container is being ignored and since we are assuming negligible heat exchange with the environment, the principle of conservation of energy applies in the following form: **heat gained equals heat lost.** In reaching equilibrium the colder aluminum gains heat in warming to 0.0 °C, and the warmer water loses heat in cooling to 0.0 °C. In either case, the heat Q that must be supplied or removed to change the temperature of a substance of mass m by an amount ΔT is given by Equation 12.4 as $Q = cm\Delta T$, where c is the specific heat capacity. In using this equation as we apply the energy-conservation principle, we must remember to express the change in temperature ΔT as the higher minus the lower temperature. The water that freezes into ice also loses heat. The heat Q lost when a mass m of water freezes is given by Equation 12.5 as $Q = mL_{\text{f}}$, where L_{f} is the latent heat of fusion. By including this amount of lost heat in the energy-conservation equation, we will be able to calculate the mass of water that is frozen.

SOLUTION Using the energy-conservation principle and Equations 12.4 and 12.5 gives

$$\underbrace{c_{\text{Aluminum}} m_{\text{Aluminum}} \Delta T_{\text{Aluminum}}}_{\text{Heat gained by aluminum}} = \underbrace{c_{\text{Water}} m_{\text{Water}} \Delta T_{\text{Water}}}_{\text{Heat lost by water}} + \underbrace{m_{\text{Ice}} L_{\text{f, Water}}}_{\text{Heat lost by water that freezes}}$$

Solving for m_{Ice}, taking values for the specific heat capacities from Table 12.2, and taking the latent heat for water from Table 12.3, we find that

$$m_{\text{Ice}} = \frac{c_{\text{Aluminum}} m_{\text{Aluminum}} \Delta T_{\text{Aluminum}} - c_{\text{Water}} m_{\text{Water}} \Delta T_{\text{Water}}}{L_{\text{f, Water}}}$$

$$= \frac{\left[9.00\times 10^2 \text{ J/}(\text{kg}\cdot \text{C°})\right](0.200 \text{ kg})\left[0.0 \text{ °C} - (-155 \text{ °C})\right]}{33.5\times 10^4 \text{ J/kg}}$$

$$- \frac{\left[4186 \text{J/}(\text{kg}\cdot \text{C°})\right](1.5 \text{ kg})(3.0 \text{ °C} - 0.0 \text{ °C})}{33.5\times 10^4 \text{ J/kg}} = \boxed{0.027 \text{ kg}}$$

---

59. ***REASONING*** The amount Q of heat required to melt an iceberg at 0 °C is equal to mL_f, where m is its mass and L_f is the latent heat of fusion for water (see Table 12.3). The mass is related to the density ρ and the volume V of the ice by Equation 11.1, $m = \rho V$.

SOLUTION

a. The amount of heat required to melt the iceberg is

$$Q = mL_f = \rho V L_f \qquad (12.5)$$

$$= \left(917 \text{ kg/m}^3\right) \underbrace{\left(120\times 10^3 \text{ m}\right)\left(35\times 10^3 \text{ m}\right)(230 \text{ m})}_{\text{Volume}} \left(3.35\times 10^5 \text{ J/kg}\right)$$

$$= \boxed{3.0\times 10^{20} \text{ J}}$$

b. The number of years it would take to melt the iceberg is equal to the energy required to melt it divided by the energy consumed per year by the U.S.

$$\text{Number of years} = \frac{3.0\times 10^{20} \text{ J}}{9.3\times 10^{19} \text{ J/y}} = \boxed{3.2 \text{ years}}$$

---

60. ***REASONING AND SOLUTION*** Using the value given for the specific heat c of water given in Table 12.2, we find that the energy released by the water is

$$Q = (cm\Delta T)_{\text{water}} + mL_f$$

$$Q = [4186 \text{ J/(kg·C°)}](840 \text{ kg})(10.0 \text{ C°}) + (840 \text{ kg})(3.35 \times 10^5 \text{ J/kg}) = 3.2 \times 10^8 \text{ J}$$

The 2.0-kW heater provides 2.0×10^3 J/s, so that the time is

$$\left(\frac{3.2 \times 10^8 \text{ J}}{2.0 \times 10^3 \text{ J/s}}\right)\left(\frac{1 \text{ hr}}{3600 \text{ s}}\right) = \boxed{44 \text{ hr}}$$

---

61. **SSM** **WWW** ***REASONING*** The system is comprised of the unknown material, the glycerin, and the aluminum calorimeter. From the principle of energy conservation, the heat gained by the unknown material is equal to the heat lost by the glycerin and the calorimeter. The heat gained by the unknown material is used to melt the material and then raise its temperature from the initial value of –25.0 °C to the final equilibrium temperature of $T_{\text{eq}} = 20.0$ °C.

SOLUTION

$$Q_{\substack{\text{gained by}\\ \text{unknown}}} = Q_{\substack{\text{lost by}\\ \text{glycerine}}} + Q_{\substack{\text{lost by}\\ \text{calorimeter}}}$$

$$m_{\text{u}}L_{\text{f}} + c_{\text{u}}m_{\text{u}}\Delta T_{\text{u}} = c_{\text{gl}}m_{\text{gl}}\Delta T_{\text{gl}} + c_{\text{al}}m_{\text{al}}\Delta T_{\text{al}}$$

Taking values for the specific heat capacities of glycerin and aluminum from Table 12.2, we have

$$(0.10 \text{ kg})L_{\text{f}} + [160 \text{ J/(kg·C°)}](0.10 \text{ kg})(45.0 \text{ C°}) = [2410 \text{ J/(kg·C°)}](0.100 \text{ kg})(7.0 \text{ C°}) + [9.0 \times 10^2 \text{ J/(kg·C°)}](0.150 \text{ kg})(7.0 \text{ C°})$$

Solving for L_{f} yields,

$$L_{\text{f}} = \boxed{1.9 \times 10^4 \text{ J/kg}}$$

---

62. ***REASONING*** The mass $m_{\text{remaining}}$ of the liquid water that remains at 100 °C is equal to the original mass m minus the mass $m_{\text{vaporized}}$ of the liquid water that has been vaporized. The heat Q required to vaporize this mass of liquid is given by Equation 12.5 as $Q = m\,L_{\text{v}}$, where L_{v} is the latent heat of vaporization for water. Thus, we have

$$m_{\text{remaining}} = m - m_{\text{vaporized}} = m - \frac{Q}{L_{\text{v}}}$$

The heat required to vaporize the water comes from the heat that is removed from the water at 0 °C when it changes phase from the liquid state to ice. This heat is also given by Equation 12.5 as $Q = mL_f$, where L_f is the latent heat of fusion for water. Thus, the remaining mass of liquid water can be written as

$$m_{\text{remaining}} = m - \frac{Q}{L_v} = m - \frac{mL_f}{L_v} = m\left(1 - \frac{L_f}{L_v}\right)$$

SOLUTION Using the values of L_f and L_v from Table 12.3, we find that the mass of liquid water that remains at 100 °C is

$$m_{\text{remaining}} = m\left(1 - \frac{L_f}{L_v}\right) = (2.00\text{ g})\left(1 - \frac{3.35\times10^5\text{ J/kg}}{2.26\times10^6\text{ J/kg}}\right) = \boxed{1.70\text{ g}}$$

---

63. ***REASONING*** Since all of the heat generated by friction goes into the block of ice, only this heat provides the heat needed to melt some of the ice. Since the surface on which the block slides is horizontal, the gravitational potential energy does not change, and energy conservation dictates that the heat generated by friction equals the amount by which the kinetic energy decreases or $Q_{\text{Friction}} = \frac{1}{2}Mv_0^2 - \frac{1}{2}Mv^2$, where v_0 and v are, respectively, the initial and final speeds and M is the mass of the block. In reality, the mass of the block decreases as the melting proceeds. However, only a very small amount of ice melts, so we may consider M to be essentially constant at its initial value. The heat Q needed to melt a mass m of water is given by Equation 12.5 as $Q = mL_f$, where L_f is the latent heat of fusion. Thus, by equating Q_{Friction} to mL_f and solving for m, we can determine the mass of ice that melts.

SOLUTION Equating Q_{Friction} to mL_f and solving for m gives

$$Q_{\text{Friction}} = \tfrac{1}{2}Mv_0^2 - \tfrac{1}{2}Mv^2 = mL_f$$

$$m = \frac{M\left(v_0^2 - v^2\right)}{2L_f} = \frac{(42\text{ kg})\left[(7.3\text{ m/s})^2 - (3.5\text{ m/s})^2\right]}{2\left(33.5\times10^4\text{ J/kg}\right)} = \boxed{2.6\times10^{-3}\text{ kg}}$$

We have taken the value for the latent heat of fusion for water from Table 12.3.

---

64. ***REASONING*** To freeze either liquid, heat must be removed to cool the liquid to its freezing point. In either case, the heat Q that must be removed to lower the temperature of a substance of mass m by an amount ΔT is given by Equation 12.4 as $Q = cm\Delta T$, where c is the

specific heat capacity. The amount ΔT by which the temperature is lowered is the initial temperature T_0 minus the freezing point temperature T. Once the liquid has been cooled to its freezing point, additional heat must be removed to convert the liquid into a solid at the freezing point. The heat Q that must be removed to freeze a mass m of liquid into a solid is given by Equation 12.5 as $Q = mL_f$, where L_f is the latent heat of fusion. The total heat to be removed, then, is the sum of that specified by Equation 12.4 and that specified by Equation 12.5, or $Q_{Total} = cm\,(T_0 - T) + mL_f$. Since we know that same amount of heat is removed from each liquid, we can set Q_{Total} for liquid A equal to Q_{Total} for liquid B and solve the resulting equation for $L_{f,\,A} - L_{f,\,B}$.

SOLUTION Setting Q_{Total} for liquid A equal to Q_{Total} for liquid B gives

$$c_A m\,(T_0 - T_A) + mL_{f,\,A} = c_B m\,(T_0 - T_B) + mL_{f,\,B}$$

Noting that the mass m can be eliminated algebraically from this result and solving for $L_{f,\,A} - L_{f,\,B}$, we find

$$L_{f,\,A} - L_{f,\,B} = c_B\left(T_0 - T_B\right) - c_A\left(T_0 - T_A\right)$$

$$= \left[2670\ \text{J}/\left(\text{kg}\cdot\text{C}^\circ\right)\right]\left[25.0\ ^\circ\text{C} - \left(-96.0\ ^\circ\text{C}\right)\right]$$

$$- \left[1850\ \text{J}/\left(\text{kg}\cdot\text{C}^\circ\right)\right]\left[25.0\ ^\circ\text{C} - \left(-68.0\ ^\circ\text{C}\right)\right] = \boxed{1.51\times10^5\ \text{J/kg}}$$

---

65. **SSM** ***REASONING*** In order to melt, the bullet must first heat up to 327.3 °C (its melting point) and then undergo a phase change. According to Equation 12.4, the amount of heat necessary to raise the temperature of the bullet to 327.3 °C is $Q = cm(327.3\ ^\circ\text{C} - 30.0\ ^\circ\text{C})$, where m is the mass of the bullet. The amount of heat required to melt the bullet is given by $Q_{melt} = mL_f$, where L_f is the latent heat of fusion of lead.

The lead bullet melts completely when it comes to a sudden halt; all of the kinetic energy of the bullet is converted into heat; therefore,

$$\text{KE} = Q + Q_{melt}$$

$$\tfrac{1}{2}mv^2 = cm(327.3\ ^\circ\text{C} - 30.0\ ^\circ\text{C}) + mL_f$$

The value for the specific heat c of lead is given in Table 12.2, and the value for the latent heat of fusion L_f of lead is given in Table 12.3. This expression can be solved for v, the minimum speed of the bullet for such an event to occur.

SOLUTION Solving for v, we find that the minimum speed of the lead bullet is

$$v = \sqrt{2L_f + 2c\,(327.3\ ^\circ\text{C} - 30.0\ ^\circ\text{C})}$$

$$v = \sqrt{2(2.32 \times 10^4\ \text{J/kg}) + 2[128\ \text{J/(kg} \cdot \text{C}^\circ)]\,(327.3\ ^\circ\text{C} - 30.0\ ^\circ\text{C})} = \boxed{3.50 \times 10^2\ \text{m/s}}$$

---

66. ***REASONING AND SOLUTION*** The steel band must be heated so that it can expand to fit the wheel. The diameter of the band must increase in length by an amount $\Delta L = 6.00 \times 10^{-4}$ m. Also, $\Delta L = \alpha L_0 \Delta T$, where the coefficient of thermal expansion α of steel is given in Table 12.1, so that

$$\Delta T = \Delta L/\alpha\, L_0 = (6.00 \times 10^{-4}\ \text{m})/[(12 \times 10^{-6}/\text{C}^\circ)(1.00\ \text{m})] = 5.0 \times 10^1\ \text{C}^\circ$$

The heat to expand the steel band comes from the heat released from the steam as it changes to water. Therefore $Q_{stm} = Q_{sb}$, or

$$m_{stm}L_v + (cm_{stm}\Delta T)_{water} = (cm_{sb}\Delta T)$$

Substituting the values for the latent heat of vaporization of water (see Table 12.3), the specific heat capacity of water (see Table 12.2), and the specific heat capacity of steel (see Table 12.2) gives

$$m_{stm}(22.6 \times 10^5\ \text{J/kg}) + [4186\ \text{J/(kg}\cdot\text{C}^\circ)]m_{stm}(100.0\ ^\circ\text{C} - 70.0\ ^\circ\text{C})$$

$$= [452\ \text{J/(kg}\cdot\text{C}^\circ)](25.0\ \text{kg})(50.0\ \text{C}^\circ)$$

Solving for the mass of the steam, we obtain $\boxed{m_{stm} = 0.237\ \text{kg}}$.

---

67. SSM ***REASONING AND SOLUTION*** From inspection of the graph that accompanies this problem, a pressure of 3.5×10^6 Pa corresponds to a temperature of 0 °C. Therefore, liquid carbon dioxide exists in equilibrium with its vapor phase at $\boxed{0\ ^\circ\text{C}}$ when the vapor pressure is 3.5×10^6 Pa.

---

68. ***REASONING*** The definition of percent relative humidity is given by Equation 12.6 as follows:

$$\text{Percent relative humidity} = \frac{\text{Partial pressure of water vapor}}{\text{Equilibrium vapor pressure of water at the existing temperature}} \times 100$$

Using R to denote the percent relative humidity, P to denote the partial pressure of water vapor, and P_V to denote the equilibrium vapor pressure of water at the existing temperature, we can write Equation 12.6 as

$$R = \frac{P}{P_V} \times 100$$

The partial pressure of water vapor P is the same at the two given temperatures. The relative humidity is not the same at the two temperatures, however, because the equilibrium vapor pressure P_V is different at each temperature, with values that are available from the vapor pressure curve given with the problem statement. To determine the ratio R_{10}/R_{40}, we will apply Equation 12.6 at each temperature.

SOLUTION Using Equation 12.6 and reading the values of R_{10} and R_{40} from the vapor pressure curve given with the problem statement, we find

$$\frac{R_{10}}{R_{40}} = \frac{P / P_{V,\,10}}{P / P_{V,\,40}} = \frac{P_{V,\,40}}{P_{V,\,10}} = \frac{7200 \text{ Pa}}{1300 \text{ Pa}} = \boxed{5.5}$$

---

69. **SSM** ***REASONING AND SOLUTION*** From the vapor pressure curve that accompanies Problem 68, it is seen that the partial pressure of water vapor in the atmosphere at 10 °C is about 1400 Pa, and that the equilibrium vapor pressure at 30 °C is about 4200 Pa. The relative humidity is, from Equation 12.6,

$$\begin{array}{c}\text{Percent}\\ \text{relative}\\ \text{humidity}\end{array} = \left(\frac{1400 \text{ Pa}}{4200 \text{ Pa}}\right) \times 100 = \boxed{33\%}$$

---

70. ***REASONING*** To bring the water to the point where it just begins to boil, its temperature must be increased to the boiling point. The heat Q that must be added to raise the temperature of a substance of mass m by an amount ΔT is given by Equation 12.4 as $Q = cm\Delta T$, where c is the specific heat capacity. The amount ΔT by which the temperature changes is the boiling temperature minus the initial temperature of 100.0 °C. The boiling temperature is the temperature at which the vapor pressure of the water equals the external pressure of 3.0×10^5 Pa and can be read from the vapor pressure curve for water given in Figure 12.32.

SOLUTION Using Equation 12.4, with T_{BP} being the boiling temperature and T_0 being the initial temperature, we have

$$Q = cm\Delta T = cm\left(T_{BP} - T_0\right)$$

According to Figure 12.32, an external pressure of 3.0×10^5 Pa corresponds to a boiling point temperature of $T_{BP} = 134$ °C. Using this value in Equation 12.4 and taking the specific heat capacity for water from Table 12.2, we determine the heat to be

$$Q = cm\left(T_{BP} - T_0\right) = \left[4186 \text{ J/(kg}\cdot\text{C°)}\right](2.0 \text{ kg})(134 \text{ °C} - 100.0 \text{ °C}) = \boxed{2.8\times10^5 \text{ J}}$$

---

71. ***REASONING AND SOLUTION*** At a temperature of 30.0 °C the equilibrium vapor pressure is 4200 Pa (as seen from the vapor pressure curve for water that accompanies Problem 68).

$$\text{Percent relative humidity} = \left(\frac{\text{Partial pressure}}{\text{Equilibrium vapor pressure}}\right) \times 100 \qquad (12.6)$$

The partial pressure is equal to

$$\text{Partial pressure} = \frac{(\text{Percent relative humidity})(\text{Equilibrium vapor pressure})}{100}$$

$$= \frac{(56.0\%)(4200 \text{ Pa})}{100} = \boxed{2400 \text{ Pa}}$$

---

72. ***REASONING AND SOLUTION*** The pressure inside the container is due to the weight on the piston in addition to the pressure of the atmosphere. The 120-kg mass produces a pressure of

$$P = \frac{F}{A} = \frac{mg}{\pi r^2} = \frac{(120 \text{ kg})(9.80 \text{ m/s}^2)}{\pi(0.061 \text{ m})^2} = 1.0 \times 10^5 \text{ Pa}$$

If we include atmospheric pressure (1.01×10^5 Pa), the total pressure inside the container is

$$P_{total} = P + P_{atm} = 2.0 \times 10^5 \text{ Pa}$$

Examination of the vaporization curve for water in Figure 12.32 shows that the temperature corresponding to this pressure at equilibrium is $\boxed{T = 120 \text{ °C}}$.

---

73. **SSM** ***REASONING*** We must first find the equilibrium temperature T_{eq} of the iced tea. Once this is known, we can use the vapor pressure curve that accompanies Problem 68 to find the partial pressure of water vapor at that temperature and then estimate the relative humidity using Equation 12.6.

According to the principle of energy conservation, when the ice is mixed with the tea, the heat lost by the tea is gained by the ice, or $Q_{tea} = Q_{ice}$. The heat gained by the ice is used to melt the ice at 0.0 °C; the remainder of the heat is used to bring the water at 0.0 °C up to the final equilibrium temperature T_{eq}.

SOLUTION

$$Q_{tea} = Q_{ice}$$

$$c_{water} m_{tea} (30.0\ °C - T_{eq}) = m_{ice} L_f + c_{water} m_{ice} (T_{eq} - 0.00\ °C)$$

The specific heat capacity of water is given in Table 12.2, and the latent heat of fusion L_f of water is given in Table 12.3. Solving for T_{eq}, we have

$$T_{eq} = \frac{c_{water} m_{tea} (30.0\ °C) - m_{ice} L_f}{c_{water} (m_{tea} + m_{ice})}$$

$$= \frac{[4186\ J/(kg \cdot C°)]\ (0.300\ kg)(30.0\ °C) - (0.0670\ kg)(33.5 \times 10^4\ J/kg)}{[4186\ J/(kg \cdot C°)]\ (0.300\ kg + 0.0670\ kg)} = 9.91\ °C$$

According to the vapor pressure curve that accompanies Problem 68, at a temperature of 9.91 °C, the equilibrium vapor pressure is approximately 1250 Pa. At 30 °C, the equilibrium vapor pressure is approximately 4400 Pa. Therefore, according to Equation 12.6, the percent relative humidity is approximately

$$\text{Percent relative humidity} = \left(\frac{1250\ Pa}{4400\ Pa}\right) \times 100 = \boxed{28\%}$$

---

74. ***REASONING AND SOLUTION*** Equation 12.6 defines the relative humidity as

$$\text{Percent relative humidity} = \frac{\text{Partial pressure of water vapor}}{\text{Equilibrium vapor pressure of water at the existing temeprature}} \times 100$$

According to the vapor pressure curve that accompanies Problem 68, the equilibrium vapor pressure of water at 36°C is 5.8 x 10^3 Pa. Since the water condenses on the coils when the temperature of the coils is 30 °C, the relative humidity at 30 °C is 100 percent. From Equation 12.6 this implies that the partial pressure of water vapor in the air must be equal to the equilibrium vapor pressure of water at 30 °C. From the vapor pressure curve, this pressure is 4.4 x 10^3 Pa. Thus, the relative humidity in the room is

$$\text{Percent relative humidity} = \frac{4.4 \times 10^3\ Pa}{5.8 \times 10^3\ Pa} \times 100 = \boxed{76\%}$$

---

75. ***REASONING AND SOLUTION*** At 10 °C the equilibrium vapor pressure is 1250 Pa. At 25 °C, the equilibrium vapor pressure is 3200 Pa. To get the smallest possible value for the relative humidity assume that the air at 10 °C is saturated. That is, take the partial pressure to be 1250 Pa (see the vapor pressure curve for water that accompanies Problem 68). Then we have,

$$\text{Percent relative humidity} = \left(\frac{\text{Partial pressure}}{\text{Equilibrium vapor pressure}}\right) \times 100 = \left(\frac{1250\text{ Pa}}{3200\text{ Pa}}\right) \times 100 = \boxed{39\%}$$

---

76. ***REASONING AND SOLUTION*** The water will boil if the vapor pressure of the water is equal to the ambient pressure. The pressure at a depth h in the water can be determined from Equation 11.4: $P_2 = P_1 + \rho gh$. When h = 10.3 m,

$$P_2 = (1.01 \times 10^5\text{ Pa}) + [(1.000 \times 10^3\text{ kg/m}^3)(9.80\text{ m/s}^2)(10.3\text{ m})] = 2.02 \times 10^5\text{ Pa}$$

The vapor pressure curve in Figure 12.32 shows that the vapor pressure of water is equal to 2.02×10^5 Pa at a temperature of 123 °C. Thus, the water at that depth has a temperature of $\boxed{T = 123\text{ °C}}$.

---

77. ***REASONING*** The change in length ΔL of the pipe is proportional to the coefficient of linear expansion α for steel, the original length L_0 of the pipe, and the change in temperature ΔT. The coefficient of linear expansion for steel can be found in Table 12.1.

SOLUTION The change in length of the pipe is

$$\Delta L = \alpha L_0 \Delta T = \left[1.2 \times 10^{-5}\,(\text{C°})^{-1}\right](65\text{ m})\left[18\text{ °C} - (-45\text{ °C})\right] = \boxed{4.9 \times 10^{-2}\text{ m}} \qquad (12.2)$$

---

78. ***REASONING AND SOLUTION*** The block of ice must undergo a change in temperature, followed by a change in phase, and then another change in temperature,

$$Q = Q_{\text{ice}} + Q_{\text{ice-to-water}} + Q_{\text{water}}$$

$$Q = (cm\Delta T)_{\text{ice}} + mL_f + (cm\Delta T)_{\text{water}}$$

Table 12.2 gives values for the specific heats of ice and water. Table 12.3 gives the latent heat of fusion of water. Using these values, we have

$$4.11 \times 10^6\text{ J} = [2.00 \times 10^3\text{ J/(kg·C°)}](10.0\text{ kg})(10.0\text{ C°}) + (10.0\text{ kg})(33.5 \times 10^4\text{ J/kg})$$

$$+ [4186\text{ J/(kg·C°)}](10.0\text{ kg})T_f$$

Solving for T_f we obtain, $\boxed{T_f = 13\text{ °C}}$.

---

79. **SSM** ***REASONING*** From the principle of conservation of energy, the heat lost by the coin must be equal to the heat gained by the liquid nitrogen. The heat lost by the silver coin is, from Equation 12.4, $Q = c_{\text{coin}} m_{\text{coin}} \Delta T_{\text{coin}}$ (see Table 12.2 for the specific heat capacity of silver). If the liquid nitrogen is at its boiling point, –195.8 °C, then the heat gained by the nitrogen will cause it to change phase from a liquid to a vapor. The heat gained by the liquid nitrogen is $Q = m_{\text{nitrogen}} L_{\text{v}}$, where m_{nitrogen} is the mass of liquid nitrogen that vaporizes, and L_{v} is the latent heat of vaporization for nitrogen (see Table 12.3).

SOLUTION

$$Q_{\substack{\text{lost by}\\ \text{coin}}} = Q_{\substack{\text{gained by}\\ \text{nitrogen}}}$$

$$c_{\text{coin}} m_{\text{coin}} \Delta T_{\text{coin}} = m_{\text{nitrogen}} L_{\text{v}}$$

Solving for the mass of the nitrogen that vaporizes

$$m_{\text{nitrogen}} = \frac{c_{\text{coin}} m_{\text{coin}} \Delta T_{\text{coin}}}{L_{\text{v}}}$$

$$= \frac{[235\ \text{J/(kg}\cdot\text{C°)}](1.5\times10^{-2}\ \text{kg})[25\ \text{°C} - (-195.8\ \text{°C})]}{2.00\times10^{5}\ \text{J/kg}} = \boxed{3.9\times10^{-3}\ \text{kg}}$$

---

80. ***REASONING*** Since there is no heat lost or gained by the system, the heat lost by the coffee in cooling down must be equal to the heat gained by the ice as it melts plus the heat gained by the melted water as it subsequently heats up. The heat Q lost or gained by a substance is given by Equation 12.4 as $Q = cm\Delta T$, where c is the specific heat capacity (see Table 12.2), m is the mass, and ΔT is the change in temperature. The heat that is required to change ice at 0 °C into liquid water at 0 °C is given by Equation 12.5 as $Q = m_{\text{ice}} L_{\text{f}}$, where m_{ice} is the mass of ice and L_{f} is the latent heat of fusion for water (see Table 12.3). Thus, we have that

$$\underbrace{m_{\text{coffee}} c_{\text{coffee}} \Delta T_{\text{coffee}}}_{\text{Heat lost by coffee}} = \underbrace{m_{\text{ice}} L_{\text{f}} + m_{\text{ice}} c_{\text{water}} \Delta T_{\text{water}}}_{\substack{\text{Heat gained by ice}\\ \text{and liquid water}}}$$

The mass of the coffee can be expressed in terms of its density as $m_{\text{coffee}} = \rho_{\text{coffee}} V_{\text{coffee}}$ (Equation 11.1). The change in temperature of the coffee is $\Delta T_{\text{coffee}} = 85\ \text{°C} - T$, where T is the final temperature of the coffee. The change in temperature of the water is $\Delta T_{\text{water}} = T - 0\ \text{°C}$. With these substitutions, the equation above becomes

$$\rho_{\text{coffee}} V_{\text{coffee}} c_{\text{coffee}} (85\ \text{°C} - T) = m_{\text{ice}} L_{\text{f}} + m_{\text{ice}} c_{\text{water}} (T - 0\ \text{°C})$$

Solving this equation for the final temperature gives

$$T = \frac{-m_{\text{ice}}L_{\text{f}} + \rho_{\text{coffee}}V_{\text{coffee}}c_{\text{coffee}}(85\ ^\circ\text{C})}{\rho_{\text{coffee}}V_{\text{coffee}}c_{\text{coffee}} + m_{\text{ice}}c_{\text{water}}}$$

$$= \frac{-(2)(11\times10^{-3}\text{ kg})(3.35\times10^{5}\text{ J/kg}) + (1.0\times10^{3}\text{ kg/m}^3)(150\times10^{-6}\text{ m}^3)[4186\text{ J/(kg}\cdot\text{C}^\circ)](85\ ^\circ\text{C})}{(1.0\times10^{3}\text{ kg/m}^3)(150\times10^{-6}\text{ m}^3)[4186\text{ J/(kg}\cdot\text{C}^\circ)] + (2)(11\times10^{-3}\text{ kg})[4186\text{ J/(kg}\cdot\text{C}^\circ)]}$$

$$= \boxed{64\ ^\circ\text{C}}$$

---

81. ***REASONING AND SOLUTION*** The fractional change in the length of the beam is given by $\Delta L = \alpha L_0 \Delta T$ to be $\Delta L/L_0 = \alpha\Delta T$. Taking the value for the coefficient of thermal expansion α for steel from Table 12.1, we find that

$$\frac{\Delta L}{L_0} = (12\times10^{-6}\ \text{C}^{\circ -1})[105\ ^\circ\text{F} - (-15\ ^\circ\text{F})]\left(\frac{1\ \text{C}^\circ}{\frac{9}{5}\ \text{F}^\circ}\right) = \boxed{8.0\times10^{-4}}$$

---

82. ***REASONING AND SOLUTION*** The change in the coin's diameter is $\Delta d = \alpha d_0 \Delta T$, according to Equation 12.2. Solving for α gives

$$\alpha = \frac{\Delta d}{d_0\,\Delta T} = \frac{2.3\times10^{-5}\text{ m}}{(1.8\times10^{-2}\text{ m})(75\ \text{C}^\circ)} = \boxed{1.7\times10^{-5}\ (\text{C}^\circ)^{-1}} \qquad (12.2)$$

---

83. **SSM** ***REASONING AND SOLUTION*** The cider will expand according to Equation 12.3; therefore, the change in volume of the cider is

$$\Delta V = \beta V_0 \Delta T = [280\times10^{-6}\ (\text{C}^\circ)^{-1}](1.0\text{ gal})(22\ \text{C}^\circ) = 6.2\times10^{-3}\text{ gal}$$

At a cost of two dollars per gallon, this amounts to

$$(6.2\times10^{-3}\text{ gal})\left(\frac{\$\ 2.00}{1\text{ gal}}\right) = \$\ 0.01 \quad \text{or} \quad \boxed{\text{one penny}}$$

---

84. ***REASONING*** We assume that no heat is lost through the chest to the outside. Then, energy conservation dictates that the heat gained by the soda is equal to the heat lost by the watermelon in reaching the final temperature T_f. Each quantity of heat is given by Equation 12.4, $Q = cm\Delta T$, where we write the change in temperature ΔT as the higher temperature minus the lower temperature.

SOLUTION Starting with the statement of energy conservation, we have

$$\text{Heat gained by soda} = \text{Heat lost by watermelon}$$

$$(cm\Delta T)_{\text{soda}} = (cm\Delta T)_{\text{watermelon}}$$

Since the watermelon is being treated like water, we take the specific heat capacity of water from Table 12.2. Thus, the above equation becomes

$$[3800\ \text{J/(kg}\cdot\text{C°)}](12\times 0.35\ \text{kg})(T_f - 5.0\ \text{°C}) = [4186\ \text{J/(kg}\cdot\text{C°)}](6.5\ \text{kg})(27\ \text{°C} - T_f)$$

Suppressing units for convenience and algebraically simplifying, we have

$$1.6\times 10^4\ T_f - 8.0\times 10^4 = 7.3\times 10^5 - 2.7\times 10^4\ T_f$$

Solving for T_f, we obtain

$$T_f = \frac{8.1\times 10^5}{4.3\times 10^4} = \boxed{19\ \text{°C}}$$

---

85. ***REASONING AND SOLUTION***

$$\text{Percent relative humidity} = \left(\frac{\text{Partial pressure}}{\text{Equilibrium vapor pressure}}\right)\times 100 = 35\%$$

Using the vaporization curve from Problem 68, we see that at 27 °C, the equilibrium vapor pressure is 3800 Pa. The partial pressure is, therefore,

$$\text{Partial pressure} = \frac{(\text{Percent relative humidity})(\text{Equilibrium vapor pressure})}{100}$$

$$= \frac{(35\%)(3800\ \text{Pa})}{100} = 1300\ \text{Pa}$$

From the vaporization curve accompanying Problem 68 we see that the dew point (100% relative humidity at equilibrium temperature) is

$$\boxed{\text{Dew Point} = 10\ \text{°C}}$$

---

86. ***REASONING AND SOLUTION*** The Rankine and Fahrenheit degrees are the same size, since the difference between the steam point and ice point temperatures is the same for both. The difference in the ice points of the two scales is 491.67 − 32.00 = 459.67. To get Rankine from Fahrenheit this amount must be added, so $\boxed{T_R = T_F + 459.67}$

---

87. **SSM** ***REASONING*** According to the statement of the problem, the initial state of the system is comprised of the ice and the steam. From the principle of energy conservation, the heat lost by the steam equals the heat gained by the ice, or $Q_{steam} = Q_{ice}$. When the ice and the steam are brought together, the steam immediately begins losing heat to the ice. An amount $Q_{1(lost)}$ is released as the temperature of the steam drops from 130 °C to 100 °C, the boiling point of water. Then an amount of heat $Q_{2(lost)}$ is released as the steam condenses into liquid water at 100 °C. The remainder of the heat lost by the "steam" $Q_{3(lost)}$ is the heat that is released as the water at 100 °C cools to the equilibrium temperature of $T_{eq} = 50.0\ °C$. According to Equation 12.4, $Q_{1(lost)}$ and $Q_{3(lost)}$ are given by

$$Q_{1(lost)} = c_{steam} m_{steam} (T_{steam} - 100.0\ °C) \quad \text{and} \quad Q_{3(lost)} = c_{water} m_{steam} (100.0\ °C - T_{eq})$$

$Q_{2(lost)}$ is given by $Q_{2(lost)} = m_{steam} L_v$, where L_v is the latent heat of vaporization of water. The total heat lost by the steam has three effects on the ice. First, a portion of this heat $Q_{1(gained)}$ is used to raise the temperature of the ice to its melting point at 0.00 °C. Then, an amount of heat $Q_{2(gained)}$ is used to melt the ice completely (we know this because the problem states that after thermal equilibrium is reached the liquid phase is present at 50.0 °C). The remainder of the heat $Q_{3(gained)}$ gained by the "ice" is used to raise the temperature of the resulting liquid at 0.0 °C to the final equilibrium temperature. According to Equation 12.4, $Q_{1(gained)}$ and $Q_{3(gained)}$ are given by

$$Q_{1(gained)} = c_{ice} m_{ice} (0.00\ °C - T_{ice}) \quad \text{and} \quad Q_{3(gained)} = c_{water} m_{ice} (T_{eq} - 0.00\ °C)$$

$Q_{2(gained)}$ is given by $Q_{2(gained)} = m_{ice} L_f$, where L_f is the latent heat of fusion of ice.

SOLUTION According to the principle of energy conservation, we have

$$Q_{steam} = Q_{ice}$$

$$Q_{1(lost)} + Q_{2(lost)} + Q_{3(lost)} = Q_{1(gained)} + Q_{2(gained)} + Q_{3(gained)}$$

or

$$c_{\text{steam}} m_{\text{steam}} (T_{\text{steam}} - 100.0\ ^\circ\text{C}) + m_{\text{steam}} L_{\text{v}} + c_{\text{water}} m_{\text{steam}} (100.0\ ^\circ\text{C} - T_{\text{eq}})$$

$$= c_{\text{ice}} m_{\text{ice}} (0.00\ ^\circ\text{C} - T_{\text{ice}}) + m_{\text{ice}} L_{\text{f}} + c_{\text{water}} m_{\text{ice}} (T_{\text{eq}} - 0.00\ ^\circ\text{C})$$

Values for specific heats are given in Table 12.2, and values for the latent heats are given in Table 12.3. Solving for the ratio of the masses gives

$$\frac{m_{\text{steam}}}{m_{\text{ice}}} = \frac{c_{\text{ice}} (0.00\ ^\circ\text{C} - T_{\text{ice}}) + L_{\text{f}} + c_{\text{water}} (T_{\text{eq}} - 0.00\ ^\circ\text{C})}{c_{\text{steam}} (T_{\text{steam}} - 100.0\ ^\circ\text{C}) + L_{\text{v}} + c_{\text{water}} (100.0\ ^\circ\text{C} - T_{\text{eq}})}$$

$$= \frac{\left[2.00\times10^3\ \text{J}/(\text{kg}\cdot\text{C}^\circ)\right]\left[0.0\ ^\circ\text{C}-(-10.0^\circ\text{C})\right] + 33.5\times10^4\ \text{J/kg} + \left[4186\ \text{J}/(\text{kg}\cdot\text{C}^\circ))\right](50.0\ ^\circ\text{C}-0.0\ ^\circ\text{C})}{\left[2020\ \text{J}/(\text{kg}\cdot\text{C}^\circ)\right](130\ ^\circ\text{C}-100.0\ ^\circ\text{C}) + 22.6\times10^5\ \text{J/kg} + \left[4186\ \text{J}/(\text{kg}\cdot\text{C}^\circ)\right](100.0\ ^\circ\text{C}-50.0\ ^\circ\text{C})}$$

or

$$\frac{m_{\text{steam}}}{m_{\text{ice}}} = \boxed{0.223}$$

---

88. ***REASONING AND SOLUTION***

a. According to Equation 12.5 and the definition of the density ρ given in Equation 11.1, it follows that $Q = mL_{\text{V}} = \rho V L_{\text{V}}$. Taking the latent heat of vaporization of water from Table 12.3 and the density of water from Table 11.1, we find that

$$Q = (1.000 \times 10^3\ \text{kg/m}^3)(2.59 \times 10^6\ \text{m}^2)(0.0254\ \text{m})(22.6 \times 10^5\ \text{J/kg}) = \boxed{1.49 \times 10^{14}\ \text{J}}$$

b. $$\text{Number of homes} = (1.49 \times 10^{14}\ \text{J})\left(\frac{1\ \text{home}}{1.50 \times 10^{11}\ \text{J}}\right) = \boxed{993\ \text{homes}}$$

---

89. ***REASONING*** According to Equation 6.10b, the average power is the change in energy divided by the time. The change in energy in this problem is the heat supplied to the water and the coffee mug to raise their temperature from 15 to 100 °C, which is the boiling point of water. The time is given as three minutes (180 s). The heat Q that must be added to raise the temperature of a substance of mass m by an amount ΔT is given by Equation 12.4 as $Q = cm\Delta T$, where c is the specific heat capacity. This equation will be used for the water and the material of which the mug is made.

SOLUTION Using Equation 6.10b, we write the average power $\overline{P}$ as

$$\overline{P} = \frac{\text{Change in energy}}{\text{Time}} = \frac{Q_{\text{Water}} + Q_{\text{Mug}}}{\text{Time}}$$

The heats Q_{Water} and Q_{Mug} each can be expressed with the aid of Equation 12.4, so that we obtain

$$\overline{P} = \frac{Q_{\text{Water}} + Q_{\text{Mug}}}{\text{Time}} = \frac{c_{\text{Water}} m_{\text{Water}} \Delta T + c_{\text{Mug}} m_{\text{Mug}} \Delta T}{\text{Time}}$$

$$= \frac{\begin{array}{c}\left[4186\ \text{J/}(\text{kg}\cdot\text{C}°)\right](0.25\ \text{kg})(100.0\ °\text{C} - 15\ °\text{C}) \\ + \left[920\ \text{J/}(\text{kg}\cdot\text{C}°)\right](0.35\ \text{kg})(100.0\ °\text{C} - 15\ °\text{C})\end{array}}{180\ \text{s}} = \boxed{650\ \text{W}}$$

The specific heat of water has been taken from Table 12.2.

---

90. ***REASONING*** The force F_{applied} to compress a spring by an amount ΔL is given by Equation 10.1 as $F_{\text{applied}} = k\Delta L$, where k is the spring constant of the spring. The change in length of the spring is related to its initial length L_0 and the change in temperature ΔT by Equation 12.2, $\Delta L = \alpha L_0 \Delta T$, where α is the coefficient of linear expansion for brass. Thus, the applied force is

$$F_{\text{Applied}} = k\,\Delta L = k\,\alpha\,L_0\,\Delta T$$

SOLUTION Using the value of α for brass from Table 12.1, we have

$$F_{\text{Applied}} = k\,\alpha\,L_0\,\Delta T = \left(1.3\times 10^4\ \text{N/m}\right)\left[1.9\times 10^{-5}\ (\text{C}°)^{-1}\right](0.18\ \text{m})(135\ °\text{C} - 21\ °\text{C}) = \boxed{5.1\ \text{N}}$$

---

91. SSM ***REASONING AND SOLUTION*** According to Equation 12.4, the total heat per kilogram required to raise the temperature of the water is

$$\frac{Q}{m} = c\Delta T = [4186\ \text{J/(kg}\cdot\text{C}°)]\ (32.0\ °\text{C}) = 1.34\times 10^5\ \text{J/kg}$$

The mass flow rate, $\Delta m/\Delta t$, is given by Equations 11.7 and 11.10 as $\Delta m/\Delta t = \rho A v = \rho Q_V$, where ρ and Q_V are the density and volume flow rate, respectively. We have, therefore,

$$\frac{\Delta m}{\Delta t} = (1.000\times 10^3\ \text{kg/m}^3)(5.0\times 10^{-6}\ \text{m}^3/\text{s}) = 5.0\times 10^{-3}\ \text{kg/s}$$

Therefore, the minimum power rating of the heater must be

$$(1.34\times 10^5\ \text{J/kg})(5.0\times 10^{-3}\ \text{kg/s}) = 6.7\times 10^2\ \text{J/s} = \boxed{6.7\times 10^2\ \text{W}}$$

---

92. ***REASONING AND SOLUTION***
a. As the wheel heats up, it will expand. Its radius, and therefore, its moment of inertia, will increase. Since no net external torque acts on the wheel, conservation of angular momentum applies where, according to Equation 9.10, the angular momentum is given by: $L = I\omega$, where I is the moment of inertia and ω is the angular velocity. When the moment of inertia increases at the higher temperature, the angular speed must decrease in order for the angular momentum to remain the same.

Thus, $\boxed{\text{the angular speed of the wheel decreases as the wheel heats up}}$.

b. According to the principle of conservation of angular momentum,

$$\underbrace{I_0\omega_0}_{\substack{\text{Initial angular}\\ \text{momentum, } L_0}} = \underbrace{I_f\omega_f}_{\substack{\text{Final angular}\\ \text{momentum, } L_f}}$$

Solving for ω_f, we have, treating the bicycle wheel as a thin-walled hollow hoop ($I = MR^2$, see Table 9.1)

$$\omega_f = \omega_0\left(\frac{I_0}{I_f}\right) = \omega_0\left(\frac{MR_0^2}{MR_f^2}\right) = \omega_0\left(\frac{R_0}{R_f}\right)^2$$

According to Equation 12.2, $\Delta R = \alpha R_0 \Delta T$, and the final radius of the wheel at the higher temperature is,

$$R_f = R_0 + \Delta R = R_0 + \alpha R_0 \Delta T = R_0(1 + \alpha\Delta T)$$

Therefore, taking the coefficient of thermal expansion α for steel from Table 12.1, we find that the angular speed of the wheel at the higher temperature is

$$\omega_f = \omega_0\left[\frac{R_0}{R_0(1+\alpha\Delta T)}\right]^2 = \omega_0\left(\frac{1}{1+\alpha\Delta T}\right)^2$$

$$= (18.00 \text{ rad/s})\left\{\frac{1}{1+[12\times 10^{-6}\ (\text{C°})^{-1}][300.0\ \text{°C} - (-100.0\ \text{°C})]}\right\}^2 = \boxed{17.83 \text{ rad/s}}$$

---

93. ***REASONING AND SOLUTION*** The steel ball and aluminum plate expand when heat is added to the system. Once the correct amount of heat has been added, the two have the same radius and the ball fits into the plate. Upon heating, the change in length is given by $\Delta L = \alpha L_0 \Delta T$, where $\Delta L = (L_f - L_0)$.

We require that the final lengths of the steel and aluminum be equal, i.e., $(L_f)_{st} = (L_f)_{al}$. Substituting for ΔL and equating the final lengths yields

$$(L_f)_{st} = (L_0)_{st}(1 + \alpha_{st}\Delta T) = (L_f)_{al} = (L_0)_{al}(1 + \alpha_{al}\Delta T)$$

Solving for ΔT and using values for the coefficients of thermal expansion for aluminum and steel from Table 12.1, we find that

$$\Delta T = (0.0010)/(\alpha_{al} - 1.0010\ \alpha_{st}) = 91\ C°$$

Taking the values for the specific heat capacities for steel and aluminum from Table 12.2, we find that the amount of heat that must be added is, therefore,

$$Q = (cm\Delta T)_{st} + (cm\Delta T)_{al}$$

$$Q = \{[452\ J/(kg{\cdot}C°)](1.5\ kg) + [9.00 \times 10^2\ J/(kg{\cdot}C°)](0.85\ kg)\}(91\ C°) = \boxed{1.3 \times 10^5\ J}$$

---

94. ***REASONING AND SOLUTION*** Let L_0 be the length of the wire before heating. After heating the wire will have stretched an amount $\Delta L_W = \alpha_W\ L_0\Delta T$, while the gap in the concrete "stretches" an amount $\Delta L_C = \alpha_C\ L_0\Delta T$.

The net change in length of the wire is then $\Delta L_C - \Delta L_W = (\alpha_C - \alpha_W)\ L_0\Delta T$. The additional stress needed to produce this change is

$$\text{Stress} = Y(\Delta L_C - \Delta L_W)/\ L_0 = Y\ (\alpha_C - \alpha_W)\Delta T$$

Taking values for the coefficients of thermal expansion of concrete and steel from Table 12.1 and the value for Young's modulus Y of aluminum from Table 10.1, we find that the additional tension in the wire is

$$T = (\text{Stress})(\pi R^2) = Y(\alpha_c - \alpha_w)\Delta T(\pi R^2)$$

$$= (6.9 \times 10^{10}\ N/m^2)[12 \times 10^{-6}\ C°^{-1} - 23 \times 10^{-6}\ C°^{-1}](185\ °C - 35\ °C)\pi\,(3.0 \times 10^{-4}\ m)^2$$

$$= -32\ N$$

The new tension in the wire is, then, $50.0\ N - 32\ N = \boxed{18\ N}$

---

95. ***CONCEPT QUESTIONS***

a. 1 A° is larger than 1 B°, because there are only 90 A° between the ice and boiling points of water, while there are 110 B° between these points.

b. +20 °A is hotter than +20 °B, because +20 °A is 50 A° above the ice point of water, while +20 °B is at the ice point.

SOLUTION

a. Since there are 90.0 A° and 110.0 B° between the ice and boiling points of water, we have that

$$\boxed{1\text{ A}° = \left(\frac{110.0}{90.0}\right)\text{B}° = 1.22\text{ B}°}$$

b. +40.0 °A is 70.0 A° above the ice point of water. On the B thermometer, this is

$$(70.0\text{ A}°)\left(\frac{1.22\text{ B}°}{1\text{ A}°}\right) = 85.4\text{ B}° \text{ above the ice point.}$$

The temperature on the B scale is

$$T = +20.0\text{ °B} + 85.4\text{ B}° = \boxed{105.4\text{ °B}}$$

---

96. ***CONCEPT QUESTIONS***

a. According to Equation 12.2, the factors that determine the amount ΔL by which the length of a rod changes are the coefficient of linear expansion α, its initial length L_0, and the change in temperature ΔT.

b. Since the materials from which the rods are made are different, the coefficients of linear expansion are different. The change in length is the same for each rod when the change in temperature is the same, however. Therefore, the initial lengths must be different to compensate for the fact that the expansion coefficients are different.

SOLUTION The change in length of the lead rod is (from Equation 12.2)

$$\Delta L_{\text{L}} = \alpha_{\text{L}} L_{\text{L}} \Delta T \qquad (1)$$

Similarly, the change in length of the quartz rod is

$$\Delta L_{\text{Q}} = \alpha_{\text{Q}} L_{\text{Q}} \Delta T \qquad (2)$$

where the temperature change ΔT is the same for both. Since both rods change length by the same amount, $\Delta L_{\text{L}} = \Delta L_{\text{Q}}$. Equating Equations (1) and (2) and solving for L_{Q} yields.

$$L_Q = \left(\frac{\alpha_L}{\alpha_Q}\right) L_L = \left[\frac{29\times10^{-6}\,(\text{C}°)^{-1}}{0.50\times10^{-6}\,(\text{C}°)^{-1}}\right](0.10\text{ m}) = \boxed{5.8\text{ m}}$$

Values for the coefficients of thermal expansion for lead and quartz have been taken from Table 12.1.

---

97. ***CONCEPT QUESTIONS***

a. According to Equation 12.3, the change ΔV in volume depends on the coefficient of volume expansion β, the initial volume V_0, and the change in temperature ΔT.

b. The liquid expands more, because its coefficient of volume expansion is larger and the change in volume is directly proportional to that coefficient.

c. The volume of liquid that spills over is equal to the change in volume of the liquid minus the change in volume of the can.

SOLUTION The volume of liquid that spills over the can is the difference between the increase in the volume of the liquid and that of the aluminum can:

$$\Delta V = \beta_{\text{liquid}} V_0\,\Delta T - \beta_{\text{aluminum}} V_0\,\Delta T$$

Therefore,

$$\beta_{\text{liquid}} = \frac{\Delta V}{V_0\,\Delta T} + \beta_{\text{aluminum}}$$

$$= \frac{3.6\times10^{-6}\text{ m}^3}{\left(3.5\times10^{-4}\text{ m}^3\right)\left(78\text{ °C} - 5\text{ °C}\right)} + 69\times10^{-6}\,(\text{C}°)^{-1} = \boxed{2.1\times10^{-4}\,(\text{C}°)^{-1}}$$

---

98. ***CONCEPT QUESTIONS***

a. The change in temperature is determined by the amount Q of heat added, the specific heat capacity c and mass m of the material (see Equation 12.4).

b. The heat and the mass are the same, but the changes in temperature are different. The only factor that can account for the different temperature changes is the specific heat capacities, which must be different.

SOLUTION The identity of the second bar can be made by determining its specific heat capacity and making a comparison with the values in Table 12.2. The heat supplied to each bar is given by Equation 12.4: $Q = cm\Delta T$. Since identical amounts of heat are supplied to each bar, we have

$$Q_1 = Q_2 \quad \text{or} \quad c_1\, m\,\Delta T_1 = c_2\, m\,\Delta T_2$$

Assigning c_1 to glass, we have c_1 = 840 J/(kg·C°) from Table 12.2. Solving for c_2, we have

$$c_2 = c_1\left(\frac{\Delta T_1}{\Delta T_2}\right) = \left[840\ \text{J/(kg}\cdot\text{C°)}\right]\left(\frac{88\ \text{°C} - 25\ \text{°C}}{250.0\ \text{°C} - 25\ \text{°C}}\right) = 235\ \text{J/(kg}\cdot\text{C°)}$$

Inspection of Table 12.2 shows that the second bar is made of $\boxed{\text{silver}}$.

---

99. ***CONCEPT QUESTIONS***

a. The amount of heat required to melt an object is directly proportional to the latent heat of fusion, according to Equation 12.5. Since each has the same mass and more heat is required to melt *B*, it has the larger latent heat of fusion.

b. The amount of heat required to melt an object is directly proportional to its mass, according to Equation 12.5. If the mass is doubled, the heat required to melt the object also doubles.

SOLUTION

a. According to Equation 12.5, the latent heats of fusion for *A* and *B* are

$$L_{\text{f},A} = \frac{Q_\text{A}}{m} = \frac{3.0\times10^4\ \text{J}}{3.0\ \text{kg}} = \boxed{1.0\times10^4\ \text{J/kg}}$$

$$L_{\text{f},B} = \frac{Q_\text{B}}{m} = \frac{9.0\times10^4\ \text{J}}{3.0\ \text{kg}} = \boxed{3.0\times10^4\ \text{J/kg}}$$

b. The amount of heat required to melt object *A* when its mass is 6.0 kg is

$$Q = m_\text{A}L_{\text{f},A} = (6.0\ \text{kg})\left(1.0\times10^4\ \text{J/kg}\right) = \boxed{6.0\times10^4\ \text{J}} \qquad (12.5)$$

---

100. ***CONCEPT QUESTIONS***

a. It does not mean that the partial pressure of water vapor in the air equals atmospheric pressure. The partial pressure is less than atmospheric pressure.

b. No, the humidity is not 100%. The vapor pressure at the higher temperature is greater than that at the lower temperature, but the partial pressure of the water vapor in the air has remained the same. According to Equation 12.6, the humidity has fallen below 100%.

SOLUTION

a. The percentage of atmospheric pressure is

$$\left(\frac{2500 \text{ Pa}}{1.01\times10^5 \text{ Pa}}\right)\times 100\% = \boxed{2.5\%}$$

b. The percentage is $\boxed{2.5\%}$.

c. The relative humidity at 35 °C is

$$\left(\frac{2500 \text{ Pa}}{5500 \text{ Pa}}\right)\times 100\% = \boxed{45\%} \qquad (12.6)$$

---

101. ***CONCEPT QUESTIONS***

a. When the ball and the plate are both heated to a higher common temperature, the ball passes through the hole. Since the ball's diameter is greater than the hole's diameter to start with, this must mean that the hole expands more than the ball for the same temperature change. The hole expands as if it were filled with the material that surrounds it. We conclude, therefore, that the coefficient of linear expansion for the plate is greater than that for the ball.

b. In each arrangement, the ball's diameter exceeds the hole's diameter by the same amount, the diameters of the holes are the same, the diameters of the balls are the same, and the initial temperatures are the same. The only difference between the various arrangements, then, is in the coefficients of linear expansion. The ball and the hole are both expanding. However, the hole is expanding more than the ball, to the extent that the coefficient of linear expansion of the material of the plate exceeds that of the ball. Thus, we need to examine the difference between the two coefficients, in order to decide the order in which the balls fall through the holes as the temperature increases. Referring to Table 12.1, we find the following differences for each of the arrangements:

Arrangement I $\alpha_{\text{Lead}} - \alpha_{\text{Gold}} = 29 \times 10^{-6}\,(\text{C}^\circ)^{-1} - 14 \times 10^{-6}\,(\text{C}^\circ)^{-1} = 15 \times 10^{-6}\,(\text{C}^\circ)^{-1}$

Arrangement II $\alpha_{\text{Aluminum}} - \alpha_{\text{Steel}} = 23 \times 10^{-6}\,(\text{C}^\circ)^{-1} - 12 \times 10^{-6}\,(\text{C}^\circ)^{-1} = 11 \times 10^{-6}\,(\text{C}^\circ)^{-1}$

Arrangement III $\alpha_{\text{Silver}} - \alpha_{\text{Quartz}} = 19 \times 10^{-6}\,(\text{C}^\circ)^{-1} - 0.50 \times 10^{-6}\,(\text{C}^\circ)^{-1} = 18.5 \times 10^{-6}\,(\text{C}^\circ)^{-1}$

In Arrangement III the coefficient of linear expansion of the plate-material exceeds that of the ball-material by the greatest amount. Therefore, the quartz ball will fall through the hole first. Next in sequence is Arrangement I, so the gold ball will fall second. Last is Arrangement II, so the steel ball will be the last to fall.

SOLUTION According to Equation 12.2, the diameter increases by an amount $\Delta D = \alpha D_0 \Delta T$ when the temperature increases by an amount ΔT, where D_0 is the initial diameter and α is the coefficient of linear expansion. Thus, we can write the final diameter as $D = D_0 + \alpha D_0 \Delta T$. Since the diameters of the ball and the hole are the same when the ball falls through the hole, we have

$$\underbrace{D_{0,\text{ Ball}} + \alpha_{\text{Ball}} D_{0,\text{ Ball}} \Delta T}_{\text{Final diameter of ball}} = \underbrace{D_{0,\text{ Hole}} + \alpha_{\text{Hole}} D_{0,\text{ Hole}} \Delta T}_{\text{Final diameter of hole}}$$

Solving for the change in temperature, we obtain

$$\Delta T = \frac{D_{0,\text{ Ball}} - D_{0,\text{ Hole}}}{\alpha_{\text{Hole}} D_{0,\text{ Hole}} - \alpha_{\text{Ball}} D_{0,\text{ Ball}}}$$

Arrangement I

$$\Delta T = \frac{D_{0,\text{ Gold}} - D_{0,\text{ Lead}}}{\alpha_{\text{Lead}} D_{0,\text{ Lead}} - \alpha_{\text{Gold}} D_{0,\text{ Gold}}}$$

$$= \frac{1.0\times10^{-5}\text{ m}}{\left[29\times10^{-6}\left(\text{C}^\circ\right)^{-1}\right]\left(0.10\text{ m}\right) - \left[14\times10^{-6}\left(\text{C}^\circ\right)^{-1}\right]\left(0.10\text{ m} + 1.0\times10^{-5}\text{ m}\right)} = 6.7\text{ C}^\circ$$

Arrangement II

$$\Delta T = \frac{D_{0,\text{ Steel}} - D_{0,\text{ Aluminum}}}{\alpha_{\text{Aluminum}} D_{0,\text{ Aluminum}} - \alpha_{\text{Steel}} D_{0,\text{ Steel}}}$$

$$= \frac{1.0\times10^{-5}\text{ m}}{\left[23\times10^{-6}\left(\text{C}^\circ\right)^{-1}\right]\left(0.10\text{ m}\right) - \left[12\times10^{-6}\left(\text{C}^\circ\right)^{-1}\right]\left(0.10\text{ m} + 1.0\times10^{-5}\text{ m}\right)} = 9.1\text{ C}^\circ$$

Arrangement III

$$\Delta T = \frac{D_{0,\text{ Quartz}} - D_{0,\text{ Silver}}}{\alpha_{\text{Silver}} D_{0,\text{ Silver}} - \alpha_{\text{Quartz}} D_{0,\text{ Quartz}}}$$

$$= \frac{1.0\times10^{-5}\text{ m}}{\left[19\times10^{-6}\left(\text{C}^\circ\right)^{-1}\right]\left(0.10\text{ m}\right) - \left[0.50\times10^{-6}\left(\text{C}^\circ\right)^{-1}\right]\left(0.10\text{ m} + 1.0\times10^{-5}\text{ m}\right)} = 5.4\text{ C}^\circ$$

Since each arrangement has an initial temperature of 25.0 °C, the temperatures at which the balls fall through the holes are as follows:

Arrangement I $T = 25.0\text{ °C} + 6.7\text{ C°} = 31.7\text{ °C}$

Arrangement II $T = 25.0\text{ °C} + 9.1\text{ C°} = 34.1\text{ °C}$

Arrangement III $T = 25.0\text{ °C} + 5.4\text{ C°} = 30.4\text{ °C}$

These results are consistent with our answer to Concept Question (b).

---

102. ***CONCEPT QUESTIONS***

a. Two portions of the same liquid that have the same mass, but different initial temperatures, when mixed together, will yield an equilibrium mixture that has a temperature lying exactly midway between the two initial temperatures. That is, the final temperature is $T = \frac{1}{2}\left(T_{0\text{A}} + T_{0\text{B}}\right)$. This occurs only when the mixing occurs without any exchange of heat with the surroundings. Then, all of the heat lost from the warmer portion is gained by the cooler portion. Since each portion is identical except for temperature, the warmer portion cools down by the same number of degrees that the cooler portion warms up, yielding a mixture whose final equilibrium temperature is midway between the two initial temperatures.

b. There are two ways to apply the logic described in Concept Question (a). Portions A and B have the same mass m, so they will yield a combined mass of $2m$. Thus, we can imagine portions A and B mixed together to yield an equilibrium temperature of $\frac{1}{2}\left(T_{0\text{A}} + T_{0\text{B}}\right) = \frac{1}{2}(94.0\text{ °C} + 78.0\text{ °C}) = 86.0\text{ °C}$. This mixture has a mass $2m$, just like the mass of portion C, so when it is mixed with portion C, the final equilibrium temperature will be $\frac{1}{2}(86.0\text{ °C} + 34.0\text{ °C}) = 60.0\text{ °C}$. The other way to apply the logic from Concept Question (a) is to mix half of portion C with portion A and half with portion B. This will produce two mixtures with different temperatures and each having a mass $2m$, which can then be combined. The final equilibrium temperature of this combination is again 60.0 °C.

SOLUTION Since we are assuming negligible heat exchange with the surroundings, the principle of conservation of energy applies in the following form: **heat lost equals heat gained**. In reaching equilibrium the warmer portions lose heat and the cooler portions gain heat. In either case, the heat Q that must be supplied or removed to change the temperature of a substance of mass m by an amount ΔT is given by Equation 12.4 as $Q = cm\Delta T$, where c is the specific heat capacity. We will assume that the final temperature is between 94.0 °C and 78.0 °C. We could also assume that the final temperature is between 78.0 °C and 34.0 °C. With either assumption we would obtain the same answer for the final temperature, provided that we remember to express the change in temperature ΔT as the higher minus the lower temperature in applying the energy-conservation principle. We have, then,

$$\underbrace{c_{\text{C}} m_{\text{C}} \left(T_{\text{Final}} - 34.0\text{ °C}\right)}_{\text{Heat gained}} = \underbrace{c_{\text{A}} m_{\text{A}} \left(94.0\text{ °C} - T_{\text{Final}}\right) + c_{\text{B}} m_{\text{B}} \left(78.0\text{ °C} - T_{\text{Final}}\right)}_{\text{Heat lost}}$$

The specific heat capacities for each portion have the same value c, while $m_A = m_B = m$ and $m_C = 2m$. With these substitutions, we find

$$c2m\left(T_{\text{Final}} - 34.0\ ^\circ\text{C}\right) = cm\left(94.0\ ^\circ\text{C} - T_{\text{Final}}\right) + cm\left(78.0\ ^\circ\text{C} - T_{\text{Final}}\right)$$

$$2\left(T_{\text{Final}} - 34.0\ ^\circ\text{C}\right) = \left(94.0\ ^\circ\text{C} - T_{\text{Final}}\right) + \left(78.0\ ^\circ\text{C} - T_{\text{Final}}\right)$$

$$4T_{\text{Final}} = 94.0\ ^\circ\text{C} + 78.0\ ^\circ\text{C} + 2\left(34.0\ ^\circ\text{C}\right)$$

$$T_{\text{Final}} = \frac{94.0\ ^\circ\text{C} + 78.0\ ^\circ\text{C} + 2\left(34.0\ ^\circ\text{C}\right)}{4} = \boxed{60.0\ ^\circ\text{C}}$$

This answer agrees with our answer to Concept Question (b).

CHAPTER 13 | *THE TRANSFER OF HEAT*

1. SSM ***REASONING AND SOLUTION*** According to Equation 13.1, the heat per second lost is

$$\frac{Q}{t}=\frac{kA\,\Delta T}{L}=\frac{[0.040\text{ J/(s}\cdot\text{m}\cdot\text{C}^\circ)]\,(1.6\text{ m}^2)(25\text{ C}^\circ)}{2.0\times10^{-3}\text{ m}}=\boxed{8.0\times10^2\text{ J/s}}$$

---

2. ***REASONING AND SOLUTION***

a. The heat lost by the oven is

$$Q=\frac{(kA\Delta T)t}{L}=\frac{[0.045\text{ J/(s}\cdot\text{m}\cdot\text{C}^\circ)](1.6\text{ m}^2)(160\,^\circ\text{C}-50\,^\circ\text{C})(6.0\text{ h})\left(\frac{3600\text{ s}}{1\text{ h}}\right)}{0.020\text{ m}}$$

$$=\boxed{8.6\times10^6\text{ J}}$$

b. Now 1 J = 2.78×10^{-7} kWh, so Q = 2.4 kWh. At \$ 0.10 per kWh, the cost is $\boxed{\$\ 0.24}$.

---

3. ***REASONING AND SOLUTION*** Since 1 kWh of energy costs \$ 0.10, we know that $Q = 10.0\times10^3$ W·h of energy can be purchased with \$ 1.00. Using Equation 13.1, we find that the time required is

$$t=\frac{QL}{kA\,\Delta T}=\frac{\left(10.0\times10^3\text{ W}\cdot\text{h}\right)(0.10\text{ m})}{[1.1\text{ J/(s}\cdot\text{m}\cdot\text{C}^\circ)]\left(9.0\text{ m}^2\right)\left(20.0\text{ C}^\circ-12.8\text{ C}^\circ\right)}=\boxed{14\text{ h}}$$

---

4. ***REASONING*** Since heat Q is conducted from the blood capillaries to the skin, we can use the relation $Q=\frac{(kA\Delta T)t}{L}$ (Equation 13.1) to describe how the conduction process depends on the various factors. We can determine the temperature difference between the capillaries and the skin by solving this equation for ΔT and noting that the heat conducted per second is Q/t.

SOLUTION Solving Equation 13.1 for the temperature difference, and using the fact that $Q/t = 240$ J/s, yields

$$\Delta T = \frac{(Q/t)L}{kA} = \frac{(240\ \text{J/s})(2.0\times10^{-3}\,\text{m})}{[0.20\ \text{J/(s}\cdot\text{m}\cdot\text{C°)}](1.6\ \text{m}^2)} = \boxed{1.5\ \text{C°}}$$

We have taken the thermal conductivity of body fat from Table 13.1.

---

5. SSM ***REASONING*** The heat transferred in a time t is given by Equation 13.1, $Q = (kA\,\Delta T)t/L$. If the same amount of heat per second is conducted through the two plates, then $(Q/t)_{\text{al}} = (Q/t)_{\text{st}}$. Using Equation 13.1, this becomes

$$\frac{k_{\text{al}} A \Delta T}{L_{\text{al}}} = \frac{k_{\text{st}} A \Delta T}{L_{\text{st}}}$$

This expression can be solved for L_{st}.

SOLUTION Solving for L_{st} gives

$$L_{\text{st}} = \frac{k_{\text{st}}}{k_{\text{al}}} L_{\text{al}} = \frac{14\ \text{J/(s}\cdot\text{m}\cdot\text{C°)}}{240\ \text{J/(s}\cdot\text{m}\cdot\text{C°)}}(0.035\ \text{m}) = \boxed{2.0\times10^{-3}\ \text{m}}$$

---

6. ***REASONING AND SOLUTION*** The heat lost in each case is given by Q = (kAΔT)t/L. For the goose down jacket

$$Q_{\text{g}} = \frac{[0.025\ \text{J/(s}\cdot\text{m}\cdot\text{C°)}]A\Delta T t}{1.5\times10^{-2}\ \text{m}}$$

For the wool jacket

$$Q_{\text{w}} = \frac{[0.040\ \text{J/(s}\cdot\text{m}\cdot\text{C°)}]A\Delta T t}{5.0\times10^{-3}\ \text{m}}$$

Now

$$Q_{\text{w}}/Q_{\text{g}} = \boxed{4.8}$$

---

7. SSM WWW ***REASONING AND SOLUTION*** The conductance of an 0.080 mm thick sample of Styrofoam of cross-sectional area A is

$$\frac{k_{\text{s}}A}{L_{\text{s}}} = \frac{[0.010\ \text{J/(s}\cdot\text{m}\cdot\text{C°)}]\,A}{0.080\times10^{-3}\ \text{m}} = [125\ \text{J/(s}\cdot\text{m}^2\cdot\text{C°)}]\,A$$

The conductance of a 3.5 mm thick sample of air of cross-sectional area A is

$$\frac{k_a A}{L_a} = \frac{[0.0256\ \text{J}/(\text{s}\cdot\text{m}\cdot\text{C}^\circ)]\,A}{3.5\times 10^{-3}\ \text{m}} = [7.3\ \text{J}/(\text{s}\cdot\text{m}^2\cdot\text{C}^\circ)]\,A$$

Dividing the conductance of Styrofoam by the conductance of air for samples of the same cross-sectional area A, gives

$$\frac{[125\ \text{J}/(\text{s}\cdot\text{m}^2\cdot\text{C}^\circ)]A}{[7.3\ \text{J}/(\text{s}\cdot\text{m}^2\cdot\text{C}^\circ)]\,A} = 17$$

Therefore, the body can adjust the conductance of the tissues beneath the skin by **a factor of 17**.

---

8. ***REASONING*** To find the total heat conducted, we will apply Equation 13.1 to the steel portion and the iron portion of the rod. In so doing, we use the area of a square for the cross section of the steel. The area of the iron is the area of the circle minus the area of the square. The radius of the circle is one half the length of the diagonal of the square.

SOLUTION In preparation for applying Equation 13.1, we need the area of the steel and the area of the iron. For the steel, the area is simply $A_{\text{Steel}} = L^2$, where L is the length of a side of the square. For the iron, the area is $A_{\text{Iron}} = \pi R^2 - L^2$. To find the radius R, we use the Pythagorean theorem, which indicates that the length D of the diagonal is related to the length of the sides according to $D^2 = L^2 + L^2$. Therefore, the radius of the circle is $R = D/2 = \sqrt{2}L/2$. For the iron, then, the area is

$$A_{\text{Iron}} = \pi R^2 - L^2 = \pi\left(\frac{\sqrt{2}L}{2}\right)^2 - L^2 = \left(\frac{\pi}{2} - 1\right)L^2$$

Taking values for the thermal conductivities of steel and iron from Table 13.1 and applying Equation 13.1, we find

$$Q_{\text{Total}} = Q_{\text{Steel}} + Q_{\text{Iron}}$$

$$= \left[\frac{(kA\Delta T)t}{L}\right]_{\text{Steel}} + \left[\frac{(kA\Delta T)t}{L}\right]_{\text{Iron}} = \left[k_{\text{Steel}} L^2 + k_{\text{Iron}}\left(\frac{\pi}{2}-1\right)L^2\right]\frac{(\Delta T)t}{L}$$

$$= \left[\left(14 \frac{\text{J}}{\text{s}\cdot\text{m}\cdot\text{C}^\circ}\right)(0.010\text{ m})^2 + \left(79 \frac{\text{J}}{\text{s}\cdot\text{m}\cdot\text{C}^\circ}\right)\left(\frac{\pi}{2}-1\right)(0.010\text{ m})^2\right]$$

$$\times\frac{(78\text{ °C}-18\text{ °C})(120\text{ s})}{0.50\text{ m}} = \boxed{85\text{ J}}$$

---

9. ***REASONING AND SOLUTION*** The rate of heat transfer is the same for all three materials so

$$Q/t = k_p A\Delta T_p/L = k_b A\Delta T_b/L = k_w A\Delta T_w/L$$

Let T_i be the inside temperature, T_1 be the temperature at the plasterboard-brick interface, T_2 be the temperature at the brick-wood interface, and T_o be the outside temperature. Then

$$k_p T_i - k_p T_1 = k_b T_1 - k_b T_2 \tag{1}$$

and

$$k_b T_1 - k_b T_2 = k_w T_2 - k_w T_o \tag{2}$$

Solving (1) for T_2 gives

$$T_2 = (k_p + k_b)T_1/k_b - (k_p/k_b)T_i$$

a. Substituting this into (2) and solving for T_1 yields

$$T_1 = \frac{(k_p/k_b)(1+k_w/k_b)T_i + (k_w/k_b)T_o}{(1+k_w/k_b)(1+k_p/k_b) - 1} = \boxed{21\text{ °C}}$$

b. Using this value in (1) yields

$$\boxed{T_2 = 18\text{ °C}}$$

---

10. ***REASONING*** The heat lost per second due to conduction through the glass is given by Equation 13.1 as $Q/t = (kA\Delta T)/L$. In this expression, we have no information for the thermal conductivity k, the cross-sectional area A, or the length L. Nevertheless, we can apply the equation to the initial situation and again to the situation where the outside temperature has fallen. This will allow us to eliminate the unknown variables from the calculation.

SOLUTION Applying Equation 13.1 to the initial situation and to the situation after the outside temperature has fallen, we obtain

$$\left(\frac{Q}{t}\right)_{\text{Initial}}=\frac{kA\left(T_{\text{In}}-T_{\text{Out, initial}}\right)}{L}\quad\text{and}\quad\left(\frac{Q}{t}\right)_{\text{Colder}}=\frac{kA\left(T_{\text{In}}-T_{\text{Out, colder}}\right)}{L}$$

Dividing these two equations to eliminate the common variables gives

$$\frac{(Q/t)_{\text{Colder}}}{(Q/t)_{\text{Initial}}}=\frac{\dfrac{kA\left(T_{\text{In}}-T_{\text{Out, colder}}\right)}{L}}{\dfrac{kA\left(T_{\text{In}}-T_{\text{Out, initial}}\right)}{L}}=\frac{T_{\text{In}}-T_{\text{Out, colder}}}{T_{\text{In}}-T_{\text{Out, initial}}}$$

Remembering that twice as much heat is lost per second when the outside is colder, we find

$$\frac{2(Q/t)_{\text{Initial}}}{(Q/t)_{\text{Initial}}}=2=\frac{T_{\text{In}}-T_{\text{Out, colder}}}{T_{\text{In}}-T_{\text{Out, initial}}}$$

Solving for the colder outside temperature gives

$$T_{\text{Out, colder}}=2T_{\text{Out, initial}}-T_{\text{In}}=2(5.0\ ^\circ\text{C})-(25\ ^\circ\text{C})=\boxed{-15\ ^\circ\text{C}}$$

---

11. SSM ***REASONING AND SOLUTION***

a. If we ignore the loss of heat through the sides of the rod and assume that heat does not accumulate at any point, then, from energy conservation, the rate at which heat is conducted through the two-rod combination must be the same at all points. In particular, at the interface of the two rods, $Q_a/t=Q_c/t$. Let T represent the temperature at the interface. Using Equation 13.1, we have

$$\frac{k_a A(T_a-T)}{L}=\frac{k_c A(T-T_c)}{L}$$

Solving for T, we find

$$T=\frac{k_a T_a+k_c T_c}{k_a+k_c}=\frac{[240\ \text{J}/(\text{s}\cdot\text{m}\cdot\text{C}^\circ)](302\ ^\circ\text{C})+[390\ \text{J}/(\text{s}\cdot\text{m}\cdot\text{C}^\circ)](25\ ^\circ\text{C})}{240\ \text{J}/(\text{s}\cdot\text{m}\cdot\text{C}^\circ)+390\ \text{J}/(\text{s}\cdot\text{m}\cdot\text{C}^\circ)}=\boxed{130\ ^\circ\text{C}}$$

b. Now that the temperature of the interface is known, Equation 13.1 can be used to calculate the amount of heat Q that flows through either section of the unit in a time t. Since the rate of heat flow is the same at all points throughout the unit as discussed above, we need only calculate Q for one of the rods that make up the unit. Using the aluminum rod, and considering the heat flow from its free end to the interface, we have

$$Q=\frac{k_a A(T_a-T)}{L}t=\frac{[240\ \mathrm{J/(s\cdot m\cdot C^\circ)}](4.0\times10^{-4}\ \mathrm{m}^2)(302\ ^\circ\mathrm{C}-130\ ^\circ\mathrm{C})}{4.0\times10^{-2}\ \mathrm{m}}(2.0\ \mathrm{s})=\boxed{830\ \mathrm{J}}$$

c. According to Equation 13.1, the temperature T_d at a distance d from the hot end of the aluminum rod is

$$T_a-T_d=\frac{Qd}{k_a At}=\frac{(Q/t)d}{k_a A}$$

where, from the answer to part (b), we know that $Q/t=(830\ \mathrm{J})/(2.0\ \mathrm{s})=415\ \mathrm{J/s}$. Solving for T_d, we have

$$T_d=T_a-\frac{(Q/t)d}{k_a A}=302\ ^\circ\mathrm{C}-\frac{(415\ \mathrm{J/s})(1.5\times10^{-2}\ \mathrm{m})}{[240\ \mathrm{J/(s\cdot m\cdot C^\circ)}](4.0\times10^{-4}\ \mathrm{m}^2)}=\boxed{237\ ^\circ\mathrm{C}}$$

---

12. ***REASONING*** The heat Q required to melt ice at 0 °C into water at 0 °C is given by the relation $Q=mL_f$ (Equation 12.5), where m is the mass of the ice and L_f is the latent heat of fusion. We divide both sides of this equation by the time t and solve for the mass of ice per second (m/t) that melts:

$$\frac{m}{t}=\frac{\left(\frac{Q}{t}\right)}{L_f} \tag{1}$$

The heat needed to melt the ice is conducted through the copper bar, from the hot end to the cool end. The amount of heat conducted in a time t is given by $Q=\frac{(kA\Delta T)t}{L}$ (Equation 13.1), where k is the thermal conductivity of the bar, A and L are its cross-sectional area and length, and ΔT is the temperature difference between the ends. We will use these two relations to find the mass of ice per second that melts.

SOLUTION Solving Equation 13.1 for Q/t and substituting the result into Equation (1) gives

$$\frac{m}{t}=\frac{\frac{kA\Delta T}{L}}{L_f}=\frac{kA\Delta T}{LL_f}$$

The thermal conductivity of copper can be found in Table 13.1, and the latent heat of fusion for water can be found in Table 12.3. The temperature difference between the ends of the rod is $\Delta T = 100$ C°, since the hot end is in boiling water (100 °C) and the cool end is in ice (0 °C). Thus,

$$\frac{m}{t} = \frac{kA\Delta T}{LL_f} = \frac{\left[390 \text{ J/}(\text{s}\cdot\text{m}\cdot\text{C}°)\right]\left(4.0\times10^{-4}\text{m}^2\right)(100 \text{ C}°)}{(1.5 \text{ m})\left(33.5\times10^4 \text{ J/kg}\right)} = \boxed{3.1\times10^{-5} \text{ kg/s}}$$

---

13. ***REASONING*** The heat Q conducted along the bar is given by the relation $Q = \frac{(kA\Delta T)t}{L}$ (Equation 13.1). We can determine the temperature difference between the hot end of the bar and a point 0.15 m from that end by solving this equation for ΔT and noting that the heat conducted per second is Q/t and that $L = 0.15$ m.

SOLUTION Solving Equation 13.1 for the temperature difference, using the fact that $Q/t = 3.6$ J/s, and taking the thermal conductivity of brass from Table 13.1, yield

$$\Delta T = \frac{(Q/t)L}{kA} = \frac{(3.6 \text{ J/s})(0.15 \text{ m})}{\left[110 \text{ J/}(\text{s}\cdot\text{m}\cdot\text{C}°)\right]\left(2.6\times10^{-4} \text{ m}^2\right)} = 19 \text{ C}°$$

The temperature at a distance of 0.15 m from the hot end of the bar is

$$T = 306 \text{ °C} - 19 \text{ C}° = \boxed{287 \text{ °C}}$$

---

14. ***REASONING AND SOLUTION*** The heat which must be removed to form a volume V of ice is

$$Q = mL_f = \rho VL_f = \rho AhL_f$$

The heat is conducted through the ice to the air, so Q is $Q = kA(\Delta T)t/L$. Thus, we have

$$h = \frac{k\Delta T t}{\rho L_f L} = \frac{[2.2 \text{ J/}(\text{s}\cdot\text{m}\cdot\text{C}°)](15 \text{ C}°)\left(3.0\times10^2 \text{ s}\right)}{\left(917 \text{ kg/m}^3\right)\left(3.35\times10^5 \text{ J/kg}\right)(0.30 \text{ m})} = 1.1\times10^{-4} \text{ m} = \boxed{0.11 \text{ mm}}$$

---

15. SSM WWW ***REASONING*** The rate at which heat is conducted along either rod is given by Equation 13.1, $Q/t = (kA\ \Delta T)/L$. Since both rods conduct the same amount of heat per second, we have

$$\frac{k_s A_s\ \Delta T}{L_s} = \frac{k_i A_i\ \Delta T}{L_i} \tag{1}$$

Since the same temperature difference is maintained across both rods, we can algebraically cancel the ΔTs. Because both rods have the same mass, $m_s = m_i$; in terms of the densities of silver and iron, the statement about the equality of the masses becomes $\rho_s(L_s A_s) = \rho_i(L_i A_i)$, or

$$\frac{A_s}{A_i} = \frac{\rho_i L_i}{\rho_s L_s} \quad (2)$$

Equations (1) and (2) may be combined to find the ratio of the lengths of the rods. Once the ratio of the lengths is known, Equation (2) can be used to find the ratio of the cross-sectional areas of the rods. If we assume that the rods have circular cross sections, then each has an area of $A = \pi r^2$. Hence, the ratio of the cross-sectional areas can be used to find the ratio of the radii of the rods.

SOLUTION
a. Solving Equation (1) for the ratio of the lengths and substituting the right hand side of Equation (2) for the ratio of the areas, we have

$$\frac{L_s}{L_i} = \frac{k_s A_s}{k_i A_i} = \frac{k_s(\rho_i L_i)}{k_i(\rho_s L_s)} \quad \text{or} \quad \left(\frac{L_s}{L_i}\right)^2 = \frac{k_s \rho_i}{k_i \rho_s}$$

Solving for the ratio of the lengths, we have

$$\frac{L_s}{L_i} = \sqrt{\frac{k_s \rho_i}{k_i \rho_s}} = \sqrt{\frac{[420\ \text{J}/(\text{s}\cdot\text{m}\cdot\text{C}^\circ)](7860\ \text{kg}/\text{m}^3)}{[79\ \text{J}/(\text{s}\cdot\text{m}\cdot\text{C}^\circ)](10\ 500\ \text{kg}/\text{m}^3)}} = \boxed{2.0}$$

b. From Equation (2) we have

$$\frac{\pi r_s^2}{\pi r_i^2} = \frac{\rho_i L_i}{\rho_s L_s} \quad \text{or} \quad \left(\frac{r_s}{r_i}\right)^2 = \frac{\rho_i L_i}{\rho_s L_s}$$

Solving for the ratio of the radii, we have

$$\frac{r_s}{r_i} = \sqrt{\frac{\rho_i}{\rho_s}\left(\frac{L_i}{L_s}\right)} = \sqrt{\frac{7860\ \text{kg}/\text{m}^3}{10\ 500\ \text{kg}/\text{m}^3}\left(\frac{1}{2.0}\right)} = \boxed{0.61}$$

---

16. ***REASONING*** The radiant energy Q absorbed by the person's head is given by $Q = e\sigma T^4 A t$ (Equation 13.2), where e is the emissivity, σ is the Stefan-Boltzmann constant, T is the Kelvin temperature of the environment surrounding the person (T = 28 °C + 273 = 301 K), A is the area of the head that is absorbing the energy, and t is the time. The radiant energy absorbed per second is $Q/t = e\sigma T^4 A$.

SOLUTION

a. The radiant energy absorbed per second by the person's head when it is covered with hair ($e = 0.85$) is

$$\frac{Q}{t} = e\sigma T^4 A = (0.85)\left[5.67\times10^{-8}\ \text{J}/\left(\text{s}\cdot\text{m}^2\cdot\text{K}^4\right)\right](301\ \text{K})^4\left(160\times10^{-4}\ \text{m}^2\right) = \boxed{6.3\ \text{J/s}}$$

b. The radiant energy absorbed per second by a bald person's head ($e = 0.65$) is

$$\frac{Q}{t} = e\sigma T^4 A = (0.65)\left[5.67\times10^{-8}\ \text{J}/\left(\text{s}\cdot\text{m}^2\cdot\text{K}^4\right)\right](301\ \text{K})^4\left(160\times10^{-4}\ \text{m}^2\right) = \boxed{4.8\ \text{J/s}}$$

---

17. **SSM** **WWW** ***REASONING AND SOLUTION*** Solving the Stefan-Boltzmann law, Equation 13.2, for the time t, and using the fact that $Q_{\text{blackbody}} = Q_{\text{bulb}}$, we have

$$t_{\text{blackbody}} = \frac{Q_{\text{blackbody}}}{\sigma T^4 A} = \frac{Q_{\text{bulb}}}{\sigma T^4 A} = \frac{P_{\text{bulb}}\, t_{\text{bulb}}}{\sigma T^4 A}$$

where P_{bulb} is the power rating of the light bulb. Therefore,

$$t_{\text{blackbody}} = \frac{(100.0\ \text{J/s})\ (3600\ \text{s})}{\left[5.67\times10^{-8}\ \text{J}/(\text{s}\cdot\text{m}^2\cdot\text{K}^4)\right]\ (303\ \text{K})^4\ \left[(6\ \text{sides})(0.0100\ \text{m})^2/\text{side}\right]}$$

$$\times\left(\frac{1\ \text{h}}{3600\ \text{s}}\right)\left(\frac{1\ \text{da}}{24\ \text{h}}\right) = \boxed{14.5\ \text{da}}$$

---

18. ***REASONING AND SOLUTION*** We know from Equation 13.2 that

$$A = \frac{Q/t}{e\sigma T^4} = \frac{6.0\times10^1\ \text{W}}{(0.36)\left[5.67\times10^{-8}\ \text{J}/(\text{s}\cdot\text{m}^2\cdot\text{K}^4)\right](3273\ \text{K})^4} = \boxed{2.6\times10^{-5}\ \text{m}^2}$$

---

19. ***REASONING AND SOLUTION*** For a blackbody $P_b = \sigma T^4 A$, and for the object $P_0 = e\sigma T^4 A$. Division of the above yields $P_0/P_b = e$. Thus,

$$e = \frac{P_0}{P_b} = \frac{30\ \text{W}}{90\ \text{W}} = \boxed{0.3}$$

---

20. ***REASONING*** The radiant energy Q radiated by the sun is given by $Q = e\sigma T^4 At$ (Equation 13.2), where e is the emissivity, σ is the Stefan-Boltzmann constant, T is its temperature (in Kelvins), A is the surface area of the sun, and t is the time. The radiant energy emitted per second is $Q/t = e\sigma T^4 A$. Solving this equation for T gives the surface temperature of the sun.

SOLUTION The radiant power produced by the sun is $Q/t = 3.9 \times 10^{26}$ W. The surface area of a sphere of radius r is $A = 4\pi r^2$. Since the sun is a perfect blackbody, $e = 1$. Solving Equation 13.2 for the surface temperature of the sun gives

$$T = \sqrt[4]{\frac{Q/t}{e\sigma 4\pi r^2}} = \sqrt[4]{\frac{3.9\times 10^{26}\ \text{W}}{(1)\left[5.67\times 10^{-8}\ \text{J}/\left(\text{s}\cdot\text{m}^2\cdot\text{K}^4\right)\right]4\pi\left(6.96\times 10^8\ \text{m}\right)^2}} = \boxed{5800\ \text{K}}$$

---

21. SSM ***REASONING AND SOLUTION*** The power radiated per square meter by the car when it has reached a temperature T is given by the Stefan-Boltzmann law, Equation 13.2, $P_{\text{radiated}}/A = e\sigma T^4$, where $P_{\text{radiated}} = Q/t$. Solving for T we have

$$T = \left[\frac{(P_{\text{radiated}}/A)}{e\sigma}\right]^{1/4} = \left\{\frac{560\ \text{W}/\text{m}^2}{(1.00)\left[5.67\times 10^{-8}\ \text{J}/(\text{s}\cdot\text{m}^2\cdot\text{K}^4)\right]}\right\}^{1/4} = \boxed{320\ \text{K}}$$

---

22. ***REASONING AND SOLUTION***

a. The radiant power lost by the body is

$$P_L = e\sigma T^4 A = (0.80)[5.67\times 10^{-8}\ \text{J/(s·m}^2\text{·K}^4)](307\ \text{K})^4(1.5\ \text{m}^2) = 604\ \text{W}$$

The radiant power gained by the body from the room is

$$P_g = (0.80)[5.67\times 10^{-8}\ \text{J/(s·m}^2\text{·K}^4)](298\ \text{K})^4(1.5\ \text{m}^2) = 537\ \text{W}$$

The net loss of radiant power is $P = P_L - P_g = \boxed{67\ \text{W}}$

b. The net energy lost by the body is

$$Q = Pt = (67\ \text{W})(3600\ \text{s})\left(\frac{1\ \text{Calorie}}{4186\ \text{J}}\right) = \boxed{58\ \text{Calories}}$$

---

23. ***REASONING AND SOLUTION*** The heat Q conducted during a time t through a wall of thickness L and cross sectional area A is given by Equation 13.1:

$$Q = \frac{kA\,\Delta Tt}{L}$$

The radiant energy Q, emitted in a time t by a wall that has a Kelvin temperature T, surface area A, and emissivity e is given by Equation (13.2):

$$Q = e\sigma T^4 At$$

If the amount of radiant energy emitted per second per square meter at 0 °C is the same as the heat lost per second per square meter due to conduction, then

$$\left(\frac{Q}{tA}\right)_{\text{conduction}} = \left(\frac{Q}{tA}\right)_{\text{radiation}}$$

Making use of Equations 13.1 and 13.2, the equation above becomes

$$\frac{k\Delta T}{L} = e\sigma T^4$$

Solving for the emissivity e gives:

$$e = \frac{k\Delta T}{L\sigma T^4} = \frac{[1.1\ \text{J/(s}\cdot\text{m}\cdot\text{K)}](293.0\ \text{K} - 273.0\ \text{K})}{(0.10\ \text{m})[5.67\times10^{-8}\ \text{J/(s}\cdot\text{m}^2\cdot\text{K}^4)]\,(273.0\ \text{K})^4} = \boxed{0.70}$$

Remark on units: Notice that the units for the thermal conductivity were expressed as J/(s·m·K) even though they are given in Table 13.1 as J/(s·m·C°). The two units are equivalent since the "size" of a Celsius degree is the same as the "size" of a Kelvin; that is, 1 C° = 1 K. Kelvins were used, rather than Celsius degrees, to ensure consistency of units. However, Kelvins must be used in Equation 13.2 or any equation that is derived from it.

---

24. ***REASONING AND SOLUTION*** According to Equation 13.2, for the sphere we have $Q/t = e\sigma A_S T_S{}^4$, and for the cube $Q/t = e\sigma A_C T_C{}^4$. Equating and solving we get

$$T_C{}^4 = (A_S/A_C)T_S{}^4$$

Now

$$A_S/A_C = (4\pi R^2)/(6L^2)$$

The volume of the sphere and the cube are the same, $(4/3)\,\pi R^3 = L^3$, so $R = \left(\frac{3}{4\pi}\right)^{1/3} L$.

The ratio of the areas is $\frac{A_s}{A_c} = \frac{4\pi R^2}{6L^2} = \frac{4\pi}{6}\left(\frac{3}{4\pi}\right)^{2/3} = 0.806$. The temperature of the cube is, then

$$T_c = \left(\frac{A_s}{A_c}\right)^{1/4} T_s = (0.806)^{1/4}(773\text{ K}) = \boxed{732\text{ K}}$$

---

25. **SSM** ***REASONING*** The total radiant power emitted by an object that has a Kelvin temperature *T*, surface area *A*, and emissivity *e* can be found by rearranging Equation 13.2, the Stefan-Boltzmann law: $Q = e\sigma T^4 At$. The emitted power is $P = Q/t = e\sigma T^4 A$. Therefore, when the original cylinder is cut perpendicular to its axis into *N* smaller cylinders, the ratio of the power radiated by the pieces to that radiated by the original cylinder is

$$\frac{P_{\text{pieces}}}{P_{\text{original}}} = \frac{e\sigma T^4 A_2}{e\sigma T^4 A_1} \qquad (1)$$

where A_1 is the surface area of the original cylinder, and A_2 is the sum of the surface areas of all *N* smaller cylinders. The surface area of the original cylinder is the sum of the surface area of the ends and the surface area of the cylinder body; therefore, if *L* and *r* represent the length and cross-sectional radius of the original cylinder, with $L = 10r$,

$$A_1 = (\text{area of ends}) + (\text{area of cylinder body})$$
$$= 2(\pi r^2) + (2\pi r)L = 2(\pi r^2) + (2\pi r)(10r) = 22\pi r^2$$

When the original cylinder is cut perpendicular to its axis into *N* smaller cylinders, the total surface area A_2 is

$$A_2 = N2(\pi r^2) + (2\pi r)L = N2(\pi r^2) + (2\pi r)(10r) = (2N + 20)\pi r^2$$

Substituting the expressions for A_1 and A_2 into Equation (1), we obtain the following expression for the ratio of the power radiated by the *N* pieces to that radiated by the original cylinder

$$\frac{P_{\text{pieces}}}{P_{\text{original}}} = \frac{e\sigma T^4 A_2}{e\sigma T^4 A_1} = \frac{(2N+20)\pi r^2}{22\pi r^2} = \frac{N+10}{11}$$

SOLUTION Since the total radiant power emitted by the N pieces is twice that emitted by the original cylinder, $P_{\text{pieces}} / P_{\text{original}} = 2$, we have $(N + 10)/11 = 2$. Solving this expression for N gives $N = 12$. Therefore, $\boxed{\text{there are 12 smaller cylinders}}$.

---

26. ***REASONING*** The drawing shows a cross-sectional view of the small sphere inside the larger spherical asbestos shell. The small sphere produces a net radiant energy, because its temperature (800.0 °C) is greater than that of its environment (600.0 °C). This energy is then conducted through the thin asbestos shell (thickness = L). By setting the net radiant energy produced by the small sphere equal to the energy conducted through the asbestos shell, we will be able to obtain the temperature T_2 of the outer surface of the shell.

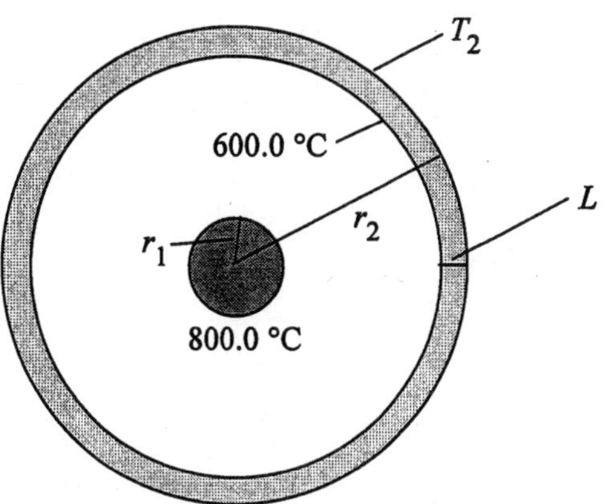

SOLUTION The heat Q conducted during a time through the thin asbestos shell is given by Equation 13.1 as $Q = \frac{\left(k_{\text{asbestos}} A_2 \,\Delta T\right)t}{L}$, where k_{asbestos} is the thermal conductivity of asbestos, A_2 is the area of the spherical shell$\left(A_2 = \pi r_2^2\right)$, ΔT is the temperature difference between the inner and outer surfaces of the shell (ΔT = 600.0 °C − T_2), and L is the thickness of the shell. Solving this equation for the T_2 yields

$$T_2 = \frac{QL}{k_{\text{asbestos}}\left(\pi r_2^2\right)t} + 600.0\ °\text{C}$$

The heat Q is produced by the net radiant energy generated by the small sphere inside the asbestos shell. According to Equation 13.3, the net radiant energy is $Q = P_{\text{net}}t = e\sigma A_1\left(T^4 - T_0^4\right)t$, where e is the emissivity, σ is the Stefan-Boltzmann constant, A_1 is the spherical area of the sphere $\left(A_1 = \pi r_1^2\right)$, T is the temperature of the sphere (T = 800.0 °C = 1073.2 K) and T_0 is the temperature of the environment that surrounds the sphere (T_0 = 600.0 °C = 873.2 K). Substituting this expression for Q into the expression above for T_2, and algebraically eliminating the time t and the factors of π, gives

$$T_2 = 600.0\ °\text{C} - \frac{\left[e\sigma\left(T^4 - T_0^4\right)\right]L}{k_{\text{asbestos}}\left(\dfrac{r_2}{r_1}\right)^2}$$

$$= 600.0\ °\text{C} - \frac{\left\{(0.90)\left(5.67\times10^{-8}\ \dfrac{\text{J}}{\text{s}\cdot\text{m}^2\cdot\text{K}^4}\right)\left[(1073.2\ \text{K})^4 - (873.2\ \text{K})^4\right]\right\}\left(1.00\times10^{-2}\,\text{m}\right)}{\left[0.090\ \dfrac{\text{J}}{\text{s}\cdot\text{m}\cdot\text{C°}}\right](10)^2}$$

$$= \boxed{557.7\ °\text{C}}$$

---

27. SSM ***REASONING*** The heat conducted through the iron poker is given by Equation 13.1, $Q = (kA\,\Delta T)t / L$. If we assume that the poker has a circular cross-section, then its cross-sectional area is $A = \pi r^2$. Table 13.1 gives the thermal conductivity of iron as $79\ \text{J}/(\text{s}\cdot\text{m}\cdot\text{C°})$.

SOLUTION The amount of heat conducted from one end of the poker to the other in 5.0 s is, therefore,

$$Q = \frac{(k\,A\,\Delta T)t}{L} = \frac{\left[79\ \text{J}/(\text{s}\cdot\text{m}\cdot\text{C°})\right]\pi\left(5.0\times10^{-3}\ \text{m}\right)^2(502\ °\text{C} - 26\ °\text{C})(5.0\ \text{s})}{1.2\ \text{m}} = \boxed{12\ \text{J}}$$

---

28. ***REASONING AND SOLUTION*** The rate at which energy is gained through the refrigerator walls is

$$\frac{Q}{t} = \frac{kA\,\Delta T}{L} = \frac{\left[0.030\ \text{J}/(\text{s}\cdot\text{m}\cdot\text{C°})\right]\left(5.3\ \text{m}^2\right)\left(2.0\times10^1\ \text{C°}\right)}{0.075\ \text{m}} = 42\ \text{J}/\text{s}$$

Therefore, the amount of heat per second that must be removed from the unit to keep it cool is $\boxed{42/\text{J}\ \ \text{s}}$.

---

29. ***REASONING*** According to the discussion in Section 13.3, the net power P_{net} radiated by the person is $P_{\text{net}} = e\sigma A\left(T^4 - T_0^4\right)$, where e is the emissivity, σ is the Stefan-Boltzmann constant, A is the surface area, and T and T_0 are the temperatures of the person and the environment, respectively. Since power is the change in energy per unit time (see

Equation 6.10b), the time t required for the person to emit the energy Q contained in the dessert is $t = Q/P_{net}$.

SOLUTION The time required to emit the energy from the dessert is

$$t = \frac{Q}{P_{net}} = \frac{Q}{e\sigma A\left(T^4 - T_0^4\right)}$$

The energy is $Q = (260 \text{ Calories})\left(\frac{4186 \text{ J}}{1 \text{ Calorie}}\right)$, and the Kelvin temperatures are $T = 36\ °\text{C} + 273 = 309\ \text{K}$ and $T_0 = 21\ °\text{C} + 273 = 294\ \text{K}$. The time is

$$t = \frac{(260 \text{ Calories})\left(\frac{4186 \text{ J}}{1 \text{ Calorie}}\right)}{(0.75)\left[5.67\times10^{-8} \text{ J/}\left(\text{s}\cdot\text{m}^2\cdot\text{K}^4\right)\right]\left(1.3 \text{ m}^2\right)\left[(309 \text{ K})^4 - (294 \text{ K})^4\right]} = \boxed{1.2\times10^4 \text{ s}}$$

---

30. ***REASONING*** The net rate at which energy is being lost via radiation can not exceed the production rate of 115 J/s, if the body temperature is to remain constant. The net rate at which an object at temperature T radiates energy in a room where the temperature is T_0 is given by Equation 13.3 as $P_{net} = e\sigma A(T^4 - T_0^4)$. P_{net} is the net energy per second radiated. We need only set P_{net} equal to 115 J/s and solve for T_0. We note that the temperatures in this equation must be expressed in Kelvins, not degrees Celsius.

SOLUTION According to Equation 13.3, we have

$$P_{net} = e\sigma A\left(T^4 - T_0^4\right) \quad \text{or} \quad T_0^4 = T^4 - \frac{P_{net}}{e\sigma A}$$

Using Equation 12.1 to convert from degrees Celsius to Kelvins, we have $T = 34 + 273 = 307$ K. Using this value, it follows that

$$T_0 = \sqrt[4]{T^4 - \frac{P_{net}}{e\sigma A}}$$

$$= \sqrt[4]{(307 \text{ K})^4 - \frac{115 \text{ J/s}}{0.700\left[5.67\times10^{-8} \text{ J/}\left(\text{s}\cdot\text{m}^2\cdot\text{K}^4\right)\right]\left(1.40 \text{ m}^2\right)}} = \boxed{287 \text{ K } (14\ °\text{C})}$$

---

31. SSM ***REASONING AND SOLUTION*** The net power generated by the stove is given by Equation 13.3, $P_{net} = e\sigma A(T^4 - T_0^4)$. Solving for T gives

$$T = \left(\frac{P_{net}}{e\sigma A} + T_0^4\right)^{1/4}$$

$$= \left\{\frac{7300 \text{ W}}{(0.900)[5.67\times10^{-8} \text{ J}/(\text{s}\cdot\text{m}^2\cdot\text{K}^4)](2.00 \text{ m}^2)} + (302 \text{ K})^4\right\}^{1/4} = \boxed{532 \text{ K}}$$

---

32. ***REASONING*** The heat Q necessary to vaporize a mass m of any substance at its boiling point is $Q = mL_V$ where L_V is the latent heat of vaporization. Therefore, the mass vaporized by an amount of heat Q is $m = Q/L_V$.

For the liquid helium system in question, it continually absorbs heat through radiation. The net power absorbed is given by Equation 13.3, $P_{net} = e\sigma A(T^4 - T_0^4)$ where T_0 is the temperature of the liquid helium, and T is the temperature maintained by the shield. Since the container is a perfect blackbody radiator, $e = 1$. Thus, the rate at which the mass of liquid helium boils away through the venting value is

$$\frac{m}{t} = \frac{(Q/t)}{L_v} = \frac{P_{net}}{L_v} = \frac{e\sigma A(T^4 - T_0^4)}{L_v}$$

This expression can be multiplied by the time t to determine the mass vaporized during that time.

SOLUTION The rate at which liquid helium mass boils away is

$$\frac{m}{t} = \frac{(1)[5.67\times10^{-8} \text{ J}/(\text{s}\cdot\text{m}^2\cdot\text{K}^4)]4\pi(0.30 \text{ m})^2[(77 \text{ K})^4 - (4.2 \text{ K})^4]}{2.1\times10^4 \text{ J/kg}} = 1.07\times10^{-4} \text{ kg/s}$$

The mass of liquid helium that boils away in one hour is, therefore,

$$\left(1.07\times10^{-4} \frac{\text{kg}}{\text{s}}\right)(1.0 \text{ h})\left(\frac{3600 \text{ s}}{1.0 \text{ h}}\right) = \boxed{0.39 \text{ kg}}$$

---

33. ***REASONING*** The heat Q required to change liquid water at 100.0 °C into steam at 100.0 °C is given by the relation $Q = mL_V$ (Equation 12.5), where m is the mass of the water and L_V is the latent heat of vaporization. The heat required to vaporize the water is conducted through the bottom of the pot and the stainless steel plate. The amount of heat

conducted in a time t is given by $Q=\dfrac{(kA\,\Delta T)t}{L}$ (Equation 13.1), where k is the thermal conductivity, A and L are the cross-sectional area and length, and ΔT is the temperature difference. We will use these two relations to find the temperatures at the aluminum-steel interface and at the steel surface in contact with the heating elèment.

SOLUTION

a. Substituting Equation 12.5 into Equation 13.1 and solving for ΔT, we have

$$\Delta T=\frac{QL}{kAt}=\frac{(mL_{\mathrm{v}})L}{kAt}$$

The thermal conductivity k_{Al} of aluminum can be found in Table 13.1, and the latent heat of vaporization for water can be found in Table 12.3. The temperature difference ΔT_{Al} between the aluminum surfaces is

$$\Delta T_{\mathrm{Al}}=\frac{(mL_{\mathrm{v}})L}{k_{\mathrm{Al}}At}=\frac{(0.15\text{ kg})\left(22.6\times10^{5}\text{ J/kg}\right)\left(3.1\times10^{-3}\text{m}\right)}{\left[240\text{ J/}(\text{s}\cdot\text{m}\cdot\text{C}^{\circ})\right]\left(0.015\text{ m}^{2}\right)(240\text{ s})}=1.2\text{ C}^{\circ}$$

The temperature at the aluminum-steel interface is $T_{\text{Al-Steel}}=100.0\text{ °C}+\Delta T_{\mathrm{Al}}=\boxed{101.2\text{ °C}}$.

b. Using the thermal conductivity k_{ss} of stainless steel from Table 13.1, we find that the temperature difference ΔT_{ss} between the stainless steel surfaces is

$$\Delta T_{\mathrm{ss}}=\frac{(mL_{\mathrm{v}})L}{k_{\mathrm{ss}}At}=\frac{(0.15\text{ kg})\left(22.6\times10^{5}\text{ J/kg}\right)\left(1.4\times10^{-3}\text{m}\right)}{\left[14\text{ J/}(\text{s}\cdot\text{m}\cdot\text{C}^{\circ})\right]\left(0.015\text{ m}^{2}\right)(240\text{ s})}=9.4\text{ C}^{\circ}$$

The temperature at the steel-burner interface is $T=101.2\text{ °C}+\Delta T_{\mathrm{ss}}=\boxed{110.6\text{ °C}}$.

---

34. ***REASONING AND SOLUTION*** Using Equation 13.1, $Q=(kA\,\Delta T)t/L$, we obtain

$$\left(\frac{Q}{At}\right)=\frac{k\Delta T}{L} \tag{1}$$

Before Equation (1) can be applied to the ice-aluminum combination, the temperature T at the interface must be determined. We find the temperature at the interface by noting that the heat conducted through the ice must be equal to the heat conducted through the aluminum: $Q_{\text{ice}}=Q_{\text{aluminum}}$. Applying Equation 13.1 to this condition, we have

$$\left(\frac{kA\Delta Tt}{L}\right)_{\text{ice}}=\left(\frac{kA\Delta Tt}{L}\right)_{\text{aluminum}} \tag{2}$$

or

$$\frac{[2.2\text{ J}/(\text{s}\cdot\text{m}\cdot\text{C}°)]A[(-10.0\text{ °C})-\text{T}]t}{0.0050\text{ m}}=\frac{[240\text{ J}/(\text{s}\cdot\text{m}\cdot\text{C}°)]A[T-(-25.0\text{ °C})]t}{0.0015\text{ m}}$$

The factors A and t can be eliminated algebraically. Solving for T gives $T = -24.959$ °C for the temperature at the interface.

a. Applying Equation (1) to the ice leads to

$$\left(\frac{Q}{At}\right)_{\text{ice}}=\frac{[2.2\text{ J}/(\text{s}\cdot\text{m}\cdot\text{C}°)][(-10.0\text{ °C})-(-24.959\text{ °C})]}{0.0050\text{ m}}=\boxed{6.58\times10^{3}\text{ J}/(\text{s}\cdot\text{m}^2)}$$

Since heat is not building up in the materials, the rate of heat transfer per unit area is the same throughout the ice-aluminum combination. Thus, this must be the heat per second per square meter that is conducted through the ice-aluminum combination.

b. Applying Equation (1) to the aluminum in the absence of any ice gives:

$$\left(\frac{Q}{At}\right)_{\text{Al}}=\frac{[240\text{ J}/(\text{s}\cdot\text{m}\cdot\text{C}°)][(-10.0\text{ °C})-(-25.0\text{ °C})]}{0.0015\text{ m}}=\boxed{2.40\times10^{6}\text{ J}/(\text{s}\cdot\text{m}^2)}$$

---

35. SSM WWW ***REASONING*** If the cylindrical rod were made of solid copper, the amount of heat it would conduct in a time t is, according to Equation 13.1, $Q_{\text{copper}} = (k_{\text{copper}}A_2\,\Delta T/L)t$. Similarly, the amount of heat conducted by the lead-copper combination is the sum of the heat conducted through the copper portion of the rod and the heat conducted through the lead portion:

$$Q_{\text{combination}}=\left[k_{\text{copper}}(A_2-A_1)\Delta T/L+k_{\text{lead}}A_1\Delta T/L\right]t.$$

Since the lead-copper combination conducts one-half the amount of heat than does the solid copper rod, $Q_{\text{combination}} = \frac{1}{2}Q_{\text{copper}}$, or

$$\frac{k_{\text{copper}}(A_2-A_1)\Delta T}{L}+\frac{k_{\text{lead}}A_1\Delta T}{L}=\frac{1}{2}\frac{k_{\text{copper}}A_2\,\Delta T}{L}$$

This expression can be solved for A_1 / A_2, the ratio of the cross-sectional areas. Since the cross-sectional area of a cylinder is circular, $A = \pi r^2$. Thus, once the ratio of the areas is known, the ratio of the radii can be determined.

SOLUTION Solving for the ratio of the areas, we have

$$\frac{A_1}{A_2} = \frac{k_{\text{copper}}}{2\left(k_{\text{copper}} - k_{\text{lead}}\right)}$$

The cross-sectional areas are circular so that $A_1 / A_2 = (\pi r_1^2)/(\pi r_2^2) = (r_1 / r_2)^2$; therefore,

$$\frac{r_1}{r_2} = \sqrt{\frac{k_{\text{copper}}}{2(k_{\text{copper}} - k_{\text{lead}})}} = \sqrt{\frac{390\ \text{J}/(\text{s}\cdot\text{m}\cdot\text{C°})}{2[390\ \text{J}/(\text{s}\cdot\text{m}\cdot\text{C°}) - 35\ \text{J}/(\text{s}\cdot\text{m}\cdot\text{C°})]}} = \boxed{0.74}$$

---

36. ***REASONING AND SOLUTION*** The heat conducted through the layered slab can be found from Equation 13.1:

$$Q = \frac{kA\Delta Tt}{L}$$

We must first find the temperature at the interface between materials 1 and 2. The temperature at the interface is found by noting that since there are no sources or sinks for heat in the slab, $Q_1 = Q_2$. Applying Equation 13.1 to this condition gives

$$\frac{k_1 A\Delta T_1 t}{L_1} = \frac{k_2 A\Delta T_2 t}{L_2} \tag{1}$$

If T represents the temperature at the interface between the two materials, then Equation (1) becomes

$$\frac{k_1}{L_1}(T_{\text{H}} - T) = \frac{k_2}{L_2}(T - T_{\text{C}})$$

Solving for T gives

$$T = \frac{\dfrac{k_1}{L_1}T_{\text{H}} + \dfrac{k_2}{L_2}T_{\text{C}}}{\dfrac{k_1}{L_1} + \dfrac{k_2}{L_2}} \tag{2}$$

Applying Equation 13.1 to material 1, we obtain

$$Q = \frac{k_1 A(T_{\text{H}} - T)t}{L_1} \tag{3}$$

Substituting the right hand side of Equation (2) into Equation (3) gives

$$Q = \frac{k_1 A t}{L_1}\left[T_H - \left(\frac{\frac{k_1}{L_1}T_H + \frac{k_2}{L_2}T_C}{\frac{k_1}{L_1} + \frac{k_2}{L_2}}\right)\right]$$

$$Q = At\left[\frac{k_1}{L_1}T_H - \left(\frac{\frac{k_1^2}{L_1^2}T_H + \frac{k_1 k_2}{L_1 L_2}T_C}{\frac{k_1}{L_1} + \frac{k_2}{L_2}}\right)\right]$$

$$Q = At\left[\frac{\frac{L_1 k_2 k_1}{L_1}T_H + \frac{L_2 k_1^2}{L_1}T_H - \frac{L_1 L_2 k_1^2}{L_1^2}T_H - k_1 k_2 T_C}{L_1 k_2 + L_2 k_1}\right]$$

$$Q = At\left[\frac{k_1 k_2 T_H - k_1 k_2 T_C}{L_1 k_2 + L_2 k_1}\right] = At(T_H - T_C)\left(\frac{k_1 k_2}{L_1 k_2 + L_2 k_1}\right)$$

$$Q = At(T_H - T_C)\left(\frac{L_1 k_2 + L_2 k_1}{k_1 k_2}\right)^{-1} = At(T_H - T_C)\left(\frac{L_1}{k_1} + \frac{L_2}{k_2}\right)^{-1}$$

Therefore,

$$\boxed{Q = \frac{A(T_H - T_C)t}{\left(\frac{L_1}{k_1} + \frac{L_2}{k_2}\right)}}$$

---

37. ***CONCEPT QUESTIONS*** a. The temperature is 100.0 °C, because water boils at 100.0 °C under one atmosphere of pressure. The temperature remains at 100.0 °C until all the water is gone.

b. When water boils, it changes from the liquid to the vapor phase. The heat needed to make the water change phase is $Q = mL_V$, according to Equation 12.5, where m is the mass and L_V is the latent heat of vaporization of water.

c. The temperature of the heating element must be greater than 100.0 °C. This is because heat flows via conduction from a higher to a lower temperature and the temperature of the boiling water is 100.0 °C.

SOLUTION Applying Equation 13.1 to the heat conduction and using Equation 12.5 to express the heat needed to boil away the water, we have

$$Q = \frac{\left(k_{\text{Copper}} A \Delta T\right) t}{L} = mL_{\text{v}}$$

The thermal conductivity of copper can be found in Table 13.1 $\left[k_{\text{copper}} = 390 \text{ J/}(\text{s} \cdot \text{m} \cdot \text{C}°)\right]$, and the latent heat of vaporization for water can be found in Table 12.4 ($L_{\text{v}} = 22.6 \times 10^5$ J/kg). The area A is the area of a circle or $A = \pi R^2$. Finally, the temperature difference is $\Delta T = T_{\text{E}} - 100.0$ °C. Using this expression for ΔT in the heat-conduction equation and solving for T_{E} gives

$$\frac{k_{\text{Copper}} A\left(T_{\text{E}} - 100.0\text{ °C}\right) t}{L} = mL_{\text{v}} \quad \text{or} \quad T_{\text{E}} = 100.0\text{ °C} + \frac{LmL_{\text{v}}}{k_{\text{Copper}} At}$$

$$T_{\text{E}} = 100.0\text{ °C} + \frac{\left(2.0 \times 10^{-3} \text{ m}\right)\left(0.45 \text{ kg}\right)\left(22.6 \times 10^5 \text{ J/kg}\right)}{\left[390 \text{ J/}(\text{s} \cdot \text{m} \cdot \text{C°})\right] \pi \left(0.065 \text{ m}\right)^2 \left(120 \text{ s}\right)} = \boxed{103.3 \text{ °C}}$$

---

38. ***CONCEPT QUESTIONS*** a. According to Equation 13.1 less heat is lost when the area through which the heat flows is smaller. Since the window has the smaller area, it would lose less heat than the wall, other things being equal.

b. According to Equation 13.1 more heat is lost when the thickness through which the heat flows is smaller. Since the window has the smaller thickness, it would lose more heat than the wall, other things being equal.

c. According to Equation 13.1 more heat is lost when the thermal conductivity of the material through which the heat flows is greater. According to Table 13.1 the thermal conductivity of glass is $k_{\text{G}} = 0.80 \text{ J/}(\text{s} \cdot \text{m} \cdot \text{C}°)$, while the value for Styrofoam is $k_{\text{S}} = 0.010 \text{ J/}(\text{s} \cdot \text{m} \cdot \text{C}°)$. Therefore, the window would lose more heat than the wall, other things being equal.

SOLUTION The percentage of the heat lost by the window is

$$\text{Percentage} = \left(\frac{Q_{\text{window}}}{Q_{\text{wall}} + Q_{\text{window}}}\right) \times 100$$

$$= \left[\frac{\dfrac{k_G A_G (\Delta T)t}{L_G}}{\dfrac{k_S A_S (\Delta T)t}{L_S} + \dfrac{k_G A_G (\Delta T)t}{L_G}}\right] \times 100 = \left(\frac{\dfrac{k_G A_G}{L_G}}{\dfrac{k_S A_S}{L_S} + \dfrac{k_G A_G}{L_G}}\right) \times 100$$

Here, we algebraically eliminated the time t and the temperature difference ΔT, since they are the same in each term. The percentage is

$$\text{Percentage} = \left(\frac{\dfrac{k_G A_G}{L_G}}{\dfrac{k_S A_S}{L_S} + \dfrac{k_G A_G}{L_G}}\right) \times 100$$

$$= \left\{\frac{\dfrac{[0.80\ \text{J/(s}\cdot\text{m}\cdot\text{C}^\circ)](0.16\ \text{m}^2)}{2.0\times10^{-3}\ \text{m}}}{\dfrac{[0.010\ \text{J/(s}\cdot\text{m}\cdot\text{C}^\circ)](18\ \text{m}^2)}{0.10\ \text{m}} + \dfrac{[0.80\ \text{J/(s}\cdot\text{m}\cdot\text{C}^\circ)](0.16\ \text{m}^2)}{2.0\times10^{-3}\ \text{m}}}\right\} \times 100$$

$$= \boxed{97\ \%}$$

---

39. ***CONCEPT QUESTIONS*** a. The cross-sectional area A through which the heat flows is greater for arrangement b; the cross-sectional area in b is twice that in a.

b. The thickness L of the material through which the heat flows is greater for arrangement a; the thickness in a is twice that in b.

c. For two reasons, Q_a is less than Q_b. First, the area in arrangement a is smaller, and the heat flows in direct proportion to the area. A smaller area means less heat. Second, the thickness in arrangement a is greater, and the heat flows in inverse proportion to the thickness. A greater thickness means less heat.

SOLUTION Applying Equation 13.1 for the conduction of heat to both arrangements gives

$$Q_a = \frac{kA_a(\Delta T)t}{L_a} \quad \text{and} \quad Q_b = \frac{kA_b(\Delta T)t}{L_b}$$

Note that the thermal conductivity k, the temperature difference ΔT, and the time t are the same in both arrangements. Dividing Q_a by Q_b gives

$$\frac{Q_a}{Q_b} = \frac{\dfrac{kA_a(\Delta T)t}{L_a}}{\dfrac{kA_b(\Delta T)t}{L_b}} = \frac{A_a L_b}{A_b L_a}$$

Remember that $A_b = 2A_a$ and that $L_a = 2L_b$. As expected then, we find that

$$\frac{Q_a}{Q_b} = \frac{A_a L_b}{A_b L_a} = \frac{A_a L_b}{(2A_a)(2L_b)} = \boxed{\frac{1}{4}}$$

---

40. ***CONCEPT QUESTIONS*** a. According to Equation 6.10b, power is the change in energy divided by the time during which the change occurs. In this case, then, the power is $P = Q/t$.

b. According to the Stefan-Boltzmann law (Equation 13.2), the power radiated is $Q/t = e\sigma T^4 A$. The power is proportional to the fourth power of the temperature T (in Kelvins). Thus, a higher temperature promotes more radiated power.

c. According to the Stefan-Boltzmann law (Equation 13.2), the power radiated is $Q/t = e\sigma T^4 A$. The power is proportional to the area A. Thus, a smaller area generates less radiated power.

d. The higher temperature of bulb #1 promotes a greater radiated power. The only way for both bulbs to radiate the same power, then, is for the filament area of bulb #1 to be smaller than that of bulb #2, in order to offset the effect of the higher temperature.

SOLUTION Using the Stefan-Boltzmann law (Equation 13.2) for both bulbs, we have

$$P_1 = Q_1/t_1 = e\sigma T_1^4 A_1 \quad \text{and} \quad P_2 = Q_2/t_2 = e\sigma T_2^4 A_2$$

Since $P_1 = P_2$, we see that

$$e\sigma T_1^4 A_1 = e\sigma T_2^4 A_2 \quad \text{or} \quad T_1^4 A_1 = T_2^4 A_2$$

As expected, solving for the ratio A_1/A_2 gives a value less than one:

$$\frac{A_1}{A_2} = \frac{T_2^4}{T_1^4} = \frac{(2100\text{ K})^4}{(2700\text{ K})^4} = \boxed{0.37}$$

---

41. ***CONCEPT QUESTIONS*** a. The power absorbed from the room is given by $Q/t = e\sigma T_0^{\ 4}A$. Except for the temperature T_0 of the room, this expression has the same form as that for the power radiated by the object. Note especially that the area A is the radiation area for the ***object, not the room***. Review part b of the solution to Example 6 in the text to understand this important point.

b. The power emitted by the object is proportional to the fourth power of the its temperature, and the power absorbed is proportional to the fourth power of the room's temperature. Since the object emits more power than it absorbs, its temperature T must be greater than the room's temperature T_0.

SOLUTION The object emits three times more power than it absorbs from the room, so it follow that $(Q/t)_{\text{emit}} = 3(Q/t)_{\text{absorb}}$. Using the Stefan-Boltzmann law for each of the powers, we find

$$e\sigma T^4 A = 3e\sigma T_0^4 A$$

Solving for T gives

$$T = \sqrt[4]{3}T_0 = \sqrt[4]{3}(293\text{ K}) = \boxed{386\text{ K}}$$

---

42. ***CONCEPT QUESTIONS*** a. According to the Stefan-Boltzmann law, the power radiated by an object is $Q/t = e\sigma T^4 A$, where A is the area from which the radiation is emitted. The power radiated is proportional to the fourth power of the temperature T. Therefore, other things being equal, the greater surface temperature of Sirius B would imply that its radiated power is greater than that of our sun.

b. The fact that Sirius B radiates less power than our sun, means that something is offsetting the effect of the greater surface temperature in the Stefan-Boltzmann law. This can only be the surface area A. The power radiated is proportional to A, according to the law. A smaller area means a smaller radiated power. Therefore, the surface area of Sirius B must be less than the surface area of our sun.

c. The surface area of a sphere is $4\pi R^2$, where R is the radius. Therefore, having less surface area, Sirius B must also have a radius that is less than the radius of our sun.

SOLUTION Writing the Stefan-Boltzmann law (Equation 13.2) for both stars, we have

$$Q_{\text{Sirius}}/t_{\text{Sirius}} = e\sigma T_{\text{Sirius}}^4 A_{\text{Sirius}} \quad \text{and} \quad Q_{\text{Sun}}/t_{\text{Sun}} = e\sigma T_{\text{Sun}}^4 A_{\text{Sun}}$$

Dividing the equation for Sirius B by the equation for our sun and remembering that $Q_{\text{Sirius}}/t_{\text{Sirius}} = 0.04\ Q_{\text{Sun}}/t_{\text{Sun}}$, we obtain

$$\frac{Q_{\text{Sirius}}/t_{\text{Sirius}}}{Q_{\text{Sun}}/t_{\text{Sun}}}=\frac{e\sigma T_{\text{Sirius}}^4 A_{\text{Sirius}}}{e\sigma T_{\text{Sun}}^4 A_{\text{Sun}}} \quad \text{or} \quad \frac{0.04Q_{\text{Sun}}/t_{\text{Sun}}}{Q_{\text{Sun}}/t_{\text{Sun}}}=\frac{T_{\text{Sirius}}^4 A_{\text{Sirius}}}{T_{\text{Sun}}^4 A_{\text{Sun}}}$$

Simplifying this result and using the fact that the surface area of a sphere is $4\pi R^2$ gives

$$0.04=\frac{T_{\text{Sirius}}^4 \pi R_{\text{Sirius}}^2}{T_{\text{Sun}}^4 \pi R_{\text{Sun}}^2}$$

Solving for the radius of Sirius B gives

$$R_{\text{Sirius}}=\sqrt{0.04}\left(\frac{T_{\text{Sun}}}{T_{\text{Sirius}}}\right)^2 R_{\text{Sun}}=\sqrt{0.04}\left(\frac{T_{\text{Sun}}}{4T_{\text{Sun}}}\right)^2\left(6.96\times10^8\text{ m}\right)=\boxed{9\times10^6\text{ m}}$$

As expected, the radius of Sirius B is less than that of our sun, so much so that it is called a white dwarf star.

---

43. ***CONCEPT QUESTIONS*** a. The heat Q conducted during a time t through a bar of length L and cross-sectional area A is $Q=\frac{(kA\Delta T)t}{L}$ (Equation 13.1). The heat depends on two geometrical factors, the cross-sectional area and the length. Even though the cross-sectional area for heat conduction through the block in C is greater than that in A, it does not necessarily mean that more heat is conducted in C, because the lengths of the conduction paths are different.

b. Even though the length of material through which heat is conducted in block A is greater than that in B, it does not necessarily follow that less heat is conducted in A, because the blocks have different cross-sectional areas.

c. The heat Q conducted during a time t through a bar of length L and cross-sectional area A is $Q=\frac{(kA\Delta T)t}{L}$ (Equation 13.1). The cross-sectional area and length of each block are: $A_{\text{A}}=2L_0^2$ and $L_{\text{A}}=3L_0$, $A_{\text{B}}=3L_0^2$ and $L_{\text{B}}=2L_0$, $A_{\text{C}}=6L_0^2$ and $L_{\text{C}}=L_0$. The heat conducted through each block is

$$Q_{\text{A}}=\left(\tfrac{2}{3}L_0\right)k\Delta Tt \qquad Q_{\text{B}}=\left(\tfrac{3}{2}L_0\right)k\Delta Tt \qquad Q_{\text{C}}=\left(6L_0\right)k\Delta Tt$$

Therefore, the ranking of the heat conduction is (highest to lowest): C, B, A

SOLUTION From the result of part c in the Concept Questions, the heat conducted in each case is:

Case A

$$Q_A = \left(\tfrac{2}{3}L_0\right)k\Delta Tt = \tfrac{2}{3}(0.30 \text{ m})\left[250 \text{ J/}(\text{s}\cdot\text{m}\cdot\text{C}^\circ)\right](35\ ^\circ\text{C} - 19\ ^\circ\text{C})(5.0 \text{ s}) = \boxed{4.0\times10^3 \text{ J}}$$

Case B

$$Q_B = \left(\tfrac{3}{2}L_0\right)k\Delta Tt = \tfrac{3}{2}(0.30 \text{ m})\left[250 \text{ J/}(\text{s}\cdot\text{m}\cdot\text{C}^\circ)\right](35\ ^\circ\text{C} - 19\ ^\circ\text{C})(5.0 \text{ s}) = \boxed{9.0\times10^3 \text{ J}}$$

Case C

$$Q_C = \left(6L_0\right)k\Delta Tt = 6(0.30 \text{ m})\left[250 \text{ J/}(\text{s}\cdot\text{m}\cdot\text{C}^\circ)\right](35\ ^\circ\text{C} - 19\ ^\circ\text{C})(5.0 \text{ s}) = \boxed{3.6\times10^4 \text{ J}}$$

---

44. ***CONCEPT QUESTIONS*** a. The net radiant power emitted by the bar in part (*a*) of the drawing is zero. The reason is that the temperature of the bar is the same as that of the room, and this temperature does not change. Therefore, the bar emits the same power into the room as it absorbs from the room, so the net radiant power emitted by the bar is zero.

b. The two bars in part (*b*) of the drawing emit more power. According to Equation 13.2, the radiant power (or energy per unit time) emitted by an object is $Q/t = e\sigma T^4 A$, which is directly proportional to its surface area *A*. The two bars in part (*b*) have a greater total surface area than the single bar in part (*a*).

c. The two bars in part (*b*) of the drawing also absorb more power. From the results of Concept Question b, we know that the two bars emit more power because of their greater surface area. However, since their temperature does not change, the two bars must absorb as much power as they emit. Thus, they absorb more power from the room than the single bar in part (*a*).

SOLUTION

a. The power (or energy per unit time) absorbed by the two bars in part (*b*) of the drawing is given by $Q/t = e\sigma T^4 A_2$, where A_2 is the total surface area of the two bars: $A_2 = 28L_0^2$. The power absorbed by the single bar in A is $Q/t = e\sigma T^4 A_1$, where A_1 is the total surface area of the single bar: $A_1 = 22L_0^2$.The ratio of the power P_2 absorbed by the two bars in part (*b*) to the power P_1absorbed by the single bar in part (*a*) is

$$\frac{P_2}{P_1} = \frac{e\sigma T^4\left(28\ L_0^2\right)}{e\sigma T^4\left(22\ L_0^2\right)} = \boxed{1.3}$$

b. If the power absorbed by the two bars in part (*b*) of the drawing is the same as that absorbed by the single bar in part (*a*), then

$$\underbrace{e\sigma T_1^4 A_1}_{\substack{\text{Power absorbed} \\ \text{by single bar} \\ \text{in part } (a)}} = \underbrace{e\sigma T_2^4 A_2}_{\substack{\text{Power absorbed} \\ \text{by two bars} \\ \text{in part } (b)}}$$

Solving for the temperature of the room and the bars in part (*b*) gives

$$T_2 = T_1 \sqrt[4]{\frac{A_1}{A_2}} = (450\ \text{K}) \sqrt[4]{\frac{22L_0^2}{28L_0^2}} = \boxed{420\ \text{K}}$$

CHAPTER 14 | *THE IDEAL GAS LAW AND KINETIC THEORY*

PROBLEMS

---

1. SSM ***REASONING AND SOLUTION*** The number n of moles contained in a sample is equal to the number N of atoms in the sample divided by the number N_A of atoms per mole (Avogadro's number):

$$n = \frac{N}{N_A} = \frac{30.1\times10^{23}}{6.022\times10^{23}\ \text{mol}^{-1}} = 5.00\ \text{mol}$$

Since the sample has a mass of 135 g, the mass per mole is

$$\frac{135\ \text{g}}{5.00\ \text{mol}} = 27.0\ \text{g/mol}$$

The mass per mole (in g/mol) of a substance has the same numerical value as the atomic mass of the substance. Therefore, the atomic mass is 27.0 u. The periodic table of the elements reveals that the unknown element is $\boxed{\text{aluminum}}$.

---

2. ***REASONING*** The number of molecules in a known mass of material is the number n of moles of the material times the number N_A of molecules per mole (Avogadro's number). We can find the number of moles by dividing the known mass m by the mass per mole.

SOLUTION Using the periodic table on the inside of the text's back cover, we find that the molecular mass of Tylenol ($C_8H_9NO_2$) is

$$\begin{array}{c}\text{Molecular mass}\\ \text{of Tylenol}\end{array} = \underbrace{8\ (12.011\ \text{u})}_{\substack{\text{Mass of 8}\\ \text{carbon atoms}}} + \underbrace{9\ (1.00794\ \text{u})}_{\substack{\text{Mass of 9}\\ \text{hydrogen atoms}}}$$

$$+\underbrace{14.0067\ \text{u}}_{\substack{\text{Mass of}\\ \text{nitrogen atom}}} + \underbrace{2\ (15.9994)}_{\substack{\text{Mass of 2}\\ \text{oxygen atoms}}} = 151.165\ \text{u}$$

The molecular mass of Advil ($C_{13}H_{18}O_2$) is

$$\begin{array}{c}\text{Molecular mass}\\ \text{of Advil}\end{array} = \underbrace{13\ (12.011\ \text{u})}_{\substack{\text{Mass of 13}\\ \text{carbon atoms}}} + \underbrace{18\ (1.00794\ \text{u})}_{\substack{\text{Mass of 18}\\ \text{hydrogen atoms}}} + \underbrace{2\ (15.9994)}_{\substack{\text{Mass of 2}\\ \text{oxygen atoms}}} = 206.285\ \text{u}$$

a. Therefore, the number of molecules of pain reliever in the standard dose of Tylenol is

$$\text{Number of molecules} = n\,N_\text{A} = \left(\frac{m}{\text{Mass per mole}}\right)N_\text{A}$$

$$= \left(\frac{325\ \times 10^{-3}\ \text{g}}{151.165\ \text{g/mol}}\right)\left(6.022\times10^{23}\ \text{mol}^{-1}\right) = \boxed{1.29\times10^{21}}$$

b. Similarly, the number of molecules of pain reliever in the standard dose of Advil is

$$\text{Number of molecules} = n\,N_\text{A} = \left(\frac{m}{\text{Mass per mole}}\right)N_\text{A}$$

$$= \left(\frac{2.00\ \times 10^{-1}\ \text{g}}{206.285\ \text{g/mol}}\right)\left(6.022\times10^{23}\ \text{mol}^{-1}\right) = \boxed{5.84\times10^{20}}$$

---

3. ***REASONING AND SOLUTION***

a. The molecular mass of a molecule is the sum of the atomic masses of its atoms. Thus, the molecular mass of aspartame ($C_{14}H_{18}N_2O_5$) is (see the periodic table on the inside of the text's back cover)

$$\text{Molecular mass} = \underbrace{14\ (12.011\ \text{u})}_{\substack{\text{Mass of 14}\\ \text{carbon atoms}}} + \underbrace{18\ (1.00794\ \text{u})}_{\substack{\text{Mass of 18}\\ \text{hydrogen atoms}}}$$

$$+\underbrace{2\ (14.0067\ \text{u})}_{\substack{\text{Mass of 2}\\ \text{nitrogen atoms}}} + \underbrace{5\ (15.9994)}_{\substack{\text{Mass of 5}\\ \text{oxygen atoms}}} = \boxed{294.307\ \text{u}}$$

b. The mass per mole of aspartame is 294.307 g/mol. The number of aspartame molecules per mole is Avogadro's number, or 6.022×10^{23} mol^{-1}. Therefore, the mass of one aspartame molecule (in kg) is

$$\left(\frac{294.307\ \text{g/mol}}{6.022\times10^{23}\ \text{mol}^{-1}}\right)\left(\frac{1\ \text{kg}}{1000\ \text{g}}\right) = \boxed{4.887\times10^{-25}\ \text{kg}}$$

---

4. ***REASONING*** The mass (in grams) of the active ingredient in the standard dosage is the number of molecules in the dosage times the mass per molecule (in grams per molecule). The mass per molecule can be obtained by dividing the molecular mass (in grams per mole) by Avogadro's number. The molecular mass is the sum of the atomic masses of the molecule's atomic constituents.

SOLUTION Using N to denote the number of molecules in the standard dosage and m_{molecule} to denote the mass of one molecule, the mass (in grams) of the active ingredient in the standard dosage can be written as follows:

$$m = Nm_{\text{molecule}}$$

Using M to denote the molecular mass (in grams per mole) and recognizing that $m_{\text{molecule}} = \frac{M}{N_A}$, where N_A is Avogadro's number and is the number of molecules per mole, we have

$$m = Nm_{\text{molecule}} = N\left(\frac{M}{N_A}\right)$$

M (in grams per mole) is equal to the molecular mass in atomic mass units, and we can obtain this quantity by referring to the periodic table on the inside of the back cover of the text to find the molecular masses of the constituent atoms in the active ingredient. Thus, we have

$$\text{Molecular mass} = \underbrace{22(12.011\text{ u})}_{\text{Carbon}} + \underbrace{23(1.00794\text{ u})}_{\text{Hydrogen}} + \underbrace{1(35.453\text{ u})}_{\text{Chlorine}} + \underbrace{2(14.0067\text{ u})}_{\text{Nitrogen}} + \underbrace{2(15.9994\text{ u})}_{\text{Oxygen}}$$

$$= 382.89\text{ u}$$

The mass of the active ingredient in the standard dosage is

$$m = N\left(\frac{M}{N_A}\right) = \left(1.572\times10^{19}\text{ molecules}\right)\left(\frac{382.89\text{ g/mol}}{6.022\times10^{23}\text{ molecules/mol}}\right) = \boxed{1.00\times10^{-2}\text{ g}}$$

---

5. **SSM** ***REASONING AND SOLUTION*** Since hemoglobin has a molecular mass of 64 500 u, the mass per mole of hemoglobin is 64 500 g/mol. The number of hemoglobin molecules per mol is Avogadro's number, or 6.022×10^{23} mol^{-1}. Therefore, one molecule of hemoglobin has a mass (in kg) of

$$\left(\frac{64\ 500\text{ g/mol}}{6.022\times10^{23}\text{ mol}^{-1}}\right)\left(\frac{1\text{ kg}}{1000\text{ g}}\right) = \boxed{1.07\times10^{-22}\text{ kg}}$$

---

6. ***REASONING*** The number n of moles of water molecules in the glass is equal to the mass m of water divided by the mass per mole. According to Equation 11.1, the mass of water is equal to its density ρ times its volume V. Thus, we have

$$n = \frac{m}{\text{Mass per mole}} = \frac{\rho V}{\text{Mass per mole}}$$

The volume of the cylindrical glass is $V = \pi r^2 h$, where r is the radius of the cylinder and h is its height. The number of moles of water can be written as

$$n = \frac{\rho V}{\text{Mass per mole}} = \frac{\rho\left(\pi r^2 h\right)}{\text{Mass per mole}}$$

SOLUTION The molecular mass of water (H_2O) is 2(1.00794 u) + (15.9994 u) = 18.0 u. The mass per mole of H_2O is 18.0 g/mol. The density of water (see Table 11.1) is 1.00 × 10^3 kg/m^3 or 1.00 g/cm^3.

$$n = \frac{\rho\left(\pi r^2 h\right)}{\text{Mass per mole}} = \frac{\left(1.00\text{ g/cm}^3\right)(\pi)(4.50\text{ cm})^2(12.0\text{ cm})}{18.0\text{ g/mol}} = \boxed{42.4\text{ mol}}$$

---

7. SSM ***REASONING AND SOLUTION*** Since the density of aluminum is 2700 kg/m^3, the number of atoms of aluminum per cubic meter is, using the given data,

$$N/V = \left(\frac{2700 \times 10^3\text{ g/m}^3}{26.9815\text{ g/mol}}\right)\left(6.022 \times 10^{23}\text{ mol}^{-1}\right) = 6.0 \times 10^{28}\text{ atoms/m}^3$$

Assuming that the volume of the solid contains many small cubes, with one atom at the center of each, then there are $(6.0 \times 10^{28})^{1/3}$ atoms/m = 3.9×10^9 atoms/m along each edge of a 1.00-m^3 cube. Therefore, the spacing between the centers of neighboring atoms is

$$d = \frac{1}{3.9 \times 10^9\text{ /m}} = \boxed{2.6 \times 10^{-10}\text{ m}}$$

---

8. ***REASONING AND SOLUTION*** The molecular mass of C_2H_5OH is

$$2(12.011\text{ u}) + 6(1.00794\text{ u}) + 15.9994\text{ u} = 46.069\text{ u}$$

a. (46.069 g/mol)/(6.022 × 10^{23}/mol) = 7.65 × 10^{-23} g = $\boxed{7.65 \times 10^{-26}\text{ kg}}$

b. 2.00 × 10^{-3} m^3 of this substance (density = 806 kg/m^3) has a mass m = ρV = 1.61 kg. The number of molecules is, therefore, equal to

$$N = (1.61 \text{ kg})/(7.65 \times 10^{-26} \text{ kg}) = \boxed{2.11 \times 10^{25}}$$

---

9. ***REASONING*** We can use the ideal gas law, Equation 14.1 ($PV = nRT$) to find the number of moles of helium in the Goodyear blimp, since the pressure, volume, and temperature are known. Once the number of moles are known, we can find the mass of helium in the blimp.

SOLUTION The number n of moles of helium in the blimp is, according to Equation 14.1,

$$n = \frac{PV}{RT} = \frac{(1.1\times10^5 \text{ Pa})(5400 \text{ m}^3)}{[8.31 \text{ J/(mol}\cdot\text{K)}](280 \text{ K})} = 2.55\times10^5 \text{ mol}$$

According to the periodic table on the inside of the text's back cover, the atomic mass of helium is 4.002 60 u. Therefore, the mass per mole is 4.002 60 g/mol. The mass m of helium in the blimp is, then,

$$m = \left(2.55\times10^5 \text{ mol}\right)\left(4.002\,60 \text{ g/mol}\right)\left(\frac{1 \text{ kg}}{1000 \text{ g}}\right) = \boxed{1.0\times10^3 \text{ kg}}$$

---

10. ***REASONING AND SOLUTION*** To find the temperature T_2, use the ideal gas law with n and V constant. Thus, $P_1/T_1 = P_2/T_2$. Then,

$$T_2 = T_1(P_2/P_1) = (284 \text{ K})[(3.01 \times 10^5 \text{ Pa})/(2.81 \times 10^5 \text{ Pa})] = \boxed{304 \text{ K}}$$

---

11. SSM ***REASONING AND SOLUTION*** The number of moles of air that is pumped into the tire is $\Delta n = n_f - n_i$. According to the ideal gas law (Equation 14.1), $n = PV/(RT)$; therefore,

$$\Delta n = n_f - n_i = \frac{P_f V_f}{RT_f} - \frac{P_i V_i}{RT_i} = \frac{V}{RT}(P_f - P_i)$$

Substitution of the data given in the problem leads to

$$\Delta n = \frac{4.1\times10^{-4} \text{ m}^3}{\left[8.31 \text{ J/(mol}\cdot\text{K)}\right](296 \text{ K})}\left[\left(6.2\times10^5 \text{ Pa}\right) - \left(4.8\times10^5 \text{ Pa}\right)\right] = \boxed{2.3\times10^{-2} \text{ mol}}$$

---

12. ***REASONING*** The maximum number of balloons that can be filled is the volume of helium available at the pressure in the balloons divided by the volume per balloon. The volume of helium available at the pressure in the balloons can be determined using the ideal gas law. Since the temperature remains constant, the ideal gas law indicates that $PV = nRT =$ constant,

and we can apply it in the form of Boyle's law, $P_iV_i = P_fV_f$. In this expression V_f is the final volume at the pressure in the balloons, V_i is the volume of the cylinder, P_i is the initial pressure in the cylinder, and P_f is the pressure in the balloons. However, we need to remember that a volume of helium equal to the volume of the cylinder will remain in the cylinder when its pressure is reduced to atmospheric pressure at the point when balloons can no longer be filled.

SOLUTION Using Boyle's law we find that

$$V_f = \frac{P_iV_i}{P_f}$$

The volume of helium available for filling balloons is

$$V_f - 0.0031\ \text{m}^3 = \frac{P_iV_i}{P_f} - 0.0031\ \text{m}^3$$

The maximum number of balloons that can be filled is

$$N_{\text{Balloons}} = \frac{\dfrac{P_iV_i}{P_f} - 0.0031\ \text{m}^3}{V_{\text{Balloon}}} = \frac{\dfrac{\left(1.6\times10^7\ \text{Pa}\right)\left(0.0031\ \text{m}^3\right)}{1.2\times10^5\ \text{Pa}} - 0.0031\ \text{m}^3}{0.034\ \text{m}^3} = \boxed{12}$$

---

13. ***REASONING*** Since the absolute pressure, volume, and temperature are known, we may use the ideal gas law in the form of Equation 14.1 to find the number of moles of gas. When the volume and temperature are raised, the new pressure can also be determined by using the ideal gas law.

SOLUTION
a. The number of moles of gas is

$$n = \frac{PV}{RT} = \frac{\left(1.72\times10^5\ \text{Pa}\right)\left(2.81\ \text{m}^3\right)}{\left[8.31\ \text{J}/(\text{mol}\cdot\text{K})\right]\left[(273.15+15.5)\ \text{K}\right]} = \boxed{201\ \text{mol}} \qquad (14.1)$$

b. When the volume is raised to 4.16 m^3 and the temperature raised to 28.2 °C, the pressure of the gas is

$$P = \frac{nRT}{V} = \frac{(201\ \text{mol})\left[8.31\ \text{J}/(\text{mol}\cdot\text{K})\right]\left[(273.15+28.2)\ \text{K}\right]}{\left(4.16\ \text{m}^3\right)} = \boxed{1.21\times10^5\ \text{Pa}} \qquad (14.1)$$

---

14. ***REASONING AND SOLUTION*** The ideal gas law gives

$$V_2 = (P_1/P_2)(T_2/T_1)\ V_1 = [(65.0\text{ atm})/(1.00\text{ atm})][(297\text{ K})/(288\text{ K})](1.00\text{ m}^3) = \boxed{67.0\text{ m}^3}$$

---

15. SSM ***REASONING AND SOLUTION*** According to the ideal gas law (Equation 14.1), the total number of moles n of fresh air in the sample is

$$n = \frac{PV}{RT} = \frac{(1.0\times10^5\text{ Pa})(5.0\times10^{-4}\text{ m}^3)}{[8.31\text{ J/(mole}\cdot\text{K)}]\ (310\text{ K})} = 1.94\times10^{-2}\text{ mol}$$

The total number of molecules in the sample is nN_A, where N_A is Avogadro's number. Since the sample contains approximately 21% oxygen, the total number of oxygen molecules in the sample is $(0.21)nN_A$ or

$$(0.21)(1.94\times10^{-2}\text{ mol})(6.022\times10^{23}\text{ mol}^{-1}) = \boxed{2.5\times10^{21}}$$

---

16. ***REASONING AND SOLUTION*** From the drawing in the text we can see that the volume of the cylinder is $V = Ah$, where A is the cross-sectional area of the piston. Assuming that pressure is constant, $V_1/T_1 = V_2/T_2$ or $A_1h_1/T_1 = A_2h_2/T_2$. Since $A_1 = A_2$, we have

$$h_2 = h_1(T_2/T_1) = (0.120\text{ m})[(318\text{ K})/(273\text{ K})] = \boxed{0.140\text{ m}}$$

---

17. ***REASONING*** According to the ideal gas law (Equation 14.1) $PV = nRT$. We need to put this in terms of the mass density, $\rho = m/V$, of the ideal gas. We can then set $P_{He} = P_{Ne}$, and solve the resulting expression for T_{Ne}, the temperature of the neon.

SOLUTION We begin by writing Equation 14.1 in terms of the mass density ρ for an ideal gas. Recall that the number n of moles of a substance is equal to its mass m in grams, divided by its mass per mole M. Therefore, $PV = mRT/M$, and we have $P = (m/V)RT/M = \rho RT/M$, where ρ is the same for each gas. Since the two gases have the same absolute pressures, $P_{He} = P_{Ne}$, or

$$\frac{\rho RT_{He}}{M_{He}} = \frac{\rho RT_{Ne}}{M_{Ne}}$$

The term ρR can be eliminated algebraically from this result. Solving for the temperature of the neon T_{Ne} and using the mass per mole for helium (4.0026 g/mol) and neon (20.179 g/mol), we find

$$T_{\text{Ne}} = T_{\text{He}}\left(\frac{M_{\text{Ne}}}{M_{\text{He}}}\right) = (175\text{ K})\left(\frac{20.179\text{ g/mol}}{4.0026\text{ g/mol}}\right) = \boxed{882\text{ K}}$$

---

18. ***REASONING*** According to Equation 14.2, $PV = NkT$, where P is the pressure, V is the volume, N is the number of molecules in the sample, k is Boltzmann's constant, and T is the Kelvin temperature. The number of gas molecules per unit volume in the atmosphere is $N/V = P/(kT)$. This can be used to find the desired ratio for the two planets.

SOLUTION We have

$$\frac{(N/V)_{\text{Venus}}}{(N/V)_{\text{Earth}}} = \frac{P_{\text{Venus}}/(kT_{\text{Venus}})}{P_{\text{Earth}}/(kT_{\text{Earth}})} = \left(\frac{P_{\text{Venus}}}{P_{\text{Earth}}}\right)\left(\frac{T_{\text{Earth}}}{T_{\text{Venus}}}\right) = \left(\frac{9.0\times10^{6}\text{ Pa}}{1.0\times10^{5}\text{ Pa}}\right)\left(\frac{320\text{ K}}{740\text{ K}}\right) = \boxed{39}$$

Thus, we can conclude that the atmosphere of Venus is 39 times "thicker" than that of Earth.

---

19. **SSM** ***REASONING*** The graph that accompanies Problem 68 in Chapter 12 can be used to determine the equilibrium vapor pressure of water in the air when the temperature is 30.0 °C (303 K). Equation 12.6 can then be used to find the partial pressure of water in the air at this temperature. Using this pressure, the ideal gas law can then be used to find the number of moles of water vapor per cubic meter.

SOLUTION According to the graph that accompanies Problem 68 in Chapter 12, the equilibrium vapor pressure of water vapor at 30.0 °C is approximately 4250 Pa. According to Equation 12.6,

$$\begin{array}{c}\text{Partial}\\ \text{pressure of}\\ \text{water vapor}\end{array} = \left(\begin{array}{c}\text{Percent relative}\\ \text{humidity}\end{array}\right)\left(\begin{array}{c}\text{Equilibrium vapor pressure of}\\ \text{water at the existing temperature}\end{array}\right)\times\frac{1}{100\%}$$

$$= \frac{(55\%)(4250\text{ Pa})}{100\%} = 2.34\times10^{3}\text{ Pa}$$

The ideal gas law then gives the number of moles of water vapor per cubic meter of air as

$$\frac{n}{V} = \frac{P}{RT} = \frac{(2.34\times10^{3}\text{ Pa})}{[8.31\text{ J/(mol}\cdot\text{K)}](303\text{ K})} = \boxed{0.93\text{ mol/m}^{3}} \qquad (14.1)$$

---

20. ***REASONING*** The desired percentage is the volume the atoms themselves occupy divided by the total volume that the gas occupies, multiplied by the usual factor of 100. The volume V_{Atoms} that the atoms themselves occupy is the volume of an atomic sphere ($\frac{4}{3}\pi r^3$, where r

is the atomic radius), times the number of atoms present, which is the number n of moles times Avogadro's number N_A. The total volume V_{Gas} that the gas occupies can be taken to be that calculated from the ideal gas law, because the atoms themselves occupy such a small volume.

SOLUTION The total volume V_{Gas} that the gas occupies is given by the ideal gas law as $V_{Gas} = nRT/P$, where the temperature and pressure at STP conditions are 273 K and 1.01×10^5 Pa. Thus, we can write the desired percentage as

$$\text{Percentage} = \frac{V_{\text{Atoms}}}{V_{\text{Gas}}} \times 100 = \frac{\left(\frac{4}{3}\pi r^3\right) n N_A}{\frac{nRT}{P}} \times 100 = \frac{\left(\frac{4}{3}\pi r^3\right) N_A P}{RT} \times 100$$

$$= \frac{\frac{4}{3}\pi\left(2.0\times10^{-10}\text{ m}\right)^3\left(6.022\times10^{23}\text{ mol}^{-1}\right)\left(1.01\times10^5\text{ Pa}\right)}{\left[8.31\text{ J/}\left(\text{mol}\cdot\text{K}\right)\right]\left(273\text{ K}\right)} \times 100 = \boxed{0.090\ \%}$$

---

21. ***REASONING*** The mass (in grams) of the air in the room is the mass of the nitrogen plus the mass of the oxygen. The mass of the nitrogen is the number of moles of nitrogen times the molecular mass (in grams/mol) of nitrogen. The mass of the oxygen can be obtained in a similar way. The number of moles of each species can be found using the given percentages and the total number of moles. To obtain the total number of moles, we apply the ideal gas law. If we substitute the Kelvin temperature T, the pressure P, and the volume V (length × width × height) of the room in the ideal gas law, we can obtain the total number n_{Total} of moles of gas as $n_{\text{Total}} = \frac{PV}{RT}$, because the ideal gas law does not distinguish between types of ideal gases.

SOLUTION Using m to denote the mass (in grams), n to denote the number of moles, and M to denote the molecular mass (in grams/mol), we can write the mass of the air in the room as follows:

$$m = m_{\text{Nitrogen}} + m_{\text{Oxygen}} = n_{\text{Nitrogen}} M_{\text{Nitrogen}} + n_{\text{Oxygen}} M_{\text{Oxygen}}$$

We can now express this result using f to denote the fraction of a species that is present and n_{Total} to denote the total number of moles:

$$m = n_{\text{Nitrogen}} M_{\text{Nitrogen}} + n_{\text{Oxygen}} M_{\text{Oxygen}} = f_{\text{Nitrogen}} n_{\text{Total}} M_{\text{Nitrogen}} + f_{\text{Oxygen}} n_{\text{Total}} M_{\text{Oxygen}}$$

According to the ideal gas law, we have $n_{\text{Total}} = \frac{PV}{RT}$. With this substitution, the mass of the air becomes

$$m = \left(f_{\text{Nitrogen}}M_{\text{Nitrogen}} + f_{\text{Oxygen}}M_{\text{Oxygen}}\right)n_{\text{Total}} = \left(f_{\text{Nitrogen}}M_{\text{Nitrogen}} + f_{\text{Oxygen}}M_{\text{Oxygen}}\right)\frac{PV}{RT}$$

The mass per mole for nitrogen (N_2) and for oxygen (O_2) can be obtained from the periodic table on the inside of the back cover of the text. They are, respectively, 28.0 and 32.0 g/mol. The temperature of 22 °C must be expressed on the Kelvin scale as 295 K (see Equation 12.1). The mass of the air is, then,

$$m = \left(f_{\text{Nitrogen}}M_{\text{Nitrogen}} + f_{\text{Oxygen}}M_{\text{Oxygen}}\right)\frac{PV}{RT}$$

$$= \left[0.79(28.0\text{ g/mol}) + 0.21(32.0\text{ g/mol})\right]\frac{\left(1.01\times10^5\text{ Pa}\right)\left[(2.5\text{ m})(4.0\text{ m})(5.0\text{ m})\right]}{\left[8.31\text{ J/}(\text{mol}\cdot\text{K})\right](295\text{ K})}$$

$$= \boxed{5.9\times10^4\text{ g}}$$

---

22. ***REASONING AND SOLUTION*** If the pressure at the surface is P_1 and the pressure at a depth h is P_2, we have that $P_2 = P_1 + \rho gh$. We also know that $P_1V_1 = P_2V_2$. Then,

$$\frac{V_1}{V_2} = \frac{P_2}{P_1} = \frac{P_1 + \rho gh}{P_1} = 1 + \frac{\rho gh}{P_1}$$

Therefore,

$$\frac{V_1}{V_2} = 1 + \frac{(1.000\times10^3\text{ kg/m}^3)(9.80\text{ m/s}^2)(0.200\text{ m})}{(1.01\times10^5\text{ Pa})} = \boxed{1.02}$$

---

23. SSM WWW ***REASONING*** Since the temperature of the confined air is constant, Boyle's law applies, and $P_{\text{surface}}V_{\text{surface}} = P_hV_h$, where P_{surface} and V_{surface} are the pressure and volume of the air in the tank when the tank is at the surface of the water, and P_h and V_h are the pressure and volume of the trapped air after the tank has been lowered a distance h below the surface of the water. Since the tank is completely filled with air at the surface, V_{surface} is equal to the volume V_{tank} of the tank. Therefore, the fraction of the tank's volume that is filled with air when the tank is a distance h below the water's surface is

$$\frac{V_h}{V_{\text{tank}}} = \frac{V_h}{V_{\text{surface}}} = \frac{P_{\text{surface}}}{P_h}$$

We can find the absolute pressure at a depth *h* using Equation 11.4. Once the absolute pressure is known at a depth *h*, we can determine the ratio of the pressure at the surface to the pressure at the depth *h*.

SOLUTION According to Equation 11.4, the trapped air pressure at a depth $h = 40.0$ m is

$$P_h = P_{\text{surface}} + \rho g h = (1.01\times10^5\ \text{Pa}) + \left[(1.00\times10^3\ \text{kg/m}^3)(9.80\ \text{m/s}^2)(40.0\ \text{m})\right]$$

$$= 4.93\times10^5\ \text{Pa}$$

where we have used a value of $\rho = 1.00\times10^3\ \text{kg/m}^3$ for the density of water. The desired volume fraction is

$$\frac{V_h}{V_{\text{tank}}} = \frac{P_{\text{surface}}}{P_h} = \frac{1.01\times10^5\ \text{Pa}}{4.93\times10^5\ \text{Pa}} = \boxed{0.205}$$

---

24. ***REASONING AND SOLUTION***

a. Since the heat gained by the gas in one tank is equal to the heat lost by the gas in the other tank, $Q_1 = Q_2$, or (letting the subscript 1 correspond to the neon in the left tank, and letting 2 correspond to the neon in the right tank) $cm_1\Delta T_1 = cm_2\Delta T_2$,

$$cm_1(T - T_1) = cm_2(T_2 - T)$$

$$m_1(T - T_1) = m_2(T_2 - T)$$

Solving for T gives

$$T = \frac{m_2T_2 + m_1T_1}{m_2 + m_1} \qquad (1)$$

The masses m_1 and m_2 can be found by first finding the number of moles n_1 and n_2. From the ideal gas law, $PV = nRT$, so

$$n_1 = \frac{P_1V_1}{RT_1} = \frac{(5.0\times10^5\ \text{Pa})(2.0\ \text{m}^3)}{[8.31\ \text{J/(mol}\cdot\text{K)}](220\ \text{K})} = 5.5\times10^2\ \text{mol}$$

This corresponds to a mass $m_1 = (5.5\times10^2\ \text{mol})(20.179\ \text{g/mol}) = 1.1\times10^4\ \text{g} = 1.1\times10^1\ \text{kg}$. Similarly, $n_2 = 2.4\times10^2$ mol and $m_2 = 4.9\times10^3$ g $=$ 4.9 kg. Substituting these mass values into Equation (1) yields

$$T = \frac{(4.9\ \text{kg})(580\ \text{K}) + (1.1\times10^1\ \text{kg})(220\ \text{K})}{(4.9\ \text{kg}) + (1.1\times10^1\ \text{kg})} = \boxed{3.3\times10^2\ \text{K}}$$

b. From the ideal gas law,

$$P=\frac{nRT}{V}=\frac{[(5.5\times 10^{2}\text{ mol})+(2.4\times 10^{2}\text{ mol})][8.31\text{ J/(mol}\cdot\text{K)}](3.3\times 10^{2}\text{ K})}{(2.0\text{ m}^{3}+5.8\text{ m}^{3})}=\boxed{2.8\times 10^{5}\text{Pa}}$$

---

25. ***REASONING AND SOLUTION*** We need to determine the amount of He inside the balloon. Begin by using Archimedes' principle; the balloon is being buoyed up by a force equal to the weight of the air displaced. The buoyant force, F_b, therefore, is equal to

$$F_b = mg = \rho Vg = (1.19\text{ kg/m}^3)(4/3)\pi(1.50\text{ m})^3(9.80\text{ m/s}^2) = 164.9\text{ N}$$

Since the balloon has a mass of 3.00 kg (weight = 29.4 N), the He inside the balloon weighs 164.9 N − 29.4 N = 135.5 N. Hence, the mass of the helium present in the balloon is m = 13.8 kg. Now we can determine the number of moles of He present in the balloon:

$$n = m/M = (13.8\text{ kg})/(4.0026\times 10^{-3}\text{ kg/mol}) = 3450\text{ mol}$$

Using the ideal gas law to find the pressure, we have

$$P = nRT/V = (3450\text{ mol})[8.31\text{ J/(mol}\cdot\text{K)}](305\text{ K})/[\,(4/3)\pi(1.50\text{ m})^3] = \boxed{6.19\times 10^{5}\text{ Pa}}$$

---

26. ***REASONING AND SOLUTION*** The volume of the cylinder is $V = AL$ where A is the cross- sectional area of the piston and L is the length. We know $P_1V_1 = P_2V_2$ so that the new pressure, P_2 can be found. We have

$$P_2 = P_1(V_1/V_2) = P_1A_1L_1/A_2L_2 = P_1L_1/L_2 \quad (\text{since } A_1 = A_2)$$

$$P_2 = (1.01\times 10^5\text{ Pa})(L/2L) = 5.05\times 10^4\text{ Pa}$$

The force on the piston and spring is, therefore,

$$F = P_2A = (5.05\times 10^4\text{ Pa})\pi(0.0500\text{ m})^2 = 397\text{ N}$$

The spring constant is $k = F/x$ so

$$k = F/x = (397\text{ N})/(0.200\text{ m}) = \boxed{1.98\times 10^3\text{ N/m}}$$

---

27. SSM ***REASONING*** According to the ideal gas law (Equation 14.1), $PV = nRT$. Since n, the number of moles of the gas, is constant, $n_1R = n_2R$. Therefore, $P_1V_1/T_1 = P_2V_2/T_2$, where T_1 = 273 K and T_2 is the temperature we seek. Since the beaker is cylindrical, the

volume V of the gas is equal to Ad, where A is the cross-sectional area of the cylindrical volume and d is the height of the region occupied by the gas, as measured from the bottom of the beaker. With this substitution for the volume, the expression obtained from the ideal gas law becomes

$$\frac{P_1 d_1}{T_1} = \frac{P_2 d_2}{T_2} \qquad (1)$$

where the pressures P_1 and P_2 are equal to the sum of the atmospheric pressure and the pressure caused by the mercury in each case. These pressures can be determined using Equation 11.4. Once the pressures are known, Equation (1) can be solved for T_2.

SOLUTION Using Equation 11.4, we obtain the following values for the pressures P_1 and P_2. Note that the initial height of the mercury is $h_1 = (1.520 \text{ m})/2 = 0.760 \text{ m}$, while the final height of the mercury is $h_2 = (1.520 \text{ m})/4 = 0.380 \text{ m}$.

$$P_1 = P_0 + \rho g h_1 = (1.01 \times 10^5 \text{ Pa}) + \left[(1.36 \times 10^4 \text{ kg/m}^3)(9.80 \text{ m/s}^2)(0.760 \text{ m})\right] = 2.02 \times 10^5 \text{ Pa}$$

$$P_2 = P_0 + \rho g h_2 = (1.01 \times 10^5 \text{ Pa}) + \left[(1.36 \times 10^4 \text{ kg/m}^3)(9.80 \text{ m/s}^2)(0.380 \text{ m})\right] = 1.52 \times 10^5 \text{ Pa}$$

In these pressure calculations, the density of mercury is $\rho = 1.36 \times 10^4 \text{ kg/m}^3$. In Equation (1) we note that $d_1 = 0.760 \text{ m}$ and $d_2 = 1.14 \text{ m}$. Solving Equation (1) for T_2 and substituting values, we obtain

$$T_2 = \left(\frac{P_2 d_2}{P_1 d_1}\right) T_1 = \left[\frac{(1.52 \times 10^5 \text{ Pa})(1.14 \text{ m})}{(2.02 \times 10^5 \text{ Pa})(0.760 \text{ m})}\right](273 \text{ K}) = \boxed{308 \text{ K}}$$

---

28. ***REASONING AND SOLUTION*** To find the rms-speed of the CO_2 we need to first find the temperature. We can do this since we know the rms-speed of the H_2O molecules. Using $(1/2)\,mv^2_{rms} = (3/2)kT$, we can solve for the temperature to get $T = mv^2_{rms}/(3k)$ where the mass of an H_2O molecule is $(18.015 \text{ g/mol})/(6.022 \times 10^{23} \text{ mol}^{-1}) = 2.99 \times 10^{-23}$ g.
Thus, $T = (2.99 \times 10^{-26} \text{ kg})(648 \text{ m/s})^2/[3(1.38 \times 10^{-23} \text{ J/K})] = 303$ K. The molecular mass of CO_2 is 44.01 u, hence the mass of a CO_2 molecule is 7.31×10^{-26} kg.

The rms-speed for CO_2 is

$$v_{rms} = \sqrt{\frac{3kT}{m}} = \sqrt{\frac{3(1.38 \times 10^{-23} \text{ J/K})(303 \text{ K})}{7.31 \times 10^{-26} \text{ kg}}} = \boxed{414 \text{ m/s}}$$

---

29. SSM ***REASONING AND SOLUTION*** Using the expressions for $\overline{v^2}$ and $(\bar{v})^2$ given in the statement of the problem, we obtain:

a. $$\overline{v^2} = \frac{1}{3}(v_1^2 + v_2^2 + v_3^2) = \frac{1}{3}\left[(3.0 \text{ m/s})^2 + (7.0 \text{ m/s})^2 + (9.0 \text{ m/s})^2\right] = \boxed{46.3 \text{ m}^2/\text{s}^2}$$

b. $$(\bar{v})^2 = \left[\frac{1}{3}(v_1 + v_2 + v_3)\right]^2 = \left[\frac{1}{3}(3.0 \text{ m/s} + 7.0 \text{ m/s} + 9.0 \text{ m/s})\right]^2 = \boxed{40.1 \text{ m}^2/\text{s}^2}$$

$\overline{v^2}$ and $(\bar{v})^2$ are *not* equal, because they are two different physical quantities.

---

30. ***REASONING*** The average kinetic energy per molecule is proportional to the Kelvin temperature of the carbon dioxide gas. This relation is expressed by Equation 14.6 as $\frac{1}{2}mv_{\text{rms}}^2 = \frac{3}{2}kT$, where m is the mass of a carbon dioxide molecule. The mass m is equal to the molecular mass of carbon dioxide (44.0 u), expressed in kilograms.

SOLUTION Solving Equation 14.6 for the temperature of the gas, we have

$$T = \frac{mv_{\text{rms}}^2}{3k} = \frac{(44.0 \text{ u})\left(\dfrac{1.66\times10^{-27} \text{ kg}}{1 \text{ u}}\right)(650 \text{ m/s})^2}{3\left(1.38\times10^{-23} \text{ J/K}\right)} = \boxed{750 \text{ K}}$$

---

31. ***REASONING AND SOLUTION*** Using $\overline{\text{KE}}$ = (1/2) $\text{mv}_{\text{rms}}{}^2$ = (3/2) kT, we can solve for v_{rms}.

$$\text{v}_{\text{rms}} = \sqrt{\frac{3\text{kT}}{\text{m}}} = \sqrt{\frac{3(1.38 \text{ x } 10^{-23} \text{ J/K})(6.0 \text{ x } 10^{3} \text{ K})}{1.67 \text{ x } 10^{-27} \text{ kg}}} = \boxed{1.2\times10^{4} \text{ m/s}} \qquad (14.6)$$

---

32. ***REASONING*** According to the kinetic theory of gases, the average kinetic energy of an atom is $\overline{\text{KE}} = \frac{3}{2}kT$ (Equation 14.6), where k is Boltzmann's constant and T is the Kelvin temperature. Therefore, the ratio of the average kinetic energies is equal to the ratio of the Kelvin temperatures of the gases. We are given no direct information about the temperatures. However, we do know that the temperature of an ideal gas is related to the pressure P, the volume V, and the number n of moles of the gas via the ideal gas law, $PV = nRT$. Thus, the ideal gas law can be solved for the temperature, and the ratio of the temperatures can be related to the other properties of the gas. In this way, we will obtain the desired kinetic-energy ratio.

SOLUTION Using Equation 14.6 and the ideal gas law in the form $T = \frac{PV}{nR}$, we find that the desired ratio is

$$\frac{\overline{\mathrm{KE}}_{\text{Krypton}}}{\overline{\mathrm{KE}}_{\text{Argon}}} = \frac{\frac{3}{2}kT_{\text{Krypton}}}{\frac{3}{2}kT_{\text{Argon}}} = \frac{\frac{3}{2}k\left(\dfrac{PV}{n_{\text{Krypton}}R}\right)}{\frac{3}{2}k\left(\dfrac{PV}{n_{\text{Argon}}R}\right)} = \frac{n_{\text{Argon}}}{n_{\text{Krypton}}}$$

Here, we have taken advantage of the fact that the pressure and volume of each gas are the same. While we are not given direct information about the number of moles of each gas, we do know that their masses are the same. Furthermore, the number of moles can be calculated from the mass m (in grams) and the mass per mole M (in grams per mole), according to $n = \frac{m}{M}$. Substituting this expression into our result for the kinetic-energy ratio gives

$$\frac{\overline{\mathrm{KE}}_{\text{Krypton}}}{\overline{\mathrm{KE}}_{\text{Argon}}} = \frac{n_{\text{Argon}}}{n_{\text{Krypton}}} = \frac{\dfrac{m}{M_{\text{Argon}}}}{\dfrac{m}{M_{\text{Krypton}}}} = \frac{M_{\text{Krypton}}}{M_{\text{Argon}}}$$

Taking the masses per mole from the periodic table on the inside of the back cover of the text, we find

$$\frac{\overline{\mathrm{KE}}_{\text{Krypton}}}{\overline{\mathrm{KE}}_{\text{Argon}}} = \frac{M_{\text{Krypton}}}{M_{\text{Argon}}} = \frac{83.80 \text{ g/mol}}{39.948 \text{ g/mol}} = \boxed{2.098}$$

---

33. SSM ***REASONING*** The internal energy of the neon at any Kelvin temperature T is given by Equation 14.7, $U = (3/2)nRT$. Therefore, when the temperature of the neon increases from an initial temperature T_i to a final temperature T_f, the internal energy of the neon increases by an amount

$$\Delta U = U_f - U_i = \frac{3}{2}nR\,(T_f - T_i)$$

In order to use this equation, we must first determine n. Since the neon is confined to a tank, the number of moles n is constant, and we can use the information given concerning the initial conditions to determine an expression for the quantity nR. According to the ideal gas law (Equation 14.1),

$$nR = \frac{P_i V_i}{T_i}$$

These two expressions can be combined to obtain an equation in terms of the variables that correspond to the data given in the problem statement.

SOLUTION Combining the two expressions and substituting the given values yields

$$\Delta U = \frac{3}{2}\left(\frac{P_i V_i}{T_i}\right)(T_f - T_i)$$

$$= \frac{3}{2}\left[\frac{(1.01\times 10^5 \text{ Pa})(680 \text{ m}^3)}{(293.2 \text{ K})}\right](294.3 \text{ K} - 293.2 \text{ K}) = \boxed{3.9\times 10^5 \text{ J}}$$

---

34. ***REASONING*** The behavior of the molecules is described by Equation 14.5: $PV = \frac{2}{3}N(\frac{1}{2}mv_{\text{rms}}^2)$. Since the pressure and volume of the gas are kept constant, while the number of molecules is doubled, we can write $P_1V_2 = P_2V_2$, where the subscript 1 refers to the initial condition, and the subscript 2 refers to the conditions after the number of molecules is doubled. Thus,

$$\tfrac{2}{3}N_1\left[\tfrac{1}{2}m(v_{\text{rms}})_1^2\right] = \tfrac{2}{3}N_2\left[\tfrac{1}{2}m(v_{\text{rms}})_2^2\right] \quad \text{or} \quad N_1(v_{\text{rms}})_1^2 = N_2(v_{\text{rms}})_2^2$$

The last expression can be solved for $(v_{\text{rms}})_2$, the final translational rms speed.

SOLUTION Since the number of molecules is doubled, $N_2 = 2N_1$. Solving the last expression above for $(v_{\text{rms}})_2$, we find

$$(v_{\text{rms}})_2 = (v_{\text{rms}})_1\sqrt{\frac{N_1}{N_2}} = (463 \text{ m/s})\sqrt{\frac{N_1}{2N_1}} = \frac{463 \text{ m/s}}{\sqrt{2}} = \boxed{327 \text{ m/s}}$$

---

35. ***REASONING*** Since the xenon atom does not interact with any other atoms or molecules on its way up, we can apply the principle of conservation of mechanical energy (see Section 6.5) and set the final kinetic plus potential energy equal to the initial kinetic plus potential energy. Thus, during the rise, the atom's initial kinetic energy is converted entirely into gravitational potential energy, because the atom comes to a momentary halt at the top of its trajectory. The initial kinetic energy $\frac{1}{2}mv_0^2$ is equal to the average translational kinetic energy. Therefore, $\frac{1}{2}mv_0^2 = \overline{\text{KE}} = \frac{3}{2}kT$, according to Equation 14.6, where k is Boltzmann's constant and T is the Kelvin temperature. The gravitational potential energy is *mgh*, according to Equation 6.5.

SOLUTION Equation 6.9b gives the principle of conservation of mechanical energy:

$$\underbrace{\tfrac{1}{2}mv_f^2 + mgh_f}_{\text{Final mechanical energy}} = \underbrace{\tfrac{1}{2}mv_0^2 + mgh_0}_{\text{Initial mechanical energy}}$$

In this expression, we know that $\frac{1}{2}mv_0^2 = \overline{\text{KE}} = \frac{3}{2}kT$ and that $\frac{1}{2}mv_f^2 = 0$ J (since the atom comes to a halt at the top of its trajectory). Furthermore, we can take the height at the earth's surface to be $h_0 = 0$ m. Taking this information into account, we can write the energy-conservation equation as follows:

$$mgh_f = \tfrac{3}{2}kT \quad \text{or} \quad h_f = \frac{3kT}{2mg}$$

Using M to denote the molecular mass (in kilograms per mole) and recognizing that $m = \frac{M}{N_A}$, where N_A is Avogadro's number and is the number of xenon atoms per mole, we have

$$h_f = \frac{3kT}{2mg} = \frac{3kT}{2\left(\frac{M}{N_A}\right)g} = \frac{3kN_AT}{2Mg}$$

Recognizing that $kN_A = R$ and that $M = 131.29$ g/mol $= 131.29 \times 10^{-3}$ kg/mol, we find

$$h_f = \frac{3kN_AT}{2Mg} = \frac{3RT}{2Mg} = \frac{3[8.31\text{ J/(mol}\cdot\text{K)}](291\text{ K})}{2(131.29\times10^{-3}\text{ kg/mol})(9.80\text{ m/s}^2)} = \boxed{2820\text{ m}}$$

---

36. ***REASONING AND SOLUTION*** We know from the ideal gas law that PV = nRT and U = (3/2)nRT = (3/2)PV = 9300 J. A 0.25-hp engine provides (0.25 hp)(746 W)/(1 hp) = 187 J/s. In order to equal the internal energy, therefore, the engine must run for a time of

$$(9300\text{ J})/(187\text{ J/s}) = \boxed{5.0\times10^1\text{ s}}$$

---

37. SSM WWW ***REASONING AND SOLUTION*** The average force exerted by one electron on the screen is, from the impulse-momentum theorem (Equation 7.4), $\overline{F} = \Delta p/\Delta t = m\Delta v/\Delta t$. Therefore, in a time Δt, N electrons exert an average force $\overline{F} = Nm\Delta v/\Delta t = (N/\Delta t)m\Delta v$. Since the pressure on the screen is the average force per unit area (Equation 10.19), we have

$$P = \frac{\overline{F}}{A} = \frac{(N/\Delta t)m\Delta v}{A}$$

$$= \frac{(6.2\times10^{16}\text{ electrons/s})(9.11\times10^{-31}\text{ kg})(8.4\times10^7\text{ m/s} - 0\text{ m/s})}{1.2\times10^{-7}\text{ m}^2} = \boxed{4.0\times10^1\text{ Pa}}$$

---

38. ***REASONING*** When perspiration absorbs heat from the body, the perspiration vaporizes. The amount Q of heat required to vaporize a mass $m_{\text{perspiration}}$ of perspiration is given by Equation 12.5 as $Q = m_{\text{perspiration}}L_{\text{v}}$, where L_{v} is the latent heat of vaporization for water at body temperature. The average energy $\overline{E}$ given to a single water molecule is equal to the heat Q divided by the number N of water molecules.

SOLUTION Since $\overline{E} = Q/N$ and $Q = m_{\text{perspiration}}L_{\text{v}}$, we have

$$\overline{E} = \frac{Q}{N} = \frac{m_{\text{perspiration}}L_{\text{v}}}{N}$$

But the mass of perspiration is equal to the mass $m_{\text{H}_2\text{O molecule}}$ of a single water molecule times the number N of water molecules. The mass of a single water molecule is equal to its molecular mass (18.0 u), converted into kilograms. The average energy given to a single water molecule is

$$\overline{E} = \frac{m_{\text{perspiration}}L_{\text{v}}}{N} = \frac{m_{\text{H}_2\text{O molecule}}\,N\,L_{\text{v}}}{N}$$

$$\overline{E} = m_{\text{H}_2\text{O molecule}}L_{\text{v}} = (18.0\text{ u})\left(\frac{1.66\times10^{-27}\text{ kg}}{1\text{ u}}\right)\left(2.42\times10^{6}\text{ J/kg}\right) = \boxed{7.23\times10^{-20}\text{ J}}$$

---

39. ***REASONING AND SOLUTION*** According to Fick's law, we have

$$t = \frac{mL}{DA\Delta C} = \frac{\left(8.0\times10^{-13}\text{ kg}\right)(0.015\text{ m})}{\left(5.0\times10^{-10}\text{ m}^2/\text{s}\right)\left(7.0\times10^{-4}\text{ m}^2\right)\left(3.0\times10^{-3}\text{ kg/m}^3\right)} = \boxed{11\text{ s}}$$

---

40. ***REASONING*** The concentration difference between the ends of the channel is $\Delta C = C_{\text{higher}} - C_{\text{lower}}$. Therefore, we can determine the lower concentration as $C_{\text{lower}} = C_{\text{higher}} - \Delta C$, provided that we can obtain a value for ΔC. We can find ΔC by using Fick's law of diffusion. According to this law, the mass rate of diffusion is given by $\frac{m}{t} = \frac{DA\Delta C}{L}$ (Equation 14.8), where D is the diffusion constant, A is the cross-sectional area of the diffusion channel, and L is the length of the channel.

SOLUTION The lower concentration is $C_{\text{lower}} = C_{\text{higher}} - \Delta C$. Solving Fick's law for ΔC, and substituting the result into this expression gives

$$C_{\text{lower}} = C_{\text{higher}} - \frac{\left(\frac{m}{t}\right)L}{DA}$$

$$= 8.3\times10^{-3}\ \text{kg/m}^3 - \frac{\left(4.2\times10^{-14}\ \text{kg/s}\right)\left(0.020\ \text{m}\right)}{\left(1.06\times10^{-9}\ \text{m}^2/\text{s}\right)\left(1.5\times10^{-4}\ \text{m}^2\right)} = \boxed{3.0\times10^{-3}\ \text{kg/m}^3}$$

---

41. ***REASONING AND SOLUTION***

a. As stated, the time required for the first solute molecule to traverse a channel of length L is $t = L^2/(2D)$. Therefore, for water vapor in air at 293 K, where the diffusion constant is $D = 2.4\times10^{-5}\ \text{m}^2/\text{s}$, the time t required for the first water molecule to travel $L = 0.010$ m is

$$t = \frac{L^2}{2D} = \frac{(0.010\ \text{m})^2}{2(2.4\times10^{-5}\ \text{m}^2/\text{s})} = \boxed{2.1\ \text{s}}$$

b. If a water molecule were traveling at the translational rms speed v_{rms} for water, the time t it would take to travel the distance $L = 0.010$ m would be given by $t = L/v_{\text{rms}}$, where, according to Equation 14.6 ($\overline{\text{KE}} = \frac{1}{2}mv_{\text{rms}}^2$), $v_{\text{rms}} = \sqrt{2\left(\overline{\text{KE}}\right)/m}$. Before we can use the last expression for the translation rms speed v_{rms}, we must determine the mass m of a water molecule and the average translational kinetic energy $\overline{\text{KE}}$.

Using the periodic table on the inside of the text's back cover, we find that the molecular mass of a water molecule is

$$\underbrace{2(1.00794\ \text{u})}_{\text{Mass of two hydrogen atoms}} + \underbrace{15.9994\ \text{u}}_{\text{Mass of one oxygen atom}} = 18.0153\ \text{u}$$

The mass of a single molecule is

$$m = \frac{18.0153\times10^{-3}\ \text{kg/mol}}{6.022\times10^{23}\ \text{mol}^{-1}} = 2.99\times10^{-26}\ \text{kg}$$

The average translational kinetic energy of water molecules at 293 K is, according to Equation 14.6,

$$\overline{\text{KE}} = \tfrac{3}{2}kT = \tfrac{3}{2}(1.38\times10^{-23}\ \text{J/K})(293\ \text{K}) = 6.07\times10^{-21}\ \text{J}$$

Therefore, the translational rms speed of water molecules is

$$v_{\text{rms}} = \sqrt{\frac{2(\overline{\text{KE}})}{m}} = \sqrt{\frac{2(6.07 \times 10^{-21}\ \text{J})}{2.99 \times 10^{-26}\ \text{kg}}} = 637\ \text{m/s}$$

Thus, the time t required for a water molecule to travel the distance $L = 0.010$ m at this speed is

$$t = \frac{L}{v_{\text{rms}}} = \frac{0.010\ \text{m}}{637\ \text{m/s}} = \boxed{1.6 \times 10^{-5}\ \text{s}}$$

c. In part (a), when a water molecule diffuses through air, it makes millions of collisions each second with air molecules. The speed and direction changes abruptly as a result of each collision. Between collisions, the water molecules move in a straight line at constant speed. Although a water molecule does move very quickly between collisions, it wanders only very slowly in a zigzag path from one end of the channel to the other. In contrast, a water molecule traveling unobstructed at its translational rms speed [as in part (b)], will have a larger displacement over a much shorter time. Therefore, the answer to part (a) is much longer than the answer to part (b).

---

42. ***REASONING AND SOLUTION*** Fick's law of diffusion gives

$$v = \frac{L}{t} = \frac{DA\Delta C}{m} = \frac{(4.2 \times 10^{-5}\ \text{m}^2/\text{s})(4.0 \times 10^{-4}\ \text{m}^2)(3.5 \times 10^{-2}\ \text{kg/m}^3)}{8.4 \times 10^{-8}\ \text{kg}} = \boxed{7.0 \times 10^{-3}\ \text{m/s}}$$

---

43. SSM WWW ***REASONING*** Since mass is conserved, the mass flow rate is the same at all points, as described by the equation of continuity (Equation 11.8). Therefore, the mass flow rate at which CCl_4 enters the tube is the same as that at point A. The concentration difference of CCl_4 between point A and the left end of the tube, ΔC, can be calculated by using Fick's law of diffusion (Equation 14.8). The concentration of CCl_4 at point A can be found from $C_{\text{A}} = C_{\text{left end}} - \Delta C$.

SOLUTION
a. As discussed above in the reasoning, the mass flow rate of CCl_4 as it passes point A is the same as the mass flow rate at which CCl_4 enters the left end of the tube; therefore, the mass flow rate of CCl_4 at point A is $\boxed{5.00 \times 10^{-13}\ \text{kg/s}}$.

b. Solving Fick's law for ΔC, we obtain

$$\Delta C = \frac{mL}{DAt} = \frac{(m/t)L}{DA}$$

$$= \frac{(5.00\times10^{-13}\ \text{kg/s})(5.00\times10^{-3}\ \text{m})}{(20.0\times10^{-10}\ \text{m}^2/\text{s})(3.00\times10^{-4}\ \text{m}^2)} = 4.2\times10^{-3}\ \text{kg/m}^3$$

Then,

$$C_A = C_{\text{left end}} - \Delta C = (1.00\times10^{-2}\ \text{kg/m}^3) - (4.2\times10^{-3}\ \text{kg/m}^3) = \boxed{5.8\times10^{-3}\ \text{kg/m}^3}$$

---

44. ***REASONING AND SOLUTION***
a. The average concentration is $C_{av} = (1/2)\ (C_1 + C_2) = (1/2)C_2 = m/V = m/(AL)$, so that $C_2 = 2m/(AL)$. Fick's law then becomes $m = DAC_2t/L = DA(2m/AL)t/L = 2Dmt/L^2$. Solving for t yields

$$\boxed{t = L^2/(2D)}$$

b. Substituting into this expression yields

$$t = (2.5\times10^{-2}\ \text{m})^2/[2(1.0\times10^{-5}\ \text{m}^2/\text{s})] = \boxed{31\ \text{s}}$$

---

45. ***REASONING AND SOLUTION*** Equation 14.8, gives Fick's law of diffusion: $m = \dfrac{DA\Delta Ct}{L}$.
Solving for the time t gives

$$t = \frac{mL}{DA\Delta C} \qquad (1)$$

The time required for the water to evaporate is equal to the time it takes for 2.0 grams of water vapor to traverse the tube and can be calculated from Equation (1) above. Since air in the tube is completely dry at the right end, the concentration of water vapor is zero, $C_1 = 0\ \text{kg/m}^3$, and $\Delta C = C_2 - C_1 = C_2$. The concentration at the left end of the tube, C_2, is equal to the density of the water vapor above the water. This can be found from the ideal gas law:

$$PV = nRT \quad \Rightarrow \quad P = \frac{\rho RT}{M}$$

where M = 0.0180152 kg/mol is the mass per mole for water (H_2O). From the figure that accompanies Problem 68 in Chapter 12, the equilibrium vapor pressure of water at 20 °C is 2.4×10^3 Pa.

$$\therefore C_2 = \rho = \frac{PM}{RT} = \frac{(2.4\times10^3\ \text{Pa})(.0180152\ \text{kg/mol})}{[8.31\ \text{J/(mol}\cdot\text{K)}](293\ \text{K})} = 1.8\times10^{-2}\ \text{kg/m}^3$$

Substituting values into Equation (1) gives

$$t = \frac{(2.0 \times 10^{-3}\text{ kg})(0.15\text{ m})}{(2.4 \times 10^{-5}\text{ m}^2/\text{s})(3.0 \times 10^{-4}\text{ m}^2)(1.8 \times 10^{-2}\text{ kg/m}^3)} = \boxed{2.3 \times 10^6\text{ s}}$$

This is about 27 days!

---

46. ***REASONING*** Both gases fill the balloon to the same pressure P, volume V, and temperature T. Assuming that both gases are ideal, we can apply the ideal gas law $PV = nRT$ to each and conclude that the same number of moles n of each gas is needed to fill the balloon. Furthermore, the number of moles can be calculated from the mass m (in grams) and the mass per mole M (in grams per mole), according to $n = \frac{m}{M}$. Using this expression in the equation $n_{\text{Helium}} = n_{\text{Nitrogen}}$ will allow us to obtain the desired mass of nitrogen.

SOLUTION Since the number of moles of helium equals the number of moles of nitrogen, we have

$$\underbrace{\frac{m_{\text{Helium}}}{M_{\text{Helium}}}}_{\substack{\text{Number of moles}\\ \text{of helium}}} = \underbrace{\frac{m_{\text{Nitrogen}}}{M_{\text{Nitrogen}}}}_{\substack{\text{Number of moles}\\ \text{of nitrogen}}}$$

Solving for m_{Nitrogen} and taking the values of mass per mole for helium (He) and nitrogen (N_2) from the periodic table on the inside of the back cover of the text, we find

$$m_{\text{Nitrogen}} = \frac{M_{\text{Nitrogen}}\, m_{\text{Helium}}}{M_{\text{Helium}}} = \frac{(28.0\text{ g/mol})(0.16\text{ g})}{4.00\text{ g/mol}} = \boxed{1.1\text{ g}}$$

---

47. SSM ***REASONING*** According to the ideal gas law (Equation 14.1), $PV = nRT$. Since n, the number of moles, is constant, $n_1 R = n_2 R$. Thus, according to Equation 14.1, we have

$$\frac{P_1 V_1}{T_1} = \frac{P_2 V_2}{T_2}$$

SOLUTION Solving for T_2, we have

$$T_2 = \left(\frac{P_2}{P_1}\right)\left(\frac{V_2}{V_1}\right) T_1 = \left(\frac{48.5\, P_1}{P_1}\right)\left[\frac{V_1/16}{V_1}\right](305\text{ K}) = \boxed{925\text{ K}}$$

---

48. ***REASONING AND SOLUTION*** Assume the ideal gas law holds for O_2. Then

$$n=\frac{PV}{RT}=\frac{\left(1.5\times10^{7}\text{ Pa}\right)\left(2.8\times10^{-3}\text{ m}^{3}\right)}{\left[8.31\text{ J/}\left(\text{mol}\cdot\text{K}\right)\right]\left(296\text{ K}\right)}=1.71\text{ mol}$$

The mass of O_2 is (31.9988 g/mol)(1.71 mol) = 550 g = $\boxed{0.550\text{ kg}}$.

---

49. SSM ***REASONING AND SOLUTION*** According to Fick's law of diffusion (Equation 14.8), the mass of ethanol that diffuses through the cylinder in one hour (3600 s) is

$$m=\frac{(DA\,\Delta C)t}{L}=\frac{(12.4\times10^{-10}\text{ m}^2\text{/s})(4.00\times10^{-4}\text{ m}^2)(1.50\text{ kg/m}^3)(3600\text{ s})}{(0.0200\text{ m})}=\boxed{1.34\times10^{-7}\text{ kg}}$$

---

50. ***REASONING AND SOLUTION*** From Equation 14.5 $\left[PV=\frac{2}{3}N(\overline{KE})\right]$, we obtain

$$\overline{KE}=\frac{3PV}{2N}=\frac{3(4.5\times10^{5}\text{ Pa})(8.5\times10^{-3}\text{ m}^{3})}{2\left[(2.0\text{ mol})(6.022\times10^{23}\text{ mol}^{-1})\right]}=\boxed{4.8\times10^{-21}\text{ J}}$$

---

51. ***REASONING*** The smoke particles have the same average translational kinetic energy as the air molecules, namely, $\frac{1}{2}mv_{\text{rms}}^2=\frac{3}{2}kT$, according to Equation 14.6. In this expression m is the mass of a smoke particle, v_{rms} is the rms speed of a particle, k is Boltzmann's constant, and T is the Kelvin temperature. We can obtain the mass directly from this equation.

SOLUTION Solving Equation 14.6 for the mass m, we find

$$m=\frac{3kT}{v_{\text{rms}}^2}=\frac{3\left(1.38\times10^{-23}\text{ J/K}\right)\left(301\text{ K}\right)}{\left(2.8\times10^{-3}\text{ m/s}\right)^2}=\boxed{1.6\times10^{-15}\text{ kg}}$$

---

52. ***REASONING AND SOLUTION*** Assuming constant temperature we have $P_1V_1=P_2V_2$. The volume of the cylinder is $V=Ah$ where A is the cross-sectional area of the tank and h the height of the air column. Since the cross-sectional area doesn't change, we have $P_1h_1=P_2h_2$. Thus,

$$h_2=h_1(P_1/P_2)=(0.80\text{ m})(2.0\text{ atm})/(6.0\text{ atm})=\boxed{0.27\text{ m}}$$

---

53. SSM ***REASONING AND SOLUTION*** The number of moles of water in the drop is $n = m/M$, where m is the mass of the drop and M is the mass per mole of water. The mass of a drop of volume V and density ρ can be obtained from Equation 11.1 as $m = \rho V$. Thus, the number of molecules in the drop is

$$N = N_A n = \frac{N_A \rho V}{M}$$

where $N_A = 6.022 \times 10^{23}\ \text{mol}^{-1}$ is Avogadro's number. Since the drop is spherical, its volume is given by $V = \frac{4}{3}\pi r^3$, where r is the radius of the drop. Substituting the values for the data given in the problem, using a density of $\rho = 1.00 \times 10^3\ \text{kg/m}^3$ for water, and using $M = 18.015 \times 10^{-3}\ \text{kg/mol}$ for water yields

$$N = \frac{\left(6.022 \times 10^{23}\ \text{mol}^{-1}\right)\left(1.00 \times 10^3\ \frac{\text{kg}}{\text{m}^3}\right)\frac{4}{3}\pi(9.00 \times 10^{-4}\text{m})^3}{18.015 \times 10^{-3}\ \frac{\text{kg}}{\text{mol}}} = \boxed{1.02 \times 10^{20}}$$

---

54. ***REASONING AND SOLUTION*** Since we are treating the air as an ideal gas, PV = nRT, so that $U = \frac{5}{2}nRT = \frac{5}{2}PV = \frac{5}{2}(7.7 \times 10^6\ \text{Pa})(5.6 \times 10^5\ \text{m}^3) = 1.1 \times 10^{13}\ \text{J}$. The number of joules of energy consumed per day by one house is

$$30.0\ \text{kW} \cdot \text{h} = \left(30.0 \times 10^3\ \frac{\text{J} \cdot \text{h}}{\text{s}}\right)\left(\frac{3600\ \text{s}}{1\ \text{h}}\right) = 1.08 \times 10^8\ \text{J}$$

The number of homes that could be served for one day by 1.1×10^{13} J of energy is

$$\left(1.1 \times 10^{13}\ \text{J}\right)\left(\frac{1\ \text{home}}{1.08 \times 10^8\ \text{J}}\right) = \boxed{1.0 \times 10^5\ \text{homes}}$$

---

55. SSM WWW ***REASONING AND SOLUTION*** For a given mass of the liquid phase and the vapor phase, $m_{\text{liquid}} = m_{\text{vapor}}$, or using Equation 11.1, $\rho_{\text{liquid}} V_{\text{liquid}} = \rho_{\text{vapor}} V_{\text{vapor}}$. In this result, ρ and V stand for density and volume, respectively. The ratio of the volumes, is therefore,

$$\frac{V_{\text{vapor}}}{V_{\text{liquid}}} = \frac{\rho_{\text{liquid}}}{\rho_{\text{vapor}}}$$

If we assume that the volume of each phase is filled with N cubes, with one molecule at the center of each cube, then the volume is $V = Nd^3$, where d is the length of one side of a cube. The number N is the same for each phase, since each has the same mass. Since each cube contains one molecule at its center, d is also the distance between neighboring molecules. It follows that, for either phase, $d = \sqrt[3]{V/N}$. Therefore,

$$\frac{d_{\text{vapor}}}{d_{\text{liquid}}} = \left(\frac{V_{\text{vapor}}/N}{V_{\text{liquid}}/N}\right)^{1/3} = \left(\frac{\rho_{\text{liquid}}}{\rho_{\text{vapor}}}\right)^{1/3} = \left(\frac{958\ \text{kg/m}^3}{0.598\ \text{kg/m}^3}\right)^{1/3} = \boxed{11.7}$$

---

56. ***REASONING AND SOLUTION***

a. Assuming that the direction of travel of the bullets is positive, the average change in momentum per second is

$$\Delta p/\Delta t = m\Delta v/\Delta t = (200)(0.0050\ \text{kg})[(0\ \text{m/s}) - 1200\ \text{m/s})]/(10.0\ \text{s}) = \boxed{-120\ \text{N}}$$

b. The average force exerted on the bullets is $\bar{F} = \Delta p/\Delta t$. According to Newton's third law, the average force exerted on the wall is $-\bar{F} = \boxed{120\ \text{N}}$.

c. The pressure P is the magnitude of the force on the wall per unit area, so

$$P = (120\ \text{N})/(3.0 \times 10^{-4}\ \text{m}^2) = \boxed{4.0 \times 10^5\ \text{Pa}}$$

---

57. ***REASONING AND SOLUTION*** At the instant just before the balloon lifts off, the buoyant force from the outside air has a magnitude that equals the magnitude of the total weight. According to Archimedes' principle, the buoyant force is the weight of the displaced outside air (density $\rho_0 = 1.29 \times 10^3\ \text{kg/m}^3$). The mass of the displaced outside air is $\rho_0 V$, where $V = 650\ \text{m}^3$. The corresponding weight is the mass times the magnitude g of the acceleration due to gravity. Thus, we have

$$\underbrace{(\rho_0 V)g}_{\text{Buoyant force}} = \underbrace{m_{\text{total}}g}_{\substack{\text{Total weight}\\ \text{of balloon}}} \qquad (1)$$

The total mass of the balloon is $m_{\text{total}} = m_{\text{load}} + m_{\text{air}}$, where $m_{\text{load}} = 320$ kg and m_{air} is the mass of the hot air within the balloon. The mass of the hot air can be calculated from the ideal gas law by using it to obtain the number of moles n of air and multiplying n by the mass per mole of air, $M = 29 \times 10^{-3}$ kg/mol:

$$m_{\text{air}} = nM = \left(\frac{PV}{RT}\right)M$$

Thus, the total mass of the balloon is $m_{\text{total}} = m_{\text{load}} + PVM/(RT)$ and Equation (1) becomes

$$\rho_0 V = m_{\text{load}} + \left(\frac{PV}{RT}\right)M$$

Solving for T gives

$$T = \frac{PVM}{\left(\rho_0 V - m_{\text{load}}\right)R}$$

$$= \frac{\left(1.01\times 10^5 \text{ Pa}\right)\left(650 \text{ m}^3\right)\left(29\times 10^{-3} \text{ kg/mol}\right)}{\left[\left(1.29\times 10^3 \text{ kg/m}^3\right)\left(650 \text{ m}^3\right) - 320 \text{ kg}\right]\left[8.31 \text{ J}/\left(\text{mol}\cdot\text{K}\right)\right]} = \boxed{440 \text{ K}}$$

---

58. ***CONCEPT QUESTIONS***

a. The mass of one of its atoms (in atomic mass units) has the same numerical value as the mass per mole (in units of g/mol).

b. Dividing the mass of the sample by the mass per mole gives the number of moles of atoms in the sample.

SOLUTION

a. Gold has a mass per mole of 196.967 g/mol. Since the mass of one of its atoms (in atomic mass units) has the same numerical value as the mass per mole, the mass of a single gold atom is $m = \boxed{196.967 \text{ u}}$.

b. We can convert the mass from atomic mass units to kilograms by noting that 1 u = 1.6605×10^{-27} kg:

$$m = \left(196.967 \text{ u}\right)\left(\frac{1.6605\times 10^{-27} \text{ kg}}{1 \text{ u}}\right) = \boxed{3.2706\times 10^{-25} \text{ kg}}$$

c. The number of moles of gold atoms is equal to the mass of the gold divided by the mass per mole:

$$n = \frac{m}{\text{Mass per mole}} = \frac{285 \text{ g}}{196.967 \text{ g/mol}} = \boxed{1.45 \text{ mol}}$$

---

59. ***CONCEPT QUESTIONS***

a. According to the ideal gas law, $PV = nRT$, the absolute pressure P is directly proportional to the temperature T, provided the temperature is measured on the Kelvin scale. Therefore, if the temperature on the Kelvin scale doubles, the pressure also doubles.

b. The pressure is not proportional to the temperature when it is measured on the Celsius scale. (The pressure is proportional to the temperature when measured on the Kelvin scale.) Thus, the pressure does not double when the temperature on the Celsius scale doubles.

SOLUTION
a. According to Equation 14.1, the pressures at the two temperatures are

$$P_1 = \frac{nRT_1}{V} \quad \text{and} \quad P_2 = \frac{nRT_2}{V}$$

Taking the ratio P_2/P_1 of the final pressure to the initial pressure gives

$$\frac{P_2}{P_1} = \frac{\dfrac{nRT_2}{V}}{\dfrac{nRT_1}{V}} = \frac{T_2}{T_1} = \frac{70.0\text{ K}}{35.0\text{ K}} = \boxed{2.00}$$

b. The ratio of the pressures at the temperatures of 35.0 °C and 70.0 °C is

$$\frac{P_2}{P_1} = \frac{T_2}{T_1} = \frac{(273.15 + 70.0)\text{ K}}{(273.15 + 35.0)\text{ K}} = \boxed{1.11}$$

---

60. ***CONCEPT QUESTION*** According to the ideal gas law, $PV = nRT$, the temperature T is directly proportional to the product PV, for a fixed number n of moles. Therefore, tanks with equal values of PV have the same temperature. Using the data in the table given with the problem statement, we see that the values of PV for each tank are (starting with tank A): 100 J, 150 J, 100 J, and 150 J. Tanks A and C have the same temperature, while B and D have the same temperature.

SOLUTION The temperature of each gas can be found from the ideal gas law, Equation 14.1:

$$T_\text{A} = \frac{P_\text{A}V_\text{A}}{nR} = \frac{(25.0\text{ Pa})(4.0\text{ m}^3)}{(0.10\text{ mol})[8.31\text{ J/(mol}\cdot\text{K)}]} = \boxed{120\text{ K}}$$

$$T_\text{B} = \frac{P_\text{B}V_\text{B}}{nR} = \frac{(30.0\text{ Pa})(5.0\text{ m}^3)}{(0.10\text{ mol})[8.31\text{ J/(mol}\cdot\text{K)}]} = \boxed{180\text{ K}}$$

$$T_\text{C} = \frac{P_\text{C}V_\text{C}}{nR} = \frac{(20.0\text{ Pa})(5.0\text{ m}^3)}{(0.10\text{ mol})[8.31\text{ J/(mol}\cdot\text{K)}]} = \boxed{120\text{ K}}$$

$$T_\text{D} = \frac{P_\text{D}V_\text{D}}{nR} = \frac{(2.0\text{ Pa})(75\text{ m}^3)}{(0.10\text{ mol})[8.31\text{ J/(mol}\cdot\text{K)}]} = \boxed{180\text{ K}}$$

---

61. ***CONCEPT QUESTION*** According to the kinetic theory of gases, the average kinetic energy of an atom is related to the temperature of the gas by $\frac{1}{2}mv_{\text{rms}}^2 = \frac{3}{2}kT$. We see that the temperature is proportional to the product of the mass and the square of the rms-speed. Therefore, the tank with the greatest value of mv_{rms}^2 has the greatest temperature. Using the information from the table given with the problem statement, we see that the values of mv_{rms}^2 for each tank are:

Tank	
A	mv_{rms}^2
B	$m(2v_{\text{rms}})^2 = 4mv_{\text{rms}}^2$
C	$(2m)v_{\text{rms}}^2 = 2mv_{\text{rms}}^2$
D	$2m(2v_{\text{rms}})^2 = 8mv_{\text{rms}}^2$

Thus, tank D has the greatest temperature, followed by tanks B, C, and A.

SOLUTION The temperature of the gas in each tank can be determined from Equation 14.6:

$$T_{\text{A}} = \frac{mv_{\text{rms}}^2}{3k} = \frac{(3.32\times10^{-26}\text{ kg})(1223\text{ m/s})^2}{3(1.38\times10^{-23}\text{ J/K})} = \boxed{1200\text{ K}}$$

$$T_{\text{B}} = \frac{m(2v_{\text{rms}})^2}{3k} = \frac{(3.32\times10^{-26}\text{ kg})(2\times1223\text{ m/s})^2}{3(1.38\times10^{-23}\text{ J/K})} = \boxed{4800\text{ K}}$$

$$T_{\text{C}} = \frac{(2m)v_{\text{rms}}^2}{3k} = \frac{2(3.32\times10^{-26}\text{ kg})(1223\text{ m/s})^2}{3(1.38\times10^{-23}\text{ J/K})} = \boxed{2400\text{ K}}$$

$$T_{\text{D}} = \frac{2m(2v_{\text{rms}})^2}{3k} = \frac{2(3.32\times10^{-26}\text{ kg})(2\times1223\text{ m/s})^2}{3(1.38\times10^{-23}\text{ J/K})} = \boxed{9600\text{ K}}$$

---

62. ***CONCEPT QUESTIONS***

a. The diffusion rate is proportional to the rms-speed of the atoms in a gas. According to the kinetic theory of gases, the average kinetic energy of an atom is related to the temperature of the gas by $\frac{1}{2}mv_{\text{rms}}^2 = \frac{3}{2}kT$. As the temperature of the gas is raised, the rms-speed of the atoms also increases. Therefore, the diffusion rate increases with temperature.

b. Both helium and neon gases are at the same temperature, so the average kinetic energy of an atom is the same in each species. However, helium has a smaller mass, so it has a greater rms-speed. Therefore, helium has the greater diffusion rate.

SOLUTION
a. Since the diffusion rate is proportional to the rms-speed of the atoms in a gas, the ratio R_{580}/R_{290} of the diffusion rates at the two temperatures is equal to the ratio $v_{\text{rms},580}/v_{\text{rms},290}$ of the rms-speeds at those temperatures. The rms-speed is related to the temperature by Equation 14.6 as $v_{\text{rms}} = \sqrt{3kT/m}$. The ratio of the diffusion rates for helium at the two temperatures is

$$\frac{R_{580}}{R_{290}} = \frac{v_{\text{rms},580}}{v_{\text{rms},290}} = \frac{\sqrt{\dfrac{3k(580\text{ K})}{m_{\text{He}}}}}{\sqrt{\dfrac{3k(290\text{ K})}{m_{\text{He}}}}} = \sqrt{\frac{580\text{ K}}{290\text{ K}}} = \boxed{1.4}$$

b. For a fixed temperature, the ratio $R_{\text{He}}/R_{\text{Ne}}$ of the diffusion rates for helium and neon is equal to the ratio $v_{\text{rms,He}}/v_{\text{rms,Ne}}$ of the rms-speeds.

$$\frac{R_{\text{He}}}{R_{\text{Ne}}} = \frac{v_{\text{rms,He}}}{v_{\text{rms,Ne}}} = \frac{\sqrt{\dfrac{3kT}{m_{\text{He}}}}}{\sqrt{\dfrac{3kT}{m_{\text{Ne}}}}} = \sqrt{\frac{m_{\text{Ne}}}{m_{\text{He}}}} = \sqrt{\frac{20.179\text{ u}}{4.00260\text{ u}}} = \boxed{2.2}$$

---

63. ***CONCEPT QUESTIONS***
a. The number n of moles of a species can be calculated from the mass m (in grams) of the species and its molecular mass, or mass per mole M (in grams per mole), according to $n = \dfrac{m}{M}$.

b. To calculate the percentage, we divide the number n of moles of a species by the total number n_{Total} of moles in the mixture and multiply that fraction by 100. The total number of moles is the sum of the numbers of moles for each component.

c. The component with the greatest number of moles has the greatest percentage. For the three components described, this would be helium, because it has the greatest mass and the smallest mass per mole.

d. The component with the smallest number of moles has the smallest percentage. For the three components described, this would be argon, because it has the smallest mass and the greatest mass per mole.

SOLUTION The percentage p_{Argon} of argon is

$$p_{\text{Argon}} = \frac{n_{\text{Argon}}}{n_{\text{Argon}} + n_{\text{Neon}} + n_{\text{Helium}}} \times 100 = \frac{\dfrac{m_{\text{Argon}}}{M_{\text{Argon}}}}{\dfrac{m_{\text{Argon}}}{M_{\text{Argon}}} + \dfrac{m_{\text{Neon}}}{M_{\text{Neon}}} + \dfrac{m_{\text{Helium}}}{M_{\text{Helium}}}} \times 100$$

$$= \frac{\dfrac{1.20\text{ g}}{39.948\text{ g/mol}}}{\dfrac{1.20\text{ g}}{39.948\text{ g/mol}} + \dfrac{2.60\text{ g}}{20.179\text{ g/mol}} + \dfrac{3.20\text{ g}}{4.0026\text{ g/mol}}} \times 100 = \boxed{3.1\,\%}$$

The percentage of neon is

$$p_{\text{Neon}} = \frac{n_{\text{Neon}}}{n_{\text{Argon}} + n_{\text{Neon}} + n_{\text{Helium}}} \times 100 = \frac{\dfrac{m_{\text{Neon}}}{M_{\text{Neon}}}}{\dfrac{m_{\text{Argon}}}{M_{\text{Argon}}} + \dfrac{m_{\text{Neon}}}{M_{\text{Neon}}} + \dfrac{m_{\text{Helium}}}{M_{\text{Helium}}}} \times 100$$

$$= \frac{\dfrac{2.60\text{ g}}{20.179\text{ g/mol}}}{\dfrac{1.20\text{ g}}{39.948\text{ g/mol}} + \dfrac{2.60\text{ g}}{20.179\text{ g/mol}} + \dfrac{3.20\text{ g}}{4.0026\text{ g/mol}}} \times 100 = \boxed{13.5\,\%}$$

The percentage of helium is

$$p_{\text{Helium}} = \frac{n_{\text{Helium}}}{n_{\text{Argon}} + n_{\text{Neon}} + n_{\text{Helium}}} \times 100 = \frac{\dfrac{m_{\text{Helium}}}{M_{\text{Helium}}}}{\dfrac{m_{\text{Argon}}}{M_{\text{Argon}}} + \dfrac{m_{\text{Neon}}}{M_{\text{Neon}}} + \dfrac{m_{\text{Helium}}}{M_{\text{Helium}}}} \times 100$$

$$= \frac{\dfrac{3.20\text{ g}}{4.0026\text{ g/mol}}}{\dfrac{1.20\text{ g}}{39.948\text{ g/mol}} + \dfrac{2.60\text{ g}}{20.179\text{ g/mol}} + \dfrac{3.20\text{ g}}{4.0026\text{ g/mol}}} \times 100 = \boxed{83.4\,\%}$$

These results are consistent with our answers to Concept Questions (c) and (d).

64. ***CONCEPT QUESTIONS***

a. The force F_{Applied} that must be applied to stretch an ideal spring by an amount x with respect to its unstrained length is given by Equation 10.1 as

$$F_{\text{Applied}} = kx \tag{1}$$

where k is the spring constant.

b. Pressure is the magnitude of the force applied perpendicularly to a surface divided by the area of the surface. Thus, the magnitudes of the forces that the initial and final pressures apply to the piston (and, therefore, to the spring) are given by Equation 11.3 as

$$\underbrace{F_0 = P_0 A}_{\substack{\text{Force applied by}\\ \text{initial pressure}}} \quad \text{and} \quad \underbrace{F_f = P_f A}_{\substack{\text{Force applied by}\\ \text{final pressure}}} \tag{2}$$

c. The ideal gas law is $PV = nRT$. Since the number of moles is constant, this equation can be written as $\frac{PV}{T} = nR = \text{constant}$. Thus, the value of $\frac{PV}{T}$ is the same initially and finally, and we can write

$$\frac{P_0 V_0}{T_0} = \frac{P_f V_f}{T_f} \tag{3}$$

d. The final volume is the initial volume plus the amount by which the volume increases as the spring stretches. The increased volume due to the additional stretching is $A\left(x_f - x_0\right)$. Therefore, we have

$$V_f = V_0 + A\left(x_f - x_0\right) \tag{4}$$

SOLUTION The final temperature can be obtained by rearranging Equation (3) to show that

$$T_f = \left(\frac{P_f V_f}{P_0 V_0}\right) T_0 \tag{5}$$

Into this result we can now substitute expressions for P_0 and P_f. These expressions can be obtained by using Equations (2) in Equation (1) as follows:

$$\underbrace{P_0 A}_{\substack{\text{Force applied to}\\ \text{spring by initial}\\ \text{pressure}}} = kx_0 \quad \text{and} \quad \underbrace{P_f A}_{\substack{\text{Force applied to}\\ \text{spring by final}\\ \text{pressure}}} = kx_f \tag{6}$$

In addition, we can substitute Equation (4) for the final volume into Equation (5). With these substitutions Equation (5) becomes

$$T_{\mathrm{f}}=\left(\frac{P_{\mathrm{f}}V_{\mathrm{f}}}{P_0V_0}\right)T_0=\frac{\left(\dfrac{kx_{\mathrm{f}}}{A}\right)\left[V_0+A\left(x_{\mathrm{f}}-x_0\right)\right]T_0}{\left(\dfrac{kx_0}{A}\right)V_0}=\frac{x_{\mathrm{f}}\left[V_0+A\left(x_{\mathrm{f}}-x_0\right)\right]T_0}{x_0V_0}$$

$$=\frac{(0.1000\text{ m})\left[6.00\times10^{-4}\text{ m}^3+\left(2.50\times10^{-3}\text{ m}^2\right)(0.1000\text{ m}-0.0800\text{ m})\right](273\text{ K})}{(0.0800\text{ m})\left(6.00\times10^{-4}\text{ m}^3\right)}$$

$$=\boxed{3.70\times10^2\text{ K}}$$

CHAPTER 15 | *THERMODYNAMICS*

---

1. SSM ***REASONING AND SOLUTION*** In going from the initial state to the intermediate state, the internal energy of the system changes according to Equation 15.1:

$$\Delta U_1 = U_f - U_i = Q - W = +165\text{ J} - (+312\text{ J}) = -147\text{ J}$$

Since internal energy is a function of state, the change in the internal energy in going from the initial state to the intermediate state and back to the initial state is zero. That is, $\Delta U_{\text{total}} = \Delta U_1 + \Delta U_2 = 0$. Therefore, in returning to the initial state, the change in the internal energy is $\Delta U_2 = -\Delta U_1 = +147\text{ J}$.

a. Thus, the work involved is

$$W = Q - \Delta U_2 = -114\text{ J} - (+147\text{ J}) = \boxed{-261\text{ J}}$$

b. Since the work is negative, $\boxed{\text{work is done on the system}}$

---

2. ***REASONING*** Since the student does work, W is positive, according to our convention. Since his internal energy decreases, the change ΔU in the internal energy is negative. The first law of thermodynamics will allow us to determine the heat Q.

SOLUTION

a. The work is $\boxed{W = +1.6 \times 10^4\text{ J}}$.

b. The change in internal energy is $\boxed{\Delta U = -4.2 \times 10^4\text{ J}}$.

c. Applying the first law of thermodynamics from Equation 15.1, we find that

$$Q = \Delta U + W = \left(-4.2 \times 10^4\text{ J}\right) + \left(1.6 \times 10^4\text{ J}\right) = \boxed{-2.6 \times 10^4\text{ J}}$$

---

3. ***REASONING*** Energy in the form of work leaves the system, while energy in the form of heat enters. More energy leaves than enters, so we expect the internal energy of the system to decrease, that is, we expect the change ΔU in the internal energy to be negative. The first law of thermodynamics will confirm our expectation. As far as the environment is concerned, we note that when the system loses energy, the environment gains it, and when the system gains energy the environment loses it. Therefore, the change in the internal energy of the environment must be opposite to that of the system.

SOLUTION
a. The system gains heat so Q is positive, according to our convention. The system does work, so W is also positive, according to our convention. Applying the first law of thermodynamics from Equation 15.1, we find for the system that

$$\Delta U = Q - W = (77\text{ J}) - (164\text{ J}) = \boxed{-87\text{ J}}$$

As expected, this value is negative, indicating a decrease.

b. The change in the internal energy of the environment is opposite to that of the system, so that $\boxed{\Delta U_{\text{environment}} = +87\text{ J}}$.

---

4. ***REASONING*** According to the discussion in Section 14.3, the internal energy U of a monatomic ideal gas is given by $U = \frac{3}{2}nRT$ (Equation 14.7), where n is the number of moles, R is the universal gas constant, and T is the Kelvin temperature. When the temperature changes to a final value of T_f from an initial value of T_i, the internal energy changes by an amount

$$\underbrace{U_f - U_i}_{\Delta U} = \tfrac{3}{2}nR(T_f - T_i)$$

Solving this equation for the final temperature yields $T_f = \left(\dfrac{2}{3nR}\right)\Delta U + T_i$. We are given n and T_i, but must determine ΔU. The change ΔU in the internal energy of the gas is related to the heat Q and the work W by the first law of thermodynamics, $\Delta U = Q - W$ (Equation 15.1). Using these two relations will allow us to find the final temperature of the gas.

SOLUTION Substituting $\Delta U = Q - W$ into the expression for the final temperature gives

$$T_f = \left(\frac{2}{3nR}\right)(Q - W) + T_i$$

$$= \left\{\frac{2}{3(3.00\text{ mol})[8.31\text{ J/(mol}\cdot\text{K)}]}\right\}[+2438\text{ J} - (-962\text{ J})] + 345\text{ K} = \boxed{436\text{ K}}$$

Note that the heat is positive ($Q = +2438$ J) since the system (the gas) gains heat, and the work is negative ($W = -962$ J), since it is done on the system.

---

5. SSM ***REASONING*** Since the change in the internal energy and the heat released in the process are given, the first law of thermodynamics (Equation 15.1) can be used to find the work done. Since we are told how much work is required to make the car go one mile, we can determine how far the car can travel. When the gasoline burns, its internal energy decreases and heat flows into the surroundings; therefore, both ΔU and Q are negative.

SOLUTION According to the first law of thermodynamics, the work that is done when one gallon of gasoline is burned in the engine is

$$W = Q - \Delta U = -1.00 \times 10^8 \text{ J} - (-1.19 \times 10^8 \text{ J}) = 0.19 \times 10^8 \text{ J}$$

Since 6.0×10^5 J of work is required to make the car go one mile, the car can travel

$$0.19 \times 10^8 \text{ J} \left(\frac{1 \text{ mile}}{6.0 \times 10^5 \text{ J}} \right) = \boxed{32 \text{ miles}}$$

---

6. ***REASONING AND SOLUTION***

a. For the weight lifter

$$\Delta U = Q - W$$

$$= -mL_V - W = -(0.150 \text{ kg})(2.42 \times 10^6 \text{ J/kg}) - 1.40 \times 10^5 \text{ J} = \boxed{-5.03 \times 10^5 \text{ J}}$$

b. Since 1 nutritional calorie = 4186 J, the number of nutritional calories is

$$\left(5.03 \times 10^5 \text{ J}\right)\left(\frac{1 \text{ Calorie}}{4186 \text{ J}} \right) = \boxed{1.20 \times 10^2 \text{ nutritional calories}}$$

---

7. ***REASONING AND SOLUTION*** The change in the internal energy of the material is

$$\Delta U = Q = cm\Delta T = [1100 \text{ J/(kg·C°)}](2.0 \text{ kg})(6.0 \text{ C°}) = \boxed{13\,000 \text{ J}}$$

---

8. ***REASONING*** When a gas expands under isobaric conditions, its pressure remains constant. The work W done by the expanding gas is $W = P(V_f - V_i)$, Equation 15.2, where P is the pressure and V_f and V_i are the final and initial volumes. Since all the variables in this relation are known, we can solve for the final volume.

SOLUTION Solving $W = P(V_f - V_i)$ for the final volume gives

$$V_f = \frac{W}{P} + V_i = \frac{480\text{ J}}{1.6\times10^5\text{ Pa}} + 1.5\times10^{-3}\text{ m}^3 = \boxed{4.5\times10^{-3}\text{ m}^3}$$

---

9. SSM ***REASONING*** The work done in the process is equal to the "area" under the curved line between A and B in the drawing. From the graph, we find that there are about 78 "squares" under the curve. Each square has an "area" of

$$(2.0\times10^4\text{ Pa})(2.0\times10^{-3}\text{ m}^3) = 4.0\times10^1\text{ J}$$

SOLUTION

a. The work done in the process has a magnitude of

$$W = (78)(4.0\times10^1\text{ J}) = \boxed{3100\text{ J}}$$

b. The final volume is smaller than the initial volume, so the gas is compressed. Therefore, work is done on the gas so the work is $\boxed{\text{negative}}$.

---

10. ***REASONING*** For segment AB, there is no work, since the volume is constant. For segment BC the process is isobaric and Equation 15.2 applies. For segment CA, the work can be obtained as the area under the line CA in the graph.

SOLUTION

a. For segment AB, the process is isochoric, that is, the volume is constant. For a process in which the volume is constant, no work is done, so $\boxed{W = 0\text{ J}}$.

b. For segment BC, the process is isobaric, that is, the pressure is constant. Here, the volume is increasing, so the gas is expanding against the outside environment. As a result, the gas does work, which is positive according to our convention. Using Equation 15.2 and the data in the drawing, we obtain

$$W = P(V_f - V_i)$$

$$= (7.0\times10^5\text{ Pa})\left[(5.0\times10^{-3}\text{ m}^3) - (2.0\times10^{-3}\text{ m}^3)\right] = \boxed{+2.1\times10^3\text{ J}}$$

c. For segment CA, the volume of the gas is decreasing, so the gas is being compressed and work is being done on it. Therefore, the work is negative, according to our convention. The magnitude of the work is the area under the segment CA. We estimate that this area is 15 of the squares in the graphical grid. The area of each square is

$$(1.0\times10^5\text{ Pa})(1.0\times10^{-3}\text{ m}^3) = 1.0\times10^2\text{ J}$$

The work, then, is

$$W = -15\,(1.0 \times 10^2\ \text{J}) = \boxed{-1.5 \times 10^3\ \text{J}}$$

---

11. ***REASONING AND SOLUTION*** The work done by the expanding gas is

$$W = Q - \Delta U = 2050\ \text{J} - 1730\ \text{J} = 320\ \text{J}$$

The work, according to Equation 6.1, is also the magnitude F of the force exerted on the piston times the magnitude s of its displacement. But the force is equal to the weight mg of the block and piston, so that the work is $W = Fs = mgs$. Thus, we have

$$s = \frac{W}{mg} = \frac{320\ \text{J}}{(135\ \text{kg})(9.80\ \text{m/s}^2)} = \boxed{0.24\ \text{m}}$$

---

12. ***REASONING AND SOLUTION***

a. Starting at point A, the work done during the first (vertical) straight-line segment is

$$W_1 = P_1 \Delta V_1 = P_1(0\ \text{m}^3) = 0\ \text{J}$$

For the second (horizontal) straight-line segment, the work is

$$W_2 = P_2 \Delta V_2 = 10(1.0 \times 10^4\ \text{Pa})6(2.0 \times 10^{-3}\ \text{m}^3) = 1200\ \text{J}$$

For the third (vertical) straight-line segment the work is

$$W_3 = P_3 \Delta V_3 = P_3(0\ \text{m}^3) = 0\ \text{J}$$

For the fourth (horizontal) straight-line segment the work is

$$W_4 = P_4 \Delta V_4 = 15(1.0 \times 10^4\ \text{Pa})6(2.0 \times 10^{-3}\ \text{m}^3) = 1800\ \text{J}$$

The total work done is

$$W = W_1 + W_2 + W_3 + W_4 = \boxed{+3.0 \times 10^3\ \text{J}}$$

b. Since the total work is positive, work is done $\boxed{\text{by the system}}$.

---

13. $\boxed{\text{SSM}}$ ***REASONING*** The work done in an isobaric process is given by Equation 15.2, $W = P\Delta V$; therefore, the pressure is equal to $P = W/\Delta V$. In order to use this expression, we must first determine a numerical value for the work done; this can be calculated using the first law of thermodynamics (Equation 15.1), $\Delta U = Q - W$.

SOLUTION Solving Equation 15.1 for the work *W*, we find

$$W = Q - \Delta U = 1500 \text{ J} - (+4500 \text{ J}) = -3.0 \times 10^3 \text{ J}$$

Therefore, the pressure is

$$P = \frac{W}{\Delta V} = \frac{-3.0 \times 10^3 \text{ J}}{-0.010 \text{ m}^3} = \boxed{3.0 \times 10^5 \text{ Pa}}$$

The change in volume ΔV, which is the final volume minus the initial volume, is negative because the final volume is 0.010 m^3 *less* than the initial volume.

---

14. ***REASONING AND SOLUTION***
a. The work is the area under the path ACB. There are 48 "squares" under the path, so that

$$W = -48\left(2.0 \times 10^4 \text{ Pa}\right)\left(2.0 \times 10^{-3} \text{ m}^3\right) = -1900 \text{ J}$$

The minus sign is included because the gas is compressed, so that work is done on it. Since there is no temperature change between A and B (the line AB is an isotherm) and the gas is ideal, $\Delta U = 0$, so

$$Q = \Delta U + W = W = \boxed{-1900 \text{ J}}$$

b. The negative answer for *W* means that heat flows $\boxed{\text{out}}$ of the gas.

---

15. $\boxed{\text{SSM}}$ $\boxed{\text{WWW}}$ ***REASONING AND SOLUTION*** The first law of thermodynamics states that $\Delta U = Q - W$. The work W involved in an isobaric process is, according to Equation 15.2, $W = P\Delta V$. Combining these two expressions leads to $\Delta U = Q - P\Delta V$. Solving for Q gives

$$Q = \Delta U + P\Delta V \qquad (1)$$

Since this is an expansion, $\Delta V > 0$, so $P\Delta V > 0$. From the ideal gas law, $PV = nRT$, we have $P\Delta V = nR\Delta T$. Since $P\Delta V > 0$, it follows that $nR\Delta T > 0$. The internal energy of an ideal gas is directly proportional to its Kelvin temperature *T*. Therefore, since $nR\Delta T > 0$, it follows that $\Delta U > 0$. Since both terms on the right hand side of Equation (1) are positive, the left hand side of Equation (1) must also be positive. Thus, Q is positive. By the convention described in the text, this means that heat can only flow into an ideal gas during an isobaric expansion

---

16. ***REASONING AND SOLUTION*** According to the first law of thermodynamics, the change in internal energy is $\Delta U = Q - W$. The work can be obtained from the area under the graph. There are sixty squares of area under the graph, so the positive work of expansion is

$$W = 60\left(1.0\times10^{4}\ \text{Pa}\right)\left(2.0\times10^{-3}\ \text{m}^3\right) = 1200\ \text{J}$$

Since $Q = 2700$ J, the change in internal energy is

$$\Delta U = Q - W = 2700\ \text{J} - 1200\ \text{J} = \boxed{1500\ \text{J}}$$

---

17. ***REASONING AND SOLUTION*** Since the pan is open, the process takes place at constant (atmospheric) pressure P_0. The work involved in an isobaric process is given by Equation 15.2: $W = P_0\Delta V$. The change in volume of the liquid as it is heated is given according to Equation 12.3 as $\Delta V = \beta V_0 \Delta T$, where β is the coefficient of volume expansion. Table 12.1 gives $\beta = 207\times10^{-6}\ (\text{C}°)^{-1}$ for water. The heat absorbed by the water is given by Equation 12.4 as $Q = cm\Delta T$, where $c = 4186\ \text{J}/(\text{kg}\cdot\text{C}°)$ is the specific heat capacity of liquid water according to Table 12.2. Therefore,

$$\frac{W}{Q} = \frac{P_0\Delta V}{cm\Delta T} = \frac{P_0\beta V_0\Delta T}{cm\Delta T} = \frac{P_0\beta}{c(m/V_0)} = \frac{P_0\beta}{c\rho}$$

where $\rho = 1.00\times10^{3}\ \text{kg/m}^3$ is the density of the water. Thus, we find

$$\frac{W}{Q} = \frac{P_0\beta}{c\rho} = \frac{\left(1.01\times10^{5}\ \text{Pa}\right)\left(207\times10^{-6}\ \text{C}°^{-1}\right)}{\left[4186\ \text{J}/(\text{kg}\cdot\text{C}°)\right]\left(1.00\times10^{3}\ \text{kg/m}^3\right)} = \boxed{4.99\times10^{-6}}$$

---

18. ***REASONING*** We can use the first law of thermodynamics, $\Delta U = Q - W$ (Equation 15.1) to find the work W. The heat is $Q = -4700$ J, where the minus sign denotes that the system (the gas) loses heat. The internal energy U of a monatomic ideal gas is given by $U = \frac{3}{2}nRT$ (Equation 14.7), where n is the number of moles, R is the universal gas constant, and T is the Kelvin temperature. If the temperature remains constant during the process, the internal energy does not change, so $\Delta U = 0$ J.

SOLUTION The work done during the isothermal process is

$$W = Q - \Delta U = -4700\text{ J} + 0\text{ J} = \boxed{-4700\text{ J}}$$

The negative sign indicates that work is done on the system.

---

19. $\boxed{\text{SSM}}$ ***REASONING AND SOLUTION***

a. Since the temperature of the gas is kept constant at all times, the process is isothermal; therefore, the internal energy of an ideal gas does not change and $\boxed{\Delta U = 0}$.

b. From the first law of thermodynamics (Equation 15.1), $\Delta U = Q - W$. But $\Delta U = 0$, so that $Q = W$. Since work is done on the gas, the work is negative, and $\boxed{Q = -6.1 \times 10^3\text{ J}}$.

c. The work done in an isothermal compression is given by Equation 15.3:

$$W = nRT \ln\left(\frac{V_f}{V_i}\right)$$

Therefore, the temperature of the gas is

$$T = \frac{W}{nR\ln\left(V_f / V_i\right)} = \frac{-6.1 \times 10^3\text{ J}}{(3.0\text{ mol})[8.31\text{ J}/(\text{mol}\cdot\text{K})]\ \ln\left[(2.5 \times 10^{-2}\text{ m}^3)/(5.5 \times 10^{-2}\text{ m}^3)\right]} = \boxed{310\text{ K}}$$

---

20. ***REASONING*** Since the gas is expanding adiabatically, the work done is given by Equation 15.4 as $W = \frac{3}{2}nR\left(T_i - T_f\right)$. Once the work is known, we can use the first law of thermodynamics to find the change in the internal energy of the gas.

SOLUTION

a. The work done by the expanding gas is

$$W = \tfrac{3}{2}nR\left(T_i - T_f\right) = \tfrac{3}{2}(5.0\text{ mol})\left[8.31\text{ J}/(\text{mol}\cdot\text{K})\right](370\text{ K} - 290\text{ K}) = \boxed{+5.0 \times 10^3\text{ J}}$$

b. Since the process is adiabatic, Q = 0, and the change in the internal energy is

$$\Delta U = Q - W = 0 - 5.0 \times 10^3\text{ J} = \boxed{-5.0 \times 10^3\text{ J}}$$

---

21. ***REASONING*** During an adiabatic process, no heat flows into or out of the gas ($Q = 0$ J). For an ideas gas, the final pressure and volume (P_f and V_f) are related to the initial pressure and volume (P_i and V_i) by $P_i V_i^{\gamma} = P_f V_f^{\gamma}$ (Equation 15.5), where γ is the ratio of the specific

heat capacities at constant pressure and constant volume ($\gamma = \frac{5}{3}$ in this problem). We will use this relation to find V_f/V_i.

SOLUTION Solving $P_i V_i^\gamma = P_f V_f^\gamma$ for V_f/V_i and noting that the pressure doubles ($P_f/P_i = 2.0$) during the compression, we have

$$\frac{V_f}{V_i} = \left(\frac{P_i}{P_f}\right)^{\frac{1}{\gamma}} = \left(\frac{1}{2.0}\right)^{\frac{1}{(5/3)}} = \boxed{0.66}$$

---

22. ***REASONING*** An adiabatic process is one for which no heat enters or leaves the system, so $Q = 0$ J. The work is given as $W = +610$ J, where the plus sign denotes that the gas does work, according to our convention. Knowing the heat and the work, we can use the first law of thermodynamics to find the change ΔU in internal energy as $\Delta U = Q - W$ (Equation 15.1). Knowing the change in the internal energy, we can find the change in the temperature by recalling that the internal energy of a monatomic ideal gas is $U = \frac{3}{2}nRT$, according to Equation 14.7. As a result, it follows that $\Delta U = \frac{3}{2}nR\Delta T$.

SOLUTION Using the first law from Equation 15.5 and the change in internal energy from Equation 14.7, we have

$$\Delta U = Q - W \quad \text{or} \quad \tfrac{3}{2}nR\Delta T = Q - W$$

Therefore, we find

$$\Delta T = \frac{2(Q-W)}{3nR} = \frac{2[(0\text{ J})-(610\text{ J})]}{3(0.50\text{ mol})[8.31\text{ J}/(\text{mol}\cdot\text{K})]} = \boxed{-98\text{ K}}$$

The change in temperature is a decrease.

---

23. **SSM** ***REASONING*** When the expansion is isothermal, the work done can be calculated from Equation (15.3): $W = nRT\ln(V_f/V_i)$. When the expansion is adiabatic, the work done can be calculated from Equation 15.4: $W = \frac{3}{2}nR(T_i - T_f)$.

Since the gas does the same amount of work whether it expands adiabatically or isothermally, we can equate the right hand sides of these two equations. We also note that since the initial temperature is the same for both cases, the temperature T in the isothermal expansion is the same as the initial temperature T_i for the adiabatic expansion. We then have

$$nRT_i \ln\left(\frac{V_f}{V_i}\right) = \tfrac{3}{2}nR(T_i - T_f)$$

or

$$\ln\left(\frac{V_{\text{f}}}{V_{\text{i}}}\right)=\frac{\frac{3}{2}(T_{\text{i}}-T_{\text{f}})}{T_{\text{i}}}$$

SOLUTION Solving for the ratio of the volumes gives

$$\frac{V_{\text{f}}}{V_{\text{i}}}=e^{\frac{3}{2}(T_{\text{i}}-T_{\text{f}})/T_{\text{i}}}=e^{\frac{3}{2}(405\text{ K}-245\text{ K})/(405\text{ K})}=\boxed{1.81}$$

---

24. ***REASONING AND SOLUTION***

Step A → B

The internal energy of a monatomic ideal gas is $U=(3/2)nRT$. Thus, the change is

$$\Delta U=\tfrac{3}{2}nR\,\Delta T=\tfrac{3}{2}(1.00\text{ mol})\left[8.31\text{ J/}(\text{mol}\cdot\text{K})\right](800.0\text{ K}-400.0\text{ K})=\boxed{4990\text{ J}}$$

The work for this constant pressure step is $W=P\Delta V$. But the ideal gas law applies, so

$$W=P\Delta V=nR\,\Delta T=(1.00\text{ mol})\left[8.31\text{ J/}(\text{mol}\cdot\text{K})\right](800.0\text{ K}-400.0\text{ K})=\boxed{3320\text{ J}}$$

The first law of thermodynamics indicates that the heat is

$$Q=\Delta U+W=\tfrac{3}{2}nR\,\Delta T+nR\,\Delta T$$

$$=\tfrac{5}{2}(1.00\text{ mol})\left[8.31\text{ J/}(\text{mol}\cdot\text{K})\right](800.0\text{ K}-400.0\text{ K})=\boxed{8310\text{ J}}$$

Step B → C

The internal energy of a monatomic ideal gas is $U=(3/2)nRT$. Thus, the change is

$$\Delta U=\tfrac{3}{2}nR\,\Delta T=\tfrac{3}{2}(1.00\text{ mol})\left[8.31\text{ J/}(\text{mol}\cdot\text{K})\right](400.0\text{ K}-800.0\text{ K})=\boxed{-4990\text{ J}}$$

The volume is constant in this step, so the work done by the gas is $\boxed{\text{W}=0\text{ J}}$.

The first law of thermodynamics indicates that the heat is

$$Q=\Delta U+W=\Delta U=\boxed{-4990\text{ J}}$$

Step C → D

The internal energy of a monatomic ideal gas is $U=(3/2)nRT$. Thus, the change is

$$\Delta U = \tfrac{3}{2} nR\,\Delta T = \tfrac{3}{2}(1.00 \text{ mol})\left[8.31 \text{ J/}(\text{mol}\cdot\text{K})\right](200.0 \text{ K} - 400.0 \text{ K}) = \boxed{-2490 \text{ J}}$$

The work for this constant pressure step is $W = P\Delta V$. But the ideal gas law applies, so

$$W = P\Delta V = nR\,\Delta T = (1.00 \text{ mol})\left[8.31 \text{ J/}(\text{mol}\cdot\text{K})\right](200.0 \text{ K} - 400.0 \text{ K}) = \boxed{-1660 \text{ J}}$$

The first law of thermodynamics indicates that the heat is

$$Q = \Delta U + W = \tfrac{3}{2} nR\,\Delta T + nR\,\Delta T$$

$$= \tfrac{5}{2}(1.00 \text{ mol})\left[8.31 \text{ J/}(\text{mol}\cdot\text{K})\right](200.0 \text{ K} - 400.0 \text{ K}) = \boxed{-4150 \text{ J}}$$

Step D → A
The internal energy of a monatomic ideal gas is $U = (3/2)nRT$. Thus, the change is

$$\Delta U = \tfrac{3}{2} nR\,\Delta T = \tfrac{3}{2}(1.00 \text{ mol})\left[8.31 \text{ J/}(\text{mol}\cdot\text{K})\right](400.0 \text{ K} - 200.0 \text{ K}) = \boxed{2490 \text{ J}}$$

The volume is constant in this step, so the work done by the gas is $\boxed{W = 0 \text{ J}}$

The first law of thermodynamics indicates that the heat is

$$Q = \Delta U + W = \Delta U = \boxed{2490 \text{ J}}$$

---

25. ***REASONING AND SOLUTION***
a. Since the curved line between A and C is an isotherm, the initial and final temperatures are the same. Since the internal energy of an ideal monatomic gas is $U = (3/2)nRT$, the initial and final energies are also the same, and the change in the internal energy is $\Delta U = 0$. The first law of thermodynamics, then, indicates that for the process A→B→C, we have

$$\Delta U = 0 = Q - W \quad \text{or} \quad Q = W$$

The heat is equal to the work. Determining the work from the area beneath the straight line segments AB and BC, we find that

$$Q = W = -\left(4.00 \times 10^5 \text{ Pa}\right)\left(0.400 \text{ m}^3 - 0.200 \text{ m}^3\right) = \boxed{-8.00 \times 10^4 \text{ J}}$$

b. The minus sign is included because the gas is compressed, so that work is done on the gas. Since the answer for Q is negative, we conclude that $\boxed{\text{heat flows out of the gas}}$.

---

26. ***REASONING***

a. The work done by the gas is equal to the area under the pressure-versus-volume curve. We will measure this area by using the graph given with the problem.

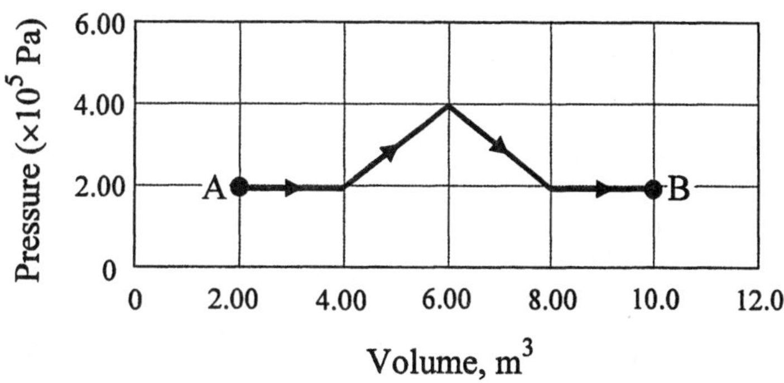

b. Since the gas is an ideal gas, it obeys the ideal gas law, $PV = nRT$ (Equation 14.1). This implies that $P_A V_A / T_A = P_B V_B / T_B$. All the variables except for T_B in this relation are known. Therefore, we can use this expression to find the temperature at point B.

c. The heat Q that has been added to or removed from the gas can be obtained from the first law of thermodynamics, $Q = \Delta U + W$ (Equation 15.1), where ΔU is the change in the internal energy of the gas and W is the work done by the gas. The work W is known from part (a) of the problem. The change ΔU in the internal energy of the gas can be obtained from Equation 14.7, $\Delta U = U_B - U_A = \frac{3}{2} nR\left(T_B - T_A\right)$, where n is the number of moles, R is the universal gas constant, and T_B and T_A are the final and initial Kelvin temperatures. We do not know n, but we can use the ideal gas law ($PV = nRT$) to replace nRT_B by $P_B V_B$ and to replace nRT_A by $P_A V_A$.

SOLUTION

a. From the drawing we see that the area under the curve is 5.00 "squares," where each square has an area of $\left(2.00\times10^5 \text{ Pa}\right)\left(2.00 \text{ m}^3\right) = 4.00\times10^5$ J. Therefore, the work W done by the gas is

$$W = (5.00 \text{ squares})\left(4.00\times10^5 \text{ J/square}\right) = \boxed{2.00\times10^6 \text{ J}}$$

b. In the Reasoning section, we have seen that $P_A V_A / T_A = P_B V_B / T_B$. Solving this relation for the temperature T_B at point B, using the fact that $P_A = P_B$ (see the graph), and taking the values for V_B and V_A from the graph, we have that

$$T_B = \left(\frac{P_B V_B}{P_A V_A}\right) T_A = \left(\frac{V_B}{V_A}\right) T_A = \left(\frac{10.0 \text{ m}^3}{2.00 \text{ m}^3}\right)(185 \text{ K}) = \boxed{925 \text{ K}}$$

c. From the Reasoning section we know that the heat Q that has been added to or removed from the gas is given by $Q = \Delta U + W$. The change ΔU in the internal energy of the gas is $\Delta U = U_B - U_A = \frac{3}{2}nR(T_B - T_A)$. Thus, the heat can be expressed as

$$Q = \Delta U + W = \tfrac{3}{2}nR(T_B - T_A) + W$$

We now use the ideal gas law ($PV = nRT$) to replace nRT_B by P_BV_B and nRT_A by P_AV_A. The result is

$$Q = \tfrac{3}{2}(P_BV_B - P_AV_A) + W$$

Taking the values for P_B, V_B, P_A, and V_A from the graph and using the result from part a that $W = 2.00 \times 10^6$ J, we find that the heat is

$$\begin{aligned} Q &= \tfrac{3}{2}(P_BV_B - P_AV_A) + W \\ &= \tfrac{3}{2}\left[(2.00\times10^5 \text{ Pa})(10.0 \text{ m}^3) - (2.00\times10^5 \text{ Pa})(2.00 \text{ m}^3)\right] + 2.00\times10^6 \text{ J} = \boxed{4.40\times10^6 \text{ J}} \end{aligned}$$

---

27. ***REASONING*** During an adiabatic process, no heat flows into or out of the gas ($Q = 0$ J). For an ideas gas, the final pressure and volume (P_f and V_f) are related to the initial pressure and volume (P_i and V_i) by $P_iV_i^\gamma = P_fV_f^\gamma$ (Equation 15.5), where γ is the ratio of the specific heat capacities at constant pressure and constant volume ($\gamma = \frac{7}{5}$ in this problem). The initial and final pressures are not given. However, the initial and final temperatures are known, so we can use the ideal gas law, $PV = nRT$ (Equation 14.1) to relate the temperatures to the pressures. We will then be able to find V_i/V_f in terms of the initial and final temperatures.

SOLUTION Substituting the ideal gas law, $PV = nRT$, into $P_iV_i^\gamma = P_fV_f^\gamma$ gives

$$\left(\frac{nRT_i}{V_i}\right)V_i^\gamma = \left(\frac{nRT_f}{V_f}\right)V_f^\gamma \quad \text{or} \quad T_iV_i^{\gamma-1} = T_fV_f^{\gamma-1}$$

Solving this expression for the ratio of the initial volume to the final volume yields

$$\frac{V_i}{V_f} = \left(\frac{T_f}{T_i}\right)^{\frac{1}{\gamma-1}}$$

The initial and final Kelvin temperatures are $T_i = (21\ ^\circ\text{C} + 273) = 294$ K and $T_f = (688\ ^\circ\text{C} + 273) = 961$ K. The ratio of the volumes is

$$\frac{V_i}{V_f} = \left(\frac{T_f}{T_i}\right)^{\frac{1}{\gamma-1}} = \left(\frac{961\text{ K}}{294\text{ K}}\right)^{\frac{1}{\left(\frac{7}{5}-1\right)}} = \boxed{19.3}$$

---

28. ***REASONING AND SOLUTION*** The three-step process is shown on the *P-V* diagram at the right. From the first law of thermodynamics,

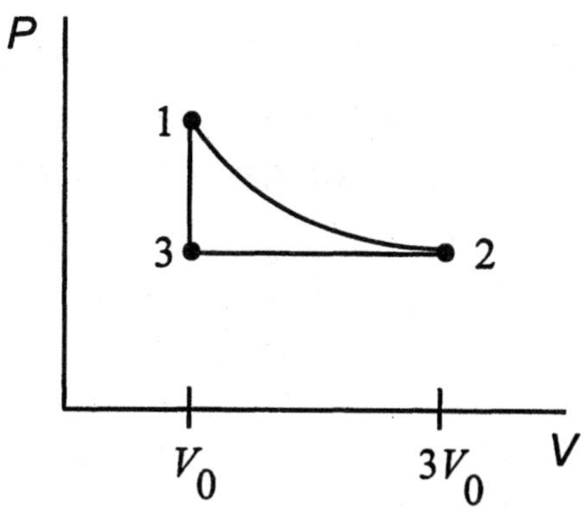

$$Q = \Delta U + W \qquad (1)$$

However, the ideal gas is back in its initial state at the end of the three-step process, so that $\Delta U = 0$ overall. With this value for ΔU, Equation (1) becomes $Q = W$, and we conclude that

$$Q = \underbrace{W_{1\to 2}}_{\text{isothermal}} + \underbrace{W_{2\to 3}}_{\text{isobaric}} + \underbrace{W_{3\to 1}}_{\text{isochoric}}$$

Using Equations 15.3 for the isothermal work and Equation 15.2 for the isobaric work, and remembering that there is no work done in an isochoric process, we find that

$$Q = nRT\ \ln\left(\frac{3V_0}{V_0}\right) + P\left(V_0 - 3V_0\right) = nRT\ \ln 3 - 2PV_0$$

In this result, $T = 438$ K, P is the pressure for step $2 \rightarrow 3$, and $2PV_0 = 2(P3V_0)/3 = 2nRT/3$. In addition, we know that $n = 1$ mol. Therefore,

$$Q = nRT\ \ln 3 - 2PV_0 = nRT\ \ln 3 - \tfrac{2}{3}nRT = nRT\left(\ln 3 - \tfrac{2}{3}\right)$$

$$= (1\text{ mol})\left[8.31\text{ J}(\text{mol}\cdot\text{K})\right](438\text{ K})\left(\ln 3 - \tfrac{2}{3}\right) = \boxed{1.57\times 10^{3}\text{ J}}$$

Since this answer is positive, $\boxed{\text{heat is absorbed}}$ by the gas.

---

29. SSM ***REASONING AND SOLUTION***

a. The final temperature of the adiabatic process is given by solving Equation 15.4 for T_f.

$$T_f = T_i - \frac{W}{\frac{3}{2}nR} = 393\text{ K} - \frac{825\text{ J}}{\frac{3}{2}(1.00\text{ mol})[8.31\text{ J}/(\text{mol}\cdot\text{K})]} = \boxed{327\text{ K}}$$

b. According to Equation 15.5 for the adiabatic expansion of an ideal gas, $P_i V_i^\gamma = P_f V_f^\gamma$. Therefore,

$$V_f^\gamma = V_i^\gamma \left(\frac{P_i}{P_f}\right)$$

From the ideal gas law, $PV = nRT$; therefore, the ratio of the pressures is given by

$$\frac{P_i}{P_f} = \left(\frac{T_i}{T_f}\right)\left(\frac{V_f}{V_i}\right)$$

Combining the previous two equations gives

$$V_f^\gamma = V_i^\gamma \left(\frac{T_i}{T_f}\right)\left(\frac{V_f}{V_i}\right)$$

Solving for V_f we obtain

$$\frac{V_f^\gamma}{V_f} = \left(\frac{T_i}{T_f}\right)\left(\frac{V_i^\gamma}{V_i}\right) \quad \text{or} \quad V_f^{(\gamma-1)} = \left(\frac{T_i}{T_f}\right) V_i^{(\gamma-1)}$$

$$V_f = \left[\left(\frac{T_i}{T_f}\right) V_i^{(\gamma-1)}\right]^{1/(\gamma-1)} = V_i \left(\frac{T_i}{T_f}\right)^{1/(\gamma-1)}$$

Therefore,

$$V_f = V_i \left(\frac{T_i}{T_f}\right)^{1/(\gamma-1)} = (0.100\text{ m}^3)\left(\frac{393\text{ K}}{327\text{ K}}\right)^{1/(2/3)} = (0.100\text{ m}^3)\left(\frac{393\text{ K}}{327\text{ K}}\right)^{3/2} = \boxed{0.132\text{ m}^3}$$

---

30. ***REASONING*** When the temperature of a gas changes as a result of heat Q being added, the change ΔT in temperature is related to the amount of heat according to $Q = Cn\Delta T$ (Equation 15.6), where C is the molar specific heat capacity, and n is the number of moles. The heat Q_V added under conditions of constant volume is $Q_V = C_V n \Delta T_V$, where C_V is the specific heat capacity at constant volume and is given by $C_V = \frac{3}{2}R$ (Equation 15.8) and R is the universal gas constant. The heat Q_P added under conditions of constant pressure is $Q_P = C_P n \Delta T_P$, where C_P is the specific heat capacity at constant pressure and is given by $C_P = \frac{5}{2}R$ (Equation 15.7). It is given that $Q_V = Q_P$, and

this fact will allow us to find the change in temperature of the gas whose pressure remains constant.

SOLUTION Setting $Q_V = Q_P$, gives

$$\underbrace{C_V n \Delta T_V}_{Q_V} = \underbrace{C_P n \Delta T_P}_{Q_P}$$

Algebraically eliminating n and solving for ΔT_P, we obtain

$$\Delta T_P = \left(\frac{C_V}{C_P}\right)\Delta T_V = \left(\frac{\frac{3}{2}R}{\frac{5}{2}R}\right)(75\text{ K}) = \boxed{45\text{ K}}$$

---

31. SSM ***REASONING AND SOLUTION*** The amount of heat required to change the temperature of the gas is given by Equation 15.6, where C_P is given by Equation 15.7.

$$Q = C_P n\Delta T = \tfrac{5}{2}Rn\Delta T = \tfrac{5}{2}[8.31\text{ J/(mol}\cdot\text{K)}]\ (1.5\text{ mol})\ (77\text{ K}) = \boxed{2400\text{ J}}$$

---

32. ***REASONING AND SOLUTION*** The heat required for an isobaric process is

$$Q = C_p n\,\Delta T = \left(\tfrac{5}{2}R\right)n\Delta T$$

$$= \tfrac{5}{2}\left[8.31\text{ J/(mol}\cdot\text{K)}\right]\left(\frac{8.0\text{ g}}{39.9\text{ g/mol}}\right)(75\text{ K}) = \boxed{310\text{ J}}$$

---

33. SSM ***REASONING AND SOLUTION*** According to the first law of thermodynamics (Equation 15.1), $\Delta U = U_f - U_i = Q - W$. Since the internal energy of this gas is doubled by the addition of heat, the initial and final internal energies are U and $2U$, respectively. Therefore,

$$\Delta U = U_f - U_i = 2U - U = U$$

Equation 15.1 for this situation then becomes $U = Q - W$. Solving for Q gives

$$Q = U + W \qquad (1)$$

The initial internal energy of the gas can be calculated from Equation 14.7:

$$U=\frac{3}{2}nRT=\frac{3}{2}(2.5\text{ mol})[8.31\text{ J/(mol}\cdot\text{K)}](350\text{ K}) = 1.1\times10^4\text{ J}$$

a. If the process is carried out isochorically (i.e., at constant volume), then $W=0$, and the heat required to double the internal energy is

$$Q=U+W=U+0=\boxed{1.1\times10^4\text{ J}}$$

b. If the process is carried out isobarically (i.e., at constant pressure), then $W=P\Delta V$, and Equation (1) above becomes

$$Q=U+W=U+P\Delta V \tag{2}$$

From the ideal gas law, $PV=nRT$, we have that $P\Delta V=nR\Delta T$, and Equation (2) becomes

$$Q=U+nR\Delta T \tag{3}$$

The internal energy of an ideal gas is directly proportional to its Kelvin temperature. Since the internal energy of the gas is doubled, the final Kelvin temperature will be twice the initial Kelvin temperature, or $\Delta T=350$ K. Substituting values into Equation (3) gives

$$Q=1.1\times10^4\text{ J}+(2.5\text{ mol})[8.31\text{ J/(mol}\cdot\text{K)}](350\text{ K})=\boxed{1.8\times10^4\text{ J}}$$

---

34. ***REASONING AND SOLUTION*** The amount of heat removed at constant volume is

$$Q=nC_V\Delta T=(2.5\text{ mol})\left(\tfrac{3}{2}R\right)(35\text{ K})=\boxed{1100\text{ J}}$$

---

35. ***REASONING AND SOLUTION*** The heat added at constant pressure is $Q=C_p n\Delta T=$ $(5R/2)\,n\Delta T$. The work done during the process is $W=P\Delta V$. The ideal gas law requires that $\Delta V=nR\Delta T/P$, so $W=nR\Delta T$. The required ratio is then

$$Q/W=\boxed{5/2}$$

---

36. ***REASONING AND SOLUTION*** The total heat generated by the students is

$$Q=(200)(130\text{ W})(3000\text{ s})=7.8\times10^7\text{ J}$$

For the isochoric process.

$$Q=C_V n\Delta T=(5R/2)n\,\Delta T$$

The number of moles of air in the room is found from the ideal gas law to be

$$n = \frac{PV}{RT} = \frac{(1.01\times10^5\ \text{Pa})(1200\ \text{m}^3)}{[8.31\ \text{J/(mol}\cdot\text{K)}](294\ \text{K})} = 5.0\times10^4\ \text{mol}$$

Now

$$\Delta T = \frac{Q}{\frac{5}{2}Rn} = \frac{7.8\times10^7\ \text{J}}{\frac{5}{2}[8.31\ \text{J/(mol}\cdot\text{K)}](5.0\times10^4\ \text{mol})} = \boxed{75\ \text{K}}$$

---

37. **SSM** ***REASONING*** According to Equations 15.6 and 15.7, the heat supplied to a monatomic ideal gas at constant pressure is $Q = C_P n\,\Delta T$, with $C_P = \frac{5}{2}R$. Thus, $Q = \frac{5}{2}nR\Delta T$. The percentage of this heat used to increase the internal energy by an amount ΔU is

$$\text{Percentage} = \left(\frac{\Delta U}{Q}\right)\times 100\ \% = \left(\frac{\Delta U}{\frac{5}{2}nR\,\Delta T}\right)\times 100\ \% \qquad (1)$$

But according to the first law of thermodynamics, $\Delta U = Q - W$. The work W is $W = P\Delta V$, and for an ideal gas $P\Delta V = nR\Delta T$. Therefore, the work W becomes $W = P\Delta V = nR\Delta T$ and the change in the internal energy is $\Delta U = Q - W = \frac{5}{2}nR\Delta T - nR\Delta T = \frac{3}{2}nR\Delta T$. Combining this expression for ΔU with Equation (1) above yields a numerical value for the percentage of heat being supplied to the gas that is used to increase its internal energy.

SOLUTION

a. The percentage is

$$\text{Percentage} = \left(\frac{\Delta U}{\frac{5}{2}nR\Delta T}\right)\times 100\ \% = \left(\frac{\frac{3}{2}nR\,\Delta T}{\frac{5}{2}nR\,\Delta T}\right)\times 100\ \% = \left(\frac{3}{5}\right)\times 100\ \% = \boxed{60.0\ \%}$$

b. The remainder of the heat, or $\boxed{40.0\%}$, is used for the work of expansion.

---

38. ***REASONING AND SOLUTION*** The change in volume is $\Delta V = -\ sA$, where s is the distance through which the piston drops and A is the piston area. The minus sign is included because the volume decreases. Thus,

$$s = \frac{-\Delta V}{A}$$

The ideal gas law states that $\Delta V = nR\Delta T/P$. But $Q = C_p n\Delta T = \frac{5}{2}Rn\Delta T$. Thus, $\Delta T = Q/\left(\frac{5}{2}Rn\right)$. Using these expressions for ΔV and ΔT, we find that

$$s = \frac{-nR\,\Delta T / P}{A} = \frac{-nR\left[Q/\left(\frac{5}{2}Rn\right)\right]}{PA} = \frac{-Q}{\frac{5}{2}PA}$$

$$= \frac{-(-2093\text{ J})}{\frac{5}{2}\left(1.01\times10^{5}\text{ Pa}\right)\left(3.14\times10^{-2}\text{ m}^2\right)} = \boxed{0.264\text{ m}}$$

---

39. ***REASONING AND SOLUTION*** Let P, V, and T represent the initial values of pressure, volume, and temperature. The first process is isochoric, so

$$Q_1 = C_V n\,\Delta T_1 = (3R/2)n\,\Delta T_1$$

The ideal gas law for this process gives $\Delta T_1 = 2PV/(nR)$, so $Q_1 = 3PV$.

The second process is isobaric, so

$$Q_2 = C_P n\,\Delta T_2 = (5R/2)n\,\Delta T_2$$

The ideal gas law for this process gives $\Delta T_2 = 3PV/nR$, so $Q_2 = (15/2)\,PV$. The total heat is $Q = Q_1 + Q_2 = (21/2)\,PV$.

But at conditions of standard temperature and pressure (see Section 14.2), $P = 1.01 \times 10^5$ Pa and $V = 22.4$ liters $= 22.4 \times 10^{-3}$ m^3, so

$$Q = \tfrac{21}{2}PV = \tfrac{21}{2}\left(1.01\times10^{5}\text{ Pa}\right)\left(22.4\times10^{-3}\text{ m}^3\right) = \boxed{2.38\times10^{4}\text{J}}$$

---

40. ***REASONING*** According to Equation 15.11, the efficiency of a heat engine is $e = W/Q_H$, where W is the work and Q_H is the input heat. Thus, the work is $W = eQ_H$. We can apply this result before and after the tune-up to compute the extra work produced.

SOLUTION Using Equation 15.11, we find the work before and after the tune-up as follows:

$$W_{\text{Before}} = e_{\text{Before}}Q_H \quad \text{and} \quad W_{\text{After}} = e_{\text{After}}Q_H$$

Subtracting the "before" equation from the "after" equation gives

$$W_{\text{After}} - W_{\text{Before}} = e_{\text{After}}Q_H - e_{\text{Before}}Q_H = \left(e_{\text{After}} - e_{\text{Before}}\right)Q_H$$

$$= 0.050(1300\text{ J}) = \boxed{65\text{ J}}$$

---

41. SSM ***REASONING AND SOLUTION*** The efficiency of a heat engine is defined by Equation 15.11 as $e = W / Q_H$, where W is the work done and Q_H is the heat input. The principle of energy conservation requires that $Q_H = W + Q_C$, where Q_C is the heat rejected to the cold reservoir (Equation 15.12). Combining Equations 15.11 and 15.12 gives

$$e = \frac{W}{W + Q_C} = \frac{16\ 600\ \text{J}}{16\ 600\ \text{J} + 9700\ \text{J}} = \boxed{0.631}$$

---

42. ***REASONING AND SOLUTION*** For engine A: $Q_C = 0.72 Q_H$, which means that the efficiency of engine A is

$$e_A = 1 - (Q_C/Q_H)_A = 1 - 0.72 = 0.28$$

Since engine B has twice this efficiency, we have $e_B = 1 - (Q_C/Q_H)_B = 0.56$. Solving gives

$$(Q_C/Q_H)_B = 1 - 0.56 = 0.44$$

So engine B discards $\boxed{44\%}$ of its input heat.

---

43. ***REASONING AND SOLUTION***
a. The efficiency is $e = W/Q_H$, so that

$$Q_H = W/e = (5500\ \text{J})/(0.64) = \boxed{8600\ \text{J}}$$

b. The rejected heat is

$$Q_C = Q_H - W = 8600\ \text{J} - 5500\ \text{J} = \boxed{3100\ \text{J}}$$

---

44. ***REASONING*** The efficiency e of an engine can be expressed as (see Equation 15.13) $e = 1 - (Q_C/Q_H)$, where Q_C is the heat delivered to the cold reservoir and Q_H is the heat supplied to the engine from the hot reservoir. Solving this equation for Q_C gives $Q_C = (1 - e)\ Q_H$. We will use this expression twice, once for the improved engine and once for the original engine. Taking the ratio of these expressions will give us the answer that we seek.

SOLUTION Taking the ratio of the heat rejected to the cold reservoir by the improved engine to that for the original engine gives

$$\frac{Q_{\text{C, improved}}}{Q_{\text{C, original}}} = \frac{\left(1-e_{\text{improved}}\right)Q_{\text{H, improved}}}{\left(1-e_{\text{original}}\right)Q_{\text{H, original}}}$$

But the input heat to both engines is the same, so $Q_{\text{H, improved}} = Q_{\text{H, original}}$. Thus, the ratio becomes

$$\frac{Q_{\text{C, improved}}}{Q_{\text{C, original}}} = \frac{1-e_{\text{improved}}}{1-e_{\text{original}}} = \frac{1-0.42}{1-0.23} = \boxed{0.75}$$

---

45. SSM ***REASONING AND SOLUTION*** We wish to find an expression for the overall efficiency e in terms of the efficiencies e_1 and e_2. From the problem statement, the overall efficiency of the two-engine device is

$$e = \frac{W_1 + W_2}{Q_{\text{H}}} \tag{1}$$

where Q_{H} is the input heat to engine 1. The efficiency of a heat engine is defined by Equation 15.11, $e = W / Q_{\text{H}}$, so we can write

$$W_1 = e_1 Q_{\text{H}} \tag{2}$$

and

$$W_2 = e_2 Q_{\text{H2}}$$

Since the heat rejected by engine 1 is used as input heat for the second engine, $Q_{\text{H2}} = Q_{\text{C1}}$, and the expression above for W_2 can be written as

$$W_2 = e_2 Q_{\text{C1}} \tag{3}$$

According to Equation 15.12, we have $Q_{\text{C1}} = Q_{\text{H}} - W_1$, so that Equation (3) becomes

$$W_2 = e_2 \left(Q_{\text{H}} - W_1\right) \tag{4}$$

Substituting Equations (2) and (4) into Equation (1) gives

$$e = \frac{e_1 Q_{\text{H}} + e_2 \left(Q_{\text{H}} - W_1\right)}{Q_{\text{H}}} = \frac{e_1 Q_{\text{H}} + e_2 \left(Q_{\text{H}} - e_1 Q_{\text{H}}\right)}{Q_{\text{H}}}$$

Algebraically canceling the Q_{H}'s in the right hand side of the last expression gives the desired result:

$$\boxed{e = e_1 + e_2 - e_1 e_2}$$

---

46. ***REASONING AND SOLUTION*** The efficiency is given by

$$e = 1 - (T_C/T_H) = 1 - [(200\text{ K})/(500\text{ K})] = 0.6 = W/Q_H$$

The work is

$$W = e\,Q_H = (0.6)(5000\text{ J}) = \boxed{3000\text{ J}}$$

---

47. ***REASONING AND SOLUTION*** The maximum efficiency is given by

$$e_{max} = 1 - (T_C/T_H) = 1 - [(620\text{ K})/(950\text{ K})] = 0.35$$

Since the engine operates at three-fifths maximum efficiency, its efficiency is

$$e = 0.60\, e_{max} = (0.60)(0.35) = \boxed{0.21}$$

---

48. ***REASONING*** We will use the subscript "27" to denote the engine whose efficiency is 27.0% (e_{27} = 0.270) and the subscript "32" to denote the engine whose efficiency is 32.0% (e_{32} = 0.320). In general, the efficiency e_{Carnot} of a Carnot engine depends on the Kelvin temperatures, T_C and T_H, of its cold and hot reservoirs through the relation (see Equation 15.15) $e_{Carnot} = 1 - (T_C/T_H)$. Solving this equation for the temperature $T_{C,\,32}$ of the engine whose efficiency is e_{32} gives $T_{C,\,32} = (1 - e_{32})T_{H,\,32}$. We are given e_{32}, but do not know the temperature $T_{H,\,32}$. However, we are told that this temperature is the same as that of the hot reservoir of the engine whose efficiency is e_{27}, so $T_{H,\,32} = T_{H,\,27}$. The temperature $T_{H,\,27}$ can be determined since we know the efficiency and cold reservoir temperature of this engine.

SOLUTION The temperature of the cold reservoir for engine whose efficiency is e_{32} is $T_{C,\,32} = (1 - e_{32})T_{H,\,32}$. Since $T_{H,\,32} = T_{H,\,27}$, we have that

$$T_{C,\,32} = (1 - e_{32})T_{H,\,27} \tag{1}$$

The efficiency e_{27} is given by Equation 15.15 as $e_{27} = 1 - (T_{C,\,27}/T_{H,\,27})$. Solving this equation for the temperature $T_{H,\,27}$ of the hot reservoir and substituting the result into Equation 1 yields

$$T_{C,32} = \left(\frac{1-e_{32}}{1-e_{27}}\right)T_{C,27} = \left(\frac{1-0.320}{1-0.270}\right)(275\text{ K}) = \boxed{256\text{ K}}$$

---

49. **SSM** ***REASONING*** The efficiency e of a Carnot engine is given by Equation 15.15, $e = 1-(T_C/T_H)$, where, according to Equation 15.14, $(Q_C/Q_H) = (T_C/T_H)$. Since the efficiency is given along with T_C and Q_C, Equation 15.15 can be used to calculate T_H. Once T_H is known, the ratio T_C/T_H is thus known, and Equation 15.14 can be used to calculate Q_H.

SOLUTION
a. Solving Equation 15.15 for T_H gives

$$T_H = \frac{T_C}{1-e} = \frac{378\text{ K}}{1-0.700} = \boxed{1260\text{ K}}$$

b. Solving Equation 15.14 for Q_H gives

$$Q_H = Q_C\left(\frac{T_H}{T_C}\right) = (5230\text{ J})\left(\frac{1260\text{ K}}{378\text{ K}}\right) = \boxed{1.74\times10^4\text{ J}}$$

---

50. ***REASONING*** The smallest possible temperature of the hot reservoir would occur when the engine is a Carnot engine, since it has the greatest efficiency of any engine operating between the same hot and cold reservoirs. The efficiency e_{Carnot} of a Carnot engine is (see Equation 15.15) $e_{Carnot} = 1 - (T_C/T_H)$, where T_C and T_H are the Kelvin temperatures of its cold and hot reservoirs. Solving this equation for T_H gives $T_H = T_C/(1 - e_{Carnot})$. We are given T_C, but do not know e_{Carnot}. However the efficiency is defined as the work W done by the engine divided by the input heat Q_H from the hot reservoir, so $e_{Carnot} = W/Q_H$ (Equation 15.11). Furthermore, the conservation of energy requires that the input heat Q_H equals the sum of the work W done by the engine and the heat Q_C it rejects to the cold reservoir, $Q_H = W + Q_C$. By combining these relations, we will be able to find the temperature of the hot reservoir of the Carnot engine.

SOLUTION From the Reasoning section, the temperature of the hot reservoir is $T_H = T_C/(1 - e_{Carnot})$. Writing the efficiency of the engine as $e_{Carnot} = W/Q_H$, the expression for the temperature becomes

$$T_H = \frac{T_C}{1-e_{Carnot}} = \frac{T_C}{1-\dfrac{W}{Q_H}}$$

From the conservation of energy, we have that $Q_H = W + Q_C$. Substituting this expression for Q_H into the one above for T_H gives

$$T_H = \frac{T_C}{1-\dfrac{W}{Q_H}} = \frac{T_C}{1-\dfrac{W}{W+Q_C}} = \frac{285\text{ K}}{1-\dfrac{18\ 500\text{ J}}{18\ 500\text{ J}+6550\text{ J}}} = \boxed{1090\text{ K}}$$

---

51. ***REASONING AND SOLUTION*** The efficiency of the engine is $e = 1 - (T_C/T_H)$ so

(i) Increase T_H by 40 K; $e = 1 - [(350\text{ K})/(690\text{ K})] = 0.493$

(ii) Decrease T_C by 40 K; $e = 1 - [(310\text{ K})/(650\text{ K})] = 0.523$

The greatest improvement is made by $\boxed{\text{lowering}}$ the temperature of the cold reservoir.

---

52. ***REASONING AND SOLUTION*** From Equation 15.14 we know that the input heat Q_H and the exhaust heat Q_C of a Carnot engine are related to the Kelvin temperatures of the hot and cold reservoirs according to $Q_H/Q_C = T_H/T_C$. We also know that $Q_C = Q_H - W$, according to Equation 15.12.

a. Therefore, we find T_1 as follows:

$$\frac{Q_H}{Q_C} = \frac{5550\text{ J}}{5550\text{ J} - 1750\text{ J}} = \frac{T_1}{503\text{ K}} \quad \text{or} \quad \boxed{T_1 = 735\text{ K}}$$

b. Similarly, we find T_2 as follows:

$$\frac{Q_H}{Q_C} = \frac{5550\text{ J} - 1750\text{ J}}{5550\text{ J} - 1750\text{ J} - 1750\text{ J}} = \frac{503\text{ K}}{T_2} \quad \text{or} \quad \boxed{T_2 = 271\text{ K}}$$

---

53. SSM ***REASONING*** The maximum efficiency e at which the power plant can operate is given by Equation 15.15, $e = 1-(T_C/T_H)$. The power output is given; it can be used to find the work output W for a 24 hour period. With the efficiency and W known, Equation 15.11, $e = W/Q_H$, can be used to find Q_H. Q_C can then be found from Equation 15.12, $Q_H = W + Q_C$.

SOLUTION

a. The maximum efficiency is

$$e = 1 - \frac{T_C}{T_H} = 1 - \frac{323\ \text{K}}{505\ \text{K}} = \boxed{0.360}$$

b. Since the power output of the power plant is $P = 84\ 000$ kW, the required heat input Q_H for a 24 hour period is

$$Q_H = \frac{W}{e} = \frac{Pt}{e} = \frac{(8.4 \times 10^7\ \text{J/s})(24\ \text{h})}{0.360}\left(\frac{3600\ \text{s}}{1\ \text{h}}\right) = 2.02 \times 10^{13}\ \text{J}$$

Therefore, solving Equation 15.12 for Q_C, we have

$$Q_C = Q_H - W = 2.02 \times 10^{13}\ \text{J} - 7.3 \times 10^{12}\ \text{J} = \boxed{1.3 \times 10^{13}\ \text{J}}$$

---

54. ***REASONING AND SOLUTION*** The temperature of the gasoline engine input is $T_1 = 904$ K, the exhaust temperature is $T_2 = 412$ K, and the air temperature is $T_3 = 300$ K. The efficiency of the engine/exhaust is

$$e_1 = 1 - (T_2/T_1) = 0.544$$

The efficiency of the second engine is

$$e_2 = 1 - (T_3/T_2) = 0.272$$

The work done by each segment is $W_1 = e_1 Q_{H1}$ and $W_2 = e_2 Q_{H2} = e_2 Q_{C1}$ since

$$Q_{H2} = Q_{C1}$$

Now examine $(W_1 + W_2)/W_1$ to find the ratio of the total work produced by both engines to that produced by the first engine alone.

$$(W_1 + W_2)/W_1 = (e_1 Q_{H1} + e_2 Q_{C1})/(e_1 Q_{H1}) = 1 + (e_2/e_1)(Q_{C1}/Q_{H1})$$

But, $e_1 = 1 - (Q_{C1}/Q_{H1})$, so that $(Q_{C1}/Q_{H1}) = 1 - e_1$. Therefore,

$$\frac{W_1+W_2}{W_1} = 1 + \frac{e_2}{e_1}(1-e_1)$$

$$= 1 + \frac{e_2}{e_1} - e_2 = 1 + 0.500 - 0.272 = \boxed{1.23}$$

---

55. SSM ***REASONING AND SOLUTION*** The efficiency e of the power plant is three-fourths its Carnot efficiency so, according to Equation 15.15,

$$e = 0.75\left(1 - \frac{T_C}{T_H}\right) = 0.75\left(1 - \frac{40\text{ K} + 273\text{ K}}{285\text{ K} + 273\text{ K}}\right) = 0.33$$

The power output of the plant is 1.2×10^9 watts. According to Equation 15.11, $e = W/Q_H = (\text{Power} \cdot t)/Q_H$. Therefore, at 33% efficiency, the heat input per unit time is

$$\frac{Q_H}{t} = \frac{\text{Power}}{e} = 3.6 \times 10^9 \text{ J/s}$$

From the principle of conservation of energy, the heat output per unit time must be

$$\frac{Q_C}{t} = \frac{Q_H}{t} - \text{Power} = 2.4 \times 10^9 \text{ J/s}$$

The rejected heat is carried away by the flowing water and, according to Equation 12.4, $Q_C = cm\Delta T$. Therefore,

$$\frac{Q_C}{t} = \frac{cm\Delta T}{t} \quad \text{or} \quad \frac{Q_C}{t} = c(m/t)\Delta T$$

Solving the last equation for ΔT, we have

$$\Delta T = \frac{Q_C}{tc(m/t)} = \frac{(Q_C/t)}{c(m/t)} = \frac{2.4 \times 10^9 \text{ J/s}}{[4186 \text{ J/(kg} \cdot \text{C}^\circ)](1.0 \times 10^5 \text{ kg/s})} = \boxed{5.7\text{ C}^\circ}$$

---

56. ***REASONING AND SOLUTION*** The engine performs work to lift the block where,

$$W = (1/2)mv^2 + mgh = (1/2)(15.0\text{ kg})(8.50\text{ m/s})^2 + (15.0\text{ kg})(9.80\text{ m/s}^2)(5.00\text{ m}) = 1280\text{ J}$$

The efficiency is $e = 1 - (T_C/T_H) = W/Q_H$, so that

$$Q_H = \frac{W}{1-\frac{T_C}{T_H}} = \frac{1280\text{ J}}{1-\frac{395\text{ K}}{845\text{ K}}} = \boxed{2.40\times10^3\text{ J}}$$

---

57. ***REASONING AND SOLUTION*** The heat delivered to the kitchen is Q_H, where

$$Q_H = Q_C(T_H/T_C) = (3.00\times10^4\text{ J})(298\text{ K})/(276\text{ K}) = \boxed{3.24\times10^4\text{ J}}$$

---

58. ***REASONING*** For any refrigerator, the first law of thermodynamics (Equation 15.12) indicates that $W = Q_H - Q_C$. In this expression, we know that $W = 2500$ J and wish to find Q_C. To do so, we need information about Q_H. But the refrigerator is a Carnot device, so we know in addition that $Q_C/Q_H = T_C/T_H$ (Equation 15.14). With this additional equation, we can solve for Q_H and substitute into the first law, obtaining in the process an equation that contains only Q_C as an unknown.

SOLUTION From Equation 15.14 we have

$$\frac{Q_C}{Q_H} = \frac{T_C}{T_H} \quad \text{or} \quad Q_H = Q_C\left(\frac{T_H}{T_C}\right)$$

Substituting this expression for Q_H into the first law of thermodynamics gives

$$W = Q_H - Q_C = Q_C\left(\frac{T_H}{T_C}\right) - Q_C$$

Solving for Q_C, we find

$$Q_C = \frac{W}{\frac{T_H}{T_C}-1} = \frac{2500\text{ J}}{\frac{299\text{ K}}{277\text{ K}}-1} = \boxed{3.1\times10^4\text{ J}}$$

---

59. SSM WWW ***REASONING AND SOLUTION*** Equation 15.14 holds for a Carnot air conditioner as well as a Carnot engine. Therefore, solving Equation 15.14 for Q_C, we have

$$Q_C = Q_H\left(\frac{T_C}{T_H}\right) = (6.12\times10^5\text{ J})\left(\frac{299\text{ K}}{312\text{ K}}\right) = \boxed{5.86\times10^5\text{ J}}$$

---

60. ***REASONING*** The coefficient of performance COP is defined as COP = Q_C/W (Equation 15.16), where Q_C is the heat removed from the cold reservoir and W is the work done on the refrigerator. The work is related to the heat Q_H deposited into the hot reservoir and the heat Q_C taken from the cold reservoir by the conservation of energy, $W = Q_H - Q_C$. Thus, the coefficient of performance can be written as (after some algebraic manipulations)

$$\text{COP} = \frac{1}{\dfrac{Q_H}{Q_C} - 1}$$

The maximum coefficient of performance occurs when the refrigerator is a Carnot refrigerator. For a Carnot refrigerator, the ratio Q_H/Q_C is equal to the ratio T_H/T_C of the Kelvin temperatures of the hot and cold reservoirs, $Q_H/Q_C = T_H/T_C$ (Equation 15.14).

SOLUTION Substituting $Q_H/Q_C = T_H/T_C$ into the expression above for the COP gives

$$\text{COP} = \frac{1}{\dfrac{T_H}{T_C} - 1} = \frac{1}{\dfrac{296\text{ K}}{275\text{ K}} - 1} = \boxed{13}$$

---

61. ***REASONING AND SOLUTION*** We know that the efficiency is given by $e = 1 - (T_C/T_H) = 1 - [(265\text{ K})/(298\text{ K})] = 0.111$. The coefficient of performance is

$$\text{CP} = \frac{Q_H}{W} = \frac{1}{e} = \frac{1}{0.111} = \boxed{9.03}$$

---

62. ***REASONING*** The coefficient of performance of an air conditioner is Q_C/W, according to Equation 15.16, where Q_C is the heat removed from the house and W is the work required for the removal. In addition, we know that the first law of thermodynamics (energy conservation) applies, so that $W = Q_H - Q_C$, according to Equation 15.12. In this equation Q_H is the heat discarded outside. While we have no direct information about the heats Q_C and Q_H, we do know that the air conditioner is a Carnot device. This means that Equation 15.14 applies: $Q_C/Q_H = T_C/T_H$. Thus, the given temperatures will allow us to calculate the coefficient of performance.

SOLUTION Using Equation 15.16 for the definition of the coefficient of performance and Equation 15.12 for the fact that $W = Q_H - Q_C$, we have

$$\text{Coefficient of performance} = \frac{Q_C}{W} = \frac{Q_C}{Q_H - Q_C} = \frac{Q_C / Q_H}{1 - Q_C / Q_H}$$

Equation 15.14 applies, so that $Q_C/Q_H = T_C/T_H$. With this substitution, we find

$$\text{Coefficient of performance} = \frac{Q_C / Q_H}{1 - Q_C / Q_H} = \frac{T_C / T_H}{1 - T_C / T_H}$$

$$= \frac{T_C}{T_H - T_C} = \frac{297 \text{ K}}{(311 \text{ K}) - (297 \text{ K})} = \boxed{21}$$

---

63. ***REASONING AND SOLUTION*** Let CP denote the coefficient of performance. By definition (Equation 15.16), CP $= Q_C/W$, so that

$$W = \frac{Q_C}{\text{CP}} = \frac{7.6 \times 10^4 \text{ J}}{2.0} = 3.8 \times 10^4 \text{ J}$$

Thus, the amount of heat that is pumped out of the back of the air conditioner is

$$Q_H = W + Q_C = 3.8 \times 10^4 \text{ J} + 7.6 \times 10^4 \text{ J} = 1.14 \times 10^5 \text{ J}$$

The temperature rise in the room can be found as follows:

$$Q_H = C_V n \Delta T$$

Solving for ΔT gives

$$\Delta T = \frac{Q_H}{C_V n} = \frac{Q_H}{\left(\frac{5}{2}R\right)n} = \frac{1.14 \times 10^5 \text{ J}}{\frac{5}{2}[8.31 \text{ J/(mol}\cdot\text{K)}](3800 \text{ mol})} = \boxed{1.4 \text{ K}}$$

---

64. ***REASONING AND SOLUTION*** The amount of heat removed from the ice Q_C is

$$Q_C = mL_f = (2.0 \text{ kg})(33.5 \times 10^4 \text{ J/kg}) = 6.7 \times 10^5 \text{ J}$$

The amount of heat leaving the refrigerator Q_H is therefore,

$$Q_H = Q_C(T_H/T_C) = (6.7 \times 10^5 \text{ J})(300 \text{ K})/(258 \text{ K}) = 7.8 \times 10^5 \text{ J}$$

The amount of work done by the refrigerator is therefore,

$$W = Q_H - Q_C = 1.1 \times 10^5\ \text{J}$$

At \$0.10 per kWh (or \$0.10 per 3.6×10^6 J), the cost is

$$\frac{\$0.10}{3.6 \times 10^6\ \text{J}}\left(1.1 \times 10^5\ \text{J}\right) = \boxed{\$3.0 \times 10^{-3} = 0.30\ \text{cents}}$$

---

65. SSM WWW ***REASONING*** Let the coefficient of performance be represented by the symbol CP. Then according to Equation 15.16, $\text{CP} = Q_C / W$. From the statement of energy conservation for a Carnot refrigerator (Equation 15.12), $W = Q_H - Q_C$. Combining Equations 15.16 and 15.12 leads to

$$\text{CP} = \frac{Q_C}{Q_H - Q_C} = \frac{Q_C / Q_C}{(Q_H - Q_C)/Q_C} = \frac{1}{(Q_H / Q_C) - 1}$$

Replacing the ratio of the heats with the ratio of the Kelvin temperatures, according to Equation 15.14, leads to

$$\text{CP} = \frac{1}{(T_H / T_C) - 1} \qquad (1)$$

The heat Q_C that must be removed from the refrigerator when the water is cooled can be calculated using Equation 12.4, $Q_C = cm\Delta T$; therefore,

$$W = \frac{Q_C}{\text{CP}} = \frac{cm\Delta T}{\text{CP}} \qquad (2)$$

SOLUTION

a. Substituting values into Equation (1) gives

$$\text{CP} = \frac{1}{\frac{T_H}{T_C} - 1} = \frac{1}{\frac{(20.0+273.15)\ \text{K}}{(6.0+273.15)\ \text{K}} - 1} = \boxed{2.0 \times 10^1}$$

b. Substituting values into Equation (2) gives

$$W = \frac{cm\Delta T}{\text{CP}} = \frac{[4186\ \text{J/(kg} \cdot \text{C}^\circ)](5.00\ \text{kg})(14.0\ \text{C}^\circ)}{2.0 \times 10^1} = \boxed{1.5 \times 10^4\ \text{J}}$$

---

66. ***REASONING*** According to the conservation of energy, the work W done by the electrical energy is $W = Q_H - Q_C$, where Q_H is the heat delivered to the outside (the hot reservoir) and

Q_C is the heat removed from the house (the cold reservoir). Dividing both sides of this relation by the time t, we have

$$\frac{W}{t} = \frac{Q_H}{t} - \frac{Q_C}{t}$$

The term W/t is the work per second that must be done by the electrical energy, and the terms Q_H/t and Q_C/t are, respectively, the heat per second delivered to the outside and removed from the house. Since the air conditioner is a Carnot air conditioner, we know that Q_H/Q_C is equal to the ratio T_H/T_C of the Kelvin temperatures of the hot and cold reservoirs, $Q_H/Q_C = T_H/T_C$ (Equation 15.14). This expression, along with the one above for W/t, will allow us to determine the work per second done by the electrical energy.

SOLUTION Solving the expression $Q_H/Q_C = T_H/T_C$ for Q_H, substituting the result into the relation $\frac{W}{t} = \frac{Q_H}{t} - \frac{Q_C}{t}$, and recognizing that $Q_C/t = 10\ 500$ J/s, give

$$\frac{W}{t} = \frac{Q_H}{t} - \frac{Q_C}{t} = \frac{\frac{Q_C T_H}{T_C}}{t} - \frac{Q_C}{t}$$

$$= \left(\frac{Q_C}{t}\right)\left(\frac{T_H}{T_C} - 1\right) = (10\ 500 \text{ J/s})\left(\frac{306.15 \text{ K}}{292.15 \text{ K}} - 1\right) = \boxed{5.0 \times 10^2 \text{ J/s}}$$

In this result we have used the fact that $T_H = 273.15 + 33.0\ °\text{C} = 306.15$ K and $T_C = 273.15 + 19.0\ °\text{C} = 292.15$ K.

---

67. **SSM** ***REASONING*** The efficiency of the Carnot engine is, according to Equation 15.15,

$$e = 1 - \frac{T_C}{T_H} = 1 - \frac{842 \text{ K}}{1684 \text{ K}} = \frac{1}{2}$$

Therefore, the work delivered by the engine is, according to Equation 15.11,

$$W = e\, Q_H = \tfrac{1}{2} Q_H$$

The heat pump removes an amount of heat Q_H from the cold reservoir. Thus, the amount of heat Q' delivered to the hot reservoir of the heat pump is

$$Q' = Q_H + W = Q_H + \tfrac{1}{2} Q_H = \tfrac{3}{2} Q_H$$

Therefore, $Q'/Q_H = 3/2$. According to Equation 15.14, $Q'/Q_H = T'/T_C$, so $T'/T_C = 3/2$.

SOLUTION Solving for T' gives

$$T' = \frac{3}{2}T_C = \frac{3}{2}(842 \text{ K}) = \boxed{1.26\times10^3 \text{ K}}$$

---

68. ***REASONING AND SOLUTION*** The change in entropy is $\Delta S = Q/T$, where

$$Q = mL_s = (4.00 \text{ kg})(5.77 \times 10^5 \text{ J/kg}) = 2.31 \times 10^6 \text{ J}$$

Thus,

$$\Delta S = Q/T = (2.31 \times 10^6 \text{ J})/(194.7 \text{ K}) = \boxed{1.19\times10^4 \text{ J/K}}$$

---

69. ***REASONING*** According to the discussion on Section 15.11, the change $\Delta S_{universe}$ in the entropy of the universe is the sum of the change in entropy ΔS_C of the cold reservoir and the change in entropy ΔS_H of the hot reservoir, or $\Delta S_{universe} = \Delta S_C + \Delta S_H$. The change in entropy of each reservoir is given by Equation 15.18 as $\Delta S = (Q/T)_R$, where Q is the heat removed from or delivered to the reservoir and T is the Kelvin temperature of the reservoir. In applying this equation we imagine a process in which the heat is lost by the house and gained by the outside in a reversible fashion.

SOLUTION Since heat is lost from the hot reservoir (inside the house), the change in entropy is negative: $\Delta S_H = -Q_H/T_H$. Since heat is gained by the cold reservoir (the outdoors), the change in entropy is positive: $\Delta S_C = +Q_C/T_C$. Here we are using the symbols Q_H and Q_C to denote the magnitudes of the heats. The change in the entropy of the universe is

$$\Delta S_{universe} = \Delta S_C + \Delta S_H = \frac{Q_C}{T_C} - \frac{Q_H}{T_H} = \frac{24\,500 \text{ J}}{258 \text{ K}} - \frac{24\,500 \text{ J}}{294 \text{ K}} = \boxed{11.6 \text{ J/K}}$$

In this calculation we have used the fact that $T_C = 273 - 15\ °\text{C} = 258$ K and $T_H = 273 + 21\ °\text{C} = 294$ K.

---

70. ***REASONING AND SOLUTION*** Equation 15.19 gives the unavailable work as

$$W_{unavailable} = T_0\Delta S \tag{1}$$

where $T_0 = 248\text{ K}$. We also know that $W_{\text{unavailable}} = 0.300\,Q$. Furthermore, we can apply Equation 15.18 to the heat lost from the 394-K reservoir and the heat gained by the reservoir at temperature T, with the result that

$$\Delta S = \frac{-Q}{394\text{ K}} + \frac{Q}{T}$$

With these substitutions for T_0, $W_{\text{unavailable}}$, and ΔS, Equation (1) becomes

$$0.300\,Q = (248\text{ K})\left(\frac{-Q}{394\text{ K}} + \frac{Q}{T}\right) \quad \text{or} \quad \boxed{T = 267\text{ K}}$$

---

71. SSM ***REASONING AND SOLUTION*** The change in entropy ΔS of a system for a process in which heat Q enters or leaves the system reversibly at a constant temperature T is given by Equation 15.18, $\Delta S = (Q/T)_\text{R}$. For a phase change, $Q = mL$, where L is the latent heat (see Section 12.8).

a. If we imagine a reversible process in which 3.00 kg of ice melts into water at 273 K, the change in entropy of the water molecules is

$$\Delta S = \left(\frac{Q}{T}\right)_\text{R} = \left(\frac{mL_\text{f}}{T}\right)_\text{R} = \frac{(3.00\text{ kg})(3.35\times10^5\text{ J/kg})}{273\text{ K}} = \boxed{3.68\times10^3\text{ J/K}}$$

b. Similarly, if we imagine a reversible process in which 3.00 kg of water changes into steam at 373 K, the change in entropy of the water molecules is

$$\Delta S = \left(\frac{Q}{T}\right)_\text{R} = \left(\frac{mL_\text{v}}{T}\right)_\text{R} = \frac{(3.00\text{ kg})(2.26\times10^6\text{ J/kg})}{373\text{ K}} = \boxed{1.82\times10^4\text{ J/K}}$$

c. Since the change in entropy is greater for the vaporization process than for the fusion process, the **vaporization process creates more disorder** in the collection of water molecules.

---

72. ***REASONING AND SOLUTION***

a. We know that the hot and cold waters exchange equal amounts of heat, i.e.,

$\Delta Q_\text{hw} = \Delta Q_\text{cw}$, so that $(mc\Delta T)_\text{hw} = (mc\Delta T)_\text{cw}$, or

$$(1.00 \text{ kg})[4186 \text{ J/(kg}\cdot\text{C}^\circ)](373 \text{ K} - T_f) = (2.00 \text{ kg})[4186 \text{ J/(kg}\cdot\text{C}^\circ)](T_f - 283 \text{ K})$$

Solving for T_f, we obtain $\boxed{T_f = 313\text{K}}$.

b. Since $\Delta S = mc \ln(T_f/T_i)$:

$$\Delta S_{hw} = m_{hw}c \ln[(313 \text{ K})/(373 \text{ K})] = -734 \text{ J/K}$$

$$\Delta S_{cw} = m_{cw}c \ln[(313 \text{ K})/(283 \text{ K})] = +844 \text{ J/K}$$

Therefore,

$$\Delta S_{universe} = \Delta S_{hw} + \Delta S_{cw} = \boxed{1.10 \times 10^2 \text{ J/K}}$$

c. The energy unavailable for doing work is, therefore,

$$W_{unavailable} = T_0 \, \Delta S_{universe} = (273 \text{ K})(1.10 \times 10^2 \text{ J/K}) = \boxed{3.00 \times 10^4 \text{ J}}$$

---

73. ***REASONING*** The change $\Delta S_{universe}$ in entropy of the universe for this process is the sum of the entropy changes for (1) the warm water (ΔS_{water}) as it cools down from its initial temperature of 85.0 °C to its final temperature T_f, (2) the ice (ΔS_{ice}) as it melts at 0 °C, and (3) the ice water ($\Delta S_{ice\ water}$) as it warms up from 0 °C to the final temperature T_f: $\Delta S_{universe} = \Delta S_{water} + \Delta S_{ice} + \Delta S_{ice\ water}$.

To find the final temperature T_f, we will follow the procedure outlined in Sections 12.7 and 12.8, where we set the heat lost by the warm water as it cools down equal to the heat gained by the melting ice and the resulting ice water as it warms up. The heat Q that must be supplied or removed to change the temperature of a substance of mass m by an amount ΔT is $Q = cm\Delta T$ (Equation 12.4), where c is the specific heat capacity. The heat that must be supplied to melt a mass m of a substance is $Q = mL_f$ (Equation 12.5), where L_f is the latent heat of fusion.

SOLUTION

a. We begin by finding the final temperature T_f of the water. Setting the heat lost equal to the heat gained gives

$$\underbrace{cm_{water}\left(85.0\ ^\circ\text{C} - T_f\right)}_{\text{Heat lost by water}} = \underbrace{m_{ice}L_f}_{\substack{\text{Heat gained}\\ \text{by melting ice}}} + \underbrace{m_{ice}c\left(T_f - 0.0\ ^\circ\text{C}\right)}_{\text{Heat gained by ice water}}$$

Solving this relation for the final temperature T_f yields

$$T_f = \frac{cm_{\text{water}}(85.0\ °\text{C}) - m_{\text{ice}}L_f}{c(m_{\text{ice}} + m_{\text{water}})}$$

$$= \frac{[4186\ \text{J}/(\text{kg}\cdot\text{C}°)](6.00\ \text{kg})(85.0\ °\text{C}) - (3.00\ \text{kg})(33.5\times10^4\ \text{J/kg})}{[4186\ \text{J}/(\text{kg}\cdot\text{C}°)](3.00\ \text{kg} + 6.00\ \text{kg})} = 30.0\ °\text{C}$$

We have taken the specific heat capacity of $4186\ \text{J}/(\text{kg}\cdot\text{C}°)$ for water from Table 12.2 and the latent heat of 33.5×10^4 J/kg from Table 12.3. This temperature is equivalent to $T_f = (273 + 30.0\ °\text{C}) = 303\ \text{K}$.

The change $\Delta S_{\text{universe}}$ in the entropy of the universe is the sum of three contributions:

[Contribution 1]

$$\Delta S_{\text{water}} = m_{\text{water}}c \ln\left(\frac{T_f}{T_i}\right) = (6.00\ \text{kg})[4186\ \text{J}/(\text{kg}\cdot\text{C}°)] \ln\left(\frac{303\ \text{K}}{358\ \text{K}}\right) = -4190\ \text{J/K}$$

where $T_i = 273 + 85.0\ °\text{C} = 358\ \text{K}$.

[Contribution 2]

$$\Delta S_{\text{ice}} = \frac{Q}{T} = \frac{mL_f}{T} = \frac{(3.00\ \text{kg})(33.5\times10^4\ \text{J/kg})}{273\ \text{K}} = +3680\ \text{J/K}$$

[Contribution 3]

$$\Delta S_{\text{ice water}} = m_{\text{ice}}c \ln\left(\frac{T_f}{T_i}\right) = (3.00\ \text{kg})[4186\ \text{J}/(\text{kg}\cdot\text{C}°)] \ln\left(\frac{303\ \text{K}}{273\ \text{K}}\right) = +1310\ \text{J/K}$$

The change in the entropy of the universe is

$$\Delta S_{\text{universe}} = \Delta S_{\text{water}} + \Delta S_{\text{ice}} + \Delta S_{\text{ice water}} = \boxed{+8.0\times10^2\ \text{J/K}}$$

b. The entropy of the universe **increases**, because the mixing process is irreversible.

---

74. ***REASONING AND SOLUTION*** As Section 14.1 discusses, the number of moles n is given by the mass m divided by the mass per mole:

$$n = \frac{m}{\text{Mass per mole}} = \frac{6.0\ \text{g}}{4.0\ \text{g/mol}} = 1.5\ \text{mol}$$

For an isothermal process we have (see Equation 15.3)

$$\ln\left(\frac{V_f}{V_i}\right) = \frac{W}{nRT} = \frac{9600\ \text{J}}{(1.5\ \text{mol})\left[8.31\ \text{J}/(\text{mol}\cdot\text{K})\right](370\ \text{K})} = 2.08$$

Therefore, $V_f/V_i = e^{2.08} = \boxed{8.0}$.

---

75. SSM ***REASONING*** According to the first law of thermodynamics (Equation 15.1), $\Delta U = Q - W$. For a monatomic ideal gas (Equation 14.7), $U = \frac{3}{2}nRT$. Therefore, for the process in question, the change in the internal energy is $\Delta U = \frac{3}{2}nR\Delta T$. Combining the last expression for ΔU with Equation 15.1 yields

$$\tfrac{3}{2}nR\Delta T = Q - W$$

This expression can be solved for ΔT.

SOLUTION
a. The heat is $Q = +1200\ \text{J}$, since it is absorbed by the system. The work is $W = +2500\ \text{J}$, since it is done *by* the system. Solving the above expression for ΔT and substituting the values for the data given in the problem statement, we have

$$\Delta T = \frac{Q-W}{\frac{3}{2}nR} = \frac{1200\ \text{J} - 2500\ \text{J}}{\frac{3}{2}(0.50\ \text{mol})[8.31\ \text{J}/(\text{mol}\cdot\text{K})]} = \boxed{-2.1\times10^{2}\ \text{K}}$$

b. Since $\Delta T = T_{\text{final}} - T_{\text{initial}}$ is negative, T_{initial} must be greater than T_{final}; this change represents a $\boxed{\text{decrease}}$ in temperature.

Alternatively, one could deduce that the temperature decreases from the following physical argument. Since the system loses more energy in doing work than it gains in the form of heat, the internal energy of the system decreases. Since the internal energy of an ideal gas depends only on the temperature, a decrease in the internal energy must correspond to a decrease in the temperature.

---

76. ***REASONING AND SOLUTION*** We know that $Q_H = W + Q_C$, so that

$$W = Q_H - Q_C = 2.41\times10^{4}\ \text{J} - 5.86\times10^{3}\ \text{J} = \boxed{1.82\times10^{4}\ \text{J}}$$

---

77. ***REASONING*** We will apply the first law of thermodynamics as given in Equation 15.1 ($\Delta U = Q - W$) to the overall process. First, however, we add the changes in the internal energy to obtain the overall change ΔU and add the work values to get the overall work W.

SOLUTION In both steps the internal energy increases, so overall we have $\Delta U = 228\text{ J} + 115\text{ J} = +343\text{ J}$. In both steps the work is negative according to our convention, since it is done on the system. Overall, then, we have $W = -166\text{ J} - 177\text{ J} = -343\text{ J}$. Using the first law of thermodynamics from Equation 15.1, we find

$$\Delta U = Q - W \quad \text{or} \quad Q = \Delta U + W = (+343\text{ J}) + (-343\text{ J}) = \boxed{0\text{ J}}$$

Since the heat is zero, the overall process is $\boxed{\text{adiabatic}}$.

---

78. ***REASONING AND SOLUTION*** We know that

$$Q_C = Q_H - W = 14\,200\text{ J} - 800\text{ J} = 13\,400\text{ J}$$

Therefore,

$$T_C = T_H(Q_C/Q_H) = (301\text{ K})[(13\,400\text{ J})/(14\,200\text{ J})] = \boxed{284\text{ K}}$$

---

79. $\boxed{\text{SSM}}$ ***REASONING AND SOLUTION***

a. Since the energy that becomes unavailable for doing work is zero for the process, we have from Equation 15.19, $W_{\text{unavailable}} = T_0 \Delta S_{\text{universe}} = 0$. Therefore, $\Delta S_{\text{universe}} = 0$ and according to the discussion in Section 15.11, the process is $\boxed{\text{reversible}}$.

b. Since the process is reversible, we have (see Section 15.11)

$$\Delta S_{\text{universe}} = \Delta S_{\text{system}} + \Delta S_{\text{surroundings}} = 0$$

Therefore,

$$\Delta S_{\text{surroundings}} = -\Delta S_{\text{system}} = \boxed{-125\text{ J/K}}$$

---

80. ***REASONING AND SOLUTION***

a. The amount of heat needed to raise the temperature of the gas at constant volume is given by Equations 15.6 and 15.8, $Q = n\,C_V\,\Delta T$. Solving for ΔT yields

$$\Delta T = \frac{Q}{nC_v} = \frac{5.24 \times 10^3\text{ J}}{(3.00\text{ mol})(\frac{3}{2}R)} = \boxed{1.40 \times 10^2\text{ K}}$$

b. The change in the internal energy of the gas is given by the first law of thermodynamics with $W = 0$, since the gas is heated at constant volume:

$$\Delta U = Q - W = 5.24 \times 10^3 \text{ J} - 0 = \boxed{5.24 \times 10^3 \text{ J}}$$

c. The change in pressure can be obtained from the ideal gas law,

$$\Delta P = \frac{nR\Delta T}{V} = \frac{(3.00 \text{ mol})R(1.40 \times 10^2 \text{ K})}{1.50 \text{ m}^3} = \boxed{2.33 \times 10^3 \text{ Pa}}$$

---

81. ***REASONING AND SOLUTION***
a. The work is

$$W = Q_H - Q_C = 3140 \text{ J} - 2090 \text{ J} = \boxed{1050 \text{ J}}$$

b. The coefficient of performance (CP) is

$$\text{CP} = Q_H/W = (3140 \text{ J})/(1050 \text{ J}) = \boxed{2.99}$$

---

82. ***REASONING AND SOLUTION*** In order to find out how many kilograms of ice in the tub are melted, we must determine Q_C, the amount of exhaust heat delivered to the cold reservoir. Since the hot reservoir consists of boiling water (T_H = 373.0 K) and the cold reservoir consists of ice and water (T_C = 273.0 K), the efficiency of this engine is

$$e = 1 - \frac{T_C}{T_H} = 1 - \frac{273.0 \text{ K}}{373.0 \text{ K}} = 0.2681$$

The work done by the engine is

$$W = eQ_H = (0.2681)(6800 \text{ J}) = 1823 \text{ J}$$

Therefore, the amount of heat delivered to the cold reservoir is

$$Q_C = Q_H - W = 6800 \text{ J} - 1823 \text{ J} = 4977 \text{ J}$$

Using the definition of the latent heat of fusion (melting) L_f, we find that the amount of ice that melts is

$$m = \frac{Q_C}{L_f} = \frac{4977 \text{ J}}{33.5 \times 10^4 \text{ J/kg}} = \boxed{0.015 \text{ kg}}$$

---

83. SSM ***REASONING*** According to Equation 15.2, $W = P\Delta V$, the average pressure $\overline{P}$ of the expanding gas is equal to $\overline{P} = W / \Delta V$, where the work W done by the gas on the bullet can be found from the work-energy theorem (Equation 6.3). Assuming that the barrel of the gun is cylindrical with radius r, the volume of the barrel is equal to its length L multiplied by the area (πr^2) of its cross section. Thus, the change in volume of the expanding gas is $\Delta V = L\pi r^2$.

SOLUTION The work done by the gas on the bullet is given by Equation 6.3 as

$$W = \tfrac{1}{2}m(v_{\text{final}}^2 - v_{\text{initial}}^2) = \tfrac{1}{2}(2.6 \times 10^{-3}\ \text{kg})[(370\ \text{m/s})^2 - 0] = 180\ \text{J}$$

The average pressure of the expanding gas is, therefore,

$$\overline{P} = \frac{W}{\Delta V} = \frac{180\ \text{J}}{(0.61\ \text{m})\pi(2.8 \times 10^{-3}\ \text{m})^2} = \boxed{1.2 \times 10^7\ \text{Pa}}$$

---

84. ***REASONING AND SOLUTION*** The rod's volume increases by an amount $\Delta V = \beta V_0 \Delta T$, according to Equation 12.3. The work done by the expanding aluminum is, from Equation 15.2,

$$W = P\Delta V = P\beta V_0 \Delta T = (1.01 \times 10^5\ \text{Pa})(69 \times 10^{-6}/\text{C}°)(1.4 \times 10^{-3}\ \text{m}^3)(3.0 \times 10^2\ \text{C}°) = \boxed{2.9\ \text{J}}$$

---

85. ***REASONING*** The power rating $\overline{P}$ of the heater is equal to the heat Q supplied to the gas divided by the time t the heater is on, $\overline{P} = Q/t$ (Equation 6.10b). Therefore, $t = Q/\overline{P}$. The heat required to change the temperature of a gas under conditions of constant pressure is given by $Q = C_P n \Delta T$ (Equation 15.6), where C_P is the molar specific heat capacity at constant pressure, n is the number of moles, and $\Delta T = T_f - T_i$ is the change in temperature. For a monatomic ideal gas, the specific heat capacity at constant pressure is $C_P = \frac{5}{2}R$, Equation (15.7), where R is the universal gas constant. We do not know n, T_f and T_i, but we can use the ideal gas law, $PV = nRT$, (Equation 14.1) to replace nRT_f by $P_f V_f$ and to replace nRT_i by $P_i V_i$, quantities that we do know.

SOLUTION Substituting $Q = C_P n \Delta T = C_P n (T_f - T_i)$ into $t = Q/\overline{P}$ and using the fact that $C_P = \frac{5}{2}R$ give

$$t = \frac{Q}{\overline{P}} = \frac{C_P n(T_f - T_i)}{\overline{P}} = \frac{\frac{5}{2}Rn(T_f - T_i)}{\overline{P}}$$

Replacing RnT_f by P_fV_f and RnT_i by P_iV_i and remembering that $P_i = P_f$, we find

$$t = \frac{\frac{5}{2}P_i\left(V_f - V_i\right)}{\overline{P}}$$

Since the volume of the gas increases by 25.0%, $V_f = 1.250V_i$. The time that the heater is on is

$$t = \frac{\frac{5}{2}P_i\left(V_f - V_i\right)}{\overline{P}} = \frac{\frac{5}{2}P_i\left(1.250V_i - V_i\right)}{\overline{P}} = \frac{\frac{5}{2}P_i\left(0.250\right)V_i}{\overline{P}}$$

$$= \frac{\frac{5}{2}\left(7.60\times10^5\ \text{Pa}\right)\left(0.250\right)\left(1.40\times10^{-3}\,\text{m}^3\right)}{15.0\ \text{W}} = \boxed{44.3\ \text{s}}$$

---

86. ***REASONING AND SOLUTION*** The amount of heat removed when the ice freezes is

$$Q_C = cm\Delta T + mL_f$$

$$= [4186\ \text{J/(kg}\cdot\text{C}°)](1.50\ \text{kg})(20.0\ \text{C}°) + (1.50\ \text{kg})(33.5\times10^4\ \text{J/kg}) = 6.28\times10^5\ \text{J}$$

Since the coefficient of performance is $\text{CP} = Q_C/W$, the work done by the refrigerator is

$$W = Q_C/\text{CP} = (6.28\times10^5\ \text{J})/3.00 = 2.09\times10^5\ \text{J}$$

The heat delivered to the kitchen is

$$Q_H = Q_C + W = 8.37\times10^5\ \text{J}$$

The space heater has a power output P of

$$P = Q_H/t = 3.00\times10^3\ \text{J/s}$$

Therefore,

$$t = Q_H/P = (8.37\times10^5\ \text{J})/(3.00\times10^3\ \text{J/s}) = \boxed{279\ \text{s}}$$

---

87. SSM ***REASONING*** According to Equation 15.5, $P_iV_i^{\gamma} = P_fV_f^{\gamma}$. The ideal gas law states that $V = nRT/P$ for both the initial and final conditions. Thus, we have

$$P_i\left(\frac{nRT_i}{P_i}\right)^{\gamma} = P_f\left(\frac{nRT_f}{P_f}\right)^{\gamma} \quad \text{or} \quad \frac{P_f}{P_i} = \left(\frac{T_i}{T_f}\right)^{\gamma/(1-\gamma)}$$

Since the ratio of the temperatures is known, the last expression can be solved for the final pressure P_f.

SOLUTION Since $T_i / T_f = 1/2$, and $\gamma = 5/3$, we find that

$$P_f = P_i \left(\frac{T_i}{T_f} \right)^{\gamma/(1-\gamma)} = \left(1.50 \times 10^5 \text{ Pa}\right)\left(\frac{1}{2}\right)^{(5/3)/[1-(5/3)]} = \boxed{8.49 \times 10^5 \text{ Pa}}$$

---

88. ***REASONING*** The change in the internal energy of the gas can be found using the first law of thermodynamics, since the heat added to the gas is known and the work can be calculated by using Equation 15.2, $W = P\,\Delta V$. The molar specific heat capacity at constant pressure can be evaluated by using Equation 15.6 and the ideal gas law.

SOLUTION
a. The change in the internal energy is

$$\Delta U = Q - W = Q - P\Delta V$$
$$= 31.4 \text{ J} - \left(1.40 \times 10^4 \text{ Pa}\right)\left(8.00 \times 10^{-4} \text{ m}^3 - 3.00 \times 10^{-4} \text{ m}^3\right) = \boxed{24.4 \text{ J}}$$

b. According to Equation 15.6, the molar specific heat capacity at constant pressure is $C_p = Q/(n\,\Delta T)$. The term $n\,\Delta T$ can be expressed in terms of the pressure and change in volume by using the ideal gas law:

$$P\,\Delta V = n\,R\,\Delta T \quad \text{or} \quad n\,\Delta T = P\,\Delta V/R$$

Substituting this relation for $n\,\Delta T$ into $C_p = Q/(n\,\Delta T)$, we obtain

$$C_p = \frac{Q}{\dfrac{P\Delta V}{R}} = \frac{31.4 \text{ J}}{\dfrac{\left(1.40\times10^4 \text{ Pa}\right)\left(5.00\times10^{-4} \text{ m}^3\right)}{R}} = \boxed{37.3 \text{ J/(mol}\cdot\text{K)}}$$

---

89. ***REASONING*** The heat added is given by Equation 15.6 as $Q = C_V n\Delta T$, where C_V is the molar specific heat capacity at constant volume, n is the number of moles, and ΔT is the change in temperature. But the heat is supplied by the heater at a rate of ten watts, or ten joules per second, so $Q = (10.0 \text{ W})t$, where t is the on-time for the heater. In addition, we know that the ideal gas law applies: $PV = nRT$ (Equation 14.1). Since the volume is constant while the temperature changes by an amount ΔT, the amount by which the pressure

changes is ΔP. This change in pressure is given by the ideal gas law in the form $(\Delta P)V = nR(\Delta T)$.

SOLUTION Using Equation 15.6 and the expression $Q = (10.0\text{ W})t$ for the heat delivered by the heater, we have

$$Q = C_V n\Delta T \quad \text{or} \quad (10.0\text{ W})t = C_V n\Delta T \quad \text{or} \quad t = \frac{C_V n\Delta T}{10.0\text{ W}}$$

Using the ideal gas law in the form $(\Delta P)V = nR(\Delta T)$, we can express the change in temperature as $\Delta T = (\Delta P)V/nR$. With this substitution for ΔT, the expression for the time becomes

$$t = \frac{C_V n(\Delta P)V}{(10.0\text{ W})nR}$$

According to Equation 15.8, $C_V = \frac{3}{2}R$ for a monatomic ideal gas, so we find

$$t = \frac{3Rn(\Delta P)V}{2(10.0\text{ W})nR} = \frac{3(\Delta P)V}{2(10.0\text{ W})} = \frac{3(5.0\times10^4\text{ Pa})(1.00\times10^{-3}\text{ m}^3)}{2(10.0\text{ W})} = \boxed{7.5\text{ s}}$$

---

90. ***REASONING AND SOLUTION*** The amount of work delivered by the engines can be determined from Equation 15.12, $Q_H = W + Q_C$. Solving for W for each engine gives:

$$W_1 = (Q_H)_1 - (Q_C)_1 \qquad \text{and} \qquad W_2 = (Q_H)_2 - (Q_C)_2 .$$

The total work delivered by the two engines is

$$W = W_1 + W_2 = [(Q_H)_1 - (Q_C)_1] + [(Q_H)_2 - (Q_C)_2]$$

But we know that $(Q_H)_2 = (Q_C)_1$, so that

$$W = [(Q_H)_1 - (Q_C)_1] + [(Q_C)_1 - (Q_C)_2] = (Q_H)_1 - (Q_C)_2 \qquad (1)$$

Since these are Carnot engines,

$$\frac{(Q_C)_1}{(Q_H)_1} = \frac{(T_C)_1}{(T_H)_1} \quad \Rightarrow \quad (Q_C)_1 = (Q_H)_1\frac{(T_C)_1}{(T_H)_1} = (4800\text{ J})\frac{670\text{ K}}{890\text{ K}} = 3.61\times10^3\text{ J}$$

Similarly, noting that $(Q_H)_2 = (Q_C)_1$ and that $(T_H)_2 = (T_C)_1$, we have

$$(Q_C)_2 = (Q_H)_2 \frac{(T_C)_2}{(T_H)_2} = (Q_C)_1 \frac{(T_C)_2}{(T_C)_1} = (3.61\times10^3 \text{ J})\frac{420 \text{ K}}{670 \text{ K}} = 2.26\times10^3 \text{ J}$$

Substituting into Equation (1) gives

$$W = 4800 \text{ J} - 2.26\times10^3 \text{ J} = \boxed{2.5\times10^3 \text{ J}}$$

---

91. SSM WWW ***REASONING*** The efficiency of either engine is given by Equation 15.13, $e = 1-(Q_C/Q_H)$. Since engine A receives three times more input heat, produces five times more work, and rejects two times more heat than engine B, it follows that $Q_{HA} = 3Q_{HB}$, $W_A = 5W_B$, and $Q_{CA} = 2Q_{CB}$. As required by the principle of energy conservation for engine A (Equation 15.12),

$$\underbrace{Q_{HA}}_{3Q_{HB}} = \underbrace{Q_{CA}}_{2Q_{CB}} + \underbrace{W_A}_{5W_B}$$

Thus,

$$3Q_{HB} = 2Q_{CB} + 5W_B \qquad (1)$$

Since engine B also obeys the principle of energy conservation (Equation 15.12),

$$Q_{HB} = Q_{CB} + W_B \qquad (2)$$

Substituting Q_{HB} from Equation (2) into Equation (1) yields

$$3(Q_{CB} + W_B) = 2Q_{CB} + 5W_B$$

Solving for W_B gives

$$W_B = \tfrac{1}{2}Q_{CB}$$

Therefore, Equation (2) predicts for engine B that

$$Q_{HB} = Q_{CB} + W_B = \frac{3}{2}Q_{CB}$$

SOLUTION

a. Substituting $Q_{CA} = 2Q_{CB}$ and $Q_{HA} = 3Q_{HB}$ into Equation 15.13 for engine A, we have

$$e_A = 1 - \frac{Q_{CA}}{Q_{HA}} = 1 - \frac{2Q_{CB}}{3Q_{HB}} = 1 - \frac{2Q_{CB}}{3(\frac{3}{2}Q_{CB})} = 1 - \frac{4}{9} = \boxed{\frac{5}{9}}$$

b. Substituting $Q_{HB} = \frac{3}{2}Q_{CB}$ into Equation 15.13 for engine B, we have

$$e_B = 1 - \frac{Q_{CB}}{Q_{HB}} = 1 - \frac{Q_{CB}}{\frac{3}{2}Q_{CB}} = 1 - \frac{2}{3} = \boxed{\frac{1}{3}}$$

---

92. ***REASONING AND SOLUTION*** Let the left be side 1 and the right be side 2. Since the partition moves to the right, side 1 does work on side 2, so that the work values involved satisfy the relation $W_1 = -W_2$. Using Equation 15.4 for each work value, we find that

$$\frac{3}{2}nR\left(T_{1i}-T_{1f}\right) = -\frac{3}{2}nR\left(T_{2i}-T_{2f}\right) \quad \text{or}$$

$$T_{1f} + T_{2f} = T_{1i} + T_{2i} = 525\text{ K} + 275\text{ K} = 8.00 \times 10^2\text{ K}$$

We now seek a second equation for the two unknowns T_{1f} and T_{2f}. Equation 15.5 for an adiabatic process indicates that $P_{1i}V_{1i}^{\gamma} = P_{1f}V_{1f}^{\gamma}$ and $P_{2i}V_{2i}^{\gamma} = P_{2f}V_{2f}^{\gamma}$. Dividing these two equations and using the facts that $V_{1i} = V_{2i}$ and $P_{1f} = P_{2f}$, gives

$$\frac{P_{1i}V_{1i}^{\gamma}}{P_{2i}V_{2i}^{\gamma}} = \frac{P_{1f}V_{1f}^{\gamma}}{P_{2f}V_{2f}^{\gamma}} \quad \text{or} \quad \frac{P_{1i}}{P_{2i}} = \left(\frac{V_{1f}}{V_{2f}}\right)^{\gamma}$$

Using the ideal gas law, we find that

$$\frac{P_{1i}}{P_{2i}} = \left(\frac{V_{1f}}{V_{2f}}\right)^{\gamma} \quad \text{becomes} \quad \frac{nRT_{1i}/V_{1i}}{nRT_{2i}/V_{2i}} = \left(\frac{nRT_{1f}/P_{1f}}{nRT_{2f}/P_{2f}}\right)^{\gamma}$$

Since $V_{1i} = V_{2i}$ and $P_{1f} = P_{2f}$, the result above reduces to

$$\frac{T_{1i}}{T_{2i}} = \left(\frac{T_{1f}}{T_{2f}}\right)^{\gamma} \quad \text{or} \quad \frac{T_{1f}}{T_{2f}} = \left(\frac{T_{1i}}{T_{2i}}\right)^{1/\gamma} = \left(\frac{525\text{ K}}{275\text{ K}}\right)^{1/\gamma} = 1.474$$

Using this expression for the ratio of the final temperatures in $T_{1f} + T_{2f} = 8.00 \times 10^2\text{ K}$, we find that

a. $\boxed{T_{1f} = 477\text{ K}}$ and b. $\boxed{T_{2f} = 323\text{ K}}$

---

93. ***CONCEPT QUESTIONS***

a. By itself, the work would decrease the internal energy of the system. This is because the system does work and would use some of its internal energy in the process.

b. By itself, the heat would increase the internal energy of the system, because it flows into the system. Thus, it would add to the supply of internal energy that the system already has.

c. The conservation-of-energy principle indicates that energy can neither be created nor destroyed, but can only be converted from one form to another. Therefore, the internal energy of the system increases, because more energy enters the system as heat than leaves the system as work.

SOLUTION Using the first law of thermodynamics (conservation of energy) from Equation 15.1, we obtain

$$\Delta U = Q - W = \left(7.6\times10^4\ \text{J}\right) - \left(4.8\times10^4\ \text{J}\right) = \boxed{+2.8\times10^4\ \text{J}}$$

The plus sign indicates that the internal energy increases, as expected.

---

94. ***CONCEPT QUESTIONS***

a. The energy gain in the form of heat means that the internal energy of the system would increase by an equal amount in the absence of work. This follows from the first law of thermodynamics:$\Delta U = Q - W$ (Equation 15.1). But the internal energy increases by an even greater amount, which means that energy also enters the system because work is being done on it.

b. According to our convention, work done on the system is negative.

c. The volume of the system decreases. This is because work is done on the system. In other words, the environment is pushing inward on the system, compressing it. Alternatively, Equation 15.2 indicates that work W done at constant pressure P is $W = P\Delta V$, where ΔV is the change in volume. Since we know that the work is negative, the change in volume must also be negative. But $\Delta V = V_f - V_i$, so the final volume V_f is less than the initial volume V_i.

SOLUTION Using Equations 15.1 ($\Delta U = Q - W$) and 15.2 ($W = P\Delta V$), we get

$$\Delta U = Q - W = Q - P\Delta V$$

Solving for ΔV gives

$$\Delta V = \frac{Q - \Delta U}{P} = \frac{(2780\ \text{J}) - (3990\ \text{J})}{1.26\times10^5\ \text{Pa}} = \boxed{-9.60\times10^{-3}\ \text{m}^3}$$

As expected, ΔV is negative, reflecting a decrease in volume.

---

95. ***CONCEPT QUESTIONS***

a. The internal energy of an ideal gas remains the same during an isothermal process. The temperature is constant in an isothermal process, and the internal energy of an ideal gas is proportional to the kelvin temperature, as Section 14.3 discusses. Since the temperature is constant, the internal energy is constant.

b. The work done is equal to the heat that flows into the gas. According to the first law of thermodynamics, the change ΔU in the internal energy is given by Equations 15.1 as $\Delta U = Q - W$. Since the internal energy U is constant, it follows that $\Delta U = 0 = Q - W$, or $W = Q$.

SOLUTION According to Equation 15.3, the work done in an isothermal process involving an ideal gas is

$$W = nRT \ln\left(\frac{V_\text{f}}{V_\text{i}}\right) \quad \text{or} \quad T = \frac{W}{nR \ln\left(\dfrac{V_\text{f}}{V_\text{i}}\right)}$$

Since $W = Q$ for an isothermal process utilizing an ideal gas, we find

$$T = \frac{Q}{nR \ln\left(\dfrac{V_\text{f}}{V_\text{i}}\right)} = \frac{4.75 \times 10^3 \text{ J}}{(3.00 \text{ mol})[8.31 \text{ J}/(\text{mol}\cdot\text{K})] \ln\left(\dfrac{0.250 \text{ m}^3}{0.100 \text{ m}^3}\right)} = \boxed{208 \text{ K}}$$

---

96. ***CONCEPT QUESTIONS***

a. The 4.1×10^6 J of energy equals the input heat Q_H. It makes no sense for this energy to be the work W, because then the work would be independent of the climb. Common sense indicates that more work is done when the climb is through a greater versus a smaller height.

b. The work done in climbing upward is related to the vertical height of the climb via the work-energy theorem (Equation 6.8), which is

$$W_\text{nc} = \underbrace{\text{KE}_\text{f} + \text{PE}_\text{f}}_{\substack{\text{Final total} \\ \text{mechanical energy}}} - \underbrace{(\text{KE}_0 + \text{PE}_0)}_{\substack{\text{Initial total} \\ \text{mechanical energy}}}$$

Here, W_nc is the net work done by nonconservative forces, in this case the work done by the climber in going upward. Since the climber starts at rest and finishes at rest, the final kinetic energy KE_f and the initial kinetic energy KE_0 are zero. As a result, we have

$W_{nc} = PE_f - PE_0$, where PE_f and PE_0 are the final and initial gravitational potential energies, respectively. Equation 6.5 gives the gravitational potential energy as PE = mgh, where h is the vertical height. Taking the height at her starting point to be zero, we then have $W_{nc} = mgh$.

SOLUTION Using Equation 15.11 for the efficiency and relating the work to the height via the work-energy theorem, we find

$$e = \frac{W}{Q_H} = \frac{mgh}{Q_H} = \frac{(52\text{ kg})(9.80\text{ m/s}^2)(730\text{ m})}{4.1\times10^6\text{ J}} = \boxed{0.091}$$

---

97. ***CONCEPT QUESTIONS***

a. According to Equation 15.11, the efficiency of a heat engine is $e = W/Q_H$, where W is the work and Q_H is the input heat. Thus, the work is $W = eQ_H$. The engine that delivers more work for a given heat input is the engine with the higher efficiency. In this case, that is engine B.

b. The efficiency of a Carnot engine is given by Equation 15.15 as $e_{Carnot} = 1 - T_C/T_H$. Smaller values of the cold-reservoir temperature T_C mean greater efficiencies for a given value of the hot-reservoir temperature T_H. Thus, the engine with the greater efficiency has the lower cold-reservoir temperature. In this case, that is engine B.

SOLUTION Using Equation 15.11, we find the work delivered by each engine as follows:

Engine A $\quad W = eQ_H = (0.60)(1200\text{ J}) = \boxed{720\text{ J}}$

Engine B $\quad W = eQ_H = (0.80)(1200\text{ J}) = \boxed{960\text{ J}}$

Equation 15.15 for the efficiency e_{Carnot} of a Carnot engine can be solved for the temperature of the cold reservoir:

$$e_{Carnot} = 1 - \frac{T_C}{T_H} \quad \text{or} \quad T_C = (1 - e_{Carnot})T_H$$

Applying this result to each engine gives

Engine A $\quad T_C = (1 - 0.60)(650\text{ K}) = \boxed{260\text{ K}}$

Engine B $\quad T_C = (1 - 0.80)(650\text{ K}) = \boxed{130\text{ K}}$

As expected, engine B delivers more work and has the lower cold-reservoir temperature.

98. ***CONCEPT QUESTIONS***

a. An air conditioner removes heat from a room by doing work to make the heat flow up the temperature "hill" from cold to hot. More work is required to remove a given amount of heat when the temperature difference against which the air conditioner is working is greater. Here the hot temperature outside is the same for each unit, but the room serviced by unit A is kept colder. Thus, unit A must work against the greater temperature difference and uses more work than unit B.

b. The heat deposited outside is equal to the heat removed plus the work done. Since both units remove the same amount of heat, the unit that deposits more heat outside is the unit that uses the greater amount of work. That is unit A.

SOLUTION According to the first law of thermodynamics (Equation 15.12), we know that $W = Q_H - Q_C$. In addition, since the air conditioners are Carnot devices, we know that the ratio of the heat Q_C removed from the cold reservoir to the heat Q_H deposited in the hot reservoir is equal to the ratio of the reservoir temperatures or $Q_C/Q_H = T_C/T_H$ (Equation 15.14). Using these two equations, we have

$$W = Q_H - Q_C = \left(\frac{Q_H}{Q_C} - 1\right)Q_C = \left(\frac{T_H}{T_C} - 1\right)Q_C$$

Applying this result to each air conditioner gives

Unit A $$W = \left(\frac{309.0\text{ K}}{294.0\text{ K}} - 1\right)(4330\text{ J}) = \boxed{220\text{ J}}$$

Unit B $$W = \left(\frac{309.0\text{ K}}{301.0\text{ K}} - 1\right)(4330\text{ J}) = \boxed{120\text{ J}}$$

We can find the heat deposited outside directly from Equation 15.14 by solving it for Q_H.

$$\frac{Q_C}{Q_H} = \frac{T_C}{T_H} \quad \text{or} \quad Q_H = \left(\frac{T_H}{T_C}\right)Q_C$$

Applying this result to each air conditioner gives

Unit A $$Q_H = \left(\frac{309.0\text{ K}}{294.0\text{ K}}\right)(4330\text{ J}) = \boxed{4550\text{ J}}$$

Unit B $$Q_H = \left(\frac{309.0\text{ K}}{301.0\text{ K}}\right)(4330\text{ J}) = \boxed{4450\text{ J}}$$

As expected, unit A uses more work and deposits more heat outside.

---

99. ***CONCEPT QUESTIONS***

a. There is no work done for the process A→B. The reason is that the volume is constant (see the drawing), which means that the change ΔV in the volume is zero. In other words, the area under the plot of pressure versus volume is zero, and we see that $W_{A\to B} = 0$ J.

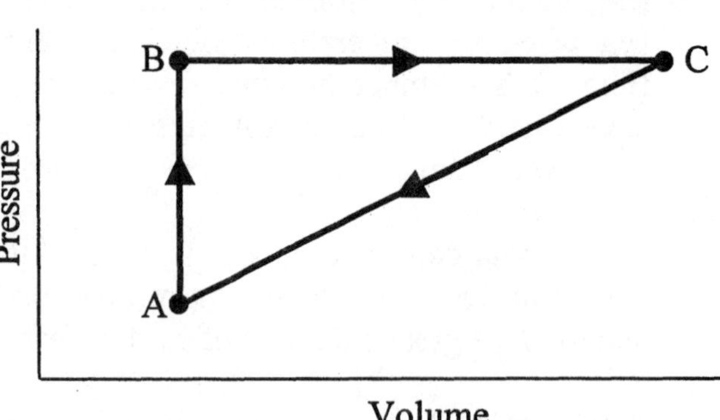

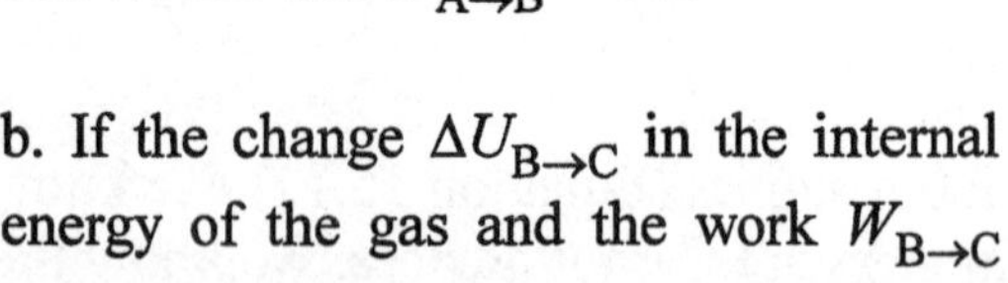

b. If the change $\Delta U_{B\to C}$ in the internal energy of the gas and the work $W_{B\to C}$ are known for the process B→C, the heat $Q_{B\to C}$ can be determined by using the first law of thermodynamics as $Q_{B\to C} = \Delta U_{B\to C} + W_{B\to C}$ (Equation 15.1).

c. Yes. For the process C→A it is possible to find the change in the internal energy of the gas if the change in the internal energies for the processes A→B and B→C are known. The total change ΔU_{total} in the internal energy for the three processes is $\Delta U_{\text{total}} = \Delta U_{A\to B} + \Delta U_{B\to C} + U_{C\to A}$. We can use this equation to find $U_{C\to A}$. The quantities $\Delta U_{A\to B}$ and $\Delta U_{B\to C}$ are known. We also know ΔU_{total}, which is the change in the internal energy for the total process A→B→C→A. This process begins and ends at the same place on the pressure-versus-volume plot. Therefore, the value of the internal energy U is the same at the start and the end, with the result that $\Delta U_{\text{total}} = 0$ J.

SOLUTION

a. Since the change in volume is zero ($\Delta V = 0\ \text{m}^3$), the area under the plot of pressure versus volume is zero, with the result that the work is $W_{A\to B} = \boxed{0\ \text{J}}$.

b. The change in the internal energy of the gas for the process A→B is $\Delta U_{A\to B} = Q_{A\to B} - W_{A\to B} = +561\ \text{J} - 0\ \text{J} = \boxed{+561\ \text{J}}$.

c. According to the first law of thermodynamics, the change in the internal energy of the gas for the process B→C is $\Delta U_{B\to C} = Q_{B\to C} - W_{B\to C}$. Thus, the heat added to the gas is $Q_{B\to C} = \Delta U_{B\to C} + W_{B\to C} = +4303\ \text{J} + 2867\ \text{J} = \boxed{+7170\ \text{J}}$.

d. According to the first law of thermodynamics, the change ΔU_{total} in the total internal energy for the three processes is $\Delta U_{\text{total}} = \Delta U_{A\to B} + \Delta U_{B\to C} + \Delta U_{C\to A}$. Solving this equation for $\Delta U_{C\to A}$ gives $\Delta U_{C\to A} = \Delta U_{\text{total}} - \Delta U_{A\to B} - \Delta U_{B\to C}$. As discussed in part c of the Concept Questions, $\Delta U_{\text{total}} = 0$ J, since the third process ends at point A, which is the start of the first process. Therefore, $\Delta U_{\text{C}\to\text{A}} = 0\text{ J} - 561\text{ J} - 4303\text{ J} = \boxed{-4864\text{ J}}$.

e. According to the first law of thermodynamics, the change in the internal energy of the gas for the process C→A is $\Delta U_{\text{C}\to\text{A}} = Q_{\text{C}\to\text{A}} - W_{\text{C}\to\text{A}}$. The heat removed during this process is $Q_{\text{C}\to\text{A}} = \Delta U_{\text{C}\to\text{A}} + W_{\text{C}\to\text{A}} = -4864\text{ J} - 3740\text{ J} = \boxed{-8604\text{ J}}$.

---

100. ***CONCEPT QUESTIONS***

a. According to the discussion in Section 15.11, the change $\Delta S_{\text{universe}}$ in the entropy of the universe is the sum of the change in entropy ΔS_{H} of the hot reservoir and the change in entropy ΔS_{C} of the cold reservoir, so $\Delta S_{\text{universe}} = \Delta S_{\text{H}} + \Delta S_{\text{C}}$. The change in entropy of each reservoir is given by Equation 15.18 as $\Delta S = (Q/T)_{\text{R}}$. The engine is irreversible, so we must imagine a process in which the heat Q is added to or removed from the reservoirs reversibly. T is the Kelvin temperature of a reservoir. Since heat is lost from the hot reservoir, the change in entropy is negative: $\Delta S_{\text{H}} = -Q_{\text{H}}/T_{\text{H}}$. Since heat is gained by the cold reservoir, the change in entropy is positive: $\Delta S_{\text{C}} = +Q_{\text{C}}/T_{\text{C}}$. The change in entropy of the universe is

$$\Delta S_{\text{universe}} = \Delta S_{\text{H}} + \Delta S_{\text{C}} = -\frac{Q_{\text{H}}}{T_{\text{H}}} + \frac{Q_{\text{C}}}{T_{\text{C}}}.$$

b. The change in the entropy of the universe is greater than zero, $\Delta S_{\text{universe}} > 0$ J/K, as it must be for any irreversible process. (This is the second law of thermodynamics stated in terms of entropy.)

c. When a reversible engine (a Carnot engine) operates between the same hot and cold temperatures as the irreversible engine, the reversible engine produces more work, assuming that the input heat Q_{H} to both engines is the same. This is because the reversible engine is more efficient, according to the second law of thermodynamics.

d. The difference in the work produced by the two engines is labeled $W_{\text{unavailable}}$ in Section15.11, where $W_{\text{unavailable}} = W_{\text{reversible}} - W_{\text{irreversible}}$. The difference in the work is related to the change in the entropy of the universe by $W_{\text{unavailable}} = T_0\, \Delta S_{\text{universe}}$ (Equation 15.19), where T_0 is the Kelvin temperature of the coldest heat reservoir. In this case $T_0 = T_{\text{C}}$.

SOLUTION

a. From part a of the Concept Questions, the change in entropy of the universe is

$$\Delta S_{\text{universe}} = -\frac{Q_{\text{H}}}{T_{\text{H}}} + \frac{Q_{\text{C}}}{T_{\text{C}}}$$

The magnitude Q_C of the heat rejected to the cold reservoir is related to the magnitude Q_H of the heat supplied to the engine from the hot reservoir and the work W done by the engine via $Q_C = Q_H - W$ (Equation 15.1). Thus, $\Delta S_{\text{universe}}$ becomes

$$\Delta S_{\text{universe}} = -\frac{Q_H}{T_H} + \frac{Q_H - W}{T_C} = -\frac{1285\text{ J}}{852\text{ K}} + \frac{1285\text{ J} - 264\text{ J}}{314\text{ K}} = \boxed{+1.74\text{ J/K}}$$

As expected, the entropy of the universe increases when an irreversible process occurs.

b. The work W done by any engine depends on its efficiency e and input heat Q_H via $W = eQ_H$ (Equation 15.11). For a reversible engine, the efficiency is related to the temperatures of its hot and cold reservoirs by $e = 1 - (T_C/T_H)$, Equation 15.15. The work done by the reversible engine is

$$W_{\text{reversible}} = eQ_H = \left(1 - \frac{T_C}{T_H}\right)Q_H = \left(1 - \frac{314\text{ K}}{852\text{ K}}\right)(1285\text{ J}) = \boxed{811\text{ J}}$$

c. According to the discussion in part d of the Concept Questions, the difference between the work done by the reversible and irreversible engines is

$$\underbrace{\text{W}_{\text{reversible}} - \text{W}_{\text{irreversible}}}_{\text{W}_{\text{unavailable}}} = \text{T}_\text{C}\Delta\text{S}_{\text{universe}} = (314\text{ K})(1.74\text{ J/K}) = \boxed{547\text{ J}}$$

CHAPTER 16 | *WAVES AND SOUND*

PROBLEMS

---

1. SSM ***REASONING AND SOLUTION*** The frequency f (in hertz) is the number of wave cycles that pass an observer and is the reciprocal of the period T (in seconds). The period T is the time required for one complete cycle of the wave to pass a fixed point. From the data given in the problem, $T = (\text{time for } n \text{ cycles})/n = (120\text{ s})/10 = 12\text{ s}$. Therefore,

$$f = \frac{1}{T} = \frac{1}{12\text{ s}} = \boxed{0.083\text{ Hz}}$$

---

2. ***REASONING*** Since light behaves as a wave, its speed v, frequency f, and wavelength λ are related to according to $v = f\lambda$ (Equation 16.1). We can solve this equation for the frequency in terms of the speed and the wavelength.

SOLUTION Solving Equation 16.1 for the frequency, we find that

$$f = \frac{v}{\lambda} = \frac{3.00\times10^{8}\text{ m/s}}{5.45\times10^{-7}\text{ m}} = \boxed{5.50\times10^{14}\text{ Hz}}$$

---

3. ***REASONING AND SOLUTION*** From Equation 16.1, we have $\lambda = v/f$. But $v = x/t$, so we find

$$\lambda = \frac{v}{f} = \frac{x}{tf} = \frac{2.5\text{ m}}{(1.7\text{ s})(3.0\text{ Hz})} = \boxed{0.49\text{ m}}$$

---

4. ***REASONING AND SOLUTION***

a. Since 15 boxcars pass by in 12.0 s, the boxcars pass by with a frequency of

$$f = \frac{15}{12.0\text{ s}} = \boxed{1.25\text{ Hz}}$$

b. Since the length of a boxcar corresponds to the wavelength λ of a wave, we have, from Equation 16.1, that

$$v = f\lambda = (1.25\text{ Hz})(14.0\text{ m}) = \boxed{17.5\text{ m/s}}$$

---

5. SSM ***REASONING*** When the end of the Slinky is moved up and down continuously, a transverse wave is produced. The distance between two adjacent crests on the wave, is, by definition, one wavelength. The wavelength λ is related to the speed and frequency of a periodic wave by Equation 16.1, $\lambda = v/f$. In order to use Equation 16.1, we must first determine the frequency of the wave. The wave on the Slinky will have the same frequency as the simple harmonic motion of the hand. According to the data given in the problem statement, the frequency is $f = (2.00 \text{ cycles})/(1 \text{ s}) = 2.00 \text{ Hz}$.

SOLUTION Substituting the values for λ and f, we find that the distance between crests is

$$\lambda = \frac{v}{f} = \frac{0.50 \text{ m/s}}{2.00 \text{ Hz}} = \boxed{0.25 \text{ m}}$$

---

6. ***REASONING AND SOLUTION*** The period of the wave is the same as the period of the person, so $T = 5.00$ s.

a. $$f = 1/T = \boxed{0.200 \text{ Hz}} \qquad (10.5)$$

b. $$v = \lambda f = (20.0 \text{ m})(0.200 \text{ Hz}) = \boxed{4.00 \text{ m/s}} \qquad (16.1)$$

---

7. SSM ***REASONING*** As the transverse wave propagates, the colored dot moves up and down in simple harmonic motion with a frequency of 5.0 Hz. The amplitude (1.3 cm) is the magnitude of the maximum displacement of the dot from its equilibrium position.

SOLUTION The period T of the simple harmonic motion of the dot is $T = 1/f = 1/(5.0 \text{ Hz}) = 0.20 \text{ s}$. In one period the dot travels through one complete cycle of its motion, and covers a vertical distance of $4 \times (1.3 \text{ cm}) = 5.2 \text{ cm}$. Therefore, in 3.0 s the dot will have traveled a *total vertical distance* of

$$\left(\frac{3.0 \text{ s}}{0.20 \text{ s}}\right)(5.2 \text{ cm}) = \boxed{78 \text{ cm}}$$

---

8. ***REASONING*** The wave is moving with a speed v (unknown), and the jetskier is moving in the same direction with a speed v_{jetskier} (known). To find v, we proceed as follows. If the jetskier were traveling with the same speed as the wave, his speed relative to the wave would be zero and he would never experience any "bumps." He is traveling faster than the wave, however, so his speed relative to the wave is $v_{\text{jetskier}} - v$. This relative speed times the time between "bumps" must, then, equal the wavelength of the water wave, because the distance between adjacent crests is the wavelength. Moreover, the time between "bumps" or crests is

the reciprocal of the given "bump" frequency. In this fashion, we can obtain an equation that can be solved for the wave speed.

SOLUTION The speed of the jetskier relative to the wave times the time t between "bumps" is the wavelength λ of the water wave, so that we have

$$\underbrace{\left(v_{\text{jetskier}} - v\right)}_{\substack{\text{Speed of jetskier} \\ \text{relative to wave}}} t = \lambda$$

The times between bumps is the reciprocal of the "bump" frequency f or $t = 1/f$. With this substitution we find

$$\left(v_{\text{jetskier}} - v\right)\left(\frac{1}{f}\right) = \lambda \quad \text{or} \quad v_{\text{jetskier}} - v = f\lambda$$

Solving for the speed v of the wave gives

$$v = v_{\text{jetskier}} - f\lambda = 8.4 \text{ m/s} - (1.2 \text{ Hz})(5.8 \text{ m}) = \boxed{1.4 \text{ m/s}}$$

---

9. ***REASONING*** During each cycle of the wave, a particle of the string moves through a total distance that equals $4A$, where A is the amplitude of the wave. The number of wave cycles per second is the frequency f of the wave. Therefore, the distance moved per second by a string particle is $4Af$. The time to move through a total distance L, then, is $t = L/(4Af)$. According to Equation 16.1, however, we have that $f = v/\lambda$, so that

$$t = \frac{L}{4Af} = \frac{L}{4A(v/\lambda)}$$

SOLUTION Using the result obtained above, we find that

$$t = \frac{L\lambda}{4Av} = \frac{\left(1.0 \times 10^3 \text{ m}\right)(0.18 \text{ m})}{4\left(2.0 \times 10^{-3} \text{ m}\right)(450 \text{ m/s})} = \boxed{5.0 \times 10^1 \text{ s}}$$

---

10. ***REASONING AND SOLUTION*** First find the speed of the record at a distance of 0.100 m from the center:

$$v = r\omega = (0.100 \text{ m})(3.49 \text{ rad/s}) = 0.349 \text{ m/s} \tag{8.9}$$

The wavelength is, then,

$$\lambda = v/f = (0.349 \text{ m/s})/(5.00 \times 10^3 \text{ Hz}) = \boxed{6.98 \times 10^{-5} \text{ m}} \tag{16.1}$$

---

11. ***REASONING AND SOLUTION*** When traveling with the waves, the skier "sees" the waves to be traveling with a velocity of $v_w - v_s$ and with a period of $T_1 = 0.600$ s. Suppressing the units for convenience, we find that the distance between crests is (in meters)

$$\lambda = (v_s - v_w)T_1 = (12.0 - v_w)(0.600) = 7.20 - 0.600\, v_w \qquad (1)$$

Similarly, for the skier traveling opposite the waves

$$\lambda = (v_s + v_w)T_2 = 6.00 + 0.500\, v_w \qquad (2)$$

a. Subtracting Equation (2) from Equation(1) and solving for v_w gives $v_w = \boxed{1.1 \text{ m/s}}$.

b. Substituting the value for v_w into Equation (1) gives $\lambda = \boxed{6.55 \text{ m}}$.

---

12. ***REASONING*** The length L of the string is one of the factors that affects the speed of a wave traveling on it, in so far as the speed v depends on the mass per unit length m/L according to $v = \sqrt{\dfrac{F}{m/L}}$ (Equation 16.2). The other factor affecting the speed is the tension F. The speed is not directly given here. However, the frequency f and the wavelength λ are given, and the speed is related to them according to $v = f\lambda$ (Equation 16.1). Substituting Equation 16.1 into Equation 16.2 will give us an equation that can be solved for the length L.

SOLUTION Substituting Equation 16.1 into Equation 16.2 gives

$$v = f\lambda = \sqrt{\frac{F}{m/L}}$$

Solving for the length L, we find that

$$L = \frac{f^2\lambda^2 m}{F} = \frac{(260 \text{ Hz})^2 (0.60 \text{ m})^2 \left(5.0\times10^{-3} \text{ kg}\right)}{180 \text{ N}} = \boxed{0.68 \text{ m}}$$

---

13. **SSM** ***REASONING*** The tension F in the violin string can be found by solving Equation 16.2 for F to obtain $F = mv^2/L$, where v is the speed of waves on the string and can be found from Equation 16.1 as $v = f\lambda$.

SOLUTION Combining Equations 16.2 and 16.1 and using the given data, we obtain

$$F = \frac{mv^2}{L} = (m/L)f^2\lambda^2 = \left(7.8\times10^{-4} \text{ kg/m}\right)(440 \text{ Hz})^2\left(65\times10^{-2} \text{ m}\right)^2 = \boxed{64 \text{ N}}$$

---

14. ***REASONING AND SOLUTION*** Initially, the speed of the wave is $v_0 = \lambda f_0$. After the adjustment the speed of the wave is $v = 2\lambda f_0$. Dividing, we obtain $v/v_0 = 2$. Now for each case

$$F_0 = (m/L)v_0^2 \quad \text{and} \quad F = (m/L)v^2$$

Division yields

$$F/F_0 = v^2/v_0^2 = 4$$

Hence,

$$F = 4F_0 = 4(58\text{ N}) = \boxed{230\text{ N}}$$

---

15. ***REASONING AND SOLUTION*** The speed of a wave on a string is proportional to the square root of the tension, Equation 16.2. If the tension increases by a factor of four then the speed will increase by a factor of two. The new speed will then be $\boxed{600\text{ m/s}}$.

---

16. ***REASONING*** Each pulse travels a distance that is given by vt, where v is the wave speed and t is the travel time up to the point when they pass each other. The sum of the distances traveled by each pulse must equal the 50.0-m length of the wire, since each pulse starts out from opposite ends of the wires.

SOLUTION Using v_A and v_B to denote the speeds on either wire, we have

$$v_A t + v_B t = 50.0\text{ m}$$

Solving for the time t and using Equation 16.2 $\left(v = \sqrt{\dfrac{F}{m/L}}\right)$, we find

$$t = \frac{50.0\text{ m}}{v_A + v_B} = \frac{50.0\text{ m}}{\sqrt{\dfrac{F_A}{m/L}} + \sqrt{\dfrac{F_B}{m/L}}} = \frac{50.0\text{ m}}{\sqrt{\dfrac{6.00\times10^2\text{ N}}{0.020\text{ kg/m}}} + \sqrt{\dfrac{3.00\times10^2\text{ N}}{0.020\text{ kg/m}}}} = \boxed{0.17\text{ s}}$$

---

17. ***REASONING*** The speed v of the transverse pulse on the wire is determined by the tension F in the wire and the mass per unit length m/L of the wire, according to $v = \sqrt{\dfrac{F}{m/L}}$ (Equation 16.2). The ball has a mass M. Since the wire supports the weight Mg of the ball and since the weight of the wire is negligible, it is only the ball's weight that determines the tension in the wire, $F = Mg$. Therefore, we can use Equation 16.2 with this value of the tension and solve it for the acceleration g due to gravity. The speed of the transverse pulse is not given, but we know that the pulse travels the length L of the wire in a time t and that the speed is $v = L/t$.

SOLUTION Substituting the tension $F = Mg$ and the speed $v = L/t$ into Equation 16.2 for the speed of the pulse on the string gives

$$v = \sqrt{\frac{F}{m/L}} \quad \text{or} \quad \frac{L}{t} = \sqrt{\frac{Mg}{m/L}}$$

Solving for the acceleration g due to gravity, we obtain

$$g = \frac{\left(\frac{L}{t}\right)^2 (m/L)}{M} = \frac{\left(\frac{0.95\text{ m}}{0.016\text{ s}}\right)^2 \left(1.2\times10^{-4}\text{ kg/m}\right)}{0.055\text{ kg}} = \boxed{7.7\text{ m/s}^2}$$

---

18. ***REASONING*** A particle of the string is moving in simple harmonic motion. The maximum speed of the particle is given by Equation 10.8 as $v_{max} = A\omega$, where A is the amplitude of the wave and ω is the angular frequency. The angular frequency is related to the frequency f by Equation 10.6, $\omega = 2\pi f$, so the maximum speed can be written as $v_{max} = 2\pi f A$. The speed v of a wave on a string is related to the frequency f and wavelength λ by Equation 16.1, $v = f\lambda$. The ratio of the maximum particle speed to the speed of the wave is

$$\frac{v_{max}}{v} = \frac{2\pi f A}{f\lambda} = \frac{2\pi A}{\lambda}$$

The equation can be used to find the wavelength of the wave.

SOLUTION Solving the equation above for the wavelength, we have

$$\lambda = \frac{2\pi A}{\left(\frac{v_{max}}{v}\right)} = \frac{2\pi(4.5\text{ cm})}{3.1} = \boxed{9.1\text{ cm}}$$

---

19. SSM WWW ***REASONING*** Using the procedures developed in Chapter 4 for using Newton's second law to analyze the motion of bodies and neglecting the weight of the wire relative to the tension in the wire lead to the following equations of motion for the two blocks:

$$\sum F_x = F - m_1 g\ (\sin 30.0°) = 0 \qquad (1)$$

$$\sum F_y = F - m_2 g = 0 \qquad (2)$$

where F is the tension in the wire. In Equation (1) we have taken the direction of the $+x$ axis for block 1 to be parallel to and up the incline. In Equation (2) we have taken the direction of the $+y$ axis to be upward for block 2. This set of equations consists of two equations in three unknowns, m_1, m_2, and F. Thus, a third equation is needed in order to solve for any of the unknowns. A useful third equation can be obtained by solving Equation 16.2 for F:

$$F = (m/L)v^2 \tag{3}$$

Combining Equation (3) with Equations (1) and (2) leads to

$$(m/L)v^2 - m_1 g \sin 30.0° = 0 \tag{4}$$

$$(m/L)v^2 - m_2 g = 0 \tag{5}$$

Equations (4) and (5) can be solved directly for the masses m_1 and m_2.

SOLUTION Substituting values into Equation (4), we obtain

$$m_1 = \frac{(m/L)v^2}{g \sin 30.0°} = \frac{(0.0250 \text{ kg/m})(75.0 \text{ m/s})^2}{(9.80 \text{ m/s}^2) \sin 30.0°} = \boxed{28.7 \text{ kg}}$$

Similarly, substituting values into Equation (5), we obtain

$$m_2 = \frac{(m/L)v^2}{g} = \frac{(0.0250 \text{ kg/m})(75.0 \text{ m/s})^2}{(9.80 \text{ m/s}^2)} = \boxed{14.3 \text{ kg}}$$

---

20. ***REASONING AND SOLUTION*** Initially, the tension in the wire is

$$F_0 = (m/L)v^2 = (7.0 \times 10^{-3} \text{ kg/m})(46 \text{ m/s})^2$$

As the temperature is lowered, the wire will attempt to shrink by an amount $\Delta L = \alpha L \Delta T$, where α is the coefficient of thermal expansion. Since the wire cannot shrink, a stress will develop, according to

$$\text{stress} = Y\Delta L/L = Y\alpha\Delta T$$

where Y is Young's modulus. This stress corresponds to an additional tension:

$$F' = (\text{stress})A = YA\alpha\Delta T = (1.1 \times 10^{11} \text{ Pa})(1.1 \times 10^{-6} \text{ m}^2)(17 \times 10^{-6}/\text{C°})(14 \text{ C°})$$

The total tension in the wire at the lower temperature is now $F = F_0 + F'$, so that the new speed of the waves on the wire is

$$v = \sqrt{\frac{F_0 + F'}{m/L}} = \boxed{79\ \text{m/s}}$$

---

21. ***REASONING AND SOLUTION*** If the string has length L, the time required for a wave on the string to travel from the center of the circle to the ball is

$$t = \frac{L}{v_{wave}} \qquad (1)$$

The speed of the wave is given by text Equation 16.2

$$v_{wave} = \sqrt{\frac{F}{m_{string}/L}} \qquad (2)$$

The tension F in the string acts as a centripetal force on the ball, so that

$$F = m_{ball}\omega^2 r = m_{ball}\omega^2 L \qquad (3)$$

Eliminating the tension F from Equations (2) and (3) above yields

$$v_{wave} = \sqrt{\frac{m_{ball}\omega^2 L}{m_{string}/L}} = \sqrt{\frac{m_{ball}\omega^2 L^2}{m_{string}}} = L\sqrt{\frac{m_{ball}\omega^2}{m_{string}}}$$

Substituting this expression for v_{wave} into Equation (1) gives

$$t = \frac{L}{L\sqrt{\dfrac{m_{ball}\omega^2}{m_{string}}}} = \sqrt{\frac{m_{string}}{m_{ball}\omega^2}} = \sqrt{\frac{0.0230\ \text{kg}}{(15.0\ \text{kg})(12.0\ \text{rad/s})^2}} = \boxed{3.26 \times 10^{-3}\ \text{s}}$$

---

22. ***REASONING*** Since the wave is traveling in the +*x* direction, its form is given by Equation 16.3 as

$$y = A\sin\left(2\pi ft - \frac{2\pi x}{\lambda}\right)$$

We are given that the amplitude is $A = 0.35$ m. However, we need to evaluate $2\pi f$ and $\frac{2\pi}{\lambda}$. Although the wavelength λ is not stated directly, it can be obtained from the values for the speed v and the frequency f, since we know that $v = f\lambda$ (Equation 16.1).

SOLUTION Since the frequency is $f = 14$ Hz, we have

$$2\pi f = 2\pi(14 \text{ Hz}) = 88 \text{ rad/s}$$

It follows from Equation 16.1 that

$$\frac{2\pi}{\lambda} = \frac{2\pi f}{v} = \frac{2\pi(14 \text{ Hz})}{5.2 \text{ m/s}} = 17 \text{ m}^{-1}$$

Using these values for $2\pi f$ and $\frac{2\pi}{\lambda}$ in Equation 16.3, we have

$$y = A\sin\left(2\pi ft - \frac{2\pi x}{\lambda}\right)$$

$$\boxed{y = (0.35 \text{ m})\sin\left[(88 \text{ rad/s})t - \left(17 \text{ m}^{-1}\right)x\right]}$$

---

23. SSM ***REASONING*** The mathematical form for the displacement of a wave traveling in the $-x$ direction is given by Equation 16.4: $y = A\sin\left(2\pi ft + \frac{2\pi x}{\lambda}\right)$.

SOLUTION Using Equation 16.1 and the fact that $f = 1/T$, we obtain the following numerical values for f and λ: $f = 1/T = 1.3$ Hz, and $\lambda = v/f = 9.2$ m. Omitting units and substituting these values for f and λ into Equation 16.4 gives

$$\boxed{y = (0.37 \text{ m})\sin(2.6\pi t + 0.22\pi x)}$$

---

24. ***REASONING AND SOLUTION*** We find from the upper graph that $\lambda = 0.060$ m $-$ 0.020 m $= 0.040$ m and $A = 0.010$ m. From the lower graph we find that $T = 0.30$ s $-$ 0.10 s $= 0.20$ s. Then, $f = 1/(0.20 \text{ s}) = 5.0$ Hz. Substituting these into Equation 16.3 we get

$$y = A\sin\left(2\pi ft - \frac{2\pi x}{\lambda}\right) \quad \text{and} \quad \boxed{y = (0.010 \text{ m})\sin(10\pi t - 50\pi x)}$$

---

25. ***REASONING AND SOLUTION***

a. Comparing with Equation 16.4, we find $A = \boxed{0.45\text{ m}}$. $2f = 8.0$ Hz, so $f = \boxed{4.0\text{ Hz}}$. $2/\lambda = 1.0\text{ m}^{-1}$, so $\lambda = \boxed{2.0\text{ m}}$. Using Equation 16.1, we find

$$v = f\lambda = (2.0\text{ m})(4.0\text{ Hz}) = \boxed{8.0\text{ m/s}}$$

b. Equation 16.4 applies to a wave traveling in the $\boxed{-x\text{ direction}}$.

---

26. ***REASONING*** The speed of a wave on the string is given by Equation 16.2 as $v = \sqrt{\dfrac{F}{m/L}}$, where F is the tension in the string and m/L is the mass per unit length (or linear density) of the string. The wavelength λ is the speed of the wave divided by its frequency f (Equation 16.1).

SOLUTION

a. The speed of the wave on the string is

$$v = \sqrt{\frac{F}{(m/L)}} = \sqrt{\frac{15\text{ N}}{0.85\text{ kg/m}}} = \boxed{4.2\text{ m/s}}$$

b. The wavelength is

$$\lambda = \frac{v}{f} = \frac{4.2\text{ m/s}}{12\text{ Hz}} = \boxed{0.35\text{ m}}$$

c. The amplitude of the wave is $A = 3.6\text{ cm} = 3.6 \times 10^{-2}$ m. Since the wave is moving along the $-x$ direction, the mathematical expression for the wave is given by Equation 16.4 as

$$y = A\sin\left(2\pi f t + \frac{2\pi x}{\lambda}\right)$$

Substituting in the numbers for A, f, and λ, we have

$$y = A\sin\left(2\pi f t + \frac{2\pi x}{\lambda}\right) = \left(3.6\times10^{-2}\text{ m}\right)\sin\left[2\pi\left(12\text{ Hz}\right)t + \frac{2\pi x}{0.35\text{ m}}\right]$$

$$= \boxed{\left(3.6\times10^{-2}\text{ m}\right)\sin\left[\left(75\text{ s}^{-1}\right)t + \left(18\text{ m}^{-1}\right)x\right]}$$

---

27. SSM ***REASONING*** According to Equation 16.2, the tension F in the string is given by $F = v^2(m/L)$. Since $v = \lambda f$ from Equation 16.1, the expression for F can be written

$$F = (\lambda f)^2\left(\frac{m}{L}\right) \qquad (1)$$

where the quantity m/L is the linear density of the string. In order to use Equation (1), we must first obtain values for f and λ; these quantities can be found by analyzing the expression for the displacement of a string particle.

SOLUTION The displacement is given by $y = (0.021\text{ m})\sin(25t - 2.0x)$. Inspection of this equation and comparison with Equation 16.3, $y = A\sin\left(2\pi f t - \frac{2\pi x}{\lambda}\right)$, gives

$$2\pi f = 25\text{ rad/s} \quad \text{or} \quad f = \frac{25}{2\pi}\text{ Hz}$$

and

$$\frac{2\pi}{\lambda} = 2.0\text{ m}^{-1} \quad \text{or} \quad \lambda = \frac{2\pi}{2.0}\text{ m}$$

Substituting these values f and λ into Equation (1) gives

$$F = (\lambda f)^2\left(\frac{m}{L}\right) = \left[\left(\frac{2\pi}{2.0}\text{ m}\right)\left(\frac{25\text{ Hz}}{2\pi}\right)\right]^2 (1.6\times10^{-2}\text{ kg/m}) = \boxed{2.5\text{ N}}$$

---

28. ***REASONING*** Using either Equation 16.3 or 16.4 with $x = 0$ m, we obtain

$$y = A\sin(2\pi f t)$$

Applying this equation with $f = 175$ Hz, we can find the times at which $y = 0.10$ m.

SOLUTION Using the given values for y and A, we obtain

$$0.10\text{ m} = (0.20\text{ m})\sin(2\pi f t) \quad \text{or} \quad 2\pi f t = \sin^{-1}\left(\frac{0.10\text{ m}}{0.20\text{ m}}\right) = \sin^{-1}(0.50)$$

The smallest two angles for which the sine function is 0.50 are 30° and 150°. However, the values for $2\pi ft$ must be expressed in radians. Thus, we find that the smallest two angles for which the sine function is 0.50 are expressed in radians as follows:

$$(30°)\left(\frac{2\pi}{360°}\right) \quad \text{and} \quad (150°)\left(\frac{2\pi}{360°}\right)$$

The times corresponding to these angles can be obtained as follows:

$$2\pi f t_1 = (30°)\left(\frac{2\pi}{360°}\right) \qquad 2\pi f t_2 = (150°)\left(\frac{2\pi}{360°}\right)$$

$$t_1 = \frac{30°}{f(360°)} \qquad t_2 = \frac{150°}{f(360°)}$$

Subtracting and using $f = 175$ Hz gives

$$t_2 - t_1 = \frac{150°}{f(360°)} - \frac{30°}{f(360°)} = \frac{(150° - 30°)}{(175 \text{ Hz})(360°)} = \boxed{1.9 \times 10^{-3} \text{ s}}$$

---

29. SSM ***REASONING AND SOLUTION*** The speed of sound in an ideal gas is given by Equation 16.5, $v = \sqrt{\gamma kT/m}$. The ratio of the speed of sound v_2 in the container (after the temperature change) to the speed v_1 (before the temperature change) is

$$\frac{v_2}{v_1} = \sqrt{\frac{T_2}{T_1}}$$

Thus, the new speed is

$$v_2 = v_1\sqrt{\frac{T_2}{T_1}} = (1220 \text{ m/s})\sqrt{\frac{405 \text{ K}}{201 \text{ K}}} = \boxed{1730 \text{ m/s}}$$

---

30. ***REASONING*** The speed v, frequency f, and wavelength λ of the sound are related according to $v = f\lambda$ (Equation 16.1). This expression can be solved for the wavelength in terms of the speed and the frequency. The speed of sound in seawater is 1522 m/s, as given in Table 16.1. While the frequency is not given directly, the period T is known and is related to the frequency according to $f = 1/T$ (Equation 10.5).

SOLUTION Substituting the frequency from Equation 10.5 into Equation 16.1 gives

$$v = f\lambda = \left(\frac{1}{T}\right)\lambda$$

Solving this result for the wavelength yields

$$\lambda = vT = (1522 \text{ m/s})\left(71 \times 10^{-3} \text{ s}\right) = \boxed{110 \text{ m}}$$

---

31. ***REASONING AND SOLUTION***
a. The travel time is

$$t = x/v = (2.70\text{ m})/(343\text{ m/s}) = \boxed{7.87 \times 10^{-3}\text{ s}}$$

b. The wavelength is

$$\lambda = v/f = (343\text{ m/s})/(523\text{ Hz}) = 0.656\text{ m} \qquad (16.1)$$

The number of wavelengths in 2.70 m is, therefore,

$$\text{Number of wavelengths} = (2.70\text{ m})/(0.656\text{ m}) = \boxed{4.12}$$

---

32. ***REASONING*** A rail can be approximated as a long slender bar, so the speed of sound in the rail is given by Equation 16.7. With this equation and the given data for Young's modulus and the density of steel, we can determine the speed of sound in the rail. Then, we will be able to compare this speed to the speed of sound in air at 20 °C, which is 343 m/s.

SOLUTION The speed of sound in the rail is

$$v = \sqrt{\frac{Y}{\rho}} = \sqrt{\frac{2.0 \times 10^{11}\text{ N/m}^2}{7860\text{ kg/m}^3}} = 5.0 \times 10^3\text{ m/s}$$

This speed is greater than the speed of sound in air at 20 °C by a factor of

$$\frac{v_{\text{rail}}}{v_{\text{air}}} = \frac{5.0 \times 10^3\text{ m/s}}{343\text{ m/s}} = \boxed{15}$$

---

33. SSM ***REASONING AND SOLUTION*** The speed of sound in a liquid is given by Equation 16.6, $v = \sqrt{B_{\text{ad}}/\rho}$, where B_{ad} is the adiabatic bulk modulus and ρ is the density of the liquid. Solving for B_{ad}, we obtain $B_{\text{ad}} = v^2\rho$. Values for the speed of sound in fresh water and in ethyl alcohol are given in Table 16.1. The ratio of the adiabatic bulk modulus of fresh water to that of ethyl alcohol at 20°C is, therefore,

$$\frac{(B_{\text{ad}})_{\text{water}}}{(B_{\text{ad}})_{\text{ethyl alcohol}}} = \frac{v^2_{\text{water}}\rho_{\text{water}}}{v^2_{\text{ethyl alcohol}}\rho_{\text{ethyl alcohol}}} = \frac{(1482\text{ m/s})^2(998\text{ kg/m}^3)}{(1162\text{ m/s})^2(789\text{ kg/m}^3)} = \boxed{2.06}$$

---

34. ***REASONING AND SOLUTION*** Since $f = v/\lambda$ and the frequency of the sound remains unchanged in water, we have

$$v_a/\lambda_a = v_w/\lambda_w$$

Taking the speed of sound in fresh water from Table 16.1, we find that

$$\lambda_w = \lambda_a(v_w/v_a) = (2.74 \text{ m})[(1482 \text{ m/s})/(343 \text{ m/s})] = \boxed{11.8 \text{ m}}$$

---

35. ***REASONING AND SOLUTION***

a. In order to determine the order of arrival of the three waves, we need to know the speeds of each wave. The speeds for air, water and the metal are

$$v_a = 343 \text{ m/s},\ v_w = 1482 \text{ m/s},\ v_m = 5040 \text{ m/s}$$

The order of arrival is $\boxed{\text{metal wave first, water wave second, air wave third}}$

b. Calculate the length of time each wave takes to travel 125 m.

$$t_m = (125 \text{ m})/(5040 \text{ m/s}) = 0.025 \text{ s}$$
$$t_w = (125 \text{ m})/(1482 \text{ m/s}) = 0.084 \text{ s}$$
$$t_a = (125 \text{ m})/(343 \text{ m/s}) = 0.364 \text{ s}$$

Therefore, the delay times are

$$\Delta t_{12} = t_w - t_m = 0.084 \text{ s} - 0.025 \text{ s} = \boxed{0.059 \text{ s}}$$
$$\Delta t_{13} = t_a - t_m = 0.364 \text{ s} - 0.025 \text{ s} = \boxed{0.339 \text{ s}}$$

---

36. ***REASONING AND SOLUTION*** The speed of sound in an ideal gas is given by text Equation 16.5

$$v = \sqrt{\frac{\gamma kT}{m}}$$

where *m* is the mass of a single gas particle (atom or molecule). Solving for *T* gives

$$T = \frac{mv^2}{\gamma k} \qquad (1)$$

The mass of a single helium atom is

$$\frac{4.003 \text{ g/mol}}{6.022\times10^{23} \text{ mol}^{-1}}\left(\frac{1 \text{ kg}}{1000 \text{ g}}\right) = 6.650\times10^{-27} \text{ kg}$$

The speed of sound in oxygen at 0 °C is 316 m/s. Since helium is a monatomic gas, $\gamma = 1.67$. Then, substituting into Equation (1) gives

$$T = \frac{\left(6.65\times10^{-27}\text{ kg}\right)\left(316\text{ m/s}\right)^2}{1.67\left(1.38\times10^{-23}\text{ J/K}\right)} = \boxed{28.8\text{ K}}$$

---

37. ***REASONING AND SOLUTION*** The wheel must rotate at a frequency of $f = (2200\text{ Hz})/20 = 110\text{ Hz}$. The angular speed ω of the wheel is

$$\omega = 2\pi f = 2\pi(110\text{ Hz}) = \boxed{690\text{ rad/s}}$$

---

38. ***REASONING*** Two facts allow us to solve this problem. First, the sound reaches the microphones at different times, because the distances between the microphones and the source of sound are different. Since sound travels at a speed $v = 343$ m/s and the sound arrives at microphone 2 later by an interval of $\Delta t = 1.46\times10^{-3}$ s, it follows that

$$L_2 - L_1 = v\Delta t \tag{1}$$

Second, the microphones and the source of sound are located at the corners of a right triangle. Therefore, the Pythagorean theorem applies:

$$L_2^2 = D^2 + L_1^2 \tag{2}$$

Since v, Δt, and D are known, these two equations may be solved simultaneously for the distances L_1 and L_2.

SOLUTION Solving Equation (1) for L_2 gives

$$L_2 = L_1 + v\Delta t$$

Substituting this result into Equation (2) gives

$$\left(L_1 + v\Delta t\right)^2 = D^2 + L_1^2$$

$$L_1^2 + 2L_1 v\Delta t + \left(v\Delta t\right)^2 = D^2 + L_1^2$$

$$2L_1 v\Delta t + \left(v\Delta t\right)^2 = D^2$$

Solving for L_1, we find that

$$L_1 = \frac{D^2 - (v\Delta t)^2}{2v\Delta t} = \frac{(1.50\text{ m})^2 - (343\text{ m/s})^2(1.46\times10^{-3}\text{ s})^2}{2(343\text{ m/s})(1.46\times10^{-3}\text{ s})} = \boxed{2.00\text{ m}}$$

Solving Equation (2) for L_2 and substituting the result for L_1 reveal that

$$L_2 = \sqrt{D^2 + L_1^2} = \sqrt{(1.50\text{ m})^2 + (2.00\text{ m})^2} = \boxed{2.50\text{ m}}$$

---

39. SSM ***REASONING*** Equation 16.7 relates the Young's modulus Y, the mass density ρ, and the speed of sound v in a long slender solid bar. According to Equation 16.7, the Young's modulus is given by $Y = \rho v^2$. The data given in the problem can be used to compute values for both ρ and v.

SOLUTION Using the values of the data given in the problem statement, we find that the speed of sound in the bar is

$$v = \frac{L}{t} = \frac{0.83\text{ m}}{1.9\times10^{-4}\text{ s}} = 4.4\times10^3\text{ m/s}$$

where L is the length of the rod and t is the time required for the wave to travel the length of the rod. The mass density of the bar is, from Equation 11.1, $\rho = m/V = m/(LA)$, where m and A are, respectively, the mass and the cross-sectional area of the rod. The density of the rod is, therefore,

$$\rho = \frac{m}{LA} = \frac{2.1\text{ kg}}{(0.83\text{ m})(1.3\times10^{-4}\text{ m}^2)} = 1.9\times10^4\text{ kg/m}^3$$

Using these values, we find that the bulk modulus for the rod is

$$Y = \rho v^2 = (1.9\times10^4\text{ kg/m}^3)(4.4\times10^3\text{ m/s})^2 = 3.7\times10^{11}\text{ N/m}^2$$

Comparing this value to those given in Table 10.1, we conclude that the bar is most likely made of $\boxed{\text{tungsten}}$.

---

40. ***REASONING*** Let v_P represent the speed of the primary wave and v_S the speed of the secondary wave. The travel times for the primary and secondary waves are t_P and t_S, respectively. If x is the distance from the earthquake to the seismograph, then $t_P = x/v_P$ and $t_S = x/v_S$. The difference in the arrival times is

$$t_S - t_P = \frac{x}{v_S} - \frac{x}{v_P} = x\left(\frac{1}{v_S} - \frac{1}{v_P}\right)$$

We can use this equation to find the distance from the earthquake to the seismograph.

SOLUTION Solving the equation above for x gives

$$x = \frac{t_S - t_P}{\frac{1}{v_S} - \frac{1}{v_P}} = \frac{78\text{ s}}{\frac{1}{4.5\times10^3\text{ m/s}} - \frac{1}{8.0\times10^3\text{ m/s}}} = \boxed{8.0\times10^5\text{ m}}$$

---

41. **SSM** ***REASONING*** The sound will spread out uniformly in all directions. For the purposes of counting the echoes, we will consider only the sound that travels in a straight line parallel to the ground and reflects from the vertical walls of the cliff.

Let the distance between the hunter and the closer cliff be x_1 and the distance from the hunter to the further cliff be x_2.

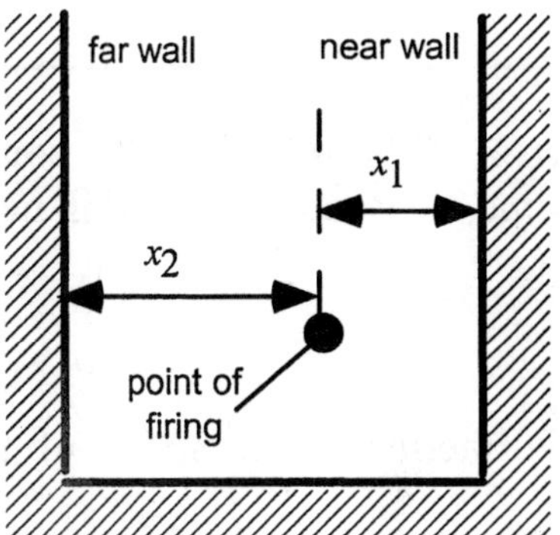

The first echo arrives at the location of the hunter after traveling a total distance $2x_1$ in a time t_1, so that, if v_s is the speed of sound, $t_1 = 2x_1/v_s$. Similarly, the second echo arrives after reflection from the far wall and in an amount of time t_2 after the firing of the gun. The quantity t_2 is related to the distance x_2 and the speed of sound v_s according to $t_2 = 2x_2/v_s$. The time difference between the first and second echo is, therefore

$$\Delta t = t_2 - t_1 = \frac{2}{v_s}(x_2 - x_1) \tag{1}$$

The third echo arrives in a time t_3 after the second echo. It arises from the sound of the second echo that is reflected from the closer cliff wall. Thus, $t_3 = 2x_1/v_s$, or, solving for x_1, we have

$$x_1 = \frac{v_s t_3}{2} \tag{2}$$

Combining Equations (1) and (2), we obtain

$$\Delta t = t_2 - t_1 = \frac{2}{v_s}\left(x_2 - \frac{v_s t_3}{2}\right)$$

Solving for x_2, we have

$$x_2 = \frac{v_s}{2}(\Delta t + t_3) \tag{3}$$

The distance between the cliffs can be found from $d = x_1 + x_2$, where x_1 and x_2 can be determined from Equations (2) and (3), respectively.

SOLUTION According to Equation (2), the distance x_1 is

$$x_1 = \frac{(343\text{ m/s})(1.1\text{ s})}{2} = 190\text{ m}$$

According to Equation (3), the distance x_2 is

$$x_2 = \frac{(343\text{ m/s})}{2}(1.6\text{ s} + 1.1\text{ s}) = 460\text{ m}$$

Therefore, the distance between the cliffs is

$$d = x_1 + x_2 = 190\text{ m} + 460\text{ m} = \boxed{650\text{ m}}$$

---

42. ***REASONING*** Since the sound wave travels twice as far in neon as in krypton in the same time, the speed of sound in neon must be twice that in krypton:

$$v_{\text{neon}} = 2v_{\text{krypton}} \tag{1}$$

Furthermore, the speed of sound in an ideal gas is given by $v = \sqrt{\frac{\gamma kT}{m}}$, according to Equation 16.5. In this expression γ is the ratio of the specific heat capacities at constant pressure and constant volume and is the same for either gas (see Section 15.6), k is Boltzmann's constant, T is the Kelvin temperature, and m is the mass of an atom. This expression for the speed can be used for both gases in Equation (1) and the result solved for the temperature of the neon.

SOLUTION Using Equation 16.5 in Equation (1), we have

$$\underbrace{\sqrt{\frac{\gamma kT_{\text{neon}}}{m_{\text{neon}}}}}_{\text{Speed in neon}} = 2\underbrace{\sqrt{\frac{\gamma kT_{\text{krypton}}}{m_{\text{krypton}}}}}_{\text{Speed in krypton}}$$

Squaring this result and solving for the temperature of the neon give

$$\frac{\gamma kT_{\text{neon}}}{m_{\text{neon}}} = 4\left(\frac{\gamma kT_{\text{krypton}}}{m_{\text{krypton}}}\right) \quad \text{or} \quad T_{\text{neon}} = 4\left(\frac{m_{\text{neon}}}{m_{\text{krypton}}}\right)T_{\text{krypton}}$$

We note here that the mass m of an atom is proportional to its atomic mass in atomic mass units (u). As a result, the temperature of the neon is

$$T_{\text{neon}} = 4\left(\frac{m_{\text{neon}}}{m_{\text{krypton}}}\right)T_{\text{krypton}} = 4\left(\frac{20.2\ \text{u}}{83.8\ \text{u}}\right)(293\ \text{K}) = \boxed{283\ \text{K}}$$

---

43. ***REASONING AND SOLUTION*** The speed of sound in an ideal gas is given by text Equation 16.5:

$$v = \sqrt{\frac{\gamma kT}{m}}$$

where $\gamma = c_p/c_v$, k is Boltzmann's constant, T is the Kelvin temperature of the gas, and m is the mass of a single gas molecule. If m_{TOTAL} is the mass of the gas sample and N is the total number of molecules in the sample, then the above equation can be written as

$$v = \sqrt{\frac{\gamma kT}{(m_{\text{TOTAL}}/N)}} = \sqrt{\frac{\gamma NkT}{m_{\text{TOTAL}}}} \qquad (1)$$

For an ideal gas, $PV = NkT$, so that Equation (1) becomes,

$$v = \sqrt{\frac{\gamma PV}{m_{\text{TOTAL}}}} = \sqrt{\frac{(1.67)(3.5\times10^5\ \text{Pa})(2.5\ \text{m}^3)}{2.3\ \text{kg}}} = \boxed{8.0\times10^2\ \text{m/s}}$$

---

44. ***REASONING AND SOLUTION*** If y_a is the one-way distance the sound travels in air ($y_a = 10.0$ m) and y_w is the distance the sound travels in water, then the total travel time, t, is given by

$$t = 2y_a/v_a + 2y_w/v_w$$

where the speeds can be obtained from Table 16.1. We can, therefore, solve for the depth of the lake.

$$y_w = \left[t - 2\left(\frac{y_a}{v_a}\right)\right]\left(\frac{v_w}{2}\right) = \left[(0.140\ \text{s}) - 2\left(\frac{10.0\ \text{m}}{343\ \text{m/s}}\right)\left(\frac{1482\ \text{m/s}}{2}\right)\right] = \boxed{61\ \text{m}}$$

---

45. ***REASONING AND SOLUTION*** The sound emitted by the plane at A reaches the person after a time t. This time, t, required for the sound wave at A to reach the person is the same as the time required for the plane to fly from A to B.

The figure at the right shows the relevant geometry. During the time t, the plane travels the distance x, while the sound wave travels the distance d. The sound wave travels with constant speed. The plane has a speed of v_0 at A and a speed v at B and travels with constant acceleration. Thus,

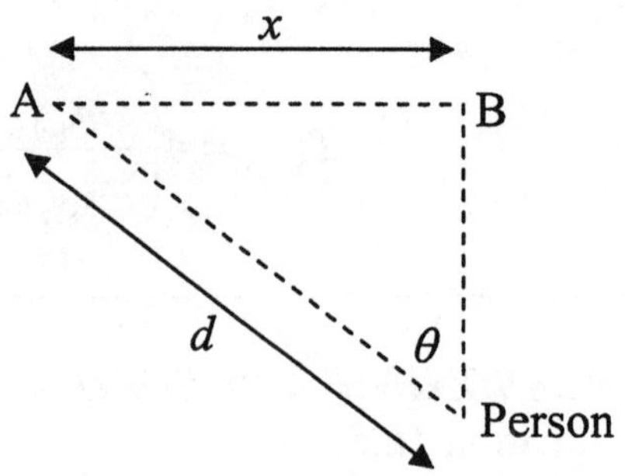

$$d = v_{\text{sound}} t$$

and

$$x = \tfrac{1}{2}(v + v_0)t$$

From the drawing, we have

$$\sin\theta = \frac{x}{d} = \frac{\tfrac{1}{2}(v+v_0)t}{v_{\text{sound}}t} = \frac{v+v_0}{2v_{\text{sound}}}$$

Solving for v gives

$$v = 2v_{\text{sound}}\sin\theta - v_0 = 2(343 \text{ m/s})\sin 36.0° - 164 \text{ m/s} = \boxed{239 \text{ m/s}}$$

---

46. ***REASONING AND SOLUTION*** The speed of sound in fresh water is

$$v_{\text{fresh}} = \sqrt{\frac{B_{\text{ad}}}{\rho}} = \sqrt{\frac{2.20\times10^9 \text{ Pa}}{1.00\times10^3 \text{ kg/m}^3}} = 1.48\times10^3 \text{ m/s}$$

The speed of sound in salt water is

$$v_{\text{salt}} = \sqrt{\frac{B_{\text{ad}}}{\rho}} = \sqrt{\frac{2.37\times10^9 \text{ Pa}}{1025 \text{ kg/m}^3}} = 1.52\times10^3 \text{ m/s}$$

The ratio of these two speeds is

$$\frac{v_{\text{salt}}}{v_{\text{fresh}}} = \frac{1.52\times10^3 \text{ m/s}}{1.48\times10^3 \text{ m/s}} = 1.03$$

Thus, the sonar unit erroneously calculates that the objects are 1.03 times the distance measured in salt water:

$$d_{\text{actual}} = (1.03)(10.0 \text{ m}) = \boxed{10.3 \text{ m}}$$

---

47. SSM WWW ***REASONING*** We must determine the time t for the warning to travel the vertical distance $h = 10.0$ m from the prankster to the ears of the man when he is just under the window. The desired distance above the man's ears is the distance that the balloon would travel in this time and can be found with the aid of the equations of kinematics.

SOLUTION Since sound travels with constant speed v_s, the distance h and the time t are related by $h = v_s t$. Therefore, the time t required for the warning to reach the ground is

$$t = \frac{h}{v_s} = \frac{10.0 \text{ m}}{343 \text{ m/s}} = 0.0292 \text{ s}$$

We now proceed to find the distance that the balloon travels in this time. To this end, we must find the balloon's final speed v_y after falling from rest for 10.0 m. Since the balloon is dropped from rest, we use Equation 3.6*b* ($v_y^2 = v_{0y}^2 + 2a_y y$) with $v_{0y} = 0$ m/s:

$$v_y = \sqrt{v_{0y}^2 + 2a_y y} = \sqrt{(0 \text{ m/s})^2 + 2(9.80 \text{ m/s}^2)(10.0 \text{ m})} = 14.0 \text{ m/s}$$

Using this result, we can find the balloon's speed 0.0292 seconds before it hits the man by solving Equation 3.3b ($v_y = v_{0y} + a_y t$) for v_{0y}:

$$v_{0y} = v_y - a_y t = \left[(14.0 \text{ m/s}) - (9.80 \text{ m/s}^2)(0.0292 \text{ s})\right] = 13.7 \text{ m/s}$$

Finally, we can find the desired distance y above the man's head from Equation 3.5b:

$$y = v_{0y} t + \tfrac{1}{2} a_y t^2 = (13.7 \text{ m/s})(0.0292 \text{ s}) + \tfrac{1}{2}(9.80 \text{ m/s}^2)(0.0292 \text{ s})^2 = \boxed{0.404 \text{ m}}$$

---

48. ***REASONING AND SOLUTION*** We have $I = P/A$. Therefore,

$$P = IA = (3.2 \times 10^{-6} \text{ W/m}^2)(2.1 \times 10^{-3} \text{ m}^2) = \boxed{6.7 \times 10^{-9} \text{ W}} \qquad (16.8)$$

---

49. SSM ***REASONING AND SOLUTION*** According to Equation 16.8, the power radiated by the speaker is $P = IA = I\pi r^2$, where r is the radius of the circular opening. Thus, the radiated power is

$$P = (17.5 \text{ W/m}^2)(\pi)(0.0950 \text{ m})^2 = 0.496 \text{ W}$$

As a percentage of the electrical power, this is

$$\frac{0.496\text{ W}}{25.0\text{ W}} \times 100\ \% = \boxed{1.98\ \%}$$

---

50. ***REASONING AND SOLUTION*** The energy carried by the sound into the ear is

$$\text{Energy} = IAt = (3.2 \times 10^{-5}\text{ W/m}^2)(2.1 \times 10^{-3}\text{ m}^2)(3600\text{ s}) = \boxed{2.4 \times 10^{-4}\text{ J}}$$

---

51. ***REASONING AND SOLUTION*** The intensity of the sound falls off according to the square of the distance according to Equation 16.9. Therefore, the intensity at the new location will be

$$I_{78} = \left(\frac{22\text{ m}}{78\text{ m}}\right)^2 \left(3.0 \times 10^{-4}\text{ W/m}^2\right) = \boxed{2.4 \times 10^{-5}\text{ W/m}^2}$$

---

52. ***REASONING AND SOLUTION*** The intensity of the "direct" sound is given by text Equation 16.9,

$$I_{\text{DIRECT}} = \frac{P}{4\pi r^2}$$

The total intensity at the point in question is

$$\begin{aligned} I_{\text{TOTAL}} &= I_{\text{DIRECT}} + I_{\text{REFLECTED}} \\ &= \left[\frac{1.1 \times 10^{-3}\text{ W}}{4\pi\,(3.0\text{ m})^2}\right] + 4.4 \times 10^{-6}\text{ W/m}^2 = \boxed{1.4 \times 10^{-5}\text{ W/m}^2} \end{aligned}$$

---

53. **SSM** **WWW** ***REASONING AND SOLUTION*** Since the sound radiates uniformly in all directions, at a distance *r* from the source, the energy of the sound wave is distributed over the area of a sphere of radius *r*. Therefore, according to Equation 16.9 $[I = P/(4\pi r^2)]$ with $r = 3.8$ m, the power radiated from the source is

$$P = 4\pi I r^2 = 4\pi(3.6 \times 10^{-2}\text{ W/m}^2)(3.8\text{ m})^2 = \boxed{6.5\text{ W}}$$

---

54. ***REASONING*** According to Equation 16.8, the intensity I of a sound wave is equal to the sound power P divided by the area A through which the power passes; $I = P/A$. The area is that of a circle, so $A = \pi r^2$, where r is the radius. The power, on the other hand, is the energy E per unit time, $P = E/t$, according to Equation 6.10b. Thus, we have

$$I = \frac{P}{A} = \frac{\frac{E}{t}}{\pi r^2}$$

All the variables in this equation are known except for the time.

SOLUTION Solving the expression above for the time yields

$$t = \frac{E}{I\pi r^2} = \frac{4800 \text{ J}}{\left(5.9 \times 10^3 \text{ W/m}^2\right)\pi\left(1.8 \times 10^{-2} \text{ m}\right)^2} = \boxed{8.0 \times 10^2 \text{ s}}$$

---

55. ***REASONING*** Intensity I is power P divided by the area A, or $I = \frac{P}{A}$, according to Equation 16.8. The area is given directly, but the power is not. Therefore, we need to recast this expression in terms of the data given in the problem. Power is the change in energy per unit time, according to Equation 6.10b. In this case the energy is the heat Q that causes the temperature of the lasagna to increase. Thus, the power is $P = \frac{Q}{t}$, where t denotes the time. As a result, Equation 16.8 for the intensity becomes

$$I = \frac{P}{A} = \frac{Q}{tA} \tag{1}$$

According to Equation 12.4, the heat that must be supplied to increase the temperature of a substance of mass m by an amount ΔT is $Q = cm\Delta T$, where c is the specific heat capacity. Substituting this expression into Equation (1) gives

$$I = \frac{Q}{tA} = \frac{cm\Delta T}{tA} \tag{2}$$

SOLUTION Equation (2) reveals that the intensity of the microwaves is

$$I = \frac{cm\Delta T}{tA} = \frac{\left[3200 \text{ J/}\left(\text{kg}\cdot\text{C}^\circ\right)\right]\left(0.35 \text{ kg}\right)\left(72 \text{ C}^\circ\right)}{\left(480 \text{ s}\right)\left(2.2\times10^{-2} \text{ m}^2\right)} = \boxed{7.6\times10^3 \text{ W/m}^2}$$

---

56. ***REASONING AND SOLUTION*** The power radiated by the helicopter is

$$P = IA = I\,(4\pi r^2)$$

At the overhead position, the intensity is

$$I_1 = P/\left(4\pi r_1^2\right)$$

At some distance away, the new intensity is

$$I_2 = P/\left(4\pi r_2^2\right)$$

Since $I_1 = 4I_2$, we can write

$$P/\left(4\pi r_1^2\right) = P/\left(4\pi r_2^2\right)$$

This result indicates that $r_2 = 2r_1$. The helicopter must therefore travel a distance d where

$$r_1^2 + d^2 = r_2^2 = 4r_1^2$$

So,

$$d = \sqrt{3}\, r_1 = \sqrt{3}\,(1450 \text{ m}) = \boxed{2510 \text{ m}}$$

---

57. **SSM** ***REASONING*** Since the sound is emitted from the rocket uniformly in all directions, the energy carried by the sound wave spreads out uniformly over concentric spheres of increasing radii $r_1, r_2, r_3, \ldots$ as the wave propagates. Let r_1 represent the radius when the measured intensity at the ground is I (position 1 for the rocket) and r_2 represent the radius when the measured intensity at the ground is $\frac{1}{3}I$ (position 2 of the rocket). The time required for the energy to spread out over the sphere of radius r_1 is $t_1 = r_1/v_s$, where v_s is the speed of sound. Similarly, the time required for the energy to spread out over a sphere of radius r_2 is $t_2 = r_2/v_s$. As the sound wave emitted at position 1 spreads out uniformly, the rocket continues to accelerate upward to position 2 with acceleration a_y for a time t_{12}. Therefore, the time that elapses between the two intensity measurements is

$$t = t_2 - t_1 + t_{12} \tag{1}$$

From Equation 3.3*b*, the time t_{12} is

$$t_{12} = \frac{\left(v_2 - v_1\right)}{a_y} \tag{2}$$

where v_1 and v_2 are the speeds of the rocket at positions 1 and 2, respectively. Combining Equations (1) and (2), we obtain

$$t = t_2 - t_1 + \frac{(v_2 - v_1)}{a_y} \tag{3}$$

The respective values of v_1 and v_2 can be found from Equation 3.6b ($v_y^2 = v_{0y}^2 + 2a_y y$) with $v_{0y} = 0$ m/s, and $y = r$. Once values for v_1 and v_2 are known, Equation (3) can be solved directly.

SOLUTION From the data given in the problem statement, $r_1 = 562$ m. We can find r_2 using the following reasoning: from the definition of intensity, $I_1 = P/(4\pi r_1^2)$ and $I_2 = P/(4\pi r_2^2)$. Since $I_1 = 3I_2$, we have

$$P/(4\pi r_1^2) = 3P/(4\pi r_2^2) \qquad \text{or} \qquad r_2 = r_1\sqrt{3} = 973 \text{ m}$$

The times t_1 and t_2 are

$$t_1 = \frac{r_1}{v_s} = \frac{562 \text{ m}}{343 \text{ m/s}} = 1.64 \text{ s}$$

and

$$t_2 = \frac{r_2}{v_s} = \frac{973 \text{ m}}{343 \text{ m/s}} = 2.84 \text{ s}$$

Then, taking up as the positive direction, we have from Equation 3.6b,

$$v_1 = \sqrt{2a_y r_1} = \sqrt{2(58.0 \text{ m/s}^2)(562 \text{ m})} = 255 \text{ m/s}$$

and

$$v_2 = \sqrt{2a_y r_2} = \sqrt{2(58.0 \text{ m/s}^2)(973 \text{ m})} = 336 \text{ m/s}$$

Substituting these values into Equation (3), we have

$$t = 2.84 \text{ s} - 1.64 \text{ s} + \frac{(336 \text{ m/s} - 255 \text{ m/s})}{58.0 \text{ m/s}^2} = \boxed{2.6 \text{ s}}$$

---

58. ***REASONING*** This is a situation in which the intensities I_{man} and I_{woman} (in watts per square meter) detected by the man and the woman are compared using the intensity level β, expressed in decibels. This comparison is based on Equation 16.10, which we rewrite as follows:

$$\beta = (10\ \text{dB})\log\left(\frac{I_{\text{man}}}{I_{\text{woman}}}\right)$$

SOLUTION Using Equation 16.10, we have

$$\beta = 7.8\ \text{dB} = (10\ \text{dB})\log\left(\frac{I_{\text{man}}}{I_{\text{woman}}}\right) \quad \text{or} \quad \log\left(\frac{I_{\text{man}}}{I_{\text{woman}}}\right) = \frac{7.8\ \text{dB}}{10\ \text{dB}} = 0.78$$

Solving for the intensity ratio gives

$$\frac{I_{\text{man}}}{I_{\text{woman}}} = 10^{0.78} = \boxed{6.0}$$

---

59. ***REASONING*** Knowing that the threshold of hearing corresponds to an intensity of $I_0 = 1.00 \times 10^{-12}\ \text{W/m}^2$, we can solve Equation 16.10 directly for the desired intensity.

SOLUTION Using Equation 16.10, we find

$$\beta = 115\ \text{dB} = (10\ \text{dB})\log\left(\frac{I}{I_0}\right) \quad \text{or} \quad \frac{I}{I_0} = 10^{(115\ \text{dB})/(10\ \text{dB})} = 10^{11.5}$$

Solving for I gives

$$I = I_0 10^{11.5} = \left(1.00 \times 10^{-12}\ \text{W/m}^2\right)10^{11.5} = \boxed{0.316\ \text{W/m}^2}$$

---

60. ***REASONING*** The sound intensity level outside the room is

$$\beta_{\text{outside}} = (10\ \text{dB})\log\left(\frac{I_{\text{outside}}}{I_0}\right) \tag{16.10}$$

where I_0 is the threshold of hearing. Solving for the intensity I_{outside} gives

$$I_{\text{outside}} = I_0 10^{\frac{\beta_{\text{outside}}}{10\ \text{dB}}}$$

These two relations will allow us to find the sound intensity outside the room.

SOLUTION From the problem statement we know that $\beta_{\text{outside}} = \beta_{\text{inside}} + 44.0$ dB. We can evaluate β_{inside} by applying Equation 16.10 to the inside of the room:

$$\beta_{\text{inside}} = (10\text{ dB})\log\left(\frac{I_{\text{inside}}}{I_0}\right) = (10\text{ dB})\log\left(\frac{1.20\times10^{-10}\text{ W/m}^2}{1.00\times10^{-12}\text{ W/m}^2}\right) = 20.8\text{ dB}$$

Thus, the sound intensity level outside the room is $\beta_{\text{outside}} = 20.8\text{ dB} + 44.0\text{ dB} = 64.8\text{ dB}$. The sound intensity outside the room is

$$I_{\text{outside}} = I_0 10^{\frac{\beta_{\text{outside}}}{10\text{ dB}}} = (1.00\times10^{-12}\text{ W/m}^2)10^{\frac{64.8\text{ dB}}{10\text{ dB}}} = \boxed{3.02\times10^{-6}\text{ W/m}^2}$$

---

61. SSM ***REASONING AND SOLUTION*** The sound intensity level β in decibels (dB) is related to the sound intensity I according to Equation 16.10, $\beta = (10\text{ dB})\log(I/I_0)$, where I_0 is the reference intensity. If β_1 and β_2 represent two different sound level intensities, then,

$$\beta_2 - \beta_1 = (10\text{ dB})\log\left(\frac{I_2}{I_0}\right) - (10\text{ dB})\log\left(\frac{I_1}{I_0}\right) = (10\text{ dB})\log\left(\frac{I_2/I_0}{I_1/I_0}\right) = (10\text{ dB})\log\left(\frac{I_2}{I_1}\right)$$

Therefore,

$$\frac{I_2}{I_1} = 10^{(\beta_2-\beta_1)/(10\text{ dB})}$$

When the difference in sound intensity levels is $\beta_2 - \beta_1 = 1.0\text{ dB}$, the ratio of the sound intensities is

$$\frac{I_2}{I_1} = 10^{(1\text{ dB})/(10\text{ dB})} = \boxed{1.3}$$

---

62. ***REASONING AND SOLUTION***

a. $\beta = (10\text{ dB})\log(P_A/P_B) = (10\text{ dB})\log[(250\text{ W})/(45\text{ W}) = \boxed{7.4\text{ dB}}$

b. No , A will not be twice as loud as B since it requires an increase of 10 dB to double the loudness.

---

63. ***REASONING AND SOLUTION*** From the information given in Problem 62 we have

$$\beta = (10\text{ dB})\log(P/P_0)$$

where $P_0 = 1.00$ W. Solving for P in the above equation yields

$$P = P_0 10^{\beta/(10\text{ dB})} = (1.00\text{ W})10^{1.75} = \boxed{56.2\text{ W}}$$

---

64. SSM **_REASONING_** We must first find the intensities that correspond to the given sound intensity levels (in decibels). The total intensity is the sum of the two intensities. Once the total intensity is known, Equation 16.10 can be used to find the total sound intensity level in decibels.

SOLUTION Since, according to Equation 16.10, $\beta = (10\text{ dB})\log(I/I_0)$, where I_0 is the reference intensity corresponding to the threshold of hearing $\left(I_0 = 1.00\times10^{-12}\text{ W/m}^2\right)$, it follows that $I = I_0\ 10^{\beta/(10\text{ dB})}$. Therefore, if $\beta_1 = 75.0$ dB and $\beta_2 = 72.0$ dB at the point in question, the corresponding intensities are

$$I_1 = I_0\ 10^{\beta_1/(10\text{ dB})} = (1.00\times10^{-12}\text{ W/m}^2)\ 10^{(75.0\text{ dB})/(10\text{ dB})} = 3.16\times10^{-5}\text{ W/m}^2$$

$$I_2 = I_0\ 10^{\beta_2/(10\text{ dB})} = (1.00\times10^{-12}\text{ W/m}^2)\ 10^{(72.0\text{ dB})/(10\text{ dB})} = 1.58\times10^{-5}\text{ W/m}^2$$

Therefore, the total intensity I_{total} at the point in question is

$$I_{\text{total}} = I_1 + I_2 = \left(3.16\times10^{-5}\text{ W/m}^2\right) + \left(1.58\times10^{-5}\text{ W/m}^2\right) = 4.74\times10^{-5}\text{ W/m}^2$$

and the corresponding intensity level β_{total} is

$$\beta_{\text{total}} = (10\text{ dB})\log\left(\frac{I_{\text{total}}}{I_0}\right) = (10\text{ dB})\log\left(\frac{4.74\times10^{-5}\text{ W/m}^2}{1.00\times10^{-12}\text{ W/m}^2}\right) = \boxed{76.8\text{ dB}}$$

---

65. **_REASONING_** Suppose we knew the sound intensity ratio $\frac{I_A}{I_B}$. In addition, suppose we knew the sound intensity ratio $\frac{I_C}{I_A}$. Then, all that would be necessary to obtain the desired ratio $\frac{I_C}{I_B}$ would be to multiply the two known ratios:

$$\left(\frac{I_A}{I_B}\right)\left(\frac{I_C}{I_A}\right)=\frac{I_C}{I_B}$$

This is exactly the procedure we will follow, except that first we will obtain the intensity ratios from the given intensity levels (expressed in decibels). The intensity level β in decibels is given by Equation 16.10.

SOLUTION Applying Equation 16.10 to persons A and B gives

$$\beta_{A/B}=1.5\text{ dB}=(10\text{ dB})\log\left(\frac{I_A}{I_B}\right)$$

$$\log\left(\frac{I_A}{I_B}\right)=\frac{1.5\text{ dB}}{10\text{ dB}}=0.15 \quad \text{or} \quad \frac{I_A}{I_B}=10^{0.15}$$

In a similar fashion for persons C and A we obtain

$$\beta_{C/A}=2.7\text{ dB}=(10\text{ dB})\log\left(\frac{I_C}{I_A}\right)$$

$$\log\left(\frac{I_C}{I_A}\right)=\frac{2.7\text{ dB}}{10\text{ dB}}=0.27 \quad \text{or} \quad \frac{I_C}{I_A}=10^{0.27}$$

Multiplying the two intensity ratios reveals that

$$\frac{I_C}{I_B}=\left(\frac{I_A}{I_B}\right)\left(\frac{I_C}{I_A}\right)=\left(10^{0.15}\right)\left(10^{0.27}\right)=10^{0.42}=\boxed{2.6}$$

---

66. ***REASONING AND SOLUTION*** We can write

$$\beta_2-\beta_1=(10\text{ dB})\log(I_2/I_1)$$

so

$$I_2/I_1=10^{(\beta_2-\beta_1)/(10\text{ dB})}=10^{(78\text{ dB}-65\text{ dB})/(10\text{ dB})}=20$$

Since the 78 dB level is 20 times more intense than the 65 dB level, this means that the number of people speaking at the 65 dB level needed to reach the 78 dB level is $\boxed{20}$.

---

67. ***REASONING*** The sound intensity level heard by the gardener increases by 10.0 dB because distance between him and the radio decreases. The intensity of the sound is greater when the radio is closer than when it is further away. Since the unit is emitting sound uniformly (neglecting any reflections), the intensity is inversely proportional to the square of the distance from the radio, according to Equation 16.9. Combining this equation with Equation 16.10, which relates the intensity level in decibels to the intensity I, we can use the 10.0-dB change in the intensity level to find the final vertical position of the radio. Once that position is known, we can then use kinematics to determine the fall time.

SOLUTION Let the sound intensity levels at the initial and final positions of the radio be β_i and β_f, respectively. Using Equation 16.10 for each, we have

$$\beta_f - \beta_i = (10\text{ dB})\log\left(\frac{I_f}{I_0}\right) - (10\text{ dB})\log\left(\frac{I_i}{I_0}\right) = (10\text{ dB})\log\left(\frac{I_f/I_0}{I_i/I_0}\right) = (10\text{ dB})\log\left(\frac{I_f}{I_i}\right) \quad (1)$$

Since the radiation is uniform, Equation 16.9 can be used to substitute for the intensities I_i and I_f, so that Equation (1) becomes

$$\beta_f - \beta_i = (10\text{ dB})\log\left(\frac{I_f}{I_i}\right) = (10\text{ dB})\log\left[\frac{P/(4\pi h_f)^2}{P/(4\pi h_i)^2}\right] = (10\text{ dB})\log\left(\frac{h_i}{h_f}\right)^2 = (20\text{ dB})\log\left(\frac{h_i}{h_f}\right) \quad (2)$$

Since $\beta_f - \beta_i = 10.0$ dB, Equation (2) gives

$$\beta_f - \beta_i = 10.0\text{ dB} = (20\text{ dB})\log\left(\frac{h_i}{h_f}\right) \quad \text{or} \quad \frac{h_i}{h_f} = 10^{(10.0\text{ dB})/(20\text{ dB})} = 10^{0.500} = 3.16$$

We can now determine the final position of the radio as follows:

$$\frac{h_i}{h_f} = \frac{5.1\text{ m}}{h_f} = 3.16 \quad \text{or} \quad h_f = \frac{5.1\text{ m}}{3.16} = 1.61\text{ m}$$

It follows, then, that the radio falls through a distance of 5.1 m − 1.61 m = 3.49 m. Taking upward as the positive direction and noting that the radio falls from rest ($v_0 = 0$ m/s), we can solve Equation 2.8 $\left(y = v_0 t + \frac{1}{2}at^2\right)$ from the equations of kinematics for the fall time t:

$$t = \sqrt{\frac{2y}{a}} = \sqrt{\frac{2(-3.49\text{ m})}{-9.80\text{ m/s}^2}} = \boxed{0.84\text{ s}}$$

68. ***REASONING AND SOLUTION*** The intensity level at each point is given by

$$I = \frac{P}{4\pi r^2}$$

Therefore,

$$\frac{I_1}{I_2} = \left(\frac{r_2}{r_1}\right)^2$$

Since the two intensity levels differ by 2.00 dB, the intensity ratio is

$$\frac{I_1}{I_2} = 10^{0.200} = 1.58$$

Thus,

$$\left(\frac{r_2}{r_1}\right)^2 = 1.58$$

We also know that $r_2 - r_1 = 1.00$ m. We can then solve the two equations simultaneously by substituting, i.e., $r_2 = r_1\sqrt{1.58}$ gives

$$r_1\sqrt{1.58} - r_1 = 1.00 \text{ m}$$

so that

$$r_1 = (1.00 \text{ m})/[\sqrt{1.58} - 1] = \boxed{3.9 \text{ m}} \qquad \text{and} \qquad r_2 = 1.00 \text{ m} + r_1 = \boxed{4.9 \text{ m}}$$

---

69. SSM ***REASONING AND SOLUTION*** The sound intensity level β in decibels (dB) is related to the sound intensity I according to Equation 16.10, $\beta = (10 \text{ dB}) \log (I/I_0)$, where the quantity I_0 is the reference intensity. According to the problem statement, when the sound intensity level triples, the sound intensity also triples; therefore,

$$3\beta = (10 \text{ dB}) \log\left(\frac{3I}{I_0}\right)$$

Then,

$$3\beta - \beta = (10 \text{ dB}) \log\left(\frac{3I}{I_0}\right) - (10 \text{ dB}) \log\left(\frac{I}{I_0}\right) = (10 \text{ dB}) \log\left(\frac{3I/I_0}{I/I_0}\right)$$

Thus, $2\beta = (10 \text{ dB}) \log 3$ and

$$\beta = (5 \text{ dB}) \log 3 = \boxed{2.39 \text{ dB}}$$

---

70. ***REASONING*** You hear a frequency f_o that is 1.0% lower than the frequency f_s emitted by the source. This means that the frequency you observe is 99.0% of the emitted frequency, so that $f_o = 0.990 f_s$. You are an observer who is moving away from a stationary source of sound. Therefore, the Doppler-shifted frequency that you observe is specified by Equation 16.14, which can be solved for the bicycle speed v_o.

SOLUTION Equation 16.14, in which v denotes the speed of sound, states that

$$f_o = f_s\left(1 - \frac{v_o}{v}\right)$$

Solving for v_o and using the fact that $f_o = 0.990 f_s$ reveal that

$$v_o = v\left(1 - \frac{f_o}{f_s}\right) = (343\ \text{m/s})\left(1 - \frac{0.990 f_s}{f_s}\right) = \boxed{3.4\ \text{m/s}}$$

---

71. ***REASONING*** The observed frequency changes because of the Doppler effect. As you drive toward the parked car (a stationary source of sound), the Doppler effect is that given by Equation 16.13. As you drive away from the parked car, Equation 16.14 applies.

SOLUTION Equations 16.13 and 16.14 give the observed frequency f_o in each case:

$$\underbrace{f_{o,\ \text{toward}} = f_s\left(1 + v_o/v\right)}_{\text{Driving toward parked car}} \qquad \text{and} \qquad \underbrace{f_{o,\ \text{away}} = f_s\left(1 - v_o/v\right)}_{\text{Driving away from parked car}}$$

Subtracting the equation on the right from the one on the left gives the change in the observed frequency:

$$f_{o,\ \text{toward}} - f_{o,\ \text{away}} = 2 f_s v_o / v$$

Solving for the observer's speed (which is your speed), we obtain

$$v_o = \frac{v\left(f_{o,\ \text{toward}} - f_{o,\ \text{away}}\right)}{2 f_s} = \frac{(343\ \text{m/s})(95\ \text{Hz})}{2(960\ \text{Hz})} = \boxed{17\ \text{m/s}}$$

---

72. ***REASONING AND SOLUTION*** The frequency of the sound reaching your ears is

$$f_o = f_s\left(\frac{1}{1-v_s/v}\right) = (955\text{ Hz})\left[\frac{1}{1-(18\text{ m/s})/(343\text{ m/s})}\right] = 1010\text{ Hz} \quad (16.11)$$

This corresponds to a wavelength of

$$\lambda = \frac{v}{f_o} = \frac{343\text{ m/s}}{1010\text{ Hz}} = \boxed{0.340\text{ m}} \quad (16.1)$$

---

73. SSM ***REASONING AND SOLUTION: METHOD 1*** Since the police car is matching the speed of the speeder, there is no relative motion between the police car and the listener (the speeder). Therefore, the frequency that the speeder hears when the siren is turned on is the same as if the police car and the listener are stationary. Thus, the speeder hears a frequency of $\boxed{860\text{ Hz}}$.

REASONING AND SOLUTION: METHOD 2 In this situation, both the source (the siren) and the observer (the speeder) are moving through the air. Therefore, the frequency of the siren heard by the speeder is given by Equation 16.15 using the minus signs in both the numerator and the denominator.

$$f_o = f_s\left(\frac{1-\dfrac{v_o}{v}}{1-\dfrac{v_s}{v}}\right) = (860\text{ Hz})\left(\frac{1-\dfrac{38\text{ m/s}}{343\text{ m/s}}}{1-\dfrac{38\text{ m/s}}{343\text{ m/s}}}\right) = \boxed{860\text{ Hz}}$$

---

74. ***REASONING*** The observer of the sound (the bird-watcher) is stationary, while the source (the bird) is moving toward the observer. Therefore, the Doppler-shifted observed frequency is given by Equation 16.11. This expression can be solved to give the ratio of the bird's speed to the speed of sound, from which the desired percentage follows directly.

SOLUTION According to Equation 16.11, the observed frequency f_o is related to the frequency f_s of the source, and the ratio of the speed of the source v_s to the speed of sound v by

$$f_o = f_s\left(\frac{1}{1-v_s/v}\right) \quad \text{or} \quad \frac{f_o}{f_s} = \frac{1}{1-v_s/v} \quad \text{or} \quad \frac{f_s}{f_o} = 1-\frac{v_s}{v}$$

Solving for v_s/v gives

$$\frac{v_s}{v} = 1-\frac{f_s}{f_o} = 1-\frac{1250\text{ Hz}}{1290\text{ Hz}} = 0.031$$

This ratio corresponds to $\boxed{3.1\%}$.

---

75. SSM ***REASONING AND SOLUTION*** In this situation, both the source and the observer are moving through the air. Therefore, the frequency of sound heard by the crew is given by Equation 16.15 using the plus signs in both the numerator and the denominator.

$$f_o = f_s \left(\frac{1 + \frac{v_o}{v}}{1 + \frac{v_s}{v}} \right) = (1550 \text{ Hz}) \left(\frac{1 + \frac{13.0 \text{ m/s}}{343 \text{ m/s}}}{1 + \frac{67.0 \text{ m/s}}{343 \text{ m/s}}} \right) = \boxed{1350 \text{ Hz}}$$

---

76. ***REASONING AND SOLUTION*** The speed of the Bungee jumper, v_s, after she has fallen a distance y is given by

$$v_s^2 = v_0^2 + 2ay$$

Since she falls from rest, $v_0 = 0$ m/s, and

$$v_s = \sqrt{2ay} = \sqrt{2\left(-9.80 \text{ m/s}^2\right)\left(-11.0 \text{ m}\right)} = 14.7 \text{ m/s}$$

Then, from Equation 16.11

$$f_o = f_s \left(\frac{1}{1 - v_s / v} \right) = (589 \text{ Hz}) \left[\frac{1}{1 - (14.7 \text{ m/s})/(343 \text{ m/s})} \right] = \boxed{615 \text{ Hz}}.$$

---

77. ***REASONING*** The Doppler shift that occurs here arises because both the source and the observer of the sound are moving. Therefore, the expression for the Doppler-shifted observed frequency f_o is given by Equation 16.15 as

$$f_o = f_s \left(\frac{1 \pm v_o / v}{1 \mp v_s / v} \right)$$

where f_s is the frequency emitted by the source, v_o is the speed of the observer, v_s is the speed of the source, and v is the speed of sound. The observer is moving toward the source, so we use the plus sign in the numerator. The source is moving toward the observer, so we use the minus sign in the denominator. Thus, Equation 16.15 becomes

$$f_o = f_s \left(\frac{1 + v_o / v}{1 - v_s / v} \right)$$

Recognizing that both trucks move at the same speed, we can substitute $v_o = v_s = v_{Truck}$ and solve for v_{Truck}.

SOLUTION Using Equation 16.15 as described in the ***REASONING*** and substituting $v_o = v_s = v_{Truck}$, we have

$$f_o = f_s\left(\frac{1+v_{Truck}/v}{1-v_{Truck}/v}\right) \quad \text{or} \quad \frac{f_o}{f_s} - \left(\frac{f_o}{f_s}\right)\left(\frac{v_{Truck}}{v}\right) = 1 + \frac{v_{Truck}}{v}$$

Rearranging, with a view toward solving for v_{Truck}/v, gives

$$\frac{f_o}{f_s} - 1 = \frac{v_{Truck}}{v} + \left(\frac{f_o}{f_s}\right)\left(\frac{v_{Truck}}{v}\right) \quad \text{or} \quad \frac{v_{Truck}}{v}\left(1+\frac{f_o}{f_s}\right) = \frac{f_o}{f_s} - 1$$

Finally, we obtain

$$\frac{v_{Truck}}{v} = \frac{\dfrac{f_o}{f_s} - 1}{1+\dfrac{f_o}{f_s}} = \frac{1.14-1}{1+1.14} = \frac{0.14}{2.14} \quad \text{or} \quad v_{Truck} = \left(\frac{0.14}{2.14}\right)(343\text{ m/s}) = \boxed{22\text{ m/s}}$$

---

78. ***REASONING***

a. Since the two submarines are approaching each other head on, the frequency f_o detected by the observer (sub B) is related to the frequency f_s emitted by the source (sub A) by

$$f_o = f_s\left(\frac{1+\dfrac{v_o}{v}}{1-\dfrac{v_s}{v}}\right) \tag{16.15}$$

where v_o and v_s are the speed of the observer and source, respectively, and v is the speed of the underwater sound

b. The sound reflected from submarine B has the same frequency that it detects, namely, f_o. Now sub B becomes the source of sound and sub A is the observer. We can still use Equation 16.15 to find the frequency detected by sub A.

SOLUTION

a. The frequency f_o detected by sub B is

$$f_o = f_s \left(\frac{1 + \frac{v_o}{v}}{1 - \frac{v_s}{v}} \right) = (1550 \text{ Hz}) \left(\frac{1 + \frac{8 \text{ m/s}}{1522 \text{ m/s}}}{1 - \frac{12 \text{ m/s}}{1522 \text{ m/s}}} \right) = \boxed{1570 \text{ Hz}}$$

b. The sound reflected from submarine B has the same frequency that it detects, namely, 1570 Hz. Now sub B is the source of sound whose frequency is f_s = 1570 Hz. The speed of sub B is v_s = 8 m/s. The frequency detected by sub A (whose speed is v_o = 12 m/s) is

$$f_o = f_s \left(\frac{1 + \frac{v_o}{v}}{1 - \frac{v_s}{v}} \right) = (1570 \text{ Hz}) \left(\frac{1 + \frac{12 \text{ m/s}}{1522 \text{ m/s}}}{1 - \frac{8 \text{ m/s}}{1522 \text{ m/s}}} \right) = \boxed{1590 \text{ Hz}}$$

---

79. SSM ***REASONING*** We can use the Doppler shift formula to determine the speed of the motorcycle. Once this speed is known, the equations of kinematics (specifically, Equation 2.9) can be used to determine the distance covered by the motorcycle.

SOLUTION As the motorcycle accelerates, the frequency of the siren, as heard by the driver, decreases When the frequency of the siren heard by the driver is 90.0% of the value it has when the motorcycle is stationary, we have $f_o / f_s = 0.900$. Since the driver (the observer) is moving away from the source, we can use Equation 16.14 $\{f_o = f_s[1-(v_o/v)]\}$ to obtain the speed of the driver, and hence the speed of the motorcycle;

$$\frac{f_o}{f_s} = 1 - \frac{v_o}{v} = 0.900$$

Solving for v_o, we have

$$v_o = 0.100\ v = 0.100\ (343 \text{ m/s}) = 34.3 \text{ m/s}$$

According to Equation 2.9 ($v^2 = v_0^2 + 2ax$), since the motorcycle starts from rest so that $v_0 = 0$ m/s, the distance traveled by the motorcycle is

$$x = \frac{v^2}{2a} = \frac{(34.3 \text{ m/s})^2}{2(2.81 \text{ m/s}^2)} = \boxed{209 \text{ m}}$$

---

80. ***REASONING AND SOLUTION*** The maximum observed frequency is f_o^{max}, and the minimum observed frequency is f_o^{min}. We are given that $f_o^{\text{max}} - f_o^{\text{min}} = 2.1$ Hz, where

$$f_o^{\text{max}} = f_s[1 + (v_o/v)] \qquad (16.13)$$

and

$$f_o^{\text{min}} = f_s[1 - (v_o/v)] \qquad (16.14)$$

We have

$$f_o^{\text{max}} - f_o^{\text{min}} = f_s[1 + (v_o/v)] - f_s[1 - (v_o/v)] = 2f_s(v_o/v)$$

We can now solve for the maximum speed of the microphone, v_o:

$$v_o = (f_o^{\text{max}} - f_o^{\text{min}})v/(2f_s) = (2.1\text{ Hz})(343\text{ m/s})/[2(440\text{ Hz})] = 0.82\text{ m/s}$$

Using $v_{\text{max}} = v_o = A\omega$, Equation 10.6, where A is the amplitude of the simple harmonic motion and ω is the angular frequency, $\omega = 2\pi/T = 2\pi/(2.0\text{ s}) = 3.1$ rad/s, we have

$$A = v_o/\omega = (0.82\text{ m/s})/(3.1\text{ rad/s}) = \boxed{0.26\text{ m}}$$

---

81. ***REASONING AND SOLUTION*** The intensity level in dB is $\beta = (10\text{ dB})\log(I/I_0)$, Equation 16.10, where $\beta = 14$ dB. Therefore,

$$I/I_0 = 10^{\beta/(10\text{ dB})} = 10^{1.4} = \boxed{25}$$

---

82. ***REASONING*** Since the sonar pulse travels to the object and back, the total distance it travels is $2x$, where x is the distance from the submarine to the object. The total distance is equal to the product of the speed v_{seawater} of the sonar wave in seawater and the time t for the pulse to reach the object and return, $2x = v_{\text{seawater}}t$. We can use this relation to find the distance to the object.

SOLUTION Solving the equation above for x, and using the value of v_{seawater} from Table 16.1, we have

$$x = \frac{v_{\text{seawater}}t}{2} = \frac{(1522\text{ m/s})(1.30\text{ s})}{2} = \boxed{989\text{ m}}$$

---

83. SSM ***REASONING*** If we treat the sample of argon atoms like an ideal monatomic gas ($\gamma = 1.67$) at 298 K, Equation 14.6 $\left(\frac{1}{2}mv_{\text{rms}}^2 = \frac{3}{2}kT\right)$ can be solved for the root-mean-square speed v_{rms} of the argon atoms. The speed of sound in argon can be found from Equation 16.5: $v = \sqrt{\gamma kT/m}$.

SOLUTION We first find the mass of an argon atom. Since the molecular mass of argon is 39.9 u, argon has a mass per mole of 39.9×10^{-3} kg/mol. Thus, the mass of a single argon atom is

$$m = \frac{39.9\times10^{-3} \text{ kg/mol}}{6.022\times10^{23} \text{ mol}^{-1}} = 6.63\times10^{-26} \text{ kg}$$

a. Solving Equation 14.6 for v_{rms} and substituting the data given in the problem statement, we find

$$v_{\text{rms}} = \sqrt{\frac{3kT}{m}} = \sqrt{\frac{3\left(1.38\times10^{-23} \text{ J/K}\right)(298 \text{ K})}{6.63\times10^{-26} \text{ kg}}} = \boxed{431 \text{ m/s}}$$

b. The speed of sound in argon is, according to Equation 16.5,

$$v = \sqrt{\frac{\gamma kT}{m}} = \sqrt{\frac{(1.67)\left(1.38\times10^{-23} \text{ J/K}\right)(298 \text{ K})}{6.63\times10^{-26} \text{ kg}}} = \boxed{322 \text{ m/s}}$$

---

84. ***REASONING*** This problem deals with the Doppler effect in a situation where the source of the sound is moving and the observer is stationary. Thus, the observed frequency is given by Equation 16.11 when the car is approaching the observer and Equation 16.12 when the car is moving away from the observer. These equations relate the frequency f_o heard by the observer to the frequency f_s emitted by the source, the speed v_s of the source, and the speed v of sound. They can be used directly to calculate the desired ratio of the observed frequencies. We note that no information is given about the frequency emitted by the source. We will see, however, that none is needed, since f_s will be eliminated algebraically from the solution.

SOLUTION Equations 16.11 and 16.12 are

$$f_o^{\text{Approach}} = f_s\left(\frac{1}{1 - v_s/v}\right) \quad (16.11) \qquad f_o^{\text{Recede}} = f_s\left(\frac{1}{1 + v_s/v}\right) \quad (16.12)$$

The ratio is

$$\frac{f_o^{\text{Approach}}}{f_o^{\text{Recede}}}=\frac{f_s\left(\dfrac{1}{1-v_s/v}\right)}{f_s\left(\dfrac{1}{1+v_s/v}\right)}=\frac{1+v_s/v}{1-v_s/v}=\frac{1+\dfrac{9.00\text{ m/s}}{343\text{ m/s}}}{1-\dfrac{9.00\text{ m/s}}{343\text{ m/s}}}=\boxed{1.054}$$

As mentioned in the ***REASONING***, the unknown source frequency f_s has been eliminated algebraically from this calculation.

·85. SSM WWW ***REASONING*** According to Equation 16.2, the linear density of the string is given by $(m/L)=F/v^2$, where the speed v of waves on the middle C string is given by Equation 16.1, $v=\lambda f=\lambda/T$.

SOLUTION Combining Equations 16.2 and 16.1 and using the given data, we obtain

$$m/L=\frac{F}{v^2}=\frac{FT^2}{\lambda^2}=\frac{(944\text{ N})(3.82\times10^{-3}\text{ s})^2}{(1.26\text{ m})^2}=\boxed{8.68\times10^{-3}\text{ kg/m}}$$

---

86. ***REASONING*** The speed of a Tsunamis is equal to the distance x it travels divided by the time t it takes for the wave to travel that distance. The frequency f of the wave is equal to its speed divided by the wavelength λ, $f=v/\lambda$ (Equation 16.1). The period T of the wave is related to its frequency by Equation 10.5, $T=1/f$.

SOLUTION

a. The speed of the wave is (in m/s)

$$v=\frac{x}{t}=\frac{3700\times10^3\text{ m}}{5.3\text{ h}}\left(\frac{1\text{ h}}{3600\text{ s}}\right)=\boxed{190\text{ m/s}}$$

b. The frequency of the wave is

$$f=\frac{v}{\lambda}=\frac{190\text{ m/s}}{750\times10^3\text{ m}}=\boxed{2.5\times10^{-4}\text{ Hz}} \qquad (16.1)$$

c. The period of any wave is the reciprocal of its frequency:

$$T=\frac{1}{f}=\frac{1}{2.5\times10^{-4}\text{ Hz}}=\boxed{4.0\times10^3\text{ s}} \qquad (10.5)$$

---

87. SSM ***REASONING AND SOLUTION*** The intensity level β in decibels (dB) is related to the sound intensity I according to Equation 16.10:

$$\beta = (10\ \text{dB})\log\left(\frac{I}{I_0}\right)$$

where the quantity I_0 is the reference intensity. Therefore, we have

$$\beta_2 - \beta_1 = (10\ \text{dB})\log\left(\frac{I_2}{I_0}\right) - (10\ \text{dB})\log\left(\frac{I_1}{I_0}\right) = (10\ \text{dB})\log\left(\frac{I_2/I_0}{I_1/I_0}\right) = (10\ \text{dB})\log\left(\frac{I_2}{I_1}\right)$$

Solving for the ratio I_2/I_1, we find

$$30.0\ \text{dB} = (10\ \text{dB})\log\left(\frac{I_2}{I_1}\right) \quad \text{or} \quad \frac{I_2}{I_1} = 10^{3.0} = 1000$$

Thus, we conclude that the sound intensity $\boxed{\text{increases by a factor of 1000}}$.

---

88. ***REASONING AND SOLUTION*** Since the sound spreads out uniformly in all directions, the intensity is uniform over any sphere centered on the source. From text Equation 16.9,

$$I = \frac{P}{4\pi r^2}$$

Then,

$$\frac{I_1}{I_2} = \frac{P/(4\pi r_1^2)}{P/(4\pi r_2^2)} = \frac{r_2^2}{r_1^2}$$

Solving for r_2, we obtain

$$r_2 = r_1\sqrt{\frac{I_1}{I_2}} = (120\ \text{m})\sqrt{\frac{2.0\times10^{-6}\ \text{W/m}^2}{0.80\times10^{-6}\ \text{W/m}^2}} = \boxed{190\ \text{m}}$$

---

89. ***REASONING AND SOLUTION*** Using Equation 16.1, we find that

$$\lambda = v/f = (343\ \text{m/s})/(4185.6\ \text{Hz}) = \boxed{8.19\times10^{-2}\ \text{m}} \qquad (16.1)$$

---

90. ***REASONING AND SOLUTION***
a. Comparing with Equations 16.3 and 16.4, we see that the wave travels in the $\boxed{+x \text{ direction}}$

b. The displacement at $x = 13$ m and $t = 38$ s is

$$y = (0.26\text{ m})\sin\left[\pi(38) - (3.7)\pi(13)\right] = \boxed{-0.080\text{ m}}$$

---

91. SSM ***REASONING*** Since you detect a frequency that is smaller than that emitted by the car when the car is stationary, the car must be moving away from you. Therefore, according to Equation 16.12, the frequency f_o heard by a stationary observer from a source moving away from the observer is given by

$$f_o = f_s\left(\frac{1}{1 + \frac{v_s}{v}}\right)$$

where f_s is the frequency emitted from the source when it is stationary with respect to the observer, v is the speed of sound, and v_s is the speed of the moving source. This expression can be solved for v_s.

SOLUTION We proceed to solve for v_s and substitute the data given in the problem statement. Rearrangement gives

$$\frac{v_s}{v} = \frac{f_s}{f_o} - 1$$

Solving for v_s and noting that $f_o / f_s = 0.86$ yields

$$v_s = v\left(\frac{f_s}{f_o} - 1\right) = (343\text{ m/s})\left(\frac{1}{0.86} - 1\right) = \boxed{56\text{ m/s}}$$

---

92. ***REASONING AND SOLUTION*** The sound wave will travel at constant speed through the water (speed = v_{water}) and the copper block (speed = v_{copper}). Thus, the times it takes for the sound wave to travel downward through the water a distance d_{water} and downward through the copper a distance d_{copper} are

$$\Delta t_{water} = \frac{d_{water}}{v_{water}} \quad \text{and} \quad \Delta t_{copper} = \frac{d_{copper}}{v_{copper}}$$

The time required for the incident sound wave to reach the bottom of the copper block just before reflection is equal to the time required for the reflected sound wave to reach the surface of the water. Using values for the speed of sound in water and copper from Table 16.1 in the text, we find that the total time interval between when the sound enters and leaves the water is, then,

$$\Delta t_{\text{TOTAL}} = 2(\Delta t_{\text{water}} + \Delta t_{\text{copper}}) = 2\left[\left(\frac{d_{\text{water}}}{v_{\text{water}}}\right) + \left(\frac{d_{\text{copper}}}{v_{\text{copper}}}\right)\right]$$

$$\Delta t_{\text{TOTAL}} = 2\left[\left(\frac{0.45\text{ m}}{1482\text{ m/s}}\right) + \left(\frac{0.15\text{ m}}{5010\text{ m/s}}\right)\right] = \boxed{6.7\times10^{-4}\text{ s}}$$

---

93. SSM ***REASONING*** According to Equation 16.10, the sound intensity level β in decibels (dB) is related to the sound intensity I according to $\beta = (10\text{ dB})\log(I/I_0)$, where the quantity I_0 is the reference intensity. Since the sound is emitted uniformly in all directions, the intensity, or power per unit area, is given by $I = P/(4\pi r^2)$. Thus, the sound intensity at position 1 can be written as $I_1 = P/(4\pi r_1^2)$, while the sound intensity at position 2 can be written as $I_2 = P/(4\pi r_2^2)$. We can obtain the sound intensity levels from Equation 16.10 for these two positions by using these expressions for the intensities.

SOLUTION Using Equation 16.10 and the expressions for the intensities at the two positions, we can write the difference in the sound intensity levels β_{21} between the two positions as follows:

$$\beta_{21} = \beta_2 - \beta_1 = (10\text{ dB})\log\left(\frac{I_2}{I_0}\right) - (10\text{ dB})\log\left(\frac{I_1}{I_0}\right)$$

$$= (10\text{ dB})\log\left(\frac{I_2/I_0}{I_1/I_0}\right) = (10\text{ dB})\log\left(\frac{I_2}{I_1}\right)$$

$$\beta_{21} = (10\text{ dB})\log\left[\frac{P/(4\pi r_2^2)}{P/(4\pi r_1^2)}\right] = (10\text{ dB})\log\left(\frac{r_1^2}{r_2^2}\right) = (10\text{ dB})\log\left(\frac{r_1}{r_2}\right)^2$$

$$= (20\text{ dB})\log\left(\frac{r_1}{r_2}\right) = (20\text{ dB})\log\left(\frac{r_1}{2r_1}\right) = (20\text{ dB})\log(1/2) = \boxed{-6.0\text{ dB}}$$

The negative sign indicates that the sound intensity level decreases.

---

94. ***REASONING AND SOLUTION*** The emitted power is 2.0×10^5 J/s = 2.0×10^5 W. The intensity is, therefore,

$$I = \frac{P}{A} = \frac{P}{4\pi r^2} = \frac{2.0\times10^5\ \text{W}}{4\pi(85\ \text{m})^2} = 2.2\ \text{W/m}^2 \qquad (16.9)$$

Thus, the sound intensity level is

$$\beta = (10\ \text{dB})\log\left(\frac{I}{I_0}\right) = (10\ \text{dB})\log\left(\frac{2.2\ \text{W/m}^2}{1.00\times10^{-12}\ \text{W/m}^2}\right) = \boxed{120\ \text{dB}} \qquad (16.10)$$

---

95. ***REASONING AND SOLUTION*** Let β_2 and I_2 denote the intensity level and the intensity, respectively, when two rifles are shot. Let β_1 and I_1 denote the intensity level and intensity, respectively, when one rifle is shot. Then

$$\beta_2 = (10\ \text{dB})\log\left(\frac{I_2}{I_0}\right) = (10\ \text{dB})\log\left(\frac{2I_1}{I_0}\right)$$

where I_1 is the intensity of a single rifle. Therefore,

$$\frac{2I_1}{I_0} = 10^{\beta_2/(10\ \text{dB})}$$

$$I_1 = \left(\frac{I_0}{2}\right)10^{\beta_2/(10\ \text{dB})} = \left(\frac{1.00\times10^{-12}\ \text{W/m}^2}{2}\right)10^{8.00} = 5.00\times10^{-5}\ \text{W/m}^2$$

$$\therefore \beta_1 = (10\ \text{dB})\log\left(\frac{I_1}{I_0}\right) = (10\ \text{dB})\log\left(\frac{5.00\times10^{-5}\ \text{W/m}^2}{1.00\times10^{-12}\ \text{W/m}^2}\right) = \boxed{77.0\ \text{dB}}$$

---

96. ***REASONING AND SOLUTION*** The maximum acceleration of the dot occurs at the extreme positions and is

$$a_{\text{max}} = (2\pi f)^2 A = [2\pi(4.0\ \text{Hz})]^2(5.4\times10^{-3}\ \text{m}) = \boxed{3.4\ \text{m/s}^2} \qquad (10.10)$$

---

97. SSM ***REASONING*** Newton's second law can be used to analyze the motion of the blocks using the methods developed in Chapter 4. We can thus determine an expression that relates the magnitude P of the pulling force to the magnitude F of the tension in the wire. Equation 16.2 [$v = \sqrt{F/(m/L)}$] can then be used to find the tension in the wire.

SOLUTION The following figures show a schematic of the situation described in the problem and the free-body diagrams for each block, where $m_1 = 42.0$ kg and $m_2 = 19.0$ kg. The pulling force is **P**, and the tension in the wire gives rise to the forces **F** and –**F**, which act on m_1 and m_2, respectively.

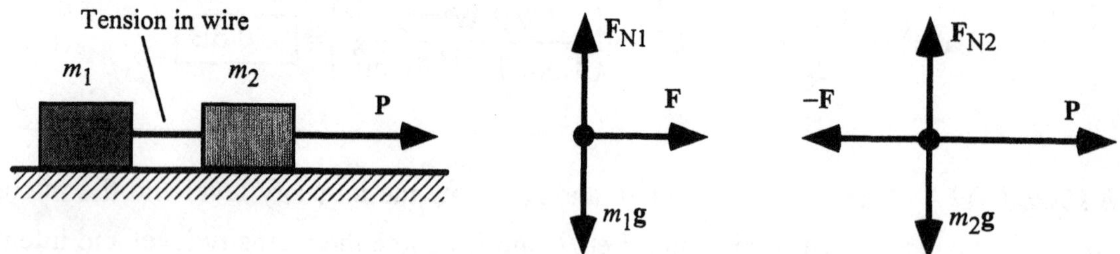

Newton's second law for block 1 is, taking forces that point to the right as positive, $F = m_1 a$, or $a = F/m_1$. For block 2, we obtain $P - F = m_2 a$. Using the expression for a obtained from the equation for block 1, we have

$$P - F = F\left(\frac{m_2}{m_1}\right) \quad \text{or} \quad P = F\left(\frac{m_2}{m_1}\right) + F = F\left(\frac{m_2}{m_1} + 1\right)$$

According to Equation 16.2, $F = v^2(m/L)$, where m/L is the mass per unit length of the wire. Combining this expression for F with the expression for P, we have

$$P = v^2(m/L)\left(\frac{m_2}{m_1} + 1\right) = (352 \text{ m/s})^2(8.50 \times 10^{-4} \text{ kg/m})\left(\frac{19.0 \text{ kg}}{42.0 \text{ kg}} + 1\right) = \boxed{153 \text{ N}}$$

---

98. ***REASONING AND SOLUTION*** The speed of the wave on the cable is

$$v = \sqrt{\frac{F}{(m/L)}} = \sqrt{\frac{F}{\rho A}} = \sqrt{\frac{1.00 \times 10^4 \text{ N}}{(7860 \text{ kg/m}^3)(2.83 \times 10^{-3} \text{ m}^2)}} = \boxed{21.2 \text{ m/s}} \qquad (16.2)$$

---

99. ***REASONING AND SOLUTION***
a. According to Equation 16.2, the speed of the wave is

$$v = \sqrt{\frac{F}{m/L}} = \sqrt{\frac{315 \; N}{6.50 \times 10^{-3} \text{ kg/m}}} = \boxed{2.20 \times 10^2 \text{ m/s}}$$

b. According to Equations 10.6 and 10.8, the maximum speed of the point on the wire is

$$v_{\text{max}} = (2\pi f)A = 2\pi(585\ \text{Hz})(2.50 \times 10^{-3}\ \text{m}) = \boxed{9.19\ \text{m/s}}$$

---

100. ***REASONING*** If I_1 is the sound intensity produced by a single person, then NI_1 is the sound intensity generated by N people. The sound intensity level generated by N people is given by Equation 16.10 as

$$\beta_{\text{N}} = (10\ \text{dB})\log\left(\frac{NI_1}{I_0}\right)$$

where I_0 is the threshold of hearing. Solving this equation for N yields

$$N = \left(\frac{I_0}{I_1}\right)10^{\frac{\beta_{\text{N}}}{10\ \text{dB}}} \qquad (1)$$

We also know that the sound intensity level for one person is

$$\beta_1 = (10\ \text{dB})\log\left(\frac{I_1}{I_0}\right) \quad \text{or} \quad I_1 = I_0\,10^{\frac{\beta_1}{10\ \text{dB}}} \qquad (2)$$

Equations (1) and (2) are all that we need in order to find the number of people at the football game.

SOLUTION Substituting the expression for I_1 from Equation (2) into Equation (1) gives the desired result.

$$N = \frac{I_0\,10^{\frac{\beta_{\text{N}}}{10\ \text{dB}}}}{I_0\,10^{\frac{\beta_1}{10\ \text{dB}}}} = \frac{10^{\frac{109\ \text{dB}}{10\ \text{dB}}}}{10^{\frac{60.0\ \text{dB}}{10\ \text{dB}}}} = \boxed{79\ 400}$$

---

101. ***REASONING AND SOLUTION*** The speed of sound in an ideal gas is

$$v = \sqrt{\frac{\gamma kT}{m}}$$

so that the mass of a gas molecule is

$$m = \gamma kT/v^2 = (1.67)(1.38 \times 10^{-23}\ \text{J/K})(3.00 \times 10^2\ \text{K})/(363\ \text{m/s})^2 = 5.25 \times 10^{-23}\ \text{g}$$

We now need to determine what fraction of the gas is argon and what fraction is neon. First find the mass of each molecule.

$$m_{ar} = (39.9 \text{ g/mol})/(6.022 \times 10^{23}/\text{mol}) = 6.63 \times 10^{-23} \text{ g}$$

$$m_{ne} = (20.2 \text{ g/mol})/(6.022 \times 10^{23}/\text{mol}) = 3.35 \times 10^{-23} \text{ g}$$

Let q be the fraction of gas that is in the form of argon and p be the fraction that is neon. We know that $q + p = 1$. Also, $qm_{ar} + pm_{ne} = m$. Substituting $p = 1 - q$ in this equation gives

$$qm_{ar} + (1 - q)m_{ne} = m$$

and we can now solve for the fraction

$$q = (m - m_{ne})/(m_{ar} - m_{ne})$$

Suppressing the units and canceling the factor of 10^{-23}, we find

$$q = (5.25 - 3.35)/(6.63 - 3.35) = 0.57 = \boxed{57\% \text{ argon}}$$

and

$$p = 1 - q = 1 - 0.57 = 0.43 = \boxed{43\% \text{ neon}}$$

---

102. ***REASONING AND SOLUTION***

a. The distance traveled is $x = vt = (343 \text{ m/s})(0.580 \text{ s}) = 199 \text{ m}$. The one-way distance is half this value or $\boxed{99.5 \text{ m}}$.

b. Since the speed is proportional to the square-root of the absolute temperature, i.e.,

$$\frac{v_2}{v_1} = \sqrt{\frac{T_2}{T_1}}$$

or

$$v_2 = v_1\sqrt{\frac{T_2}{T_1}} = (343 \text{ m/s})\sqrt{\frac{298 \text{ K}}{293 \text{ K}}} = 346 \text{ m/s}$$

The distance computed by the theodolite at the higher temperature would be $(1/2)\,vt_2$, where $v = 343$ m/s and $t_2 = (199 \text{ m})/(346 \text{ m/s})$. Then

$$(1/2)(343 \text{ m/s})(199 \text{ m})/(346 \text{ m/s}) = 98.6 \text{ m}$$

The percent error is then

$$\% \text{ error} = [(99.5 \text{ m} - 98.6 \text{ m})/(99.5 \text{ m})] \times 100\% = \boxed{0.9\ \%}$$

---

103. ***CONCEPT QUESTIONS***

a. The wavelength is the horizontal distance between two successive crests. The horizontal distance between successive crests of wave B is two times greater than that of wave A. Therefore, B has the greater wavelength.

b. The frequency f of a wave is related to its speed v and wavelength λ by Equation 16.1, $f = v/\lambda$. Since the speed is the same for both waves, the wave with the smaller wavelength has the larger frequency. Therefore wave A, having the smaller wavelength, has the larger frequency.

c. The maximum speed of a particle attached to a wave is given by Equation 10.8 as $v_{max} = A\omega$, where A is the amplitude of the wave and ω is the angular frequency, $\omega = 2\pi f$. Wave A has both a larger amplitude and frequency, so the maximum particle speed is greater for A.

SOLUTION

a. From the drawing, we determine the wavelength of each wave to be

$$\lambda_A = \boxed{2.0\text{ m}} \quad \text{and} \quad \lambda_B = \boxed{4.0\text{ m}}.$$

b. The frequency of each wave is given by Equation 16.1 as:

$$f_A = \frac{v}{\lambda_A} = \frac{12\text{ m/s}}{2.0\text{ m}} = \boxed{6.0\text{ Hz}} \quad \text{and} \quad f_B = \frac{v}{\lambda_B} = \frac{12\text{ m/s}}{4.0\text{ m}} = \boxed{3.0\text{ Hz}}$$

c. The maximum speed for a particle moving in simple harmonic motion is given by Equation 10.8 as $v_{max} = A\omega$. The amplitude of each wave can be obtained from the drawing: $A_A = 0.50$ m and $A_B = 0.25$ m.

Wave A $$v_{max} = A_A\omega_A = A_A\,2\pi f_A = (0.50\text{ m})2\pi(6.0\text{ Hz}) = \boxed{19\text{ m/s}}$$

Wave B $$v_{max} = A_B\omega_B = A_B\,2\pi f_B = (0.25\text{ m})2\pi(3.0\text{ Hz}) = \boxed{4.7\text{ m/s}}$$

---

104. ***CONCEPT QUESTIONS***

a. Generally, the speed of sound in a liquid like water is greater than in a gas. And, in fact, according to Table 16.1, the speed of sound in water is greater than the speed of sound in air.

b. Since the speed of sound in water is greater than in air, an underwater ultrasonic pulse returns to the ruler in a shorter time than a pulse in air. The ruler has been designed for use in air, not in water, so this quicker return time fools the ruler into believing that the object is much closer than it actually is. Therefore, the reading on the ruler is less than the actual distance.

SOLUTION Let x be the actual distance from the ruler to the object. The time it takes for the ultrasonic pulse to reach the object and return to the ruler, a distance of $2x$, is equal to the distance divided by the speed of sound in water v_{water}: $t = 2x/v_{\text{water}}$. The speed of sound in water is given by Equation 16.6 as $v_{\text{water}} = \sqrt{B_{\text{ad}}/\rho}$, where B_{ad} is the adiabatic bulk modulus and ρ is the density of water. Thus, the time it takes for the pulse to return is

$$t = \frac{2x}{v_{\text{water}}} = \frac{2x}{\sqrt{\dfrac{B_{\text{ad}}}{\rho}}} = \frac{2(25.0\text{ m})}{\sqrt{\dfrac{2.31\times10^{9}\text{ Pa}}{1025\text{ kg/m}^3}}} = 3.33\times10^{-2}\text{ s}$$

The ruler measures this value for the time and computes the distance to the object by using the speed of sound in air, 343 m/s. The distance x_{ruler} displayed by the ruler is equal to the speed of sound in air multiplied by the time $\frac{1}{2}t$ it takes for the pulse to go from the ruler to the object:

$$x_{\text{ruler}} = v_{\text{air}}\left(\tfrac{1}{2}t\right) = (343\text{ m/s})\left(\tfrac{1}{2}\right)\left(3.33\times10^{-2}\text{ s}\right) = \boxed{5.71\text{ m}}$$

Thus, the ruler displays a distance of $x_{\text{ruler}} = 5.71$ m. As expected, the reading on the ruler's display is less than the actual distance.

---

105. ***CONCEPT QUESTION*** You don't want to run the car any faster than you have to, so you'll try and break the sound barrier when the speed of sound in air has its smallest value. The speed of sound in an ideal gas depends on its temperature T through Equation 16.5 as $v = \sqrt{\dfrac{\gamma kT}{m}}$, where γ = 1.40 for air, k is Boltzmann's constant, and m is the mass of a molecule in the air. Thus, the speed of sound depends on the air temperature, with a lower temperature giving rise to a smaller speed of sound. Thus, you should attempt to break the sound barrier in the early morning when the temperature is lower.

SOLUTION The speed of sound in air is $v = \sqrt{\dfrac{\gamma kT}{m}}$, where the temperature T must be expressed on the Kelvin scale ($T = T_{\text{c}} + 273$, Equation 12.1). Taking the ratio of the speed of sound at 43 °C to that at 0 °C, we have

$$\frac{v_{43\,^\circ\text{C}}}{v_{0\,^\circ\text{C}}} = \frac{\sqrt{\dfrac{\gamma k(43+273)}{m}}}{\sqrt{\dfrac{\gamma k(0+273)}{m}}} = \sqrt{\frac{316\text{ K}}{273\text{ K}}}$$

The speed of sound at 43 °C is

$$v_{43\,°\text{C}} = v_{0\,°\text{C}}\sqrt{\frac{316\text{ K}}{273\text{ K}}} = (331\text{ m/s})\sqrt{\frac{316\text{ K}}{273\text{ K}}} = \boxed{356\text{ m/s}}$$

---

106. ***CONCEPT QUESTIONS***

a. The source emits sound uniformly in all directions, so the sound intensity I at any distance r is given by Equation 16.9 as $I = P/(4\pi r^2)$, where P is the sound power emitted by the source. Since patches 1 and 2 are at the same distance from the source of sound, the sound intensity at each location is the same, so $I_1 = I_2$. Patch 3 is farther from the sound source, so the intensity I_3 is smaller for points on that patch. Therefore, patches 1 and 2 have equal intensities, each of which is greater than the intensity at patch 3.

b. According to Equation 16.8, the sound intensity I is defined as the sound power P that passes perpendicularly through a surface divided by the area A of that surface, $I = P/A$. The area of the surface is, then, $A = P/I$. Since the same sound power passes through patches 1 and 2, and the intensity at each one is the same, their areas must also be the same, $A_1 = A_2$. The same sound power passes through patch 3, but the intensity at that surface is smaller than that at patches 1 and 2. Thus, the area A_3 of patch 3 is larger than that of surface 1 or 2. In summary, A_3 is the largest area, followed by A_1 and A_2, which are equal.

SOLUTION

a. The sound intensity at the inner spherical surface is given by Equation 16.9 as

$$I_A = \frac{P}{4\pi r_A^2} = \frac{2.3\text{ W}}{4\pi(0.60\text{ m})^2} = 0.51\text{ W/m}^2$$

This intensity is the same at all points on the inner surface, since all points are equidistant from the sound source. Therefore, the sound intensity at patches 1 and 2 are equal; $I_1 = I_2 =$ $\boxed{0.51\text{ W/m}^2}$.

The sound intensity at the outer spherical surface is

$$I_B = \frac{P}{4\pi r_B^2} = \frac{2.3\text{ W}}{4\pi(0.80\text{ m})^2} = 0.29\text{ W/m}^2$$

This intensity is the same at all points on outer surface. Therefore, the sound intensity at patch 3 is $I_3 = \boxed{0.29\text{ W/m}^2}$.

b. The area of a surface is given by Equation 16.8 as the sound power passing perpendicularly through that area divided by the sound intensity, $A = P/I$. The areas of the three surfaces are:

Surface 1 $$A_1 = \frac{P}{I_1} = \frac{1.8\times10^{-3}\text{ W}}{0.51\text{ W/m}^2} = \boxed{3.5\times10^{-3}\text{ m}^2}$$

Surface 2 $$A_2 = \frac{P}{I_2} = \frac{1.8\times10^{-3}\text{ W}}{0.51\text{ W/m}^2} = \boxed{3.5\times10^{-3}\text{ m}^2}$$

Surface 3 $$A_3 = \frac{P}{I_3} = \frac{1.8\times10^{-3}\text{ W}}{0.29\text{ W/m}^2} = \boxed{6.2\times10^{-3}\text{ m}^2}$$

---

107. ***CONCEPT QUESTIONS***

a. A threshold of hearing of –8.00 dB means that this individual can hear a sound whose intensity is *less* than $I_0 = 1.00\times10^{-12}\text{ W/m}^2$, which is the intensity of the reference level.

b. The person with a threshold of hearing of +12.0 dB can only hear sounds that have intensities greater than $I_0 = 1.00\times10^{-12}\text{ W/m}^2$. Thus, the person whose threshold of hearing is –8.00 dB has the better hearing, because he can hear sounds with intensities less than $1.00\times10^{-12}\text{ W/m}^2$.

SOLUTION The relation between the sound intensity level β and the sound intensity I is given by Equation 16.10:

$$\beta = (10\text{ dB})\log\left(\frac{I}{I_0}\right) \quad\text{or}\quad I = I_0\,10^{\frac{\beta}{10\text{ dB}}}$$

The threshold of hearing intensities for the two people are

$$I_1 = I_0\,10^{\frac{\beta_1}{10\text{ dB}}} \quad\text{and}\quad I_2 = I_0\,10^{\frac{\beta_2}{10\text{ dB}}}$$

Taking the ratio I_1/I_2 gives

$$\frac{I_1}{I_2} = \frac{I_0\,10^{\frac{\beta_1}{10\text{ dB}}}}{I_0\,10^{\frac{\beta_2}{10\text{ dB}}}} = \frac{10^{\frac{+12.0\text{ dB}}{10\text{ dB}}}}{10^{\frac{-8.00\text{ dB}}{10\text{ dB}}}} = \boxed{100}$$

---

108. ***CONCEPT QUESTIONS***

	Velocity of Sound Source (Toward the Observer)	Velocity of Observer (Toward the Source)	Wavelength	Frequency Heard by Observer
(a)	0 m/s	0 m/s	Remains the same	Remains the same
(b)	⟶	0 m/s	Decreases	Increases
(c)	⟶	⟵	Decreases	Increases

a. Since the sound source and the observer are stationary, there is no Doppler effect. The wavelength remains the same and the frequency of the sound heard by the observer remains the same as that emitted by the sound source.

b. When the sound source moves toward a stationary observer, the wavelength decreases (see Figure 16.30*b*). This decrease arises because the condensations "bunch-up" as the source moves toward the observer. The frequency heard by the observer increases, because, according to Equation 16.1, the frequency is inversely proportional to the wavelength; a smaller wavelength gives rise to a greater frequency.

c. The wavelength decreases for the same reason given in part (b). The increase in frequency is due to two effects; the decrease in wavelength, and the fact that the observer intercepts more wave cycles per second as she moves toward the sound source.

SOLUTION

a. The frequency of the sound is the same as that emitted by the siren; $f_o = f_s = \boxed{2450\text{ Hz}}$. The wavelength is given by Equation 16.1 as

$$\lambda = \frac{v}{f_s} = \frac{343\text{ m/s}}{2450\text{ Hz}} = \boxed{0.140\text{ m}}$$

b. According to the discussion in Section 16.9 (see the subsection "Moving source") the wavelength λ' of the sound is given by $\lambda' = \lambda - v_s T$, where v_s is the speed of the source and T is the period of the sound. However, $T = 1/f_s$ so that

$$\lambda' = \lambda - v_s T = \lambda - \frac{v_s}{f_s} = 0.140\text{ m} - \frac{26.8\text{ m/s}}{2450\text{ Hz}} = \boxed{0.129\text{ m}}$$

The frequency f_o heard by the observer is equal to the speed of sound v divided by the shortened wavelength λ':

$$f_o = \frac{v}{\lambda'} = \frac{343 \text{ m/s}}{0.129 \text{ m}} = \boxed{2660 \text{ Hz}}$$

c. The wavelength is the same as that in part (b), so $\lambda' = \boxed{0.129 \text{ m}}$. The frequency heard by the observer can be obtained from Equation 16.15, where we use the fact that the observer is moving toward the sound source:

$$f_o = f_s \left(\frac{1 + \frac{v_o}{v}}{1 - \frac{v_s}{v}} \right) = (2450 \text{ Hz}) \left(\frac{1 + \frac{14.0 \text{ m/s}}{343 \text{ m/s}}}{1 - \frac{26.8 \text{ m/s}}{343 \text{ m/s}}} \right) = \boxed{2770 \text{ Hz}}$$

---

109. ***CONCEPT QUESTIONS***

a. The tension in the rope is greater near the top than near the bottom. This is because the rope has weight. The part of the rope near the top must support more of that weight than the part of the rope near the bottom. In fact, the very top end must support all the weight of the rope beneath it. In contrast, the very bottom end supports no weight at all, since nothing hangs beneath it.

b. The speed of the wave is greater near the top of the rope. This follows directly from Equation 16.2, which indicates that the speed of the wave is proportional to the square root of the tension. Since the tension near the top of the rope is greater than the tension near the bottom, the speed is greater near the top.

c. The weight is the mass of the section of rope times the acceleration g due to gravity. Since the rope is uniform, the mass of the section is simply the total mass m of the rope times the fraction y/L, which is the length of the section divided by the total length of the rope. Thus, the weight of the section is $m\left(\frac{y}{L}\right)g$.

SOLUTION

a. According to Equation 16.2, the speed v of the wave is

$$v = \sqrt{\frac{F}{m/L}}$$

where F is the tension, m is the total mass of the rope, and L is the length of the rope. At a point y meters above the bottom end, the rope is supporting the weight of the section beneath that point, which is $m\left(\frac{y}{L}\right)g$, as discussed in Concept Question (c). The rope supports the weight by virtue of the tension in the rope. Since the rope does not accelerate upward or downward, the tension must be equal to $m\left(\frac{y}{L}\right)g$, according to Newton's second law of motion. Substituting this tension for F in Equation 16.2 reveals that the speed at a point y meters above the bottom end is

$$v = \sqrt{\frac{m\left(\frac{y}{L}\right)g}{m/L}} = \sqrt{yg}$$

b. Using the expression just derived, we find the following speeds

[y = 0.50 m] $v = \sqrt{yg} = \sqrt{(0.50\text{ m})(9.80\text{ m/s}^2)} = \boxed{2.2\text{ m/s}}$

[y = 2.0 m] $v = \sqrt{yg} = \sqrt{(2.0\text{ m})(9.80\text{ m/s}^2)} = \boxed{4.4\text{ m/s}}$

As expected, the speed is greater at the spot higher up the rope.

---

110. ***CONCEPT QUESTIONS***

a. Following its release from rest, the platform begins to move down the incline, picking up speed as it goes. Thus, the platform's velocity points down the incline, and its magnitude increases with time. The reason for the increasing velocity is gravity. The acceleration due to gravity points vertically downward and has a component along the length of the incline.

b. The changing velocity is related to the acceleration of the platform according to Equation 2.4, which gives the acceleration as the change in the velocity divided by the time interval during which the change occurs.

c. The frequency detected by the microphone at the instant the platform is released from rest is the same as the frequency broadcast by the speaker. However, as the platform begins to move away from the speaker, the microphone detects fewer wave cycles per second than the speaker broadcasts. In other words, the microphone detects a frequency that is smaller than that broadcast by the speaker. This is an example of the Doppler effect and occurs because the platform is moving in the same direction as the sound is traveling. As the platform picks up speed, the microphone detects an ever decreasing frequency.

d. The speaker is the source of the sound, and the microphone is the "observer." Since the source is stationary and the observer is moving away from the source, the Doppler-shifted observed frequency is given by Equation 16.14.

SOLUTION The acceleration a is directed down the incline and is the change in the velocity divided by the time interval during which the change occurs. The change in the velocity is the velocity v at a later time t minus the velocity v_0 at an earlier time t_0. Thus, according to Equation 2.4, the acceleration is

$$a = \frac{v - v_0}{t - t_0}$$

Equation 16.14 gives the frequency f_o detected by the microphone in terms of the frequency f_s emitted by the speaker, the speed v_{mike}, and the speed v of sound:

$$f_o = f_s\left(1 - \frac{v_{mike}}{v}\right)$$

Solving for the speed of the mike gives

$$v_{mike} = v\left(1 - \frac{f_o}{f_s}\right)$$

Using this result, we can determine the speed of the microphone platform at the two given times:

[t = 1.5 s] $$v_{mike} = (343\ \text{m/s})\left(1 - \frac{9939\ \text{Hz}}{1.000 \times 10^4\ \text{Hz}}\right) = 2.1\ \text{m/s}$$

[t = 3.5 s] $$v_{mike} = (343\ \text{m/s})\left(1 - \frac{9857\ \text{Hz}}{1.000 \times 10^4\ \text{Hz}}\right) = 4.9\ \text{m/s}$$

Using these two values for the velocity, we can now obtain the acceleration using Equation 2.4

$$a = \frac{v - v_0}{t - t_0} = \frac{4.9\ \text{m/s} - 2.1\ \text{m/s}}{3.5\ \text{s} - 1.5\ \text{s}} = \boxed{1.4\ \text{m/s}^2}$$

CHAPTER 17 | *THE PRINCIPLE OF LINEAR SUPERPOSITION AND INTERFERENCE PHENOMENA*

---

1. ***REASONING AND SOLUTION*** In a time of $t = 1$ s, the pulse on the left has moved to the right a distance of 1 cm, while the pulse on the right has moved to the left a distance of 1 cm. Adding the shapes of these two pulses when $t = 1$ s reveals that the height of the resultant pulse is

   a. $\boxed{2\text{ cm}}$ at $x = 3$ cm.

   b. $\boxed{1\text{ cm}}$ at $x = 4$ cm.

---

2. ***REASONING*** When the difference in path lengths traveled by the two sound waves is a half-integer number $\left(\frac{1}{2}, 1\frac{1}{2}, 2\frac{1}{2}, \ldots\right)$ of wavelengths, the waves are out of phase and destructive interference occurs at the listener. The smallest separation d between the speakers is when the difference in path lengths is $\frac{1}{2}$ of a wavelength, so $d = \frac{1}{2}\lambda$. The wavelength is, according to Equation 16.1, is equal to the speed v of sound divided by the frequency f; $\lambda = v/f$.

   SOLUTION Substituting $\lambda = v/f$ into $d = \frac{1}{2}\lambda$ gives

$$d = \tfrac{1}{2}\lambda = \tfrac{1}{2}\left(\frac{v}{f}\right) = \tfrac{1}{2}\left(\frac{343\text{ m/s}}{245\text{ Hz}}\right) = \boxed{0.700\text{ m}}$$

---

3. SSM ***REASONING AND SOLUTION*** According to the principle of linear superposition, when two or more waves are present simultaneously at the same place, the resultant wave is the sum of the individual waves. Therefore, the shape of the string at the indicated times looks like the following:

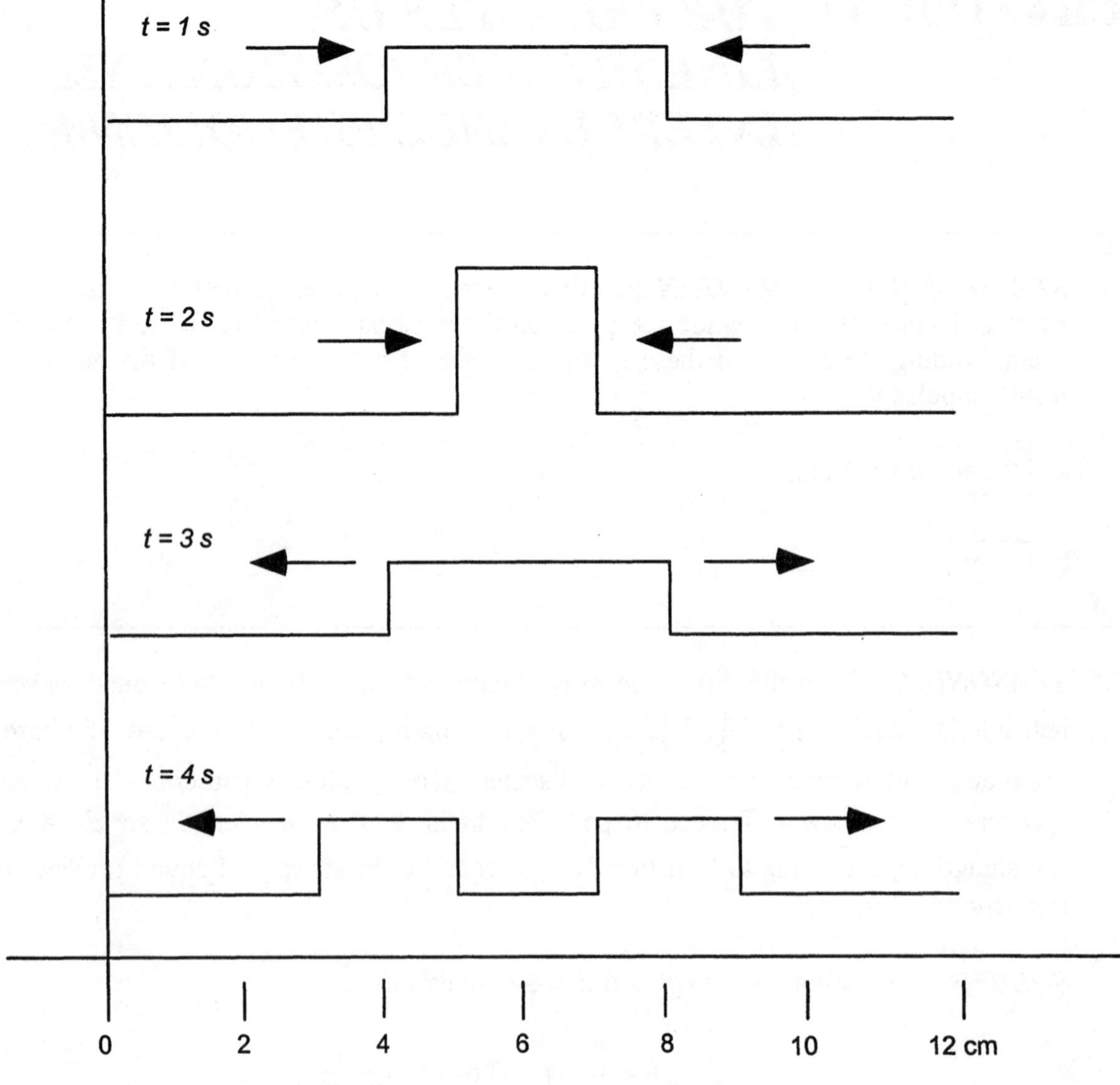

4. ***REASONING*** For destructive interference to occur, the difference in travel distances for the sound waves must be an integer number of half wavelengths. For larger and larger distances between speaker *B* and the observer at *C*, the difference in travel distances becomes smaller and smaller. Thus, the largest possible distance between speaker *B* and the observer at *C* occurs when the difference in travel distances is just one half wavelength.

SOLUTION Since the triangle *ABC* in Figure 17.7 is a right triangle, we can apply the Pythagorean theorem to obtain the distance d_{AC} as $\sqrt{(5.00\text{ m})^2 + d_{BC}^2}$. Therefore, the difference in travel distances is

$$\sqrt{(5.00\text{ m})^2 + d_{BC}^2} - d_{BC} = \frac{\lambda}{2} = \frac{v}{2f}$$

where we have used Equation 16.1 to express the wavelength λ as $\lambda = v/f$. Solving for the distance d_{BC} gives

$$\sqrt{(5.00\text{ m})^2 + d_{BC}^2} = d_{BC} + \frac{v}{2f} \quad \text{or} \quad (5.00\text{ m})^2 + d_{BC}^2 = \left(d_{BC} + \frac{v}{2f}\right)^2$$

$$(5.00\text{ m})^2 + d_{BC}^2 = d_{BC}^2 + \frac{d_{BC}v}{f} + \frac{v^2}{4f^2} \quad \text{or} \quad (5.00\text{ m})^2 = \frac{d_{BC}v}{f} + \frac{v^2}{4f^2}$$

$$d_{BC} = \frac{(5.00\text{ m})^2 - \dfrac{v^2}{4f^2}}{\dfrac{v}{f}} = \frac{(5.00\text{ m})^2 - \dfrac{(343\text{ m/s})^2}{4(125\text{ Hz})^2}}{\dfrac{343\text{ m/s}}{125\text{ Hz}}} = \boxed{8.42\text{ m}}$$

---

5. **SSM** ***REASONING*** The tones from the two speakers will produce destructive interference with the smallest frequency when the path length difference at C is one-half of a wavelength. From Figure 17.7, we see that the path length difference is $\Delta s = s_{\text{AC}} - s_{\text{BC}}$. From Example 1, we know that $s_{\text{AC}} = 4.00\text{ m}$, and from Figure 17.7, $s_{\text{BC}} = 2.40\text{ m}$. Therefore, the path length difference is $\Delta s = 4.00\text{ m} - 2.40\text{ m} = 1.60\text{ m}$.

SOLUTION Thus, destructive interference will occur when

$$\frac{\lambda}{2} = 1.60\text{ m} \quad \text{or} \quad \lambda = 3.20\text{ m}$$

This corresponds to a frequency of

$$f = \frac{v}{\lambda} = \frac{343\text{ m/s}}{3.20\text{ m}} = \boxed{107\text{ Hz}}$$

---

6. ***REASONING AND SOLUTION*** As the LYM section of the tube is pulled out, the length of the path on that side increases by $2(0.020\text{ m}) = 0.040\text{ m}$. This additional path difference between the LYM and LXM waves must correspond to $\lambda/2$, since the sound at the microphone goes from a maximum to a minimum. Hence, $\lambda/2 = 0.040$ m, or $\lambda = 0.080$ m. Then

$$v = \lambda f = (0.080\text{ m})(12\ 000\text{ Hz}) = \boxed{960\text{ m/s}}$$

---

7. **SSM** **WWW** ***REASONING*** The geometry of the positions of the loudspeakers and the listener is shown in the following drawing.

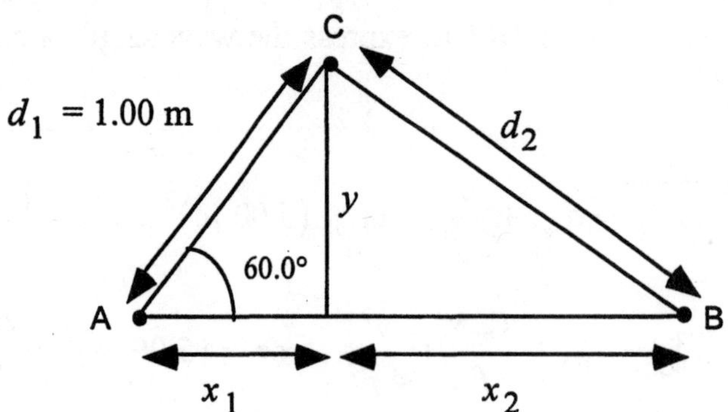

The listener at C will hear either a loud sound or no sound, depending upon whether the interference occurring at C is constructive or destructive. If the listener hears no sound, destructive interference occurs, so

$$d_2 - d_1 = \frac{n\lambda}{2} \qquad n = 1, 3, 5, \ldots \qquad (1)$$

SOLUTION Since $v = \lambda f$, according to Equation 16.1, the wavelength of the tone is

$$\lambda = \frac{v}{f} = \frac{343 \text{ m/s}}{68.6 \text{ Hz}} = 5.00 \text{ m}$$

Speaker B will be closest to Speaker A when $n = 1$ in Equation (1) above, so

$$d_2 = \frac{n\lambda}{2} + d_1 = \frac{5.00 \text{ m}}{2} + 1.00 \text{ m} = 3.50 \text{ m}$$

From the figure above we have that,

$$x_1 = (1.00 \text{ m}) \cos 60.0° = 0.500 \text{ m}$$

$$y = (1.00 \text{ m}) \sin 60.0° = 0.866 \text{ m}$$

Then

$$x_2^2 + y^2 = d_2^2 = (3.50 \text{ m})^2 \qquad \text{or} \qquad x_2 = \sqrt{(3.50 \text{ m})^2 - (0.866 \text{ m})^2} = 3.39 \text{ m}$$

Therefore, the closest that speaker A can be to speaker B so that the listener hears no sound is $x_1 + x_2 = 0.500 \text{ m} + 3.39 \text{ m} = \boxed{3.89 \text{ m}}$.

---

8. ***REASONING*** When the difference $\ell_1 - \ell_2$ in path lengths traveled by the two sound waves is a half-integer number $\left(\frac{1}{2}, 1\frac{1}{2}, 2\frac{1}{2}, \ldots\right)$ of wavelengths, destructive interference occurs at the listener. When the difference in path lengths is zero or an integer number $(1, 2, 3, \ldots)$

of wavelengths, constructive interference occurs. Therefore, we will divide the distance $\ell_1 - \ell_2$ by the wavelength of the sound to determine if constructive or destructive interference occurs. The wavelength is, according to Equation 16.1, the speed v of sound divided by the frequency f; $\lambda = v/f$.

SOLUTION

a. The distances ℓ_1 and ℓ_2 can be determined by applying the Pythagorean theorem to the two right triangles in the drawing:

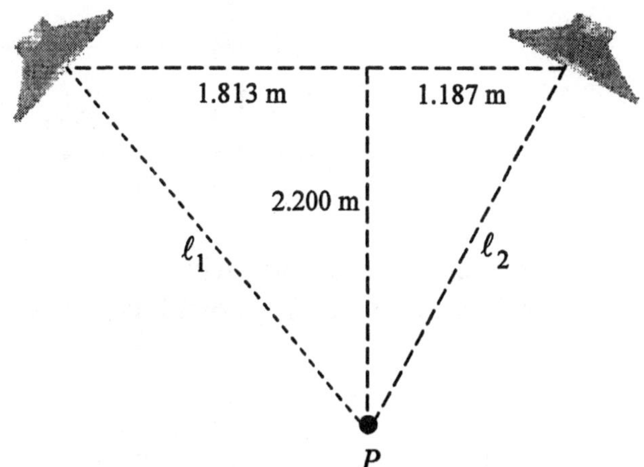

$$\ell_1 = \sqrt{(2.200\text{ m})^2 + (1.813\text{ m})^2} = 2.851\text{ m}$$

$$\ell_2 = \sqrt{(2.200\text{ m})^2 + (1.187\text{ m})^2} = 2.500\text{ m}$$

Therefore, $\ell_1 - \ell_2 =$ 0.351 m. The wavelength of the sound is $\lambda = \dfrac{v}{f} = \dfrac{343\text{ m/s}}{1466\text{ Hz}} = 0.234\text{ m}$. Dividing the distance $\ell_1 - \ell_2$ by the wavelength λ gives the number of wavelengths in this distance:

$$\text{Number of wavelengths} = \frac{\ell_1 - \ell_2}{\lambda} = \frac{0.351\text{ m}}{0.233\text{ m}} = 1.5$$

Since the number of wavelengths is a half-integer number $\left(1\frac{1}{2}\right)$, $\boxed{\text{destructive interference}}$ occurs at the listener.

b. The wavelength of the sound is now $\lambda = \dfrac{v}{f} = \dfrac{343\text{ m/s}}{977\text{ Hz}} = 0.351\text{ m}$. Dividing the distance $\ell_1 - \ell_2$ by the wavelength λ gives the number of wavelengths in that distance:

$$\text{Number of wavelengths} = \frac{\ell_1 - \ell_2}{\lambda} = \frac{0.351\text{ m}}{0.351\text{ m}} = 1$$

Since the number of wavelengths is an integer number (1), $\boxed{\text{constructive interference}}$ occurs at the listener.

---

9. ***REASONING AND SOLUTION*** Since $v = \lambda f$, the wavelength of the tone is

$$\lambda = \frac{v}{f} = \frac{343 \text{ m/s}}{73.0 \text{ Hz}} = 4.70 \text{ m}$$

The figure below shows the line between the two speakers and the distances in question.

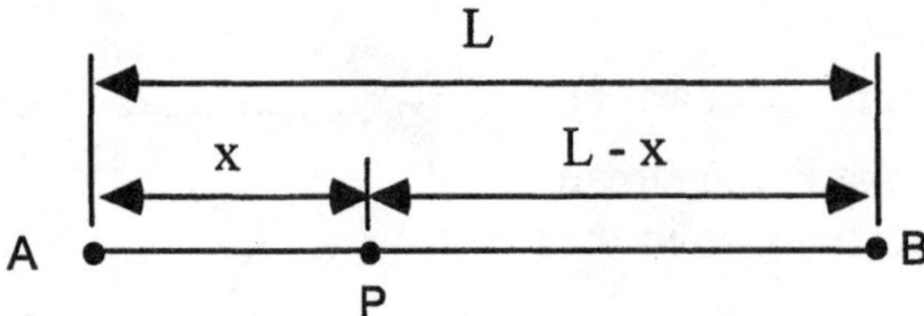

Constructive interference will occur when the difference in the distances traveled by the two sound waves in reaching point P is an integer number of wavelengths. That is, when

$$(L - x) - x = n\lambda$$

where n is an integer (or zero). Solving for x gives

$$x = \frac{L - n\lambda}{2} \tag{1}$$

When $n = 0$, $x = L/2 = (7.80 \text{ m})/2 = \boxed{3.90 \text{ m}}$. This corresponds to the point halfway between the two speakers. Clearly in this case, each wave has traveled the same distance and therefore, they will arrive in phase.

When $n = 1$,

$$x = \frac{(7.80 \text{ m}) - (4.70 \text{ m})}{2} = 1.55 \text{ m}$$

Thus, there is a point of constructive interference $\boxed{1.55 \text{ m from speaker A}}$. The points of constructive interference will occur symmetrically about the center point at $L/2$, so there is also a point of constructive interference 1.55 m from speaker B, that is at the point 7.80 m – 1.55 m = $\boxed{6.25 \text{ m from speaker A}}$.

When $n > 1$, the values of x obtained from Equation (1) will be negative. These values correspond to positions of constructive interference that lie to the left of A or to the right of C. They do not lie on the line between the speakers.

---

10. ***REASONING*** For a rectangular opening ("single slit") such as a doorway, the diffraction angle θ at which the first minimum in the sound intensity occurs is given by $\sin\theta = \frac{\lambda}{D}$ (Equation 17.1), where λ is the wavelength of the sound and D is the width of the opening.

This relation can be used to find the angle provided we realize that the wavelength λ is related to the speed v of sound and the frequency f by $\lambda = v/f$ (Equation 16.1).

SOLUTION
a. Substituting $\lambda = v/f$ into Equation 17.1 and using $D = 0.700$ m (only one door is open) gives

$$\sin\theta = \frac{\lambda}{D} = \frac{v}{f\,D} = \frac{343\text{ m/s}}{(607\text{ Hz})(0.700\text{ m})} = 0.807 \qquad \theta = \sin^{-1}(0.807) = \boxed{53.8°}$$

b. When both doors are open, $D = 2 \times 0.700$ m and the diffraction angle is

$$\sin\theta = \frac{\lambda}{D} = \frac{v}{f\,D} = \frac{343\text{ m/s}}{(607\text{ Hz})(2\times 0.700\text{ m})} = 0.404 \qquad \theta = \sin^{-1}(0.404) = \boxed{23.8°}$$

---

11. SSM ***REASONING*** The diffraction angle for the first minimum for a circular opening is given by Equation 17.2: $\sin\theta = 1.22\lambda / D$, where D is the diameter of the opening.

SOLUTION
a. Using Equation 16.1, we must first find the wavelength of the 2.0-kHz tone:

$$\lambda = \frac{v}{f} = \frac{343\text{ m/s}}{2.0\times 10^{3}\text{ Hz}} = 0.17\text{ m}$$

The diffraction angle for a 2.0-kHz tone is, therefore,

$$\theta = \sin^{-1}\left(1.22 \times \frac{0.17\text{ m}}{0.30\text{ m}}\right) = \boxed{44°}$$

b. The wavelength of a 6.0-kHz tone is

$$\lambda = \frac{v}{f} = \frac{343\text{ m/s}}{6.0\times 10^{3}\text{ Hz}} = 0.057\text{ m}$$

Therefore, if we wish to generate a 6.0-kHz tone whose diffraction angle is as wide as that for the 2.0-kHz tone in part (a), we will need a speaker of diameter D, where

$$D = \frac{1.22\ \lambda}{\sin\theta} = \frac{(1.22)(0.057\text{ m})}{\sin 44°} = \boxed{0.10\text{ m}}$$

---

12. ***REASONING*** In both cases, Equation 17.1 ($\sin\theta = \lambda/D$) offers a direct solution, since the width D of the doorway is given and we can use Equation 16.1 ($\lambda = v/f$) to determine the wavelength λ in terms of the speed v and frequency f.

SOLUTION Using Equations 17.1 and 16.1, we find

$$\sin\theta = \frac{\lambda}{D} = \frac{v}{Df}$$

Using this result we have

a. $$\theta = \sin^{-1}\left(\frac{v}{Df}\right) = \sin^{-1}\left[\frac{343 \text{ m/s}}{(0.77 \text{ m})(5.0\times10^{3} \text{ Hz})}\right] = \boxed{5.1°}$$

b. $$\theta = \sin^{-1}\left(\frac{v}{Df}\right) = \sin^{-1}\left[\frac{343 \text{ m/s}}{(0.77 \text{ m})(5.0\times10^{2} \text{ Hz})}\right] = \boxed{63°}$$

---

13. ***REASONING AND SOLUTION*** The condition for the first diffraction minimum for a single slit is given by Equation 17.1:

$$\sin\,\theta = \frac{\lambda}{D} \quad \text{or} \quad \theta = \sin^{-1}\left(\frac{\lambda}{D}\right)$$

Since $v = \lambda f$, the expression for θ becomes

$$\theta = \sin^{-1}\left(\frac{v}{fD}\right) = \sin^{-1}\left[\frac{343 \text{ m/s}}{(8100 \text{ Hz})(0.060 \text{ m})}\right] = 45°$$

Thus, at the frequency 8100 Hz, the person sitting at this angle off to the side of the diffraction horn hears no sound. If the person moves to an angle $\theta/2$, the frequency that interferes destructively (leading again to no sound) is found as follows:

$$\sin\,\theta = \frac{\lambda}{D} = \frac{v}{fD} \quad \text{or} \quad f = \frac{v}{D\sin\,\theta} = \left\{\frac{343 \text{ m/s}}{(0.060 \text{ m})[\sin\,(45°/2)]}\right\} = \boxed{1.5\times10^{4} \text{ Hz}}$$

---

14. ***REASONING AND SOLUTION***
The figure at the right shows the geometry of the situation.

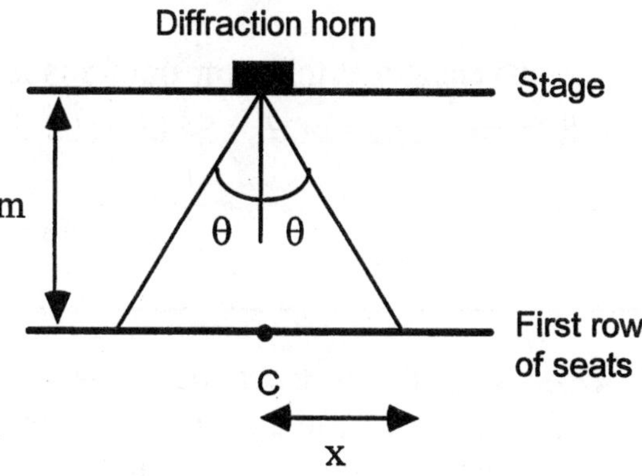

The tone will not be heard at seats located at the first diffraction minimum. This occurs when

$$\sin\theta = \frac{\lambda}{D} = \frac{v}{fD}$$

That is, the angle θ is given by

$$\theta = \sin^{-1}\left(\frac{v}{fD}\right) = \sin^{-1}\left[\frac{343\text{ m/s}}{(1.0\times10^{4}\text{ Hz})(0.075\text{ m})}\right] = 27.2°$$

From the figure at the right, we see that

$$\tan 27.2° = \frac{x}{8.7\text{ m}} \quad\Rightarrow\quad x = (8.7\text{ m})(\tan 27.2°) = 4.47\text{ m}$$

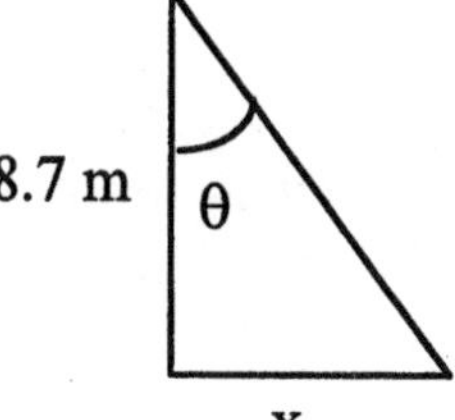

Thus, seats at which the tone cannot be heard are a distance x on either side of the center seat C. Thus, the distance between the two seats is

$$2x = 2(4.47\text{ m}) = \boxed{8.9\text{ m}}$$

---

15. ***REASONING AND SOLUTION*** At 0 °C the speed of sound is air is given as 331 m/s in Table 16.1 in the text. This corresponds to a wavelength of

$$\lambda_1 = v/f = (331\text{ m/s})/(3.00\times10^{3}\text{ Hz}) = 0.1103\text{ m}$$

The diffraction angle is given by Equation 17.2 as

$$\theta_1 = \sin^{-1}\left(\frac{1.22\,\lambda}{\text{D}}\right) = \sin^{-1}\left[\frac{1.22(0.1103\text{ m})}{0.175\text{ m}}\right] = 50.3°$$

For an ideal gas, the speed of sound is proportional to the square root of the Kelvin temperature, according to Equation 16.5. Therefore, the speed of sound at 29 °C is

$$v = (331\text{ m/s})\sqrt{\frac{302\text{ K}}{273\text{ K}}} = 348\text{ m/s}$$

The wavelength at this temperature is λ_2 = (348 m/s)/(3.00 × 10^3 Hz) = 0.116 m. This gives a diffraction angle of θ_2 = 54.0°. The change in the diffraction angle is thus

$$\Delta\theta = 54.0° - 50.3° = \boxed{3.7°}$$

---

16. ***REASONING*** When two frequencies are sounded simultaneously, the beat frequency produced is the difference between the two. Thus, knowing the beat frequency between the tuning fork and one flute tone tells us only the difference between the known frequency and the tuning-fork frequency. It does not tell us whether the tuning-fork frequency is greater or smaller than the known frequency. However, two different beat frequencies and two flute frequencies are given. Consideration of both beat frequencies will enable us to find the tuning-fork frequency.

    SOLUTION The fact that a 1-Hz beat frequency is heard when the tuning fork is sounded along with the 262-Hz tone implies that the tuning-fork frequency is either 263 Hz or 261 Hz. We can eliminate one of these values by considering the fact that a 3-Hz beat frequency is heard when the tuning fork is sounded along with the 266-Hz tone. This implies that the tuning-fork frequency is either 269 Hz or 263 Hz. Thus, the tuning-fork frequency must be $\boxed{263\text{ Hz}}$.

---

17. SSM ***REASONING*** The beat frequency of two sound waves is the difference between the two sound frequencies. From the graphs, we see that the period of the wave in the upper figure is 0.020 s, so its frequency is $f_1 = 1/T_1 = 1/(0.020\text{ s}) = 5.0\times10^1\text{ Hz}$. The frequency of the wave in the lower figure is $f_2 = 1/(0.024\text{ s}) = 4.2\times10^1\text{ Hz}$.

    SOLUTION The beat frequency of the two sound waves is

$$f_{\text{beat}} = f_1 - f_2 = 5.0\times10^1\text{ Hz} - 4.2\times10^1\text{ Hz} = \boxed{8\text{ Hz}}$$

---

18. ***REASONING*** The beat frequency is the difference between two sound frequencies. Therefore, the original frequency of the guitar string (before it was tightened) was either 3 Hz lower than that of the tuning fork (440.0 Hz – 3 Hz = 337 Hz) or 3 Hz higher (440.0 Hz + 3 Hz = 443 Hz):

443 Hz ——— } 3-Hz beat frequency
440.0 Hz ———
437 Hz ——— } 3-Hz beat frequency

To determine which of these frequencies is the correct one (437 or 443 Hz), we will use the information that the beat frequency decreases when the guitar string is tightened

SOLUTION When the guitar string is tightened, its frequency of vibration (either 437 or 443 Hz) increases. As the drawing below shows, when the 437-Hz frequency increases, it becomes closer to 440.0 Hz, so the beat frequency decreases. When the 443-Hz frequency increases, it becomes farther from 440.0 Hz, so the beat frequency increases. Since the problem states that the beat frequency decreases, the original frequency of the guitar string was $\boxed{437\text{ Hz}}$.

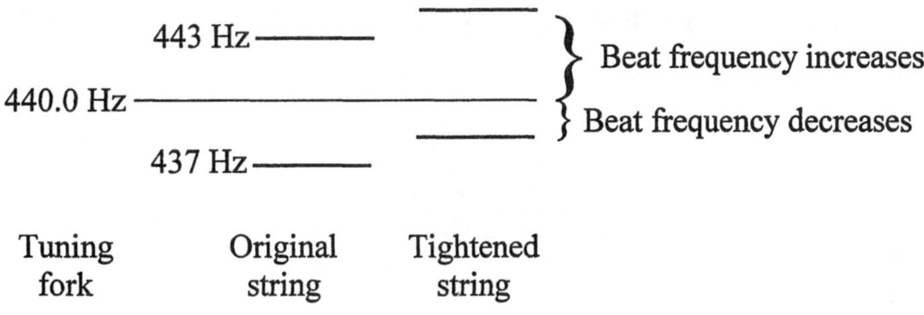

---

19. SSM ***REASONING AND SOLUTION*** Two ultrasonic sound waves combine and form a beat frequency that is in the range of human hearing. The frequency of one of the ultrasonic waves is 70 kHz. The beat frequency is the difference between the two sound frequencies. The smallest possible value for the ultrasonic frequency can be found by subtracting the upper limit of human hearing from the value of 70 kHz. The largest possible value for the ultrasonic frequency can be determined by adding the upper limit of human hearing to the value of 70 kHz. We know that the frequency range of human hearing is from 20 Hz to 20 kHz.

a. The smallest possible frequency of the other ultrasonic wave is

$$f = 70\text{ kHz} - 20\text{ kHz} = \boxed{50\text{ kHz}}$$

which results in a beat frequency of $70\text{ kHz} - 50\text{ kHz} = 20\text{ kHz}$.

b. The largest possible frequency for the other wave is

$$f = 70\text{ kHz} + 20\text{ kHz} = \boxed{90\text{ kHz}}$$

which results in a beat frequency of $90\text{ kHz} - 70\text{ kHz} = 20\text{ kHz}$.

---

20. ***REASONING AND SOLUTION*** The beat frequency is f_{beat} = 529 Hz – 524 Hz = 5 Hz. The period of the beats is then

$$T_{\text{beat}} = 1/f_{\text{beat}} = 1/(5\text{ Hz}) = \boxed{0.2\text{ s}}$$

---

21. ***REASONING*** The beat frequency is the difference between the frequency of the sound wave traveling in seawater and that generated by the 440.0-Hz tuning fork. We can use Equation 16.1, $v = f\lambda$, to find the frequency of the sound wave, provided we can determine the speed v of sound in seawater. This can be accomplished by using Equation 16.6.

SOLUTION The speed of sound in a liquid is given by Equation 16.6 as $v = \sqrt{B_{\text{ad}}/\rho}$, where B_{ad} is the adiabatic bulk modulus and ρ is the density of the liquid. The frequency of the sound traveling in sea water is

$$f = \frac{v}{\lambda} = \frac{\sqrt{\dfrac{B_{\text{ad}}}{\rho}}}{\lambda} = \frac{\sqrt{\dfrac{2.31\times 10^{9}\text{ Pa}}{1025\text{ kg/m}^{3}}}}{3.35\text{ m}} = 448\text{ Hz}$$

The beat frequency is

$$448\text{ Hz} - 440.0\text{ Hz} = \boxed{8\text{ Hz}}$$

---

22. ***REASONING AND SOLUTION*** The speed of the speakers is

$$v_{\text{s}} = 2\pi r/t = 2\pi(9.01\text{ m})/(20.0\text{ s}) = 2.83\text{ m/s}$$

The sound that an observer hears coming from the right speaker is Doppler shifted to a new frequency given by Equation 16.11 as

$$f_{\text{OR}} = \frac{f_{\text{s}}}{1 - v_{\text{s}}/v} = \frac{100.0\text{ Hz}}{1 - \left[(2.83\text{ m/s})/(343.00\text{ m/s})\right]} = 100.83\text{ Hz}$$

The sound that an observer hears coming from the left speaker is shifted to a new frequency given by Equation 16.12 as

$$f_{\text{OL}} = \frac{f_{\text{s}}}{1 + v_{\text{s}}/v} = \frac{100.0\text{ Hz}}{1 + \left[(2.83\text{ m/s})/(343.00\text{ m/s})\right]} = 99.18\text{ Hz}$$

The beat frequency heard by the observer is then

$$100.83\text{ Hz} - 99.18\text{ Hz} = \boxed{1.7\text{ Hz}}$$

---

23. SSM ***REASONING*** For standing waves on a string that is clamped at both ends, Equations 17.3 and 16.2 indicate that the standing wave frequencies are

$$f_n = n\left(\frac{v}{2L}\right) \qquad \text{where} \qquad v = \sqrt{\frac{F}{m/L}}$$

Combining these two expressions, we have, with $n = 1$ for the fundamental frequency,

$$f_1 = \frac{1}{2L}\sqrt{\frac{F}{m/L}}$$

This expression can be used to find the ratio of the two fundamental frequencies.

SOLUTION The ratio of the two fundamental frequencies is

$$\frac{f_{\text{old}}}{f_{\text{new}}} = \frac{\dfrac{1}{2L}\sqrt{\dfrac{F_{\text{old}}}{m/L}}}{\dfrac{1}{2L}\sqrt{\dfrac{F_{\text{new}}}{m/L}}} = \sqrt{\frac{F_{\text{old}}}{F_{\text{new}}}}$$

Since $F_{\text{new}} = 4F_{\text{old}}$, we have

$$f_{\text{new}} = f_{\text{old}}\sqrt{\frac{F_{\text{new}}}{F_{\text{old}}}} = f_{\text{old}}\sqrt{\frac{4\,F_{\text{old}}}{F_{\text{old}}}} = f_{\text{old}}\sqrt{4} = (55.0\text{ Hz})\,(2) = \boxed{1.10\times 10^2\text{ Hz}}$$

---

24. ***REASONING*** The frequencies f_n of the standing waves on a string fixed at both ends are given by Equation 17.3 as $f_n = n\left(\frac{v}{2L}\right)$, where n is an integer that specifies the harmonic number, v is the speed of the traveling waves that make up the standing waves, and L is the length of the string. For the second harmonic, $n = 2$.

SOLUTION The frequency f_2 of the second harmonic is

$$f_2 = n\left(\frac{v}{2L}\right) = 2\left[\frac{140\text{ m/s}}{2(0.28\text{ m})}\right] = \boxed{5.0\times 10^2\text{ Hz}}$$

---

25. ***REASONING AND SOLUTION*** Assuming that fretting the string does NOT change the tension, the speed of waves on the string will be the same in both cases. The speed of the

waves is given $v = \lambda f = 2Lf$. Applied to the first case (unfretted), this relation gives $v = 2L_1 f_1$. Applied to the second case, it gives $v = 2L_2 f_2$.

Equating the above equations and rearranging yields

$$L_2 = L_1(f_1/f_2) = (0.62\text{ m})(196\text{ Hz})/(262\text{ Hz}) = \boxed{0.46\text{ m}}$$

---

26. ***REASONING*** Equation 17.3 (with $n = 1$) gives the fundamental frequency as $f_1 = v/(2L)$, where L is the length of the wire and v is the speed of the waves traveling on the string. The speed is given by Equation 16.2 as $v = \sqrt{\dfrac{F}{m/L}}$, where F is the tension and m is the mass of the wire.

SOLUTION Using Equations 17.3 and 16.2, we obtain

$$f_1 = \frac{v}{2L} = \frac{\sqrt{\dfrac{F}{m/L}}}{2L} = \frac{1}{2}\sqrt{\frac{F}{mL}} = \frac{1}{2}\sqrt{\frac{160\text{ N}}{\left(6.0\times10^{-3}\text{ kg}\right)\left(0.41\text{ m}\right)}} = \boxed{130\text{ Hz}}$$

---

27. SSM ***REASONING*** The fundamental frequency f_1 is given by Equation 17.3 with $n = 1$: $f_1 = v/(2L)$. Since values for f_1 and L are given in the problem statement, we can use this expression to find the speed of the waves on the cello string. Once the speed is known, the tension F in the cello string can be found by using Equation 16.2, $v = \sqrt{F/(m/L)}$.

SOLUTION Combining Equations 17.3 and 16.2 yields

$$2Lf_1 = \sqrt{\frac{F}{m/L}}$$

Solving for F, we find that the tension in the cello string is

$$F = 4L^2 f_1^2 (m/L) = 4(0.800\text{ m})^2(65.4\text{ Hz})^2(1.56\times10^{-2}\text{ kg/m}) = \boxed{171\text{ N}}$$

---

28. ***REASONING*** The time it takes for a wave to travel the length L of the string is $t = L/v$, where v is the speed of the wave. The speed can be obtained since the fundamental frequency is known, and Equation 17.3 (with $n = 1$) gives the fundamental frequency as $f_1 = v/(2L)$. The length is not needed, since it can be eliminated algebraically between this expression and the expression for the time.

SOLUTION Solving Equation 17.3 for the speed gives $v = 2Lf_1$. With this result for the speed, the time for a wave to travel the length of the string is

$$t = \frac{L}{v} = \frac{L}{2Lf_1} = \frac{1}{2f_1} = \frac{1}{2(256\ \text{Hz})} = \boxed{1.95\times10^{-3}\ \text{s}}$$

---

29. ***REASONING*** A standing wave is composed of two oppositely traveling waves. The speed v of these waves is given by $v = \sqrt{\frac{F}{m/L}}$ (Equation 16.2), where F is the tension in the string and m/L is its linear density (mass per unit length). Both F and m/L are given in the statement of the problem. The wavelength λ of the waves can be obtained by visually inspecting the standing wave pattern. The frequency of the waves is related to the speed of the waves and their wavelength by $f = v/\lambda$ (Equation 16.1).

SOLUTION

a. The speed of the waves is

$$v = \sqrt{\frac{F}{m/L}} = \sqrt{\frac{280\ \text{N}}{8.5\times10^{-3}\ \text{kg/m}}} = \boxed{180\ \text{m/s}}$$

b. Two loops of any standing wave comprise one wavelength. Since the string is 1.8 m long and consists of three loops (see the drawing), the wavelength is

1.8 m

λ

$$\lambda = \tfrac{2}{3}(1.8\ \text{m}) = \boxed{1.2\ \text{m}}$$

c. The frequency of the waves is

$$f = \frac{v}{\lambda} = \frac{180\ \text{m/s}}{1.2\ \text{m}} = \boxed{150\ \text{Hz}}$$

---

30. ***REASONING*** Each string has a node at each end, so the frequency of vibration is given by Equation 17.3 as $f_n = nv/(2L)$, where n = 1, 2, 3, ... The speed v of the wave can be determined from Equation 16.2 as $v = \sqrt{F/(m/L)}$. We will use these two relations to find the lowest frequency that permits standing waves in both strings with a node at the junction.

SOLUTION

Since the frequency of the left string is equal to the frequency of the right string, we can write

$$\frac{n_{\text{left}}\sqrt{\dfrac{F}{(m/L)_{\text{left}}}}}{2L_{\text{left}}} = \frac{n_{\text{right}}\sqrt{\dfrac{F}{(m/L)_{\text{right}}}}}{2L_{\text{right}}}$$

Substituting in the data given in the problem yields

$$\frac{n_{\text{left}}\sqrt{\dfrac{190.0\text{ N}}{6.00\times10^{-2}\text{ kg/m}}}}{2(3.75\text{ m})} = \frac{n_{\text{right}}\sqrt{\dfrac{190.0\text{ N}}{1.50\times10^{-2}\text{ kg/m}}}}{2(1.25\text{ m})}$$

This expression gives $n_{\text{left}} = 6n_{\text{right}}$. Letting $n_{\text{left}} = 6$ and $n_{\text{right}} = 1$, the frequency of the left string (which is also equal to the frequency of the right string) is

$$f_6 = \frac{(6)\sqrt{\dfrac{190.0\text{ N}}{6.00\times10^{-2}\text{ kg/m}}}}{2(3.75\text{ m})} = \boxed{45.0\text{ Hz}}$$

---

31. SSM ***REASONING*** We can find the extra length that the D-tuner adds to the E-string by calculating the length of the D-string and then subtracting from it the length of the E string. For standing waves on a string that is fixed at both ends, Equation 17.3 gives the frequencies as $f_n = n(v/2L)$. The ratio of the fundamental frequency of the D-string to that of the E-string is

$$\frac{f_{\text{D}}}{f_{\text{E}}} = \frac{v/(2L_{\text{D}})}{v/(2L_{\text{E}})} = \frac{L_{\text{E}}}{L_{\text{D}}}$$

This expression can be solved for the length L_{D} of the D-string in terms of quantities given in the problem statement.

SOLUTION The length of the D-string is

$$L_{\text{D}} = L_{\text{E}}\left(\frac{f_{\text{E}}}{f_{\text{D}}}\right) = (0.628\text{ m})\left(\frac{41.2\text{ Hz}}{36.7\text{ Hz}}\right) = 0.705\text{ m}$$

The length of the E-string is extended by the D-tuner by an amount

$$L_{\text{D}} - L_{\text{E}} = 0.705\text{ m} - 0.628\text{ m} = \boxed{0.077\text{ m}}$$

---

32. ***REASONING AND SOLUTION*** We are given $f_j = \sqrt[12]{2}\, f_{j-1}$.

a. The length of the unfretted string is $L_0 = v/(2f_0)$ and the length of the string when it is pushed against fret 1 is $L_1 = v/(2f_1)$. The distance between the frets is

$$L_0 - L_1 = \frac{v}{2f_0} - \frac{v}{2f_1} = \left(\frac{v}{2f_0}\right)\left(1 - \frac{f_0}{f_1}\right) = \left(\frac{v}{2f_0}\right)\left(1 - \frac{1}{\sqrt[12]{2}}\right)$$

$$= L_0\left(1 - \frac{1}{\sqrt[12]{2}}\right) = (0.628 \text{ m})(0.0561) = \boxed{0.0352 \text{ m}}$$

b. The frequencies corresponding to the sixth and seventh frets are $f_6 = \left(\sqrt[12]{2}\right)^6 f_0$ and $f_7 = \left(\sqrt[12]{2}\right)^7 f_0$. The distance between fret 6 and fret 7 is

$$L_6 - L_7 = \frac{v}{2f_6} - \frac{v}{2f_7} = \frac{v}{2\left(\sqrt[12]{2}\right)^6 f_0} - \frac{v}{2\left(\sqrt[12]{2}\right)^7 f_0} = \left(\frac{v}{2f_0}\right)\left[\frac{1}{\left(\sqrt[12]{2}\right)^6} - \frac{1}{\left(\sqrt[12]{2}\right)^7}\right]$$

$$= L_0\left[\frac{1}{\left(\sqrt[12]{2}\right)^6} - \frac{1}{\left(\sqrt[12]{2}\right)^7}\right] = (0.628 \text{ m})(0.0397) = \boxed{0.0249 \text{ m}}$$

---

33. SSM WWW ***REASONING*** The natural frequencies of the cord are, according to Equation 17.3, $f_n = nv/(2L)$, where $n = 1, 2, 3, ...$ The speed v of the waves on the cord is, according to Equation 16.2, $v = \sqrt{F/(m/L)}$, where F is the tension in the cord. Combining these two expressions, we have

$$f_n = \frac{nv}{2L} = \frac{n}{2L}\sqrt{\frac{F}{m/L}} \qquad \text{or} \qquad \left(\frac{f_n 2L}{n}\right)^2 = \frac{F}{m/L}$$

Applying Newton's second law of motion, $\Sigma F = ma$, to the forces that act on the block and are parallel to the incline gives

$$F - Mg\sin\theta = Ma = 0 \qquad \text{or} \qquad F = Mg\sin\theta$$

where $Mg\sin\theta$ is the component of the block's weight that is parallel to the incline. Substituting this value for the tension into the equation above gives

$$\left(\frac{f_n 2L}{n}\right)^2 = \frac{Mg\sin\theta}{m/L}$$

This expression can be solved for the angle θ and evaluated at the various harmonics. The answer can be chosen from the resulting choices.

SOLUTION Solving this result for $\sin\theta$ shows that

$$\sin\theta = \frac{(m/L)}{Mg}\left(\frac{f_n 2L}{n}\right)^2 = \frac{1.20\times10^{-2}\ \text{kg/m}}{(15.0\ \text{kg})(9.80\ \text{m/s}^2)}\left[\frac{(165\ \text{Hz})2(0.600\ \text{m})}{n}\right]^2 = \frac{3.20}{n^2}$$

Thus, we have

$$\theta = \sin^{-1}\left(\frac{3.20}{n^2}\right)$$

Evaluating this for the harmonics corresponding to the range of n from $n = 2$ to $n = 4$, we have

$$\theta = \sin^{-1}\left(\frac{3.20}{2^2}\right) = 53.1^\circ \text{ for } n = 2$$

$$\theta = \sin^{-1}\left(\frac{3.20}{3^2}\right) = 20.8^\circ \text{ for } n = 3$$

$$\theta = \sin^{-1}\left(\frac{3.20}{4^2}\right) = 11.5^\circ \text{ for } n = 4$$

The angles between 15.0° and 90.0° are $\boxed{\theta = 20.8^\circ}$ and $\boxed{\theta = 53.1^\circ}$.

---

34. ***REASONING*** The frequency of a pipe open at both ends is given by Equation 17.4 as $f_n = n\left(\frac{v}{2L}\right)$, where n is an integer specifying the harmonic number, v is the speed of sound, and L is the length of the pipe. This relation can be used to find L, since all the other variables are known.

SOLUTION Solving the equation above for L, and recognizing that n = 3 for the third harmonic, we have

$$L = n\left(\frac{v}{2f_n}\right) = 3\left[\frac{343\ \text{m/s}}{2(262\ \text{Hz})}\right] = \boxed{1.96\ \text{m}}$$

---

35. ***REASONING AND SOLUTION***

a. For a string fixed at both ends the fundamental frequency is $f_1 = v/(2L)$ so $f_n = nf_1$.

$$\boxed{f_2 = 800\text{ Hz},\quad f_3 = 1200\text{ Hz},\quad f_4 = 1600\text{ Hz}}$$

b. For a pipe with both ends open the fundamental frequency is $f_1 = v/(2L)$ so $f_n = nf_1$.

$$\boxed{f_2 = 800\text{ Hz},\quad f_3 = 1200\text{ Hz},\quad f_4 = 1600\text{ Hz}}$$

c. For a pipe open at one end only the fundamental frequency is $f_1 = v/(4L)$ so $f_n = nf_1$ with n odd.

$$\boxed{f_3 = 1200\text{ Hz},\quad f_5 = 2000\text{ Hz},\quad f_7 = 2800\text{ Hz}}$$

---

36. ***REASONING*** Equation 17.5 (with n = 1) gives the fundamental frequency as $f_1 = v/(4L)$, where L is the length of the auditory canal and v is the speed of sound.

SOLUTION Using Equation 17.5, we obtain

$$f_1 = \frac{v}{4L} = \frac{343\text{ m/s}}{4(0.029\text{ m})} = \boxed{3.0\times10^3\text{ Hz}}$$

---

37. SSM ***REASONING AND SOLUTION*** The distance between one node and an adjacent antinode is $\lambda/4$. Thus, we must first determine the wavelength of the standing wave. A tube open at only one end can develop standing waves only at the odd harmonic frequencies. Thus, for a tube of length L producing sound at the third harmonic (n = 3), $L = 3(\lambda/4)$. Therefore, the wavelength of the standing wave is

$$\lambda = \tfrac{4}{3}L = \tfrac{4}{3}(1.5\text{ m}) = 2.0\text{ m}$$

and the distance between one node and the adjacent antinode is $\lambda/4 = \boxed{0.50\text{ m}}$.

---

38. ***REASONING*** The natural frequencies of a tube open at only one end are given by Equation 17.5 as $f_n = n\left(\dfrac{v}{4L}\right)$, where n is any odd integer (n = 1, 3, 5, …), v is the speed of sound, and L is the length of the tube. We can use this relation to find the value for n for the 450-Hz sound and to determine the length of the pipe.

SOLUTION

a. The frequency f_n of the 450-Hz sound is given by $450\text{ Hz} = n\left(\frac{v}{4L}\right)$. Likewise, the frequency of the next higher harmonic is $750\text{ Hz} = (n+2)\left(\frac{v}{4L}\right)$, because n is an odd integer and this means that the value of n for the next higher harmonic must be $n + 2$. Taking the ratio of these two relations gives

$$\frac{750\text{ Hz}}{450\text{ Hz}} = \frac{(n+2)\left(\frac{v}{4L}\right)}{n\left(\frac{v}{4L}\right)} = \frac{n+2}{n}$$

Solving this equation for n gives $n = \boxed{3}$.

b. Solving the equation $450\text{ Hz} = n\left(\frac{v}{4L}\right)$ for L and using $n = 3$, we find that the length of the tube is

$$L = n\left(\frac{v}{4f_n}\right) = 3\left[\frac{343\text{ m/s}}{4(450\text{ Hz})}\right] = \boxed{0.57\text{ m}}$$

---

39. SSM ***REASONING AND SOLUTION*** Since both tubes have the same length, the wavelength of the fundamental frequency is the same for both gases. From Equation 16.1 we have $f_H = v_H / \lambda$ and $f_N = v_N / \lambda$, where the subscript "H" refers to helium and "N" refers to neon. Dividing the first equation by the second gives

$$f_H / f_N = v_H / v_N$$

We know from Equation 16.5 that the speed of sound in a monatomic ideal gas is given by $v = \sqrt{\gamma kT / m}$, where $\gamma = 1.40$ and m is the mass of one atom of the gas. Therefore,

$$\frac{v_H}{v_N} = \sqrt{\frac{m_N}{m_H}}$$

Substituting the relation above into $f_H/f_N = v_H/v_N$, and using the fact that the atomic mass of neon is $m_N = 20.179$ u and that of helium is $m_H = 4.0026$ u, yields

$$f_H = f_N\sqrt{\frac{m_N}{m_H}} = (268\text{ Hz})\sqrt{\frac{20.179\text{ u}}{4.0026\text{ u}}} = \boxed{602\text{ Hz}}$$

---

40. ***REASONING AND SOLUTION*** The distance between the nodes of the standing wave is $L = \lambda/2 = v/(2f)$. The man travels this distance in a time

$$t = 1/(3.0\text{ Hz}) = 0.33\text{ s}$$

His speed is then

$$v_{\text{listener}} = L/t = v/(2ft) = (343\text{ m/s})/[2(0.33\text{ s})(440\text{ Hz})] = \boxed{1.2\text{ m/s}}$$

---

41. ***REASONING*** The well is open at the top and closed at the bottom, so it can be approximated as a column of air that is open at only one end. According to Equation 17.5, the natural frequencies for such an air column are

$$f_n = n\left(\frac{v}{4L}\right) \quad \text{where} \quad n = 1, 3, 5, \ldots$$

The depth L of the well can be calculated from the speed of sound, v = 343 m/s, and a knowledge of the natural frequencies f_n.

SOLUTION We know that two of the natural frequencies are 42 and 70.0 Hz. The ratio of these two frequencies is

$$\frac{70.0\text{ Hz}}{42\text{ Hz}} = \frac{5}{3}$$

Therefore, the value of n for each frequency is n = 3 for the 42-Hz sound, and $n = 5$ for the 70.0-Hz sound. Using $n = 3$, for example, the depth of the well is

$$L = \frac{nv}{4f_3} = \frac{3(343\text{ m/s})}{4(42\text{ Hz})} = \boxed{6.1\text{ m}}$$

---

42. ***REASONING AND SOLUTION*** According to Equation 16.5, the molecular mass is given by $m = 1.40\ kT/v^2$. The speed of waves in the tube is $v = \lambda f = 2Lf$.

Thus,

$$m = \frac{1.40\ kT}{4L^2 f^2} = \frac{(1.40)(1.38 \times 10^{-23}\text{ J/K})(293\text{ K})}{4(0.248\text{ m})^2(294\text{ Hz})^2} = \boxed{2.66 \times 10^{-25}\text{ kg}}$$

---

43. SSM WWW ***REASONING*** According to Equation 11.4, the absolute pressure at the bottom of the mercury is $P = P_{\text{atm}} + \rho g h$, where the height h of the mercury column is the original length L_0 of the air column minus the shortened length L. Hence,

$$P = P_{\text{atm}} + \rho g(L_0 - L)$$

SOLUTION From Equation 17.5, the fundamental (n = 1) frequency f_1 of the shortened tube is $f_1 = 1(v/4L)$, where L is the length of the air column in the tube. Likewise, the frequency f_3 of the third (n = 3) harmonic in the original tube is $f_3 = 3(v/4L_0)$, where L_0 is the length of the air column in the original tube. Since $f_1 = f_3$, we have that

$$1\left(\frac{v}{4L}\right) = 3\left(\frac{v}{4L_0}\right) \quad \text{or} \quad L = \tfrac{1}{3}L_0$$

The pressure at the bottom of the mercury is

$$P = P_{\text{atm}} + \rho g\left(\tfrac{2}{3}L_0\right)$$

$$= 1.01\times 10^5 \text{ Pa} + (13\,600 \text{ kg/m}^3)(9.80 \text{ m/s}^2)\left(\tfrac{2}{3}\times 0.75 \text{ m}\right) = \boxed{1.68\times 10^5 \text{ Pa}}$$

---

44. ***REASONING AND SOLUTION*** The original tube has a fundamental given by $f = v/(4L)$, so that its length is $L = v/(4f)$. The cut tube that has one end closed has a length of $L_c = v/(4f_c)$, while the cut tube that has both ends open has a length $L_o = v/(2f_o)$.

We know that $L = L_c + L_o$. Substituting the expressions for the lengths and solving for f gives

$$f = \frac{f_o f_c}{2f_c + f_o} = \frac{(425 \text{ Hz})(675 \text{ Hz})}{2(675 \text{ Hz}) + 425 \text{ Hz}} = \boxed{162 \text{ Hz}}$$

---

45. ***REASONING AND SOLUTION*** Since the wavelength is twice the distance between two successive nodes, we can use Equation 16.1 and see that

$$v = \lambda f = (2L)f = 2(0.30 \text{ m})(4.0 \text{ Hz}) = \boxed{2.4 \text{ m/s}}$$

---

46. ***REASONING AND SOLUTION*** We know that $L = v/(2f)$. For 20.0 Hz

$$L = (343 \text{ m/s})/[2(20.0 \text{ Hz})] = \boxed{8.6 \text{ m}}$$

For 20.0 kHz

$$L = (343 \text{ m/s})/[2(20.0\times 10^3 \text{ Hz})] = \boxed{8.6\times 10^{-3} \text{ m}}$$

---

47. SSM ***REASONING*** When constructive interference occurs again at point C, the path length difference is two wavelengths, or $\Delta s = 2\lambda = 3.20\text{ m}$. Therefore, we can write the expression for the path length difference as

$$s_{AC} - s_{BC} = \sqrt{s_{AB}^2 + s_{BC}^2} - s_{BC} = 3.20\text{ m}$$

This expression can be solved for s_{AB}.

SOLUTION Solving for s_{AB}, we find that

$$s_{AB} = \sqrt{(3.20\text{ m} + 2.40\text{ m})^2 - (2.40\text{ m})^2} = \boxed{5.06\text{ m}}$$

---

48. ***REASONING AND SOLUTION*** The first case requires that the frequency be either

$$440\text{ Hz} - 5\text{ Hz} = 435\text{ Hz} \quad \text{or} \quad 440\text{ Hz} + 5\text{ Hz} = 445\text{ Hz}$$

The second case requires that the frequency be either

$$436\text{ Hz} - 9\text{ Hz} = 427\text{ Hz} \quad \text{or} \quad 436\text{ Hz} + 9\text{ Hz} = 445\text{ Hz}$$

The frequency of the tuning fork is $\boxed{445\text{ Hz}}$.

---

49. ***REASONING*** The fundamental frequency f_1^A of air column A, which is open at both ends, is given by Equation 17.4 with $n = 1$: $f_1^A = (1)\left(\frac{v}{2L_A}\right)$, where v is the speed of sound in air and L_A is the length of the air column. Similarly, the fundamental frequency f_1^B of air column B, which is open at only one end, can be expressed using Equation 17.5 with $n = 1$: $f_1^B = (1)\left(\frac{v}{4L_B}\right)$. These two relations will allow us to determine the length of air column B.

SOLUTION Since the fundamental frequencies of the two air columns are the same $f_1^A = f_1^B$, so that

$$\underbrace{(1)\left(\frac{v}{2L_A}\right)}_{f_1^A} = \underbrace{(1)\left(\frac{v}{4L_B}\right)}_{f_1^B} \quad \text{or} \quad L_B = \tfrac{1}{2}L_A = \tfrac{1}{2}(0.70\text{ m}) = \boxed{0.35\text{ m}}$$

---

50. ***REASONING AND SOLUTION*** The shape of the string looks like

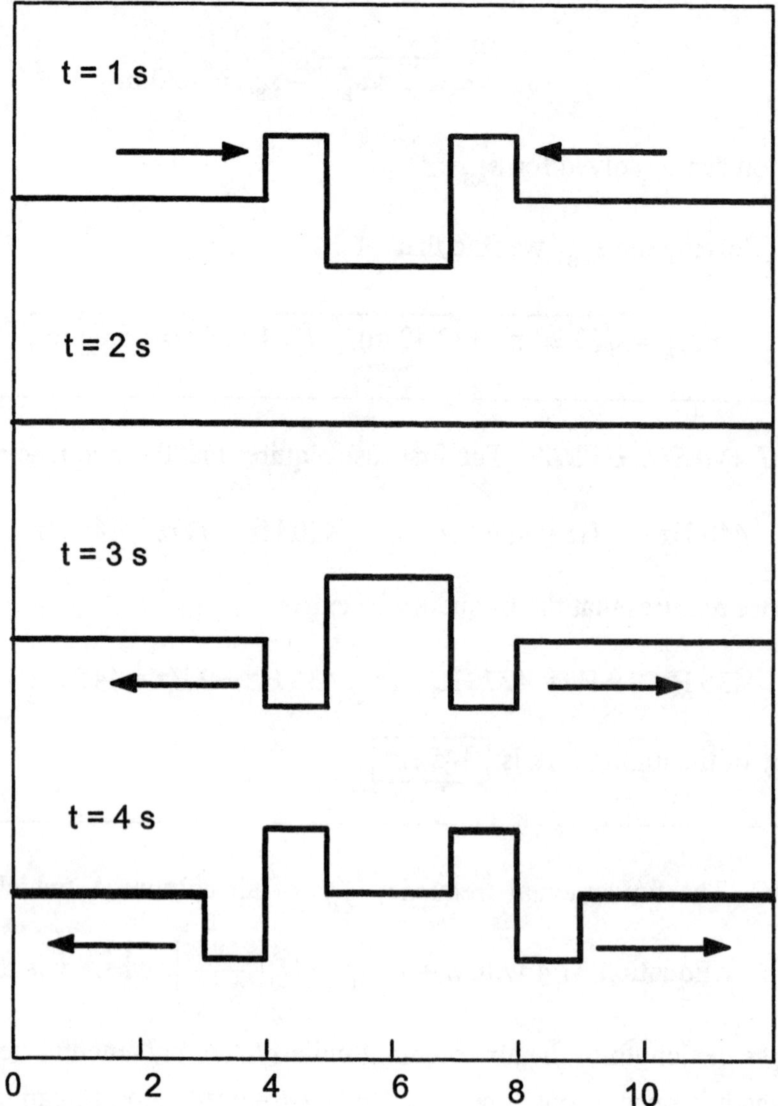

---

51. SSM ***REASONING*** According to Equation 17.3, the fundamental ($n = 1$) frequency of a string fixed at both ends is related to the wave speed v by $f_1 = v/2L$, where L is the length of the string. Thus, the speed of the wave is $v = 2Lf_1$. Combining this with Equation 16.2, $v = \sqrt{F/(m/L)}$, we have, after some rearranging,

$$\frac{F}{L^2} = 4f_1^2(m/L)$$

Since the strings have the same tension and the same lengths between their fixed ends, we have

$$f_{1E}^2(m/L)_E = f_{1G}^2(m/L)_G$$

where the symbols "E" and "G" represent the E and G strings on the violin. This equation can be solved for the linear density of the G string.

SOLUTION The linear density of the string is

$$(m/L)_G = \frac{f_{1E}^2}{f_{1G}^2}(m/L)_E = \left(\frac{f_{1E}}{f_{1G}}\right)^2 (m/L)_E$$

$$= \left(\frac{659.3\text{ Hz}}{196.0\text{ Hz}}\right)^2 (3.47\times 10^{-4}\text{ kg/m}) = \boxed{3.93\times 10^{-3}\text{ kg/m}}$$

---

52. ***REASONING*** The frequencies f_n of the standing waves allowed on a string fixed at both ends are given by Equation 17.3 as $f_n = n\left(\frac{v}{2L}\right)$, where n is an integer that specifies the harmonic number, v is the speed of the traveling waves that make up the standing waves, and L is the length of the string. The speed v is related to the tension F in the string and the linear density m/L via $v = \sqrt{\frac{F}{m/L}}$ (Equation 16.2). Therefore, the frequencies of the standing waves can be written as

$$f_n = n\left(\frac{v}{2L}\right) = n\left(\frac{\sqrt{\frac{F}{m/L}}}{2L}\right) = \frac{n}{2L}\sqrt{\frac{F}{m/L}}$$

The tension F in each string is provided by the weight W (either W_A or W_B) that hangs from the right end, so $F = W$. Thus, the expression for f_n becomes $f_n = \frac{n}{2L}\sqrt{\frac{W}{m/L}}$. We will use this relation to find the weight W_B.

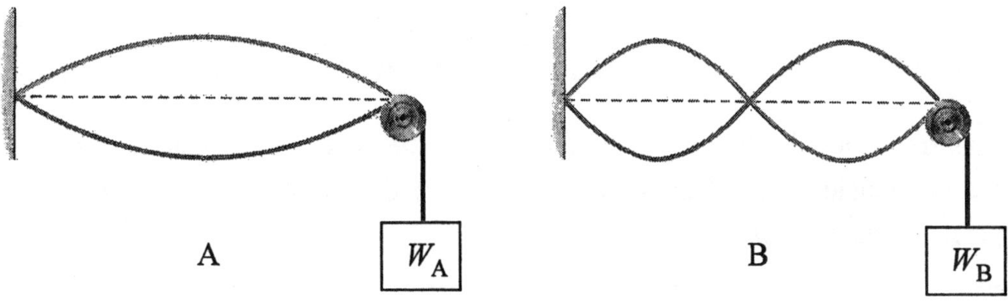

SOLUTION String A has one loop so $n = 1$, and the frequency f_1^{A} of this standing wave is $f_1^{\text{A}} = \dfrac{1}{2L}\sqrt{\dfrac{W_{\text{A}}}{m/L}}$. String B has two loops so $n = 2$, and the frequency f_2^{B} of this standing wave is $f_2^{\text{B}} = \dfrac{2}{2L}\sqrt{\dfrac{W_{\text{B}}}{m/L}}$. We are given that the two frequencies are equal, so

$$\underbrace{\frac{1}{2L}\sqrt{\frac{W_{\text{A}}}{m/L}}}_{f_1^{\text{A}}} = \underbrace{\frac{2}{2L}\sqrt{\frac{W_{\text{B}}}{m/L}}}_{f_2^{\text{B}}}$$

Solving this expression for W_{B} gives

$$W_{\text{B}} = \tfrac{1}{4}W_{\text{A}} = \tfrac{1}{4}(44\text{ N}) = \boxed{11\text{ N}}$$

53\. ***REASONING*** The fact that a loud sound is heard implies constructive interference, which occurs when the difference in path lengths is an integer number $(1, 2, 3, \ldots)$ of wavelengths. This difference is 50.5 m − 26.0 m = 24.5 m. Therefore, constructive interference occurs when $24.5\text{ m} = n\lambda$, where $n = 1, 2, 3, \ldots$. The wavelength is equal to the speed v of sound divided by the frequency f; $\lambda = v/f$ (Equation 16.1). Substituting this relation for λ into $24.5\text{ m} = n\lambda$, and solving for the frequency gives

$$f = \frac{nv}{24.5\text{ m}}$$

This relation will allow us to find the two lowest frequencies that the listener perceives as being loud due to constructive interference.

SOLUTION The lowest frequency occurs when $n = 1$:

$$f = \frac{nv}{24.5\text{ m}} = \frac{(1)(343\text{ m/s})}{24.5\text{ m}} = 14\text{ Hz}$$

This frequency lies below 20 Hz, so it cannot be heard by the listener. For $n = 2$ and $n = 3$, the frequencies are $\boxed{28\text{ and }42\text{ Hz}}$, which are the two lowest frequencies that the listener perceives as being loud.

---

54. ***REASONING AND SOLUTION***

a. We know that $f_x - f_y = 8$ Hz. Also, f_x is either 395 Hz or 389 Hz and f_y is either 397 Hz or 387 Hz. The fact that $f_x > f_y$ requires that

$$\boxed{f_y = 387\text{ Hz} \quad \text{and, therefore,} \quad f_x = 395\text{ Hz}}$$

b. Now we know that $f_y - f_x = 8$ Hz and f_x is either 395 Hz or 389 Hz and f_y is either 397 Hz or 387 Hz. The fact that $f_x < f_y$ requires that

$$\boxed{f_y = 397\text{ Hz} \quad \text{and, therefore,} \quad f_x = 389\text{ Hz}}$$

---

55. SSM WWW ***REASONING*** The beat frequency produced when the piano and the other instrument sound the note (three octaves higher than middle C) is $f_{\text{beat}} = f - f_0$, where f is the frequency of the piano and f_0 is the frequency of the other instrument ($f_0 = 2093$ Hz). We can find f by considering the temperature effects and the mechanical effects that occur when the temperature drops from 25.0 °C to 20.0 °C.

SOLUTION The fundamental frequency f_0 of the wire at 25.0 °C is related to the tension F_0 in the wire by

$$f_0 = \frac{v}{2L_0} = \frac{\sqrt{F_0/(m/L)}}{2L_0} \tag{1}$$

where Equations 17.3 and 16.2 have been combined.

The amount ΔL by which the piano wire attempts to contract is (see Equation 12.2) $\Delta L = \alpha L_0 \Delta T$, where α is the coefficient of linear expansion of the wire, L_0 is its length at 25.0 °C, and ΔT is the amount by which the temperature drops. Since the wire is prevented from contracting, there must be a stretching force exerted at each end of the wire. According to Equation 10.17, the magnitude of this force is

$$\Delta F = Y\left(\frac{\Delta L}{L_0}\right)A$$

where Y is the Young's modulus of the wire, and A is its cross-sectional area. Combining this relation with Equation 12.2, we have

$$\Delta F = Y\left(\frac{\alpha L_0 \Delta T}{L_0}\right)A = \alpha(\Delta T)YA$$

Thus, the frequency f at the lower temperature is

$$f=\frac{v}{2L_0}=\frac{\sqrt{(F_0+\Delta F)/(m/L)}}{2L_0}=\frac{\sqrt{\left[F_0+\alpha(\Delta T)YA\right]/(m/L)}}{2L_0} \quad (2)$$

Using Equations (1) and (2), we find that the frequency f is

$$f=f_0\frac{\sqrt{\left[F_0+\alpha(\Delta T)YA\right]/(m/L)}}{\sqrt{F_0/(m/L)}}=f_0\sqrt{\frac{F_0+\alpha(\Delta T)YA}{F_0}}$$

$$f=(2093\text{ Hz})\sqrt{\frac{818.0\text{ N}+(12\times10^{-6}/\text{C}^\circ)(5.0\text{ C}^\circ)(2.0\times10^{11}\text{ N/m}^2)(7.85\times10^{-7}\text{ m}^2)}{818.0\text{ N}}}$$

$$=2105\text{ Hz}$$

Therefore, the beat frequency is $2105\text{ Hz}-2093\text{ Hz}=\boxed{12\text{ Hz}}$.

---

56. ***CONCEPT QUESTIONS*** a. In drawing 1 the two speakers are equidistant from the observer. Since each wave travels the same distance in reaching the observer, the difference in travel-distances is zero, and constructive interference will occur. It will occur for any frequency. Different frequencies will correspond to different wavelengths, but the path difference will always be zero. Condensations will always meet condensations and rarefactions will always meet rarefactions at the observation point.

b. Destructive interference occurs only when the difference in travel distances for the two waves is an odd integer number n of half-wavelengths. Only certain frequencies, therefore, will be consistent with this requirement.

SOLUTION Frequency and wavelength are related by Equation 16.1 ($\lambda = v/f$). Using this equation together with the requirement for destructive interference in drawing 2, we have

$$\underbrace{\sqrt{L^2+L^2}-L}_{\text{Difference in travel distances}}=n\frac{\lambda}{2}=n\frac{v}{2f} \qquad \text{where } n=1,\ 3,\ 5,\ \ldots$$

Here we have used the Pythagorean theorem to determine the length of the diagonal of the square. Solving for the frequency f gives

$$f=\frac{nv}{2\left(\sqrt{2}-1\right)L}$$

The problem asks for the minimum frequency, so $n = 1$, and we obtain

$$f = \frac{v}{2\left(\sqrt{2}-1\right)L} = \frac{343 \text{ m/s}}{2\left(\sqrt{2}-1\right)(0.75 \text{ m})} = \boxed{550 \text{ Hz}}$$

---

57. ***CONCEPT QUESTIONS*** a. The diffraction angle θ is determined by the ratio of the wavelength λ to the diameter D, according to Equation 17.2 ($\sin\theta = 1.22\ \lambda/D$).

b. The wavelength is related to the frequency f and the speed v of the wave by Equation 16.1 ($\lambda = v/f$).

SOLUTION Using Equations 17.2 and 16.1, we have

$$\sin\theta = 1.22\frac{\lambda}{D} = 1.22\frac{v}{Df}$$

Since the speed of sound is a constant, this result indicates that the diffraction angle will be the same for each of the three speakers, provided that the diameter D times the frequency f has the same value. Thus, we pair the diameter and the frequency as follows

Diameter × Frequency
(0.050 m)(12.0 × 10^3 Hz) = 6.0 × 10^2 m/s
(0.10 m)(6.0 × 10^3 Hz) = 6.0 × 10^2 m/s
(0.15 m)(4.0 × 10^3 Hz) = 6.0 × 10^2 m/s

The common value of the diffraction angle, then, is

$$\theta = \sin^{-1}\left(1.22\frac{v}{Df}\right) = \sin^{-1}\left[\frac{1.22(343 \text{ m/s})}{6.0\times10^2 \text{ m/s}}\right] = \boxed{44°}$$

---

58. ***CONCEPT QUESTIONS*** a. The bystander hears a frequency from the moving horn that is greater than the emitted frequency f_s. This is because of the Doppler effect (Section 16.9). The bystander is the observer of the sound wave emitted by the horn. Since the horn is moving toward the observer, more condensations and rarefactions of the wave arrive at the observer's ear per second than would otherwise be the case. More cycles per second means that the observed frequency is greater than the emitted frequency.

b. The bystander hears a frequency from the stationary horn that is equal to the frequency f_s produced by the horn. Since the horn is stationary, there is no Doppler effect.

c. Yes. Because the two frequencies heard by the bystander are different, he hears a beat frequency that is the difference between the two.

SOLUTION According to Equation 16.11, the frequency that the bystander hears from the moving horn is

$$f_o = f_s\left(\frac{1}{1-v_s/v}\right)$$

where v_s is the speed of the moving horn and v is the speed of sound. The beat frequency heard by the bystander is $f_o - f_s$, so we find that

$$f_o - f_s = f_s\left(\frac{1}{1-v_s/v}\right) - f_s = f_s\left(\frac{1}{1-v_s/v} - 1\right)$$

$$= (395\ \text{Hz})\left[\frac{1}{1-(12.0\ \text{m/s})/(343\ \text{m/s})} - 1\right] = \boxed{14\ \text{Hz}}$$

---

59. ***CONCEPT QUESTIONS*** a. The harmonic frequencies are integer multiples of the fundamental frequency. Therefore, for wire A, the fundamental is one half of 660 Hz, or 330 Hz. Similarly, for wire B, the fundamental is one third of 660 Hz, or 220 Hz. Thus, the fundamental frequency of wire A is greater than that for wire B.

b. The fundamental frequency is related to the length L of the wire and the speed v at which individual waves travel back and forth on the wire by Equation 17.3 with $n = 1$: $f_1 = v/(2L)$.

c. According to Equation 17.3, the fundamental frequency is proportional to the speed. Therefore, since the fundamental frequency of wire A is greater, the speed must be greater on wire A than on wire B.

SOLUTION Using Equation 17.3 with $n = 1$, we find

$$f_1 = \frac{v}{2L} \quad \text{or} \quad v = 2Lf_1$$

Wire A $\qquad v = 2(1.2\ \text{m})(330\ \text{Hz}) = \boxed{790\ \text{m/s}}$

Wire B $\qquad v = 2(1.2\ \text{m})(220\ \text{Hz}) = \boxed{530\ \text{m/s}}$

As expected, the speed for wire A is greater.

---

60. ***CONCEPT QUESTIONS*** a. The speed v at which individual waves travel on the wire related to the tension F according to Equation 16.2: $v = \sqrt{\dfrac{F}{m/L}}$, where m/L is the mass per unit length of the wire.

b. The tension in the wire in Part 2 less than the tension in Part 1. The reason is related to Archimedes' principle (Equation 11.6). This principle indicates that when an object is immersed in a fluid, the fluid exerts an upward buoyant force on the object. In Part 2 the upward buoyant force from the mercury supports part of the block's weight, thus reducing the amount of the weight that the wire must support. As a result, the tension in the wire is less than in Part 1.

c. Since the tension F is less in Part 2, the speed v is also less. The fundamental frequency of the wire is given by Equation 17.3 with $n = 1$: $f_1 = v/(2L)$. Since v is less, the fundamental frequency of the wire is less in Part 2 than in Part 1.

SOLUTION Using Equations 17.3 and 16.2, we can obtain the fundamental frequency of the wire as follows:

$$f_1 = \frac{v}{2L} = \frac{1}{2L}\sqrt{\frac{F}{m/L}} \tag{1}$$

In Part 1 of the drawing, the tension F balances the weight of the block, keeping it from falling. The weight of the block is its mass times the acceleration due to gravity. The mass, according to Equation 11.1 is the density ρ_{copper} times the volume V of the block. Thus, the tension in Part 1 is

Part 1 tension $$F = (mass)g = \rho_{\text{copper}} Vg$$

In Part 2 of the drawing, the tension is reduced from this amount by the amount of the upward buoyant force. According to Archimedes' principle, the buoyant force is the weight of the liquid mercury displaced by the block. Since half of the block's volume is immersed, the volume of mercury displaced is $V/2$. The weight of this mercury is the mass times the acceleration due to gravity. Once again, according to Equation 11.1, the mass is the density ρ_{mercury} times the volume, which is $V/2$. Thus, the tension in Part 2 is

Part 2 tension $$F = \rho_{\text{copper}} Vg - \rho_{\text{mercury}} (V/2)g$$

With these two values for the tension we can apply Equation (1) to both parts of the drawing and obtain

Part 1
$$f_1 = \frac{1}{2L}\sqrt{\frac{\rho_{\text{copper}} V g}{m/L}}$$

Part 2
$$f_1 = \frac{1}{2L}\sqrt{\frac{\rho_{\text{copper}} V g - \rho_{\text{mercury}}(V/2)g}{m/L}}$$

Dividing the Part 2 by the Part 1 result, gives

$$\frac{f_{1,\text{ Part 2}}}{f_{1,\text{ Part 1}}} = \frac{\dfrac{1}{2L}\sqrt{\dfrac{\rho_{\text{copper}} V g - \rho_{\text{mercury}}(V/2)g}{m/L}}}{\dfrac{1}{2L}\sqrt{\dfrac{\rho_{\text{copper}} V g}{m/L}}} = \sqrt{\frac{\rho_{\text{copper}} - \frac{1}{2}\rho_{\text{mercury}}}{\rho_{\text{copper}}}}$$

$$= \sqrt{\frac{8890 \text{ kg/m}^3 - \frac{1}{2}(13\,600 \text{ kg/m}^3)}{8890 \text{ kg/m}^3}} = \boxed{0.485}$$

As expected, the fundamental frequency is less in Part 2 than Part 1.

---

61. ***CONCEPT QUESTIONS*** a. At the end of the tube where the tuning fork is, there is an antinode, because the gas molecules there are free to vibrate.

b. At the plunger, there is a node, because the gas molecules there are not free to vibrate.

c. Since there is an antinode at one end of the tube and a node at the other, the smallest value of L occurs when the length of the tube is one quarter of a wavelength.

SOLUTION Since the smallest value for L is a quarter of a wavelength, we have $L = \lambda/4$ or $\lambda = 4L$. According to Equation 16.1, the speed of sound is

$$v = f\lambda = f\,4L = (485 \text{ Hz})4(0.264 \text{ m}) = \boxed{512 \text{ m/s}}$$

---

62. ***CONCEPT QUESTIONS***

a. The phrase "the speakers are vibrating out of phase" means that when the diaphragm of one speaker is moving outward, the diaphragm of the other speaker is moving inward. In other words, when one speaker is creating a condensation, the other is creating a rarefaction.

b. The sound waves reaching the listener would exhibit destructive interference. Both sound waves travel the same distance from the speakers to the listener. Since the speakers are vibrating out of phase, whenever a condensation from one speaker reaches the listener, it is met by a rarefaction from the other, and vice versa. Therefore, the two sound waves experience destructive interference, and the listener hears no sound.

c. When the listener begins to move sideways, the distance between the listener and each speaker is no longer the same. Consequently, the sound waves no longer produce destructive interference, and the sound intensity begins to increase.

SOLUTION The two speakers are vibrating out of phase. Therefore, when the difference in path lengths $\ell_1 - \ell_2$ traveled by the two sounds is one-half a wavelength, or $\ell_1 - \ell_2 = \frac{1}{2}\lambda$, destructive interference occurs. The frequency f of the sound is equal to the speed v of sound divided by the wavelength λ; $f = v/\lambda$ (Equation 16.1). Thus, we have that

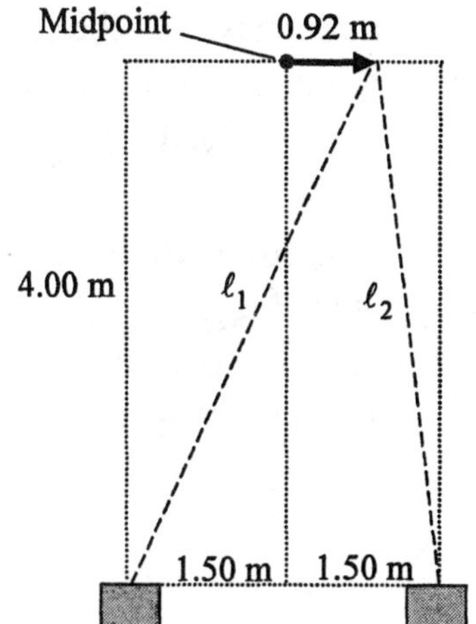

$$\ell_1 - \ell_2 = \tfrac{1}{2}\lambda = \frac{v}{2f} \qquad \text{or} \qquad f = \frac{v}{2(\ell_1 - \ell_2)}$$

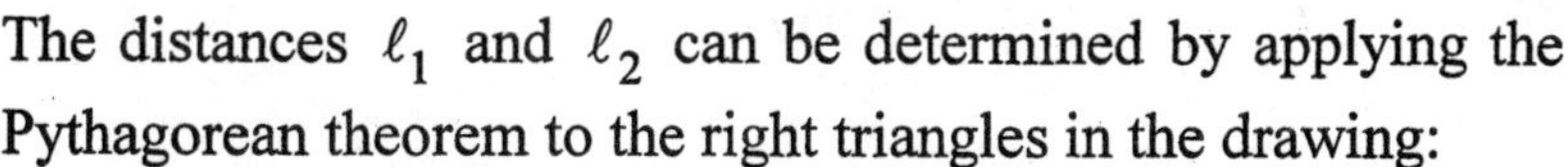

The distances ℓ_1 and ℓ_2 can be determined by applying the Pythagorean theorem to the right triangles in the drawing:

$$\ell_1 = \sqrt{(4.00\text{m})^2 + (1.50\text{ m} + 0.92\text{ m})^2} = 4.68\text{ m}$$

$$\ell_2 = \sqrt{(4.00\text{ m})^2 + (1.50\text{ m} - 0.92\text{ m})^2} = 4.04\text{ m}$$

The frequency of the sound is

$$f = \frac{v}{2(\ell_1 - \ell_2)} = \frac{343\text{ m/s}}{2(4.68\text{ m} - 4.04\text{ m})} = \boxed{270\text{ Hz}}$$

---

63. ***CONCEPT QUESTIONS***

a. The waves on the longer string have the same speed as those on the shorter string. The speed v of a transverse wave on a string is given by $v = \sqrt{F/(m/L)}$ (Equation 16.2), where F is the tension in the string and m/L is the mass per unit length (or linear density). Since F and m/L are the same for both strings, the speed of the waves is the same.

b. From the drawing it is evident that both strings are vibrating at their fundamental frequencies, and the longer string will vibrate at a lower frequency. The reason is that the fundamental frequency of vibration (n = 1) for a string fixed at each end is given by Equation 17.3 as $f_1 = v/(2L)$. Since the speed v is the same for both strings, but L is greater for the longer string, the longer string vibrates at the lower frequency.

c. The beat frequency will increase. The beat frequency is equal to the higher frequency of the shorter string minus the lower frequency of the longer string. If the longer string is increased in length, its frequency decreases even more, so the difference between the frequencies (the beat frequency) increases.

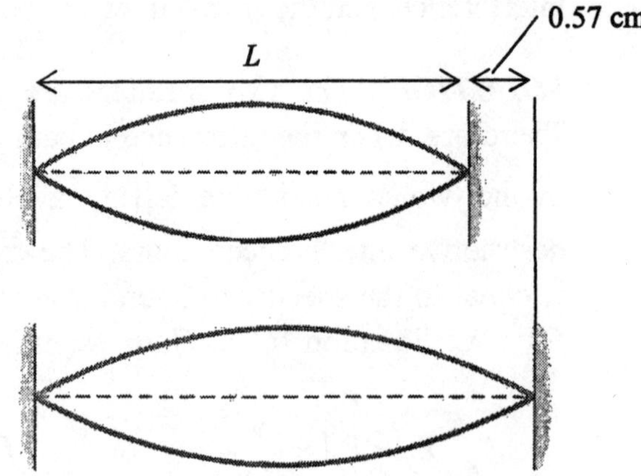

SOLUTION The beat frequency is the frequency of the shorter string minus the frequency of the longer string; $f_{\text{shorter}} - f_{\text{longer}}$. We are given that $f_{\text{shorter}} = 225$ Hz.

According to Equation 17.3 with $n = 1$, we have $f_{\text{longer}} = v/(2L_{\text{longer}})$, where L_{longer} is the length of the longer string; $L_{\text{longer}} = L + 0.0057$ m. Thus,

$$f_{\text{longer}} = \frac{v}{2L_{\text{longer}}} = \frac{v}{2(L + 0.0057\text{ m})}$$

According to our answer to Concept Question (a), the speed v of the waves on the longer string is the same as those on the shorter string, so $v = 41.8$ m/s. The length L of the shorter string can be obtained directly from Equation 17.3:

$$L = \frac{v}{2f_1} = \frac{41.8\text{ m/s}}{2(225\text{ Hz})} = 0.0929\text{ m}$$

Substituting this number back into the expression for f_{longer} yields

$$f_{\text{longer}} = \frac{v}{2(L + 0.0057\text{ m})} = \frac{41.8\text{ m/s}}{2(0.0929\text{ m} + 0.0057\text{ m})} = 212\text{ Hz}$$

The beat frequency is $f_{\text{shorter}} - f_{\text{longer}} = 225\text{ Hz} - 212\text{ Hz} = \boxed{13\text{ Hz}}$.

CHAPTER 1 | INTRODUCTION AND MATHEMATICAL CONCEPTS

CONCEPTUAL QUESTIONS

---

1. ***REASONING AND SOLUTION***

a. The SI unit for x is m. The SI units for the quantity vt are

$$\left(\frac{\text{m}}{\text{s}}\right)(\text{s}) = \text{m}\left(\frac{\text{m}}{\text{s}}\right)(\text{s}) = \text{m}$$

Therefore, the units on the left hand side of the equation are consistent with the units on the right hand side.

b. As described in part *a*, the SI units for the quantities x and vt are both m. The SI units for the quantity $\frac{1}{2}at^2$ are

$$\left(\frac{\text{m}}{\text{s}^2}\right)(\text{s}^2) = \text{m}$$

Therefore, the units on the left hand side of the equation are consistent with the units on the right hand side.

c. The SI unit for v is m/s. The SI unit for the quantity at is

$$\left(\frac{\text{m}}{\text{s}^2}\right)(\text{s}) = \frac{\text{m}}{\text{s}}$$

Therefore, the units on the left hand side of the equation are consistent with the units on the right hand side.

d. As described in part *c*, the SI units of the quantities v and at are both m/s. The SI unit of the quantity $\frac{1}{2}at^3$ is

$$\left(\frac{\text{m}}{\text{s}^2}\right)(\text{s}^3) = \text{m}\cdot\text{s}$$

Thus, the units on the left hand side are *not* consistent with the units on the right hand side. In fact, the right hand side is not a valid operation because it is not possible to add physical quantities that have different units.

e. The SI unit for the quantity v^3 is m^3/s^3. The SI unit for the quantity $2ax^3$ is

$$\left(\frac{\text{m}}{\text{s}^2}\right)(\text{m}^2) = \frac{\text{m}^3}{\text{s}^2}$$

Therefore, the units on the left hand side of the equation are *not* consistent with the units on the right hand side.

f. The SI unit for the quantity t is s. The SI unit for the quantity $\sqrt{\dfrac{2x}{a}}$ is

$$\sqrt{\frac{\text{m}}{(\text{m}/\text{s}^2)}} = \sqrt{\text{m}\left(\frac{\text{s}^2}{\text{m}}\right)} = \sqrt{\text{s}^2} = \text{s}$$

Therefore, the units on the left hand side of the equation are consistent with the units on the right hand side.

---

2. ***REASONING AND SOLUTION*** The quantity tan θ is dimensionless and has no units. The units of the ratio x/v are

$$\frac{\text{m}}{(\text{m}/\text{s})} = \text{m}\left(\frac{\text{s}}{\text{m}}\right) = \text{s}$$

Thus, the units on the left side of the equation are not consistent with those on the right side, and the equation tan $\theta = x/v$ is not a possible relationship between the variables x, v, and θ.

---

3. ***REASONING AND SOLUTION*** It is not always possible to add two numbers that have the same dimensions. In order to add any two physical quantities they must be expressed in the same *units*. Consider the two lengths: 1.00 m and 1.00 cm. Both quantities are lengths and, therefore, have the dimension [L]. Since the units are different, however, these two numbers cannot be added.

---

4. ***REASONING AND SOLUTION***

a. The dimension of a physical quantity describes the physical *nature* of the quantity and the *kind* of unit that is used to express the quantity. It is possible for two quantities to have the same dimensions but different units. All lengths, for example, have the dimension [L]. However, a length may be expressed in any length unit, such as kilometers, meters, centimeters, millimeters, inches, feet or yards.

As another illustration, the quantities 100 g and 1.5 kg are masses and have the dimensions [M]; however, they have different units.

b. All quantities with the same units must have the same dimensions. For example, all quantities expressed in kilograms have the dimension [M]; all quantities expressed in meters have the dimensions [L].

---

5. ***REASONING AND SOLUTION*** For the equation to be valid, the dimensions of the left hand side of the equation must be the same as the dimensions on the right hand side. Since the quantity *c* has no dimensions, it does not contribute to the dimensions of the right hand side, regardless of the value of *n*. Therefore, the value of *n* cannot be determined from dimensional analysis.

---

6. ***REASONING AND SOLUTION*** The following table shows the value of sin θ, cos θ, the ratio (sin θ)/(cos θ) and tan θ.

θ	sin θ	cos θ	(sin θ)/(cos θ)	tan θ
30.0	0.500	0.866	0.577	0.577
39.0	0.629	0.777	0.810	0.810
53.0	0.799	0.602	1.33	1.33
60.0	0.866	0.500	1.73	1.73

From the definitions given in Equations 1.1-1.3, we have

$$\frac{\sin\theta}{\cos\theta} = \frac{h_o / h}{h_a / h} = \frac{h_o}{h_a} = \tan\theta$$

---

7. ***REASONING AND SOLUTION***

a. The graph below shows sin θ plotted on the vertical axis and θ on the horizontal axis, with θ in 15° increments from θ = 0 to θ = 720°.

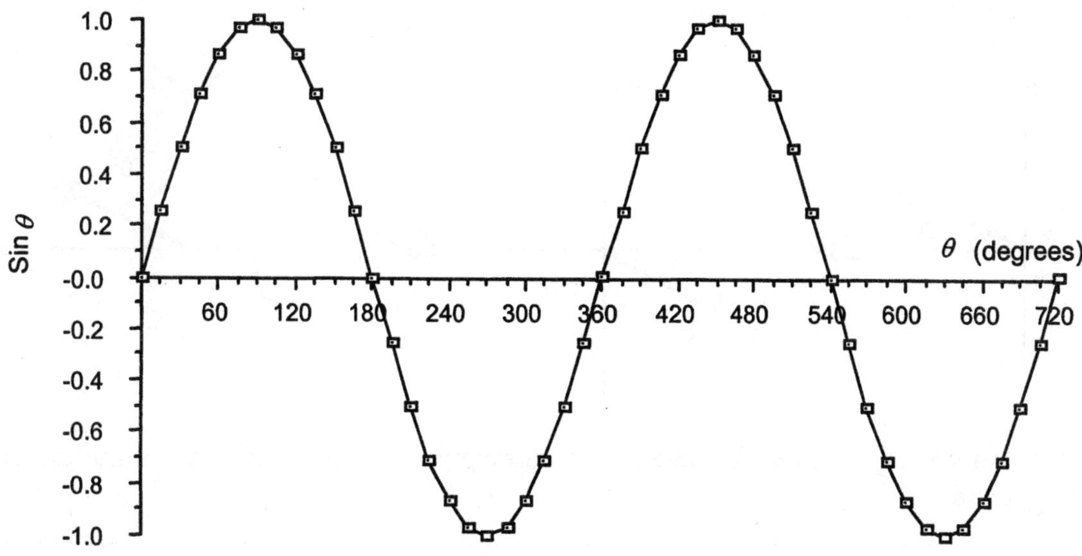

b. The graph below shows cos θ plotted on the vertical axis and θ on the horizontal axis, with θ in 15° increments from θ = 0 to θ = 720°.

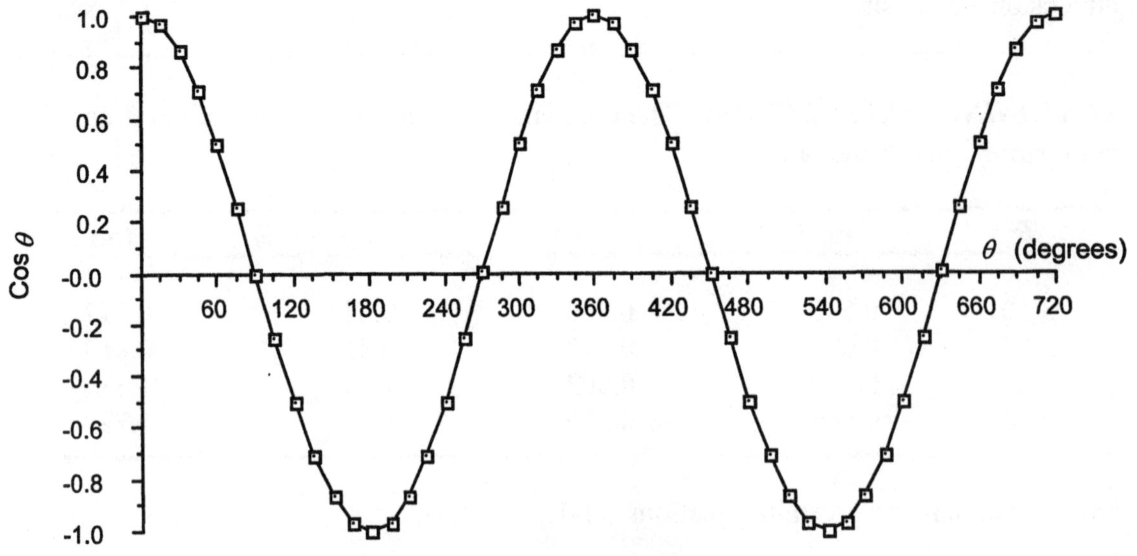

8. ***REASONING AND SOLUTION*** A vector quantity has both magnitude and direction. The number of people attending a football game, the number of days in a month, and the number of pages in a book can all be completely specified by giving a magnitude only. Hence, none of these quantities can be considered a vector.

9. ***REASONING AND SOLUTION*** For two vectors to be equal, they must be equal in magnitude and have the same direction. Only vectors **A**, **B**, and **D** have the same magnitude. These three vectors are shown below:

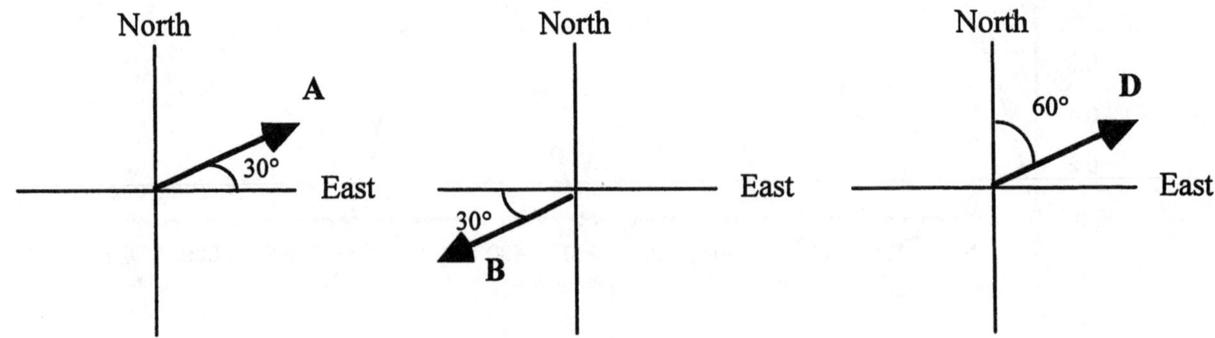

Clearly, only vectors **A** and **D** have the same magnitude and point in the same direction; hence, **A** and **D** are equal.

10. ***REASONING AND SOLUTION*** For two vectors to be equal, they must be equal in magnitude and have the same direction. Thus, two vectors with the same magnitude are not necessarily equal. They must also point in the same direction.

---

11. ***REASONING AND SOLUTION*** One can arrive back at the starting point after making eight consecutive displacements that add to zero only if one traverses three of the edges on one face and three edges on the opposite face (six displacements; the remaining two displacements occur in going from one opposite face to the other). For any particular starting point, there are four independent ways to traverse three edges on opposite faces. These are illustrated in the figures below. In (A) and (B), one traverses three consecutive edges before moving to the opposite face; in (C) and (D), one traverses two consecutive edges, moves to the opposite face, traverses three consecutive edges, then moves back to the original face and traverses the third edge on that face ending at the starting point.

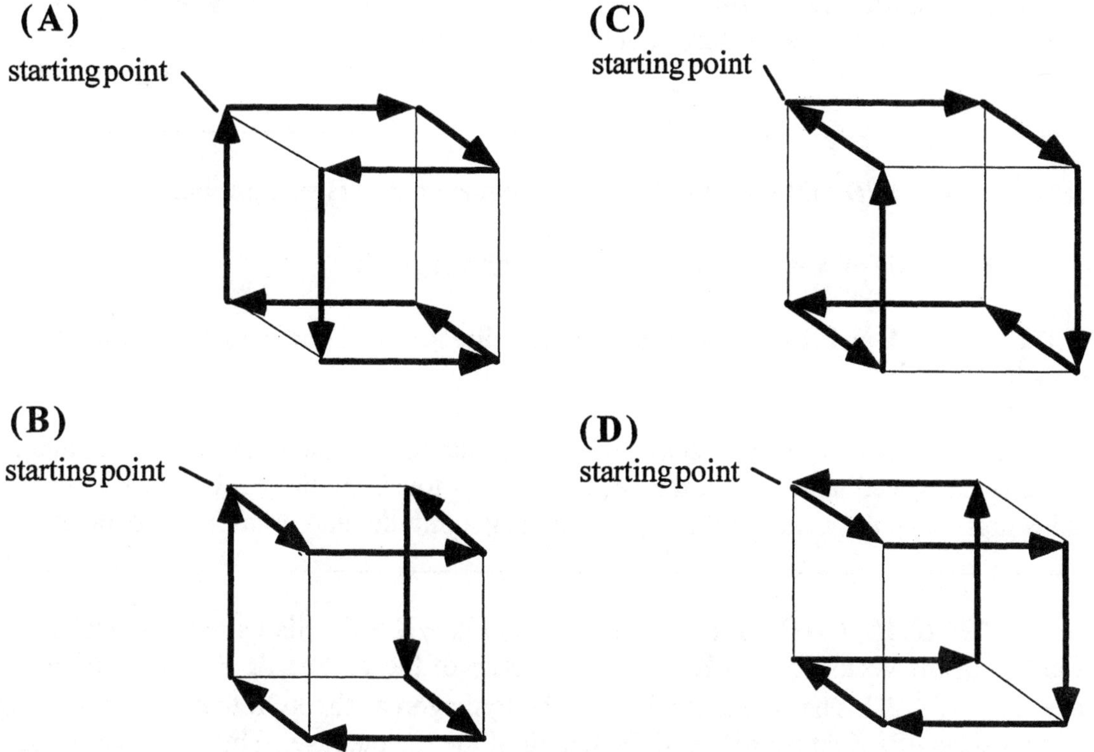

For any particular starting point, one can initially move in any one of three possible directions. Thus, there are a total of 4 (independent ways) × 3 (possible directions) or 12 ways to arrive back at the starting point that involve eight displacement vectors.

---

12. ***REASONING AND SOLUTION***
(a) Any vector has a zero component in the direction that is perpendicular to the direction of that vector. Thus, it *is* possible for one component of a vector to be zero, while the vector itself is not zero.

(b) In order for a vector to be zero, its components in all mutually perpendicular directions must also be zero. Therefore, it is *not* possible for a vector to be zero, while one of its components is not zero.

---

13. ***REASONING AND SOLUTION*** If two or more vectors have a resultant of zero, then, when they are arranged in a tail-to-head fashion the head of the last vector must touch the tail of the first vector. In order to satisfy this requirement with two vectors, the vectors must be equal in magnitude and opposite in direction. Therefore, it is *not* possible to add two *perpendicular* vectors so that the vector sum is zero.

---

14. ***REASONING AND SOLUTION*** Three or more vectors with unequal magnitudes may add together so that their vector sum is zero. To do so, they must be oriented so that, when they are placed in a tail-to-head fashion, the head of the last vector touches the tail of the first vector, as in the drawing at the right.

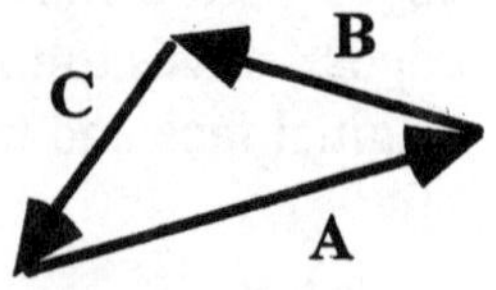

A + B + C = 0

---

15. ***REASONING AND SOLUTION*** The vector equation **A** + **B** = 0, implies that **A** = –**B**.

a. The magnitude of **A** must be equal to the magnitude of **B**.

b. The vectors **A** and **B** must point in opposite directions, as indicated by the minus sign in **A** = –**B**.

The same conclusions can be reached from a geometric argument. If **A** + **B** = 0, then, when **A** and **B** are placed tail-to-head, the head of **B** must touch the tail of **A**. Thus, the vectors **A** and **B** must have the same length (i.e., the same magnitude) and point in opposite directions.

---

16. ***REASONING AND SOLUTION*** The equation **A** + **B** = **C** tells us that the vector **C** is the resultant of the vectors **A** and **B**. The magnitudes of the vectors **A**, **B**, and **C** are related by $A^2 + B^2 = C^2$. This has the same form as the Pythagorean theorem that relates the length of the two sides of a right triangle and the length of the hypotenuse. Thus, the vectors **A** and **B** must be at right angles (or perpendicular) to each other.

---

17. ***REASONING AND SOLUTION*** The equation **A** + **B** = **C** tells us that the vector **C** is the resultant of the vectors **A** and **B**. The magnitudes of the vectors **A**, **B**, and **C** are related by $A + B = C$. In other words, the length of the vector **C** is equal to the combined lengths of vectors **A** and **B**. Therefore, the vectors **A** and **B** must point in the same direction.

---

18. ***REASONING AND SOLUTION*** If the magnitude of a vector is doubled, we may conclude that the magnitude of each component of the vector has also doubled.

This is best explained using the arbitrary vector, **A**, shown at the right. The angle θ, that specifies the direction of **A**, is given by

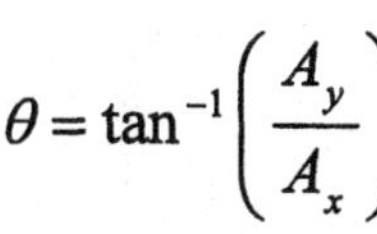

$$\theta = \tan^{-1}\left(\frac{A_y}{A_x}\right)$$

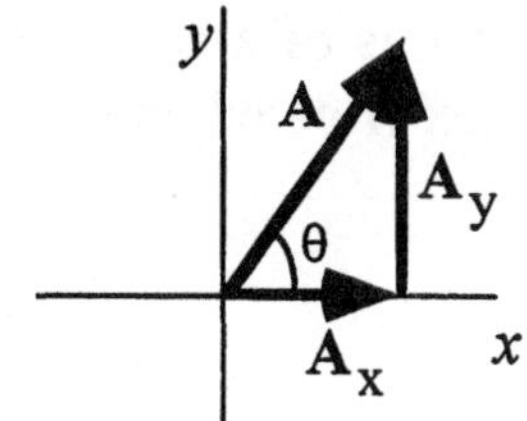

Suppose that the magnitude of **A** is doubled. The vector **A** is now twice as long, but its direction (and therefore the angle θ) remains unchanged. In order for the angle θ to remain the same after the magnitude of **A** is doubled, both $\mathbf{A_x}$ and $\mathbf{A_y}$ must double, so that the ratio A_y / A_x remains the same.

---

19. ***REASONING AND SOLUTION***

a. As the angle increases from 0° to 90°, the x component of the vector decreases in magnitude while the y component increases in magnitude. When the vector has rotated through an angle of 90°, the x component has zero magnitude and the magnitude of the y component is equal to the magnitude of the original vector (see figure below).

b. As the angle increases from 90° to 180°, the y component of the vector decreases in magnitude while the x component increases in magnitude. When the vector has rotated through an angle of 180°, the y component has zero magnitude and the magnitude of the x component is equal to the magnitude of the original vector.

c. As the angle increases from 180° to 270°, the x component of the vector decreases in magnitude while the y component increases in magnitude. When the vector has rotated through an angle of 270°, the x component has zero magnitude and the magnitude of the y component is equal to the magnitude of the original vector.

d. As the angle increases from 270° to 360°, the y component of the vector decreases in magnitude while the x component increases in magnitude. When the vector has rotated through an angle of 360°, the y component has zero magnitude and the magnitude of the x component is equal to the magnitude of the original vector.

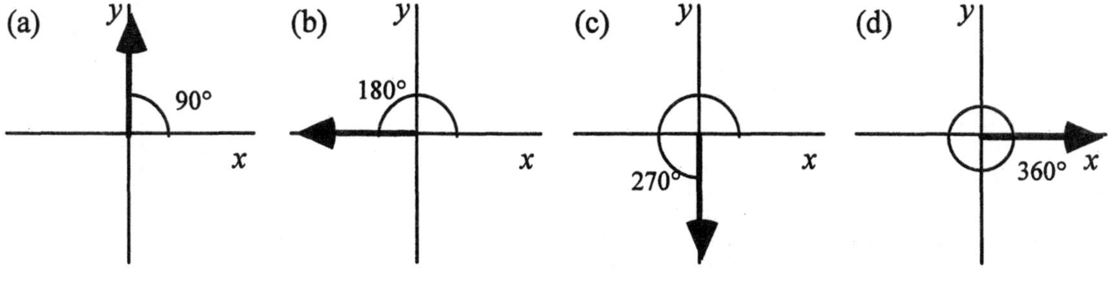

---

20. ***REASONING AND SOLUTION*** In general, a vector that has a component of zero along the x axis of a certain axes system does *not* have a component of zero along the x axis of another (rotated) axes system. Consider the vector **A** shown below.

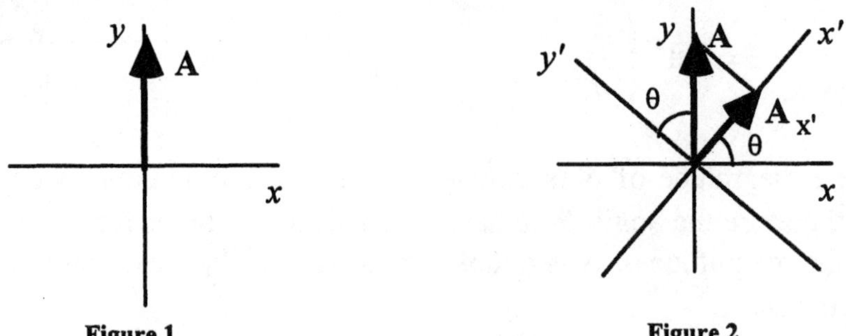

Figure 1 **Figure 2**

Relative to the x, y axes shown in Figure 1, the vector **A** has a component of zero along the x axis. Figure 2 shows the same vector along with a new axes system x', y'. The primed axes system is rotated at an arbitrary angle θ counterclockwise relative to the unprimed x, y axes system. Clearly the component of **A** along the x' axis is non zero for *most* arbitrary values of the (rotation) angle θ. The exceptions occur when $\theta = 0$, 180° and 360°.

CHAPTER 2 | *KINEMATICS IN ONE DIMENSION*

CONCEPTUAL QUESTIONS

---

1. ***REASONING AND SOLUTION*** The displacement of the honeybee for the trip is *not* the same as the distance traveled by the honeybee. As stated in the question, the distance traveled by the honeybee is 2 km. The displacement for the trip is the shortest distance between the initial and final positions of the honeybee. Since the honeybee returns to the hive, its initial position and final position are the same; therefore, the displacement of the honeybee is zero.

---

2. ***REASONING AND SOLUTION*** The buses *do not* have equal velocities. Velocity is a vector, with both magnitude and direction. In order for two vectors to be equal, they must have the same magnitude and the same direction. The direction of the velocity of each bus points in the direction of motion of the bus. Thus, the directions of the velocities of the buses are different. Therefore, the velocities are not equal, even though the speeds are the same.

---

3. ***REASONING AND SOLUTION*** The average speed of a vehicle is defined as the total distance covered by the vehicle divided by the time required for the vehicle to cover the distance. Both distance and time are scalar quantities. Since the average speed is the ratio of two scalar quantities, it is a scalar quantity.

---

4. ***REASONING AND SOLUTION*** Consider the four traffic lights **1**, **2**, **3** and **4** shown below. Let the distance between lights **1** and **2** be x_{12}, the distance between lights **2** and **3** be x_{23}, and the distance between lights **3** and **4** be x_{34}.

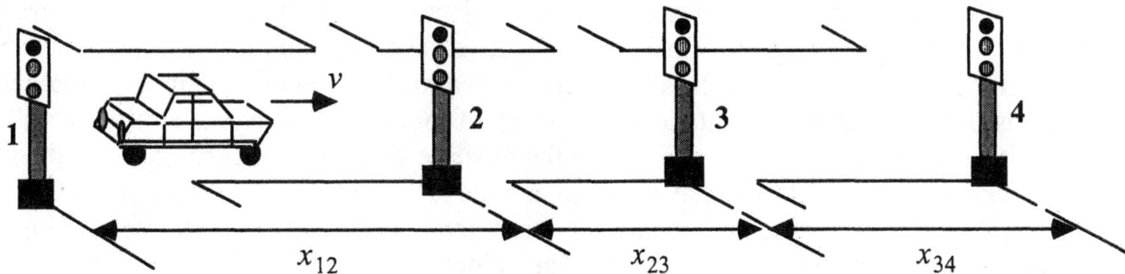

The lights can be timed so that if a car travels with a constant speed v, red lights can be avoided in the following way. Suppose that at time $t = 0$ s, light **1** turns green while the rest are red. Light **2** must then turn green in a time t_{12}, where $t_{12} = x_{12}/v$. Light **3** must turn green in a time t_{23} after light **2** turns green, where $t_{23} = x_{23}/v$. Likewise, light **4** must turn green in a time t_{34} after light **3** turns green, where $t_{34} = x_{34}/v$. Note that the timing of traffic lights is more complicated than indicated here when groups of cars are stopped at light **1**.

Then the acceleration of the cars, the reaction time of the drivers, and other factors must be considered.

---

5. ***REASONING AND SOLUTION*** The velocity of the car is a vector quantity with both magnitude and direction The speed of the car is a scalar quantity and has nothing to do with direction. It is possible for a car to drive around a track at constant speed. As the car drives around the track, however, the car must change direction. Therefore, the direction of the velocity *changes*, and the velocity cannot be constant. The *incorrect* statement is (a).

---

6. ***REASONING AND SOLUTION*** Answers will vary. One example is a ball that is thrown straight up in the air. When the ball is at its highest point, its velocity is momentarily zero. Since the ball is close to the surface of the earth, its acceleration is nearly constant and is equal to the acceleration due to gravity. Thus, the velocity of the ball is momentarily zero, but since the ball is accelerating, the velocity is changing. An instant later, the velocity of the ball is nonzero as the ball begins to fall. Another example is a swimmer in a race, reversing directions at the end of the pool.

---

7. ***REASONING AND SOLUTION*** The acceleration of an object is the rate at which its velocity is changing. No information can be gained concerning the acceleration of an object if all that is known is the velocity of the object at a single instant. No conclusion can be reached concerning the accelerations of the two vehicles, so the car does not necessarily have a greater acceleration.

---

8. ***REASONING AND SOLUTION*** It *is* possible for the instantaneous velocity at any point during a trip to have a negative value, even though the average velocity for the entire trip has a positive value. The average velocity for the trip is the displacement for the trip divided by the elapsed time. It depends only on the initial and final positions, and the time required for the trip. The average velocity contains no information concerning the actual path taken by the object. Let us assume that the object is constrained to move in a straight line with directions designated as positive or negative. The direction of the average velocity is the same as the direction of the displacement, while the direction of the instantaneous velocity is the same as the instantaneous direction of motion. The average velocity will be positive if the displacement vector points in the positive direction. At any point in the trip, the object could temporarily reverse direction and move in the negative direction. While the object is moving in the negative direction, its instantaneous velocity is negative. As long as the overall displacement is positive, the average velocity for the trip is positive.

---

9. ***REASONING AND SOLUTION*** Acceleration is the rate of change of velocity. The average velocity for an object over a time interval Δt is given by $\bar{\mathbf{v}} = \Delta \mathbf{x} / \Delta t$, where $\Delta \mathbf{x}$ is the displacement of the object during the time interval.

   A runner runs half the remaining distance to the finish line every ten seconds. In each successive ten-second interval, the distance covered by the runner, and hence the runner's displacement, becomes smaller by a factor of 2. Thus, during each successive ten-second interval, the ratio $\Delta\mathbf{x} / \Delta t$ becomes smaller by a factor of 2. The acceleration, however, is the *change* in velocity per unit time. Assume that the distance to the finish line is 100 m and consider the change in the average velocity between the first and second 10 s intervals; it is $\Delta v_{1,2}$ = (1/2)(50 m)/(10 s) – (1/2)(100 m)/(10 s) = – 2.5 m/s. Now consider the change in average velocity between the second and third 10 s intervals; it is $\Delta v_{2,3}$ = (1/2)(25 m)/(10 s) – (1/2)(50 m)/(10 s) = – 1.25 m/s. Thus, as time passes, the magnitude of the velocity changes by a decreasing amount from one 10 s interval to the next. As a result, the acceleration of the runner does not have a constant magnitude.

---

10. ***REASONING AND SOLUTION*** An object moving with a constant acceleration will slow down if the acceleration vector points in the opposite direction to the velocity vector; however, if the acceleration remains constant, the object will never come to a permanent halt. As time increases, the magnitude of the velocity will get smaller and smaller. At some time, the velocity will be instantaneously zero. If the acceleration is constant, however, the velocity vector will continue to change at the same rate. An instant after the velocity is zero, the magnitude of the velocity will begin increasing in the same direction as the acceleration. As time increases, the velocity of the object will then increase in the same direction as the acceleration. In other words, if the acceleration truly remains constant, the object will slow down, stop *for an instant*, reverse direction and then speed up.

---

11. ***REASONING AND SOLUTION*** Other than being horizontal, the motion of an experimental vehicle that slows down, comes to a momentary halt, reverses direction and then speeds up with a constant acceleration of 9.80 m/s^2, is identical to that of a ball that is thrown straight upward near the surface of the earth, comes to a halt, and falls back down. In both cases, the acceleration is constant in magnitude and direction and the objects begin their motion with the acceleration and velocity vectors pointing in opposite directions.

---

12. ***REASONING AND SOLUTION*** The first ball has an initial speed of zero, since it is dropped from rest. It picks up speed on the way down, striking the ground at a speed v_f. The second ball has a motion that is the reverse of that of the first ball. The second ball starts out with a speed v_f and loses speed on the way up. By symmetry, the second ball will come to a halt at the top of the building. Thus, in approaching the crossing point, the second ball travels faster than the first ball. Correspondingly, the second ball must travel farther on its way to the crossing point than the first ball does. Thus, the crossing point must be located in the upper half of the building.

---

13. ***REASONING AND SOLUTION*** Two objects are thrown vertically upward, first one, and then, a bit later, the other. The time required for either ball to reach its maximum height can be found from Equation 2.4: $v = v_0 + at$. At the maximum height, v = 0 m/s; solving for t yields $t = -v_0 / a$, where a is the acceleration due to gravity. Clearly, the time required to reach the maximum height depends on the initial speed with which the object was thrown. Since the second object is launched later, its initial speed must be less than the initial speed of the first object in order that both objects reach their maximum heights at the same instant.

The maximum height that each object attains can be found from Equation 2.9: $v^2 = v_0^2 + 2ay$. At the maximum height, $v = 0$ m/s; solving for y gives $y = -v_0^2 / (2a)$, where a is the acceleration due to gravity. Since the second object has a smaller initial speed v_0, it will also attain a smaller maximum height. Thus, it is *not* possible for both objects to reach the same maximum height at the same instant.

---

14. ***REASONING AND SOLUTION*** The magnitude of the muzzle velocity of the bullet can be found (to a very good approximation) by solving Equation 2.9, $v^2 = v_0^2 + 2ax$, with $v_0 = 0$ m/s; that is

$$v = \sqrt{2ax}$$

where a is the acceleration of the bullet and x is the distance traveled by the bullet before it leaves the barrel of the gun (i.e., the length of the barrel).

Since the muzzle velocity of the rifle with the shorter barrel is greater than the muzzle velocity of the rifle with the longer barrel, the product ax must be greater for the bullet in the rifle with the shorter barrel. But x is smaller for the rifle with the shorter barrel, thus the *acceleration of the bullet must be larger in the rifle with the shorter barrel.*

CHAPTER 3 | *KINEMATICS IN TWO DIMENSIONS*

CONCEPTUAL QUESTIONS

---

1. ***REASONING AND SOLUTION*** In addition to the x and y axes, the z axis would be required to completely describe motion in three dimensions. Motion along the z axis can be described in terms of the kinematic variables z, a_z, v_z, v_{0z}, and t. In analogy with the equations of kinematics for the x and y components, the following equations would be necessary to describe motion along the z axis:

$$v_z = v_{0z} + a_z t \qquad z = v_{0z} t + \tfrac{1}{2} a_z t^2$$

$$z = \tfrac{1}{2}(v_{0z} + v_z)t \qquad v_z^2 = v_{0z}^2 + 2a_z z$$

---

2. ***REASONING AND SOLUTION*** An object thrown upward at an angle θ will follow the trajectory shown below. Its acceleration is that due to gravity, and, therefore, always points downward. The acceleration is denoted by a_y in the figure. In general, the velocity of the object has two components, v_x and v_y. Since $a_x = 0$, v_x always equals its initial value. The y component of the velocity, v_y, decreases as the object rises, drops to zero when the object is at its highest point, and then increases in magnitude as the object falls downward.

a. Since $v_y = 0$ when the object is at its highest point, the velocity of the object points only in the x direction. As suggested in the figure below, the acceleration will be perpendicular to the velocity when the object is at its highest point and $v_y = 0$.

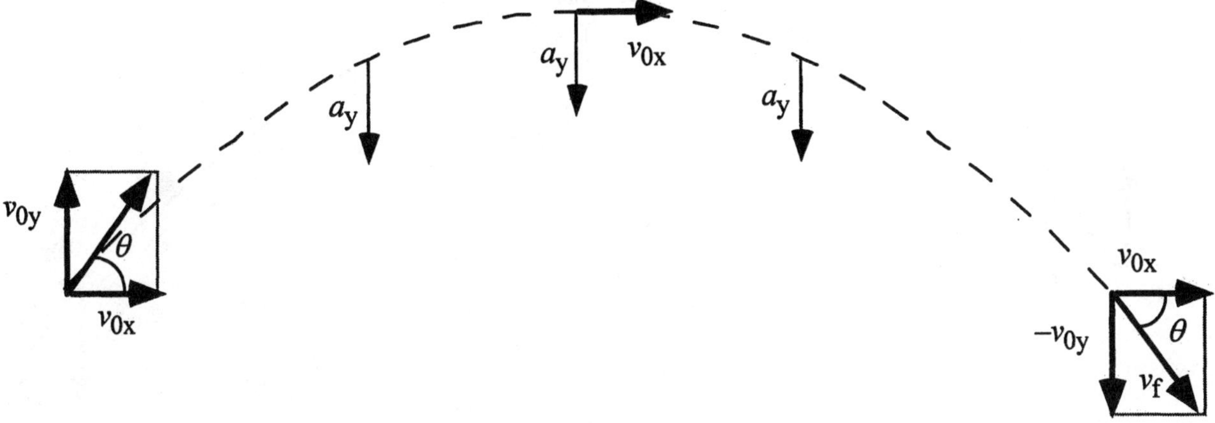

b. In order for the velocity and acceleration to be parallel, the x component of the velocity would have to drop to zero. However, v_x always remains equal to its initial value; therefore, the velocity and the acceleration can never be parallel.

---

3. ***REASONING AND SOLUTION*** As long as air resistance is negligible, the acceleration of a projectile is constant and equal to the acceleration due to gravity. The acceleration of the projectile, therefore, is the same at every point in its trajectory, and can *never* be zero.

---

4. ***REASONING AND SOLUTION*** If a baseball were pitched on the moon, it would still fall downwards as it travels toward the batter. However the acceleration due to gravity on the moon is roughly 6 times less than that on earth. Thus, in the time it takes to reach the batter, the ball will not fall as far vertically on the moon as it does on earth. Therefore, the pitcher's mound on the moon would be at a lower height than it is on earth.

---

5. ***REASONING AND SOLUTION*** The figure below shows the ball's trajectory. The velocity of the ball (along with its x and y components) are indicated at three positions. As long as air resistance is neglected, we know that $a_x = 0$ and a_y is the acceleration due to gravity. Since $a_x = 0$, the x component of the velocity remains the same and is given by v_{0x}. The initial y component of the velocity is v_{0y} and decreases as the ball approaches the highest point, where $v_y = 0$. The magnitude of the y component of the velocity then increases as the ball falls downward. Just before the ball strikes the ground, the y velocity component is equal in magnitude to v_{0y}. The speed is the magnitude of the velocity.

a. Since $v_y = 0$ when the ball is at the highest point in the trajectory, the speed is a minimum there.

b. Similarly, since v_y is a maximum at the initial and final positions of the motion, the speed is a maximum at these positions.

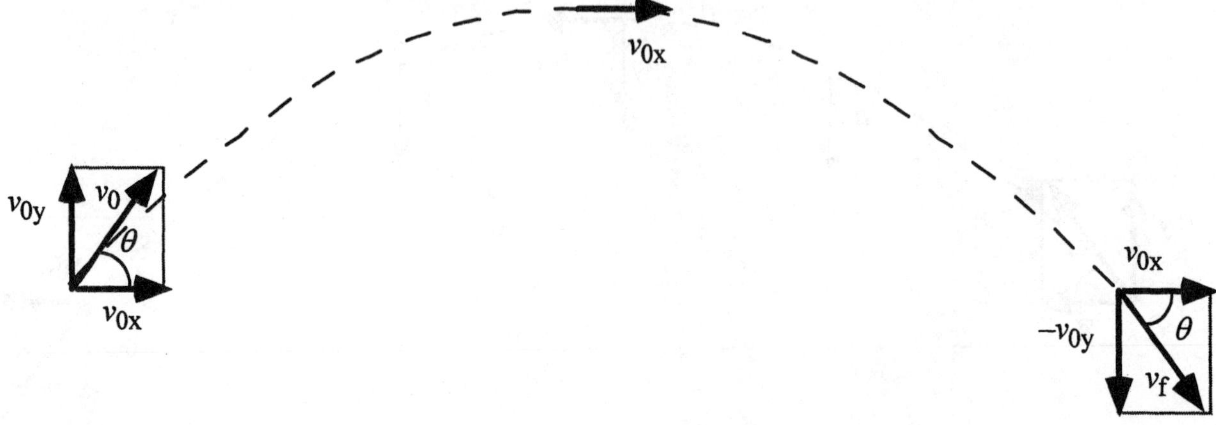

---

6. ***REASONING AND SOLUTION*** If there were a wind blowing parallel to the ground and toward the kicker in Figure 3.12, the acceleration component in the horizontal direction would not be zero. The horizontal component of the velocity would decrease as time goes on. The flight time of the ball, however, depends only on the vertical component of the initial velocity. Thus, the flight time of the ball would not be affected by the wind. The horizontal range of the football will be shortened, but the flight time will remain the same.

---

7. ***REASONING AND SOLUTION*** The wrench will hit at the same place on the deck of the ship regardless of whether the sailboat is at rest or moving with a constant velocity. If the sailboat is at rest, the wrench will fall straight down hitting the deck at some point *P*. If the sailboat is moving with a constant velocity, the motion of the wrench will be two dimensional. However, the horizontal component of the velocity of the wrench will be the same as the velocity of the sailboat. Therefore, the wrench will always remain above the same point *P* as it is falling. This situation is analogous to the falling care package which is described in Example 2 of the text.

---

8. ***REASONING AND SOLUTION*** The two bullets differ only in their horizontal motion. One bullet has $v_x = 0$, while the other bullet has $v_x = v_{0x}$. The time of flight, however, is determined only by the vertical motion, and both bullets have the same initial vertical velocity component ($v_{0y} = 0$). Both bullets, therefore, reach the ground at the same time.

---

9. ***REASONING AND SOLUTION*** Since the launch speed of projectile A is twice that of B, it follows that

$$(v_{0x})_{\text{A}} = 2(v_{0x})_{\text{B}} \quad \text{and} \quad (v_{0y})_{\text{A}} = 2(v_{0y})_{\text{B}}$$

As seen from the result of Example 5, the maximum height attained by either projectile is directly proportional to the square of v_{0y}^2; therefore, the ratio of the maximum heights is $H_{\text{A}} / H_{\text{B}} = (2)^2 = 4$. Example 7 in the text shows that the range of either projectile is directly proportional to the product of v_{0x} and *t*. Example 6 shows that *t* is proportional to v_{0y}; thus,

$$\frac{R_{\text{A}}}{R_{\text{B}}} = \frac{(v_{0x})_{\text{A}} t_{\text{A}}}{(v_{0x})_{\text{B}} t_{\text{B}}} = \frac{(v_{0x})_{\text{A}} (v_{0y})_{\text{A}}}{(v_{0x})_{\text{B}} (v_{0y})_{\text{B}}} = \frac{\left[2(v_{0x})_{\text{B}}\right]\left[2(v_{0y})_{\text{B}}\right]}{(v_{0x})_{\text{B}} (v_{0y})_{\text{B}}} = 4$$

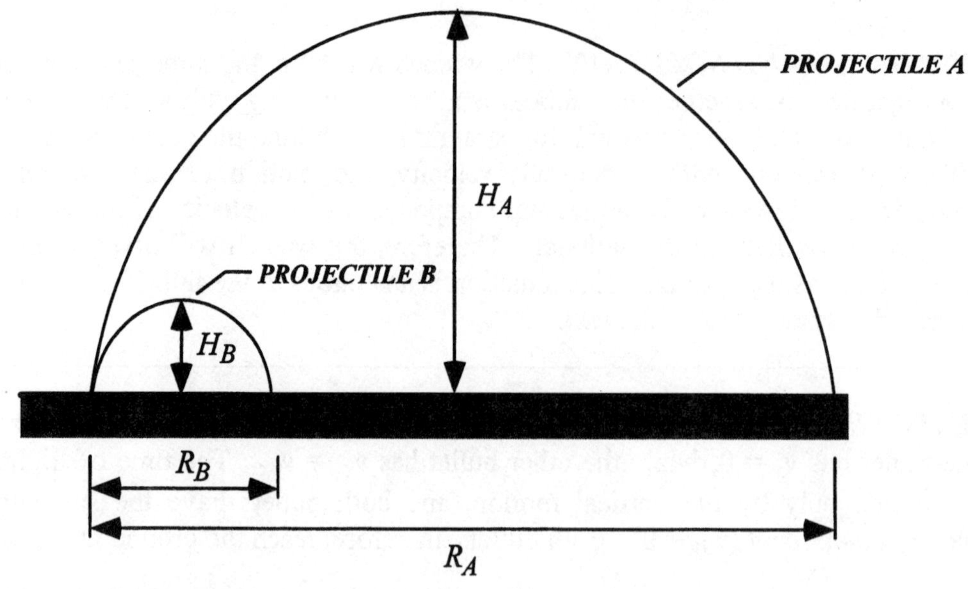

---

10. ***REASONING AND SOLUTION***

a. The displacement is greater for the stone that is thrown horizontally, because it has the same vertical component as the dropped stone and, in addition, has a horizontal component.

b. The impact speed is greater for the stone that is thrown horizontally. The reason is that it has the same vertical velocity component as the dropped stone but, in addition, also has a horizontal component that equals the throwing velocity.

c. The time of flight is the same in each case, because the vertical part of the motion for each stone is the same. That is, each stone has an initial vertical velocity component of zero and falls through the same height.

---

11. ***REASONING AND SOLUTION*** A leopard springs upward at a 45° angle and then falls back to the ground. If we neglect air resistance, the leopard will follow the usual parabolic trajectory of projectile motion. If we let v_0 denote the initial speed of the leopard, then, at any time *t*, the *x* and *y* components of the velocity of the leopard are given by

$$v_x = v_0 \cos 45° = \tfrac{\sqrt{2}}{2} v_0$$

$$v_y = v_0 \sin 45° + a_y t = \tfrac{\sqrt{2}}{2} v_0 + a_y t$$

The speed at any time *t* is then given by

$$v = \sqrt{v_x^2 + v_y^2}$$

As the leopard rises, its speed decreases reaching a minimum at the highest point in the trajectory. Then, as the leopard falls back to the ground, its speed increases. To determine the minimum speed of the leopard, we note that when the leopard is at the highest point in its trajectory, $v_y = 0$. Therefore, the minimum speed of the leopard is

$$v = \sqrt{v_x^2 + v_y^2} = \sqrt{\left(\tfrac{\sqrt{2}}{2}v_0\right)^2 + 0} = 0.707v_0$$

In other words, as the leopard moves upward, its speed decreases from its initial value of v_0 to its minimum value of 0.707 v_0 at the leopard's highest point. We can conclude, therefore, that the leopard never has a speed that is one-half of its initial value.

---

12. ***REASONING AND SOLUTION*** A football quarterback throws a pass on the run and then keeps running without changing his velocity. If the quarterback throws the ball *straight upward*, and keeps running without changing his velocity, the quarterback can catch the ball. The initial horizontal velocity component of the ball will be equal to the (horizontal) velocity of the quarterback. If air resistance is neglected, the ball is not accelerated in the horizontal direction; therefore, the ball retains its initial horizontal velocity component. As a result, the ball remains directly above the quarterback at all times during the flight. Thus, the ball will fall into the hands of the quarterback. This situation is analogous to the bullet in Conceptual Example 4, which remains directly above the gun barrel during its flight.

---

13. ***REASONING AND SOLUTION*** Since the plastic bottle moves with the current, the passenger is estimating the velocity of the boat relative to the water. Therefore, the passenger cannot conclude that the boat is moving at 5 m/s with respect to the shore.

---

14. ***REASONING AND SOLUTION*** The plane flies at the same speed with respect to the ground during the entire flight. Therefore, even though the earth is rotating, the speed of the plane relative to the earth is identical in both cases, the distances are the same, and, therefore, the two flight times are identical.

---

15. ***REASONING AND SOLUTION***

a. Since the child is at rest with respect to the floor of the RV and with respect to the cannon, there is no relative motion between the child, the floor of the RV and the cannon. Therefore, the range of the marble toward the front is the same as the range toward the rear from the point of view of the child.

b. The velocity of the marble relative to an observer standing still on the ground is $\mathbf{v}_{\text{MG}} = \mathbf{v}_{\text{MV}} + \mathbf{v}_{\text{VG}}$, where $\mathbf{v}_{\text{MV}}$ is the velocity of the marble relative to the vehicle and $\mathbf{v}_{\text{VG}}$ is the velocity of the vehicle relative to the ground. Suppose that the RV moves in the positive direction; then the quantity $\mathbf{v}_{\text{VG}}$ is positive. When the marble is shot toward the front of the RV, the x component of $\mathbf{v}_{\text{MV}}$ will also be a positive quantity. Alternatively, when the marble is shot toward the back of the RV, the x component of $\mathbf{v}_{\text{MV}}$ is a negative quantity. Thus, the magnitude of the x component of $\mathbf{v}_{\text{MG}}$ is smaller when the marble is shot toward the rear than when it is shot toward the front. The time that the marble is in the air is determined by the vertical motion which is the same in both cases. Since the magnitude of the x component of $\mathbf{v}_{MG}$ is smaller when the marble is shot toward the rear, it will cover less distance in the time it takes to hit the floor of the RV compared to when the marble is shot toward the front. Therefore, from the point of view of an observer on the ground, the range of the marble is greater when the marble is shot toward the front.

---

16. ***REASONING AND SOLUTION*** The time required for any given swimmer to cross the river is equal to the width of the river divided by the magnitude of the component of the velocity that is parallel to the width of the river. All three swimmers can swim equally fast relative to the water; however, all three swim at different angles relative to the current. Since swimmer A heads straight across the width of the river, swimmer A will have the largest velocity component parallel to the width of the river; therefore, swimmer A crosses the river in the least time.

CHAPTER 4 | *FORCES AND NEWTON'S LAWS OF MOTION*

CONCEPTUAL QUESTIONS

---

1. ***REASONING AND SOLUTION*** When the car comes to a sudden halt, the upper part of the body continues forward (as predicted by Newton's first law) if the force exerted by the lower back muscles is not great enough to give the upper body the same deceleration as the car. The lower portion of the body is held in place by the force of friction exerted by the car seat and the floor.

   When the car rapidly accelerates, the upper part of the body tries to remain at a constant velocity (again as predicted by Newton's first law). If the force provided by the lower back muscles is not great enough to give the upper body the same acceleration as the car, the upper body appears to be pressed backward against the seat as the car moves forward.

---

2. ***REASONING AND SOLUTION*** When the birdfeeder is hanging freely and no one is pulling on the dangling (lower) cord, there is a tension in the cord between the birdfeeder and the tree limb (the upper cord), because the upper cord supports the weight of the birdfeeder. When the lower cord is pulled down with a slow continuous pull, the tension in both cords increases slowly. Since the upper cord has a larger tension to begin with, it always has the greater tension as the lower cord is pulled. Thus, the upper cord snaps first.

   On the other hand, when the child gives the lower cord a sudden, downward pull, the tension in the lower cord increases suddenly. However, the tension in the upper cord does not increase as suddenly. The reason is that the birdfeeder has a large mass, so it accelerates very slowly. Thus, the upper cord is stretched slowly and, consequently, the tension in the upper cord rises slowly. Since the tension rises much faster in the lower cord, it is the first to snap.

---

3. ***REASONING AND SOLUTION*** If the net external force acting on an object is zero, it is possible for the object to be traveling with a nonzero velocity. According to Newton's second law, $\Sigma \mathbf{F} = m\mathbf{a}$, if the net external force $\Sigma \mathbf{F}$ is zero, the acceleration $\mathbf{a}$ is also zero. If the acceleration is zero, the velocity must be constant, both in magnitude and in direction. Thus, an object can move with a constant nonzero velocity when the net external force is zero.

---

4. ***REASONING AND SOLUTION*** According to Newton's second law, a net force is required to give an object a non-zero acceleration.

   a. If an object is moving with a constant acceleration of 9.80 m/s^2, we can conclude that there is a net force on the object.

b. If an object moves with a constant velocity of 9.80 m/s, its acceleration is zero; therefore, we can conclude that the net force acting on the object is zero.

---

5. ***REASONING AND SOLUTION*** An object will not necessarily accelerate when two or more forces are applied to the object simultaneously. The applied forces may cancel so the net force is zero; in such a case, the object will not accelerate. The resultant of all the forces that act on the object must be nonzero in order for the object to accelerate.

---

6. ***REASONING AND SOLUTION***
Since the father and the daughter are standing on ice skates, there is virtually no friction between their bodies and the ground. We can assume, therefore, that the only horizontal force that acts on the daughter is due to the father, and similarly, the only horizontal force that acts on the father is due to the daughter.

a. According to Newton's third law, when they push off against each other, the force exerted on the father by the daughter must be equal in magnitude and opposite in direction to the force exerted on the daughter by the father. In other words, both the father and the daughter experience pushing forces of equal magnitude.

b. According to Newton's second law, $\Sigma\mathbf{F} = m\mathbf{a}$. Therefore, $\mathbf{a} = \Sigma\mathbf{F} / m$. The magnitude of the net force on the father is the same as the magnitude of the net force on the daughter, so we can conclude that, since the daughter has the smaller mass, she will acquire the larger acceleration.

---

7. ***REASONING AND SOLUTION***

a. The force of the gymnast on the trampoline causes the elastic surface of the trampoline to deform.

b. The reaction force exerted by the trampoline on the gymnast causes the gymnast to decelerate and come to a momentary stop.

---

8. ***REASONING AND SOLUTION*** The two oppositely directed pushing forces predicted by Newton's third law act on different bodies; one force acts on the crate, while the other acts on the person who is pushing on the crate. Since the two forces act on different objects, they cannot cancel each other. Whether or not the crate moves depends on the net force that acts on the crate. If the crate does not move under the action of a single pushing force, the only reasonable conclusion is that there must be another force *acting on the crate* that cancels the pushing force. The other force is the force of static friction.

---

9. ***REASONING AND SOLUTION*** The magnitude of the gravitational force between any two of the particles is given by Newton's law of universal gravitation: $F = Gm_1 m_2 / r^2$ where m_1 and m_2 are the masses of the particles and r is the distance between them. Since the particles have equal masses, we can arrange the particles so that each one experiences a net gravitational force that has the same magnitude if we arrange the particles so that the distance between any two of the particles is the same. Therefore, the particles should be placed at the corners of an equilateral triangle with all three sides of equal length.

---

10. ***REASONING AND SOLUTION*** The mass of an object is a quantitative measure of its inertia. The mass of an object is an intrinsic property of the object and is independent of the location of the object. The weight of an object is the gravitational force exerted on the object by the earth. The gravitational force depends on the distance between the object and the center of the earth. Therefore, when an object is moved from sea level to the top of a mountain, its weight will change, while the mass of the object remains constant.

---

11. ***REASONING AND SOLUTION*** The weight of the ball always acts downward. The force of air resistance will always act in the direction that is opposite to the direction of motion of the ball. The net force on the ball is the resultant of the weight and the force of air resistance.

a. As the ball moves upward, the force of air resistance acts downward. Since air resistance and the weight of the ball act in the same direction in this case, the net force on the ball will be greater in magnitude than the weight of the ball.

b. As the ball falls downward, the force of air resistance is upward. Since air resistance and the weight of the ball act in opposite directions, the net force that acts on the ball will be smaller in magnitude than the weight of the ball.

Note that in both cases, the net force points downward since the object accelerates downward in both cases.

---

12. ***REASONING AND SOLUTION*** From Equation 4.5, $W = mg$, we know that the weight of an object is directly proportional to its mass. The proportionality constant on a given planet is g, the magnitude of the acceleration due to gravity on that planet.

If object A weighs twice as much as object B at the same spot on the earth, then the mass of object A is twice as much as the mass of object B. The mass of any object is independent of the object's location in the universe. Since the weight of an object is directly proportional to its mass, object A will weigh twice as much as object B at the same spot on any planet. While the values of the weights will differ from planet to planet, as the value of g varies from planet to planet, the ratio of the weights (which equals the ratio of their masses) will be the same on any planet.

---

13. ***REASONING AND SOLUTION*** Equations 4.4 and 4.5 may be combined to give the acceleration due to gravity at a distance r from the center of the earth: $g = GM_E / r^2$. Since this expression depends on r, the acceleration of a freely falling object does depend on its location. The acceleration due to gravity will be greater at Death Valley, California (where r is smaller) than it is on the top of Mt. Everest (where r is greater). Since r is measured from the center of the earth, however, these values will be very close.

---

14. ***REASONING AND SOLUTION*** Assuming that the accelerating mechanism remains attached to the rocket, the acceleration will be greater when the rocket is fired horizontally. The accelerating mechanism provides an acceleration that points in the initial direction of motion of the rocket. The net acceleration is the resultant of the accelerating mechanism and the acceleration due to gravity. When the rocket is fired horizontally, these accelerations will be at right angles to each other. When the rocket is fired straight up, these accelerations will be in opposite directions. The magnitude of the resultant will be greater when these two accelerations are at right angles rather than when they are in opposition. Therefore, the acceleration will be greater when the rocket is fired horizontally.

If we assume that the accelerating mechanism is not attached to the rocket, then once the rocket is fired, the only force on the rocket is that due to gravity, and the rocket has the acceleration due to gravity regardless of its orientation. In this case, the acceleration of the rocket is the same regardless of whether it is fired straight up or fired horizontally.

---

15. ***REASONING AND SOLUTION*** If the elevator were at rest, or moving with a constant velocity, the scale would read the true weight of $mg = 98$ N. When the elevator is accelerating, the scale reading will differ from 98 N and will display the apparent weight, F_N, which is given by Equation 4.6: $F_N = mg + ma$ where a, the acceleration of the elevator, is positive when the elevator accelerates upward and negative when the elevator accelerates downward.

a. When the apparent weight is $F_N = 75$ N, the apparent weight is less than the true weight ($mg = 98$ N) so a must be negative. The elevator is accelerating downward.

b. When the apparent weight is $F_N = 120$ N, the apparent weight is greater than the true weight ($mg = 98$ N) so a must be positive. The elevator is accelerating upward.

---

16. ***REASONING AND SOLUTION*** The apparent weight will differ from the true weight only in an accelerating elevator. When the scale in an elevator reads the true weight, the only conclusion that can be made is that the elevator has zero acceleration. Therefore, one cannot conclude whether the elevator is moving with any constant velocity upward, any constant velocity downward, or whether the elevator is at rest, since each of these conditions involves zero acceleration.

---

17. ***REASONING AND SOLUTION*** If the elevator were at rest, or moving with a constant velocity, the scale will read the true weight mg. When the elevator is accelerating, the scale

reading will differ from the true weight and will register the apparent weight, F_N, which is given by Equation 4.6: $F_N = mg + ma$ where a, the acceleration of the elevator, is positive when the elevator accelerates upward and negative when the elevator accelerates downward.

Since the scale registers 600 N when the elevator is moving with constant velocity, we know that the true weight is 600 N.

a. The elevator is moving upward. When the elevator slows down, its acceleration vector points downward. The term ma will be negative; therefore, the scale reading will be less than 600 N.

b. When the elevator is stopped, the scale will register the true weight; therefore, the scale reads 600 N.

c. The elevator is moving downward and speeding up. The acceleration vector points downward, and the term ma is negative. Therefore, the scale registers a value that is less than 600 N.

---

18. ***REASONING AND SOLUTION*** Since the sled moves with constant velocity, the force of kinetic friction is present. The magnitude of this force is given by $\mu_k F_N$, where μ_k is the coefficient of kinetic friction and F_N is the magnitude of the normal force that acts on the sled. Furthermore, the horizontal component of the applied force must be equal in magnitude to the force of kinetic friction, since there is no acceleration.

When the person pulls on the sled, the vertical component of the pulling force tends to decrease the magnitude of the normal force relative to that when the sled is not being pulled or pushed. On the other hand, when the person pushes on the sled, the vertical component of the pushing force tends to increase the normal force relative to that when the sled is not being pulled or pushed. Therefore, when the sled is pulled, the magnitude of the force of kinetic friction, and therefore the magnitude of the applied force, is less than when the sled is pushed.

---

19. ***REASONING AND SOLUTION*** We know that $\mu_s = 2.0\,\mu_k$ for a crate in contact with a cement floor. The maximum force of static friction is $f_s^{\text{MAX}} = \mu_s F_N$ while the force of kinetic friction is $f_k = \mu_k F_N$. As long as the crate is on the cement floor, we can conclude that the magnitude of the maximum static frictional force acting on the crate will always be twice the magnitude of the kinetic frictional force on the moving crate, once the crate has begun moving. However, the force of static friction may not have its maximum value. Thus, the magnitude of the static frictional force is not always twice the magnitude of the kinetic frictional force.

---

20. ***REASONING AND SOLUTION*** A box rests on the floor of a stationary elevator. Because of static friction, a force is required to start the box sliding across the floor of the

elevator. The magnitude of this force is given by $f_s^{MAX} = \mu_s F_N$, where F_N is the magnitude of the normal force exerted on the box by the floor of the elevator. When the elevator is stationary, the magnitude of the normal force exerted on the box is mg where m is the mass of the crate and g is the magnitude of the acceleration due to gravity. When the crate accelerates upward, the floor of the elevator will push against the box to accelerate it upward; therefore, when the crate accelerates upward, the magnitude of the normal force will be greater than mg. When the elevator accelerates downward, the normal force between the box and the floor of the elevator will be less than when the elevator is stationary; therefore, the magnitude of the normal force will be less than mg. Thus, the magnitude of the force required to start the box sliding across the floor of the elevator are ranked as follows in ascending order: (c) elevator accelerating downward; (a) elevator stationary; (b) elevator accelerating upward.

---

21. ***REASONING AND SOLUTION*** When the rope is tied to a tree and pulled by the ten people, the tension in the rope is twice as great as it was when it was used in a tug-of-war with five people on each team. Therefore, the rope is more likely to break when it is tied to the tree and pulled by ten people.

Note that when the rope is tied to the tree and pulled by ten people, the situation is equivalent to a tug-of-war with *ten* people on each team.

---

22. ***REASONING AND SOLUTION*** An object is in equilibrium when its acceleration is zero. When a stone is thrown from the top of a cliff, its acceleration is the acceleration due to gravity; therefore, the stone is not in equilibrium.

---

23. ***REASONING AND SOLUTION*** An object is in equilibrium when its acceleration is zero.

a. If a single nonzero force acts on an object, the object will accelerate according to Newton's second law. The object is not in equilibrium.

b. If two forces that point in mutually perpendicular directions act on an object, the object will experience a net force. By Newton's second law, the object will, therefore, have a nonzero acceleration. The object is not in equilibrium.

c. If two forces that point in directions that are not perpendicular act on the object, the object may or may not be in equilibrium, depending on how the forces are oriented. In general, the resultant of two such forces is nonzero, the object will accelerate, and it is not in equilibrium. In the special case where the two forces point in opposite directions and have the same magnitude, the net force is zero, the object has zero acceleration, and the object is, therefore, in equilibrium.

---

24. ***REASONING AND SOLUTION*** A circus performer hangs from a stationary rope. Since there is no acceleration, the tension in the rope must be equal in magnitude to the weight of

the performer. She then begins to climb upward by pulling herself up, hand-over-hand. Whether the tension in the rope is greater than or equal to the tension when she hangs stationary depends on whether or not she accelerates as she moves upward. When she moves upward at constant velocity, the tension in the rope will be the same. When she accelerates upward, the rope must support the net upward force in addition to her weight; therefore, in this case, the tension in the rope will be greater than when she hangs stationary.

---

25. ***REASONING AND SOLUTION*** If a sky diver with an open parachute approaches the ground with a constant velocity, the acceleration of the sky diver is zero, and the sky diver is, therefore, in equilibrium. The two forces responsible for the equilibrium are the weight of the sky diver and the force exerted on the sky diver by the parachute. These forces must be equal in magnitude and opposite in direction so that their resultant is zero.

---

26. ***REASONING AND SOLUTION*** There are three forces that act on the ring as shown in the figure below. The weight of the block, which acts downward, and two forces of tension that act along the rope away from the ring. Since the ring is at rest, the net force on the ring is zero. The weight of the block is balanced by the vertical components of the tension in the rope. Clearly, the rope can never be made horizontal, for then there would be no vertical components of the tension forces to balance the weight of the block.

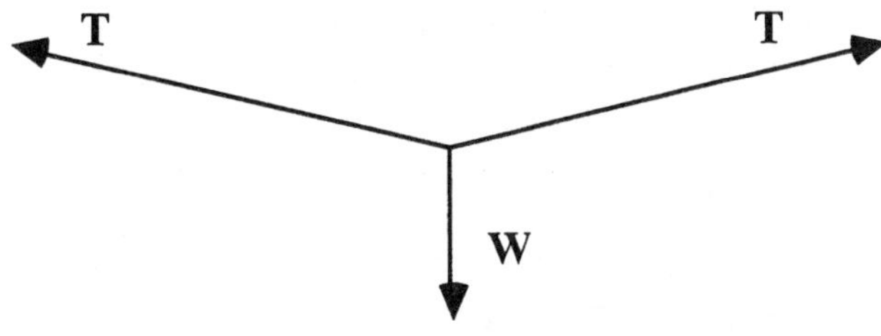

---

27. ***REASONING AND SOLUTION*** A freight train is accelerating on a level track. The tension in the coupling between the engine and the first car depends on the total mass being pulled by the engine. Therefore, other things being equal, if some of the cargo in the last car were transferred to any one of the other cars, the tension in the coupling between the engine and the first car would remain the same. This is because the transfer does not change the total mass being pulled by the engine.

CHAPTER 5 | *DYNAMICS OF UNIFORM CIRCULAR MOTION*

CONCEPTUAL QUESTIONS

---

1. ***REASONING AND SOLUTION*** The car will accelerate if its velocity changes in magnitude, in direction, or both. If a car is traveling at a constant speed of 35 m/s, it can be accelerating if its direction of motion is changing.

---

2. ***REASONING AND SOLUTION*** Consider two people, one on the earth's surface at the equator, and the other at the north pole. If we combine Equations 5.1 and 5.2, we see that the centripetal acceleration of an object moving in a circle of radius r with period T can be written as $a_c = (4\pi^2 r)/T^2$. The earth rotates about an axis that passes approximately through the north pole and is perpendicular to the plane of the equator. Since both people are moving on the earth's surface, they have the same period T. The person at the equator moves in a larger circle so that r is larger for the person at the equator. Therefore, the person at the equator has a larger centripetal acceleration than the person at the north pole.

---

3. ***REASONING AND SOLUTION*** The equations of kinematics (Equations 3.3 - 3.6) cannot be applied to uniform circular motion because an object in uniform circular motion does not have a constant acceleration. While the acceleration vector is constant in magnitude $\left(a = v^2/r\right)$, its direction changes constantly -- it always points toward the center of the circle. As the object moves around the circle the direction of the acceleration must constantly change. Because of this changing direction, the condition of constant acceleration that is required by Equations 3.3 – 3.6 is violated.

---

4. ***REASONING AND SOLUTION*** Acceleration is the rate of change of velocity. In order to have an acceleration, the velocity vector must change either in magnitude or direction, or both. Therefore, if the velocity of the object is constant, the acceleration must be zero. On the other hand, if the speed of the object is constant, the object could be accelerating if the direction of the velocity is changing.

---

5. ***REASONING AND SOLUTION*** When the car is moving at constant speed along the straight segments (i.e., *AB* and *DE*) , the acceleration is zero. Along the curved segments, the magnitude of the acceleration is given by v^2/r. Since the speed of the car is constant, the magnitude of the acceleration is largest where the radius r is smallest. Ranked from smallest to largest the magnitudes of the accelerations in each of the four sections are: *AB* or *DE, CD, BC*.

---

6. ***REASONING AND SOLUTION*** From Example 7, the maximum safe speed with which a car can round an unbanked horizontal curve of radius r is given by $v = \sqrt{\mu_s g r}$. Since the acceleration due to gravity on the moon is roughly one sixth that on earth, the safe speed for the same curve on the moon would be less than that on earth. In other words, other things being equal, it would be more difficult to drive at high speed around an unbanked curve on the moon as compared to driving around the same curve on the earth.

---

7. ***REASONING AND SOLUTION*** A bug lands on a windshield wiper. The wipers are turned on. Since the wipers move along the arc of a circle, the bug will experience a centripetal acceleration, and hence, a centripetal force must be present. The magnitude of the centripetal force is given by $F_c = mv^2 / r$. In order for the bug to remain at rest on the wiper blade, the force of static friction between the bug and the wiper blade must contribute in a major way to the centripetal force. Without the centripetal force, the bug will be dislodged. When the wipers are turned on at a higher setting, v is larger, and the centripetal force required to keep the bug moving along the arc of the circle is larger than if the wipers are turned on the low setting. Since the high setting requires a larger centripetal force to keep the bug on the wiper, the bug is more likely to be dislodged at that setting than at the low setting.

---

8. ***REASONING AND SOLUTION*** From Example 7, the maximum safe speed with which a car can round an unbanked curve of radius r is given by $v = \sqrt{\mu_s g r}$. This expression is independent of the mass (and therefore the weight) of the car. Thus, the chance of a light car safely rounding an unbanked curve on an icy road is the same as that for a heavier car (assuming that all other factors are the same).

---

9. ***REASONING AND SOLUTION*** Since the speed and radius of the circle are constant, the centripetal acceleration is constant. As the water leaks out, however, the mass of the object undergoing the uniform circular motion decreases. Centripetal force is mass times the centripetal acceleration, so that the centripetal force applied to the container must be decreasing. It is the tension in the rope that provides the centripetal force. You are holding the free end of the rope and pulling on it in order to create the tension. Therefore, you must be reducing your pull as the water leaks out. In turn, according to Newton's third law, the rope must be pulling back on your hand with a force of decreasing magnitude, and you feel this pull weakening as time passes.

---

10. ***REASONING AND SOLUTION*** As the propeller rotates faster, the centripetal acceleration of the parts of the propeller increase. As the centripetal acceleration increases, the centripetal force required to cause the various parts of the propeller to rotate in the circle also increases. When the necessary centripetal force exceeds the mechanical forces that hold the propeller together, the propeller will come apart.

---

11. ***REASONING AND SOLUTION*** The centripetal force on the penny is given by $F_c = mv^2 / r$, where $v = 2\pi r / T$ and T is the constant period of the turntable. Therefore, the centripetal force on the penny is given by $F_c = 4\pi^2 mr / T^2$. Clearly, the penny will require the largest centripetal force to remain in place when located at the largest value of r; that is, at the edge of the turntable.

---

12. ***REASONING AND SOLUTION*** A model airplane on a guideline can fly in a circle because the tension in the guideline provides the horizontal centripetal force necessary to pull the plane into a horizontal circle. A real airplane has no such horizontal forces. The air on the wings on a real plane exerts an upward lifting force that is perpendicular to the wings. The plane must bank so that a component of the lifting force can be oriented horizontally, thereby providing the required centripetal force to cause the plane to fly in a circle.

---

13. ***REASONING AND SOLUTION***

a. Referring to Figure 5.10 in the text, we can see that the centripetal force on the plane is $L \sin \theta = mv^2 / r$, where L is the magnitude of the lifting force. In addition, the vertical component of the lifting force must balance the weight of the plane, so that $L \cos \theta = mg$. Dividing these two equations reveals that $\tan \theta = v^2 / (rg)$.

b. The banking condition for a car traveling at speed v around a curve of radius r, banked at angle θ is $\tan \theta = v^2 / (rg)$, according to Equation 5.4 in the text.

c. The speed v of a satellite in a circular orbit of radius r about the earth is given by $v = \sqrt{GM_E / r}$, according to Equation 5.5 in the text.

d. The minimum speed required for a loop-the-loop trick around a loop of radius r is $v = \sqrt{rg}$, according to the discussion in Section 5.7 of the text. According to Equations 4.4 and 4.5, $g = GM_E / r^2$. Thus, any expression that depends on g also depends on M_E and would be affected by a change in the earth's mass. Such is the case for each of the four situations discussed above.

---

14. ***REASONING AND SOLUTION*** When the string is whirled in a horizontal circle, the tension in the string, F_T, provides the centripetal force which causes the stone to move in a circle. Since the speed of the stone is constant, $mv^2 / r = F_T$ and the tension in the string is constant.

When the string is whirled in a vertical circle, the tension in the string and the weight of the stone both contribute to the centripetal force, depending on where the stone is on the circle. Now, however, the tension increases and decreases as the stone traverses the vertical circle. When the stone is at the lowest point in its swing, the tension in the string pulls the stone upward, while the weight of the stone acts downward. Therefore, the centripetal force is $mv^2/r = F_T - mg$. Solving for the tension shows that $F_T = mv^2/r + mg$. This tension is larger than in the horizontal case. Therefore, the string has a greater chance of breaking when the stone is whirled in a vertical circle.

---

15. ***REASONING AND SOLUTION*** A fighter pilot pulls out of a dive on a vertical circle and begins to climb upward. As the pilot moves along the circle, all parts of his body, including the blood in his head, must experience a centripetal force in order to remain on the circle. The blood, however, is not rigidly attached to the body and does not experience the requisite centripetal force until it flows out of the head, away from the circle's center, and collects in the lower body parts, which ultimately push on it enough to keep it on the circular path.

---

16. ***REASONING AND SOLUTION*** When a car moves around an unbanked horizontal curve, the centripetal force that keeps the car on the road so that it can negotiate the curve comes from the force of static friction. If car A cannot negotiate the curve, then the force of static friction between the treads of car A's tires and the road is not great enough to provide the centripetal force. The coefficient of static friction between the tires and the road are less for car A than for car B, since car B can negotiate the turn.

CHAPTER 6 | *WORK AND ENERGY*

CONCEPTUAL QUESTIONS

---

1. ***REASONING AND SOLUTION*** The work done by $\mathbf{F}_1$ in moving the box through a displacement **s** is $W_1 = (F_1 \cos 0°)s = F_1 s$. The work done by $\mathbf{F}_2$ is $W_2 = (F_2 \cos \theta)s$. From the drawing, we see that $F_1 > F_2 \cos \theta$; therefore, the force $\mathbf{F}_1$ does more work.

---

2. ***REASONING AND SOLUTION*** The force **P** acts along the displacement; therefore, it does positive work. Both the normal force $\mathbf{F}_N$ and the weight *m***g** are perpendicular to the displacement; therefore, they do zero work. The kinetic frictional force $\mathbf{f}_k$ acts opposite to the direction of the displacement; therefore, it does negative work.

---

3. ***REASONING AND SOLUTION*** Work can be positive or negative. Work is positive when the force has a component in the same direction as the direction of the displacement, or equivalently, when the angle θ between the force **F** and the displacement **s** is less than 90°. The work is negative when the force has a component in the direction *opposite* to the displacement; that is, when θ is greater than 90°.

Since the force does positive work on a particle that has a displacement pointing in the $+x$ direction, the force must have an x component that points in the $+x$ direction. Furthermore, since the same force does negative work on a particle that has a displacement pointing in the $+y$ direction, the force has a y component that points in the *negative* y direction.

When the x and y components of this force are sketched head to tail, as in the figure at the right, we see that the force must point in the fourth quadrant.

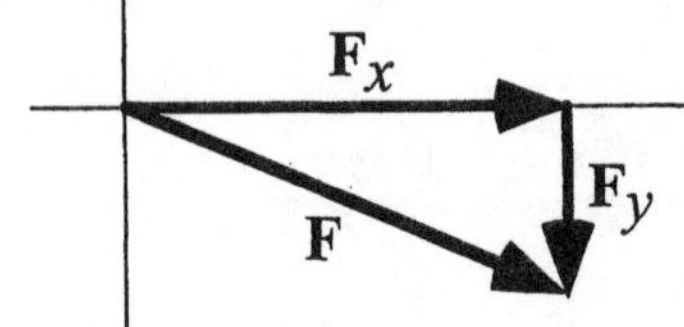

---

4. ***REASONING AND SOLUTION*** The sailboat moves at constant velocity, and, therefore, has zero acceleration. From Newton's second law, we know that the net external force on the sailboat must be zero.

a. There is no work done on the sailboat by a zero net external force.

b. Work is done by the individual forces that act on the boat; namely the wind that propels the boat forward and the water that resists the motion of the boat. Since the wind propels the boat forward, it does positive work on the boat. Since the force of the water is a resistive force, it acts opposite to the displacement of the boat, and, therefore, it does negative work. Since the total work done on the boat is zero, each force must do an equal amount of work with one quantity being positive, and the other being negative.

Note: The answer to part (a) could have been deduced from the work-energy theorem as well. Since the velocity of the boat is constant, the kinetic energy of the boat does not change and the total work done on the boat is zero.

---

5. ***REASONING AND SOLUTION*** The speed of the ball decreases; therefore, the ball is subjected to an external resistive force. A resistive force always points opposite to the direction of the displacement of the ball. Therefore, the external force does negative work.

   Using the work-energy theorem, we see that the change in the kinetic energy of the ball is negative; therefore, the total work done on the ball is negative. We can conclude, therefore, that the net force did negative work on the ball.

---

6. ***REASONING AND SOLUTION*** The kinetic energy of an object of mass m moving with speed v is given by $\mathrm{KE} = \frac{1}{2}mv^2$. The kinetic energy depends on both the mass and the speed of the object. The mass of an automobile is significantly greater than the mass of a motorcycle. Therefore, even if an automobile is moving slowly, it is possible that the product $\frac{1}{2}mv^2$ is greater for the car than it is for the faster-moving motorcycle.

---

7. ***REASONING AND SOLUTION*** A net external force acts on a particle. This net force is not zero. From Newton's second law, we can conclude that the net external force causes the particle to accelerate. Since the particle experiences an acceleration, its velocity must change. The change in velocity, however, may occur as a change in magnitude only, a change in direction only, or a change in both magnitude and direction.

   a. This is sufficient information to conclude that the velocity of the particle changes; however, there is not sufficient information to determine exactly how the velocity changes.

   b. There is not sufficient information to determine if the kinetic energy of the particle changes. In terms of the work-energy theorem, the kinetic energy will change if the net external force does work on the particle. But without knowing the direction of the net external force with respect to the particle's displacement, we cannot know if work is done.

   c. There is not sufficient information to determine if the speed of the particle changes. Kinetic energy is $\frac{1}{2}mv^2$, so the speed v will change if the kinetic energy changes. But, as explained in part (b), there is insufficient information to determine whether the kinetic energy changes.

---

8. ***REASONING AND SOLUTION*** The speed of a particle doubles and then doubles again, because a net external force acts on it. Let m and v_0 represent the mass and initial speed of the particle, respectively. During the first doubling, the change in the kinetic energy of the particle is

$$\text{KE}_\text{f} - \text{KE}_0 = \tfrac{1}{2}m(2v_0)^2 - \tfrac{1}{2}mv_0^2 = \tfrac{3}{2}mv_0^2$$

During the second doubling, the change in the kinetic energy of the particle is

$$\text{KE}_\text{f} - \text{KE}_0 = \tfrac{1}{2}m(4v_0)^2 - \tfrac{1}{2}m(2v_0)^2 = 6mv_0^2$$

From the work-energy theorem, we know that a change in kinetic energy is equal to the work done by the net external force. Therefore, more work is done by the net force during the second doubling.

---

9. ***REASONING AND SOLUTION*** A shopping bag hangs straight down from the hand of a person who walks across a horizontal floor at constant velocity.

a. The force exerted on the shopping bag by the hand is perpendicular to the direction of motion; hence, the work done by the hand is zero.

b. If the person then rides up an escalator at constant velocity, a component of the force exerted by the hand will point in the direction of the displacement and do a positive amount of work.

---

10. ***REASONING AND SOLUTION*** The change in the gravitational potential energy of the astronaut is given by $\Delta\text{PE} = mgH$, where H is the height of the ladder. The value of g is greater on earth than it is on the moon; therefore, the gravitational potential energy of the astronaut changes by a greater amount on earth.

---

11. ***REASONING AND SOLUTION*** A net external nonconservative force does positive work on a particle, and both its kinetic and potential energies change.

a. The change in the total mechanical energy of the particle is given by $\Delta E = \Delta\text{KE} + \Delta\text{PE}$. From Equation 6.7b, $W_\text{nc} = \Delta\text{KE} + \Delta\text{PE}$; in other words, the work done on the particle by the net external nonconservative force is equal to the change in the particle's total mechanical energy. Since the force does positive work on the particle, we can conclude from Equation 6.7b that the particle's total mechanical energy must increase.

b. We do not have sufficient information to reach any conclusions on the individual changes in the kinetic and potential energies. Either ΔKE or ΔPE could be positive or negative, while the sum $\Delta\text{KE} + \Delta\text{PE}$ is positive. If either the change in the kinetic energy or the change in the potential energy is negative, the only conclusion that can be reached is that the magnitude of the negative quantity must be smaller than the magnitude of the positive quantity.

---

12. ***REASONING AND SOLUTION*** If the total mechanical energy of an object is conserved, then the sum of the kinetic energy and the potential energy must be constant.

a. If the kinetic energy decreases, the gravitational potential energy must increase by the same amount that the kinetic energy decreases.

b. If the potential energy decreases, the kinetic energy must increase by the same amount that the potential energy decreases.

c. If the kinetic energy does not change, the potential energy cannot change either.

---

13. ***REASONING AND SOLUTION*** Yes, the speed of the Steel Dragon would be 3.0 m/s at the top of the second hill. As the roller coaster descends the first hill, its total mechanical energy remains constant at all times. Its gravitational potential energy at the top of the hill is converted entirely into kinetic energy at the bottom. When the roller coaster ascends the second hill, its total mechanical energy also remains constant, and is equal to what it had during the descent of the first hill. As the roller coaster moves up the hill, kinetic energy is converted into gravitational potential energy. By the time it reaches the top of the second (identical) hill, the kinetic energy gained in descending the hill has been converted back into gravitational potential energy. Therefore, its kinetic energy at the top of the second hill is the same as it was at the top of the first hill, and its speed is 3.0 m/s at the top.

---

14. ***REASONING AND SOLUTION*** As the person moves downward from the top of the Ferris wheel, his displacement points downward. Since the person's weight also points downward, the work done by gravity is positive. As the person moves upward from the bottom, his displacement points upward. Since the weight still points downward, the work done by gravity is negative. The magnitude of the work done in each half cycle is the same; therefore, the net work done in one revolution is zero.

---

15. ***REASONING AND SOLUTION*** Car A turns off its engine and coasts up the hill. Car B keeps its engine running and drives up the hill at constant speed. If air resistance and friction are negligible, then only the motion of car A is an example of the principle of conservation of mechanical energy. During the motion of car A, there are no nonconservative forces present. As car A coasts up the hill it loses kinetic energy and gains an equal amount of gravitational potential energy. The motion of car B is not described by the principle of conservation of mechanical energy, because the driving force that propels car B up the hill is a nonconservative force.

---

16. ***REASONING AND SOLUTION*** A trapeze artist, starting from rest, swings downward on the bar, lets go at the bottom of the swing, and falls freely to the net. An assistant, standing on the same platform as the trapeze artist, jumps from rest straight downward.

a. The work done by gravity, on either person, is $W = mgh$, where m is the mass of the person, and h is the magnitude of the vertical component of the person's displacement. The value of h is the same for both the trapeze artist and the assistant; however, the value of m is, in general, different for the trapeze artist and the assistant. Therefore, gravity does the most work on the more massive person.

b. Both the trapeze artist and the assistant strike the net with *the same speed.* Here's why. They both start out at rest on the platform above the net. As the trapeze artist swings downward, before letting go of the bar, she moves along the arc of a circle. The work done by the tension in the trapeze cords is zero because it points perpendicular to the circular path of motion. Thus, if air resistance is ignored, the work done by the nonconservative forces is zero, $W_{nc} = 0$ J. The total mechanical energy of each person is conserved, regardless of the path taken to the net. Since both the trapeze artist and the assistant had the same initial total mechanical energy (all potential), they must have the same total mechanical energy (all kinetic) when they reach the net. That is, for either person, $mgh = (1/2)mv^2$. Solving for v gives $v = \sqrt{2gh}$, independent of the mass of the person. The value of h is the same for both the trapeze artist and the assistant; therefore, they strike the net with the same speed.

---

17. ***REASONING AND SOLUTION*** Since each plane has the same speed, the kinetic energy of each fuel tank will be the same at the instant of release. Since each plane is at the same height above the ground, each fuel tank must fall through the same vertical displacement. Therefore, the work done by gravity on each fuel tank is the same. From the work-energy theorem, each fuel tank will gain the same amount of kinetic energy during the fall. Therefore, each fuel tank will hit the ground with the same speed.

---

18. ***REASONING AND SOLUTION*** Average power $\overline{P}$ is the average rate at which work W is done: $\overline{P} = W/t$. It is not correct to conclude that one engine is doing twice the work of another just because it is generating twice the power. If one engine generates twice the power as the other, it may be generating the same amount of work in one half of the time.

CHAPTER 7 | *IMPULSE AND MOMENTUM*

CONCEPTUAL QUESTIONS

---

1. ***REASONING AND SOLUTION*** The linear momentum **p** of an object is the product of its mass and its velocity. Since the automobiles are identical, they have the same mass; however, although the automobiles have the same speed, they have different velocities. One automobile is traveling east, while the other one is traveling west. Therefore, the automobiles do not have the same momentum. Note that both momenta have the same magnitude, however, one car has a momentum that points east, while the other car has a momentum that points west.

---

2. ***REASONING AND SOLUTION*** Since linear momentum is a vector quantity, the total linear momentum of any system is the resultant of the linear momenta of the constituents. The people who are standing around have zero momentum. Those who move randomly carry momentum randomly in all directions. Since there is such a large number of people, there is, on average, just as much linear momentum in any one direction as in any other. On average, the resultant of this random distribution is zero. Therefore, the approximate linear momentum of the Times Square system is zero.

---

3. ***REASONING AND SOLUTION***

   a. Yes. Momentum is a vector, and the two objects have the same momentum. This means that the direction of each object's momentum is the same. Momentum is mass times velocity, and the direction of the momentum is the same as the direction of the velocity. Thus, the velocity directions must be the same.

   b. No. Momentum is mass times velocity. The fact that the objects have the same momentum means that the product of the mass and the magnitude of the velocity is the same for each. Thus, the magnitude of the velocity of one object can be smaller, for example, as long as the mass of that object is proportionally greater to keep the product of mass and velocity unchanged.

---

4. ***REASONING AND SOLUTION***

   a. If a single object has kinetic energy, it must have a velocity; therefore, it must have linear momentum as well.

   b. In a system of two or more objects, the individual objects could have linear momenta that cancel each other. In this case, the linear momentum of the system would be zero. The kinetic energies of the objects, however, are scalar quantities that are always positive; thus, the total kinetic energy of the system of objects would necessarily be nonzero. Therefore, it is possible for a system of two or more objects to have a total kinetic energy that is not zero but a total momentum that is zero.

---

5. ***REASONING AND SOLUTION*** The impulse-momentum theorem, Equation 7.4, states that $(\Sigma\overline{\mathbf{F}})\Delta t = m\mathbf{v}_f - m\mathbf{v}_0$.

a. If an airplane is flying horizontally with a constant momentum during a time Δt, then from Equation 7.4, $(\Sigma\overline{\mathbf{F}})\Delta t = 0$. There is no net impulse $(\Sigma\overline{\mathbf{F}})\Delta t$ on the plane during this time interval.

b. The fact that the net impulse on the plane is zero indicates that the impulses of the two horizontal forces must cancel each other. That is, the impulse of the thrust is equal in magnitude and opposite in direction to the impulse of the resistive force.

---

6. ***REASONING AND SOLUTION*** The severity of the collision is determined by the amount of momentum transferred by the colliding object. If the child is moving twice as fast as the adult, and the mass of the child and the bicycle is one half that of the adult, the magnitude of the linear momenta (mass x speed) of the child and the adult are the same. Therefore, it does not matter whether one is struck by the fast-moving child or the slow-moving adult.

---

7. ***REASONING AND SOLUTION*** An object slides along the surface of the earth and slows down because of kinetic friction.

a. If the object and the earth are considered to be part of the system, then the force of kinetic friction arises between components of the system. Therefore, the force of kinetic friction is an internal force, not an external force.

b. A force can only change the linear momentum of a system if the force is an external force. Since the force of friction is an internal force, it cannot change the total linear momentum of the two-body system. In fact, the total linear momentum of the two-body system remains the same. The linear momentum lost by the object is equal to the linear momentum gained by the earth.

---

8. ***REASONING AND SOLUTION*** The impulse-momentum theorem, Equation 7.4, states that $(\Sigma\overline{\mathbf{F}})\Delta t = m\mathbf{v}_f - m\mathbf{v}_0$. Assuming that the golf ball is at rest when it is struck with the club, $(\Sigma\overline{\mathbf{F}})\Delta t = m\mathbf{v}_f$.

During a good "follow-through" when driving a golf ball, the club is in contact with the ball for the longest possible time. From the impulse-momentum theorem, it is clear that when the contact time Δt is a maximum, the final linear momentum $m\mathbf{v}_f$ of the ball is a maximum. In other words, during a good "follow through" the maximum amount of momentum is transferred to the ball. Therefore, the ball will travel through a larger horizontal distance.

---

9. ***REASONING AND SOLUTION***

a. Since the water leaves each nozzle with a speed that is greater than the speed inside the arm, the quantity $mv_f - mv_0$ is positive. From the impulse-momentum theorem, $(\Sigma \overline{\mathbf{F}})\Delta t = m\mathbf{v}_f - m\mathbf{v}_0$, and we can deduce that there is a net positive or outward impulse. Therefore, a net outward force is exerted on the water.

b. From Newton's third law, the water must exert a net force that is equal in magnitude, but negative and directed toward the nozzle. The nozzle and the arm, in turn, move. Since each arm is free to rotate about the vertical axis, the arm will whirl.

---

10. ***REASONING AND SOLUTION*** To throw the villain forward, Superman must exert a force on him. For a system comprised of Superman and the villain, this force is an internal force. Internal forces cannot change the total linear momentum of the two-person system. Therefore, when Superman throws the villain forward, the villain gains forward momentum, and Superman must gain an equal amount of momentum in the backward direction. Therefore, Superman could not remain stationary after throwing the villain forward.

---

11. ***REASONING AND SOLUTION*** Since the satellite explodes in outer space far from any other body, it may be considered to be an isolated system. The forces of the explosion are forces that are internal to the system. Therefore, after the explosion, the total linear momentum of all the pieces must be equal to the linear momentum of the satellite before it exploded.

---

12. ***REASONING AND SOLUTION*** Let the system be defined by the jetliner (and all the other passengers, luggage, etc.) and the walking passenger. When the passenger walks toward the front of the plane, the linear momentum of this passenger increases. Any forces exerted by this passenger on the jetliner are internal forces and cannot change the total momentum of the system. Therefore, from momentum conservation, the forward momentum of the jetliner will decrease by the same amount as the momentum gained by the passenger. This amount is small compared to the forward momentum of the jetliner and goes unnoticed.

---

13. ***REASONING AND SOLUTION***

a. No. The person overhead jumps straight down and, therefore, applies only a vertical force to the boat. Since friction and air resistance are negligible, no horizontal force is applied to the boat. According to the impulse-momentum theorem, this means that the horizontal momentum of the boat cannot change.

b. The speed of the boat decreases. Momentum is mass times velocity. The only way for the horizontal momentum of the boat to remain unchanged when the mass increases due to the presence of the jumper is for the magnitude of the boat's velocity (that is, the speed) to decrease.

---

14. ***REASONING AND SOLUTION*** Let the system be comprised of the asteroid, the catapult and the supply of stones. When the catapult is used to "throw" chunks of stones into space, the force exerted by the catapult is an internal force; therefore, the total momentum of the system must remain the same. When rocks are thrown in one direction, they carry linear momentum in that direction. From the conservation of momentum, the asteroid must carry an equal amount of momentum in the opposite direction. It will, therefore, move in that direction. Such a device could be used as a propulsion system to move the asteroid closer to earth.

---

15. ***REASONING AND SOLUTION*** For a system comprised of the three balls, there is no net external force. The forces that occur when the three balls collide are internal forces. Therefore, the total linear momentum of the system is conserved. Note, however, that the momentum of each ball is not conserved. The momentum of any given ball will change as it interacts with the other balls; the momentum of each ball will change in such a way as to conserve the momentum of the system. It is the momentum of the system of balls, not the momentum of an individual ball, that is conserved.

---

16. ***REASONING AND SOLUTION*** An elastic collision is one in which the total kinetic energy of the system after the collision is equal to the total kinetic energy before the collision. The kinetic energy of the individual objects in the system will, in general, change during a collision. They will change so that the total kinetic energy of the system remains the same before and after the collision. Even in an elastic collision, however, the kinetic energy of each object is not necessarily the same before and after the collision.

---

17. ***REASONING AND SOLUTION*** Example 7 concerns the elastic collision between two balls: one initially moving (ball 1), and one initially at rest (ball 2). The results of that example show that the final speeds of the balls are given by

$$v_{f1} = \left(\frac{m_1 - m_2}{m_1 + m_2} \right) v_{01} \tag{7.8a}$$

and

$$v_{f2} = \left(\frac{2m_1}{m_1 + m_2} \right) v_{01} \tag{7.8b}$$

If the two objects have equal mass, $m_1 = m_2$, and Equation 7.8a indicates that $v_{f1} = 0$, and Equation 7.8b indicates that $v_{f2} = v_{01}$. In other words, the first object is stopped completely and the second object takes off with the velocity the first object originally had.

---

18. ***REASONING AND SOLUTION*** Many objects have a point, a line, or a plane of symmetry. If the mass of the system is uniformly distributed, the center of mass of such an object lies at that point, on that line, or in that plane. The point of symmetry of a doughnut is at the geometric center of the hole. Thus, the center of mass of a doughnut is at the center of the hole.

---

19. ***REASONING AND SOLUTION*** Since more of the mass of the bat is located near the heavier end of the bat, the center of mass of the bat will be located nearer the heavier end.

---

20. ***REASONING AND SOLUTION*** A sunbather is lying on a floating, stationary raft. She then gets up and walks to one end of the raft. The sunbather and the raft are considered as an isolated system.

a. As the sunbather walks to one end of the raft, she exerts a force on the raft; however, the force is internal to the isolated system. Since there are no external forces acting on the system, the linear momentum of the system cannot change. Since the linear momentum of the system is initially zero, it must remain zero. Therefore, the velocity of the center of mass of the system must be zero.

b. The sunbather has linear momentum as she walks to one end of the raft. Since the linear momentum of the isolated system must remain zero, the raft must acquire a linear momentum that is equal in magnitude and opposite in direction to that of the sunbather. From the definition of linear momentum, $\mathbf{p} = m\mathbf{v}$, we know that the direction of the linear momentum of an object is the same as the direction of the velocity of the object. Thus, the raft acquires a velocity that is opposite to the direction of motion of the sunbather.

CHAPTER 8 | *ROTATIONAL KINEMATICS*

CONCEPTUAL QUESTIONS

---

1. ***REASONING AND SOLUTION*** The figures below show two axes in the plane of the paper and located so that the points *B* and *C* move in circular paths having the same radii (radius = *r*).

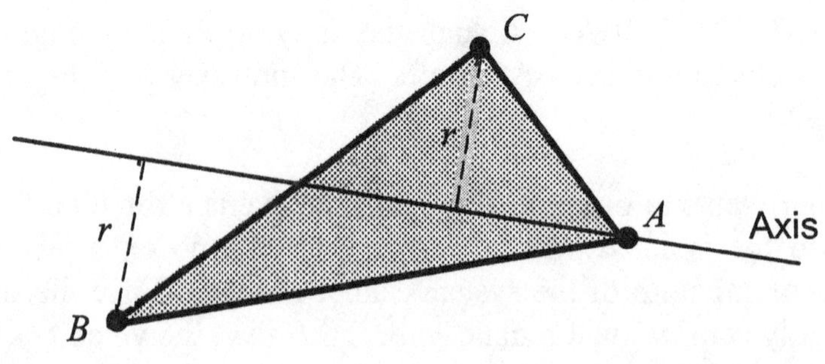

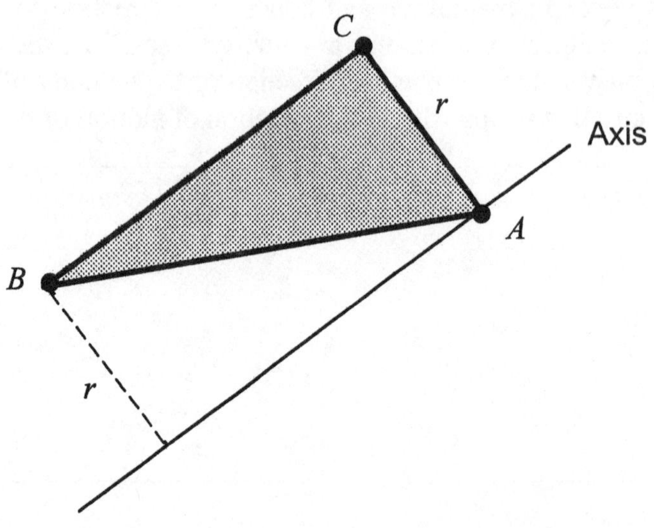

---

2. ***REASONING AND SOLUTION*** When a pair of scissors is used to cut a string, each blade of the scissors does not have the same angular velocity at a given instant during the cut. The angular speed of each blade is the same; however, each blade rotates in the opposite direction. Therefore, it is correct to conclude that the blades have opposite angular velocities at any instant during the cut.

---

3. ***REASONING AND SOLUTION*** Just before the clock is unplugged, the second hand is rotating so that its angular velocity is clockwise. The second hand moves with a constant angular velocity, so the angular acceleration is zero. When the plug is pulled, the second hand will continue to rotate with its clockwise angular velocity, but it will slow down. Therefore, the angular acceleration must be opposite to the angular velocity, or counterclockwise.

---

4. ***REASONING AND SOLUTION*** The tangential speed, v_T, of a point on the earth's surface is related to the earth's angular speed ω according to $v_T = r\omega$, Equation 8.9, where r is the perpendicular distance from the point to the earth's rotation axis. At the equator, r is equal to the earth's radius. As one moves away from the equator toward the north or south geographic pole, the distance r becomes smaller. Since the earth's rotation axis passes through the geographic poles, r is effectively zero at those locations. Therefore, your tangential speed would be a minimum if you stood as close as possible to either the north or south geographic pole.

---

5. ***REASONING AND SOLUTION***

a. A thin rod rotates at a constant angular speed about an axis of rotation that is perpendicular to the rod at its center. As the rod rotates, each point at a distance r from the center on one half of the rod has the same tangential speed as the point at a distance r from the center on the other half of the rod. This is true for all values of r for $0 < r \leq (L/2)$ where L is the length of the rod.

b. If the rod rotates about an axis that is perpendicular to the rod at one end, no two points are the same distance from the axis of rotation. Therefore, no two points on the rod have the same tangential speed.

---

6. ***REASONING AND SOLUTION*** The wheels are rotating with a constant angular velocity.

a. Since the angular velocity is constant, each wheel has zero angular acceleration, α = 0 rad/s. Since the tangential acceleration a_T is related to the angular acceleration through Equation 8.10, $a_T = r\alpha$, every point on the rim has zero tangential acceleration.

b. Since the particles on the rim of the wheels are moving along a circular path, they must have a centripetal acceleration. This can be supported by Equation 8.11, $a_c = r\omega^2$, where a_c is the magnitude of the centripetal acceleration. Since ω is nonzero, a_c is nonzero.

---

7. ***REASONING AND SOLUTION*** The wheel and the two points under consideration are shown in the figure at the right.

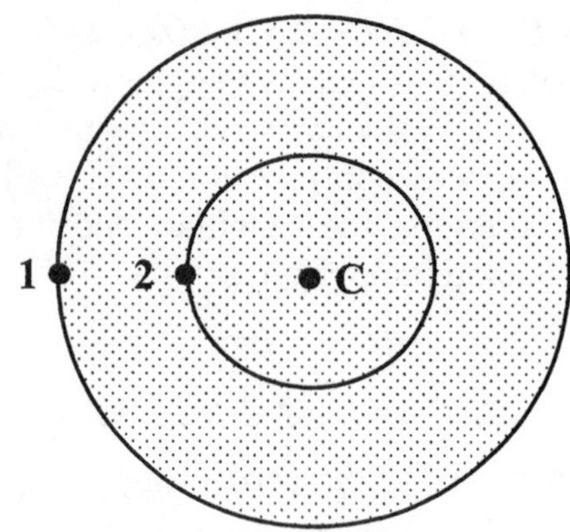

a. Each point must undergo the same angular displacement in the same time interval. Therefore, at any given instant, the angular velocity of both points is the same.

b. Each point must increase its angular velocity at the same rate; therefore, each point has the same angular acceleration.

c. The tangential speed is given by $v_T = r\omega$. Since both points have the same angular speed ω, but point **1** is further from the center than point **2**, point **1** has the larger tangential speed.

d. The tangential acceleration is given by $a_T = r\alpha$. Since both points have the same angular acceleration α, but point **1** is further from the center than point **2**, point **1** has the larger tangential acceleration.

e. The centripetal acceleration is given by $a_c = r\omega^2$. Since both points have the same angular speed ω, but point **1** is further from the center than point **2**, point **1** has the larger centripetal acceleration.

---

8. ***REASONING AND SOLUTION*** For a building that is located on the earth's equator, the points in the building rotate about the center of the earth. The tangential speed of any point is given by $v_T = r\omega$, where, in this case, r is measured from the center of the earth. The top floor of the building has a larger value of r than the other floors; therefore, the top floor has the greater tangential speed.

---

9. ***REASONING AND SOLUTION*** The centripetal acceleration of any point in the space station is given by $a_c = r\omega^2$ (Equation 8.11),where r is the distance from the point to the axis of rotation of the space station. If the centripetal acceleration is adjusted to be g at a given value of r (such as the astronaut's feet), the centripetal acceleration will be different at other values of r. Therefore, if the adjustment is made so that the centripetal acceleration at the astronaut's feet equals g, then the centripetal acceleration will not equal g at the astronaut's head.

---

10. ***REASONING AND SOLUTION*** The tangential speed for any point on the tip of a clock hand is given by $v_T = r\omega$. The angular speeds of the second hand, the minute hand, and the hour hand differ greatly, with the second hand having the largest angular speed and the hour hand having the smallest. If one desired to create a clock in which the tips of the second

hand, the minute hand, and the hour hand moved with the same tangential speed, the lengths of the arms would also have to differ greatly, with the second hand having the smallest arm and the hour hand having the largest. Such a clock would not be very practical.

---

11. ***REASONING AND SOLUTION*** Any point on a rotating object possesses a centripetal acceleration that is directed radially toward the axis of rotation. This also applies to a tire on a moving car and is true regardless of whether the car has a constant linear velocity or whether it is accelerating.

---

12. ***REASONING AND SOLUTION*** The bicycle wheel has an angular acceleration. The arrows are perpendicular to the radius of the wheel. The magnitude of the arrows increases with increasing distance from the center in accordance with $v_T = r\omega$, or $a_T = r\alpha$. The arrows in the picture could represent either the tangential velocity or the tangential acceleration. The arrows are not directed radially inward; therefore, they cannot represent the centripetal acceleration.

---

13. ***REASONING AND SOLUTION*** The speedometer of a truck uses a device that measures the angular speed of the tires. The angular speed is related to the linear speed of the truck by $v = r\omega$.

Suppose two trucks are traveling side-by-side along a highway at the same linear speed v, and one truck has larger wheels than the other. The angular speed of the larger-diameter wheels is less than that of the smaller-diameter wheels. Since the speedometer uses a device that measures the angular speed of the tires, the speedometer reading on the truck with the larger wheels will indicate a smaller linear speed than that on the truck with the smaller wheels.

---

14. ***REASONING AND SOLUTION*** When a fan is shut off, the fan blades gradually slow down until they eventually stop. The angular acceleration is never really constant because it gradually decreases, becoming zero at the instant the blades stop rotating. In solving problems, it is sometimes convenient to approximate such motion as having constant angular acceleration. This is a fairly good approximation for a short time interval just after the switch is turned off.

---

15. ***REASONING AND SOLUTION*** Three examples that involve rotation about an axis that is not fixed:

(1) The motion of a Frisbee. The axis of the rotating Frisbee moves through the air with the Frisbee.
(2) The motion of the earth in its orbit. The rotation axis of the earth changes position as the earth revolves around the sun.
(3) The motion of a twirling baton that has been thrown into the air. The rotation axis constantly changes location as the baton rises and falls.

CHAPTER 9 | *ROTATIONAL DYNAMICS*

CONCEPTUAL QUESTIONS

---

1. ***REASONING AND SOLUTION*** The magnitude of the torque produced by a force **F** is given by $\tau = F\ell$, where ℓ is the lever arm. When a long pipe is slipped over a wrench handle, the length of the wrench handle is effectively increased. The same force can be applied with a larger lever arm. Therefore, a greater torque can be used to remove the "frozen" nut.

---

2. ***REASONING AND SOLUTION*** The magnitude of the torque produced by a force **F** is given by $\tau = F\ell$ where ℓ is the lever arm.

   a. A large force can be used to produce a small torque by choosing a small lever arm. Then the product $F\ell$ is small even though the magnitude of the force, F, is large. If the line of action passes through the axis of rotation, ℓ = 0 m, and the torque is zero regardless of how large F is.

   b. A small force may be used to produce a large torque by choosing a large lever arm ℓ. Then, the product $F\ell$ can be large even though the magnitude of the force F is small.

---

3. ***REASONING AND SOLUTION*** In a cassette deck, the magnetic tape applies a torque to the supply reel. The magnitude of the torque is equal to the product of the tension F_T in the tape, assumed to be constant, and the lever arm ℓ. As shown below, the lever arm is the radius of the tape remaining on the supply reel. Figure A shows the reel and tape at an earlier time; figure B shows the tape and the reel at a later time.

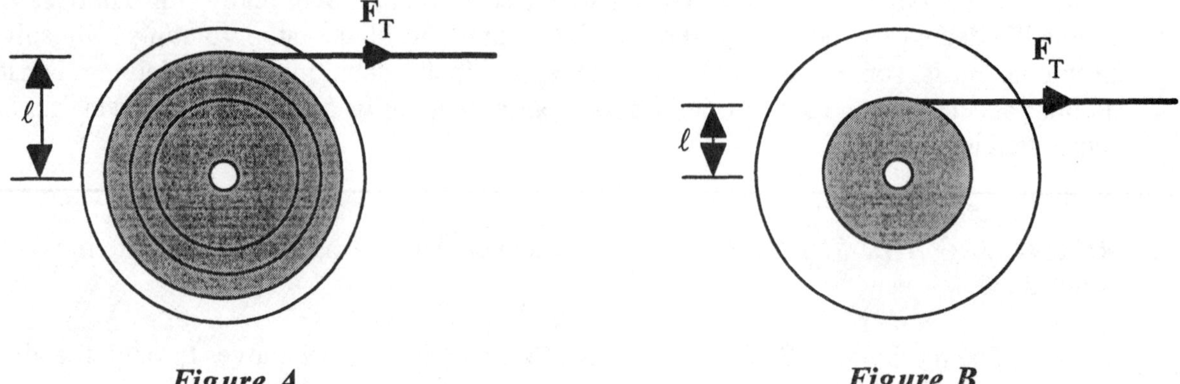

Figure A *Figure B*

As the reel becomes empty, the lever arm ℓ gets smaller, as suggested in the figures above. Therefore, the torque exerted on the reel by the tape becomes smaller.

---

4. ***REASONING AND SOLUTION*** A flat rectangular sheet of plywood is fixed so that it can rotate about an axis perpendicular to the sheet through one corner, as shown in the figure. In order to produce the largest possible torque with the force **F** (acting in the plane of the paper), the force **F** must be applied with the largest possible lever arm. This can be attained by orienting the force **F** so that it is perpendicular to the diagonal that passes through the rotation axis as shown below. The lever arm is equal to the length of the diagonal of the plywood.

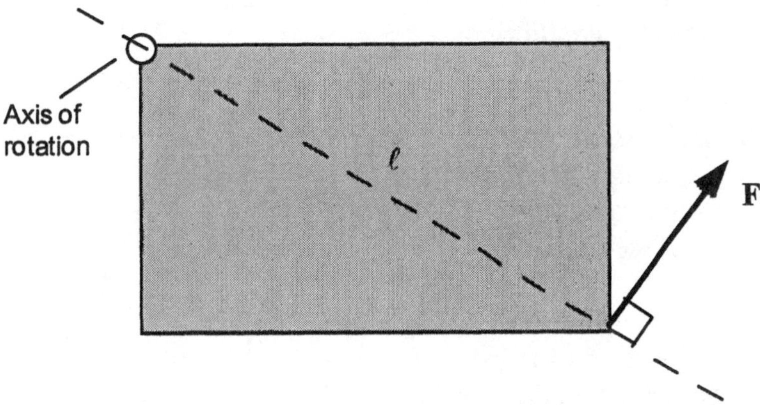

---

5. ***REASONING AND SOLUTION*** Work and torque are both the product of force and distance. Work and torque are distinctly different physical quantities, as is evident by considering the distances in the definitions. Work is defined by $W = (F\cos\theta)s$, according to Equation 6.1, where F is the magnitude of the force, θ is the angle between the force and the displacement, and s is the magnitude of the displacement. The magnitude of the torque is defined as the magnitude of the force times the lever arm, according to Equation 9.1. In the definition of work, the "distance" is the magnitude of the displacement over which the force acts. In the definition of torque, the distance is the lever arm, a "static" distance. The lever arm is not the same physical quantity as the displacement. Therefore, work and torque are different quantities.

---

6. ***REASONING AND SOLUTION*** A person stands on a train, both feet together, facing a window. The front of the train is to the person's left. When the train starts to move forward, a force of static friction is applied to the person's feet. This force tends to produce a torque about the center of gravity of the person, causing the person to fall backward. If the right foot is slid out toward the rear of the train, the normal force of the floor on the right foot produces a counter torque about the center of gravity of the person. The resultant of these two torques is zero, and the person can maintain his balance.

---

7. ***REASONING AND SOLUTION*** If we assume that the branch is fairly uniform, then in the spring, just when the fruit begins to grow, the center of gravity of the branch is near the center of the branch. As the fruit grows, the mass at the outer end of the branch increases. Therefore, as the fruit grows, the center of gravity of the fruit-growing branch moves along the branch toward the fruit.

8. ***REASONING AND SOLUTION*** For a rigid body in equilibrium, there can be no net torque acting on the object *with respect to any axis*. Consider an axis perpendicular to the plane of the paper at the right end of the rod. The lines of action of both $\mathbf{F}_1$ and $\mathbf{F}_3$ pass through this axis. Therefore, no torque is created by either of these forces. However, the force $\mathbf{F}_2$ does create a torque with respect to this axis. As a result, there is a net torque acting on the object with respect to at least one axis, and the three forces shown in the drawing cannot keep the object in equilibrium.

---

9. ***REASONING AND SOLUTION*** The forces that keep the *right section* of the ladder in equilibrium are shown in the free-body diagram at the right.

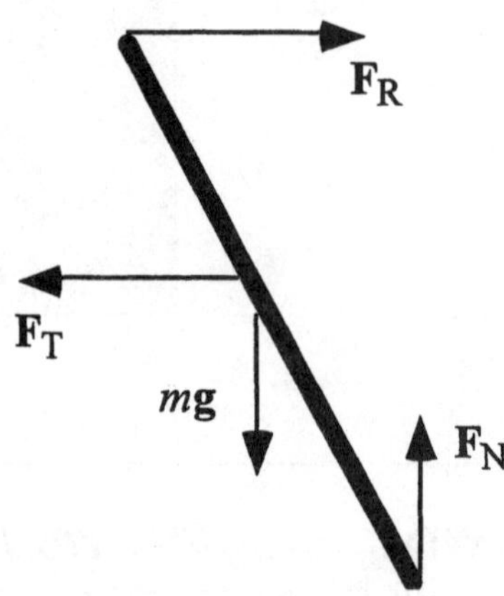

They are: the weight of the section, mg, the normal force exerted on the ladder by the floor, $\mathbf{F}_N$, the tension in the crossbar, $\mathbf{F}_T$, and the reaction force due to the left side of the ladder, $\mathbf{F}_R$.

---

10. ***REASONING AND SOLUTION*** Treating the wine rack and the bottle as a rigid body, the two external forces that keep it in equilibrium are its weight, mg, located at the center of gravity, and the normal force, $\mathbf{F}_N$, exerted on the base by the table.

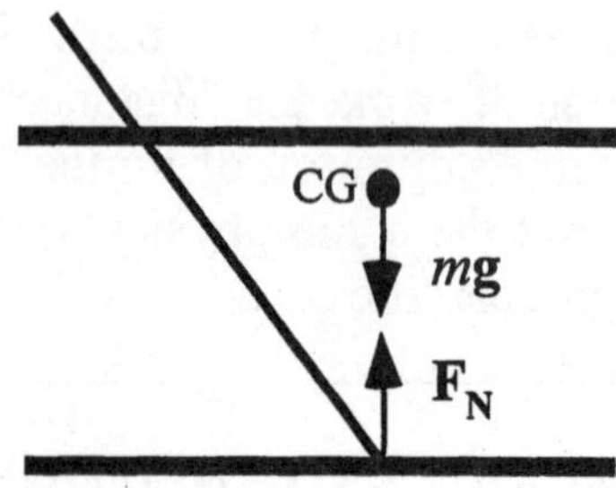

Note that the weight acts at the center of gravity (CG), which must be located exactly above the point where the normal force acts.

---

11. ***REASONING AND SOLUTION*** The torque required to give the triangular sheet an angular acceleration α is given by the rotational analog of Newton's second law for a rigid body: $\Sigma\tau = I\alpha$. The greatest net torque will be required about the axis for which the moment of inertia I is the greatest. The greatest value for I occurs when the greatest portion of the mass of the sheet is far from the axis. This will be for axis B. Therefore, for axis B, the greatest net external torque is required to bring the rectangular plate up to 10.0 rad/s in 10.0 s, starting from rest.

---

12. ***REASONING AND SOLUTION*** The object has an angular velocity and an angular acceleration due to torques that are present.

a. If additional torques are applied so as to make the net torque suddenly equal to zero, the object will continue to rotate with constant angular velocity. The angular velocity will remain constant at the value that it had at the instant that the net torque became zero.

b. If all the torques are suddenly removed, the effect is the same as when the net torque is zero as in part (a) above. That is, the object will continue to rotate with constant angular velocity. The angular velocity will remain constant at the value that it had at the instant that the torques are removed.

---

13. ***REASONING AND SOLUTION*** The space probe is initially moving with a constant translational velocity and zero angular velocity through outer space.

a. When the two engines are fired, each generates a thrust T in opposite directions; hence, the net force on the space probe is zero. Since the net force on the probe is zero, there is no translational acceleration and the translational velocity of the probe remains the same.

b. The thrust of each engine produces a torque about the center of the space probe. If R is the radius of the probe, the thrust of each engine produces a torque of magnitude TR, each in the same direction, so that the net torque on the probe has magnitude $2TR$. Since there is a net torque on the probe, there will be an angular acceleration. The angular velocity of the probe will increase.

---

14. ***REASONING AND SOLUTION*** When you are lying down with your hands behind your head, your moment of inertia I about your stomach is greater than it is when your hands are on your stomach. Sit-ups are then more difficult because, with a greater I, you must exert a greater torque about your stomach to give your upper torso the same angular acceleration, according to Equation 9.7.

---

15. ***REASONING AND SOLUTION*** The translational kinetic energy of a rigid body depends only on the mass and the speed of the body. It does not depend on how the mass is distributed. Therefore, for purposes of computing the body's translational kinetic energy, the mass of a rigid body can be considered as concentrated at its center of mass.

Unlike the translational kinetic energy, the moment of inertia of a body *does depend* on the location and orientation of the axis relative to the particles that make up the rigid body. Therefore, for purposes of computing the body's moment of inertia, the mass of a rigid body cannot be considered as concentrated at its center of mass. The distance of any individual mass particle from the axis is important.

---

16. ***REASONING AND SOLUTION*** A thin sheet of plastic is uniform and has the shape of an equilateral triangle, as shown in the figure at the right. The axes *A* and *B* are two rotation axes that are perpendicular to the plane of the triangle. If the triangle rotates about each axis with the same angular speed ω, the rotational kinetic energy of the sheet is given by $\text{KE} = \frac{1}{2} I\omega^2$. The rotational kinetic energy will be greater if the rotation occurs about the axis for which the triangle has the larger moment of inertia *I*. The value for *I* will be greater when more of the mass is located farther away from the axis, as is the case for axis *B*. Therefore, the triangle has the greater rotational kinetic energy when it rotates about axis *B*.

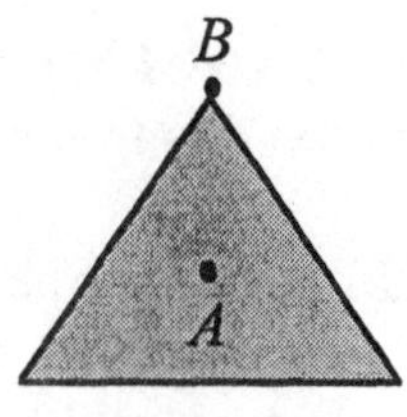

---

17. ***REASONING AND SOLUTION*** With feet flat on the floor and body straight, you can lean over until your center of gravity is above your toes. If you lean any farther, it is no longer possible to balance the torque due to your weight with a torque created by the upward reaction force that pushing downward with your toes produces. Bob's center of gravity must be closer to the ground. The reason is that, with the center of gravity closer to the ground, a greater amount of leaning can occur before it moves beyond his toes.

---

18. ***REASONING AND SOLUTION*** Let each object have mass *M* and radius *R*. The objects are initially at rest. In rolling down the incline, the objects lose gravitational potential energy Mgh , where *h* is the height of the incline. The objects gain kinetic energy equal to $(1/2)I\omega^2 + (1/2)Mv^2$, where *I* is the moment of inertia about the center of the object, *v* is the linear speed of the center of mass at the bottom of the incline, and ω is the angular speed about the center of mass at the bottom. From conservation of mechanical energy

$$Mgh = \tfrac{1}{2} I\omega^2 + \tfrac{1}{2} Mv^2$$

Since the objects roll down the incline, $\omega = v/R$. For the hoop, $I = MR^2$, hence,

$$Mgh = \tfrac{1}{2}(MR^2)\frac{v^2}{R^2} + \tfrac{1}{2} Mv^2$$

Solving for *v* gives $v = \sqrt{gh}$ for the hoop. Repeating the calculation for the solid cylinder, the spherical shell, and the solid sphere gives

$v = \sqrt{gh}$	(hoop)
$v = 1.15\sqrt{gh}$	(solid cylinder)
$v = 1.10\sqrt{gh}$	(spherical shell)
$v = 1.20\sqrt{gh}$	(solid sphere)

The solid sphere is faster at the bottom than any of the other objects. The solid cylinder is faster at the bottom than the spherical shell or the hoop. The spherical shell is faster at the bottom than the hoop. Thus, the solid sphere reaches the bottom first, followed by the solid cylinder, the spherical shell, and the hoop, in that order.

---

19. ***REASONING AND SOLUTION*** A woman sits on the spinning seat of a piano stool with her arms folded. Let the woman and the piano stool comprise the system. When the woman extends her arms, the torque that she applies is an internal torque. Since there is no net external torque acting on the system, the angular momentum of the woman and the piano stool, $I\omega$, must remain constant.

a. With her arms extended, the woman's moment of inertia I about the rotation axis is greater than it was when her arms were folded. Since the product of $I\omega$ must remain constant and I increases, ω must decrease. Therefore, the woman's angular velocity decreases.

b. As discussed above, since there is no net external torque acting on the system, the angular momentum of the system remains the same.

---

20. ***REASONING AND SOLUTION*** Consider the earth to be an isolated system. Note that the earth rotates about an axis that passes through the North and South poles and is perpendicular to the plane of the equator. If the ice cap at the South Pole melted and the water were uniformly distributed over the earth's oceans, the mass at the South Pole would be uniformly distributed and, on average, be farther from the earth's rotational axis. The moment of inertia of the earth would, therefore, increase. Since the earth is an isolated system, any torques involved in the redistribution of the water would be internal torques; therefore, the angular momentum of the earth must remain the same. If the moment of inertia increases, and the angular momentum is to remain constant, the angular velocity of the earth must decrease.

---

21. ***REASONING AND SOLUTION*** Note that the earth rotates about an axis that passes through the North and South poles and is perpendicular to the plane of the equator. When rivers like the Mississippi carry sediment toward the equator, they redistribute the mass from a more uniform distribution to a distribution with more mass concentrated around the equator. This increases the moment of inertia of the earth. If the earth is considered to be an isolated system, then any torques involved in the redistribution of mass are internal torques. Therefore, the angular momentum of the earth must remain constant. The moment of inertia increases, and the angular momentum must remain the same; therefore, the angular velocity must decrease.

---

22. ***REASONING AND SOLUTION*** Let the system be the rotating cloud of interstellar gas. The gravitational force that pulls the particles together is internal to the system; hence, the torque resulting from the gravitational force is an internal torque. Since there is no net external torque acting on the collapsing cloud, the angular momentum, $I\omega$, of the cloud must remain constant. Since the cloud is shrinking, its moment of inertia I decreases. Since the product $I\omega$ remains constant, the angular velocity ω must increase as I decreases. Therefore, the angular velocity of the formed star would be greater than the angular velocity of the original gas cloud.

---

23. ***REASONING AND SOLUTION*** A person is hanging motionless from a vertical rope over a swimming pool. She lets go of the rope and drops straight down. Consider the woman to be the system. When she lets go of the rope her angular momentum is zero. It is possible for the woman to curl into a ball; however, since her angular momentum is zero, she cannot change her angular velocity by exerting any internal torques. In order to begin spinning, an external torque must be applied to the woman. Therefore, it is not possible for the woman to curl into a ball and start spinning.

---

24. ***REASONING AND SOLUTION*** The box that creates the greatest torque is the one that appears shaded in the drawing at the right.

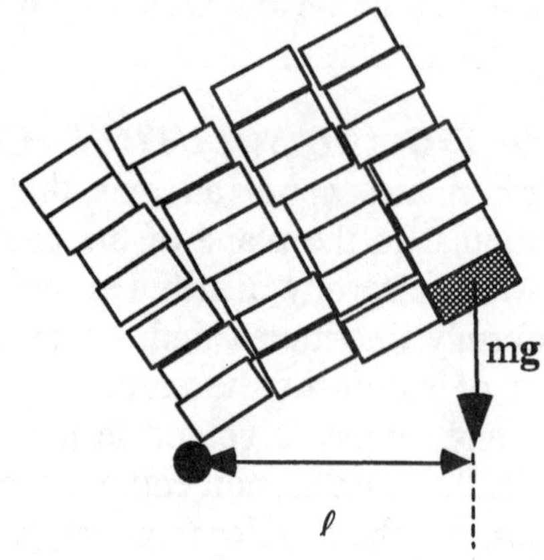

All boxes give rise to the same force *mg*. Each box produces a torque about the axle. The box that produces the greatest torque is the one for which the lever arm ℓ is the greatest. By inspection of the drawing, the lever arm relative to the axle is greatest for the shaded box.

CHAPTER 10 | *SIMPLE HARMONIC MOTION AND ELASTICITY*

CONCEPTUAL QUESTIONS

---

1. ***REASONING AND SOLUTION*** A horizontal spring is attached to an immovable wall. Two people pull on the spring. They then detach it from the wall and pull on opposite ends of the horizontal spring. They pull just as hard in each case.

Suppose that each person pulls with a force **P**. When the spring is attached to the wall, both people pull on the same end, so that they pull with a total force 2**P**. The wall exerts a force –2**P** at the other end of the spring. The tension in the spring has magnitude 2*P*. When the two people pull on opposite ends of the horizontal spring, one person exerts a force **P** at one end of the spring, and the other person exerts a force –**P** at the other end of the spring. The tension in the spring has magnitude *P*.

The tension in the spring is greater when it is attached to the immovable wall; therefore, the spring will stretch more in this case.

---

2. ***REASONING AND SOLUTION*** Identical springs are attached to a box in two different ways, as shown in the drawing in the text. Initially, the springs are unstrained. The box is then pulled to the right and released. The displacement of the box is the same in both cases.

For the box on the left, each spring is stretched and its displacement is $+\Delta x$, where the plus sign indicates a displacement to the right. According to Equation 10.2, the restoring force exerted by each spring on the box is $F = -k(+\Delta x)$. Thus, the net force due to both springs is $\sum F = -2k(+\Delta x)$.

For the box on the right, one spring is stretched and the other is compressed. However, the displacement of each spring is still $+\Delta x$. The net force exerted on the box is $\sum F = -k(+\Delta x) - k(+\Delta x) = -2k(+\Delta x)$.

Thus, both boxes experience the same net force.

---

3. ***REASONING AND SOLUTION*** Figures 10.11 and 10.14 show the velocity and the acceleration, respectively, of the shadow of a ball that undergoes uniform circular motion. The shadow undergoes simple harmonic motion.

a. The velocity of the shadow is given by Equation 10.7: $v = -A\omega \sin\theta$. The velocity of the shadow will be zero when $\theta = 0$ or π rad. From Figure 10.11, we see that the velocity equals zero each time the shadow reaches the right and left endpoints of its motion (that is, when the ball crosses the *x* axis).

b. The acceleration of the shadow is given by Equation 10.9: $a = -A\omega^2 \cos\theta$. The acceleration of the shadow will be zero when θ is $\pi/2$ or $3\pi/2$. From Figure 10.14, we see that the acceleration equals zero each time that the shadow passes through the point $x = 0$ m (that is, when the ball crosses the *y* axis).

---

4. ***REASONING AND SOLUTION*** Simple harmonic motion is the oscillatory motion that occurs when a restoring force of the form of Equation 10.2, $F = -kx$, acts on an object. The force changes continually as the displacement x changes.

A steel ball is dropped onto a concrete floor. Over and over again, it rebounds to its original height. During the time when the ball is in the air, either falling down or rebounding up, the only force acting on the ball is its weight, which is nearly constant, to the extent that the ball remains near the earth's surface. Thus, the motion of the bouncing ball is not simple harmonic motion.

---

5. ***REASONING AND SOLUTION*** The vibrational frequency f of an object in simple harmonic motion is given by (see Example 6): $f = [1/(2\pi)]\sqrt{k/m}$. The shock absorbers of a car are "springs" that obey this relation. As the number of passengers in a car increases, the total mass supported by any one of the shock absorbers increases. Therefore, the vibrational frequency of the shock absorber f must decrease.

---

6. ***REASONING AND SOLUTION*** A block is attached to a horizontal spring and slides back and forth in simple harmonic motion on a frictionless horizontal surface. A second identical block is suddenly attached to the first block when the first block is at one extreme end of the oscillation cycle.

a. Since the attachment is made at one extreme end of the oscillation cycle, where the velocity is zero, the extreme end of the oscillation cycle will remain at the same point; in other words, the amplitude remains the same.

b. The angular frequency of an object of mass m in simple harmonic motion at the end of a spring of force constant k is given by Equation 10.11: $\omega = \sqrt{k/m}$. Since the mass m is doubled while the force constant k remains the same, the angular frequency decreases by a factor of $\sqrt{2}$. The vibrational frequency f is related to ω by $f = \omega/(2\pi)$; the vibrational frequency f will also decrease by a factor of $\sqrt{2}$.

c. The maximum speed of oscillation is given by Equation 10.8: $v_{max} = A\omega$. Since the amplitude, A, remains the same and the angular frequency, ω, decreases by a factor of $\sqrt{2}$, the maximum speed of oscillation also decreases by a factor of $\sqrt{2}$.

---

7. ***REASONING AND SOLUTION*** The time required for a particle in simple harmonic motion to travel through one complete cycle (the period) is independent of the amplitude of the motion, even though at larger amplitudes the particle travels further. This is possible because, at larger amplitudes, the maximum speed of the particle is greater. Thus, even though the particle must cover larger distances at larger amplitudes, it does so with greater speeds.

---

8. ***REASONING AND SOLUTION*** The back-and-forth motion of the saw blade is simple harmonic motion for the following reason. The shadow of the ball in Figure 10.9 has a displacement x that moves in simple harmonic motion (see Equation 10.3). The horizontal back-and-forth motion of the saw blade is identical to the displacement x of the ball's shadow, since the saw blade is driven by a pin mounted on a rotating disk. Therefore, the motion of the saw blade is simple harmonic motion.

---

9. ***REASONING AND SOLUTION*** The elastic potential energy that a spring has by virtue of being stretched or compressed is given by Equation 10.13: $\mathrm{PE}_{\text{elastic}} = (1/2)kx^2$, where x is the amount by which the spring is stretched or compressed relative to its unstrained length. The amount of stretch or compression appears squared, so that the elastic potential energy is positive and independent of the sign of x. Therefore, the amount of elastic potential energy stored in a spring when it is compressed by one centimeter is the same as when it is stretched by the same amount.

---

10. ***REASONING AND SOLUTION*** We can deduce from Equation 10.16 and Example 10 that, for small angles, the period, T, of a simple pendulum is given by $T = 2\pi\sqrt{L/g}$ where L is the length of the pendulum.

If a grandfather clock is running slowly, then its period is too long, so one needs to decrease the period to make the clock keep the correct time. Since $T = 2\pi\sqrt{L/g}$, we, therefore, need to decrease the length of the pendulum.

---

11. ***REASONING AND SOLUTION*** From Equations 10.5 and 10.11, we can deduce that the period of the simple harmonic motion of an ideal spring is given by $T = 2\pi\sqrt{m/k}$, where m is the mass at the end of the ideal spring and k is the spring constant. We can deduce from Equations 10.5 and 10.16 that, for small angles, the period, T, of a simple pendulum is given by $T = 2\pi\sqrt{L/g}$ where L is the length of the pendulum.

In principle, the motion of a simple pendulum and an object on an ideal spring can both be used to provide the period of a clock. However, it is clear from the expressions for the period given above that the period of the mass-spring system depends only on the mass and the spring constant, while the period of the pendulum depends on the acceleration due to gravity. Therefore, a pendulum clock is likely to become more inaccurate when it is carried to the top of a high mountain where the value of g will be smaller than it is at sea level.

---

12. ***REASONING AND SOLUTION*** We can deduce from Equations 10.5 and 10.16 that, for small angles, the period, T, of a simple pendulum is given by $T = 2\pi\sqrt{L/g}$ where L is the length of the pendulum. This can be solved for the acceleration due to gravity to yield: $g = 4\pi^2 L/T^2$.

If you were held prisoner in a room and had only a watch and a pair of shoes with shoelaces of known length, you could determine whether this room is on earth or on the moon in the following way: You could use one of the shoelaces and one of the shoes to make a pendulum. You could then set the pendulum into oscillation and use the watch to measure the period of the pendulum. The acceleration due to gravity could then be calculated from the expression above. If the value is close to 9.80 m/s^2, then it can be concluded that the room is on earth. If the value is close to 1.6 m/s^2, then it can be concluded that the room is on the moon.

---

13. ***REASONING AND SOLUTION*** The playground swing may be treated, to a good approximation, as a simple pendulum. The period of a simple pendulum is given by $T = 2\pi\sqrt{L/g}$. This expression for the period depends only on the length of the pendulum and the acceleration due to gravity; for angles less than 10° the period is independent of the amplitude of the motion. Therefore, if one person is pulled back 4° from the vertical while another person is pulled back 8° from the vertical, they will both have the same period. If they are released simultaneously, they will both come back to the starting points at the same time.

---

14. ***REASONING AND SOLUTION*** A car travels over a road that contains a series of equally spaced bumps. If the distance between successive bumps is d and the car travels at a constant horizontal velocity **v**, then the car will encounter a bump at regular intervals separated by a time $T = d/v$. The frequency at which the car encounters the bumps is $f = 1/T = v/d$. If the oscillation frequency of the suspension system of the car is v/d, then the bumps will "drive" the suspension system into large amplitude oscillations. That is, the bumps will drive the suspension system at resonance.

---

15. ***REASONING AND SOLUTION*** The amount of force F needed to stretch a rod is given by Equation 10.17: $F = Y(\Delta L / L_0)A$, where A is the cross-sectional area of the rod, L_0 is the original length, ΔL is the change in length, and Y is Young's modulus of the material.

Since the cylinders are made of the same material, they have the same Young's modulus. However, the cross-sectional area of the material in the hollow cylinder is smaller than that of the solid cylinder since most of its cross section is empty. When identical forces of magnitude F are applied to the right end of each cylinder, each stretches by an amount $\Delta L = (FL_0)/(YA)$. Since the hollow cylinder has a smaller cross-sectional area of material compared to the solid cylinder, the hollow cylinder will stretch the most.

---

16. ***REASONING AND SOLUTION*** Young's modulus for steel is thousands of times greater than that for rubber. This means that, all other factors being equal, the force needed to stretch a steel rod by a given amount must be thousands of times greater than the force needed to stretch a rubber rod by the same amount. It does not mean that steel stretches much more easily than rubber.

---

17. ***REASONING AND SOLUTION*** A trash compactor crushes empty aluminum cans, thereby reducing the total volume of the cans by 75%. The value given in Table 10.3 for the bulk modulus of aluminum cannot be used to calculate the change in pressure generated in the trash compactor. The value of the bulk modulus in Table 10.3 pertains to solid aluminum objects. Most of the 75% reduction in the total volume of the cans arises from collapsing the cans and forcing air out of them.

---

18. ***REASONING AND SOLUTION*** Pressure is defined in terms of the magnitude of the force per unit area, where the force is applied perpendicular to the surface.

Both sides of Equation 10.18, $F = S(\Delta X / L_0)A$ can be divided by the area A to give F/A on the left side. This F/A term cannot be called a pressure, such as the pressure that appears in Equation 10.19, because the force F in the "F/A" term is a shear force that is parallel to the surface, not perpendicular to it.

---

19. ***REASONING AND SOLUTION*** When the block rests on the ground, the ground exerts an upward normal force on the block. Regardless of which face the block is resting on, the magnitude of the normal force is equal to the weight of the block. The magnitude of the force per unit area is the stress.

a. Face B has the smallest area; therefore, the stress, F/A, is largest when the block rests on face B.

b. Face C has the largest area; therefore, the stress, F/A, is smallest when the block rests on face C.

CHAPTER 11 | *FLUIDS*

CONCEPTUAL QUESTIONS

---

1. ***REASONING AND SOLUTION*** A pile of empty aluminum cans has a volume of 1.0 m^3. This volume includes the inner volume of each can which contains air. It also includes the air in the spaces between the cans. Therefore, most of the volume is occupied by air, not by aluminum. The mass of the pile of cans is *not* $\rho_{Al}V = (2700\ \text{kg/m}^3)(1.0\ \text{m}^3) = 2700\ \text{kg}$, because the volume of aluminum in the pile of cans is not 1.0 m^3.

---

2. ***REASONING AND SOLUTION*** Initially, the net force on the plunger is zero. The pressure in the bottle is equal to atmospheric pressure; therefore, the magnitude of the force per unit area pushing outward on the plunger is equal to the magnitude of the force per unit area that pushes inward on the plunger. After the first pull-and-release cycle, the force per unit area pushing outward on the plunger is less than the force per unit area pushing inward, because there is less pressure inside the bottle. Therefore, a larger external force is required to pull the plunger. On the 15th pull, the force per unit area pushing outward is very close to zero. The net force on the plunger is inward and due mostly to atmospheric pressure. Therefore, on the 15th pull, a much greater force will be required than on previous pulls.

---

3. ***REASONING AND SOLUTION*** A person could not balance her entire weight on the pointed end of a single nail, because it would penetrate her skin. According to Equation 11.3, the pressure exerted by the nail is $P = F/A$ where F represents the weight of the person, and A is the area of the tip of the nail. Since the tip of the nail has a very small radius, its area is very small; therefore, the pressure that the nail exerts on the person is large. The reason she can safely lie on a "bed of nails" is that the effective area of the nails is very large if the nails are closely spaced. Thus, the weight of the person F is distributed over all the nails so that the pressure exerted by any one nail is small.

---

4. ***REASONING AND SOLUTION*** As you climb a mountain, your ears "pop" because of the change in atmospheric pressure.

   a. As you climb up, the outside pressure becomes lower than the pressure in your inner ear. The outward force per unit area on your eardrum is greater than the inward force per unit area; therefore, your eardrum moves outward.

   b. As you climb down, the outside pressure becomes greater than the pressure in your inner ear. The outward force per unit area on your eardrum is less than the inward force per unit area on your eardrum; therefore, your eardrum moves inward.

---

5. ***REASONING AND SOLUTION*** The bottle of juice is sealed under a partial vacuum. Therefore, when the seal is intact, the button remains depressed, because the pressure inside the bottle is less than the atmospheric pressure outside of the bottle. The force per unit area pushing up on the button from the inside is significantly smaller than the force per unit area pushing down on the outside. When the seal is broken, air rushes inside the bottle. The force pushing up on the button increases and offsets the force pushing down. The natural springiness of the material from which the lid is made can then make the button "pop up."

---

6. ***REASONING AND SOLUTION*** Even though the tube is thin, pouring liquid into it increases the pressure uniformly throughout the vessel by an amount ρgh, where h is the height of the water in the tube. Apparently, before the tube is full, the outward force exerted on the walls of the container due to the fluid pressure is greater than the container can withstand. Therefore, the container bursts.

---

7. ***REASONING AND SOLUTION*** A closed tank is completely filled with water. A valve is opened at the bottom of the tank and the water begins to flow out. The water at the valve is pushed out by the force that results from the pressure exerted by the column of water above the valve. This force must be larger than the opposing inward force caused by atmospheric pressure at the valve. When these two forces are equal, the water will cease to exit at the valve. Therefore, when the water stops flowing, there will still be a noticeable amount of water in the tank.

   On the other hand, if the tank were open to the atmosphere, rather than closed, the water at the valve would experience an additional outward force due to atmospheric pressure on the top of the water column and the tank would empty completely.

---

8. ***REASONING AND SOLUTION*** When you drink through a straw, you draw the air out of the straw, and the external air pressure leads to the unbalanced force that pushes the liquid up into the straw. This action requires the presence of an atmosphere. The moon has no atmosphere, so you could not use a straw to sip a drink on the moon.

---

9. ***REASONING AND SOLUTION*** The *gauge pressure* of a fluid enclosed in a container is the amount by which the container pressure *exceeds* atmospheric pressure. The actual pressure of the fluid, taking atmospheric pressure into account, is called the absolute pressure. On earth, at sea level, the atmosphere exerts a pressure of 1.0×10^5 Pa on every surface with which it is in contact.

   An exam question asks, "Why does the cork fly out with a loud 'pop' when a bottle of champagne is opened?" To answer this question, a student replies, "Because the gas pressure in the bottle is about 0.3×10^5 Pa." The cork flies out because the absolute pressure within the bottle exceeds the atmospheric pressure on the outside. Assuming that the student recognizes this fact, the student must be referring to a gauge pressure of 0.3×10^5 Pa. Then, the absolute pressure within the bottle would be 1.0×10^5 Pa plus 0.3×10^5 Pa and would indeed exceed atmospheric pressure.

---

10. ***REASONING AND SOLUTION*** A scuba diver is below the surface of the water when a storm approaches, dropping the air pressure above the water. According to Equation 11.4, the absolute pressure at the location of the scuba diver is $P_{\text{diver}} = P_0 + \rho gh$ where P_0 is atmospheric pressure, ρ is the density of the water, and h is the depth of the diver below the surface. Since the storm causes the atmospheric pressure to drop, the absolute water pressure at the location of the diver will also drop. If the diver were wearing a sensitive gauge designed to measure the *absolute pressure*, it would register the drop in pressure.

---

11. ***REASONING AND SOLUTION*** According to Archimedes' principle, the water will apply a buoyant force to the steel beam; the magnitude of the buoyant force will be equal to the weight of the water displaced by the beam. Therefore, the buoyant force experienced by the beam is given by $(\rho_{\text{water}} V_{\text{beam}})g$ where ρ_{water} is the density of the water and V_{beam} is the volume of the beam. If we neglect any change in water density with depth, then, the quantity $\rho_{\text{water}} V_{\text{beam}}$ has the same value regardless of whether the beam is suspended vertically or horizontally under the water. Therefore, both the vertically suspended and the horizontally suspended beam experience the same buoyant force.

---

12. ***REASONING AND SOLUTION*** Archimedes' principle states that any fluid applies a buoyant force to an object that is partially or completely immersed in it; the magnitude of the buoyant force equals the weight of the fluid that the object displaces.

A glass beaker, filled to the brim with water, is resting on a scale. A block is placed in the water, causing some of it to spill over. The water that spills is wiped away, and the beaker is still filled to the brim.

a. According to Table 11.1, the density of wood is 550 kg/m^3, while the density of water at 4 °C is 1.000×10^3 kg/m^3. Since the density of the wood is less than the density of water, the block will float with part of its volume above the surface of the water. According to Archimedes' principle, the buoyant force exerted on the bottom surface of the block must be equal in magnitude to the weight of water that is displaced by the block. However, since the block floats on the surface of the water, the buoyant force must be equal in magnitude and opposite in direction to the weight of the block, as suggested in the free body diagram at the right. Hence, the weight of the block *must be equal to* the weight of the water displaced by the block. The weight of the contents of the beaker remains the same. Therefore, the initial and final scale readings are the same if the block is made of wood.

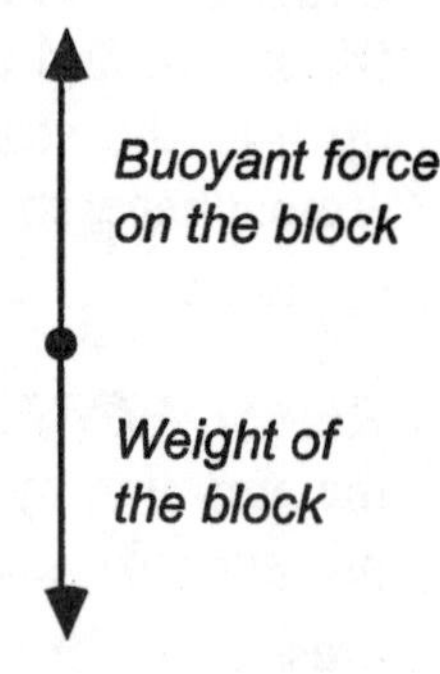

b. According to Table 11.1, the density of iron is 7860 kg/m^3, while the density of water at 4 °C is 1.000×10^3 kg/m^3. Since the density of iron is greater than the density of water, the iron block will sink to the bottom of the beaker. Let V represent the volume of water that spills over. According to the definition of mass density (Equation 11.1) the mass of water that spills over is equal to $m_{water} = \rho_{water} V$. The beaker is still filled to the brim, but the block now occupies the volume V where the water used to be. According to Equation 11.1, the mass that now occupies the volume V is given by $m_{block} = \rho_{iron} V$. Since $\rho_{iron} > \rho_{water}$, the mass of the block is greater than the mass of the water that spilled over. The weight of the contents of the beaker increases. Therefore, the final reading on the scale will be greater than the initial reading.

---

13. ***REASONING AND SOLUTION*** According to Archimedes' principle, any fluid applies a buoyant force to an object that is partially or completely immersed in it; the magnitude of the buoyant force equals the weight of the fluid that the object displaces. Therefore, the magnitude of the buoyant force exerted on an object immersed in water is given by $F_B = \rho_{water} V g$, where ρ_{water} is the density of water, V is the volume displaced by the immersed object, and g is the magnitude of the acceleration due to gravity. If the acceleration due to gravity on a distant planet is less than it is on earth, then, other factors remaining the same, the buoyant force will be less on the planet than it is on earth. However, the weight of the object will also be less than the weight of the object on earth.

When an object floats in water, the upward buoyant force exerted by the water must be equal in magnitude and opposite in direction to the weight of the object, as shown at the right. Hence, $F_B = m_{object} g$. It follows that $\rho_{water} V g = m_{object} g$. Notice that the acceleration due to gravity, g, appears on both sides of this equation. Algebraically canceling the g's we have $\rho_{water} V = m_{object}$. Therefore, the object will float so that it displaces a volume of water V, where $V = m_{object} / \rho_{water}$. This result is independent of g. It is the same on earth as it is on the distant planet. Therefore, it would be no more difficult to float in water on this planet than it would be on earth.

Buoyant force on object

Weight of object

---

14. ***REASONING AND SOLUTION*** An ice cube is placed in a glass, and the glass is filled to the brim with water. When the ice cube melts, and the temperature of the water returns to its initial value, the water level remains the same.

According to Table 11.1, the density of ice is less than that of liquid water; therefore, the ice will float in the glass with some of the ice above the level of the water. It will displace a volume of water V_{water} that is less than the volume of the ice cube, V_{ice}. According to Archimedes' principle, the water will apply a buoyant force to the ice cube; the magnitude of this force will be equal to the weight of the water that is displaced by the ice cube. Since the ice cube floats, the buoyant force must be equal in magnitude and opposite in direction to the weight of the ice cube. That is, $m_{ice} g = m_{water} g$; thus, the mass of the ice cube must be equal to the mass of the water that is displaced by the ice cube. When the ice cube melts, and the water has returned to its initial temperature, the mass of the water that originated from the ice cube occupies the volume V_{water}. In other words, the water that originated from the ice cube will occupy a volume that is exactly equal to the volume that was displaced by the ice cube. Therefore, the water level will remain the same.

---

15. ***REASONING AND SOLUTION*** As a person dives toward the bottom of a swimming pool, the pressure increases noticeably. The buoyant force, however, remains the same. According to Archimedes' principle, the water applies a buoyant force to the swimmer, the magnitude of which is equal to the weight of the water displaced by the swimmer. In other words, the buoyant force experienced by the swimmer is given by $(\rho_{\text{water}} V_{\text{swimmer}})g$, where ρ_{water} is the density of the water and V_{swimmer} is the volume of water displaced by the swimmer. If we neglect any change in water density with depth, then the quantity $\rho_{\text{water}} V_{\text{swimmer}}$ remains the same as the swimmer approaches the bottom of the pool.

---

16. ***REASONING AND SOLUTION*** According to Archimedes' principle, the magnitude of the buoyant force exerted on an object immersed in water is equal to the weight of the water displaced by the object.

As discussed in Section 5.6 of the text, objects in orbiting satellites are in uniform circular motion and "fall" with the same acceleration toward the center of the orbit. The apparent weight of an object in an orbiting satellite is, therefore, zero. Thus, the apparent weight of the water displaced by the swimmer is zero and the swimmer will not experience a buoyant force. This assumes, as stated, that there is no artificial gravity.

---

17. ***REASONING AND SOLUTION*** When the gate is not in use, the hollow, triangular-shaped gate is filled with seawater and rests on the ocean floor. As shown in the figure, the gate is anchored to the ocean floor by a hinge mechanism, about which it can rotate. When a dangerously high tide is expected, water is pumped out of the gate, and the structure rotates into a position where it can protect against flooding.

When the gate is filled with sea water, its average density (including the gate material) is greater than that of seawater, and the gate sinks to the ocean floor. When water is pumped from the gate, the density of the "gate system" decreases. Therefore, the mass, and hence the weight, of the "gate system" decreases. When the buoyant force exerted on the bottom of the "gate system" by the sea water exceeds the weight of the "gate system," the gate begins to rise. Since the left end of the gate is anchored to the hinge mechanism, the gate will rotate about the hinge as its right end rises.

---

18. ***REASONING AND SOLUTION*** In steady flow, the velocity **v** of a fluid particle at any point is constant in time. On the other hand, a fluid accelerates when it moves into a region of smaller cross-sectional area, as shown in the figure below.

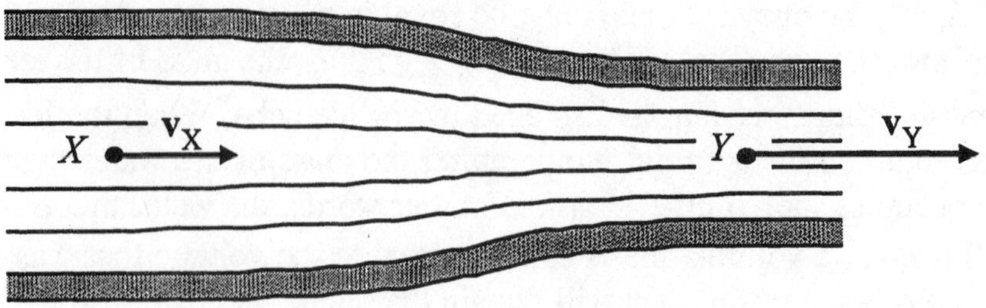

a. A fluid particle at X with speed v_X must be accelerated to the right in order to acquire the greater speed v_Y at Y. From Newton's second law, this acceleration can arise only from a net force that acts in the direction XY. If there are no other external forces acting on the fluid, this force must arise from the change in pressure within the fluid. The pressure at X must be greater than the pressure at Y.

b. The definition of steady flow makes no reference as to how the velocity of a fluid particle varies from point to point as the fluid flows. It simply states that the velocity of a fluid particle *at any particular point* is constant in time. Therefore, the condition of steady flow does not rule out the acceleration discussed in part (a).

---

19. ***REASONING AND SOLUTION*** Consider a stream of water that falls from a faucet. According to the equation of continuity, the mass flow rate ρAv must be the same at every point along the stream (ρ = water density, A = cross-sectional area of the stream, v = speed of stream). As the water falls, it is accelerated due to gravity; therefore, the speed of the water increases as it falls. Since the density of water is uniform throughout the stream, when v increases, the cross-sectional area A of the stream must decrease in order to maintain a constant mass flow rate. Therefore, the cross-sectional area of the stream becomes smaller as the water falls.

If the water is shot upward, as it is in a fountain, the velocity of the stream is upward, while the acceleration due to gravity is directed downward. Therefore, the speed of the stream decreases as the stream rises. Since v decreases, the cross-sectional area A must increase in order to maintain a constant mass flow rate ρAv. Therefore, the cross-sectional area of the stream becomes larger as the water is shot upward.

---

20. ***REASONING AND SOLUTION*** Suppose you are driving your car alongside a moving truck. You will notice that your car will be pulled toward the truck as it passes.

Consider the air at two points. One point O is on the open side of your car, that is, the side that is not adjacent to the truck. The other point B is between the car and the truck. Bernoulli's principle, as applied to a constant elevation (Equation 11.12), indicates that where the air speed is smaller, the air pressure is greater. Since the car is pulled toward the truck, the pressure at point O must be greater than at point B. Therefore, the air at point O is not moving very fast, while the air at point B is rushing more rapidly around the curved surfaces of the front of the vehicles and through the narrow space between them.

---

21. ***REASONING AND SOLUTION*** Bernoulli's equation applies to the steady flow of a nonviscous, incompressible fluid. Turbulent flow is an extreme type of unsteady flow. It occurs when there are sharp obstacles or bends in the path of a fast-moving fluid. Bernoulli's equation does not apply to turbulent flow. When water cascades down a rock-strewn spillway, the water exhibits turbulent flow. The velocity at any particular point changes erratically from one instant to another. Therefore, Bernoulli's equation could not be used to describe water flowing down the rock-strewn spillway.

---

22. ***REASONING AND SOLUTION*** Two sheets of paper are held by adjacent corners, so that they hang downward. They are oriented so that they are parallel and slightly separated by a gap through which the floor can be seen. When air is blown strongly down through the gap (see the drawing at the right), the sheets come closer together.

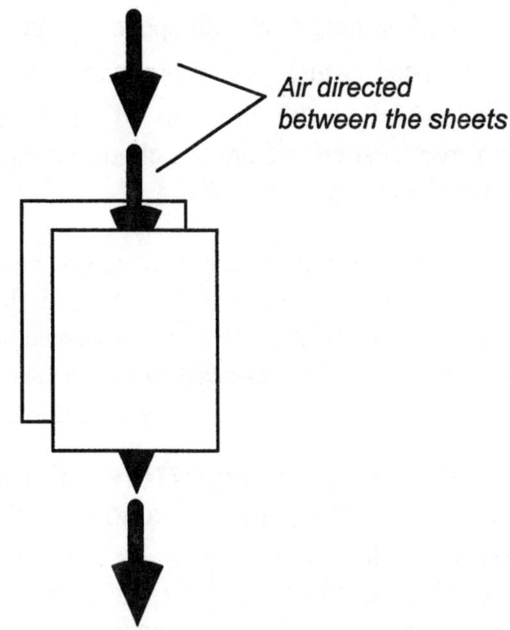

In accord with Bernoulli's equation, the moving air between the sheets has a lower pressure than the stationary air on either side of the sheets. The greater pressure on either side of the sheets generates an inward force on the outer surface of the sheets, and the sheets are pushed together.

---

23. ***REASONING AND SOLUTION*** The figure below shows a baseball, ***as viewed from the side***, moving to the right with no spin. Since the air flows with the same speed above and below the ball, the pressure is the same above and below the ball. There is no net force to cause the ball to curve in any particular direction (except for gravity which results in the usual parabolic trajectory).

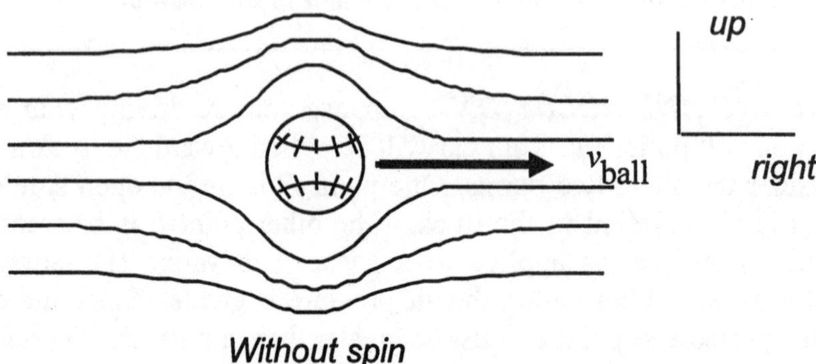

Without spin

If the ball is given a spin that is counterclockwise when viewed from the side, as shown below, the air close to the surface of the ball is dragged with the ball. In accord with Bernoulli's equation, the air on the top half of the ball is "speeded up" (lower pressure), while that on the lower half of the ball is slowed down (higher pressure).

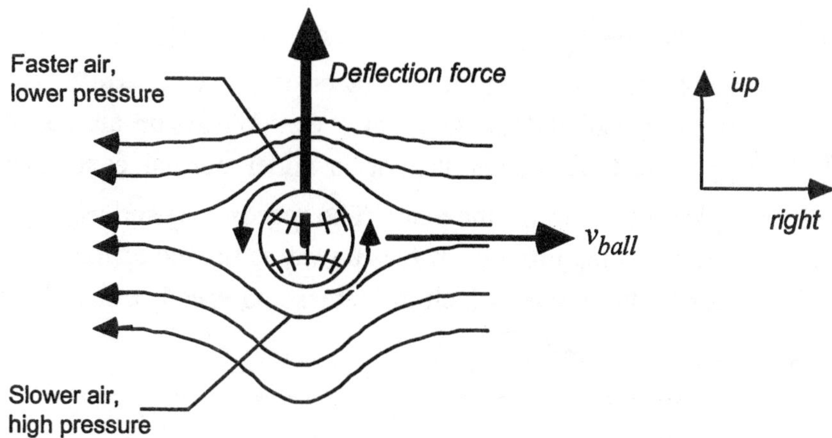

Because of the pressure difference, a deflection force is generated that is directed from the higher pressure side of the ball to the lower pressure side of the ball. Therefore, the ball curves upward on its way to the plate.

---

24. ***REASONING AND SOLUTION*** You are traveling on a train with your window open. As the train approaches its rather high operating speed, your ears "pop."

When the train is stationary, the air inside and outside the train is stationary. The pressures inside and outside, therefore, are equal. When the train moves at a high speed, the air rushes rapidly by the window. According to Bernoulli's principle as expressed in Equation 11.12, the faster moving air outside has a lower pressure than the nearly stationary air inside. Under this condition some of the inside air is pushed out the window. As a result, the pressure within the train drops. However, the pressure of the air in your inner ear is still at the higher level that it was when the train was stationary. Thus, your ears "pop" outward.

---

25. ***REASONING AND SOLUTION*** A passenger is smoking a cigarette in the back seat of a moving car. To remove the smoke, the driver opens a window just a bit. Relative to the car, the air in the car is stationary, while the air outside the car is moving. According to Bernoulli's principle, as expressed in Equation 11.12, the moving air has a lower pressure than the stationary air. Hence, the higher-pressure air inside the car will flow through the opening to the lower-pressure region on the outside. Thus, the smoke will be drawn from the higher-pressure region in the car to the lower-pressure region outside.

---

26. ***REASONING AND SOLUTION*** As discussed in the text, the dynamic lift on airplane wings occurs because the air travels faster over the curved upper surface of the wing than it does over the flatter lower surface (see Figure 11.34). According to Bernoulli's equation, the pressure above the wing is lower (higher air speed), while the pressure below the wing is higher (lower air speed). The net upward pressure on the wing causes an upward force that lifts the plane.

The airport in Phoenix, Arizona, has occasionally been closed to large planes because of unusually low air density. According to Equation 11.12, the net upward pressure on the wing of a plane is given by $P_{\text{bottom}} - P_{\text{top}} = (1/2)\rho_{\text{air}}(v_{\text{top}}^2 - v_{\text{bottom}}^2)$ where P_{bottom} and P_{top} are the air pressures on the bottom and top surfaces of the wing, while v_{top} and v_{bottom} are the air speeds at the top and bottom wing surfaces. When the air density is low, the quantity $(v_{\text{top}}^2 - v_{\text{bottom}}^2)$ must be greater than it is under normal conditions to provide the same minimum upward force to lift the plane. The speeds, v_{top} and v_{bottom}, depend on the speed of the plane relative to the ground. If the runway is not exceptionally long, the plane may not have sufficient distance over which to increase its speed so that the quantity $(v_{\text{top}}^2 - v_{\text{bottom}}^2)$ is large enough for take-off.

---

27. ***REASONING AND SOLUTION*** To change the oil in a car, you remove a plug beneath the engine and let the old oil run out. Your car has been sitting in the garage on a cold day. Before changing the oil, it is advisable to run the engine for a while.

As discussed in the text, the viscosity of liquids usually decreases as the temperature is increased. Since the car was sitting in a cold garage, the viscosity of the oil will be higher than it is at room temperature. Therefore, if the plug is removed, the oil will run out, but at a speed which is small compared to its speed of flow at room temperature. If the engine is run for a while, the oil will heat up and its viscosity will decrease. The oil will then flow more easily, and it can be drained from the car in a much shorter time than if it were at the ambient temperature of the garage.

CHAPTER 12 | *TEMPERATURE AND HEAT*

CONCEPTUAL QUESTIONS

---

1. ***REASONING AND SOLUTION*** The length L_0 of the tape rule changes by an amount ΔL when its temperature changes by an amount ΔT, where ΔL is given by $\Delta L = \alpha L_0 \Delta T$ and α is the coefficient of linear expansion. When the length of the tape rule changes, the location of the graduations changes and makes the rule inaccurate. For year-round outdoor use, the range of ΔT will be large; therefore, to minimize the thermal expansion of the rule we would want the rule to be made from a material with a small coefficient of linear expansion. Since the coefficient of linear expansion of steel is less than that of aluminum, we would choose a steel tape rule for highest accuracy.

---

2. ***REASONING AND SOLUTION*** The distance between the two fine lines engraved near the ends of the bar is unique only at a specified temperature. When the temperature of the bar is raised, the length of the bar increases, and the distance between the engraved lines increases. Since the bar was an international standard for length, it was essential that the distance between the lines always remained the same; therefore, it was important that the bar be kept at a constant temperature.

---

3. ***REASONING AND SOLUTION*** The plate is made of aluminum; the spherical ball is made of brass. The coefficient of linear expansion of aluminum is greater than the coefficient of linear expansion of brass. Therefore, if the plate and the ball are heated, both will expand; however, the diameter of the hole in the aluminum plate will expand more than the diameter of the brass ball. In order to prevent the ball from falling through the hole, the plate and the ball must be cooled. Both the diameter of the hole in the plate and the diameter of the ball will contract. The diameter of the hole will decrease more than the diameter of the ball, thereby preventing the ball from falling through the hole.

---

4. ***REASONING AND SOLUTION*** Reinforced concrete is concrete that is reinforced with embedded steel rods. For this arrangement to be practical, it is essential that the coefficient of linear expansion of concrete is the same as that of steel. When the reinforced concrete is heated or cooled, both the concrete and the steel must expand or contract by the same amount. If the coefficients of linear expansion were not the same, the concrete and steel would expand and contract at different rates when heated or cooled. If the steel expanded or contracted more than the concrete, it would place internal stresses on the concrete, thereby causing it to weaken and possibly crack.

---

5. ***REASONING AND SOLUTION*** A rod is hung from an aluminum frame such that there is a small gap between the rod and the floor. The rod and the frame are heated uniformly; both the frame and the rod will expand. The linear expansion of the vertical portion of the frame and the linear expansion of the rod can be determined from Equation 12.2: $\Delta L = \alpha L_0 \Delta T$.

a. If the rod is made of aluminum, the fractional increase in length $\Delta L / L_0$ is the same for the vertical portion of the frame and the rod. Since the rod is initially shorter than the vertical portion of the frame, the rod will never be as long as the vertical portion of the frame. Therefore, the rod will never touch the floor.

b. If the rod is made of lead, the heated rod could touch the floor. The coefficient of linear expansion of lead is greater than that of aluminum. Therefore, the fractional change in the length of the rod will be greater than the fractional change in the length of the vertical portion of the frame. If the temperature is raised high enough, the length of the rod could expand enough to fill the gap and touch the floor.

---

6. ***REASONING AND SOLUTION*** A simple pendulum is made using a long thin metal wire. From Equations 10.6 and 10.16, we know that the period of the pendulum is proportional to $\sqrt{L/g}$, where L is the length of the wire. When the temperature drops, the length of the wire decreases; therefore, the period of the pendulum decreases.

---

7. ***REASONING AND SOLUTION*** The bottom of the pot bows outward, because it is a kind of bimetallic strip. Table 12.1 shows that the coefficient of linear thermal expansion for copper is $\alpha = 17 \times 10^{-6}\ (\text{C}^\circ)^{-1}$, while for steel it is $\alpha = 12 \times 10^{-6}\ (\text{C}^\circ)^{-1}$. Thus, as the pot is heated, the copper expands more than the steel. Since the copper is on the outside, the bottom of the pot bows outward.

---

8. ***REASONING AND SOLUTION*** A hot steel ring fits snugly over a cold brass cylinder. The temperatures of the ring and the cylinder are, respectively, above and below room temperature. As the assembly equilibrates to room temperature, the diameter of the steel ring contracts while the diameter of the brass cylinder expands. The ring and the cylinder, therefore, fit much more snugly than they did initially. It becomes nearly impossible to pull the ring off the cylinder once the assembly has equilibrated at room temperature.

---

9. ***REASONING AND SOLUTION*** The coefficient of volume expansion of Pyrex glass is only about 39 percent that of common glass. Therefore, the fractional change in volume, $\Delta V / V_0$, of a Pyrex baking dish is significantly less than that of common glass when both objects are taken from a refrigerator and placed in a hot oven. The Pyrex dish, therefore, is less likely to crack from thermal stress.

---

10. ***REASONING AND SOLUTION*** An ordinary mercury-in-glass thermometer works on the following principle. When the thermometer is placed in contact with the object whose temperature is to be measured, the thermometer comes into thermal equilibrium with the object. As thermal equilibrium is reached, the glass tube and the enclosed mercury expand or contract, depending on whether the thermometer heats up or cools down. The coefficient of volume expansion of mercury is about 7 times greater than that of ordinary glass. Thus, the expansion or contraction of the mercury column relative to the scale on the glass is evident when one reads the thermometer. If the coefficients of volume expansion of mercury and glass were both the same, both would expand or contract by the same amount. The reading on the thermometer would never change.

---

11. ***REASONING AND SOLUTION*** According to Archimedes' principle, any fluid applies a buoyant force to an object that is partially or completely immersed in it; the magnitude of the buoyant force equals the weight of the fluid that the object displaces. Therefore, the magnitude of the buoyant force exerted on an object immersed in water is given by $F_B = \rho_{water} V g$, where ρ_{water} is the density of water, and V is the volume displaced by the immersed object.

As shown in Figure 12.20, the density of cold water (above 4 °C) is greater than the density of warm water (above 4 °C). Therefore, cold water provides a greater buoyant force than warm water.

---

12. ***REASONING AND SOLUTION*** When the bulb of a mercury-in-glass thermometer is inserted into boiling water, the glass is heated first. Therefore, the glass envelope expands. The cavity that contains the mercury increases in volume, and the mercury level drops slightly. Then, the mercury begins to be heated, and its volume expands. Since the coefficient of volume expansion for mercury is much greater than that of the glass, the mercury expands to a greater extent than the glass cavity, and the mercury level then begins to rise.

---

13. ***REASONING AND SOLUTION*** When an amount of heat Q is added to an object of mass m, its temperature will increase. According to Equation 12.4, the amount by which the temperature will increase is given by $\Delta T = Q / cm$, where c is the specific heat capacity of the object.

Two different objects are supplied with equal amounts of heat. The objects could have the same specific heat capacities, but they could have different masses. The objects could have the same masses, but they could have different specific heat capacities. Or, the objects could have both different masses and different specific heat capacities. In any of these cases, the temperature changes of the objects would not be the same.

---

14. ***REASONING AND SOLUTION*** Two objects are made from the same material, but have different masses and temperatures. When the two objects are placed in contact, they will come into thermal equilibrium and reach a common temperature. The heat gained by the cooler object must be equal to the heat lost by the warmer object, so that $cm_1 \Delta T_1 = cm_2 \Delta T_2$,

where one object has mass m_1 and the other object has mass m_2. Since the objects are made of the same material, they have the same specific heat capacities so that $m_1 \Delta T_1 = m_2 \Delta T_2$ or $\Delta T_2 = (m_1 / m_2) \Delta T_1$. Suppose that m_1 is greater than m_2. Then, ΔT_2 is greater than ΔT_1, so the less massive object must undergo the greater temperature change.

---

15. ***REASONING AND SOLUTION*** Two identical mugs contain hot coffee from the same pot. One mug is full, while the other is only one-quarter full. According to Equation 12.4, the amount of heat that must be removed from either cup to bring it to room temperature is $Q = cm\Delta T$, where m is the mass of coffee in the cup, c is the specific heat capacity of the coffee and ΔT is the temperature difference between the initial temperature of the coffee and room temperature. Since the mug that is full contains a larger mass m of coffee, more heat Q must be removed from the full mug than from the mug that is only one-quarter full. Therefore, the mug that is full remains warmer longer.

---

16. ***REASONING AND SOLUTION*** As heat flows from the higher-temperature air to the lower-temperature water bottle, it passes through the wet sock. Most of the heat causes the water in the sock to evaporate, and only a relatively small amount of heat reaches the water bottle. Thus, as long as the sock remains wet, the water bottle remains cool.

---

17. ***REASONING AND SOLUTION*** An alcohol rub has a cooling effect on a sick patient for the following reason. In order for the alcohol to evaporate, heat must be supplied to it. The evaporating alcohol removes heat from the skin of the sick patient. The evaporating alcohol, therefore, helps to lower the body temperature of the patient.

---

18. ***REASONING AND SOLUTION*** Your head feels colder under an air-conditioning vent when your hair is wet than when it is dry. When your hair is wet, the water evaporates. In order to evaporate, heat must be supplied to the water. Therefore, heat is removed from your head as the water evaporates, thereby causing a cooling effect on your head.

---

19. ***REASONING AND SOLUTION*** Suppose that the latent heat of vaporization of H_2O were one-tenth of its actual value.

a. Other factors being equal, it would take only one-tenth the amount of heat to "boil away" a pot of water on the stove. Assuming that the stove supplies heat at a constant rate, it would, therefore, take a shorter time for a pot of water to boil away.

b. The evaporative cooling mechanism of the human body would be less effective. When a given amount of water evaporates from the skin, it would now require only one-tenth as much heat. Therefore, only one-tenth as much heat would be removed from the skin surface for a given amount of perspiration.

---

20. ***REASONING AND SOLUTION*** Orchard owners sometimes spray a film of water over fruit blossoms to protect them when a hard freeze is expected. The thin film of water will freeze before the water in the tissues of the blossoms. When the thin film of water freezes at 0 °C, each kilogram of water releases an amount of heat equal to the latent heat of fusion. Some of this heat will warm and, therefore, protect the blossoms from reaching the damaging temperature of – 4 °C.

---

21. ***REASONING AND SOLUTION*** When a liquid is heated, its equilibrium vapor pressure increases as the temperature increases, as shown in Figure 12.32 for water. When the vapor pressure of the liquid is equal to the atmospheric pressure above the liquid, the liquid begins to boil. Therefore, at a higher altitude, where the atmospheric pressure is less than at lower altitudes, boiling is more easily achieved. At a higher altitude, the vapor pressure does not have to be raised as much to reach the boiling point, and the temperature required to boil a liquid at the higher altitude is less than that at a lower altitude.

A camping stove is used to boil water on a mountain. Since water will boil at a lower temperature at higher altitudes, it does not necessarily follow that the same stove can boil water at lower altitudes. If the stove cannot provide sufficient heat Q to raise the temperature of the water to the temperature at which the vapor pressure of the liquid is equal to the atmospheric pressure at the lower altitude, the water will not boil.

---

22. ***REASONING AND SOLUTION*** If a bowl of water is placed in a closed container and water vapor is pumped away rapidly enough, the liquid water turns to ice. As water vapor is pumped away, the pressure in the jar is reduced. The liquid attempts to reestablish the equilibrium vapor pressure, so more water molecules enter the gaseous state. As the water vaporizes, heat must be supplied in the form of the latent heat of vaporization. Because of the heat loss, the remaining liquid is cooled. If this heat loss occurs rapidly enough, the surroundings cannot replenish the heat loss, and, when sufficient heat has been removed, the remaining liquid turns to ice.

---

23. ***REASONING AND SOLUTION*** If medical instruments were sterilized in an open container of water, the water would be subjected to atmospheric pressure. Therefore, the highest possible temperature that could be attained is 100 °C. In order to raise the temperature above 100 °C, we would have to raise the pressure above the liquid, as suggested by Figure 12.31. An autoclave heats medical instruments in water under high pressure. We can see from Figure 12.32 that at pressures above atmospheric pressure, the boiling point of water is greater than 100 °C. Therefore, by sterilizing medical instruments in an autoclave, it is possible to attain temperatures much greater than 100 °C.

---

24. ***REASONING AND SOLUTION*** A bottle of carbonated soda is left outside in subfreezing temperatures. The temperature of the soda is "subfreezing," but the soda remains in the liquid state. Because the bottle is sealed, the pressure above the soda remains too high for the soda to freeze. When the soda is brought inside and opened, the pressure drops to atmospheric pressure. Because the temperature of the soda is below the freezing point at atmospheric pressure, the soda freezes.

---

25. ***REASONING AND SOLUTION*** A bowl of water is covered tightly and allowed to sit at a constant temperature of 23 °C for a long time. Since the bowl is tightly covered, none of the water vapor can escape, and since the bowl is allowed to sit for a long time at a constant temperature, the vapor above the liquid attains equilibrium with the liquid. The partial pressure of the water vapor will equal the equilibrium vapor pressure of water at 23 °C, and the relative humidity will be 100%.

---

26. ***REASONING AND SOLUTION*** Dew forms on Tuesday night and not on Monday night, even though Monday night is the cooler night. Since dew did not form on Monday night, we can conclude that the temperature on Monday night was above the dew point. The partial pressure of water in the air must have been greater on Tuesday night; then, the dew point would occur at a higher temperature. As long as the temperature on Tuesday is below the dew point, dew will form.

---

27. ***REASONING AND SOLUTION*** Two rooms in a house have the same temperature. One of the rooms contains an indoor swimming pool. The partial pressure of water vapor in the air will be greater in the room that contains the indoor pool. Therefore, the dew point will be higher in the room with the pool. On a cold day, the windows of one of the rooms are "steamed up." Water will condense on the window when the temperature of the air next to the glass falls below the dew point. Since only one of the rooms has "steamed" windows, and the dew point is higher for the room with the pool, then the outside temperature must be less than the dew point of the room with the pool, but greater than the dew point for the room without the pool. Therefore, the windows are "steamed up" in the room with the pool.

---

28. ***REASONING AND SOLUTION*** A jar is half filled with boiling water. The lid is then screwed on the jar. The temperature of the jar and the lid will be close to the boiling point of water. As the jar is cooled to room temperature, the lid and the jar contract. Since the coefficient of linear expansion of the metal jar lid is greater than the coefficient of linear expansion of the glass jar, the lid will contract more than the jar. Therefore, the lid will fit more snugly when the jar has cooled down to room temperature and be difficult to remove.

In addition, as the temperature of the water vapor inside the jar drops, the pressure inside becomes less than the outside pressure. Thus, there is a net force pushing the lid onto the jar, making the lid harder to turn.

CHAPTER 13 | *THE TRANSFER OF HEAT*

CONCEPTUAL QUESTIONS

---

1. ***REASONING AND SOLUTION*** Convection is the process in which heat is carried from one place to another by the bulk movement of the medium. In liquids and gases, the molecules are free to move; hence, convection occurs as a result of bulk molecular motion. In solids, however, the molecules are generally bound to specific locations (lattice sites). While the molecules in a solid can vibrate about their equilibrium locations, they are not free to move from place to place within the solid. Therefore, convection does not generally occur in solids.

---

2. ***REASONING AND SOLUTION*** A heavy drape, hung close to a cold window, reduces heat loss through the window by interfering with the process of convection. Without the drape, convection currents bring the warm air of the room into contact with the cold window. With the drape, convection currents are less prominent, and less room air is circulated directly past the cold surface of the window.

---

3. ***REASONING AND SOLUTION*** Forced convection plays the principal role in the wind chill factor. The wind mixes the cold ambient air with the warm layer of air that immediately surrounds the exposed portions of your body. The forced convection removes heat from your exposed body surfaces, thereby making you feel colder than you would otherwise feel if there were no wind.

---

4. ***REASONING AND SOLUTION*** A road surface is exposed to the air on its upper surface and to the earth on its lower surface. Even when the air temperature is at the freezing point, the road surface may be above this temperature as heat flows through the road from the earth. In order for a road to freeze, sufficient heat must be lost from the earth by conduction through the road surface. The temperature of the earth under the road must be reduced at least to the freezing point. A bridge is exposed to the air on both its upper and lower surfaces. It will, therefore, lose heat from both surfaces and reach thermal equilibrium with the air much more quickly than an ordinary roadbed. It is reasonable, then, that the bridge surface will usually freeze before the road surface.

---

5. ***REASONING AND SOLUTION*** A piece of Styrofoam and a piece of wood are sandwiched together to form a layered slab. The two pieces have the same thickness and cross-sectional area. The exposed surfaces have constant temperatures. The temperature of the exposed Styrofoam surface is greater than the temperature of the exposed wood surface. The rate of heat flow through either layer can be determined from Equation 13.1: $Q/t = kA\Delta T/L$, where k is the thermal conductivity of the layer, A and L are the cross-sectional area and thickness of the layer, respectively, and ΔT is the temperature difference between the ends of the layer. Since heat is not trapped within the sandwich, the rate at

which heat flows through the sandwich, Q/t, must be uniform throughout both layers. Therefore, $(kA\Delta T / L)_{\text{Styrofoam}} = (kA\Delta T / L)_{\text{wood}}$. Since both layers have the same cross-sectional area and thickness, A and L are the same for both layers. Therefore, $k_{\text{Styrofoam}} \Delta T_{\text{Styrofoam}} = k_{\text{wood}} \Delta T_{\text{wood}}$. From Table 13.1, we see that the thermal conductivity of Styrofoam is less than the thermal conductivity of wood; therefore, the temperature difference between the two ends of the wood layer must be smaller than the temperature difference between the two ends of the Styrofoam layer. From this, we can conclude that the temperature at the Styrofoam-wood interface must be closer to the lower temperature of the exposed wood surface.

---

6. ***REASONING AND SOLUTION*** When heat is transferred from place to place inside the human body by the flow of blood, the main method of heat transfer is forced convection, similar to that illustrated for the radiator fluid in Figure 13.7. The heart is analogous to the water pump in the figure.

---

7. ***REASONING AND SOLUTION*** Some animals have hair, the strands of which are hollow, air-filled tubes. Other animals have hair that is composed of solid, tubular strands. For animals that live in very cold climates, hair that is composed of hollow air-filled tubes would be advantageous for survival. Since air has a small thermal conductivity, hair shafts composed of hollow air-filled tubes would reduce the loss of body heat by conduction. Since hair shafts are small, no appreciable convection would occur within them. Thus, the hollow air-filled structure of the hair shaft inhibits the loss of heat by conduction.

---

8. ***REASONING AND SOLUTION*** A poker used in a fireplace is held at one end, while the other end is in the fire. Such pokers are made of iron rather than copper because the thermal conductivity of iron is roughly smaller by a factor of five than the thermal conductivity of copper. Therefore, the transfer of heat along the poker by conduction is considerably reduced by using iron. Hence, one end of the poker can be placed in the fire, and the other end will remain cool enough to be comfortably handled.

---

9. ***REASONING AND SOLUTION*** Snow, with air trapped within it, is a thermal insulator, because air has a relatively low thermal conductivity and the small, dead-air spaces inhibit heat transfer by convection. Therefore, a lack of snow allows the ground to freeze at depths greater than normal.

---

10. ***REASONING AND SOLUTION*** Table 13.1 indicates that the thermal conductivity of steel is 14 $\text{J/(s} \cdot \text{m} \cdot \text{C}^\circ)$, while that of concrete is 1.1 $\text{J/(s} \cdot \text{m} \cdot \text{C}^\circ)$. According to Equation 13.1, $Q = kA\Delta Tt / L$, this implies that heat will flow more readily through a volume of steel than it will through an identically shaped volume of concrete. Therefore, while steel reinforcement bars can enhance the structural stability of concrete walls, they degrade the insulating value of the concrete.

---

11. ***REASONING AND SOLUTION*** A potato will bake faster if a nail is driven into it before it is placed in the oven. Since the nail is metal, we can assume that the thermal conductivity of the nail is greater than the thermal conductivity of the potato. The nail conducts more heat from the oven to the interior of the potato than does the flesh of the potato, thereby causing the potato to bake faster.

---

12. ***REASONING AND SOLUTION*** Several days after a snowstorm, the roof on a house is uniformly covered with snow. On a neighboring house, the snow on the roof has completely melted. Since one of the houses still has snow on the roof, it is reasonable to conclude that the ambient temperature is still below the freezing point of water. Since the snow has melted from the roof of the neighboring house, we can conclude that the heat required to melt the snow must have come through the attic and the roof by conduction. Hence, the house which has the uniform layer of snow on the roof is probably better insulated. The better the insulation, the smaller is the amount of heat conducted through the roof to melt the snow.

---

13. ***REASONING AND SOLUTION*** One car has a metal body, while another car has a plastic body. On a cold winter day, these cars are parked side by side. The metal car feels colder to the touch of your bare hand even though both cars are at the same temperature. This is because your fingers are sensitive to the rate at which heat is transferred to or from them, rather than to the temperature itself. The metal car feels colder than the plastic car at the same temperature, because heat flows from your bare hand into the metal car more readily than it flows into the plastic car. The flow occurs into the metal more readily, because the thermal conductivity of the metal is greater than that of the plastic.

---

14. ***REASONING AND SOLUTION*** Many high-quality pots have copper bases and polished stainless steel sides. Since copper has a high thermal conductivity, heat can readily enter the bottom of the pot by means of conduction. Since the temperature of the pot is greater than the temperature of its environment, the pot will lose heat by means of radiation. Polished stainless steel has a low emissivity; that is, it is a poor emitter of radiant energy. Hence, by making the sides of the pot polished stainless steel, the amount of heat that would be lost by radiation is minimized. This design is optimal. If the pot were constructed entirely of copper, the bottom would efficiently conduct heat into the pan; however, heat would also be conducted efficiently into the sides of the pot, raising their temperature and increasing the loss from the sides via radiation. If, on the other hand, the pot were constructed entirely of stainless steel, the loss of heat through radiant energy would be minimized; however, since stainless steel has a low thermal conductivity, heat would not efficiently enter the bottom of the pot through conduction.

---

15. ***REASONING AND SOLUTION*** The radiant energy Q emitted in a time t by an object that has a Kelvin temperature T, a surface area A, and an emissivity e, is given by Equation 13.2, $Q = e\sigma T^4 At$, where σ is the Stefan-Boltzmann constant.

We now consider two objects that have the same size and shape. Object A has an emissivity of 0.3, and object B has an emissivity of 0.6. Since each object radiates the same power, $e_A \sigma T_A^4 A_A = e_B \sigma T_B^4 A_B$. The Stefan-Boltzmann constant is a universal constant, and since the objects have the same size and shape, $A_A = A_B$; therefore, $e_A T_A^4 = e_B T_B^4$, or $T_A / T_B = \sqrt[4]{e_B / e_A} = \sqrt[4]{2}$. Hence, the Kelvin temperature of A is $\sqrt[4]{2}$ or 1.19 times the Kelvin temperature of B, not twice the temperature of B.

---

16. ***REASONING AND SOLUTION*** A concave mirror can be used to start a fire by directing sunlight onto a small spot on a piece of paper. The mirror, being a good reflector, is a poor absorber; therefore it will not absorb much of the sun's radiant energy. Paper is a better absorber than the mirror, so the paper absorbs the radiant energy. Therefore, the mirror does not get as hot as the paper. Eventually, the paper gets hot enough to burn.

---

17. ***REASONING AND SOLUTION*** Two strips of material, A and B, are identical except that they have emissivities of 0.4 and 0.7, respectively. The strips are heated to the same temperature and have a bright glow. The emissivity is the ratio of the energy that an object actually radiates to the energy that the object would radiate if it were a perfect emitter. The strip with the higher emissivity will radiate more energy per second than the strip with the lower emissivity, other things being equal. Therefore, strip B will have the brighter glow.

---

18. ***REASONING AND SOLUTION*** The thermal conductivity of the bottom of the pot is greater than the thermal conductivity of air; therefore, the portion of the heating element beneath the pot loses heat by conduction through the bottom of the pot. The exposed portion of the heating element loses some heat through convection, but the convective process is not as efficient as the conductive process through the bottom of the pot. The exposed portion of the heating element will, therefore, lose less heat and be at a higher temperature than the portion of the heating element beneath the pot. Thus, the exposed portion glows cherry red.

---

19. ***REASONING AND SOLUTION*** If we consider a glove and a mitten, each of the same "size" and made of the same material, we can deduce that the mitten has less surface area A exposed to the cold winter air. Thus, according to Equation 13.1, $Q = kA\Delta Tt / L$, we can conclude that the mitten will conduct less heat per unit time from the hand to the winter air. Therefore, to keep your hands as warm as possible during skiing, you should wear mittens as opposed to gloves.

---

20. ***REASONING AND SOLUTION*** Two identical hot cups of cocoa are sitting on a kitchen table. One has a metal spoon in it and one does not. After five minutes, the cocoa with the metal spoon in it will be cooler. The metal spoon conducts heat from the cocoa to the handle of the spoon. Convection currents in the air and radiation then remove the heat from the spoon handle. The conduction-convection-radiation process removes heat from the cocoa, thereby cooling it faster than the cocoa that does not have a spoon in it.

---

21. ***REASONING AND SOLUTION*** The radiant energy Q emitted in a time t by an object that has a Kelvin temperature T, a surface area A, and an emissivity e, is given by Equation 13.2: $Q = e\sigma T^4 At$, where σ is the Stefan-Boltzmann constant.

a. A hot solid cube will cool more rapidly if it is cut in half, rather than if it is left intact. Since the cube is warmer than its environment, it will lose heat primarily through radiation. Convection currents will also remove some heat from the surface of the cube. When the cube has been cut in half, the surface area of the solid has been increased. If the length of one edge of the original cube is L, then cutting the cube in half increases the surface area from $6L^2$ to $8L^2$. From Equation 13.2, the amount of heat Q radiated in a time t is proportional to the surface area of the cube; therefore, the cube will radiate more rapidly and cool more rapidly if it is cut in half.

b. One pound of spaghetti noodles has a larger effective surface area than one pound of lasagna noodles. Imagine cutting many spaghetti noodles from one large lasagna noodle, in a way similar to what was done to the cube in part (a). Since the effective surface area of the spaghetti noodles is greater than that of the lasagna, heat will be radiated from the surface of the spaghetti noodles more effectively than heat will be radiated from the surface of the lasagna noodles. Therefore, the spaghetti noodles will cool more rapidly from the same initial temperature than the lasagna noodles.

---

22. ***REASONING AND SOLUTION*** The black asphalt is a better absorber than the cement; the black asphalt will absorb more of the sun's radiant energy than the cement. Since the sun has been shining all day, the asphalt will be at a higher temperature than the cement. The temperature of the asphalt is apparently above the freezing point of water, while the temperature of the cement playground is below the freezing point of water. Therefore, when snow hits the asphalt, it melts immediately, while the snow collects on the cement.

---

23. ***REASONING AND SOLUTION*** The radiant energy Q emitted in a time t by an object that has a Kelvin temperature T, a surface area A, and an emissivity e, is given by Equation 13.2: $Q = e\sigma T^4 At$, where σ is the Stefan-Boltzmann constant.

If you are stranded in the mountains in bitter cold weather, you could minimize energy losses from your body by curling up into the tightest possible ball. In doing so, you minimize your effective surface area. Therefore, A in Equation 13.2 is made smaller, and you would radiate less heat.

CHAPTER 14 | *THE IDEAL GAS LAW AND KINETIC THEORY*

CONCEPTUAL QUESTIONS

---

1. ***REASONING AND SOLUTION***

a. Avogadro's number N_A is the number of particles per mole of substance. Therefore, one mole of hydrogen gas (H_2) and one mole of oxygen gas (O_2) contain the same number (Avogadro's number) of molecules.

b. One mole of a substance has a mass in grams that is equal to the atomic or molecular mass of the substance. The molecular mass of oxygen is greater than the molecular mass of hydrogen. Therefore, one mole of oxygen has more mass than one mole of hydrogen.

---

2. ***REASONING AND SOLUTION*** Substances A and B have the same mass densities. Therefore, the mass per unit volume of substance A is equal to that of substance B.

a. The mass of one mole of a substance depends on the molecular mass of the substance. In general, the molecular masses of substances A and B will differ, and one mole of each substance will not occupy the same volume; therefore, even though substances A and B have the same mass density, one mole of substance A will not have the same mass as substance B.

b. Since the mass per unit volume of substance A is the same as the mass per unit volume of substance B, 1 m^3 of substance A has the same mass as 1 m^3 of substance B.

---

3. ***REASONING AND SOLUTION*** A tightly sealed house has a large ceiling fan that blows air out of the house and into the attic. The fan is turned on, and the owners forget to open any windows or doors. As the fan transports air molecules from the house into the attic, the number of air molecules in the house decreases. Since the house is tightly sealed, the volume of the house remains constant. If the temperature of the air inside the house remains constant, then from the ideal gas law, $PV = nRT$, the pressure in the house must decrease. The air pressure in the attic, however, increases. The fan must now blow air from a lower pressure region to a higher pressure region. Thus, it becomes harder for the fan to do its job.

---

4. ***REASONING AND SOLUTION*** According to the ideal gas law (Equation 14.1), $PV = nRT$. Therefore, the pressure of a gas confined to a fixed volume is directly proportional to the Kelvin temperature. Therefore, when the temperature is increased, the pressure increases proportionally.

The gas above the liquid in a can of hair spray is at a relatively high pressure. If the temperature of the can is increased to a sufficiently high temperature, the pressure of the confined gas could increase to a value larger than the walls of the can will sustain. The can would then explode. Therefore, the label on hair spray cans usually contains the warning "Do not store at high temperatures."

---

5. ***REASONING AND SOLUTION*** When an electric furnace in a tightly sealed house is turned on for a while, the temperature of the air in the house increases. Since the house is tightly sealed, both the volume of the air and the number of air molecules remain constant. If we assume that air behaves like an ideal gas, then, from the ideal gas law, $PV = nRT$, the pressure in the house must increase.

---

6. ***REASONING AND SOLUTION*** At the sea coast, there is a cave that can be entered only by swimming beneath the water through a submerged passage and emerging into a pocket of air within the cave. Since the cave is not vented to the external atmosphere, the air in the cave contains a fixed number of moles ***n***. As the tide comes in, the water level in the cave rises, thereby decreasing the volume of the air above the water. According to the ideal gas law (Equation 14.1), $PV = nRT$. The pressure of the air in the cave is, therefore, inversely proportional to the air volume. Since the volume of the air decreases, the pressure increases. Thus, if you are in the cave when the tide comes in, your ears will "pop" inward in a manner that is analogous to what happens when you climb down a mountain.

---

7. ***REASONING AND SOLUTION*** Atmospheric pressure decreases with increasing altitude. Helium filled weather balloons are under-inflated when they are launched. As the balloon rises, the pressure exerted on the outside of it decreases. The number of helium molecules in the balloon is fixed. If we assume that the temperature of the atmosphere remains constant, then from the ideal gas law, $PV = nRT$, we see that the volume of the helium in the balloon will increase, thereby further inflating the balloon. If the balloon is fully inflated when it is launched from earth, it would burst when it reaches an altitude where the expanded volume of the helium is greater than the maximum volume of the balloon.

---

8. ***REASONING AND SOLUTION*** A slippery cork is being pressed into a very full bottle of wine. When released, the cork slowly slides back out. If some of the wine is removed from the bottle before the cork is inserted, the cork does not slide out. When the bottle is very full, the volume of air in the bottle above the wine is relatively small. Therefore, pushing the cork in reduces the volume of that air by an appreciable fraction. As a result, the pressure of the air increases appreciably and becomes large enough to push the slippery cork back out of the bottle. If some of the wine is removed, the volume of air above the wine is much larger to begin with, and pushing the cork in reduces the volume of that air by a much smaller fraction. Consequently, the pressure of the air increases by a much smaller amount and does not become large enough to push the cork back out.

---

9. ***REASONING AND SOLUTION*** Packing material consists of "bubbles" of air trapped between bonded layers of plastic. The packing material exerts normal forces on the packed object at each place where the bubbles make contact with the object. The motion of the packed object is thereby restricted. The magnitude of the normal force that any given

bubble can exert depends on the pressure of the air trapped inside the bubble. The magnitude of the normal force is equal to the pressure of the air times the area of contact between the bubble and the object. Since the air is trapped in the bubbles, the number of air molecules and the volume of the gas are fixed. The pressure exerted by the bubbles, then, depends on the temperature according to the ideal gas law: $PV = nRT$. On colder days, T is smaller than on warmer days; therefore, on colder days, the pressure P of the air in the bubbles is less than on warmer days. When the pressure is smaller, the magnitude of the normal force that each bubble can exert on a packed object is smaller. Therefore, the packing material offers less protection on cold days.

---

10. ***REASONING AND SOLUTION*** The kinetic theory of gases assumes that a gas molecule rebounds with the same speed after colliding with the wall of a container. From the impulse-momentum theorem, Equation 7.4, we know that $\mathbf{F}\Delta t = m(\mathbf{v}_f - \mathbf{v}_0)$, where the magnitude of **F** is the magnitude of the force on the wall. This implies that in a time interval t, a gas molecule of mass m, moving with velocity **v** before the collision and –**v** after the collision, exerts a force of magnitude $F = 2mv/t$ on the walls of the container, since the magnitude of $\mathbf{v}_f - \mathbf{v}_0$ is $2v$. If the speed of the gas molecule after the collision is less than that before the collision, the force exerted on the wall of the container will be less than $F = 2mv/t$, since the magnitude of $\mathbf{v}_f - \mathbf{v}_0$ is then less than $2v$. Therefore, the pressure of the gas will be less than that predicted by kinetic theory.

---

11. ***REASONING AND SOLUTION*** The relationship between the average kinetic energy per particle in an ideal gas and the Kelvin temperature T of the gas is given by Equation 14.6: $\frac{1}{2}mv_{rms}^2 = \frac{3}{2}kT$.

    If the translational speed of each molecule in an ideal gas is tripled, then the root-mean-square speed for the gas is also tripled. From Equation 14.6, the Kelvin temperature is proportional to the square of the root-mean-square speed. Therefore, the Kelvin temperature will increase by a factor of 9.

---

12. ***REASONING AND SOLUTION*** If the temperature of an ideal gas is doubled from 50 to 100 °C, the average translational kinetic energy per particle does not double. Equation 14.6, $\frac{1}{2}mv_{rms}^2 = \frac{3}{2}kT$, relates the average kinetic energy per particle to the Kelvin temperature of the gas, not the Celsius temperature. If the Celsius temperature increases from 50 to 100 °C, the Kelvin temperature increases from 323.15 to 373.15 K. This represents a fractional increase of 1.15; therefore, the average translational kinetic energy per particle increases only by a factor of 1.15.

---

13. ***REASONING AND SOLUTION*** According to Equation 14.7, the internal energy of a sample consisting of n moles of a monatomic ideal gas at Kelvin temperature T is $U = \frac{3}{2}nRT$; therefore, the internal energy of such an ideal gas depends only on the Kelvin temperature. If the pressure and volume of this sample is changed *isothermally*, the internal

energy of the ideal gas will remain the same. Physically, this means that the experimenter would have to change the pressure and volume in such a way, that the product PV remains the same. This can be verified from the ideal gas law (Equation 14.1), $PV = nRT$. If the values of P and V are varied so that the product PV remains constant, then T will remain constant and, from Equation 14.7, the internal energy of the gas remains the same.

---

14. ***REASONING AND SOLUTION*** The atoms in a container of helium have the same translational rms speed as the molecules in a container of argon. Equation 14.6 relates the average translational kinetic energy per particle in an ideal gas to the Kelvin temperature: $\frac{1}{2}mv_{\text{rms}}^2 = \frac{3}{2}kT$. Since the mass of an argon atom is greater than the mass of a helium atom, then, from Equation 14.6, the average translational kinetic energy per atom of the argon atoms is greater than the average translational kinetic energy per atom of the helium atoms. Therefore, the temperature of the argon atoms is greater than that of the helium atoms.

---

15. ***REASONING AND SOLUTION*** When an object is placed in a warm environment, its temperature increases for two reasons: (1) the object gains electromagnetic radiation from objects in the warm environment, and (2) the object gains heat from air molecules that collide with the object.

    Suppose that an astronaut were placed in the ionosphere, where the temperature of the ionized gas is about 1000 K, and the density of the gas is on the order of 10^{11} molecules/m^3. Since the density is extremely low, the distances between gas molecules are very large. Even though the temperature of the gas is large, there is an insufficient number of molecules to radiate sufficient energy to heat the astronaut. The number of molecules that collide with the astronaut per unit time is very small; therefore, even though the average kinetic energy of the gas molecules is large, the amount of heat transferred by the small number of collisions per unit time is low. Most of the surface area of the astronaut is in contact with empty space. The tissues of the astronaut would radiate more electromagnetic radiation than they absorb. Therefore, the astronaut would not burn up; in fact, if the astronaut remained in the environment very long, he would freeze to death.

---

16. ***REASONING AND SOLUTION*** Fick's law of diffusion relates the mass m of solute that diffuses in a time t through a solvent contained in a channel of length L and cross-sectional area A: $m = (DA\Delta C)t / L$, where ΔC is the concentration difference between the ends of the channel and D is the diffusion constant.

    In the lungs, oxygen in very small sacs (alveoli) diffuses into the blood. The walls of the alveoli are thin, so the oxygen diffuses over a small distance L. Since the number of alveoli is large, the effective area A across which diffusion occurs is very large. From Fick's law, we see that the mass of oxygen that diffuses per unit time is directly proportional to the effective cross-sectional area A and inversely proportional to the diffusion distance L. Since A is large and L is small, the mass of oxygen per second that diffuses into the blood is large.

CHAPTER 15 | *THERMODYNAMICS*

CONCEPTUAL QUESTIONS

---

1. ***REASONING AND SOLUTION*** The plunger of a bicycle tire pump is pushed down rapidly with the end of the pump sealed so that no air escapes. Since the compression occurs rapidly, there is no time for heat to flow into or out of the system. Therefore, to a very good approximation, the process may be treated as an adiabatic compression that is described by Equation 15.4:

$$W = (3/2)nR(T_i - T_f)$$

The person who pushes the plunger down does work on the system, therefore W is negative. It follows that the term $(T_i - T_f)$ must also be negative. Thus, the final temperature T_f must be greater than the initial temperature T_i. This increase in temperature is evidenced by the fact that the pump becomes warm to the touch.

Alternate Explanation:
Since the compression occurs rapidly, there is no time for heat to flow into or out of the system. Therefore, to a very good approximation, the process may be treated as an adiabatic compression. According to the first law of thermodynamics, the change in the internal energy is $\Delta U = Q - W = -W$, since $Q = 0$ for adiabatic processes. Since work is done on the system, W is negative; therefore the change in the internal energy, ΔU, is positive. The work done by the person pushing the plunger is manifested as an increase in the internal energy of the air in the pump. The internal energy of an ideal gas is proportional to the Kelvin temperature. Since the internal energy of the gas increases, the temperature of the air in the pump must also increase. This increase in temperature is evidenced by the fact that the pump becomes warm to the touch.

---

2. ***REASONING AND SOLUTION*** The work done in an isobaric process is given by Equation 15.2: $W = P(V_f - V_i)$. According to the first law of thermodynamics, the change in the internal energy is $\Delta U = Q - W = Q - P(V_f - V_i)$.

One hundred joules of heat is added to a gas, and the gas expands at constant pressure (isobarically). Since the gas expands, the final volume will be greater than the initial volume. Therefore, the term $P(V_f - V_i)$ will be positive. Since $Q = +100$ J, and the term $P(V_f - V_i)$ is positive, the change in the internal energy must be less than 100 J. It is not possible that the internal energy increases by 200 J.

---

3. ***REASONING AND SOLUTION*** The internal energy of an ideal gas is proportional to its Kelvin temperature (see Equation 14.7). In an isothermal process the temperature remains constant; therefore, the internal energy of an ideal gas remains constant throughout an isothermal process. Thus, if a gas is compressed isothermally and its internal energy increases, the gas is not an ideal gas.

---

4. ***REASONING AND SOLUTION*** According to the first law of thermodynamics (Equation 15.1), the change in the internal energy is $\Delta U = Q - W$. The process is isochoric, which means that the volume is constant. Consequently, no work is done, so $W = 0$. The process is also adiabatic, which means that no heat enters or leaves the system, so $Q = 0$. According to the first law, then, $\Delta U = Q - W = 0$. There is no change in the internal energy, and the internal energy of the material at the end of the process is the same as it was at the beginning.

---

5. ***REASONING AND SOLUTION***

a. It is possible for the temperature of a substance to rise without heat flowing into the substance. Consider, for example, the adiabatic compression of an ideal gas. Since the process is an adiabatic process, $Q = 0$. The work done by the external agent increases the internal energy of the gas. Since the internal energy of an ideal gas is proportional to the Kelvin temperature, the temperature of the gas must increase.

b. The temperature of a substance does not necessarily have to change because heat flows into or out of it. Consider, for example, the isothermal expansion of an ideal gas. Since the internal energy of an ideal gas is proportional to the Kelvin temperature, the internal energy, ΔU, remains constant during an isothermal process. The first law of thermodynamics gives $\Delta U = Q - W = 0$, or $Q = W$. The heat that is added to the gas during the isothermal expansion is used by the gas to perform the work involved in the expansion. The temperature of the gas remains unchanged. Similarly, in an isothermal compression, the work done on the gas as the gas is compressed causes heat to flow out of the gas while the temperature of the gas remains constant.

---

6. ***REASONING AND SOLUTION*** The text drawing shows a pressure-volume graph in which a gas undergoes a two-step process from A to B and from B to C.

From A to B: The volume V of the gas increases at constant pressure P. According to the ideal gas law (Equation 14.1), $PV = nRT$, the temperature T of the gas must increase. According to Equation 14.7, $U = (3/2)nRT$, if T increases, then ΔU, the change in the internal energy, must be positive. Since the volume increases at constant pressure (ΔV increases), we know from Equation 15.2, $W = P\Delta V$, that the work done is positive. The first law of thermodynamics (Equation 15.1) states that $\Delta U = Q - W$; since ΔU and W are both positive, Q must also be positive.

From B to C The pressure P of the gas increases at constant volume V. According to the ideal gas law (Equation 14.1), $PV = nRT$, the temperature T of the gas must increase. According to Equation 14.7, $U = (3/2)nRT$, if T increases, then ΔU, the change in the internal energy, must be positive. Since the process occurs isochorically ($\Delta V = 0$), and according to Equation 15.2, $W = P\Delta V$, the work done is zero. The first law of thermodynamics (Equation 15.1) states that $\Delta U = Q - W$; since $W = 0$, Q is also positive since ΔU is positive.

These results are summarized in the table below:

	ΔU	Q	W
$A \rightarrow B$	+	+	+
$B \rightarrow C$	+	+	0

---

7. ***REASONING AND SOLUTION*** Since the process is an adiabatic process, $Q = 0$. Since the gas expands into chamber B under zero external pressure, the work done by the gas is $W = P\Delta V = 0$. According to the first law of thermodynamics, the change in the internal energy is, therefore, zero: $\Delta U = Q - W = 0$. The internal energy of an ideal gas is proportional to the Kelvin temperature of the gas (Equation 14.7). Since the change in the internal energy of the gas is zero, the temperature change of the gas is zero. The final temperature of the gas is the same as the initial temperature of the gas.

---

8. ***REASONING AND SOLUTION*** A material contracts when it is heated. To determine the molar specific heat capacities, we first calculate the heat Q needed to raise the temperature of the material by an amount ΔT. From the first law of thermodynamics, $Q = \Delta U + W$. When the heating occurs at constant pressure, the work done is given by Equation 15.2: $W = P\Delta V = P(V_f - V_i)$. When the volume is constant, $\Delta V = 0$. Therefore, we have:

$$Q_P = \Delta U + P(V_f - V_i) \quad \text{and} \quad Q_V = \Delta U$$

Equation 15.6 indicates that the molar heat capacities will be given by $C = Q/(n\Delta T)$. Therefore

$$C_P = \frac{\Delta U + P(V_f - V_i)}{n\Delta T} \quad \text{and} \quad C_V = \frac{\Delta U}{n\Delta T}$$

Since the material contracts when it is heated, V_f is less than V_i. Therefore the term $P(V_f - V_i)$ is negative. Hence, the numerator of C_P is smaller than the numerator of C_V. Therefore, C_V is larger than C_P.

---

9. ***REASONING AND SOLUTION*** When a solid melts at constant pressure, the volume of the resulting liquid does not differ much from the volume of the solid. According to the first law of thermodynamics, $\Delta U = Q - W = Q - P(V_f - V_i) \approx Q$. Hence, the heat that must be added to melt the solid is used primarily to increase the internal energy of the molecules. The internal energy of the liquid has increased by an amount $Q = mL_f$ compared to that of the solid, where m is the mass of the material and L_f is the latent heat of fusion.

---

10. ***REASONING AND SOLUTION*** According to Equation 14.7, the Kelvin temperature T of the gas is related to its internal energy U by $U = (3/2)nRT$. The change in the internal energy is given by the first law of thermodynamics (Equation 15.1), $\Delta U = Q - W$.

It is desired to heat a gas so that its temperature will be as high as possible. If the process occurs at constant pressure, so that the volume of the gas increases, work is done by the gas. The available heat is used to do work *and* to increase the internal energy of the gas. On the other hand, if the process is carried out at constant volume, the work done is zero, and all of the heat increases the internal energy of the gas. From Equation 14.7, the internal energy is directly proportional to the Kelvin temperature of the gas. Since the internal energy increases by a greater amount when the process occurs at constant volume, the temperature increase is greatest under conditions of constant volume. Therefore, if it is desired to heat a gas so that its temperature will be as high as possible, you should heat it under conditions of constant volume.

---

11. ***REASONING AND SOLUTION*** A hypothetical device takes 10 000 J of heat from a hot reservoir and 5000 J of heat from a cold reservoir and produces 15 000 J of work.

a. According to the first law of thermodynamics, $\Delta U = Q - W$. This is a statement of energy conservation. The hypothetical device does not violate energy conservation. It does not create or destroy energy. It converts one form of energy (15 000 J of heat) into another form of energy (15 000 J of work) with no gain or loss.

b. This hypothetical device does violate the second law of thermodynamics. It converts all of its input heat (15 000 J) into work (15 000 J). Therefore, the efficiency of this device is 1.0 or 100 %. But Equation 15.15 is a consequence of the second law of thermodynamics and sets the limits of the maximum possible efficiency of any heat device. Since T_C is greater than 0 K, the ratio T_C / T_H must be positive. Furthermore, since $T_C < T_H$, the ratio T_C / T_H must be less than one. Therefore, Equation 15.15 implies that the efficiency of any device must be less than 1 or 100%. Since the efficiency of the hypothetical device is equal to 100%, it violates the second law of thermodynamics.

---

12. ***REASONING AND SOLUTION*** According to the second law of thermodynamics, heat flows spontaneously from a substance at a higher temperature to a substance at a lower temperature and does not flow spontaneously in the reverse direction. Therefore, according to the second law of thermodynamics, work must be done to remove heat from a substance at a lower temperature and deposit it in a substance at a higher temperature. In other words, the second law requires that energy in the form of work must be supplied to an air conditioner in order for it to remove heat from a cool space and deposit the heat in a warm space. An advertisement for an automobile that claimed the same gas mileage with and without the air conditioner would be suspect. Since the car would use more energy with the air conditioner on, the car would use more gasoline. Therefore, the mileage should be less with the air conditioner running.

---

13. ***REASONING AND SOLUTION*** Carnot's principle states that the most efficient engine operating between two temperatures is a reversible engine. This means that a *reversible* engine operating between the temperatures of 600 and 400 K must be more efficient than an *irreversible* engine operating between the *same two temperatures*. No comparison can be made with an irreversible engine operating between two temperatures that are different than 600 and 400 K.

---

14. ***REASONING AND SOLUTION*** The efficiency of a Carnot engine is given by Equation 15.15: $\text{efficiency} = 1-(T_C / T_H)$. Three reversible engines A, B, and C, use the same cold reservoir for their exhaust heats. They use different hot reservoirs with the following temperatures: (A) 1000 K; (B) 1100 K; and (C) 900 K. We can rank these engines in order of increasing efficiency according to the following considerations. The ratio T_C / T_H is inversely proportional to the value of T_H. The ratio T_C / T_H will be smallest for engine B; therefore, the quantity $1 - (T_C / T_H)$ will be largest for engine B. Thus, engine B has the largest efficiency. Similarly, the ratio T_C / T_H will be largest for engine C; therefore, the quantity $1 - (T_C / T_H)$ will be smallest for engine C. Thus, engine C has the smallest efficiency. Hence, the engines are, in order of increasing efficiency: engine C, engine A, and engine B.

---

15. ***REASONING AND SOLUTION*** The efficiency of a Carnot engine is given by Equation 15.15: $\text{efficiency} = 1-(T_C / T_H)$.

a. Lowering the Kelvin temperature of the cold reservoir by a factor of four makes the ratio T_C / T_H one-fourth as great.

b. Raising the Kelvin temperature of the hot reservoir by a factor of four makes the ratio T_C / T_H one-fourth as great.

c. Cutting the Kelvin temperature of the cold reservoir in half and doubling the Kelvin temperature of the hot reservoir makes the ratio T_C / T_H one-fourth as great.

Therefore, all three possible improvements have the same effect on the efficiency of a Carnot engine.

---

16. ***REASONING AND SOLUTION*** A refrigerator is kept in a garage that is not heated in the cold winter or air-conditioned in the hot summer. In order to make ice cubes, the refrigerator uses electrical energy to provide the work to remove heat from the interior of the freezer and deposit the heat outside of the refrigerator. In the summer, the "hot" reservoir will be at a higher temperature than it is in the winter. Therefore, more work will be required to remove heat from the interior of the freezer in the summer, and the refrigerator will use more electrical energy. Hence, it will cost more for the refrigerator to make a kilogram of ice cubes in the summer.

---

17. ***REASONING AND SOLUTION*** The coefficient of performance of a heat pump is given by Equation 15.17: coefficient of performance = Q_{H}/W. From the conservation of energy, $Q_{\text{H}} = W + Q_{\text{C}}$. Thus, the ratio Q_{H}/W can be written $1+(Q_{\text{C}}/W)$. The job of a heat pump is to remove heat from a cold reservoir, and deliver it to a hot reservoir; therefore, the ratio Q_{C}/W must be nonzero and positive. Hence, the coefficient of performance, $1+(Q_{\text{C}}/W)$, must always be greater than one.

---

18. ***REASONING AND SOLUTION*** In a refrigerator, the interior of the unit is the cold reservoir, while the warmer exterior of the room is the hot reservoir. An air conditioner is like a refrigerator, except that the room being cooled is the cold reservoir, and the outdoor environment is the hot reservoir. Therefore, an air conditioner cools the inside of the house, while a refrigerator warms the interior of the house.

---

19. ***REASONING AND SOLUTION*** Heat pumps *can* deliver more energy into your house than they consume in operating. A heat pump consumes an amount of energy *W*, which it uses to make heat Q_{C} flow from the cold outdoors into the warm house. The amount of energy the heat pump delivers to the house is $Q_{\text{H}} = W + Q_{\text{C}}$. This is greater than the energy, *W*, consumed by the heat pump.

---

20. ***REASONING AND SOLUTION*** A refrigerator is advertised as being easier to "live with" during the summer, because it puts into your kitchen only the heat that it removes from the food. The advertisement is describing a refrigerator in which heat is removed from the interior of the refrigerator and deposited outside the refrigerator without requiring any work. Since no work is required, the flow must be spontaneous. This violates the second law of thermodynamics, which states that heat spontaneously flows from a higher-temperature substance to a lower-temperature substance, and does not flow spontaneously in the reverse direction. Heat can be made to flow from a cold reservoir to a hot reservoir, but only when work is done. Both the heat and the work are deposited in the hot reservoir.

---

21. ***REASONING AND SOLUTION*** On a summer day, a window air conditioner cycles on and off, according to how the temperature within the room changes. When the unit is on it will be depositing heat, along with the work required to remove the heat, to the outside. Therefore, the outside of the unit will be hotter when the unit is on. Hence, you would be more likely to fry an egg on the outside part of the unit when the unit is on.

---

22. ***REASONING AND SOLUTION*** The second law of thermodynamics states that the total entropy of the universe does not change when a reversible process occurs $(\Delta S_{\text{universe}} = 0)$ and increases when an irreversible process occurs $(\Delta S_{\text{universe}} > 0)$.

An event happens somewhere in the universe and, as a result, the entropy of an object changes by –5 J/K. If the event is a reversible process, then the entropy change for the rest of the universe must be +5 J/K; this results in a total entropy change of zero for the universe. If the process is irreversible, the only possible choice for the change in the entropy of the rest of the universe is +10 J/K; this results in a total entropy change of +5 J/K for the universe. The choices –5 J/K and 0 J/K are not possible choices for the entropy change of the rest of the universe, because they imply that the total entropy change would be negative. This would violate the second law of thermodynamics.

---

23. ***REASONING AND SOLUTION*** When water freezes from a less-ordered liquid to a more-ordered solid, its entropy decreases. This decrease in entropy does not violate the second law of thermodynamics, because it is a decrease for only one part of the universe. In terms of entropy, the second law indicates that the total change in entropy for the entire universe must be either zero (reversible process) or greater than zero (irreversible process). In the case of freezing water, heat must be removed from the water and deposited in the environment. The entropy of the environment increases as a result. If the freezing occurs reversibly, the increase in entropy of the environment will exactly match the decrease in entropy of the water, with the result that $\Delta S_{\text{universe}} = \Delta S_{\text{water}} + \Delta S_{\text{environment}} = 0$. If the freezing occurs irreversibly, then the increase in entropy of the environment will exceed the decrease in entropy of the water, with the result that $\Delta S_{\text{universe}} = \Delta S_{\text{water}} + \Delta S_{\text{environment}} > 0$.

---

24. ***REASONING AND SOLUTION*** Since we can interpret the increase of entropy as an increase in disorder, the more disordered system will have the greater entropy.

a. The popcorn that results from the kernels is more disorderly than a handful of popcorn kernels; therefore, the popcorn that results from the kernels has the greater entropy.

b. A salad has more disorder after it has been tossed; therefore, the tossed salad has the greater entropy.

c. A messy apartment is more disorderly than a neat apartment; therefore, a messy apartment has the greater entropy.

---

25. ***REASONING AND SOLUTION*** A glass of water contains a teaspoon of dissolved sugar. After a while, the water evaporates, leaving behind sugar crystals. The entropy of the sugar crystals is less than the entropy of the dissolved sugar, because the sugar crystals are in a more ordered state. However, think about the water. The entropy of the gaseous water vapor is greater than the entropy of the liquid water, because the molecules in the vapor are in a less ordered state. Since the increase in the entropy of the water is greater than the decrease in entropy of the sugar, the net change in entropy of the universe is positive. The process, therefore, does not violate the entropy version of the second law of thermodynamics.

---

26. ***REASONING AND SOLUTION*** Since we can interpret the increase of entropy as an increase in disorder, the more disordered state will have the greater entropy. The finished building is the most ordered state; therefore it has the smallest entropy. The burned-out shell of a building is the most disordered state; therefore it has the largest amount of entropy.

The states can be ranked in order of decreasing entropy (largest first) as follows: (3) the burned-out shell of a building, (1) the unused building material, and (2) the building.

CHAPTER 16 | *WAVES AND SOUND*

CONCEPTUAL QUESTIONS

---

1. ***REASONING AND SOLUTION*** As Figure 16.4 shows, in a water wave, the wave motion of the water includes both transverse and longitudinal components. The water at the surface moves on nearly circular paths. When the wave passes beneath a fishing float, the float will simultaneously bob up and down, as well as move back and forth horizontally. Thus, the float will move in a nearly circular path in the vertical plane. It is not really correct, therefore, to say that the float bobs straight "up and down."

---

2. ***REASONING AND SOLUTION*** "Domino Toppling" is an event that consists of lining up an incredible number of dominoes and then letting them topple, one after another. As the dominoes topple, their displacements contain both vertical and horizontal components. Therefore, the disturbance that propagates along the line of dominoes has both longitudinal (horizontal) and transverse (vertical) components.

---

3. ***REASONING AND SOLUTION*** A longitudinal wave moves along a Slinky at a speed of 5 m/s. We *cannot* conclude that one coil of the Slinky moves through a distance of 5 m in one second. The quantity 5 m/s is the longitudinal wave speed, v_{speed}; it specifies how fast the *disturbance* travels along the spring. The wave speed depends on the properties of the spring. Like the transverse wave speed, the longitudinal wave speed depends upon the tension F in the spring and its linear mass density m/L. As long as the tension and the linear mass density remain the same, the disturbance will travel along the spring at constant speed.

   The particles in the Slinky oscillate longitudinally in simple harmonic motion with the same amplitude and frequency as the source. As with all particles in simple harmonic motion, the particle speed is not constant. The particle speed is a maximum as the particle passes through its equilibrium position and reaches zero when the particle has reached its maximum displacement from the equilibrium position. The particle speed depends upon the amplitude and frequency of the particle's motion. Thus, the particle speed, and therefore the longitudinal speed of a single coil, depends upon the properties of the source that causes the disturbance.

---

4. ***REASONING AND SOLUTION*** A wave moves on a string with constant velocity. It is *not* correct to conclude that the particles of the string always have zero acceleration. As Conceptual Example 3 discusses, it is important to distinguish between the speed of the waves on the string, v_{wave}, and the speed of the particles in the string, v_{particle}. The wave speed v_{wave} is determined by the properties of the string; namely, the tension in the string and the linear mass density of the string. These properties determine the speed with which the disturbance travels along the string. The wave speed will remain constant as long as these properties remain unchanged.

The particles in the string oscillate transversely in simple harmonic motion with the same amplitude and frequency as the source of the disturbance. Like all particles in simple harmonic motion, the acceleration of the particles continually changes. It is zero when the particles pass through their equilibrium positions and is a maximum when the particles are at their maximum displacements from their equilibrium positions.

---

5. ***REASONING AND SOLUTION*** A wire is strung tightly between two immovable posts. The speed of a transverse wave on the wire is given by Equation 16.2: $v_{\text{wave}} = \sqrt{F/(m/L)}$. The wire will expand because of the increase in temperature. Since the length of the wire increases slightly, it will sag, and the tension in the wire will decrease. From Equation 16.2, we see that the speed of the wave is directly proportional to the square root of the tension in the wire. If we ignore any change in the mass per unit length of the wire, then we can conclude that, when the temperature is increased, the speed of waves on the wire will decrease.

---

6. ***REASONING AND SOLUTION*** A rope of mass m is hanging down from the ceiling. Nothing is attached to the loose end of the rope. A transverse wave is traveling up the rope.

   The tension in the rope is not constant. The lower portion of the rope pulls down on the higher portions of the rope. If we imagine that the rope is divided into small segments, we see that the segments near the top of the rope are being pulled down by more weight than the segments near the bottom. Therefore, the tension in the rope increases as we move up the rope. The speed of a transverse wave on the rope is given by Equation 16.2: $v_{\text{wave}} = \sqrt{F/(m/L)}$. From Equation 16.2 we see that, as the tension F in the rope increases, the speed of the wave increases. Therefore, as the transverse wave travels up the rope, the speed of the wave increases.

---

7. ***REASONING AND SOLUTION*** One end of each of two identical strings is attached to a wall. Each string is being pulled tightly by someone at the other end. A transverse pulse is sent traveling along one of the strings. A bit later, an identical pulse is sent traveling along the other string. In order for the second pulse to catch up with the first pulse, the speed of the pulse in the second string must be increased. The speed of the transverse pulse on the second string is given by Equation 16.2: $v_{\text{wave}} = \sqrt{F/(m/L)}$. This equation indicates that we can increase the speed of the second pulse by increasing the tension F in the string. Thus, the second string must be pulled more tightly.

---

8. ***REASONING AND SOLUTION*** In Section 4.10 the concept of a "massless" rope is discussed. For a truly massless rope, the linear density of the rope, m/L, is zero. From Equation 16.2, $v_{\text{wave}} = \sqrt{F/(m/L)}$, the wave speed would be infinite if m/L were zero. Therefore, if the rope were really massless, the speed of transverse waves on the rope would be infinite, and a transverse wave would be instantaneously transmitted from one end of the rope to the other. It would not take any time for a transverse wave to travel the length of a massless rope.

9. ***REASONING AND SOLUTION*** As the disturbance moves outward when a sound wave is produced, it compresses the air directly in front of it. This compression causes the air pressure to rise slightly, resulting in a condensation that travels outward. The condensation is followed by a region of decreased pressure, called a rarefaction. Both the condensation and rarefaction travel away from the speaker at the speed of sound. As the condensations and rarefactions of the sound wave move away from the disturbance, the individual air molecules are not carried with the wave. Each molecule executes simple harmonic motion about a fixed equilibrium position. As the wave passes by, all the particles in the region of the disturbance participate in this motion. There are no particles that are always at rest.

---

10. ***REASONING AND SOLUTION*** Assuming that we can treat air as an ideal gas, then the speed of sound in air is given by Equation 16.5, $v = \sqrt{\gamma kT/m}$, where γ is the ratio of the specific heats c_P / c_V, k is Boltzmann's constant, T is the Kelvin temperature, and m is the mass of a molecule of the gas.

    We can see from Equation 16.5 that the speed of sound in air is proportional to the square root of the Kelvin temperature of the gas. Therefore, on a hot day, the speed of sound in air is greater than it is on a cold day. Hence, we would expect an echo to return to us more quickly on a hot day as compared to a cold day, other things being equal.

---

11. ***REASONING AND SOLUTION*** A loudspeaker produces a sound wave. The sound wave travels from air into water. As indicated in Table 16.1, the speed of sound in water is approximately four times greater than it is in air. We are told in the hint that the frequency of the sound wave does not change as the sound enters the water. The relationship between the frequency f, the wavelength λ, and the speed v of a wave is given by Equation 16.1: $v = f\lambda$. Since the wave speed increases and the frequency remains the same as the sound enters the water, the wavelength of the sound must increase.

---

12. ***REASONING AND SOLUTION*** A person is making JELL-O. It starts out as a liquid and then sets into a gel. In general, sound travels slowest in gases, faster in liquids, and fastest in solids. When the JELL-O sets into a gel, it is more "solid;" therefore, the speed of sound should be larger in the set JELL-O than when the JELL-O is in the liquid state.

---

13. ***REASONING AND SOLUTION*** Animals that rely on an acute sense of hearing for survival often have relatively large external ear parts. Sound intensity is defined as the sound power P that passes perpendicularly through a surface divided by the area A of that surface: $I = P/A$. For low intensity sounds, the power per unit area is small. Relatively large outer ears have a greater area than smaller outer ears. Hence, large outer ears intercept and direct more sound power into the auditory system than smaller outer ears do.

---

14. ***REASONING AND SOLUTION*** A source is emitting sound uniformly in all directions. According to Equation 16.9, $I = P/(4\pi r^2)$, the intensity of such a source varies as $1/r^2$. Thus, the intensity I at a point in space depends on the distance of that point from the source. A flat surface faces the source. As suggested in the following figure, the distance between the source and the flat sheet varies, in general, from point to point on the sheet. The figure indicates that, as we move up the screen, the distance between the source and the screen increases. Therefore, the sound intensity is not the same at all points on the screen.

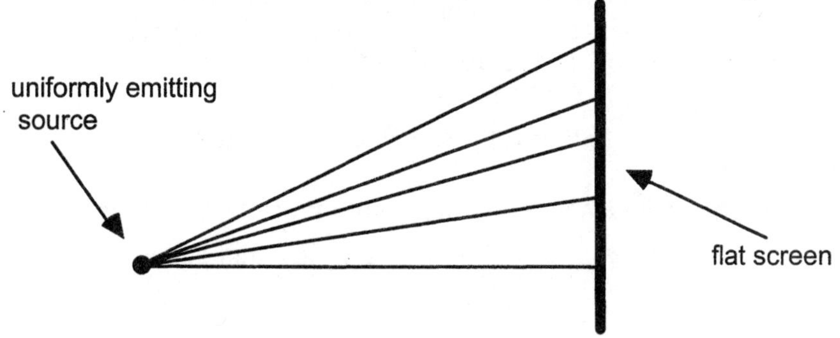

---

15. ***REASONING AND SOLUTION*** Two people talk simultaneously and each creates an intensity level of 65 dB at a certain point. The total intensity level at this point does *not* equal 130 dB. Intensity levels are defined in terms of logarithms. According to Equation 16.10, the sound intensity level in decibels is given by $\beta = (10\ \text{dB})\ \log(I/I_0)$, where I_0 is the threshold of human hearing. We can show, using Equation 16.10, that an intensity level of 65 dB corresponds to a sound intensity of $3.2 \times 10^{-6}\ \text{W/m}^2$. If two people simultaneously talk and each creates an intensity of $3.2 \times 10^{-6}\ \text{W/m}^2$ at a certain point, then the total *intensity* at that point is $6.4 \times 10^{-6}\ \text{W/m}^2$. However, since the *intensity level* is defined in terms of logarithms, the intensity level at that point is (from Equation 16.10) 68 dB. While the second person doubles the intensity at the point in question, he only increases the intensity level (or the loudness) by 3 dB.

---

16. ***REASONING AND SOLUTION*** Two cars, one behind the other, are traveling in the same direction at the same speed. The velocity of either car relative to the other one is zero. Since there is no relative motion between the cars, either driver will hear the other's horn at exactly the same frequency that would be heard if both cars were at rest.

---

17. ***REASONING AND SOLUTION*** According to Table 16.1, the speed of sound in air at 20 °C is v = 343 m/s, while its value in water at the same temperature is v = 1482 m/s. These values influence the Doppler effect, because of which an observer hears a frequency f_o that is different from the frequency f_s that is emitted by the source of sound. For purposes of this question, we assume that f_s = 1000 Hz and that the speed at which the source moves is v_s = 25 m/s. Our conclusions, however, will be valid for any values of f_s and v_s.

a. The Doppler-shifted frequency when the source approaches the observer is given by Equation 16.11 as $f_o = f_s/\left[1-(v_s/v)\right]$. Applying this equation for air and water, we find

Air $$f_o = \frac{1000\ \text{Hz}}{1-(25\ \text{m/s})/(343\ \text{m/s})} = 1079\ \text{Hz}$$

Water $$f_o = \frac{1000\ \text{Hz}}{1-(25\ \text{m/s})/(1482\ \text{m/s})} = 1017\ \text{Hz}$$

The change in frequency due to the Doppler effect in air is greater.

b. The Doppler-shifted frequency when the source moves away from the observer is given by Equation 16.12 as $f_o = f_s/\left[1+(v_s/v)\right]$. Applying this equation for air and water, we find

Air $$f_o = \frac{1000\ \text{Hz}}{1+(25\ \text{m/s})/(343\ \text{m/s})} = 932\ \text{Hz}$$

Water $$f_o = \frac{1000\ \text{Hz}}{1+(25\ \text{m/s})/(1482\ \text{m/s})} = 983\ \text{Hz}$$

The change in frequency due to the Doppler effect in air is greater.

---

18. ***REASONING AND SOLUTION*** A music fan at a swimming pool is listening to a radio on a diving platform. The radio is playing a constant frequency tone when this fellow, clutching his radio, jumps.

a. The frequency heard by a person left behind on the platform is given by Equation 16.12: $f_o = f_s/\left[1+(v_s/v)\right]$. Since the denominator, $\left[1+(v_s/v)\right]$, is necessarily greater than one, the frequency f_o is less than f_s. Thus, a person left behind on the platform will hear a frequency that is smaller than the frequency produced by the radio. As the radio falls, it accelerates, so its speed increases. As v_s increases, the quantity $\left[1+(v_s/v)\right]$ also increases, so that f_o decreases with time. Therefore, the observed frequency is not constant and decreases during the fall.

b. The frequency heard by a person down below floating on a rubber raft is given by Equation 16.11: $f_o = f_s/\left[1-(v_s/v)\right]$. We can assume that during the fall $v_s < v$ at all times so that the quantity v_s/v is less than one, and the denominator, $\left[1-(v_s/v)\right]$, is necessarily less than one. Therefore, the frequency f_o is greater than f_s. A person down below floating on a rubber raft will hear a frequency that is greater than the frequency produced by the radio. As the radio falls, v_s increases as discussed in part (a), so that the quantity $\left[1-(v_s/v)\right]$ decreases. As a result, the frequency f_o increases with time. Therefore, the observed frequency is not constant and increases during the fall.

---

19. ***REASONING AND SOLUTION*** When a car is at rest, its horn emits a frequency of 600 Hz. A person standing in the middle of the street hears the horn with a frequency of 580 Hz. As shown in Figure 16.29, when a vehicle is moving, a stationary observer in front of the vehicle will hear sound of a shorter wavelength, while an observer behind the vehicle will hear sound of a longer wavelength. We can deduce, therefore, from Equation 16.1, $v = f\lambda$, that the person standing in front of the vehicle will hear sound of a higher frequency than when the car is at rest, while a person standing behind the vehicle will hear sound of a lower frequency than when the car is at rest. Since the person standing in the middle of the street hears a frequency of 580 Hz, which is less than the stationary frequency of 600 Hz, the person is behind the car. Therefore, the person need not jump out of the way.

---

20. ***REASONING AND SOLUTION*** A car is speeding toward a large wall and sounds the horn. The sound of the horn in front of the car will have a shorter wavelength and a higher frequency than the wavelength and frequency of the sound of the horn when the car is at rest. This Doppler shift occurs because the source is moving. The echo is produced when this higher-frequency sound is reflected from the wall. The wall now acts as a stationary source of this higher-frequency sound. The driver is an observer who is moving toward a stationary source. Since the driver is moving toward the echo, he will hear a frequency that is greater than the sound heard by a stationary observer. Thus, the echo heard by the driver is Doppler-shifted twice. The sound is shifted once to a higher frequency because the source of the sound is moving. Then the sound is Doppler shifted to a higher frequency again because the driver is moving relative to the stationary source (of the echo).

CHAPTER 17 | THE PRINCIPLE OF LINEAR SUPERPOSITION AND INTERFERENCE PHENOMENA

CONCEPTUAL QUESTIONS

---

1. ***REASONING AND SOLUTION*** The principle of linear superposition states that when two or more waves are present simultaneously at the same place, the resultant wave is the sum of the individual waves. This principle does *not* imply that two sound waves, passing through the same place at the same time, always create a louder sound than either wave alone. The resultant wave pattern depends on the relative phases of the two sound waves when they meet. If two sound waves arrive at the same place at the same time, and they are exactly in phase, then the two waves will interfere constructively and create a louder sound than either wave alone. On the other hand, if two waves arrive at the same place at the same time, and they are exactly out of phase, destructive interference will occur; the net effect is a mutual cancellation of the sound. If the two sound waves have the same amplitude and frequency, they will completely cancel each other and no sound will be heard.

---

2. ***REASONING AND SOLUTION*** If you are sitting at the overlap point between the two speakers in Figure 17.4, the two sound waves reaching you are exactly out of phase. You hear no sound because of destructive interference. If one of the speakers is suddenly shut off, then only one sound wave will reach your ears. Since there is no other sound wave to interfere with this sound wave, you will hear the sound from the single speaker.

---

3. ***REASONING AND SOLUTION*** Consider the situation in Figure 17.3. If you walk along a line that is perpendicular to the line between the speakers and passes through the overlap point, you will always be equidistant from both speakers. Therefore the sound waves along this line always overlap exactly in phase. Hence, you will always hear the same loudness; you will not observe the loudness to change from loud to faint to loud.

   On the other hand, if you walk along a line that passes through the overlap point and is parallel to the line between the speakers, your distance from the two speakers will vary such that the difference in path lengths traveled by the two waves will vary. At certain points the path length difference between the two sound waves will be an integer number of wavelengths; constructive interference will occur at these points, and the sound intensity will be a maximum. At other points, the path length difference between the two sound waves will be an odd number of half-wavelengths [$(1/2)\lambda$, $(3/2)\lambda$, $(5/2)\lambda$, etc.]; destructive interference will occur and the sound intensity will be a minimum. In between the points of constructive and destructive interference, the waves will be out of phase by varying degrees. Therefore, as you walk along this line, you will observe the sound intensity to alternate between faint and loud.

---

4. ***REASONING AND SOLUTION*** If the width of the speakers is D, then sound of wavelength λ will diffract more readily if the ratio λ / D is large. Since longer wavelengths correspond to lower frequencies, we see that lower-frequency sounds diffract more readily from a given speaker than higher-frequency sounds. Since the frequencies of the sounds of the female vocalists are higher than the frequencies of the sounds of the rhythmic bass, the sounds of the female vocalists will not diffract to the same extent as the sounds of the rhythmic bass. Thus, diffraction allows the bass tones to penetrate the regions to either side of the stage, but it does not permit the same help to the sounds of the female vocalists.

---

5. ***REASONING AND SOLUTION*** When a wave encounters an obstacle or the edges of an opening, it bends or diffracts around them. The extent of the diffraction depends on the ratio of the wavelength λ to the size D of the obstacle or opening. If the ratio λ / D is small, little diffraction occurs. As the ratio λ / D is made larger, the wave diffracts to a greater extent. In Example 1 in Section 16.2, it is shown that the wavelength of AM radio waves is 244 m, while the wavelength of FM radio waves is 3.26 m. For a given obstacle, the ratio λ / D will be greater for AM radio waves; therefore AM radio waves will diffract more readily around a given obstacle than FM waves.

---

6. ***REASONING AND SOLUTION*** A tuning fork has a frequency of 440 Hz. The string of a violin and this tuning fork, when sounded together, produce a beat frequency of 1 Hz. This beat frequency is the difference between the frequency of the tuning fork and the frequency of the violin string. A violin string of frequency 439 Hz, as well as a violin string of 441 Hz, will produce a beat frequency of 1 Hz, when sounded together with the 440 Hz tuning fork. We conclude that, from these two pieces of information alone, it is *not* possible to distinguish between these two possibilities. Therefore, it is *not* possible to determine the frequency of the violin string.

---

7. ***REASONING AND SOLUTION*** Tuning forks vibrate at 438 Hz and 440 Hz. When sounded together, a listener hears a beat frequency of 2 Hz, the difference in frequency between the two forks. The forks are then vibrated underwater with the listener also underwater. The forks vibrate at 438 and 440 Hz, just as they do in air. The speed of sound, however, is four times faster in water than in air. From the relation $v = \lambda f$, we see that the wavelength of the sound in water is four times longer than it is in air. Since the speed of the waves is four times greater than it is in air, the regions of constructive and destructive interference move past the ears of the listener four times more rapidly than they do in air. However, the regions of condensation and rarefaction in water are separated by four times the distance than they are in air. Therefore, at a given point in space, the amplitude variation of the resultant wave occurs at the same rate that it does in air. Consequently, the listener hears a beat frequency of 2 Hz, just as he does in air.

---

8. ***REASONING AND SOLUTION*** The frequency of a guitar string is given by Equation 17.3, $f_n = n[v/(2L)]$, where v is the speed of the wave on the string, L is the distance between the two fixed ends of the guitar string, and $n = 1, 2, 3, 4, \ldots$ The speed v

of the wave is given by Equation 16.2, $v = \sqrt{F/(m/L)}$, where F is the tension in the guitar string, and m/L is the mass per unit length of the string.

If the tension in a guitar string is doubled, then, according to Equation 16.2, the speed of the wave on the string will increase by a factor of $\sqrt{2}$. According to Equation 17.3, the frequency of oscillation is directly proportional to the speed of the wave. Therefore, the frequency of oscillation will *not* double; rather, it will *increase by a factor of* $\sqrt{2}$.

---

9. ***REASONING AND SOLUTION*** A string is attached to a wall and vibrates back and forth as in Figure 17.18. The vibration frequency and length of the string are fixed. The tension in the string is changed. From Equation 16.2, $v = \sqrt{F/(m/L)}$, we see that increasing the tension F results in increasing the speed v of the waves on the string. From the relationship $v = \lambda f$, we see that, since the frequency remains fixed, an increase in the wave speed results in an increase in the wavelength of the wave.

   It is observed that at certain values of the tension, a standing wave pattern develops. Since the two ends of the string are "fixed," the ends of the string are nodes; thus, the length L of the string must contain an integer number of half-wavelengths. Therefore, standing waves will occur at the wavelengths $\lambda_n = 2L/n$, where n = 1, 2, 3, 4, . . . The largest possible wavelength that will result in a standing wave pattern occurs when n = 1, and the wavelength is equal to twice the length of the string. If the tension is increased beyond the value for which $\lambda = 2L$, the string cannot sustain a standing wave pattern.

---

10. ***REASONING AND SOLUTION*** A string is being vibrated back and forth as in Figure 17.18*a*. The tension in the string is decreased by a factor of four, with the frequency and the length of the string remaining the same. From Equation 16.2, $v = \sqrt{F/(m/L)}$, we see that decreasing the tension F by a factor of four results in decreasing the wave speed v by a factor of $\sqrt{4}$ or 2. From the relationship $v = \lambda f$, we see that, when the frequency f is fixed, decreasing the wave speed by a factor of 2 results in decreasing the wavelength λ by a factor of 2. Therefore, when the tension in the string is decreased by a factor of 4, the wavelength of the resulting standing wave pattern will decrease by a factor of 2. In Figure 17.18*a*, the wavelength of the standing wave is equal to twice the length of the string. Since decreasing the tension reduces the wavelength by a factor of 2, the new wavelength is equal to the length of the string. With this new wavelength, the pattern shown below results.

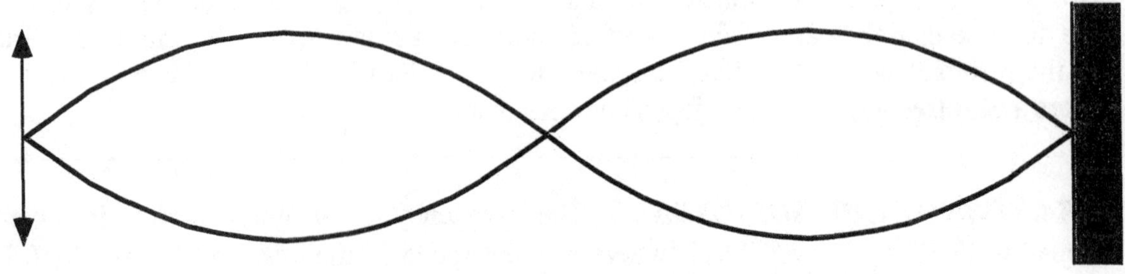

---

11. ***REASONING AND SOLUTION*** a. Since the bottom end of the rope is free to move, there is an antinode at the bottom.

b. Since the rope has mass, it has weight. The upper part of the rope supports more of the weight than the lower part does. Thus, the tension in the rope is greater near the top than near the bottom. The distance between successive nodes is one-half the wavelength of the waves traveling up and down the rope. But the wavelength is equal to the speed of the waves divided by the frequency, according to Equation 16.1. In turn, the speed is proportional to the square root of the tension, according to Equation 16.2. As a result, the wave speed is greater near the top than near the bottom and so is the wavelength. Since the separation between successive nodes is one-half the wavelength, it is also greater near the top than near the bottom.

---

12. ***REASONING AND SOLUTION*** Consider the tube in Figure 17.23. We suppose that the air in the tube is replaced with a gas in which the speed of sound is twice what it is in air. If the frequency of the tuning fork remains unchanged, we see from the relation $v = \lambda f$ that the wavelength of the sound will be twice as long in the gas as it is in air. The resulting pattern is shown in the drawing at the right.

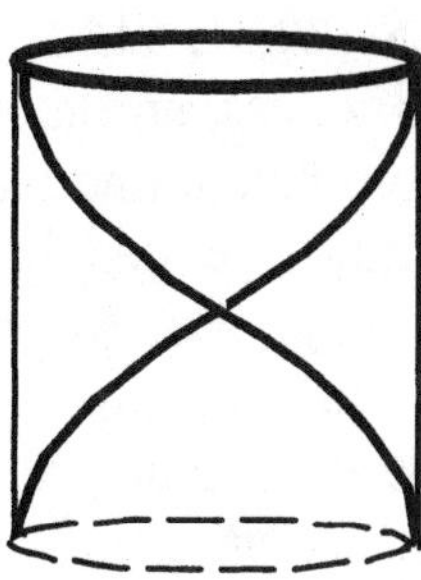

---

13. ***REASONING AND SOLUTION*** When a concert performer generates a 2093-Hz tone, the wavelength of the sound is

$$\lambda = \frac{v}{f} = \frac{343 \text{ m/s}}{2093 \text{ Hz}} = 0.164 \text{ m} \text{ or } 16.4 \text{ cm}$$

If there is excessive reflection of the sound that the performer generates, a large amplitude standing wave can result. The distance between an antinode (maximum loudness) and the next adjacent node (zero loudness) on a standing wave is one-quarter of a wavelength. For a standing sound wave with a frequency of 2093 Hz, the distance between an antinode and the adjacent node is (16.4/4) cm or 4.1 cm. It would be possible, therefore, for a listener to move a distance of only 4.1 cm and hear the loudness of the tone change from loud to faint.

---

14. ***REASONING AND SOLUTION*** The speed of sound v in an ideal gas is given by Equation 16.5, $v = \sqrt{\gamma kT/m}$, where $\gamma = C_P/C_V$ is the ratio of the specific heat capacities, k is Boltzmann's constant, T is the Kelvin temperature, and m is the mass of a molecule of the gas. Thus, the speed of sound in an ideal gas is directly proportional to the square root of the Kelvin temperature T.

If the instrument is open at both ends, then the frequency of sound produced by the instrument is given by Equation 17.4, $f_n = n[v/(2L)]$, where v is the speed of sound, L is the distance between the two open ends of the instrument, and $n = 1, 2, 3, 4, \ldots$ Similarly, if the instrument is open at only one end and closed at the other end, the frequency of sound produced by the instrument is given by Equation 17.5, $f_n = n[v/(4L)]$, where v is the speed of sound, L is the distance between the open and closed ends of the instrument, and $n = 1, 3, 5, 7, \ldots$

When a wind instrument is brought inside from the cold outdoors, its frequency will change because the temperature inside is higher than the temperature outside. If we can treat air as an ideal gas, then according to Equation 16.5, the speed of sound will increase because T increases. According to both Equations 17.4 and 17.5, the frequency produced by a wind instrument is directly proportional to v, the speed of sound; therefore, since the speed of sound increases with increasing temperature, the frequency of the sound produced by the instrument will also increase.

---

15. ***REASONING AND SOLUTION*** Tones produced by a typical orchestra are complex sound waves, and most have fundamental frequencies less than 5000 Hz. Since these tones are complex sound waves, each one consists of a mixture of the fundamental and the higher harmonic frequencies. The higher harmonics are integer multiples of the fundamental frequency. For example, in a tone with a fundamental frequency of 5000 Hz, the fourth harmonic will have a frequency of 20 000 Hz. Since most orchestra tones have fundamental frequencies that are less than 5000 Hz, a high-quality stereo system must be able to reproduce accurately all frequencies up to 20 000 Hz in order to include at least the fourth harmonic of all tones.
